Regenerative Medicine in Sports and Orthopaedics

Alberto Gobbi
Norimasa Nakamura
John G. Lane
Ignacio Dallo
Editors

Regenerative Medicine in Sports and Orthopaedics

A New Approach

Springer

ISAKOS
International Society of Arthroscopy,
Knee Surgery and Orthopaedic Sports Medicine

Editors
Alberto Gobbi
O.A.S.I. Bioresearch Foundation
Gobbi N.P.O.
Milan, Italy

Norimasa Nakamura
Department of Orthopaedic Surgery
Osaka Health Science University
Osaka, Japan

John G. Lane
Musculoskel and Joint Research
Foundation
San Diego, CA, USA

Ignacio Dallo
Department of Orthopaedic Surgery
and Sports Medicine
Unit of Biological Therapies
and Interventional Orthopaedics
SportMe Medical Center
Sevilla, Sevilla, Spain

ISBN 978-3-031-84692-2 ISBN 978-3-031-84693-9 (eBook)
https://doi.org/10.1007/978-3-031-84693-9

This Springer imprint is published by the registered company Springer Nature Switzerland AG
The registered company address is: Gewerbestrasse 11, 6330 Cham, Switzerland

Arnie, we finally made it! Who would have thought?
From mentor to friend, you have always traced my path by leaving marks that will never be forgotten. And today, those marks inspired the creation of this book. Yet, with a shred of sadness, I won't be able to share this achievement with you. Thank you for everything.

In Loving Memory of Arnold Caplan
Alberto Gobbi

I wish to dedicate this book to my family, and friends all over the world who have contributed to and are involved in scientific advancement in musculoskeletal regenerative medicine.

Norimasa Nakamura

I would like to dedicate this book to my family, colleagues and friends who share the desire to continually push boundaries in the field of regenerative medicine and further scientific discoveries in order to improve patient care.

John G. Lane

This book is dedicated to all the mentors I have encountered on my path of learning regenerative medicine. Thanks to their generosity and selfless help, I was able to incorporate a lot of knowledge. I also want to thank my family for their unconditional love and patience. I want to dedicate this book to the next generations of students and those passionate about this field of medicine.

Ignacio Dallo

Foreword

> And a great winged eagle devours your liver all day long, an evil besetment, and your liver grows by night, only to be eaten again in cruel torment.
>
> "Prometheus Bound" by Aeschylus

This passage vividly illustrates the eternal punishment of Prometheus, a Titan who defied Zeus by stealing fire and giving it to humanity. As a consequence of his transgression, Prometheus is chained to a rock, and each day an eagle—a symbol of Zeus—swoops down to devour his liver. Remarkably, his liver regenerates every night, restoring the organ only to be subjected to the same agonizing torment the next day. This relentless cycle of regeneration and destruction not only emphasizes the severity of Prometheus' punishment but also reflects an early understanding or at least a mythological representation of tissue regeneration. Through this myth, the ancient Greeks recognized the extraordinary regenerative capacity of the liver, a concept that aligns with modern biological knowledge, where the liver is known to have a remarkable ability to regenerate.

Tissue regeneration refers to the biological process by which damaged or lost tissues and organs are completely restored, both in structure and function, typically through the activation of stem cells, cellular proliferation, and tissue reorganization. In contrast, when tissue repairs itself, the result is a tissue with similar characteristics, but not identical. In some cases, it can simulate the functions of the original tissue, such as scars on the skin after an injury, but often it does not. Unlike regeneration, which seeks to fully restore the original tissue, repair often results in a partial or functional restoration of the affected structure. While tissue repair can solve the problem, it cannot replace the original tissue. Whereas tissue regeneration aims to completely restore tissues and organs through specific and complex biological processes, tissue repair focuses on solving and improving damaged structures using more immediate and pragmatic methods, though often with less complete results in terms of functionality and integral restoration.

This book, edited by Professors Alberto Gobbi, Norimasa Nakamura, John G. Lane, and Ignacio Dallo, prestigious professionals and researchers in the field of Orthopedic Surgery and Traumatology, focuses on "Regenerative Medicine in the field of sports and traumatology." The first thing to explain is what Regenerative Medicine means: it is a biomedical discipline that aims to replace or regenerate human cells, tissues, or organs that have been lost due to disease or injury, to restore their original structure and function. The goal is to restore the body's normal biological functions and potentially cure diseases that currently have limited treatment options. Regenerative medicine is characterized by being multidisciplinary, involving a variety of scientific and medical fields that collaborate to develop innovative treatments and therapies. This interdisciplinary approach combines knowledge of cellular and molecular biology, tissue engineering, genetics, biochemistry, clinical medicine, and other related disciplines. Researchers and professionals work together to understand and manipulate the biological processes that control tissue regeneration.

Thanks to Regenerative Medicine, we have managed to rejuvenate many joints using chondrocytes. Since 1996 to the present day, we have multiplied these cells in the laboratory and applied them to patients to regenerate damaged cartilage. This approach has improved joint function and relieved pain in patients with various conditions such as osteoarthritis and traumatic injuries. The technique not only seeks to alleviate symptoms but also to restore the original structure and function of the affected joints, thereby improving patients' quality of life in the long term.

In the last 100 years, life expectancy has doubled, and we have seen significant advances in prolonging human life thanks to effective medical and sanitary measures, without the need to manipulate the genome or epigenome. At the beginning of the twentieth century, the average global life expectancy was around 45 years in many countries, affected by high infant mortality rates and precarious living conditions. However, with continuous progress and the implementation of vaccines, antibiotics, potable water, food control, hygiene improvements, and access to medical care, life expectancy began to gradually increase. Today, life expectancy has significantly increased in many developed and developing countries, reaching figures that exceed 86 years in some regions. These advances have strengthened the genetic resilience of human populations against various diseases and have significantly contributed to improving the quality and duration of life.

What remains now? While effective medical and sanitary measures have significantly contributed to increasing life expectancy in the last century, scientists are now exploring the possibility of using cellular reprogramming and epigenome modulation to potentially prolong human life even further. The epigenome, which regulates gene activity without changing the DNA sequence, plays a crucial role in determining how and when genes are expressed throughout life. The ability to reprogram cells to reverse or modify these epigenetic changes opens new perspectives for therapeutic interventions aimed at mitigating the effects of aging and promoting a healthier and longer life. These advances promise not only to improve the quality of life for older people but also to provide tools to address the growing challenges associated with the aging global population. The next revolution will be to understand and modify our epigenome. And in this, we are.

All these advances in the knowledge of mechanisms that promote cellular regeneration, differentiation, and control gene expression represent a promising hope for humanity. With this deeper understanding, scientists are now capable of redirecting damaged cells towards normalization and tissue regeneration. This implies the ability to intervene in the natural course of cells to correct deviations and restore normal biological functions. Instead of merely treating symptoms, these investigations offer the possibility of addressing the underlying causes of diseases and injuries, thereby promoting improved health and prolonged quality of life for affected individuals. The cell as a drug. The cell as a therapeutic opportunity.

This publication on "Regenerative Medicine in the field of sports and traumatology" seems to us a colossal success in clarifying concepts. Research advances step by step, but until the time comes when climate change is a painful dream, diseases have disappeared, and artificial intelligence instructs and advises us, there is a long way to go. We must be careful with the possibilities of Regenerative Medicine and not negotiate with the hope of the sick. For those studying Regenerative Medicine of the musculoskeletal system, it is a highly recommended book.

Department of Traumatology and Research Unit
Clinica CEMTRO
Madrid, Spain
Pedro Guillén

Prólogo

> Y una gran águila alada devora tu hígado durante todo el día, un tormento malévolo,
> y tu hígado crece por la noche, solo para ser comido de nuevo en un tormento cruel.
>
> "Prometeo encadenado" de Esquilo

Este pasaje ilustra vívidamente el castigo eterno de Prometeo, un Titán que desafió a Zeus al robar el fuego y dárselo a la humanidad. Como consecuencia de su transgresión, Prometeo está encadenado a una roca, y cada día un águila—un símbolo de Zeus—se abalanza para devorar su hígado. De manera notable, su hígado se regenera cada noche, restaurando el órgano solo para ser sometido al mismo tormento agonizante al día siguiente. Este ciclo implacable de regeneración y destrucción no solo enfatiza la severidad del castigo de Prometeo, sino que también refleja una comprensión temprana o al menos una representación mitológica de la regeneración tisular. Los antiguos griegos, a través de este mito, reconocieron la extraordinaria capacidad regenerativa del hígado, un concepto que se alinea con el conocimiento biológico moderno, donde se sabe que el hígado tiene una notable capacidad de regenerarse.

La regeneración de tejidos se refiere al proceso biológico mediante el cual los tejidos y órganos dañados o perdidos son restaurados completamente, tanto en su estructura como en su función, generalmente a través de la activación de células madre, la proliferación celular y la reorganización tisular. Por otro lado, cuando un tejido se repara el resultado un tejido de características similares, pero no igual. En algunas ocasiones puede simular las funciones del tejido original, como el caso de las cicatrices en la piel tras una herida, pero otras muchas no. A diferencia de la regeneración, que busca restablecer completamente el tejido original, la reparación a menudo resulta en una restauración parcial o funcional de la estructura afectada. Muchas veces la reparación de tejidos puede solucionar el problema, pero es incapaz de reemplazar el tejido original. Mientras que la regeneración de tejidos apunta a restaurar completamente los tejidos y órganos mediante procesos biológicos específicos y complejos, la reparación de tejidos se enfoca en solucionar y mejorar estructuras dañadas utilizando métodos más inmediatos y pragmáticos, aunque a menudo con resultados menos completos en términos de funcionalidad y restauración integral.

Este libro, editado por los Profesores Alberto Gobbi, Norimasa Nakamura, John G. Lane e Ignacio Dallo, prestigiosos profesionales e investigadores dentro del campo de la Cirugía Ortopédica y Traumatología, está enfocado en la "Medicina Regenerativa en el campo del deporte y la traumatología". Lo primero que debemos hacer es explicar qué se entiende por Medicina Regenerativa: es una disciplina biomédica que trata de sustituir o regenerar células humanas y tejidos u órganos, que se han perdido a causa de una enfermedad o un accidente, para restablecer su estructura y función original. El objetivo es restaurar las funciones biológicas normales del cuerpo y potencialmente curar enfermedades que actualmente tienen opciones limitadas de tratamiento. La medicina regenerativa se caracteriza por ser multidisciplinaria, ya que involucra una variedad de campos científicos y médicos que colaboran para desarrollar tratamientos y terapias innovadoras. Este enfoque interdisciplinario combina conocimientos de biología celular y molecular, ingeniería de tejidos, genética, bioquímica, medicina clínica y otras disciplinas relacionadas. Los investigadores y profesionales trabajan en equipo para entender y manipular los procesos biológicos que controlan la regeneración de tejidos.

Gracias a la Medicina Regenerativa, hemos logrado rejuvenecer muchas articulaciones utilizando condrocitos. Desde 1996 hasta el día de hoy, hemos multiplicado estas células en el laboratorio y las hemos aplicado a pacientes con el objetivo de regenerar el cartílago dañado. Este enfoque ha permitido mejorar la función articular y aliviar el dolor en pacientes con diversas condiciones como osteoartritis y lesiones traumáticas. La técnica no solo busca aliviar los síntomas, sino también restaurar la estructura y función original de las articulaciones afectadas, mejorando así la calidad de vida de los pacientes a largo plazo.

En los últimos 100 años la esperanza de vida se ha duplicado y hemos visto avances significativos en la prolongación de la vida humana gracias a medidas médico-sanitarias efectivas, sin necesidad de manipular el genoma o el epigenoma. A principios del siglo XX, la esperanza de vida global promedio era de alrededor de 45 años en muchos países, afectada por altas tasas de mortalidad infantil y condiciones de vida precarias. Sin embargo, con el progreso continuo y la implementación de vacunas, antibióticos, agua potable, control de los alimentos, mejoras en la higiene y el acceso a la atención médica, la esperanza de vida comenzó a aumentar gradualmente. Hoy en día, la esperanza de vida ha experimentado un notable incremento en muchos países desarrollados y en desarrollo, alcanzando cifras que superan los 86 años en algunas regiones. Estos avances han fortalecido la resistencia genética de las poblaciones humanas frente a diversas enfermedades y han contribuido significativamente a mejorar la calidad y la duración de la vida.

¿Qué resta ahora? Mientras que las medidas médico-sanitarias efectivas han contribuido significativamente a aumentar la esperanza de vida en el último siglo, ahora los científicos están explorando la posibilidad de utilizar la reprogramación celular y la modulación del epigenoma para potencialmente prolongar aún más la vida humana. El epigenoma, que regula la actividad génica sin cambiar la secuencia del ADN, desempeña un papel crucial en la determinación de cómo y cuándo se expresan los genes a lo largo de la vida. La capacidad de reprogramar células para revertir o modificar estos cambios epigenéticos abre nuevas perspectivas para intervenciones terapéuticas destinadas a mitigar los efectos del envejecimiento y promover una vida más saludable y longeva. Estos avances prometen no solo mejorar la calidad de vida de las personas mayores, sino también proporcionar herramientas para abordar los desafíos crecientes asociados con el envejecimiento de la población mundial. La próxima revolución va a ser entender y modificar nuestro epigenoma. Y en eso, estamos.

Todos estos avances en el conocimiento de los mecanismos que promueven la regeneración celular, la diferenciación y que controlan la expresión génica representan una esperanza prometedora para la humanidad. Con esta comprensión más profunda, los científicos están ahora capacitados para reconducir la célula dañada hacia la normalización y la regeneración de tejidos. Esto implica la capacidad de intervenir en el curso natural de las células para corregir desviaciones y restaurar funciones biológicas normales. En lugar de simplemente tratar los síntomas, estas investigaciones ofrecen la posibilidad de abordar las causas subyacentes de enfermedades y lesiones, promoviendo así una salud mejorada y una calidad de vida prolongada para las personas afectadas. La célula como medicamento. La célula una oportunidad terapéutica.

Esta publicación sobre "Medicina Regenerativa en el campo del deporte y la traumatología" nos parece un acierto colosal para aclarar conceptos. La investigación avanza paso a paso, pero hasta llegar el momento que el cambio climático sea un penoso sueño, que las enfermedades han desaparecido y que la inteligencia artificial nos instruye y aconseja, resta un gran trecho. Tenemos que ser cuidadosos con las posibilidades de la Medicina Regenerativa y no negociar con la esperanza del enfermo. Para los estudiosos de la Medicina Regenerativa del sistema músculo-esquelético, es un libro muy recomendable.

Department of Traumatology and Research Unit
Clinica CEMTRO
Madrid, Spain

Pedro Guillén

Introduction to Regenerative Medicine in Sports and Orthopaedics

This mission for all providers on the Sports Medicine team is to manage the health of the athlete through Prevention, Optimizing Performance and Injury and Illness care. The most significant challenges that we face now and the future include Muscle Injury, Osteoarthritis, ACL injury and consequence and most importantly maintaining the athlete's career and their longevity. The solutions include the systematic development of new concepts, techniques and technologies. The first concept is "injury burden" which is frequency multiplied by the number of days lost. The second concept is PLAY span, which is to increase the amplitude of career and the number of years, games played or overall exposure. This book, *Orthobiologics in Sports Medicine*, is a compendium that focuses on decreasing injury burden by increasing PLAY span utilizing the family of orthobiologic and regenerative interventions. To date the applications of these interventions have followed the innovation diffusion curves progression from innovators, early adapters to early and late majority and then the laggards. The developmental progression may have some limitations as some animals such as the axolotl and the zebrafish have a highly regenerative capacity that regulates successful limb regeneration once injured. In contrast the human being does not have that same regenerative capability for muscle, tendon, bone, ligament or cartilage. As consequence in the last 15 years, we have found the applications of biological interventions have increased exponentially in Orthopaedic surgery and sports medicine. The challenge for the Orthopaedic surgeon and physician is to develop specific, resilient, disease-modifying and durable interventions for the spectrum of musculoskeletal injuries. In addition, diseases like osteoarthritis which affect over 300 million people globally are amongst the most common disabling conditions treated as a consequence of pain, symptoms and loss of function. From the regulatory perspective the Food and Drug Administration (FDA) and the Europe, the Middle East and Africa (EMEA) and other global agencies regulate human cell and tissue therapies. These innovations fall under section 361 and are considered minimally manipulated and are indicated for homologous use. Cultured and expanded bone marrow-derived adipose "Stem cells" are permitted in several countries but are regulated by the FDA and can be only used in clinical trials or utilizing an IND (Investigational New Drug). Examples of minimally, manipulated autologous therapies include Platelet Rich Plasma (PRP), Bone Marrow Concentrate (BMAC) and micro fragmented adipose tissue. Moreover, the only FDA-approved "stem cell" product is hematopoietic or blood forming cells directed towards blood disorders.

As a consequence of these nuances and complexities the definitive text *Orthobiologic in Sports Medicine* is edited by Dr Gobbi has been written. It meticulously serves as an evidence-based compendium of basic and clinical translational rationale for the family of biologic interventions. Systematically global experts present the evidence-based regenerative triad of cells, growth factors and scaffolds and the multitude requirements for success in clinical and athlete care. The text systematically focuses on PRP, then transitions to Mesenchymal Stem Cells (MSCs) that are adipose, bone marrow and blood derived. Comprehensively the text addresses clinical and surgical applications for the present and future.

Alberto Gobbi is a pioneer in orthobiologics and with his co-authors are to be congratulated as they are definitive, extremely timely, relevant and realistic as to the benefits and risks associated with the spectrum of care and the practical clinical use of biologics. This is an outstanding resource and tool for the clinician and scientist for perspective, concept and practice.

Cedars Sinai Kerlan Jobe Institute Bert R. Mandelbaum
Los Angeles, CA, USA

Contents

Part I

Overview

1 Regenerative Medicine: Fact or Fiction

Alberto Gobbi, Melanio Acosta IV, and Camilla Boiocchi

1.1 Introduction

By simply typing and searching the keywords "regenerative medicine" or "stem cells" into your search engine, you can find approximately 54 million web pages at your fingertips. Some of the top hits are web pages of popular research articles; however, most are advertisements, announcements, or flyers from companies, clinics, and hospitals offering their own brand of "stem cell therapy" that can cure, fix, and treat any condition. Do you have diabetes? Stem cell therapy can cure it. Do you want to lose weight? Stem cell therapy can help you burn fat. Are you having difficulties in the bedroom? Yes... stem cells can fix all your problems. These advertisements and claims that regenerative therapies are a miracle cure for any condition are simply false and deceptive. Many of these claims are not scientifically backed or proven and are simply used to prey on ill-informed individuals and patients.

Not all "stem cells" are the same and a red flag that may hint to a clinic being a scam is that they offer treatment for a menu worth of ailments and conditions with a single type of stem cell or a single type of therapy. The spread of information has also changed in the last few decades. If someone wanted to get information on a topic, people went from going to the library, reading newspapers, and watching or listening to the news on television and radio to simply opening their cell phone and searching keywords or going to social media platforms like Facebook, X (formerly called Twitter), Instagram, TikTok, and YouTube. With the trend of communication shifting from credible and fact-checking information outlets toward social media platforms, podcasts, and unregulated web page articles, the spread of misinformation is easy and frequent.

As clinicians and researchers, how many times have we had arguments with those outside our field on topics they claim to be experts simply because they have read a few articles on Google and seen a few videos on YouTube? Coined by Dr. Caulfield, "Scienceploitation" is when legitimate science and vulnerable people are exploited and taken advantage of to drive consumer interest in products or services [1]. By exploiting the limited knowledge of science an average person has, these products and services, which are typically unproven with no scientific backing, give false hope and expectations. "Dr. Google" and "Scienceploitation" have created a perfect cover and camouflage full of myth and fiction hiding the truths and facts of regenerative medicine which clinicians and researchers painfully work

A. Gobbi (✉) · M. Acosta IV · C. Boiocchi
O.A.S.I. Bioresearch Foundation Gobbi N.P.O., Milan, Italy
e-mail: gobbi@cartilagedoctor.it

A. Gobbi et al. (eds.), *Regenerative Medicine in Sports and Orthopaedics*,
https://doi.org/10.1007/978-3-031-84693-9_1

to uncover. Our aim is to clear up these falsehoods and misconceptions and dive into the truth and science of regenerative medicine.

1.2 Where It Started

The father of basic science and clinical therapy of adult mesenchymal stem cells who acted as the spark to ignite the research and study of regenerative medicine, Dr. Arnold Caplan, defined regenerative medicine as "an emerging discipline on the frontiers of medical science with the potential of restoring function in a large number of tissues compromised by incurable and degenerative diseases." Through the advancement of regenerative medicine, developments are being made to potentially eliminate disease, but most importantly for orthopaedics, restore normal tissue function [2, 3]. Even with the discovery of mesenchymal stem cells in the 1990s, later termed as "medicinal signaling cells," along with the many works on regenerative medicine prior to and following Dr. Caplan's introduction of the mesengenic process, there are still many misconceptions, especially among the public and in the non-medical field, on what regenerative medicine is and what it isn't.

We, as a forever evolving and advancing human species, have always been inspired by "mother nature" and the teasing glimpse that self-regeneration is possible. For example, by observing reptiles, we witness lizards possess the ability to regenerate an amputated tail or appendage in whole, complete with an ossified skeleton with surrounding neurovascular structures and supporting soft and connective tissue elements [4]. While salamanders have been shown to not only regenerate their limbs and tails (Fig. 1.1), but regenerate structures of the heart, eyes, and spinal cord [5, 6]. On the most extreme end of the spectrum, perhaps chalked up to the human interest in the concept of "immortality," this concept is frequently illustrated in modern science fiction like the famous 1984 science fiction film "The Terminator." Human beings, even in ancient stories, describe a mythological representation of an organ's capacity for regeneration as described in the stories of the Titan Prometheus, restoring his liver every night to be punished the next day. We also have been fascinated by the possible "immortality" of species like jellyfish, starfish, and flatworms, which possess whole-body regeneration, in which the whole organism can be regenerated from a single fragment [7].

Fig. 1.1 Newts and salamanders have the ability to regenerate a completely amputated tail among other structures

One important step in the regeneration process is skeletal muscle cellularization, where the muscle fibers break down into dedifferentiated, mononuclear cells that proliferate and form new muscle tissue in the renewed limb. Regeneration can be divided into two major phases—(1) the formation of a blastema that resembles the early embryonic limb bud and (2) the redevelopment of this newly formed blastema, which engages in processes of growth and re-differentiation. This type of skeletal muscle cellularization is absent in mammals hindering tissue regeneration of this type and magnitude [8].

1.3 Regenerative Medicine as a Science

The first step to unlocking the human potential of regeneration is to establish an understanding of the difference in the cellular biology and mechanisms of these species and how they are able to form this competent blastema with the ability to regenerate. By understanding these mechanisms, we could identify and characterize specific regulatory genes that may have mammalian homologs and apply this to regenerative healing practices and therapeutics in humans. Through current and future scientific studies investigating this possible genetic link between mammals and these species, though not to the same extent of reptiles or starfish, we can discover the genetic formula that enables human regenerative potential, allowing us to restore body functions lost as a result of injury, disease, or senescence [8, 9].

In the last 30+ years, investigators have studied and described various approaches to stimulating and enhancing the body's natural ability for self-restoration and healing in order to regain normal tissue and function. However, despite decades of study, we still do not have a complete understanding of the biological processes and mechanisms of the human body to achieve these goals [10]. Through various *in vitro* and *in vivo* animal model experiments, efforts have been made in establishing an increased understanding of cellular and tissue regeneration. However, success in these models does not always guarantee success and equivalent efficacy when applied to humans.

Outside of establishing the safety and efficacy of regenerative medicine practices investigated using *in vitro* and *in vivo* animal models when translating these concepts to humans, we face challenges in differentiating if these therapeutic procedures have actual regenerative potential or are simply symptomatic processes. This calls on the need for further research of these biologic processes, as well as utilizing well-designed randomized controlled trials to give us a clearer understanding.

1.4 The Misconceptions About Stem Cells and Regenerative Medicine

Aside from identifying mammalian homologs of genes which allow complete tissue regeneration over repair, we seek alternative routes by identifying cells that possess multipotency and the ability to differentiate into multiple cell types. The strongest regenerative resource in the human body with this type of potential are stem cells, which are undifferentiated cells with the unique ability to be driven down many different lineages depending on environmental stimuli which multiply into a particular cell type.

Many advancements have been made in the last three decades through extensive research on stem cells and their application in regenerative medicine. However, due to these breakthroughs, our improved understanding of stem cells, and the increased spotlight in the public and media, many clinics, laboratories, and institutions have started a controversial and fraudulent practice of claiming stem cells are the holy grail of medicine and biologics. They are exaggerating the uses, the outcomes, and the science behind stem cells and regenerative medicine, claiming it is the cure for every and all imaginable conditions. This practice of false stem cell-related articles and media coverage can distort the reality of the potential of stem cells, leading to the spreading of misinformation, inaccurate facts, and overexaggeration of expectations by the public, which, in turn, can mislead patient decisions potentially damaging the reputation of clinicians and scientists who cannot live up to these false standards [11, 12].

In fact, stem cells are not the Fountain of Youth in the Garden of Eden, a miracle treatment that can grant eternal youth or reverse aging. As mentioned by Dr. Huard "the goal is not to make you look younger: it is to make you age better." Therefore, these interventions won't transform a 50-year-old into a 25-year-old, but they can assist an older person by decreasing the rate of aging." As part of the five pillars of healthy aging (healthy diet, lifestyle, personalized medicine, regenerative medicine, and prescription medications), regenerative medicine is linked to aging through the body's regenerative potential. As the body ages, there is a higher proportion of senescent cells, cells with less regenerative potential. An increase in the proportion of senescent cells can be linked to health and longevity, as seen in a study involving mice with progeria. Muscle-derived stem/progenitor cells (MDSPCs) from young mice were injected into aged mice who had progeria (lacking DNA repair enzyme excision repair cross-complementation group 1 ($ERCC1^{-/-}$) which causes accelerated aging). Normally, ERCC1-deficient mice ($Ercc1^{-/-}$) die prematurely at about 21 days, but the mice injected with young MDSPCs lived three times as long [13–16].

With increased spotlight and awareness on stem cells, they have not been without their own criticism and opposition. Stem cell research has become one of the most debated issues in government policy, most notably after the success of animal cloning and isolation and application of embryonic stem cells [17–19]. Much of the controversy and consideration has been due to the influence of religion on political attitudes and agendas [20–22]. The view that specific stem cell research is "unethical" and against moral imperatives has caused several countries to prohibit these types of research and clinical use. As most of the controversies originate from stem cells derived from embryos, such issues have not arisen when cells are harvested from adults [23]. To bypass criticisms and policy against use of embryonic cells, many have tried to ascertain cells of similar potential. In Japan, researchers have been able to harvest adult fibroblast cells and re-program these cells into cells with embryonic stem cell characteristics, induced pluripotent stem cells (iPSCs) similar in morphology, surface antigens, proliferation capabilities, gene expression, pluripotency, and telomerase activity [24]. These cells have been of particular interest due to their regenerative potential and high safety profile.

1.5 Regenerative Medicine, Orthopaedics, and Sports

The relationship and intertwining of regenerative medicine and orthopaedics is relatively new. The use of orthobiologics started with hyaluronic acid in the 1990s, and by 2006, platelet-rich plasma emerged in sports medicine. Bone marrow concentrate and later lipoaspirate with their processing and exploitation of mesenchymal stem cells has been the most recent generation of orthobiologic therapy. With increased research and investigation of current and future orthobiologic products, improved efficacy, standardization, and possibilities are just around the horizon as clinicians apply these products in their practice as clinic-setting injections to the incorporation into surgical procedures [25].

Since the dawn of the very first Olympic games in 1896 held in Athens, Greece, to the twilight of the most recent games in 2024 held in Paris, we have seen the evolution of human capabilities. As athletes push the boundaries of performance and world records are broken each and every year, interest grows in sports technology, supplementation, and nutrition, as well as injury treatment, rehabilitation, and healing. It is an inevitable fact that injury is part of athletics and sports; however, with improved understanding of biologics, kinematics, and rehabilitation, athletes now have an increased likelihood to recover and return to the game they love.

The major goal of physicians, surgeons, and practitioners in sports medicine, when dealing with injured athletes, is to restore pre-injury function in the shortest possible time period without compromising the natural healing process among athletes. As many athletes cannot afford to undergo aggressive, risky, sometimes expensive surgical treatments that require extensive time off the field and rehabilitation, interest has turned to the application of orthobiologics as an alternative.

As sports committees around the world, most notably the International Olympic Committee and the World Anti-Doping Agency, approve the use of platelet-rich plasma as a treatment option for professional athletes, many athletes and prominent figures have utilized and benefitted from products of regenerative medicine.

Kobe Bryant, one of the most famous and well-known basketball players, had been plagued with injury throughout his career. Still in his prime, coming off of back-to-back NBA championship wins, Kobe Bryant turned to platelet-rich plasma treatments in Germany for a knee injury as surgical treatments and prolonged rehabilitation were not an option. Through these therapies, despite the inability to complete the 3-peat, Kobe Bryant was still able to be voted an NBA All-Star, where he dominated and was named the NBA All-Star Game MVP, as well as All-NBA First Team, and All-Defensive First Team.

Tiger Woods, one of the greatest of all time in the world of golf, is no stranger to PRP as well.

Coming off of a recent knee surgery, Woods described doing "everything I possibly could to heal faster so I could get back on the golf course, you know, through the PRP injections." The following years, Woods not only returned back to the golf course but became PGA Tour Player of the Year, PGA Tour Leading Money Winner, Vardon and Byron Nelson Trophy Winner, as well as FedEx Cup Champion.

Another all-time great in their sports, Rafael Nadal, holding the record of having the second most grand slam titles, has talked about how "PRP worked unbelievably on my knee before, in 2010, 2009… [PRP] worked unbelievably. I recovered 100 percent in a very short period of time during the PRP treatment." Following his PRP injections, Nadal won 15 tournament titles in 2009, 2010, and 2011.

The use of regenerative medicine and bio-orthopaedics for sports injuries harness and enhance the body's natural healing process to heal damaged cartilage, tendons, ligaments, and muscle tissue, along with augmenting other conservative and surgical modalities. Bio-orthopaedics continue to evolve, providing athletes innovative solutions for optimal recovery and sustained performance, and offering promising and personalized alternatives for treatment [26].

1.6 The Regenerative Medicine Market

As stem cells and regenerative medicine have gained popularity and worldwide awareness, the regenerative market and its financing boomed throughout the years. Before 2000, only $146 million USD was raised, which boomed after the discovery of stem cells, mesenchymal stem cells, and induced pluripotent stem cells. Financing reached $908 million USD from 2000 to 2005. This continuously increased to $3.02 billion USD and $18.93 billion USD from 2011 to 2016. As our understanding of regenerative medicine increased exponentially, the market in turn equally grew with increasing interest in new technology and new applications [27]. This market growth is expected to only increase in the next decade, as it is expected to grow from $34.6 billion USD in 2022 to $194.9 billion by 2032 (Fig. 1.2) [28].

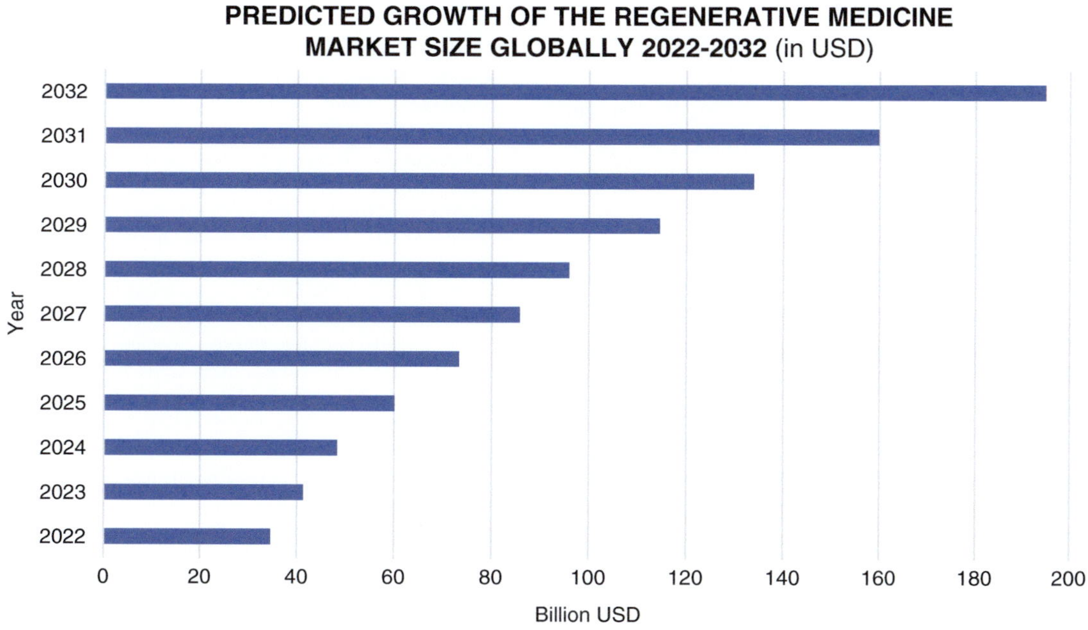

Fig. 1.2 Through private and government funding, the regenerative medicine market is expected to further increase by 2032

Regenerative medicine is typically divided by product markets such as cell therapy, gene therapy, tissue engineering, and platelet-rich plasma, which are applied to many different conditions and pathologies such as Alzheimer's and Parkinson's disease; diabetes, cardiovascular, renal, gastrointestinal, rheumatic, and oncologic diseases; and of course, orthopaedic pathologies. As stem cell and regenerative medicine application grows, we will see an increase in private and government funding toward further research and development. A boom of new companies has also arisen through partnerships, mergers, and collaborations of private investors, companies, as well as government institutions. This will lead to new product and technology, which can eventually be translated into clinical practice [28].

Despite the increasing interest and awareness of regenerative medicine, there are many constraints that prevent even further growth. Studies have shown that as high as 86% of clinical trials fail to achieve FDA approval resulting in significant economic losses. While other factors such as ethical concerns, price, timing, and setting can hinder a product or technology implementation or application despite improved outcomes [29].

1.7 Conclusion

Through observing mother nature, we, as a human species, have been fascinated with the concept of regeneration and longevity. Our interest in regenerative medicine has led to groundbreaking discoveries and improved understanding on how the body works, the biologic mechanisms of healing, and how we can apply new knowledge to improving our health, longevity, and even performance. In our excitement, we sometimes want to proclaim that we have found the "holy grail" of therapy, the one therapy that can cure all. However, it is important to decipher fact from fiction. The remainder of this book aims to review and clarify previous concepts, introduce current and future trends and research, while encouraging and inspiring new ideas on regenerative medicine in orthopaedics and sports.

References

1. Caulfield T. Blinded by science: Modern-day hucksters are cashing in on vulnerable patients. The Walrus; 2011. https://thewalrus.ca/blinded-by-science/
2. Caplan AI, West MD. Progressive approval: a proposal for a new regulatory pathway for regenerative medicine. Stem Cells Transl Med. 2014;3(5):560–3. https://doi.org/10.5966/sctm.2013-0180.
3. Wakitani S, Mera H, Nakamura N, Gobbi A. Review of Caplan (1991) on cell-based therapeutic technology using mesenchymal stem cells. J ISAKOS : joint disorders & orthopaedic sports medicine. 2024;9(3):426–30. https://doi.org/10.1016/j.jisako.2023.08.010.
4. Cox PG. Some aspects of tail regeneration in the lizard, Anolis carolinensis. I. A description based on histology and autoradiography. J Exp Zool. 1969;171:127–49. https://doi.org/10.1002/jez.1401710202.
5. Brockes JP, Kumar A. Plasticity and reprogramming of differentiated cells in amphibian regeneration. Nature reviews. Mol Cell Biol. 2002;3(8):566–74. https://doi.org/10.1038/nrm881.
6. Butler EG, Ward MB. Reconstitution of the spinal cord after ablation in adult Triturus. Dev Biol. 1967;15(5):464–86. https://doi.org/10.1016/0012-1606(67)90038-3.
7. Cary GA, Wolff A, Zueva O, Pattinato J, Hinman VF. Analysis of sea star larval regeneration reveals conserved processes of whole-body regeneration across the metazoa. BMC Biol. 2019;17(1):16. https://doi.org/10.1186/s12915-019-0633-9.
8. Kragl M, Knapp D, Nacu E, Khattak S, Maden M, Epperlein HH, Tanaka EM. Cells keep a memory of their tissue origin during axolotl limb regeneration. Nature [Internet]. 2009;460(7251):60–5. Available from: https://www.ncbi.nlm.nih.gov/pubmed/19571878
9. Sampogna G, Guraya SY, Forgione A. Regenerative medicine: historical roots and potential strategies in modern medicine. J Microsc Ultrastruct [Internet]. 2015;3(3):101–7. Available from: https://www.ncbi.nlm.nih.gov/pmc/articles/PMC6014277/
10. Stoltz JF, de Isla N, Li YP, Bensoussan D, Zhang L, Huselstein C, Chen Y, Decot V, Li N, Reppel L, He Y. Stem cells and regenerative medicine: myth or reality of the 21th century. Stem Cells Int [Internet]. 2015;2015:1–19. Available from: https://www.hindawi.com/journals/sci/2015/734731/cta/
11. Marcon AR, Murdoch B, Caulfield T. Fake news portrayals of stem cells and stem cell research. Regen Med. 2017;12(7):765–75.
12. Master Z, Matthews KRW, Abou-el-Enein M. Unproven stem cell interventions: a global public health problem requiring global deliberation. Stem Cell Rep. 2021;16(6):1435–45.

13. McCullagh KJ. Can a young muscle's stem cell secretome prolong our lives? Stem Cell Res Ther. 2012;3(3):19. https://doi.org/10.1186/scrt110.
14. Qu-Petersen Z, Deasy B, Jankowski R, Ikezawa M, Cummins J, Pruchnic R, Mytinger J, Cao B, Gates C, Wernig A, Huard J. Identification of a novel population of muscle stem cells in mice: potential for muscle regeneration. J Cell Biol. 2002;157:851–64. https://doi.org/10.1083/jcb.200108150.
15. Peault B, Rudnicki M, Torrente Y, Cossu G, Tremblay JP, Partridge T, Gussoni E, Kunkel LM, Huard J. Stem and progenitor cells in skeletal muscle development, maintenance, and therapy. Mol Ther. 2007;15:867–77. https://doi.org/10.1038/mt.sj.6300145.
16. Nicoletti K (2020, January 5) Fountain of youth. Vail Health. Retrieved August 5, 2024, from https://www.vailhealth.org/news/fountain-of-youth
17. Einsiedel E, et al. In: Bauer M, Gaskell G, editors. Brave new sheep: the clone named Dolly, in biotechnology: the making of a global controversy. Cambridge: Cambridge University Press; 2002.
18. Shepherd R, Barnett J, Cooper H, Coyle A, Moran-Ellis J, Senior V, Walton C. Towards an understanding of British public attitudes concerning human cloning. Soc Sci Med. 2007;65(2):377–92. https://doi.org/10.1016/j.socscimed.2007.03.018.
19. Reardon S. Global summit reveals divergent views on human gene editing. Nature. 2015;528(7581):173. https://doi.org/10.1038/528173a.
20. Wald KD, Calhoun-Brown A. Religion and politics in the United States. Rowman & Littlefield; 2014.
21. Haidt J. The righteous mind: why good people are divided by religion and politics. New York: Pantheon; 2012.
22. Norris P, Inglehart R. Sacred and secular: religion and politics worldwide. Cambridge University Press; 2011.
23. National Institutes of Health. NIH Guidelines for Human Stem Cell Research | STEM Cell Information [Internet]. stemcells.nih.gov. 2021. Available from: https://stemcells.nih.gov/research-policy/guidelines-for-human-stem-cell-research
24. Takahashi K, Tanabe K, Ohnuki M, Narita M, Ichisaka T, Tomoda K, Yamanaka S. Induction of pluripotent stem cells from adult human fibroblasts by defined factors. Cell. 2007;131(5):861–72.
25. Rahmani E, Rezaei M, Tavakoli R, Ghadirzadeh E, Sarnaghy FJ, Khorram R, et al. Role of regenerative medicine in the treatment of diseases. Kindle [Internet]. 2023;3(1):1–184. [cited 2024 Jul 3] Available from: https://preferpub.org/index.php/kindle/article/view/Book22
26. Neph A, Onishi K, Wang JH-C. Myths and facts of in-office regenerative procedures for tendinopathy. Am J Phys Med Rehabil. 2019;98(6):500–11.
27. Ng M, Song S, Piuzzi NS, Ng K, Gwam C, Mont MA, Muschler GF. Stem cell industry update: 2012 to 2016 reveals accelerated investment, but market capitalization and earnings lag. Cytotherapy. 2017;19(10):1131–9. https://doi.org/10.1016/j.jcyt.2017.07.006.
28. Mikulic M. Forecast of the regenerative medicine market worldwide from 2022 to 2032. Statista; 2024. Retrieved 5 August 2024, from https://www.statista.com/statistics/800676/global-market-volume-for-regenerative-medicine/#:~:text=Global%20market%20size%20for%20regenerative%20medicine%202022%2D2032&text=In%202022%2C%20the%20global%20regenerative,billion%20U.S.%20dollars%20by%202032
29. O'Donnell BT, Ives CJ, Mohiuddin OA, Bunnell BA. Beyond the present constraints that prevent a wide spread of tissue engineering and regenerative medicine approaches. Front Bioeng Biotechnol. 2019;7:95. https://doi.org/10.3389/fbioe.2019.00095.

Natural Healing: Can We Improve It?

2

Christopher J. Rogers, Mary A. Ambach, Chris T. Ha, and Alberto Gobbi

2.1 Introduction

> The doctor of the future will give no medicine, but will interest his patients in the care of the human frame, in a proper diet, and in the cause and prevention of disease
> Thomas Edison (1847–1931)

Since Edison's day, a mountain of scientific evidence has been created that supports this assertion. The benefits of a healthy diet, regular physical activity, and avoidance of substances known to promote disease have essentially become common knowledge. When it comes to using "natural" remedies for the treatment of musculoskeletal conditions, there seems to be no shortage of recommendations; however, many of these remedies remain incompletely evaluated in the medical literature. These treatments do not always assure safety or efficacy, simply because they are "natural." Regenerative medicine physicians rely upon the peer-reviewed medical literature for guidance to maximize clinical outcomes.

Recent estimates conclude that the average adult human is composed of 37.2 trillion cells [1] and, perhaps, as many bacteria [2]. The word cell derives from the Latin word cella, meaning "small room." In 1665, Robert Hooke discovered and named them for their resemblance to monastery cells inhabited by Christian monks [3]. In these "small rooms" all of the amazing properties of life are found. The cell serves such vital functions such as energy production, protein synthesis, and the repair of all the tissues of the musculoskeletal system.

The human body is in a continual state of renewal. In the average adult, more than 1.1 million new platelets and 2.4 million erythrocytes are produced every second [4, 5]. Bone and muscle cells turnover at a much slower rate, but will be completely replaced several times throughout an individual's lifespan [6]. The viability of these cells, and therefore our health, is influenced by the activities we do daily and the environment in which we do those activities. Behaviors that optimize health will naturally lead to optimal healing. Optimizing healing is therefore nothing more than the facilitation of optimizing one's overall health.

C. J. Rogers (✉)
San Diego Orthobiologics Medical Group, Carlsbad, CA, USA
e-mail: rogers@sdomg.com

M. A. Ambach
BIOEVOLVE San Diego Orthobiologics and Sports Center, Encinitas, CA, USA
e-mail: ambachmd@bioevolvesports.com

C. T. Ha
Mayo Clinic, Scottsdale, AZ, USA
e-mail: Ha.Chris@mayo.edu

A. Gobbi
O.A.S.I. Bioresearch Foundation Gobbi N.P.O., Milan, Italy
e-mail: gobbi@cartilagedoctor.it

A. Gobbi et al. (eds.), *Regenerative Medicine in Sports and Orthopaedics*,
https://doi.org/10.1007/978-3-031-84693-9_2

In the clinical context, healing can manifest in three forms: "healing better," "feeling better," and "dealing better." Healing in the truest sense implies that injured, degenerated, or diseased tissue is repaired and normal function is restored. However, very little effort extended in current medical practice achieves this high standard. Up until recently, complete tissue healing either occurs naturally or not at all. Corticosteroid injections for the treatment of chronic inflammation, analgesics for unresolving pain, and even surgical repair of damaged tendons or joints often provide only temporary relief and rarely restore normal tissue function [7–9]. Yet, these treatment options have remained the standard of care in orthopedics for decades. For many patients, simply achieving relief of their symptoms is sufficient. They simply wish to feel better. They understand the limitations of medical care and are often content with a quick fix to alleviate symptoms.

In cases where all efforts have failed to alleviate symptoms, patients are offered pain management which may include supportive counseling, behavioral modification, and education. There is no doubt that these efforts have enabled patients to better cope with their otherwise unmanageable condition. Oftentimes, simply educating the patient and taking time to listen to their concerns will result in a therapeutic benefit. Empowering the patient to control emotional responses and enhancing cognitive awareness and the impact that the environment has on their symptoms can allow these benefits to be maintained over long periods of time [10].

Regenerative medicine is the clinical translation of cellular biology and tissue engineering to replace, rejuvenate, or regenerate human cells, tissues, and organs in order to restore normal function. This emerging field of medicine is changing the face of orthopaedics at an ever-accelerating rate. Our expanding knowledge of cellular biology, human genetics, tissue engineering, and the clinical translation of this new understanding will likely continue to evolve for the foreseeable future. As clinicians and scientists, we and our patients will have an increasing number of treatment options to consider as we strive to eliminate musculoskeletal pain and dysfunction. With cellular-based therapy, patients may finally be relieved of having to deal better or merely feel better and will be enabled to actually heal better. Regenerative medicine holds great promise and has already begun to deliver on that promise.

2.2 The Phases of Tissue Healing

Healing is the process whereby tissues repair themselves in response to disease or trauma. Tissue healing is a dynamic process involving the coordinated actions of both resident and migratory cells within the extracellular matrix leading to the repair of injured tissues. The healing process is not only complex but also fragile, and it is susceptible to interruption or failure leading to chronic non-healing tissue that may cause pain or loss of function. Factors contributing to incomplete healing include diabetes, arterial disease, infection, and metabolic deficiencies [10]. The tissue healing process can be divided into four distinct phases: hemostasis, inflammation, proliferation, and remodeling (see Fig. 2.1).

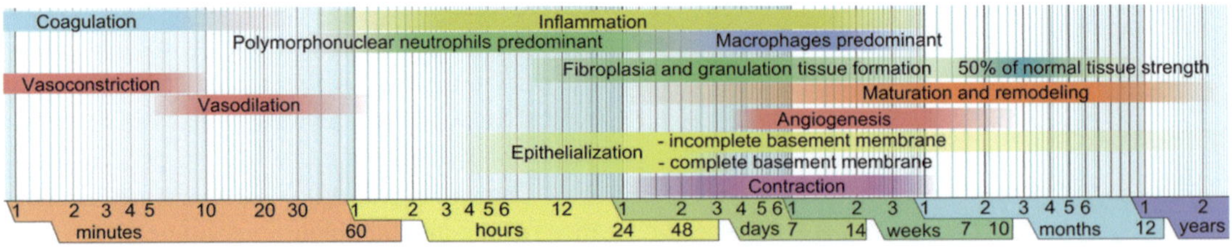

Fig. 2.1 The phases of tissue healing. Reprinted from Phases of wound healing, in *WikiJournal of Medicine*, retrieved March 30, 2024 from https://en.wikipedia.org/wiki/File:Wound_healing_phases.png#References. Copyright 2014 by Mikael Haggstrom [230]. Reprinted with permission

2.2.1 Hemostasis Phase

In the hemostasis phase, platelets in the blood become activated and adhere to the injured site within minutes following injury. Injured vessels constrict and the coagulation cascade is activated limiting blood loss. Platelets release cytokines to promote clotting by the activation of fibrin, which forms a fibrous mesh binding platelets to each other. This forms a hemostatic clot that seals the injured blood vessel and prevents further bleeding [11, 12]. The fibrin clot formed by fibrin, fibronectin, vitronectin, von Willebrand factor, and thrombospondin provides the provisional matrix for cellular migration [13]. Platelets contain alpha granules which release platelet-derived growth factor (PDGF), insulin-like growth factor-1 (IGF-1), and transforming growth factor-β (TGF-β). These proteins attract and activate fibroblasts, macrophages, and endothelial cells. The platelets also contain dense bodies that release vasoactive amines such as serotonin that increase microvascular permeability leading to tissue edema.

2.2.2 Inflammation Phase

In the inflammation phase, bacteria, damaged and dead cells, and their debris are removed via the process of leukocyte phagocytosis. The activation of complement leads to granulocyte or polymorphonuclear leukocyte infiltrates within 24–48 h. The polymorphonuclear leukocytes adhere to the endothelial cells in blood vessels by margination and exit the vessel wall by diapedesis. Bacteria in the wound are killed by the release of degrading enzymes and oxygen-derived free radical species, before being phagocytosed. Macrophages also contribute to the phagocytosis of cellular debris.

Monocytes are attracted to the wound by a variety of chemoattractants, including complement, immunoglobulin G fragments, collagen and elastin breakdown products, and the cytokines: leukotriene B4, platelet factor IV, PDGF, and TGF-β [10]. Monocytes undergo a phenotypic change and become tissue macrophages which release cytokines and growth factors recruiting fibroblasts and endothelial cells. They release collagenase to debride tissue and further stimulate inflammation with the production of other growth factors (TGF-α, heparin-binding epidermal growth factor, and basis fibroblast growth factor). After 72 h, lymphocytes are attracted to the injury by interleukin-1 (IL-1), IgG, and complement products. They may play a role in collagen and extracellular matrix remodeling [14].

2.2.3 Proliferation Phase

The proliferation phase starts about 3 days after an injury. It is the phase when angiogenesis, collagen deposition, granulation tissue formation, epithelialization, and wound contraction occur [15]. Vascular endothelial cells form new blood vessels with the stimulation of vascular endothelial growth factor (VEGF) [16]. TFG-β and PDGF, secreted by platelets, attract macrophages and granulocytes promoting early angiogenesis. The macrophages also release angiogenic substances, TNFα and bFGF. A microvascular network organizes in the fibrin clot to form granulation tissue, which is composed of proliferating fibroblasts, capillaries, tissue macrophages in a matrix of collagen, glycosaminoglycans, and glycoproteins (fibronectin and tenascin).

Fibroblasts and myofibroblasts appear between 2 and 4 days after injury. Fibroblasts are attracted by PDGF and TGF-β. They proliferate and produce extracellular matrix proteins such as fibronectin, hyaluronan (HA), collagen, and proteoglycan [10, 16]. Type I and III collagens, synthesized by the fibroblasts, provide strength and integrity to connective tissue. The extracellular matrix (ECM) supports further ingrowth of cells. Cells that are no longer producing extracellular matrix undergo apoptosis and are removed [10].

2.2.4 Remodeling Phase

In the remodeling phase the matrix matures, collagen bundles increase in diameter, and are organized along the lines of tissue tension. Collagen synthesis and the breakdown of the extracellular matrix continue until day 21 when steady state is reached. Matrix metalloproteinases (MMPs) are produced by fibroblasts, granulocytes, and macrophages to degrade collagen. The activity of tissue inhibitors of metalloproteinases (TIMPs) increases as remodeling continues. Early collagen deposition is highly disorganized but becomes more organized with wound contraction. The remodeling phase continues over a prolonged period of time leading eventually to reduced capillary growth, blood flow and metabolic activity reduction, and an acellular, avascular healed wound.

2.3 Impaired Tissue Healing

Incomplete healing may occur when any phase of the healing process is disrupted. Most chronic injuries do not progress beyond the inflammatory or proliferative phase. This leads to chronic inflammation or incomplete tissue healing causing pain and loss of function.

Altered protease activity is a biomarker of incomplete healing. In the phases of normal wound healing, the production and activity of proteases are tightly regulated, but this regulation appears to be disrupted in chronic injuries. Levels of various MMPs and serine proteases are markedly increased in fluids from chronic wounds. Levels of MMP-1 (collagenase), MMP-2 (gelatinase A), and MMP-9 (gelatinase B) have been shown to be elevated in fluid derived from pressure ulcers [17]. During granulation tissue formation in chronic ulcers, levels of MMPs have been shown to be decreased while levels of their inhibitors, TIMPs, have been found to be increased [18]. This may explain the mechanism of impaired cellular responses and extracellular matrix reorganization observed in chronic wounds.

Similarly, anabolic and catabolic imbalance in chondrocyte, subchondral bone, and extracellular matrix is seen in osteoarthritis (OA). Degradation and destruction of collagen II caused by MMP-13 is considered a key factor in the development of OA [19]. Indeed, biomarkers indicative of type II collagen (Col II) synthesis such as carboxy-peptide of Col II procollagen (CP-II), or cleaved products C2C and CTX-II, are increasingly being expressed in ratios of each other and show great promise in detecting early osteoarthritis [20].

Elevated levels of the pro-inflammatory cytokines—TNFα, interleukin-1 β (IL-1β), and TGF-β1—are found in chronic wounds [21]. Levels of these cytokines decrease as the wound heals [22]. Macrophage activation is suppressed in chronic wounds leading to an impaired inflammatory response [23]. Polymorphonuclear (PMN) cells produce more reactive oxygen species in chronic wounds [24]. The production of catabolic factors involved in both cartilage degradation and nociceptive stimulation include IL-1, IL-6, TNF-α, PGE2, FGF-2, and PKCδ [25].

Fibroblasts in chronic ulcers may not be capable of responding to growth factors such as PDGF and TGF [26], and fibroblasts isolated from chronic wounds demonstrate reduced motility or migration [27]. Free radicals are short-lived, reactive chemicals containing one or more unpaired electrons. They are classified as reactive oxygen species (ROS), reactive nitrogen species (RNS), or reactive chlorine species (RCS). In low levels, they influence cell differentiation and migration, neurotransmitters, leukocyte adhesion, thrombosis, angiogenesis, and vascular tone. But excess production can induce cellular damage by oxidizing cell components and molecules such as lipids, proteins, or DNA [28].

Free radical production results endogenously from reactions of organic compounds with oxygen or neutrophils stimulated by exposure to microbes. They are produced in the body during phagocytosis, prostaglandin synthesis, as well as in the cytochrome P450 system, respiratory chain reactions, and aerobic respiration in the mitochondria [29].

The superoxide (O_2^-) radical is one of the main reactive oxygen species in the cell deployed by the immune system to kill invading microorganisms, but is also implicated in cellular damage and aging. The superoxide dismutase enzyme catalyzes the reaction of the superoxide (O_2^-) radical to form either molecular oxygen (O_2) or hydrogen peroxide (H_2O_2) efficiently reducing oxidative stress. Modern industrialized society produces numerous exogenous sources of free radicals including ionizing radiation, lead, pesticides, cadmium, alcohol, cigarette smoke, ultraviolet light, asbestos, automobile exhaust, microwave radiation, and air pollution from fires or volcanic activity [30, 31]. Oxygen-derived free radical reactions have been associated with several diseases including Alzheimer's disease, Parkinson's disease, multiple sclerosis, amyotrophic lateral sclerosis, atherosclerosis, hypertension, ischemic heart disease, asthma, chronic obstructive pulmonary disease, rheumatoid arthritis, renal failure, and lung, breast, and hematologic cancers [32, 33].

2.4 Factors that Contribute to Impaired Healing

2.4.1 Medical Comorbidities

2.4.1.1 Diabetes

Diabetes is a chronic metabolic disease characterized by elevated levels of blood glucose and compromised insulin production and/or function. Patients with diabetes have an increased risk of comorbid conditions affecting multiple organ systems. One of the main complications of diabetes is the impairment of self-healing. This occurs through complex pathophysiological mechanisms including chronic inflammation, decreased angiogenesis and vascular maturation, a reduction of endothelial progenitor cells, and extracellular matrix dysregulation [34].

In wound healing in diabetes, inflammatory cytokines such as IL-1β and TNFα are not only elevated during the acute inflammatory phase of healing, but remain at high concentrations for a longer duration, suggesting a prolonged inflammatory response [35]. This chronic inflammatory state alters a variety of cellular and immunochemical activities implicated in delayed wound healing in patients with diabetes. Neutrophils, monocytes, macrophages, and T and B cells exhibit impaired activity which impacts their production and regulation of important cytokines and growth factors. The inhibition of various growth factors plays a critical role in the delay in progression through the phases of healing. For example, reduced levels of IGF-1 in diabetes affect the chemotaxis of endothelial cells and proliferation of keratinocytes and fibroblasts, which is implicated in impaired cell granulation and reepithelization [36].

The exposure of endothelial cells to glucose also disturbs the process of angiogenesis and vascular maturation through suppression of angiogenic factors such as VEGF and TGF-β. Dysregulation of these factors has been observed to lead to a reduction of endothelial progenitor cells in bone marrow, ultimately affecting vascular architecture in diabetic-related wounds [37]. Additionally, blood vessels stiffen in hyperglycemic environments, leading to reduced oxygenation of tissues due to slowed circulation and microvascular dysfunction [38, 39]. These vascular effects also account for reduced leukocyte migration to wounds, leaving them more vulnerable to infection [38].

Lastly, an imbalance between MMPs and tissue inhibitors of metalloproteinase (TIMPs) is observed in diabetes [40]. These proteinases play an important role in regulation of the extracellular matrix. High glucose induces the production of MMPs and reduces TIMPs, which disrupt the creation of the scaffolding required for normal wound healing.

2.4.1.2 Obesity

Obesity is defined by the World Health Organization as abnormal or excessive fat accumulation that presents a risk to health [41]. A body mass index (BMI) greater than 25 is considered overweight, and over 30 is obese. Worldwide, obesity presently affects more than 1 billion people and has more than doubled among adults between 1990 and 2022 [42]. Among the many deleterious health conditions obesity has been linked to, its role in impaired healing and musculoskeletal disease is important to discuss.

There are a variety of proposed mechanisms by which obesity leads to impaired healing. Increased adiposity has been linked to vascular insufficiency and decreased oxygen tension, decreased collagen synthesis, decreased ability to fight infection as well as nutrient-related impairment of the healing cascade [43].

Adipose tissue has been described to inherently possess less vascularity [44]. Vascular insufficiencies occur in obesity because capillary proliferation and density are not proportional to increases in adiposity [45]. Obese individuals also have a greater number of large blood vessels, which possess a greater diffusion barrier than capillaries. This relative hypoperfusion hinders the efficient delivery of oxygen to adipose tissue, thereby creating a hypoxic environment for wounds. Increased oxygen tension and hypoxia subsequently increases the risk of wound infections as well as diminishes fibroblasts' ability to make mature collagen, leading to weaker tissue and deficiencies in the healing process [46].

Obesity has also been linked to impaired angiogenesis by a variety of mechanisms. One example revolves around the anti-obesity hormone leptin, which is produced by adipocytes and leads to a reduction in caloric intake. Leptin also possesses a role in wound healing through accelerating angiogenesis and promoting keratinocyte function [47]. Obesity is associated with high levels of leptin which can cause leptin resistance and subsequently impair angiogenesis, leading to tissue ischemia [48].

Obesity has also been shown to perpetuate a chronic low-grade inflammatory process. Studies have shown that in diet-induced obesity, adipose tissue macrophages shift from an anti-inflammatory and protective M2-polarized state to a pro-inflammatory M1 state [49]. In addition to the variety of cellular and biomechanical activities affected by a chronic inflammatory state, low-level systemic inflammation has been found to play a central role in musculoskeletal diseases including osteoarthritis, tendinopathy, and osteoporosis, depicted in Fig. 2.2 [50, 51].

Excess body weight is the primary modifiable factor influencing the development of knee osteoarthritis [52, 53]. Knee pain from osteoarthritis correlates strongly with body mass index (BMI) in studies performed all over the world [54–62]. Knee osteoarthritis occurs about seven times more frequently in obese individuals than in non-obese individuals [63]. Garver et al. found that individuals in the highest BMI class had significantly more pain and poorer function than those with lower BMIs, even after controlling for radiographic osteoarthritis severity [64].

Increasing low-level activity, especially among inactive individuals, alleviates knee

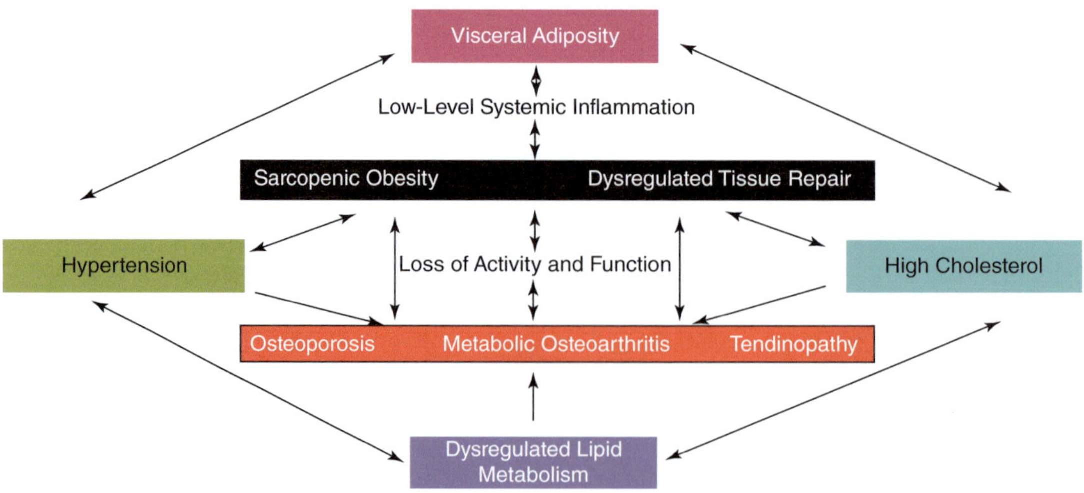

Fig. 2.2 The interface between metabolic syndrome and musculoskeletal disease. Reprinted from "Obesity, Metabolic Syndrome, and Musculoskeletal Disease: Common Inflammatory Pathways Suggest a Central Role for Loss of Muscle Integrity" by KH Collins et al., 2018, *Frontiers in physiology* vol. 9. Copyright 2018 by Kelsey Collins. Reprinted with permission

pain and reduces the risk of disability from knee osteoarthritis [52]. Randomized controlled trials show that exercise is safe and effective for overweight and obese adults with osteoarthritis [65].

Messier et al. [66] studied the effects of knee pain and body weight reduction by diet, with or without exercise in 399 older adults (>55 years) with knee osteoarthritis and body mass index of 27–41. After 18 months, lean weight loss was 11.4% for the diet/exercise group, 9.5% for the diet-alone group, and 2% for the exercise group. Pain levels and IL-6 serum concentrations were lower in the diet/exercise group when compared with the diet-alone or exercise-alone groups. The diet/exercise group also had better physical health-related quality of life scores and better function scores. Regardless of which etiological factors have the greatest effect on knee osteoarthritis, a reduction in excess body weight is the primary risk factor that can be modified to reduce knee osteoarthritis pain [52, 66].

2.4.1.3 Immunodeficiency States

The body's immune microenvironment is closely associated with tissue injury, regeneration, and healing. As such, regulation of the immune system has increasingly become a target of interest in promoting bone, cartilage, and soft tissue healing [67]. Normal wound healing requires a well-balanced and well-coordinated progression through the stages of healing. In patients with immune dysfunction or immunodeficiency, the dysregulation of immune cells, particularly in the inflammatory stage, increases the likelihood of healing failure [68].

Patients with immune diseases are known to have wounds and injuries that require a prolonged amount of time to heal compared to the general population [69]. This is the case in primary immunodeficiency disorders as well as acquired immunodeficient states such as seen in cancer, diabetes, malnutrition, stress, and aging. At their core, these diseases and factors lend to impaired healing through a shared mechanism of a blunted immune response, the specifics of which are discussed throughout various sections of this chapter.

2.4.2 Toxic Substances

2.4.2.1 Smoking

More than one billion people worldwide smoke tobacco products. Smoking is the leading cause of preventable death and causes 480,000 deaths each year in the United States alone [70]. The US Department of Health and Human Services reports that cigarettes contain at least 250 chemicals known to be harmful to humans, including hydrogen cyanide, carbon monoxide, and ammonia. Among the known harmful chemicals, at least 69 can cause cancer [70, 71].

Nicotine, present in all tobacco products, is addictive and reduces cutaneous blood flow by vasoconstriction, leading to reductions in tissue blood flow and subsequent hypoxia [72]. Nicotine also stimulates the release of proteases that may accelerate tissue destruction, suppresses the immune response, and leads to an increased risk of infection [73]. Tobacco altogether slows collagen production, weakens scar tissue, and leaves healed tissues more susceptible to the risk of recurrent injury [74]. Almost every clinical study on the effects of smoking in patients undergoing orthopedic surgery for cartilage or ligament injuries demonstrates negative outcomes in patients who smoke [75]. In patients undergoing rotator cuff repair, there is decreased tendon repair quality, decreased biomechanics, poorer clinical outcomes, and impaired healing of small to medium tears as assessed by magnetic resonance imaging in patients who smoke [76]. In cases of fracture healing, smoking tobacco is a contributory factor in delayed unions and prolonged healing times [77].

Smoking cessation is the single most effective intervention to improve healing. Heart rate, blood pressure, carbon monoxide levels, oxygenation, and blood circulation can normalize within a few weeks of quitting. Smoking cessation restores the tissue microenvironment rapidly and the inflammatory cellular functions within 4 weeks, but the proliferative response remains impaired [73]. Smoking cessation just 4 weeks prior to surgery can significantly reduce complication rates [78]. Long-term effects of quitting include decreased risk of cancer, heart disease, and pulmonary disease [79]. Assistance for patients who choose to quit smoking can be found at the National Cancer Institute's quit line, 1-800-QUIT-NOW (www.smokefree.gov).

2.4.2.2 Alcohol

Alcohol is one of the most widely used drugs in the world. The Substance Abuse and Mental Health Services Administration reported rates of alcoholism are as high as 10% in the North American population. Alcohol is a central nervous system depressant with wide-ranging and detrimental systemic effects [80]. Inappropriate alcohol consumption has been shown to have deleterious effects on responses to healing in acute and chronic injury.

Alcohol has been shown to play a role in the regulation of the immune system response. Acute alcohol consumption has inhibitory effects on the release of pro-inflammatory cytokines, whereas chronic alcohol use leads to a prolonged inflammatory response [81]. Chronic alcohol abuse is also associated with a high risk of malnutrition which results in decreased inflammatory and immune responses to tissue injury, decreased fibroblast migration, angiogenesis, and decreased type I collagen production [74]. Increased insulin resistance and higher serum sugar levels associated with alcohol consumption also impair healing [82]. Alcohol also reduces bone quality and delays fracture repairs [83, 84].

2.4.3 Drugs

2.4.3.1 NSAIDs

Nonsteroidal anti-inflammatory drugs (NSAIDs) are commonly used for their analgesic and anti-inflammatory effects. Cyclooxygenases (COX-1 and COX-2) catalyze the conversion of arachidonic acid to prostaglandins which plays a significant role in normal bone healing, osteoclastic activity, bone formation, and angiogenesis [85]. These drugs have been hypothesized to affect the healing process of bone fractures and orthopedic surgical sites. Nonsteroidal anti-inflammatory drugs (NSAIDs) have been shown to have a depressant effect on wound healing while simultaneously decreasing the granulocytic inflammatory reaction [85]. NSAIDs have an anti-proliferative effect on blood vessels and skin, resulting in a decrease in fibroblasts and thereby weakening the epithelization process and delaying wound healing rates [86, 87].

The role of NSAIDs in orthopedics remains controversial because of unclear effects on healing [88]. Delayed fracture healing and spinal fusion inhibition have been described in association with traditional NSAIDs use [89–92]. Several experimental models of bone, ligament, and tendon repair have assessed the effects of these drugs in animals with variable outcomes [93]. Celecoxib, a COX-2-specific NSAID, and indomethacin significantly inhibit tendon-to-bone healing in a rat model [94]. However, in a recent review of the literature, the short-term use of low-dose NSAIDs and COX-2 inhibitors did not appear to have a detrimental effect following soft tissue injuries [88]. Most studies on the effects of NSAIDs on healing have been completed in animal models making it difficult to translate these results to the clinical setting.

2.4.3.2 Anticoagulants

Oral anticoagulants such as warfarin and direct anticoagulants (DOACs) such as apixaban, rivaroxaban, and dabigatran are commonly prescribed for prevention of thromboembolism in a variety of disorders such as atrial fibrillation, peripheral arterial disease, and ischemic events of the central nervous and cardiac systems. Their mechanism of action centers on the inhibition of the production of coagulation factors, effectively halting the body's ability to make a clotting enzyme known as thrombin. During an initial injury, damage to endothelium precipitates the coagulation cascade through platelet activation and adhesion. The first step of wound healing, hemostasis, is mediated by platelet-derived growth factor, transforming growth factor, and other coagulation factors. Ultimately, the formation of a hemostatic plug relies on adequate amounts of thrombin [95]. Therefore, the thrombin-mediated effects of anticoagulant medication can contribute to impaired healing [96].

Nonetheless, the literature on in vivo effects of anticoagulation medication in healing remains equivocal. In a study on 264 patients with chronic leg wounds, the use of anticoagulants was found to increase the average healing time by an average of 12.5 days [87]. In a single-center retrospective study investigating patients with hip fractures, however, DOAC use was not found to influence surgical delay or length of stay compared to non-users [97]. Overall, further research is required to investigate effects on healing and an individualized approach taking into account each patient's comorbid risk factors is recommended.

2.4.3.3 Corticosteroids

Glucocorticosteroids such as prednisone and dexamethasone have been used to treat inflammation in rheumatoid and osteoarthritis since 1951. However, long-term use can delay healing by interfering with inflammation, fibroblast proliferation, collagen synthesis and degradation, deposition of connective tissue ground substances, angiogenesis, wound contraction, and re-epithelialization [98]. Subcutaneous injection of dexamethasone strongly interferes with both the synthesis and degradation of type I collagen and type III collagen [99]. Animal studies show a 30% reduction in wound tensile strength with perioperative corticosteroids at 15–40 mg/kg/day [100].

Local administration of glucocorticoid around tendons has been shown to have significant negative effects. The disruption of a normal healing response in tendons facilitates the tendon to be subject to further stress without adequate recovery time, leading to the progression of tendinopathy. At a molecular level, tendons exposed to steroids demonstrate reduced cell viability, cell proliferation, and collagen synthesis and increased apoptosis [101]. From a biomechanical level, this can cause decreased maximal load to failure and tendon stiffness [100, 101]. Frequent use of corticosteroid injections has also been associated with future tendon rupture [102].

Repeated or high-dose intra-articular steroid injections have also been shown to delay healing. Glucocorticosteroids can induce apoptosis in chondrocytes impairing cartilage homeostasis [103]. Increased chondrocyte autophagy has been described as an adaptive response to protect chondrocytes from short-term glucocorticosteroid exposure, whereas prolonged exposure decreases autophagy and increases apoptosis in vitro [103]. Yet, in vitro dexamethasone has failed to induce chondrotoxicity in low-dose exposures [104]. Recent literature has also pointed to the associated risk of intra-articular steroid injections with the accelerated progression of osteoarthritis, subchondral insufficiency fracture, osteonecrosis, as well as rapid joint destruction [105–107].

In the context of orthopedic surgery, there is some concern about the effects of preoperative corticosteroid injection and subsequent surgical healing and outcomes. One study showed that patients who had received a corticosteroid injection within 6 months prior to rotator cuff repair were much more likely to undergo a revision cuff repair within the following 3 years due to repair failure [108]. Other studies found an increased infection rate in patients who had corticosteroid injections within 1 month of arthroscopic rotator cuff repair as well as arthroscopic knee surgery [109, 110]. For overall wound healing, however, the preponderance of human literature found that high-dose corticosteroid administration for less than 10 days has no clinically important effect on wound healing, but in patients taking chronic corticosteroids for at least 30 days before surgery, wound complication rates are increased 2–5 times [100].

2.4.3.4 Immunosuppressants

Immunosuppressant drugs are often prescribed to prevent organ rejection and to treat autoimmune diseases such as lupus and other diseases such as psoriasis. Oral versions of these medications include cyclosporine, azathioprine, and mycophenolate. They work by reducing the body's immune system and primarily interfere with T cells and other inflammatory mediators, thereby impeding wound healing [111]. Immunosuppressants have also been found to delay healing by reducing fibroblast proliferation and angiogenesis [112]. An indirect contribution to the impairment of the healing process can also occur as these agents have been shown to exacerbate vasculopathic conditions such as diabetes, hypertension, and hyperlipidemia, resulting in endothelial damage and dysfunction. As such, when treating patients who are taking these medications, the associated effects on healing must be considered and counseling patients regarding potential impact on clinical outcomes is appropriate. Some have even advocated for dose reductions or avoidance of these drugs until complete healing occurs [112].

2.4.4 Stress

The effects of stress on the human body, social behavior, and healing have been well studied and documented. The pathophysiology of stress results in the deregulation of the immune system, mediated by the hypothalamic-pituitary-adrenal and sympathetic nervous systems [113]. These systems release pituitary and adrenal hormones such as cortisol and catecholamines (epinephrine and norepinephrine) which reduce the level of pro-inflammatory cytokines at the level of injury, disrupting the normal healing process [113]. Furthermore, the upregulation of cortisol and glucocorticoids influences immune cells by suppressing differentiation and proliferation, regulating gene transcription, and reducing immune cell migration [114].

Psychological stress also has a direct impact on human health behaviors. Negative emotional states such as anxiety and depression can not only directly influence endocrine and immune function, but can lead individuals to adapt unhealthy habits such as reduced sleep, poor nutrition, decreased physical activity, and an increased utilization of alcohol, tobacco, and other toxic substances [115]. As discussed in other sections of this chapter, these health behaviors can compound the impact of stress on physiological healing processes.

2.5 Factors that Improve Natural Healing

2.5.1 Sleep

Perhaps no activity influences optimal health more profoundly than sleep. In the United States, the average sleep duration has decreased steadily from 9 h per night in 1910 to 6.8 h per night [116]. The situation is even worse for patients suffering with orthopedic injuries. More than 70% of patients with lower back pain suffer from sleep disturbance when measured both subjectively and objectively [117]. Sasaki et al. measured the prevalence of sleep disturbance in patients with knee osteoarthritis finding a prevalence of 3.6%, 6.9%, 19.4%, 32.7%, and 75.0% in Kellgren-Lawrence grades 0, 1, 2, 3, and 4, respectively [118].

Sleep disruption is known to cause a range of adverse physiological and psychological effects. Poor sleep is associated with decreased cognitive performance, altered carbohydrate metabolism, and impaired endocrine function [119]. Decreased carbohydrate tolerance and increased sympathetic tone are also associated with sleep deprivation [119]. Over the long term, these effects can lead to serious health conditions such as depression, obesity, type-2 diabetes, hypertension, and coronary artery disease [120, 121].

Sleep deprivation leads to an imbalance between catabolic and anabolic hormone activity [122–124]. In humans, reduced sleep and obstructive sleep apnea adversely impact metabolic and hormonal pathways [119, 125, 126]. Poor sleep also enhances catabolic processes and is associated with increased evening cortisol levels [127, 128]. Chronic insomnia is associated with higher hypothalamic-pituitary axis (HPA) activity leading to elevated cortisol and adrenocorticotropic hormone (ACTH) levels [119, 125, 129, 130]. This in turn leads to dysregulation of the immune system and its crucial role in healing. Sleep and exercise may interact with each other by endocrine and inflammatory pathways associated with sarcopenia [131]. Exercise training in older adults possibly interacts with sleep by reducing inflammatory biomarkers, such as IL-6 and TNFα [132].

Melatonin is a neurohormone secreted by the pineal gland and it is well known for inducing and regulating sleep. Exposure to light suppresses melatonin synthesis. The primary use of melatonin as a supplement is to normalize abnormal sleep patterns. Exogenous melatonin is one of the most frequently used non-prescription sleep aids due to its role in the internal timing of biological rhythms, including promotion and regulation of sleep [133]. Melatonin may reduce the time it takes for people to fall asleep and resets the body's sleep-wake cycle [134]. Melatonin also increases total sleep time in individuals suffering from sleep restriction or altered sleep schedules and relieves daytime fatigue associated with jet lag [135].

The National Sleep Foundation's recommended dose of melatonin is 0.3–5.0 mg taken 60 min before bedtime. This supplement should only be taken as needed for occasional insomnia. Start with the lowest dose possible because synthetic melatonin is not regulated by the FDA and listed doses may not be accurate. Potential side effects include daytime drowsiness, depressed mood, irritability, abdominal pain, dizziness, and headache.

Although melatonin has important effects on sleep physiology, it may also play a role in the protection against trauma-induced cartilage degeneration. Hong et al. investigated the effects of melatonin combined with treadmill exercise on cartilage damage [136]. Intra-articular injection of collagenase in rats caused cartilage fibrillation and abnormal chondrocytes clustering and downregulated TGF-β1 mRNA levels. However, these changes were rarely seen in the melatonin alone or melatonin combined with exercise subjects. Collagenase injection caused significant serum TNF-α and IL-6 level upregulation. Melatonin, with or without exercise, significantly decreased the expression of catabolic mediators and cellular death markers with lower mineralization indicating that melatonin with exercise may have both preventive and synergistic effects in reducing cartilage degeneration.

Magnesium is an essential mineral that plays a key role in the regulation of central nervous system excitability. Magnesium deficiency is associated with poor quality sleep, increased sleep latency, predominance of light slow-wave sleep, diminution of deep slow-wave sleep, and a reduction in rapid eye movement (REM) sleep [137]. Shorter sleep time is associated with higher levels of the inflammatory markers interleukin 6 (IL-6) and C-reactive protein [138]. The highest elevations are found in patients with sleep apnea and high body mass index (BMI) [139]. Magnesium supplementation has demonstrated improved sleep quality in adults older than 51 years of age [140].

The NIH recommended daily allowance (RDA) of magnesium is 420 mg and 320 mg for men and women older than 50 years of age, respectively. It is estimated that 38% of adults have daily magnesium intakes less than the recommended allowance [141]. Magnesium deficiency is considered when serum magnesium concentrations are lower than 0.75 mmol/L. Patients with malabsorption (Crohn's, celiac disease, history of intestine surgery), type 2 diabetes, chronic alcoholism, renal disease, or those taking diuretics are at increased risk of magnesium deficiency. Green leafy vegetables, nuts, legumes, and whole grains are rich sources of magnesium. Dietary supplementation with magnesium glycinate 300 mg is preferred to other forms of supplemental magnesium due to increased bioavailability and reduced risk of side effects. Magnesium overdose can result in diarrhea, hypotension, lethargy, confusion, and cardiac rhythm abnormalities.

Sleep hygiene education is a safe and cost-effective method for managing sleep disturbance. Patients suffering from sleep disturbance should be given the following advice:

1. Get 10–15 min of morning sunlight each day. The circadian system needs bright light to reset itself. More sunlight exposure is required as you age.
2. Get a daily dose of bright sun exposure during the day to optimize melatonin production.
3. Avoid drinking caffeinated beverages several hours before bedtime. Some people do not metabolize caffeine properly and should avoid caffeine altogether.
4. Take a hot bath less than 3 h before bedtime. The core body temperature drops after a hot bath and signals that it is time to sleep.
5. After a hot bath, spend 10 min stretching or massaging muscles to minimize the likelihood of pain that triggers awakenings.
6. Avoid watching TV or using a computer at least an hour before going to bed. These devices emit blue light reducing the production of melatonin and can disrupt circadian rhythms, affecting cell cycle regulation and

metabolism. In addition, modern television rarely offers content conducive to a good night's sleep.

7. Install low-wattage yellow, orange, or red light bulbs for a light source to navigate at night. These wavelengths do not impair melatonin production to the same degree as blue light.
8. Sleep in complete darkness, cover the windows with drapes, or wear an eye mask. Light controls melatonin production in the pineal gland and even a small amount of light from a clock radio can interfere with sleep.
9. Maintain bedroom temperature to less than 70 degrees Fahrenheit (21 degrees Celsius). The optimal room temperature for sleep is between 60 to 68 degrees F (15.5–20 °C).
10. Move all electronics away from the bed at night. Electromagnetic fields found in electronic devices such as a computer or router can disrupt the pineal gland and its melatonin production.
11. Avoid using loud alarm clocks and, if possible, do not use them at all.
12. If you find that your mind is racing, consider writing your ideas down on a note pad or even spend a few minutes attending to the issue of concern.

2.5.2 Circadian Rhythms

Circadian rhythms are essentially endogenous physiological oscillations occurring at 24-h intervals. These rhythms are maintained by circadian clocks which are located in every functional cell type, including the immune system. The immune system and its regulation play a vital role in tissue healing, and disorders of immune system have a direct effect on efficacy of tissue healing. Most immune cells through their expression of circadian clock genes display profound impacts on cellular functions, including a daily rhythm in the synthesis and release of cytokines, chemokines, and cytolytic factors [142].

Alteration of circadian rhythms through clock gene mutation in many animal studies has shown altered immune response states [142]. A rat model study showed that rats with circadian rhythm disorder showed poor collagen synthesis and impaired effect of interleukin (IL)-1, IL-6, transforming growth factor α, and stress hormones on fibroblasts [143]. A molecular-level study has also shown that cardiac neutrophil oscillations have a direct effect on myocardial infarct size and healing [144]. Melatonin, a key hormone promoting circadian rhythm, has been shown to have a direct effect on skin wound healing in a mouse model. This mouse model study found transplantation of umbilical cord blood (UCB)-MSCs pre-treated with melatonin enhanced wound closure, granulation, and re-epithelialization at mouse skin wound sites [145].

In the modern era, circadian rhythm disorders have a significant impact on healing in people who travel long distances across the globe and in athletes or workers who are subjected to work night shifts. Further research in the area of human chronobiology can help in better understanding of the role of circadian rhythm and its influence on tissue healing in humans.

2.5.3 Nutrition

Nutrition has long been recognized as a keystone to proper healing. Therefore, malnutrition or specific nutrient deficiencies can have a profound impact on healing after injury or surgery.

Protein is one of the most important nutrient factors affecting the body's healing process. Decreased protein intake can negatively impact healing due to decreased collagen production, angiogenesis, and fibroblast proliferation. The essential amino acids (histidine, isoleucine, leucine, lysine, methionine, phenylalanine, tyrosine, threonine, tryptophan, valine) are not produced in the body and are only supplied by the diet. They serve crucial roles in the production of enzymes, hormones, cytokines, growth factors, and components of antibodies. Inadequate protein intake therefore impedes tissue maintenance and healing as well as the immune system, leading to increased susceptibility to infection. Insufficient protein intake can be assessed by utilizing serum albumin, pre-albumin, or total lymphocyte count. The recommended daily allowance for protein in healthy adults more than 20 years of age is 80 mg/kg of body weight. Malnourished patients may require up to 1.5 gm/kg of body weight [74]. Excellent sources of digestible protein include eggs, milk, cheese, meat, fish, rice, wheat, oatmeal, peanut butter, and beans.

Patients with decreased albumin or **zinc** levels demonstrate delayed healing and greater incidence of complications following orthopedic surgery. Zorrilla demonstrated that patients with zinc levels less than 95 μg/dL had an almost 12-fold increased risk for wound complications in patients undergoing hip hemiarthroplasty for fracture [146]. The daily recommended dose for zinc is 11 and 8 mg/day in men and women older than 50 years, respectively. Greene reported that patients with albumin level less than 3.5 g/dL showed a sevenfold increase in the risk of major wound complications following hip or knee arthroplasty [147]. Low albumin levels may be due to kidney or liver disease, poor nutrient absorption, or low protein diets. Good sources of zinc include meats, seafood, milk, cheese, nuts, legumes, and whole grains.

Vitamins also have many important roles in the process of healing. The active form of **vitamin D**, 1,25-dihydroxyvitamin D3, is essential for normal bone growth, maintenance and repair. It is essential for the health of most non-skeletal tissues, as the vitamin D receptor has been identified on almost every cell in the body. Vitamin D supplementation also significantly inhibits human mesenchymal stem cell (MSC) apoptosis and delays the development of replicative senescence in long-term cultures [148].

The optimal serum level of vitamin D is controversial and has been reported from 25 to 80 ng/mL with deficiency defined as less than 30 ng/mL [149]. By this standard, nearly 90% of healthy US adults are vitamin D deficient [150]. The prevalence of vitamin D insufficiency has been measured in 52.3% of sports medicine patients and 66.1% of trauma patients [151].

After total hip arthroplasty, patients with vitamin D deficiency have lower postoperative Harris hip scores [152]. Up to 83% of chronic lower back patients have abnormally low vitamin D levels and clinical improvement is seen with vitamin D supplementation [153]. Vitamin D insufficiency also plays a critical role in children undergoing hematopoietic stem cell transplantation for malignancy. Hansson et al. demonstrated a more rapid increase in neutrophil granulocyte levels, decreased rejection, and increased survival rates in children who were vitamin D sufficient [154].

Vitamin D is synthesized from 7-dehydrocholesterol in the skin through exposure to sunlight. Many individuals do not get enough sunlight exposure, and those living in colder climates are at even greater risk for vitamin D deficiency. Therefore, vitamin D supplementation is recommended for most adults. Singh et al. studied the effect of vitamin D supplementation on serum vitamin D levels. They noted serum concentrations are influenced by age, body mass index (BMI), and serum albumin concentration [155]. They recommend 5000 IU/day for 3 months to correct vitamin D deficiency. After 3 months, they advise repeating the vitamin D serum concentration test. Once the deficiency has been corrected, a maintenance dose of 2000 IU/day is recommended.

Vitamin C (L-ascorbic acid) is an essential water-soluble vitamin required for the biosynthesis of collagen, protein metabolism, and wound healing. It is also an important antioxidant shown to regenerate antioxidants in the body, including vitamin E (alpha-tocopherol) [156]. Good sources of vitamin C include citrus fruits, tomatoes, potatoes, peppers, broccoli, strawberries, Brussels sprouts, and cantaloupe. When necessary, vitamin C supplementation of 1 gm/day is recommended. At doses above 1 gm/day, absorption falls to less than 50% and unmetabolized ascorbic acid is excreted in the urine [156]. Each vitamin and mineral has a specific role to play in health, and once sufficient amounts of that nutrient to perform those roles are obtained, there is no benefit to consume more.

Manganese, vitamin B6 (pyridoxine), and vitamin E (alpha-tocopherol) are also linked to good tendon health; however, deficiency of these nutrients in well-nourished individuals is exceedingly rare. These essential nutrients are found in seafoods, nuts, whole wheat, seeds, tofu, chicken, turkey, and green leafy vegetables.

The development of various chronic and degenerative diseases, such as cancer [156], heart disease [157], and neuronal degeneration [158, 159], may also be attributed to oxidative stress. Oxidative stress has also been implicated in the process of aging [160] and in the pathogenesis of osteoarthritis [161]. Oxidative stress induces telomere genomic instability, replicative senescence, and dysfunction of chondrocytes in human osteoarthritic cartilage [161]. In tissue cultures of cartilage explants, nitrotyrosine staining, chondrocyte telomere length, and glycosaminoglycans (GAG) remaining in the cartilage tissue were decreased in cartilage exposed to reactive oxygen species (ROS). However, in cartilage treated with the anti-oxidative ascorbic acid, a tendency to maintain chondrocyte telomere length and proteoglycan was seen in the cartilage explants, suggesting that oxidative stress induces chondrocyte telomere instability and catabolic changes in cartilage matrix structure and composition [161].

Increased reactive oxygen species (ROS) inhibit mesenchymal stem cell (MSC) proliferation, increase senescence, enhance adipogenic differentiation, reduce osteogenic differentiation, and inhibit MSC immunomodulation [162, 163]. Furthermore, oxidative stress reduces their ex vivo expansion capabilities [163]. MSCs may have low innate antioxidant activity and are more sensitive to oxidative stress compared to more differentiated cell types [164, 165].

Antioxidants are substances capable of scavenging free radicals and preventing cellular damage. Increased consumption of whole foods such as fruits and vegetables containing natural antioxidants may offer health benefits [166, 167].

Non-enzymatic antioxidants are classified as metabolic or nutrient antioxidants. Metabolic antioxidants are produced by endogenous cellular metabolism and include coenzyme Q10, melatonin, uric acid, bilirubin, L-arginine, transferrin, and glutathione.

Nutrient antioxidants are provided in the diet and include selenium, manganese, zinc, flavonoids, vitamin E and C, and omega-3 and omega-6 fatty acids. The Oxygen Radical Absorption Capacity (ORAC) assay measures the degree of inhibition of peroxy-radical-induced oxidation by nutrient antioxidants. A review of the antioxidant activity in 326 selected foods reveals high activity in many fruits, vegetables, herbs, and spices [168]. Foods with higher ORAC values include blueberries, raspberries, strawberries, cranberries, blackberries, apples, pears, citrus fruits, gooseberries, figs, onions, broccoli, artichokes, eggplant, asparagus, garlic, thyme, oregano, basil, peppermint, black pepper, ginger, cinnamon, curry powder, cumin, paprika, peanuts, almonds, pecans, walnuts, pistachios, beans, and peas.

2.5.4 Dietary Supplements

The term "nutraceutical" was derived from the words "nutrition" and "pharmaceutical" and is defined as a food that offers health benefits [169]. The term, however, is not recognized by the US Food and Drug Administration (FDA) which uses the term "dietary supplement" instead. The FDA states that supplements are not intended to treat, diagnose, prevent, or cure diseases. Therefore, companies that produce them cannot make health claims.

Despite strong evidence of the health benefits of consuming a diet rich in whole fruits, vegetables, whole grains, and lean protein, there is little evidence for the multitude of dietary supplements offered to patients with orthopedic injuries. Despite this fact, approximately 62.5% of adults with arthritis reported taking at least one dietary supplement. Glucosamine and chondroitin sulfate are the most common supplements used by patients with osteoarthritis [170].

Eriksen et al. performed a systematic review and stratified meta-analysis of randomized placebo-controlled trials on the efficacy of glucosamine in the treatment of painful arthritis [171]. Glucosamine was shown to moderately reduce pain although a high level of between-trial inconsistency was observed. The single most important explanation was product brand. Twelve trials using the Rottapharm/Madaus product resulted in significant pain reduction although a sensitivity analysis showed less promising results.

A large National Institutes of Health (NIH) study compared glucosamine hydrochloride, chondroitin, both supplements together, celecoxib, or a placebo in patients with mild knee osteoarthritis. Those who received the celecoxib had better short-term pain relief at 6 months than those who received the placebo. Those who received glucosamine and chondroitin supplements did not receive significant improvement in knee pain or function, although the investigators saw evidence of improvement in a small subgroup of patients with moderate-to-severe pain who took glucosamine and chondroitin together [172, 173]. The American College of Rheumatology indicates that there is insufficient evidence of efficacy and does not recommend glucosamine or chondroitin sulfate for patients with hand, hip, or knee osteoarthritis [174].

Fusini et al. performed a critical review of nutraceutical supplements in the management of tendinopathy [175]. They concluded that results were encouraging but were unable to confirm a benefit from glucosamine, chondroitin sulfate, vitamin C, type 1 collagen, arginine alpha-ketoglutarate, bromelain, curcumin, boswellic acid, or methyl-sulfonyl-methane (MSM). Numerous supplements have been proposed to reduce chronic inflammation associated with joint diseases such as osteoarthritis and rheumatoid arthritis. In a review by Ghasemian et al., ginger, rosemary, borage, *Boswellia serrata*, and *Urtica dioica* (stinging nettle) have demonstrated preclinical or clinical evidence of efficacy and safety [176].

Osteoarthritis is a multifactorial process that includes articular cartilage catabolism, synovial tissue inflammation, and subchondral bone resorption. These pathological changes are associated with an excessive production of pro-inflammatory molecules such as interleukin 1-β (IL-β) and tumor necrosis factor alpha (TNFα). A review of current nutraceuticals in the management of osteoarthritis summarized the biologic effects of several neutraceuticals consumed by patients [177].

Pomegranate juice (*Punica granatum*) has strong antioxidant properties due to its high content of hydrolyzable tannin and anthocyanins, a polyphenolic compound that exhibits antioxidant and anti-inflammatory capabilities. In osteoarthritic chondrocytes, pomegranate fruit extract (PFE) has been shown to suppress COX-2 enzyme activity and IL-1β-induced prostaglandin E2 (PGE2) and nitric oxide (NO) production [178]. Interleukin 1-β (IL-1β)-induced expression of matrix metalloproteinases (MMP) 1,3, and 13 has been reduced by PFE in osteoarthritic chondrocytes as well [179]. Clinical studies of PFE efficacy in treating OA inflammation have not yet been performed.

Green tea (*Camellia sinensis*) contains catechins which have demonstrated anticancer, anti-inflammatory, antibacterial, and neuroprotective effects [177]. In particular, epigallocatechin-3-gallate (EGCG) decreases reactive oxygen species (ROS)-mediated cytotoxicity in human chondrocytes [180]. Several inflammatory mediators of inflammatory joint disease are inhibited by EGCG as well including IL-1β, TNFα, IL-6, nitric oxide, and PGE2 in human chondrocytes [181, 182]. Encouraging evidence from in vitro and animal studies demonstrates the potential anti-inflammatory and anti-arthritic effects of green tea; however, controlled clinical trials have not yet been completed.

Frankincense oil is an aromatic resin obtained from trees of the genus *Boswellia*. *Boswellia serrata* (*B. serrata*) in vitro has been reported to inhibit leukotriene biosynthesis [183], collagen IV and elastin hydrolysis [184], IL-1β-induced cartilage matrix breakdown, and nuclear factor kappa β (NFKB) in human chondrocytes [185]. Pain reduction, swelling, and disability have been demonstrated in human clinical trials using a herbomineral formulation containing *B. serrata* [186, 187], *B. serrata* extract with Aflapin or Laxin [188], or *B. serrata* 1000 mg daily [189] when compared with placebo. Bromelain is derived from the pineapple plant stems and immature fruit. Bromelain influences inflammation by decreasing PGE2 and thromboxane A2 and modulation of immune cell surface adhesion molecules [190]. Phlogenzym (which contains

90 mg bromelain, trypsin, and rutin) compared favorably to diclofenac 100–150 mg by reducing pain 80% in patients with knee osteoarthritis during a 4-week trial [191]. No serious adverse events were reported. Phlogenzym (which contains 540 mg Bromelain) showed significantly better pain relief in patients with moderate-to-severe knee osteoarthritis than diclofenac 100–150 mg during a 12-week trial [192]. Again, no serious adverse events were noted. Although bromelain has demonstrated clinical efficacy, the optimum dosage remains unclear.

Cat's claw (*Uncaria tomentosa* and *Uncaria guianensis*) is a vine found near the Amazon River with anti-inflammatory properties. *U. tomentosa* inhibits COX-1 and COX-2, TNFα, PGE2, nitric oxide production, and lipopolysaccharide (LPS)-induced iNOS gene expression [193, 194]. Pain with daily activities due to Kellgren-Lawrence grade II-III knee osteoarthritis was significantly reduced in a 4-week placebo-controlled trial of cat's claw bark extract; however, night pain was not improved [195].

Devil's claw (*Harpagophytum procumbens*) is a desert plant found in southern Africa which claims benefits for digestion, joint, and lower back pain. A dose-dependent analgesic and anti-inflammatory effect has been demonstrated in rats [196]. Devil's claw inhibits PGE2 production, TNFα, IL-1β, IL-6, MMPs, and nitric oxide [197–199]. Significant pain relief has been achieved with harpagoside 360 mg daily [200], harpagoside 57 mg daily [201], harpagoside 100 mg daily [202] when compared to placebo in patients with osteoarthritis or nonspecific lower back pain.

Turmeric (*Curcuma longa*) comes from the *C. longa* root and contains curcuminoids (curcumin, demethoxycurcumin, bis-demethoxycurcumin) which are natural polyphenols with strong antioxidant capacity. A randomized double-blind placebo-controlled trial investigated the ability of curcuminoids to reduce systemic oxidative burden in patients suffering from knee osteoarthritis [203]. Forty mild-to moderate knee osteoarthritis patients were given curcuminoid capsules 1500 mg/day in three divided doses or placebo for 6 weeks. Curcuminoids were co-administered with piperine (15 mg/day black pepper) in order to improve the bioavailability. Investigators measured the levels of superoxide dismutase (SOD) and malondialdehyde (MDA). Malondialdehyde results from lipid peroxidation of polyunsaturated fatty acids and is a product in thromboxane A2 synthesis. Serum superoxide dismutase (SOD) activity was significantly elevated and malonedialdehyde (MDA) concentration was significantly reduced in the treatment group when compared with the placebo group suggesting that short-term supplementation with curcuminoids attenuates systemic oxidative stress in patients with mild-to-moderate osteoarthritis.

Comblain performed a review of dietary supplementation in dogs with osteoarthritis. They included studies of chondroitin sulfate, glucosamine sulfate, collagen hydrolysate, avocado–soybean unsaponifiables (ASU), curcumin, and polyunsaturated fatty acids (PUFAs) [204]. They conclude that dietary supplements are promising and associated with only minor adverse effects; however, the majority of the existing studies suffered from small sample size and low statistical power.

Some physicians recommend consumption of various bovine collagen hydrolysate preparations to improve joint collagen and proteoglycan metabolism. Schadow et al. have demonstrated that various collagen hydrolysates differ with respect to their chemical composition of collagen fragments but do not stimulate the biosynthesis of collagen in humans [205]. There is evidence, however, that a leucine-rich diet combined with physical exercise stimulates collage synthesis in tendons [206].

The best evidence to date suggests that no single nutrient or supplement is sufficient to improve tendon or joint healing. Rather, a nutrient-rich diet that includes fruits, vegetables, essential fats, and proteins is recommended for optimal health.

2.5.5 Physical Activity

Optimal healing is facilitated by normalized cardiovascular health, muscle strength, and joint flexibility. According to the American College of Sports Medicine (ACSM), a program of regular exercise beyond activities of daily living is essential for most adults [207]. Adults should perform at least 150 min of moderate-intensity exercise per week. Exercise recommendations can be met through 30–60 min of moderate-intensity exercise 5 days/week or 20–60 min of vigorous-intensity exercise 3 days/week in order to maximize cardiovascular health. Gradual progression of exercise time, frequency, and intensity is recommended for best adherence and lowest risk of injury. People unable to meet these minimum requirements can still benefit from some physical activity.

Adults are also advised to engage in resistance strengthening exercises 2–3 days each week. Two to four sets of each exercise will help improve strength and power. For each exercise, 8–12 repetitions per set improve strength and power, 10–15 repetitions improve strength in middle-age and older people starting to exercise, and 15–20 repetitions improve muscular endurance. Adults should wait at least 48 h between resistance training sessions to allow for adequate recovery [207]. Static, dynamic, and ballistic stretches are all effective to maintain muscle and joint flexibility when performed 2–3 days a week. Flexibility exercises are most effective when performed after light aerobic activities to warm the muscles. Stretches should be held for 10–30 s to the point of slight discomfort for maximum effect.

Sedentary behavior, such as sitting at a computer or watching television for prolonged periods of time, has been shown to be an independent health risk. Meeting the recommended physical activity guidelines does not make up for a sedentary lifestyle.

It has been suggested that eccentric exercise programs promote collagen fiber cross-link formation and facilitate tendon remodeling [208]. The efficacy of eccentric exercise programs has been demonstrated in athletes and non-athletes with chronic Achilles tendinopathy [208, 209]. Pushing eccentric exercises through pain in patients with chronic Achilles tendinopathy was popularized in the 1990s with success rates as high as 100% [210]. Imaging of the Achilles tendon after a 12-week eccentric training protocol has demonstrated tendon structure normalization on ultrasound and MRI [211, 212]. Better results are noted with midsubstance Achilles tendinopathy when compared with insertional tendinopathy [213]. Rotator cuff tendinopathy and lateral epicondylitis also respond to strength training [214, 215].

Whether these exercise programs induce tendon healing ("healing better") or merely modify intratendinous pain sensation ("feeling better") is uncertain [216]. Evidence of histological changes following eccentric exercise is lacking and the mechanisms by which eccentric exercises help to relieve tendinopathy pain remain unclear. The best evidence to date does demonstrate that eccentric exercises are a useful management tool for tendinopathy, but this evidence is insufficient to suggest it is superior to other forms of therapeutic exercise [217].

People recovering from orthopedic injuries will have difficulty complying with all the ACSM recommendations; however, participation in regular physical activity will benefit these patients as well. The American College of Rheumatology (ACR) recommends that patients with knee osteoarthritis should consider participation in cardiovascular, resistance land-based, aquatic-based, manual therapy, or Tai Chi exercise programs [174]. Furthermore, a recent meta-analysis of 48 randomized controlled trials with a total of more than 4000 patients found that exercise therapy programs focusing on a single type of exercise are more efficacious in reducing pain and disability than those mixing several types of exercises within the same session [218]. An increased number of sessions enhances the benefits of aerobic exercise. The amount of exercise seems to be important for pain relief and disability reduction, yet greater exercise intensity did not improve the benefits of exercise. Exercise programs that only focus on knee extensor muscle strength may increase the benefits of resistance training. Exercise seems to be an effective therapy for knee osteoarthritis, regardless of age, sex, body mass index (BMI), radiographic grade, or baseline pain. For best results, the exercise program should be supervised and completed three times a week for at least 4 weeks [218]. These conclusions are supported by the Osteoarthritis Research Society International (OARSI) guidelines for non-surgical management of knee osteoarthritis [219], American Academy of Orthopedic Surgeons (AAOS) evidence-based guidelines, and European League Against Rheumatism (EULAR) recommendations for the non-pharmacological management of hip and knee osteoarthritis [220].

2.5.6 Oxygenation

Oxygen plays an essential role in the healing process of injuries and wounds. It is a requirement for various processes including collagen deposition, epithelialization, fibroplasia, angiogenesis, and resistance to infection [221]. Whether due to limited vascular supply of oxygenation, an increased oxygen demand of the tissue, or increased production of reactive species, inadequate tissue oxygenation is at the crux of injuries that do not adequately heal [222]. Several systemic conditions including advancing age and diabetes can also contribute to tissue hypoxia, leading to a microenvironment depleted of oxygen.

Hyperbaric oxygen therapy (HBOT) is a well-established adjunctive treatment for certain pathologies such as diabetic lower extremity wounds and chronic ischemic wounds. More recently, the use of HBOT has received growing interest in the field of orthopedic and sports medicine due to its potential ability to accelerate recovery in orthopedic injuries. HBOT is administered by delivering pure oxygen intermittently in a pressure vessel at higher than atmospheric pressure. This results in many physiologic changes in the body with one of the most pronounced effects being an increase of oxygen concentration in the plasma. Studies have shown HBOT has the ability to enhance oxygen delivery, reduce edema and pathologic inflammation, mitigate ischemia-reperfusion injury, stimulate collagen synthesis and deposition, and induce neovascularization and neoangiogenesis [223]. All of these effects have been proposed to enhance the recovery and rehabilitation of orthopedic injuries, especially during the inflammatory and proliferative phases of healing. Nonetheless, human trial studies remain scarce, and more rigorous research will be needed to determine the effects of HBOT on orthopedic healing.

2.5.7 Cognitive Control/Biofeedback

Restoration of lost neural or cognitive control is essential for optimizing the tissue healing process and reinjury prevention. Biofeedback is a process through which one can gain awareness of bodily functions using instruments to train, manipulate, or control the biological systems. Over the past several years, biofeedback and its methodologies were primarily applied by psychotherapists, who would use this system to treat cognitive illnesses or control a patient's behavioral patterns. More recently, this method has made its way into the area of physical rehabilitation for treating various musculoskeletal pathologies. The application of surface electromyography for biofeedback has shown promising results in treating chronic low back ache, improving gait of stroke patients, and enhancing ankle muscle strength and balance in patients with chronic ankle instability [224–226]. The use of biofeedback systems has shown significant gait improvement during postoperative periods in patients who have undergone knee and hip arthroplasty [227, 228]. In the field of sports-related injuries, biofeedback-assisted rehabilitation has shown better limb balancing after anterior cruciate ligament reconstruction [229].

2.6 Conclusion

Cells and cell products are being utilized at an accelerated pace to facilitate the regeneration of bone, tendon, and cartilage in patients with orthopedic conditions. There is strong evidence that a diet rich in fruits, vegetables, high-quality protein, and low in simple carbohydrates contributes to good health and healing. Maintenance of ideal body weight is associated with pain reduction in osteoarthritis. Consistent, gentle physical activity in all forms stimulates circulation, strengthening, and tissue healing. There is little evidence to support the use of oral glucosamine, chondroitin, bovine collagen, vitamin C, or magnesium supplementation. Herbs and spices may provide additional benefits in inflammatory conditions, but further studies are warranted.

Physicians have the opportunity to optimize their patient's health by discussing the roles of appropriate healthy behavior. Avoiding smoking, alcohol, and certain drugs and minimizing stress can help optimize healing. The specific roles of nutrition, nutrient supplementation, and oxygenation in clinical cellular-based therapies will require further examination. The information produced by these studies will allow physicians and patients to make educated decisions in order to optimize natural healing.

References

1. Bianconi E, Piovesan A, Facchin F, Beraudi A, Casadei R. Flavia Frabetti an estimation of the number of cells in the human body. Ann Hum Biol. 2013;40:6.
2. Sender R, Fuchs S, Milo R. Revised estimates for the number of human and bacteria cells in the body. PLoS Biol. 2016;14(8):e1002533. https://doi.org/10.1371/journal.pbio1002533.
3. Tero A. Achiever's biology. Allied Publishers; 1990. p. 36. ISBN 9788184243697.
4. Sackmann E. In: Lipowsky R, Sackmann E, editors. Biological membranes architecture and function., handbook of biological physics, vol. 1. Elsevier; 1995.
5. Harker LA, Roskos LK, Marzec UM, Carter RA, et al. Effects of megakaryocyte growth and development factor on platelet production, platelet life span, and platelet function in healthy human volunteers. Blood. 2000;95(8):2514–22.
6. Tate ML, Adamson JR, Tami AE, Bauer TW. Cells in focus, the osteocyte. Int J Biochem Cell Biol. 2004;36:1–8.
7. Jüni P, Hari R, Rutjes AW, Fischer R, Silletta MG, Reichenbach S, da Costa BR. Intra-articular corticosteroid for knee osteoarthritis. Cochrane Database Syst Rev. 2015;10:CD005328.
8. Ramiro S, Radner H, van der Heijde DM, Buchbinder R, Aletaha D, Landewé RB. Combination therapy for pain management in inflammatory arthritis: a Cochrane systematic review. J Rheumatol Suppl. 2012;90:47–55.
9. McElvany MD, McGoldrick E, Gee AO, Neradilek MB, Matsen FA 3rd. Rotator cuff repair: published evidence on factors associated with repair integrity and clinical outcome. Am J Sports Med. 2015;43(2):491–500.
10. Enoch S Price P. Cellular, molecular and biochemical differences in the pathophysiology of healing between acute wounds, chronic wounds and wounds in the elderly 2004.
11. Rasche H. Haemostasis and thrombosis: an overview. Eur Heart J Suppl. 2001;3(Supplement Q):Q3–7.
12. Versteeg H, Heemskerk J, Levi M, Reitsma P. New fundamentals in hemostasis. Physiol Rev. 2013;93(1):327–58.
13. Clark RA, Lanigan JM, DellaPelle P, Manseau E, Dvorak HF, Colvin RB. Fibronectin and fibrin provide a provisional matrix for epidermal cell migration during wound reepithelialization. J Invest Dermatol. 1982;79(5):264–9.
14. Sonnenberg GF, Artis D. Innate lymphoid cells in the initiation, regulation and resolution of inflammation. Nat Med. 2015;21(7):698–708.
15. Midwood KS, Williams LV, Schwarzbauer JE. Tissue repair and the dynamics of the extracellular matrix. Int J Biochem Cell Biol. 2004;36(6):1031–7. https://doi.org/10.1016/j.biocel.2003.12.003. PMID 15094118
16. Chang HY, Sneddon JB, Alizadeh AA, Sood R, West RB, Montgomery K, Chi JT, Van De Rijn M, et al. Gene expression signature of fibroblast serum response predicts human cancer progression: similarities between tumors and wounds. PLoS Biol. 2004;2(2):E7. https://doi.org/10.1371/journal.pbio.0020007.
17. Yager DR, Zhang LY, Liang HX, Diegelmann RF, Cohen IK. Wound fluids from human pressure ulcers contain elevated matrix metalloproteinase levels and activity compared to surgical wound fluids. J Invest Dermatol. 1996;107(5):743–8.
18. Cook H, Davies KJ, Harding KG, Thomas DW. Defective extracellular matrix reorganization by chronic wound fibroblasts is associated with alterations in TIMP-1, TIMP-2, and MMP-2 activity. J Invest Dermatol. 2000;115(2):225–33.
19. Chen WX, Shan FJ, Jin HT, Wang PE, Xiao LW, Tong PJ. Research on application of determination of MMP-13 in osteoarthritis. Zhongguo Gu Shang. 2014;27(7):617–20.
20. Ishijima M, Watari T, Naito K, et al. Relationships between biomarkers of cartilage, bone, synovial metabolism and knee pain provide insights into the origins of pain in early knee osteoarthritis. Arthritis Res Ther. 2011;13:R22.
21. Harris IR, Yee KC, Walters CE, Cunliffe WJ, Kearney JN, Wood EJ, Ingham E. Cytokine and protease levels in healing and non-healing chronic venous leg ulcers. Exp Dermatol. 1995;4(6):342–9.

22. Trengove N, Bielefeldt-Ohmann H, Stacey MC. Mitogenic activity and cytokine levels in nonhealing and healing chronic leg ulcers. Wound Repair Regen. 2000;8(1):13–25.
23. Moore K, Ruge F, Harding KG. T lymphocytes and the lack of activated macrophages in wound margin biopsies from chronic leg ulcers. Br J Dermatol. 1997;137(2):188–94.
24. Whiston RJ, Hallett MB, Davies EV, Harding KG, Lane IF. Inappropriate neutrophil activation in venous disease. Br J Surg. 1994;81(5):695–8.
25. Lee AS, Ellman MB, Yan D, Kroin JS, Cole BJ, van Wijnen AJ, Im HJ. A current review of molecular mechanisms regarding osteoarthritis and pain. Gene. 2013;527(2):440–7.
26. Agren MS, Steenfos HH, Dabelsteen S, Hansen JB, Dabelsteen E. Proliferation and mitogenic response to PDGF-BB of fibroblasts isolated from chronic venous leg ulcers is ulcer-age dependent. J Invest Dermatol. 1999;112(4):463–9.
27. Raffetto JD, Mendez MV, Marien BJ, Byers HR, Phillips TJ, Park HY, Menzoian JO. Changes in cellular motility and cytoskeletal actin in fibroblasts from patients with chronic venous insufficiency and in neonatal fibroblasts in the presence of chronic wound fluid. J Vasc Surg. 2001;33(6):1233–41.
28. Sen S, et al. Free radicals, antioxidants, diseases and phytomedicines: current status and future prospect. Int J Pharm Sci Rev Res. 2010;3(1):91–100.
29. Halliwell B. Biochemistry of oxidative stress. Biochem Soc Trans. 2007;35:1147–50.
30. Bagchi K, Puri S. Free radicals and antioxidants in health and disease. East Mediterr Health J. 1998;4:350–60.
31. Cadenas E. Biochemistry of oxygen toxicity. Annu Rev Biochem. 1989;58:79–110.
32. Kehrer JP, Smith CV. Free radicals in biology: sources, reactivities, and roles in the etiology of human diseases. In: Frei B, editor. Natural antioxidants in human health and disease. Academic Press: San Diego; 1994. p. 25–62.
33. Valko M, Leibfritz D, Moncol J, Cronin MTD, Mazur M, Telser J. Free radicals and antioxidants in normal physiological functions and human disease. Int J Biochem Cell Biol. 2007;39:44–84.
34. Spampinato SF, Caruso GI, De Pasquale R, Sortino MA, Merlo S. The treatment of impaired wound healing in diabetes: looking among old drugs. Pharmaceuticals. 2020;13(4):60.
35. Wetzler C, et al. Large and sustained induction of chemokines during impaired wound healing in the genetically diabetic mouse: prolonged persistence of neutrophils and macrophages during the late phase of repair. J Invest Dermatol. 2000;115(2):245–53.
36. Patel S, et al. Mechanistic insight into diabetic wounds: Pathogenesis, molecular targets and treatment strategies to pace wound healing. Biomed Pharmacother = Biomedecine & pharmacotherapie. 2019;112:108615.
37. Drela E, et al. Endothelial progenitor cells in diabetic foot syndrome. Adv Clini Exp Med. 2012;21(2):249–54.
38. Greenhalgh DG. Wound healing and diabetes mellitus. Clin Plast Surg. 2003;30:37–45.
39. Stehouwer CDA, Henry RMA, Ferreira I. Arterial stiffness in diabetes and the metabolic syndrome: a pathway to cardiovascular disease. Diabetologia. 2008;51:527–39.
40. Lobmann R, Zemlin C, Motzkau M, Reschke K, Lehnert H. Expression of matrix metalloproteinases and growth factors in diabetic foot wounds treated with a protease absorbent dressing. J Diabetes Complicat. 2006;20:329–35.
41. Obesity (2024) World Health Organization. https://www.who.int/health-topics/obesity
42. Zhou B, Chaturvedi H, Gonçalves M, et al. Worldwide trends in underweight and obesity from 1990 to 2022: a pooled analysis of 3663 population-representative studies with 222 million children, adolescents, and adults. Lancet. 2024;
43. Pierpont YN, et al. Obesity and surgical wound healing: a current review. ISRN Obesity. 2014;638936:20.
44. Wilson JA, Clark JJ. Obesity: impediment to post-surgical wound healing. Adv Skin Wound Care. 2004;17(8):426–35.
45. Khan T, Muise ES, Iyengar P, et al. Metabolic dysregulation and adipose tissue fibrosis: role of collagen VI. Mol Cell Biol. 2009;29(6):1575–91.
46. Hausberger FX, Widelitz MM. Distribution of labeled erythrocytes in adipose tissue and muscle in the rat. Am J Phys. 1963;204:649–52.
47. Tadokoro S, Ide S, Tokuyama R, Umeki H, Tatehara S, Kataoka S, Satomura K. Leptin promotes wound healing in the skin. PLoS One. 2015;10:e0121242.
48. Brown LA, Lee DE, Patton JF, Perry RA, Brown JL, Baum JI, et al. Diet-induced obesity alters anabolic signalling in mice at the onset of skeletal muscle regeneration. Acta Physiol. 2015;215:46–57. https://doi.org/10.1111/apha.12537.
49. Harford KA, Reynolds CM, McGillicuddy FC, Roche HM. Fats, inflammation and insulin resistance: insights to the role of macrophage and T-cell accumulation in adipose tissue. Proc Nutr Soc. 2011;70(4):408–17.
50. Collins KH, et al. Obesity, metabolic syndrome, and musculoskeletal disease: common inflammatory pathways suggest a central role for loss of muscle integrity. Front Physiol. 2018;9(112):23.
51. Lui PPY, Yung PSH. Inflammatory mechanisms linking obesity and tendinopathy. J Orthopaed Transl. 2021;31(80–90):9.
52. Lee J, Song J, Hootman JM. Obesity and other modifiable factors for physical inactivity measured by accelerometer in adults with knee osteoarthritis: data from the osteoarthritis initiative (OAI). Arthritis Care Res. 2013;65(1):53–61.
53. Peltonen M, Lindroos A, Torgerson J. Musculoskeletal pain in the obese: a comparison

with a general population and long-term changes after conventional and surgical obesity treatment. Pain. 2003;104(3):549–57.
54. Mork PJ, Holtermann A, Nilsen TIL. Effect of body mass index and physical exercise on risk of knee and hip osteoarthritis: longitudinal data from the Norwegian HUNT study. J Epidemiol Community Health. 2012;66:678.
55. Ho-Pham LT, Lai TQ, Mai LD. Prevalence of radiographic osteoarthritis of the knee and its relationship to self-reported pain. PLoS One. 2014;9(4):e94563.
56. Chavadaki RH, Neelkanta S. Association of obesity with osteoarthritis of knee joint in females. J Evol Med Dent Sci. 2013;2(38):7272–6.
57. Ackerman IN, Osborne RH. Obesity and increased burden of hip and knee joint disease in Australia: results from a national survey. BMC Musculoskelet Disord. 2012;13:254.
58. Jinks C, Jordan K, Croft P. Disabling knee pain—another consequence of obesity: results from a prospective cohort study. BMC Public Health. 2006;6:258.
59. Forestier R, Francon A, Briole V, et al. Prevalence of generalized osteoarthritis in a population with knee osteoarthritis. Joint Bone Spine. 2011;78(3):275–8.
60. Holla JF, van der Leeden M, Knol DL. The association of body-mass index and depressed mood with knee pain and activity limitations in knee osteoarthritis: results from the Amsterdam osteoarthritis cohort. BMC Musculoskelet Disord. 2013;14:296.
61. El Ayoubi N, Chaaya M, Mahfoud Z, et al. Risk factors for incident symptomatic knee osteoarthritis: a population based case control study in Lebanon. Int J Rheum Dis. 2013;16(2):211–8.
62. Penserga EG, Tanque AAG. Predictors of disability among Filipinos with knee osteoarthritis. Int J Rheum Dis. 2008;11(3):251–6.
63. Coggon D, Reading I, Croft P, et al. Knee osteoarthritis and obesity. Int J Obes Relat Metab Disord. 2001;25(5):622–7.
64. Garver MJ, Focht BC, Dials J, et al. Weight status and differences in mobility performance, pain symptoms, and physical activity in older, knee osteoarthritis patients. Arthritis. 2014;2014:375909.
65. Messier SP, Loeser RF, Miller GD, et al. Exercise and dietary weight loss in overweight and obese older adults with knee osteoarthritis: the arthritis, diet, and activity promotion trial. Arthritis Rheum. 2004;50:1501–10.
66. Messier SP, Mihalko SL, Legault C, et al. Effects of intensive diet and exercise on knee joint loads, inflammation, and clinical outcomes among overweight and obese adults with knee osteoarthritis: the IDEA randomized clinical. JAMA. 2013;310(12):1263–73.
67. Xiong Y, Mi BB, Lin Z, et al. The role of the immune microenvironment in bone, cartilage, and soft tissue regeneration: from mechanism to therapeutic opportunity. Military Med Res. 2022;9:65.
68. Avishai E, et al. Impaired wound healing: facts and hypotheses for multi-professional considerations in predictive, preventive and personalised medicine. EPMA J. 2017;8(1):23–33.
69. Shanmugam VK, Schilling A, Germinario A, Mete M, Kim P, Steinberg J, et al. Prevalence of immune disease in patients with wounds presenting to a tertiary wound healing center. Int Wound J. 2012;9(4):403–11.
70. U.S. Department of Health and Human Services. How tobacco smoke causes disease: the biology and behavioral basis for smoking-attributable disease: a report of the surgeon general. Atlanta, GA: U.S. Department of Health and Human Services, Centers for Disease Control and Prevention, National Center for Chronic Disease Prevention and Health Promotion, Office on Smoking and Health; 2010.
71. U.S. Department of Health and Human Services. The health consequences of smoking—50 years of progress: a report of the surgeon general, 2014. Atlanta, GA: U.S. Department of Health and Human Services, Centers for Disease Control and Prevention, National Center for Chronic Disease Prevention and Health Promotion, Office on Smoking and Health; 2014.
72. Balaji SM. Tobacco smoking and surgical healing of oral tissues: a review. Indian J Dent Res. 2008;19(4):344–8.
73. Sorensen L. Wound healing and infection in surgery: the pathophysiological impact of smoking, smoking cessation, and nicotine replacement therapy: a systematic review. Ann Surg. 2012;255(6):1069–79.
74. Anderson K, Hamm R. Factors that impair wound healing. J Am Coll Clin Wound Spec. 2012;4(4):84–91.
75. Kanneganti P, Harris J, Brophy R, Carey J, Lattermann C, Flanigan D. The effect of smoking on ligament and cartilage surgery in the knee: a systematic review. Am J Sports Med. 2012;40(12):2872–8.
76. Santiago-Torres J, Flanigan D, Butler R, Bishop J. The effect of smoking on rotator cuff and glenoid labrum surgery: a systematic review. Am J Sports Med. 2015;43(3):745–51.
77. Sloan A, Hussain I, Maqsood M, Eremin O, El-Sheemy M. The effects of smoking on fracture healing. Surgeon. 2010;8(2):111–6.
78. Truntzer J, Vopat B, Feldstein M, Matityahu A. Smoking cessation and bone healing: optimal cessation timing. Eur J Ortho Surg Traumatol. 2015;25(2):211–5.
79. U.S. Department of Health and Human Services. The health consequences of smoking: a report of the surgeon general. Atlanta, GA: U.S. Department of Health and Human Services, Centers for Disease Control and Prevention, National Center for Chronic

Disease Prevention and Health Promotion, Office on Smoking and Health; 2004.
80. Mukherjee S. Alcoholism and its effects on the central nervous system. Curr Neurovasc Res. 2013;10(3):256–62.
81. Szabo G, Mandrekar P. A recent perspective on alcohol, immunity, and host defense. Alcohol Clin Exp Res. 2009;33(2):220–32.
82. Markuson M, Hanson D, Anderson J. The relationship between hemoglobin A1c values and healing time for lower extremity ulcers in individuals with diabetes. Adv Skin Wound Care. 2009;22(8):365–72.
83. de Deco C, da Silva MA, Marchini L, da Rocha R. Extended periods of alcohol intake negatively affects osseointegration in rats. J Oral Implantol. 2015;41(3):e44–9.
84. Klein RF, Fausti KR, Carlos AS. Ethanol inhibits human osteoblastic cell proliferation. Alcohol Clin Exp Res. 1996;20(3):572–8.
85. Jones M, Wang H, Peskar B, Levin E, Itani R, Sarfeh I, et al. Inhibition of angiogenesis by nonsteroidal anti-inflammatory drugs: insight into mechanisms and implications for cancer growth and ulcer healing. Nat Med. 1999;5(12):1418–23.
86. Krischak G, Augat P, Claes L, Kinzl L, Beck A. The effects of non-steroidal anti-inflammatory drug application on incisional wound healing in rats. J Wound Care. 2007;16(2):76–8.
87. Khalil H, et al. Medications affecting healing: an evidence-based analysis. Int Wound J. 2017;14(6):1340–5.
88. Chen M, Dragoo J. The effect of nonsteroidal anti-inflammatory drugs on tissue healing. Knee Surg Sports Traumatol Arthrosc. 2013;21(3):540–9.
89. Geusens P, Emans P, de Jong JJ, van den Bergh J. NSAIDs and fracture healing. Curr Opin Rheumatol. 2013;25(4):524–31.
90. Simon A, Manigrasso M, O'Connor J. Cyclo-oxygense-2 function is essential for bone fracture healing. J Bone Miner Res. 2002;17(6):963–76.
91. Glassman S, Rose S, Dimar J, Puno R, Campbell M, Johnson J. The effect of postoperative nonsteroidal anti-inflammatory drug administration on spinal fusion. Spine. 1998;23(7):834–8.
92. Beck A, Krischak G, Sorg T, Augat P, Farker K, Merkel U, et al. Influence of diclofenac (group of nonsteroidal anti-inflammatory drugs) on fracture healing. Arch Orthop Trauma Surg. 2003;123(7):327–32.
93. Radi Z, Khan N. Effects of cyclooxygenase inhibition on bone, tendon, and ligament healing. Inflamm Res. 2005;54(9):358–66.
94. Cohen D, Kawamura S, Ehteshami J, Rodeo SA. Indomethacin and celecoxib impair rotator cuff tendon-to-bone healing. Am J Sports Med. 2006;34(3):362–9.
95. Cole W. The impact of oral anticoagulants on wound healing and development in an aging population. Podiatry Today, 2021., http://www.hmpgloballearningnetwork.com/site/podiatry/case-study/impact-oral-anticoagulants-wound-healing-and-development-aging-population.
96. Levine JM. The effect of Oral medication on wound healing. Adv Skin Wound Care. 2017;30(3):137–42.
97. Leer-Salvesen S, Dybvik E, Ranhoff AH, et al. Do direct oral anticoagulants (DOACs) cause delayed surgery, longer length of hospital stay, and poorer outcome for hip fracture patients? Eur Geriatr Med. 2020;11:563–9.
98. Anstead G. Steroids, retinoids and wound healing. Adv Wound Care. 1998;11(6):277–85.
99. Oishi Y, Fu Z, Ohnuki Y, Kato H, Noguchi T. Molecular basis of the alteration in skin collagen metabolism in response to in vivo dexamethasone treatment: effects on the synthesis of collagen type I and III, collagenase, and tissue inhibitors of metalloproteinases. Br J Dermatol. 2002;147(5):859–68.
100. Wang A, Armstrong E, Armstrong A. Corticosteroids and wound healing: clinical considerations in the perioperative period. Am J Surg. 2013;206(3):410–7.
101. Puzzitiello RN, et al. Adverse impact of corticosteroids on rotator cuff tendon health and repair: a systematic review of basic science studies. Arthrosc Sports Med Rehabil. 2020;2(2):e161–9.
102. Astrom M. Partial rupture in chronic Achilles tendinopathy: a retrospective analysis of 342 cases. Acta Orthop Scand. 1998;69(4):404–7.
103. Liu N, Wang W, Zhao Z, Zhang T, Song Y. Autophagy in human articular chondrocytes is cytoprotective following glucocorticoid stimulation. Mol Med Rep. 2014;9(6):2166–72.
104. Stueber T, Karsten J, Stoetzer C, Leffler A. Differential cytotoxic properties of drugs used for intraarticular injection on human chondrocytes: an experimental in-vitro study. Eur J Anaesthesiol. 2014;31(11):640–5.
105. Kompel A, et al. Intra-articular corticosteroid injections in the hip and knee: perhaps not as safe as we thought? Radiology. 2019;293(3):656–63.
106. Sanguino RA, et al. Prevalence of rapidly progressive osteoarthritis of the hip following intra-articular steroid injections. PM R. 2023;15(3):259–64.
107. Yasir M, Goyal A, Sonthalia S. Corticosteroid adverse effects. [Updated 2023 Jul 3]. In: StatPearls [Internet]. Treasure Island (FL): StatPearls Publishing; 2024. Available from: https://www.ncbi.nlm.nih.gov/books/NBK531462/.
108. Traven SA, et al. Preoperative shoulder injections are associated with increased risk of revision rotator cuff repair. Arthroscopy. 2019;35(3):706–13.
109. Forsythe B, Agarwalla A, Puzzitiello RN, et al. The timing of injections prior to arthroscopic rotator cuff repair impacts the risk of surgical site infection. J Bone Joint Surg Am. 2019;101:682–7.
110. Cancienne JM, Kew ME, Smith MK, et al. The timing of corticosteroid injections following simple knee arthroscopy is associated with infection risk. Arthroscopy. 2019;35:1688–94.
111. Bootun R. Effects of immunosuppressive therapy on wound healing. Int Wound J. 2013;10(1):98–104.

112. Kalluir HV, Hardinger KL. Current state of renal transplant immunosuppression: present and future. World J Transplant. 2012;2(4):51–68.
113. Godbout JP, Glaser R. Stress-induced immune dysregulation: implications for wound healing, infectious disease and cancer. J Neuroimmune Pharmacol. 2006;1:421–7.
114. Sternberg EM. Neural regulation of innate immunity: a coordinated nonspecific host response to pathogens. Nat Rev Immunol. 2006;6:318–28.
115. Gouin J-P, Kiecolt-Glaser JK. The impact of psychological stress on wound healing: methods and mechanisms. Immunol Allergy Clin N Am. 2011;31(1):81–93.
116. Jones JM. In U.S., 40% get less than recommended amount of sleep. Well Being. 2013;
117. O'Donoghue GM, Fox N, Heneghan C, Hurley D. Objective and subjective assessment of sleep inmchronic low back pain patients compared with healthy age and gender matched controls: a pilot study. BMC Musculoskelet Disord. 2009;10:122.
118. Sasaki E, Tsuda E, Yamamoto Y, Maeda S, Inoue R, Chiba D, Okubo N, Takahashi I, Nakaji S, Ishibashi Y. Nocturnal knee pain increases with the severity of knee osteoarthritis, disturbing patient sleep quality. Arthritis Care Res (Hoboken). 2014;66(7):1027–32.
119. Spiegel K, Leproult R, Van Cauter E. Impact of sleep debt on metabolic and endocrine function. Lancet. 1999;354:1435.
120. Yaggi HK, Araujo AB, McKinlay JB. Sleep duration as a risk factor for the development of type 2 diabetes. Diabetes Care. 2006;29:657–61.
121. Ayas NT, White DP, Manson JE, Stampfer MJ, Speizer FE, Malhotra A, Hu FB. A prospective study of sleep duration and coronary heart disease in women. Arch Intern Med. 2003;163:205–9.
122. Everson CA, Crowley WR. Reductions in circulating anabolic hormones induced by sustained sleep deprivation in rats. Am J Physiol Endocrinol Metab. 2004;286:E1060–70.
123. Andersen M, Martins P, D'Almeida V, Bignotto M, Tufik S. Endocrinological and catecholaminergic alterations during sleep deprivation and recovery in male rats. J Sleep Res. 2005;14:83–90.
124. Hipolide DC, Suchecki D, De Carvalho P, Pinto A, Chiconelli Faria E, Tufik S, Luz J. Paradoxical sleep deprivation and sleep recovery: effects on the hypothalamic-pituitary-adrenal axis activity, energy balance and body composition of rats. J Neuroendocrinol. 2006;18:231–8.
125. Von Treuer K, Norman T, Armstrong S. Overnight human plasma melatonin, cortisol, prolactin, TSH, under conditions of normal sleep, sleep deprivation, and sleep recovery. J Pineal Res. 1996;20:7–14.
126. Meerlo P, Sgoifo A, Suchecki D. Restricted and disrupted sleep: effects on autonomic function, neuroendocrine stress systems and stress responsivity. Sleep Med Rev. 2008;12:197–210.
127. Balbo M, Leproult R, Van Cauter E. Impact of sleep and its disturbances on hypothalamopituitary-adrenal axis activity. Int J Endocrinol. 2010;2010:759234.
128. Blackman M, Sorkin J, Munzer T, Bellantoni M, Busby-Whitehead J, Stevens T, Jayme J, O'Connor K, Christmas C, Tobin J, Stewart K, Cottrell E, Clair S, Pabst K, Harman S. Growth hormone and sex steroid administration in healthy aged women and men: a randomized controlled trial. JAMA. 2002;288:2282–92.
129. Vgontzas N, Liao D, Pejovic S, Calhoun S, Karataraki M, Bixler E. Insomnia with objective short sleep duration is associated with type 2 diabetes: a population-based study. Diabetes Care. 2009;32:1980–5.
130. Guyon A, Balbo M, Morselli L, Tasali E, Leproult R, L'hermite-Baleriaux M, Van Cauter E, Spiegel K. Adverse effects of two nights of sleep restriction on the hypothalamic-pituitary-adrenal axis in healthy men. J Clin Endocrinol Metab. 2014;99:2861–8.
131. Piovezan R, Abucham J, Thomatieli dos Santos R, Mello M, Tufik S, Poyares D. The impact of sleep on age-related sarcopenia: possible connections and clinical implications. Ageing Res Rev. 2015;23:210–20.
132. Santos R, Viana V, Boscolo R, Marques V, Santana M, Lira F, Tufik S, De Mello M. Moderate exercise training modulates cytokine profile and sleep in elderly people. Cytokine. 2012;60:731–5.
133. Wagner J, Wagner M, Hening W. Beyond benzodiazepines: alternative pharmacologic agents for the treatment of insomnia. Ann Pharmacother. 1998;32:680–91.
134. Pandi-Perumal S, Srinivasan V, Spence D, Cardinali D. Role of the melatonin system in the control of sleep: therapeutic implications. CNS Drugs. 2007;21:995–1018.
135. Costello R, Lentino C, Boyd C, O'Connell M, Crawford C, Sprengel M, Deuster P. The effectiveness of melatonin for promoting healthy sleep: a rapid evidence assessment of the literature. Nutr J. 2014;13 https://doi.org/10.1186/1475-2891-13-106.
136. Hong Y, Kim H, Lee Y, Lee S, Kim K, Jin Y, Lee SR, Chang KT, Hong Y. Salutary effects of melatonin combined with treadmill exercise on cartilage damage. J Pineal Res. 2014;57(1):53–66.
137. Popoviciu L, Babathai J, Buksa C, Delast-Popoviciu D, Bicher G, Delast-Popoviciu R, et al. Clinical and polysomnographic research in patietns with sleep disorders associated with magnesium deficiencies. In: Lasserre B, Durlach J, editors. Magnesium—a relevant ion. Eastleigh: John Libbey; 1991. p. 353–65.
138. Ferrie J, Kivimaki M, Akbaraly T, Singh-Manoux A, Milleer M, Gimeno D, et al. Associations between change in sleep duration and inflammation: findings on C-reactive protein and interleukin-6 in the Whitehall II study. Am J Epidemiol. 2013;178:956–61.
139. Vgontzas N, Tan T, Bixler E, Martin L, Shubert D, Kales A. Sleep apnea and sleep disruption in obese patients. Arch Intern Med. 1994;154:1705–11.
140. Nielsen F, Johnson L, Zeng H. Magnesium supplementation improves indicators of low magnesium

status and inflammatory stress in adults older than 51 years with poor quality sleep. Magnes Res. 2010;23:158–68.
141. Nielsen F, Lukaski H, Johnson L, Roughead Z. Reported zinc, but not copper, intakes influence whole-body bone density, mineral content and T score responses to zinc and copper supplementation in healthy postmenopausal women. Br J Nutr. 2011;106:1872–9.
142. Labrecque N, Cermakian N. Circadian clocks in the immune system. J Biol Rhythm. 2015;30(4):277–90. https://doi.org/10.1177/0748730415577723. Epub 2015 Apr 20
143. Sipahi M, Zengin K, Tanik S, Arslan E, Çubukçu A. Effects of circadian rhythm disorders on wound healing and strength of bowel anastomosis in rats. Wounds. 2014;26(11):317–22.
144. Schloss MJ, Horckmans M, Nitz K, Duchene J, Drechsler M, Bidzhekov K, Scheiermann C, Weber C, Soehnlein O, Steffens S. The time-of-day of myocardial infarction onset affects healing through oscillations in cardiac neutrophil recruitment. EMBO Mol Med. 2016;8(8):937–48. https://doi.org/10.15252/emmm.201506083.
145. Lee SJ, Jung YH, Oh SY, Yun SP, Han HJ. Melatonin enhances the human mesenchymal stem cells motility via melatonin receptor 2 coupling with Gαq in skin wound healing. J Pineal Res. 2014;57(4):393–407. https://doi.org/10.1111/jpi.12179. Epub 2014 Oct 7
146. Zorrilla P, Salido JA, Lopez-Alonso A, Silva A. Serum zinc as a prognostic tool for wound healing in hip hemiarthroplasty. Clin Orthop Relat Res. 2004;420:304–8.
147. Greene K, Wilde A, Stulberg B. Preoperative nutrional status of total joint patients. Relationship to postoperative wound complications. J Arthroplast. 1991;6(4):321–5.
148. Klotz B, Mentrup B, Regensburger M, Zeck S, Schneidereit J, Schupp N, Linden C, Merz C, Ebert R, Jakob F. 1,25-dihydroxyvitamin D3 treatment delays cellular aging in human mesenchymal stem cells while maintaining their multipotent capacity. PLoS One. 2012;7(1):e29959.
149. Kennel K, Drake M, Hurley D. Vitamin D deficiency in adults: when to test and how to treat. Mayo Clinic Proc. 2010;85(8):752–7. quiz 757-8
150. Singh G. Prevalence and implications of vitamin D deficiency in applicants for insurance. J Insur Med. 2014;44(2):103–9.
151. Bogunovic L, Kim A, Beamer B, Nguyen J, Lane J. Hypovitaminosis D in patietns scheduled to undergo orthopaedic surgery: a single-center analysis. J Bone Joint Surg Am. 2010;92(13):2300–4.
152. Nawabi D, Chin K, Keen R, Haddad F. Vitamin D deficiency in patients with osteoarthritis undergoing total hip replacement: a cause for concern? J Bone Joint Surg Br. 2010;92(4):496–9.
153. Al Faraj S, Al MK. Vitamin D deficiency and chronic low back pain in Saudi Arabia. Spine (Phila Pa 1976). 2003;28(2):177–9.
154. Hansson M, Norlin A, Omazic B, Wikstrom AC, Bergman P, Winiarski J, Remberger M, Sundin M. Vitamin D levels affect outcome in pediatric hematopoietic stem cell transplantation. 2014;20(10):1537–43.
155. Singh G, Bonham A. A predictive equation to guide vitamin D replacement dose in patients. J Am Board Fam Med. 2014;27:495–509.
156. Jacob R, Sotoudeh G. Vitamin C function and status in chronic disease. Nutr Clin Care. 2002;5:66–74.
157. Willcox B, Curb J, Rodriguez B. Antioxidants in cardiovascular health and disease: key lessons from epidemiological studies. Am J Cardiol. 2008;101(suppl):75D–86D.
158. Engelhart M, Geerlings M, Ruitenberg A, van Swieten J, Hofman A, Witteman J, Breteler M. Dietary intake of antioxidants and risk of Alzheimer disease. JAMA. 2002;287:3223–9.
159. Lang A, Lozano A. Parkinson's disease. First of two parts. New Eng J Med. 1998;339:111–4.
160. Ames B, Shigenaga M, Hagen T. Oxidants, antioxidants, and the degenerative diseases of aging. Proc Natl Acad Sci USA. 1993;90:7915–22.
161. Yudoh K, Nguyen T, Nakamura H, Hongo-Masuko K, Kato T, Nishioka K. Potential involvement of oxidative stress in cartilage senescence and development of osteoarthritis: oxidative stress induces chondrocyte telomere instability and downregulation of chondrocyte function. Arthritis Res Ther. 2005;7(2):R380–R91.
162. Choo K, Tai L, Hymavathee K, Wong C, Nguyen P, Huang C, Cheong S, Kamarul T. Oxidative stress-induced premature senescence in Wharton's jelly-derived mesenchymal stem cells. Int J Med Sci. 2014;11(11):1201–7.
163. Denu R, Hematti P. Effects of oxidative stress on mesenchymal stem cell biology. Oxid Med Cell Longev. 2016;2989076:9.
164. Orciani M, Gorbi S, Benedetti M, et al. Oxidative stress defense in human-skin-derived mesenchymal stem cells versus human keratinocytes: different mechanisms of protection and cell selection. Free Radic Biol Med. 2010;49(5):830–8.
165. Ko E, Lee K, Hwang D. Human umbilical cord blood-derived mesenchymal stem cells undergo cellular senescence in response to oxidative stress. Stem Cells Dev. 2012;21(11):1877–86.
166. Wokcik M, Burzynska-Pedziwiatr I, Wozniak L. A review of natural and synthetic antioxidants important for health and longevity. Curr Med Chem. 2010;17(28):3262–88.
167. Bielli A, Scioli M, Mazzaglia D, Doldo E, Orlandi A. Antioxidants and vascular health. Life Sci. 2015;143(15):209–16.
168. Haytowitz D, Bhagwat S. USDA database for the oxygen radical absorbance capacity (ORACO of selected foods, release 2. USDA 2010. http://www.ars.usda.gov/nutrientdata
169. Kalra E. Nutraceutical-definition and introduction. AAPS Pharm Sci. 5:E25.

170. Wilson PB. Dietary supplementation is more prevalent among adults with arthritis in the UnitedStates population. Complement Ther Med. 2016;29:152–7.
171. Eriksen P, Bartels EM, Altman RD, Bliddal H, Juhl C, Christensen R. Risk of bias and brand explain the observed inconsistency in trials on glucosamine for symptomatic relief of osteoarthritis: a meta analysis of placebo-controlled trials. Arthritis Care Res. 2014;66(12):1844–55.
172. Sawitzke AD, Shi H, Finco MF, et al. The effect of glucosamine and/or chondroitin sulfate on the progression of knee osteoarthritis: a report from the glucosamine/chondroitin arthritis intervention trial. Arthritis Rheum. 2008;58(10):3183–91.
173. Sawitzke AD, Shi H, Finco MF, et al. Clinical efficacy and safety of glucosamine, chondroitinsulphate, their combination, celecoxib or placebo taken to treat osteoarthritis of the knee: 2-year results from GAIT. Ann Rheum Dis. 2010;69(8):1459–64.
174. Hochberg MC, Altman RD, April KT, et al. American College of Rheumatology 2012 recommendations for the use of nonpharmacologic and pharmacologic therapies in osteoarthritis of the hand, hip, and knee. Arthritis Care Res. 2012;64(4):465–74.
175. Fusini F, Bisicchia S, Bottegoni C, Gigante A, Zanchini F, Busilacchi A. Nutraceutical supplement in the management of tendinopathies: a systematic review. Muscles Ligaments Tendons J. 2016;6(1):48–57.
176. Ghasemian M, Owlia S, Bagher Owlia M. Review of anti-inflammatory herbal medicines. Adv Pharmacol Sci. 2016;2016:9130979. 11
177. Akhtar N, Haqqi T. Current neutraceuticals in the management of osteoarthritis: a review. Ther Adv Musculoskelet Dis. 2012;4(3):181–207.
178. Shukla M, Gupta K, Rasheed Z, Khan K, Haqqi T. Bioavailable constituents/metabolites of pomegranate (Punica granatum L) preferentially inhibit COX2 activity ex vivo and IL-1beta-induced PGE production in human chondrocytes in vitro. J Inflamm (Lond). 2008a;5:9.
179. Ahmed S, Wang N, Hafeez B, Cheruvu V, Haqqi T. Punica granatum L. extract inhibits IL-1betainduced expression of matrix metalloproteinases by inhibiting the activation of MAP kinases and NFkappaB in human chondrocytes in vitro. J Nutr. 2005;135:2096–102.
180. Lo Y, Wong J, Cruz T. Reactive oxygen species mediate cytokine activation of c-Jun NH2-terminal kinases. J Biol Chem. 1996;271:15703–7.
181. Akhtar N, Haqqi T. Epigallocatechin-3-gallate suppresses the global interleukin-1beta-induced inflammatory response in human chondrocytes. Arthritis Res Ther. 2011;13:R93.
182. Ahmed S, Rahman A, Hasnain A, Lalonde M, Goldberg V, Haqqi T. Green tea polyphenol epigallocatechin-3-gallate inhibits the IL-1 beta-induced activity and expression of cyclooxygenase-2 and nitric oxide synthase-2 in human chondrocytes. Free Radic Biol Med. 2002;33:1097–105.
183. Gupta I, Parihar A, Malhotra P, Singh G, Lüdtke R, Safayhi H. Effects of Boswellia serrata gum resin in patients with ulcerative colitis. Eur J Med Res. 1997;2:37–43.
184. Safayhi H, Ammon H. Pharmakologische Aspecte von Weihrauch und Boswelliasäuren. Pharmazeutisch Zeitung. 1997;142:3277–80.
185. Akhtar N, Mark J, Haqqi T. Effect of a herbal-leucine mix on the IL-1β-induced cartilagedegradation and inflammatory gene expression in human chondrocytes. BMC Complement Altern Med. 2011;11:66.
186. Kulkarni R, Patki P, Jog V, Gandage S, Patwardhan B. Treatment of osteoarthritis with a herbomineral formulation: a double-blind, placebo-controlled, cross-over study. J Ethnopharmacol. 1991;33:91–5.
187. Chopra A, Lavin P, Patwardhan B, Chitre D. A 32-week randomized, placebo-controlled clinical evaluation of RA-11, an Ayurvedic drug, on osteoarthritis of the knees. J Clin Rheumatol. 2004;10:236–45.
188. Sengupta K, Krishnaraju A, Vishal A, Mishra A, Trimurtulu G, Sarma K. Comparative efficacy and tolerability of 5-Loxin® and Aflapin® against osteoarthritis of the knee: a double blind, randomized, placebo controlled clinical study. Int J Med Sci. 2010;7:366–77.
189. Kimmatkar N, Thawani V, Hingorani L, Khiyani R. Efficacy and tolerability of Boswellia serrata extract in treatment of osteoarthritis of knee—a randomized double blind placebo controlled trial. Phytomedicine. 2003;10:3–7.
190. Hale L, Greer P, Sempowski G. Bromelain treatment alters leukocyte expression of cell surface molecules involved in cellular adhesion and activation. Clin Immunol. 2002;104:183–90.
191. Klein G, Kullich W. Short-term treatment of painful osteoarthritis of the knee with oral enzymes: a randomised, double-blind study versus Diclofenac. Clin Drug Investig. 2000;19:15–23.
192. Singer F, Singer C, Oberleitner H. Phlyoenzym versus diclofenac in the treatment of activated osteoarthritis of the knee. Int J Immunother. 2001;17:135–41.
193. Aguilar J, Rojas P, Marcelo A, Plaza A, Bauer R, Reininger E. Anti-inflammatory activity of twodifferent extracts of Uncaria tomentosa (Rubiaceae). J Ethnopharmacol. 2002;81:271–6.
194. Sandoval-Chacón M, Thompson J, Zhang X, Liu X, Mannick E, Sadowska-Krowicka H. Antiinflammatory actions of cat's claw: the role of NF-kappaB. Aliment Pharmacol Ther. 1998;12:1279–89.
195. Piscoya J, Rodriguez Z, Bustamante S, Okuhama N, Miller M, Sandoval M. Efficacy and safety of freeze-dried cat's claw in osteoarthritis of the knee: mechanisms of action of the species Uncaria guianensis. Inflamm Res. 2001;50:442–8.
196. Lanhers M, Fleurentin J, Mortier F, Vinche A, Younos C. Antiinflammatory and analgesic effects

of an aqueous extract of Harpagophytum procumbens. Planta Med. 1992;58:117–23.
197. Fiebich B, Heinrich M, Hiller K, Kammerer N. Inhibition of TNF-alpha synthesis in LPS stimulated primary human monocytes by Harpagophytum extract SteiHap 69. Phytomedicine. 2001;8:28–30.
198. Huang T, Tran V, Duke R, Tan S, Chrubasik S, Roufogalis B. Harpagoside suppresses lipopolysaccharide-induced iNOS and COX-2 expression through inhibition of NF-kB activation. J Ethnopharmacol. 2005;13:11–5.
199. Schulze-Tanzil G, Hansen C, Shakibaei M. Effect of a Harpagophytum procumbens DC extract on matrix metalloproteinases in human chondrocytes in vitro. Arzneimittel-Forsch. 2004;54:213–20.
200. Chrubasik S, Thanner J, Künzel O, Conradt C, Black A, Pollak S. Comparison of outcome measures during treatment with the proprietary Harpagophytum extract Doloteffin in patients with pain in the lower back, knee, or hip. Phytomedicine. 2002;9:181–94.
201. Leblan D, Chantre P, Fournié B. Harpagophytum procumbens in the treatment of knee and hip osteoarthritis. Four-month results of a prospective, multicenter, double-blind trial versus diacerhein. Joint Bone Spine. 2000;67:462–7.
202. Brien S, Lewith G, McGregor G. Devil's claw (Harpagophytum procumbens) as a treatment for osteoarthritis: a review of efficacy and safety. J Altern Complement Med. 2006;12:981–93.
203. Panahi Y, Alishiri GH, Parvin S, Sahebkar A. Mitigation of systemic oxidative stress by curcuminoids in osteoarthritis: results of a randomized controlled trial. J Diet Suppl. 2016;13(2):209–20.
204. Comblain F, Serisier S, Barthelemy N, Balligand M, Henrotin Y. Review of dietary supplements for the management of osteoarthritis in dogs in studies from 2004 to 2014. J Vet Pharm Ther. 2016;1:1–15.
205. Schadow S, Siebert H, Lochnit G, Kordelle J, Rickert M, Steinmeyer J. Collagen metabolism of human osteoarthritic articular cartilage as modulated by bovine collagen hydrolysates. PLoS One. 2013;8(1):e53955. https://doi.org/10.1371/journal.pone.0053955. Epub 2013 Jan 16
206. Barbosa A, Benevides G, Alferes L, Salomao E, Gomes-Marcondes M, Gomes L. A leucine-rich diet and exercise affect the biomechanical characteristics of the digital flexor tendon in rats after nutritional recovery. Amino Acids. 2012;42(1):329–36.
207. Garber C, Blissman B, Deschenes M, Franklin B, Lamonte M, Lee I, Nieman D, Swain D. Quantity and quality of exercise for developing and maintaining cardiorespiratory, musculoskeletal and neuromotor fitness in apparently healthy adults: guidance for prescribing exercise. Med Sci Sports Exerc. 2011;7:1334–59.
208. Maffulli N, Walley G, Sayana M, Longo U, Denaro V. Eccentric calf muscle training in athletic patients with Achilles tendinopathy. Disabil Rehabil. 2008;30:20–2.
209. Sayana M, Maffuli N. Eccentric calf muscle training in non-athletic patients with Achilles tendinopathy. J Sci Med Sport. 2007;10(1):52–8.
210. Alfredson H, Pietila T, Jonsson P, Lorentzon R. Heavy-load eccentric calf muscle training for the treatment of chronic Achilles tendinosis. Am J Sports Med. 1998;26:360–6.
211. Ohberg L, Lorentzon R, Alfredson H. Eccentric training in patients with chronic Achilles tendinosis: normalised tendon structure and decreased thickness at follow up. Br J Sports Med. 2004;38:8–11.
212. Shalabi A, Kristoffersen-Wilberg M, Svensson L, Aspelin P, Movin T. Eccentric training of the gastrocnemius-soleus complex in chronic Achilles tendinopathy results in decreased tendon volume and intratendinous signal as evaluated by MRI. Am J Sports Med. 2004;32:1286–96.
213. Fahlstrom M, Jonsson P, Lorentzon R, Alfredson H. Chronic Achilles tendon pain treated with eccentric calf muscle training. Knee Surg Sports Traumatol Arthrosc. 2003;11:327–33.
214. Croisier J, Foidart-Dessalle M, Tinant F, Crielaard J, Forthomme B. An isokinetic eccentric programme for the management of chronic lateral epicondylar tendinopathy. Br J Sports Med. 2007;41:269–75.
215. Brox J, Staff P, Ljunggren A, Brevik J. Arthroscopic surgery compared with supervised exercises in patients with rotator cuff disease (stage II impingement syndrome). BMJ. 1993;307:899–903.
216. Maffulli N, Florio A, Osti L, Del Buono A. Achilles tendinopathy. JBJS Rev. 2014;2:9.
217. Rompe J, Nafe B, Furia J, Maffulli N. Eccentric loading, shock-wave treatment, or a wait-and-see policy for tendinopathy of the main body of tendo Achillis: a randomized controlled trial. Am J Sports Med. 2007;35:374–83.
218. Juhl C, Christensen R, Roos E, Zhang W, Lund H. Impact of exercise type and dose on pain and disability in knee osteoarthritis: a systematic review and meta-regression analysis of randomized controlled trials. Arthritis Rheumatol. 2014;66(3):622–36.
219. McAlindon T, Bannuru R, Sullivan M, Arden N, Berenbaum S, Bierma-Zeinstra S, Hawker G, Henrotin Y, Hunter D, Kawaguchi H, Kwoh K, Lohmander S, Rannou F, Roos E, Underwood M. OARSI guidelines for the non-surgical management of knee osteoarthritis. Osteoarthr Cartil. 2014;22:363–88.
220. Fernandes L, Hagen K, Bijlsma J, Andreassen O, Christensen P, Conaghan P, et al. EULAR recommendations for the non-pharmacological core management of hip and knee osteoarthritis. Ann Rheum Dis. 2013;72:1125. https://doi.org/10.1136/annrheumdis-2012-202745.
221. Castilla DM, et al. Oxygen: implications for wound healing. Adv Wound Care. 2012;1(6):225–30.
222. Sen CK. Wound healing essentials: let there be oxygen. Wound Repair Regen. 2009;17(1):1–18.

223. Moghadam N, et al. Hyperbaric oxygen therapy in sports musculoskeletal injuries. Med Sci Sports Exerc. 2020;52(6):1420–6.
224. Neblett R. Surface electromyographic (SEMG) biofeedback for chronic low back pain. Healthcare (Basel). 2016;4(2):E27. https://doi.org/10.3390/healthcare4020027.
225. Kim SJ, Cho HY, Kim KH, Lee SM. Effects of ankle biofeedback training on strength, balance, and gait in patients with stroke. J Phys Ther Sci. 2016;28(9):2596–600. Epub 2016 Sep 29
226. Donovan L, Feger MA, Hart JM, Saliba S, Park J, Hertel J. Effects of an auditory biofeedback device on plantar pressure in patients with chronic ankle instability. Gait Posture. 2016;44:29–36. https://doi.org/10.1016/j.gaitpost.2015.10.013. Epub 2015 Oct 27
227. Christiansen CL, Bade MJ, Davidson BS, Dayton MR, Stevens-Lapsley JE. Effects of weight-bearing biofeedback training on functional movement patterns following total knee arthroplasty: a randomized controlled trial. J Orthop Sports Phys Ther. 2015;45(9):647–55. https://doi.org/10.2519/jospt.2015.5593. Epub 2015 Jul 24
228. Koneva ES. The effectiveness of gait rehabilitation in the patients following endo -prosthetic hip replacement by means of the biofeedback-based hardware video reconstruction of the walking stereotype. Vopr Kurortol Fizioter Lech Fiz Kult. 2015;92(6):23–9.
229. Molka AZ, Lisiński P, Huber J. Visual biofeedback exercises for improving body balance control after anterior cruciate ligament reconstruction. J Phys Ther Sci. 2015;27(7):2357–60. https://doi.org/10.1589/jpts.27.2357. Epub 2015 Jul 22
230. Häggström M. Medical gallery of Mikael Häggström 2014. Wiki J Med. 2014;1:2. https://doi.org/10.15347/wjm/2014.008. ISSN 2002-4436

3 MSCs: A New Approach

Alberto Gobbi, Konrad Slynarski, Melanio Acosta IV, and Arnold Caplan

3.1 Introduction

MSCs or "mesenchymal stem cells" were introduced back in 1991 by Arnold Caplan. The initial theory was that MSCs did not just exist within embryos but also were found throughout adulthood. These autologous MSCs with their self-repair potential could then be the cornerstone of a new type of regenerative medicine through cell-based technology. MSCs were seen as multipotent stem cells whose progeny can give rise to different mesodermal tissues such as bone, muscle, fat, tendons, ligaments, cartilage, etc. This multipotent potential in which these cells can be driven down different tissue lineages was the initial reason Caplan termed these cells as "stem cells" [1].

Caplan saw that MSCs could have the potential to repair bone and cartilage as he observed a subset of osteochondral progenitor cells, once committed down a specific lineage, could give rise to these tissue types.

Since the introduction of MSCs, MSC-related cell-based therapies have been applied to many different areas and clinical targets such as acute kidney failure, acute myocardial infarction, amyotrophic lateral sclerosis, autism, bone marrow transplantation, burn and wound healing, graft-versus-host disease (GVHD), inflammatory bowel disease, juvenile diabetes, liver fibrosis, lung (asthma, chronic obstructive pulmonary disease), multiple sclerosis, osteoarthritis (OA), radiation syndrome, rheumatoid arthritis, sepsis, spinal cord injury, stroke, systemic lupus erythematosus, tendinitis, and urinary incontinence. In current practice, there are many tissue-derived MSCs; however, their individual function and characteristics differ depending on the origin tissue in which they were obtained. The different functions and activities of different tissue-derived MSCs can be exploited and used in various clinical applications depending on the target organ, pathology, or disease.

3.2 History of Caplan and MSCs

In 1991, Caplan introduced the world to "mesenchymal stem cells" or "MSCs" for short, where he theorized that these cells, during the embryonic stage, were involved in bone and cartilage

Arnold Caplan Passed away on January 11, 2024.

A. Gobbi (✉) · M. Acosta IV
O.A.S.I. Bioresearch Foundation Gobbi N.P.O., Milan, Italy
e-mail: gobbi@cartilagedoctor.it

K. Slynarski
Słynarski Knee Clinic, Warsaw, Poland

A. Caplan (Deceased)
Department of Biology, Case Western Reserve University School of Medicine, Cleveland, OH, USA

A. Gobbi et al. (eds.), *Regenerative Medicine in Sports and Orthopaedics*,
https://doi.org/10.1007/978-3-031-84693-9_3

formation, while being involved in repair and cell turnover in adults [1]. He proposed that these MSCs would divide, and their succeeding cells would become committed to a distinctive phenotype pathway or lineage, with steps that eventually lead to cells involved in the production of individual and specific tissue types. Through intrinsic and extrinsic factors at each step down a lineage can influence the distinct phenotype of the developing tissue, a hypothesis he later called "The Mesengenic Process" (Fig. 3.1) [2]. Through this discovery, Caplan predicted the boom of a new industry in biologic therapeutic technology and regenerative medicine through MSC's self-cell repair.

The popular concepts of MSCs by Caplan had roots in the experiments of many other researchers. As early as the 1950s and 1960s studies were documenting the osteogenic potential and the formation of bone after transplantation of bone marrow [3, 4]. Later on, it was shown that bone marrow contained osteogenic precursor cells that had high potential for differentiation and multipotency [5, 6]. These cells were identified as a sub-population of non-hematopoietic stromal cells [5]. Within these cells derived from seeding bone marrow cell suspensions, there were distinct colonies initiated by single cells, later termed as colony-forming units (CFU) [7]. *In vivo* transplantation of the progeny from these single bone marrow stromal cells was able to generate multiple tissue types (adipose tissue, bone, cartilage, and fibrous tissue) [8]. It was later theorized that there was a clinical correlation between the amount of CFUs generated and used and clinical outcomes; however, this was later proved to be not the case [9].

Although the hypothesis of the mesengenic process was decisively established and known worldwide, there were initial doubts on the concept of non-hematopoietic stem cells in bone marrow and their clinical uses. It was only

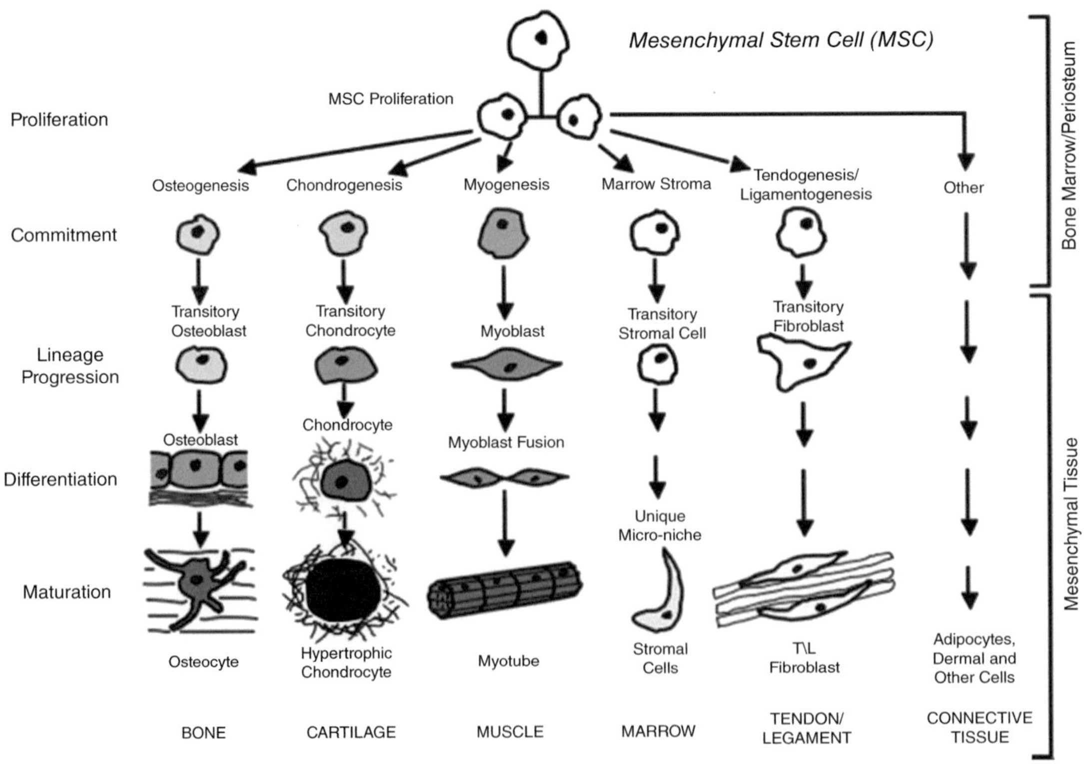

Fig. 3.1 The Mesengenic Process originated as a hypothesis to depict lineage pathways of hematopoiesis with the bone lineage. Reproduced with permission of Elsevier from [2]

through further research and work by Caplan and Pittenger and their work under Osiris Therapeutics that these cells could be the cornerstone of a new biologic technology [10]. As knowledge on MSCs grew, clinical uses were investigated. Through the use of bone marrow, heterotropic osteogenesis in porous ceramics could be achieved and was applied to the repair of bone defects in rat models [11, 12]. Autologous MSCs harvested from bone marrow and bone periosteum were then shown to be successful in rabbits which drove studies showing that autologous MSCs could be applied and used in the repair of chondral defects in humans [9, 13–15].

Further studies revealed the differentiation potential of MSCs into different types of cells, such as muscle cells, adipocytes, chondrocytes, and osteocytes [10, 16–18]. It was later shown that MSCs could be derived from tissues other than bone marrow, most notably adipose tissue, which still contained multipotency and phenotypic plasticity [19, 20]. However, the multipotent capacity of MSCs was only half the story. It was later discovered that MSCs had a trophic effect where they secrete bioactive factors that influence the local micro-environment [21–23].

With the growing understanding of MSCs, the term "mesenchymal stem cells" fell out of favor as the proper nomenclature to describe these multipotent, plastic, and secretory cells [24, 25]. Admittedly, Caplan agreed to the criticisms for his nomenclature proposed in the infancy of his research on MSCs and proposed the term "mesenchymal stem cells" be amended to "medicinal signaling cells," with emphasis in wanting to maintain the moniker "MSCs" [26, 27]. This was to better reflect the function of MSCs more accurately, especially their trophic actions, as well as their lack of "stemness" and the property of self-renewal commonly seen in stem cells. Caplan stated, "the stem cell moniker is inappropriate. Call them MSCs, but please, not stem cells" [27].

3.3 MSCs and Its Various Names

Since the introduction of Caplan's MSCs and previously named stromal cells, these culture-adherent, adult, multipotent mesenchymal cells have been referred to by many different names [27].

3.3.1 Marrow Stromal Cells

Before the term "mesenchymal stem cells," these fibroblastic cells were shown to be culturable, adhere to plastic, and possess the ability to expand. Single cells could be isolated forming colony-forming units and were shown to be osteogenic [28]. At this time, it was theorized that bone marrow stroma acted as a scaffold that could support many different derivatives of hematopoietic cell lineages; however, this theory could never be scientifically proven.

3.3.2 Multipotent Stromal Cells

This is a term used by some which describe MSCs' multipotent and plastic characteristics with the ability to differentiate into different lineages, while still having the ability to carry out other functions such as acting as sensors for damage, responding to injury, and releasing trophic factors that initiate a cascade for regeneration and reparative processes [29].

3.3.3 Mesodermal Stem Cells

This term was initially favored by Caplan and colleagues starting with their research on embryonic chick limb bug mesodermal cells (ECLBMCs) in the 1970s and 1980s. They documented that ECLBMCs had the ability to differentiate into different tissue types such as cartilage, bone, and muscle depending on the conditions when cultured [30–34]. "Mesodermal stem cells" seemed appropriate as all of the induced or bioactive factor-treated cells and tissue formed in culture were of mesodermal origin.

3.3.4 Mesenchymal Stem Cells

The original term for MSCs coined by Arnold Caplan in 1991 [1]. Caplan initially used this term as he defined mesenchyme as a type of tissue characterized by cells surrounded by a large extracellular matrix that lacks polarity. Due to his initial observation that these cells were multipotent and cloneable [10], Caplan originally theorized that these were of a similar nature to "stem cells." However, later he acknowledged that this term was not entirely accurate and later proposed a different nomenclature.

3.3.5 Mesenchymal Stromal Cells

Due to criticisms against Caplan's initial nomenclature of "stem cells," Horwitz and the International Society for Cellular Therapy used the term "mesenchymal stromal cells." They argued that Caplan's initial "mesenchymal stem cells" did not meet the generally accepted criteria for stem cell activity at that time, rendering his nomenclature scientifically inaccurate. The existence of these "MSCs" was acknowledged; however, suggestions to term them as "multipotent mesenchymal stromal cells" were made due to the fibroblast-like plastic-adherent nature of the cells [25].

3.3.6 Medicinal Signaling Cells

As knowledge grew on MSCs, most notably with their trophic and immunomodulatory effect and ability to secrete bioactive factors seen primarily after tissue injury, inflammation, or signs of disease, Caplan favored and advocated the term "medicinal signaling cells," as they act as "drug stores."

3.4 MSCs and Pericytes: The Important Relationship

Caplan's breakthrough understanding of MSCs' function in the body greatly improved with the work from Crisan et al. and their work on understanding the link between MSCs and perivascular cells, later referred to as "pericytes" [35]. Here, they showed that there were cells that had MSC markers while expressing markers for pericytes as well. This helped Caplan understand that all MSCs are related to pericytes, but the reverse is not always the case [36]. This was seen during cell sorting of pericytes and *in vitro* expansion done afterward, where the sorted cells or clones were multipotent for myogenic, adipogenic, osteogenic, and chondrogenic lineages. These characteristics were hallmark to MSCs as he knew it [37]. With the understanding of MSCs being derivatives from the differentiation of pericytes which hold *in vitro* multipotent characteristics which can be isolated from every type of vascularized tissue, this helped drive the understanding of their subsequent function [35, 36, 38].

Owing to their perivascular origin, MSCs act in two distinct ways—immunomodulatory and trophic [39]. In response to tissue injury, disease, or inflammation, MSCs secrete bioactive molecules that contribute to immunomodulatory functions and trophic activities. These bioactive molecules inhibit T cell exposure to antigens and T cell progenitor expansion. This protects the site of injury from the body's own immune system and prevents development of autoimmunity. Through the dampening of chronic inflammatory activity, activated MSCs inhibit apoptosis, especially responses due to ischemia. Activated MSCs also inhibit the formation of myofibroblasts which prevent scar formation, stimulate mitosis of tissue-intrinsic progenitors which are used to repair damaged tissue, and, stimulate and stabilize angiogenesis and the formation of vessels through pericyte functions [35, 36, 40]. Secretion of the bioactive molecules can establish a new regenerative microenvironment which can promote regeneration of the injured tissue. MSCs act as a multidrug dispensary or "drugstore" to promote and support natural regeneration [41].

Caplan explained the sequential activation of pericytes in response to injury (Fig. 3.2). At times of vessel damage, resident pericytes, in their perivascular location, are released from their contact with blood vessels and transform into MSCs which are eventually activated. After immune-activation, these cells mobilize, and as "medicinal MSCs," they secrete bioactive factors that organize and establish a powerful trophic field,

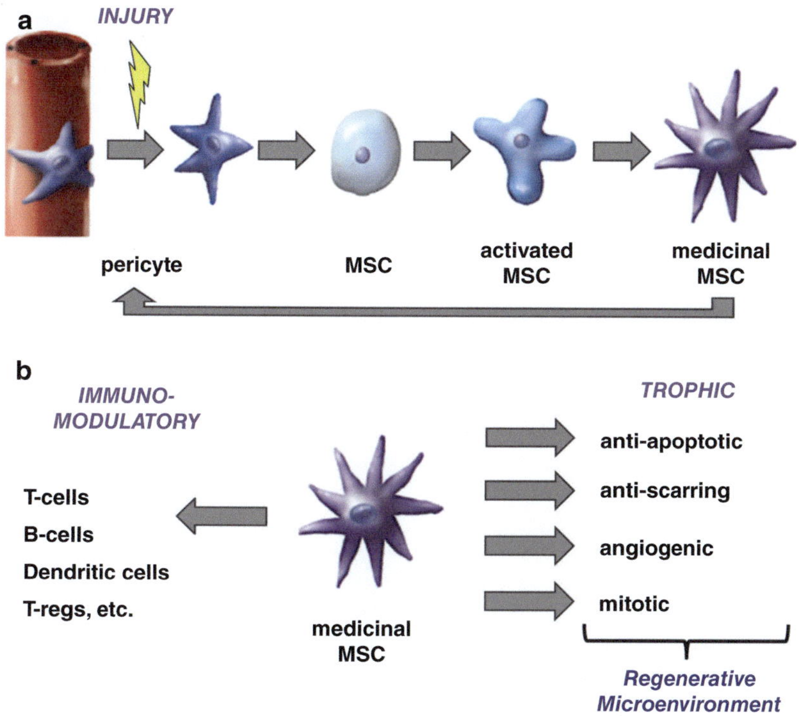

Fig. 3.2 Immunomodulatory and trophic functions of MSCs. (**a**) Sequential activation of pericytes in response to injury. (**b**) Secreted bioactive molecules by activated medicinal MSCs have both immunomodulatory and trophic effects that has action on different immune cell lineages. Reproduced with permission of Elsevier from [41]

creating a regenerative microenvironment. Subsequent repair and refabrication are strengthened when activated MSCs reacquire their stabilizing pericyte phenotype back in the abluminal space. Bioactive molecules secreted by medicinal MSCs also have an immunomodulatory effect protecting the body from self-harm [26, 41, 42].

3.5 MSCs and Immunomodulation

MSCs and their early demonstration of multipotent differentiation were the main function researchers and clinicians explored and tried to exploit in the early years of regenerative medicine with the vision to regenerate damaged tissue. However, through the decades of continuous research, emphasis has shifted to understanding and applying the MSCs' ability to produce bioactive factors and cytokines that incite tissue self-repair and modulate the body's immune responses (Table 3.1).

MSCs' immunomodulatory function is seen *in vitro* where these cells are shown to regulate the function of most effector cells involved in primary and acquired immunity [43]. Through inhibition of complement-mediated effects of mononuclear cell proliferation [44, 45], inhibiting apoptosis of activated neutrophils, and reducing neutrophils bound to endothelial cells and their ability to migrate to areas of injury [46, 47], MSCs employ their immunomodulatory effects. Activated MSCs stimulate recruitment and activation of neutrophils through the secretion of cytokines such as IL-6, IL-8, and MIF, as well as through chemoattractant mediators like anaphylotoxins and chemokines [48], While also decreasing mast cell degranulation, extension of cytoplasmic granule contents, and their secretion of pro-inflammatory cytokines [49]. When cultured with lymphocytes, MSCs caused suppression of activated CD4+ and CD8+ T cells and B-lymphocytes, reduction in the T-lymphocyte pro-inflammatory cytokines such as IFN-γ and TNF-α, and increased synthe-

Table 3.1 MicroRNA, immunomodulatory molecules, cytokine receptors, and growth factor expression of MSCs

MicroRNAs	Immunomodulatory molecules	Surface markers and cytokine receptors	Secreted growth factors and cytokines
miRNA-1224	LL37 antibac pep	SSEA-4	LIF
miRNA-486-5p	HO-1	SSEA-3	SCF
miRNA-451	Gal-9	HLA-G	GM-CSF inducible
miRNA-222-3p	Gal-1	(HLA Class II—inducible)	G-CSF
let-7a-5p	TSG-6	HLA Class I	M-CSF
miRNA-199	Inos	CD332 FGFR2	FLT-3 Ligand
miRNA-191-5p	IDO	CD331 FGFR1	FGF2
miRNA-146b	HLA-G	CD222 IGF2R	VEGF
miRNA-145	LIF	CD221 IGF1R	
miRNA-143-3p	IL10	CD166 IGF1R	
miRNA-133b	IL-6	CD146 MCAM	
miRNA-125b	IL-1RA	CD140b PDGFRB	
miRNA-29	PGE2 antibac too	CD120b TNFIIR	
miRNA-24	HGF	CD120a TNFIR	
miRNA-23b	TGF	CDw119 IFNγR	
miRNA-21		CD117 KIT	
miRNA-10b		CD106 VCAM-1	
miRNA-10a		CD105 Endoglin	**Integrins-positive**
miRNA-9-5p		CD90 Thy -1	CD104 β4
		CD73 Ectonucleotidase	CD61 β3
Hemato-negative	**Integrins-negative**	CD71 Transferrin Rec	CD29 β1
CD45	CD49d α4	CD62L L-Selectin	CD51 aα
CD34	CD18 Cβ2	CD58 LFA-3	CD49e αv
CD14	CD11a αL	CD54 ICAM-1	CD49c α3
CD11b		CD44 HA Rec	CD49b α2
CD4		CD9	CD49a α1

MSCs produce cytokines and factors that involved in innate tissue repair and immunomodulation

sis of anti-inflammatory cytokines (i.e., IL-4) [50, 51]. MSCs have the ability to inhibit the proliferation of de novo-induced NK cells, able to partially inhibit the proliferation of NK cells that are already active, and help reduce NK cells' cytotoxic action [52, 53].

3.6 Is the Future Allogenic MSCs?

When MSCs were first discovered and researched, there was excitement in the prospects of self-healing and self-regeneration. However, what if there are situations in which a patient's comorbidities cause diminished numbers of MSCs, preventing acquisition of the needed quantity or dose? Also, what if the cost of autologous MSC expansion is too great due to the need for processing and high-quality control? Patients undergoing or who had undergone chemotherapy and/or radiation therapy may have damage to their bone marrow in response to disease, malignancy, or the corresponding treatment. This may result in a reduced hematopoiesis potential and subsequently decreased ability to produce MSCs. Patients with malignancy may have reduced number of MSCs, especially those that effect the blood and bone, thus preventing clinicians from achieving the desired MSC dose [54–58].

It was initially thought that the use of allogenic MSCs would produce greater incidence of graft versus host disease in the recipient, especially if the donor is not a "perfect match" seen in identical twins or immunologically matched donors typically seen from close family members. However, this seemed to not be the case as work on allogenic MSC use showed decreased incidence or reactions of graft versus host disease [58, 59]. This may be due to the fact that MSCs do not express major histocompatibility (MHC) class II antigens on their surface, only class I. Class I molecules (HLA-A, HLA-B, HLA-C) are expressed on the surface of nucleated cells, while class II molecules (HLA-DR, HLA-DQ, HLA-DP) are expressed on immune response cells, predominantly antigen-presenting cells (APC) such as monocytes, dendritic cells, and macrophages, as well as B lymphocytes. Class II molecules present antigenic fragments to CD4+ helper T cells, while class I molecules work at the effector phase of the immune process by presenting antigens to CD8+ T cells. With lack of expression of MHC class II antigens, MSCs were shown to be non-immunogenic [60–62]. MSCs also secrete many factors that help decrease or prevent immune responses [10, 42, 59, 63–70]. In respect to macrophages and lymphocytes, MSCs incite transformation of M1 pro-inflammatory macrophages to the M2 anti-inflammatory form, reduce IgG production from B cells, reduce B lymphocyte migration through the reduction of chemokine and surface receptor expression, and prevent lymphocyte proliferation without resultant T cell apoptosis. Instead, an increased response is seen in T cells and lymphocytes when MSCs are removed [71–75].

Allogenic MSCs with their potential to be cultured and expanded can potentially be the answer. Through culture and growth in large numbers from a single donor, allogenic MSCs can be extensively tested for contamination, infection, and quality control, while allowing adequate harvest of the proper quantity of MSCs needed for management of many different patients, thus alleviating costs seen in individual processing and quality control testing when utilizing autologous MSCs.

3.7 Application of MSCs in Bio-Orthopaedics

In orthopaedics, the most common pathologies investigated in relation to treatment with MSCs include chondral defects, degenerative diseases, tendinopathies, meniscal lesions, and muscular injuries [76, 77].

With bone marrow being the primary source of MSCs, many studies have investigated their use in cartilage regeneration. MSCs derived from bone marrow aspirate concentrate (BMAC), with and without being combined with biologic scaffolds such as hyaluronic acid, proved to be an effective means of treating chondral defects with improved clinical, functional, and pain outcomes at short-term, mid-term, and long-term follow-ups. At the end of the long-term study, there was significant improvement in IKDC, KOOS, Tegner, and VAS scores at short-term follow-ups and was maintained at long-term follow-ups with a final follow-up with a mean of 8 years. Successful outcomes were seen in both patient populations <45 years old and those ≥45 years old; however, evidence may show that there is greater benefit in treatment of the younger population [78–82].

With respect to administration of MSCs, there are two general methods for MSC use—direct intra-articular injection without prior preparation or implantation of MSCs following site preparation, usually through arthroscopic means [83]. Chondral lesions and the efficacy of treatment with MSCs, especially around the knee joint, have been well established through the years [84–87]. Recent evidence has shown that there may be no correlation between clinical outcomes and the number of colony-forming units used from bone marrow aspirate concentrate when treating full-thickness cartilage lesions. Not only does this support the rationale for the feasibility and success of one-stage cartilage treatment, but it helps us further our understanding of MSCs and CFUs as a whole [9]. The safety and efficacy of MSC use have been well established; however, we are still yet to fully understand how to optimize the utilization of MSC—from identifying the best source for harvest to identifying if factors such as the quantity or functional activity of MSCs and CFUs yield different results [9, 88].

With breakthroughs in the understanding of MSCs and their relation to pericytes, it has been shown that MSCs can be harvested from almost any type of vascularized tissue due to their perivascular location. Considering factors such as ease of harvest, minimal invasiveness, and high cell concentration, adipose tissue emerged as an ideal MSC source. In fact, 1 gram of adipose tissue can yield as high as 5000 MSCs, which is 500 times more than amounts from bone marrow [89]. The use of adipose-derived MSCs alone or combined with other cartilage repair techniques such as debridement, microfracture, and biologic scaffolds has proven effective and useful in improving functional, clinical, and pain outcomes [90–96].

MSCs have been shown to be effective in other pathologies not related to cartilage. Primary ACL repair combined with MSCs has been found equally effective or superior to repair techniques without biologic augmentation in healing, function, and clinical outcomes [97–101]. There is still a scarcity of research in the efficacy of MSC use in meniscal injury and regeneration. Some studies have shown that use of MSCs is safe; however, there still is a lack high-powered studies (randomized controlled trials and cohorts) to establish a clear correlation if there are significantly improved outcomes [102–104]. MSCs have also been investigated with respect to tendinopathies, where it has been shown that there is significantly lower degrees of tendon degeneration with improved and reduced type III collagen to type I collagen ratios in Achilles tendinopathy [105] as well as decreased fatty degeneration and muscle atrophy in rotator cuff tendon injuries [106], both in rat models. Recently, MSCs have also been investigated in humans establishing their safe use; however, efficacy outcomes are still to be determined [107].

3.8 Conclusion

The emerging paradigm of MSCs as pivotal trophic mediators rather than mere progenitors marks a significant milestone in regenerative medicine. Arnold Caplan's pioneer research has been instrumental in this paradigm shift, elucidating the extensive role of MSCs in modulating the local cellular environment through the secretion of bioactive molecules. This expanded understanding emphasizes the paracrine effects of MSCs, which include immune modulation, anti-inflammatory actions, angiogenesis enhancement, and stimulation of local reparative processes, thereby highlighting their therapeutic potential beyond direct differentiation.

The concept of MSCs as trophic mediators underscores their ability to create a conducive microenvironment for healing and regeneration, making them valuable in treating various musculoskeletal conditions. This chapter has explored the multifaceted roles of MSCs in tissue repair and regeneration, illustrating how they influence cellular pathways and examining their application in various clinical contexts. If harnessed effectively, the potential of MSC-based therapies could revolutionize treatment strategies for diseases and injuries, underscoring the need for further research and standardized protocols.

The clinical application of MSCs has shown promising results in chondral lesions and osteoarthritis as evidenced by improved clinical outcomes and delayed disease progression. Studies involving intra-articular injections of MSCs, often combined with other cartilage repair techniques or agents such as platelet-rich plasma (PRP), have demonstrated significant potential in enhancing cartilage quality and function. The use of MSCs in treatment of meniscal injury, tendon repair, and the treatment of chronic tendinopathies reflects their versatility and broad therapeutic potential. Research has consistently shown that MSCs can improve structural and functional outcomes in various orthopaedic applications, supporting their potential for integration into clinical practice. Despite the encouraging results, it is crucial to acknowledge the limitations of current studies, including relatively short observation periods and a lack of extensive randomized trials. Long-term follow-up studies and comprehensive evaluations are necessary to understand the durability and efficacy of MSC-based therapies fully.

The evolving understanding of MSCs as trophic mediators has profound implications for regenerative medicine. By continuing to refine our knowledge and application of these therapies, supported by rigorous scientific research and consensus-driven guidelines, we can unlock the full potential of MSCs to transform the treatment landscape for knee osteoarthritis and other degenerative musculoskeletal conditions. The collaborative efforts of researchers, clinicians, and regulatory bodies will be essential in advancing MSC-based therapies and ensuring their safe and effective integration into clinical practice.

We would like to conclude this chapter by recalling important lessons learned and enumerated by Arnold Caplan during his 50th career anniversary celebration back in 2019 [108]:

1. The single most important human characteristic that can bring success is perseverance.
2. Perseverance requires an open mind: repeating the same wrong approach will still be wrong. An open mind sees the many solutions to the problem.
3. Curiosity beats repetition but repetition brings reliability. If you cannot repeat it, it does not exist.
4. If it works the first time, you will never be able to repeat it. Then it is up to you to make it work.
5. Dogma is the "truth" of the moment. The more that you work on the problem, the more that the dogma is challenged.
6. Money is both the source of intellectual freedom and also the bondage that muffles creativity.

This chapter was written in loving memory of Dr. Arnold Caplan, a loving husband, father, colleague, researcher, and friend

References

1. Caplan AI. Mesenchymal stem cells. J Orthop Res. 1991;9(5):641–50. https://doi.org/10.1002/jor.1100090504.
2. Caplan AI. The mesengenic process. Clin Plast Surg. 1994;21(3):429–35.
3. Denis A. Etude de l'ossification dans les graftes de moelle osseuse. 1958. Bruxelles.
4. Tavassoli M, Crosby WH. Transplantation of marrow to extramedullary sites. Science (New York, N.Y.). 1968;161(3836):54–6. https://doi.org/10.1126/science.161.3836.54.
5. Fridenshteĭn AIA. Stvolovye osteogennye kletki kostnogo mozga [Osteogenic stem cells of the bone marrow]. Ontogenez. 1991;22(2):189–97.
6. Ashton BA, Allen TD, Howlett CR, Eaglesom CC, Hattori A, Owen M. Formation of bone and cartilage by marrow stromal cells in diffusion chambers in vivo. Clin Orthop Relat Res. 1980:294–307. https://doi.org/10.1097/00003086-198009000-00040.
7. Friedenstein AJ, Chailakhjan RK, Lalykina KS. The development of fibroblast colonies in monolayer cultures of Guinea-pig bone marrow and spleen cells. Cell Tissue Kinet. 1970;3(4):393–403. https://doi.org/10.1111/j.1365-2184.1970.tb00347.x.
8. Friedenstein AJ. Osteogenic stem cells in the bone marrow. In: Heersche JNM, Kanis JA, editors. Bone and mineral research, vol. 7. New York: Elsevier Science; 1990. p. 243–72.
9. Gobbi A, Karnatzikos G, Scotti C, Mahajan V, Mazzucco L, Grigolo B. One-step cartilage repair with bone marrow aspirate concentrated cells and collagen matrix in full-thickness knee cartilage lesions: results at 2-year follow-up. Cartilage. 2011;2:286–99. https://doi.org/10.1177/1947603510392023.
10. Pittenger MF, Mackay AM, Beck SC, Jaiswal RK, Douglas R, Mosca JD, Moorman MA, Simonetti DW, Craig S, Marshak DR. Multilineage potential of adult human mesenchymal stem cells. Science (New York, N.Y.). 1999;284(5411):143–7. https://doi.org/10.1126/science.284.5411.143.
11. Ohgushi H, Goldberg VM, Caplan AI. Heterotopic osteogenesis in porous ceramics induced by marrow cells. J Orthop Res. 1989;7:568–78. https://doi.org/10.1002/jor.1100070415.
12. Ohgushi H, Goldberg VM, Caplan AI. Repair of bone defects with marrow cells and porous ceramic. Experiments in rats. Acta Orthop Scand. 1989;60:334–9. https://doi.org/10.3109/17453678909149289.
13. Wakitani S, Goto T, Pineda SJ, Young RG, Mansour JM, Caplan AI, et al. Mesenchymal cell-based repair of large, full-thickness defects of articular cartilage. J Bone Joint Surg Am. 1994;76:579–92. https://doi.org/10.2106/00004623-199404000-00013.
14. Wakitani S, Imoto K, Yamamoto T, Saito M, Murata N, Yoneda M. Human autologous culture expanded bone marrow mesenchymal cell transplantation for

repair of cartilage defects in osteoarthritic knees. Osteoarthr Cartil. 2002;10:199–206. https://doi.org/10.1053/joca.2001.0504.
15. Kamei N, Ochi M, Adachi N, Ishikawa M, Yanada S, Levin LS, et al. The safety and efficacy of magnetic targeting using autologous mesenchymal stem cells for cartilage repair. Knee Surg Sports Traumatol Arthrosc. 2018;26:3626–35. https://doi.org/10.1007/s00167-018-4898-2.
16. Wakitani S, Saito T, Caplan AI. Myogenic cells derived from rat bone marrow mesenchymal stem cells exposed to 5-azacytidine. Muscle Nerve. 1995;18:1417–26. https://doi.org/10.1002/mus.880181212.
17. Saito T, Dennis JE, Lennon DP, Young RG, Caplan AI. Myogenic expression of mesenchymal stem cells within myotubes of mdx mice in vitro and in vivo. Tissue Eng. 1995;1:327–43. https://doi.org/10.1089/ten.1995.1.327.
18. Johnstone B, Hering TM, Caplan AI, Goldberg VM, Yoo JU. In vitro chondrogenesis of bone marrow-derived mesenchymal progenitor cells. Exp Cell Res. 1998;238:265–72. https://doi.org/10.1006/excr.1997.3858.
19. Tallheden T, Dennis JE, Lennon DP, Sjogren-Jansson E, Caplan AI, Lindahl A. Phenotypic plasticity of human articular chondrocytes. J Bone Joint Surg Am. 2003;85 A(Suppl 2):93–100. https://doi.org/10.2106/00004623-200300002-00012.
20. Zuk PA, Zhu M, Mizuno H, Huang J, Futrell JW, Katz AJ, et al. Multilineage cells from human adipose tissue: implications for cell-based therapies. Tissue Eng. 2001;7:211–28. https://doi.org/10.1089/107632701300062859.
21. Haynesworth SE, Baber MA, Caplan AI. Cytokine expression by human marrowderived mesenchymal progenitor cells in vitro: effects of dexamethasone and IL-1 alpha. J Cell Physiol. 1996;166:585–92. https://doi.org/10.1002/(SICI)10974652(199603)166:3<585::AID-JCP13>3.0.CO;2-6.
22. Caplan AI, Dennis JE. Mesenchymal stem cells as trophic mediators. J Cell Biochem. 2006;98: 1076–84. https://doi.org/10.1002/jcb.20886.
23. Dennis JE, Caplan AI. Advances in mesenchymal stem cell biology. Curr Opin Orthop. 2004;15:341–6.
24. Dominici M, Le Blanc K, Mueller I, Slaper-Cortenbach I, Marini F, Krause D, Deans R, Keating A, Prockop DJ, Horwitz E. Minimal criteria for defining multipotent mesenchymal stromal cells. The International Society for Cellular Therapy position statement. Cytotherapy. 2006;8(4):315–7. https://doi.org/10.1080/14653240600855905.
25. Horwitz EM, Le Blanc K, Dominici M, Mueller I, Slaper-Cortenbach I, Marini FC, Deans RJ, Krause DS, Keating A, International Society for Cellular Therapy. Clarification of the nomenclature for MSC: the International Society for Cellular Therapy position statement. Cytotherapy. 2005;7(5):393–5. https://doi.org/10.1080/14653240500319234.
26. Caplan AI. What's in a name? Tissue Eng. 2010;16:2415–7. https://doi.org/10.1089/ten.TEA.2010.0216.
27. Caplan AI. Mesenchymal stem cells: time to change the name. Stem Cells Transl Med. 2017;6:1445–51. https://doi.org/10.1002/sctm.17-0051.
28. Owen M. Marrow stromal stem cells. J Cell Sci Suppl. 1988;10:63–76.
29. Soliman H, Theret M, Scott W, Hill L, Underhill TM, Hinz B, Rossi FMV. Multipotent stromal cells: One name, multiple identities. Cell Stem Cell. 2021;28(10):1690–707. https://doi.org/10.1016/j.stem.2021.09.001.
30. Caplan AI, Zwilling E, Kaplan NO. 3-acetylpyridine: effects in vitro related to teratogenic activity in chicken embryos. Science. 1968;160:1009–10.
31. Caplan AI, Rosenberg MJ. Interrelationship between poly (ADP-Rib) synthesis, intracellular NAD levels, and muscle or cartilage differentiation from mesodermal cells of embryonic chick limb. Proc Natl Acad Sci USA. 1975;72:1852–7.
32. Caplan AI. The molecular control of muscle and cartilage development. In: Subtelney S, Abbott U, Alan R, editors. 39th annual symposium of the Society for Developmental Biology. New York: Liss, Inc.; 1981. p. 37–68.
33. Caplan AI, Koutroupas S. The control of muscle and cartilage development in the chick limb: the role of differential vascularization. J Embryol Exp Morphol. 1973;29:571–83.
34. Caplan AI. Molecular basis for limb morphogenesis. In: Littlefield J, DeGrouchy J, editors. Proceedings of the 5th International Conference on Birth Defects. Amsterdam-Oxford: Excerp. Medica; 1978. p. 208–20.
35. Crisan M, Yap S, Casteilla L, Chen CW, Corselli M, Park TS, Andriolo G, Sun B, Zheng B, Zhang L, Norotte C, Teng PN, Traas J, Schugar R, Deasy BM, Badylak S, Buhring HJ, Giacobino JP, Lazzari L, Huard J, et al. A perivascular origin for mesenchymal stem cells in multiple human organs. Cell Stem Cell. 2008;3(3):301–13. https://doi.org/10.1016/j.stem.2008.07.003.
36. Caplan AI. All MSCs are pericytes? Cell Stem Cell. 2008;3(3):229–30. https://doi.org/10.1016/j.stem.2008.08.008.
37. Caplan AI. Adult mesenchymal stem cells for tissue engineering versus regenerative medicine. J Cell Physiol. 2007;213(2):341–7. https://doi.org/10.1002/jcp.21200.
38. Nombela-Arrieta C, Ritz J, Silberstein LE. The elusive nature and function of mesenchymal stem cells. Nat Rev Mol Cell Biol. 2011;12:126–31.
39. Krasnodembskaya A, Song Y, Fang X, et al. Antibacterial effect of human mesenchymal stem cells is mediated in part from secretion of the antimicrobial peptide LL-37. Stem Cells. 2010;28:2229–38.
40. da Silva Meirelles L, Caplan AI, Nardi NB. In search of the in vivo identity of mesenchymal stem cells.

Stem Cells (Dayton, Ohio). 2008;26(9):2287–99. https://doi.org/10.1634/stemcells.2007-1122.
41. Caplan AI, Correa D. The MSC: an injury drugstore. Cell Stem Cell. 2011;9(1):11–5. https://doi.org/10.1016/j.stem.2011.06.008.
42. Aggarwal S, Pittenger MF. Human mesenchymal stem cells modulate allogeneic immune cell responses. Blood. 2005;105(4):1815–22. https://doi.org/10.1182/blood-2004-04-1559.
43. Krampera M, Galipeau J, Shi Y, et al. Immunological characterization of multipotent mesenchymal stromal cells—the International Society for Cellular Therapy (ISCT) working proposal. Cytotherapy. 2013;15:1054–61.
44. Moll G, Jitschin R, von Bahr L, et al. Mesenchymal stromal cells engage complement and complement receptor bearing innate effector cells to modulate immune responses. PLoS One. 2011;6:e21703.
45. Tu Z, Li Q, Bu H, et al. Mesenchymal stem cells inhibit complement activation by secreting factor H. Stem Cells Dev. 2010;19:1803–9.
46. Munir H, Rainger GE, Nash GB, et al. Analyzing the effects of stromal cells on the recruitment of leukocytes from flow. J Vis Exp. 2015;95:e52480.
47. Cassatella MA, Mosna F, Micheletti A, et al. Toll-like receptor-3-activated human mesenchymal stromal cells significantly prolong the survival and function of neutrophils. Stem Cells. 2011;29:1001–11.
48. Brandau S, Jakob M, Bruderek K, et al. Mesenchymal stem cells augment the antibacterial activity of neutrophil granulocytes. PLoS One. 2014;9:e106903.
49. Brown JM, Nemeth K, Kushnir-Sukhov NM, et al. Bone marrow stromal cells inhibit mast cell function via a COX2-dependent mechanism. Clin Exp Allergy. 2011;41:526–34.
50. Glennie S, Soeiro I, Dyson PJ, et al. Bone marrow mesenchymal stem cells induce division arrest anergy of activated T cells. Blood. 2005;105:2821–7.
51. Yañez R, Lamana ML, García-Castro J, et al. Adipose tissue-derived mesenchymal stem cells have in vivo immunosuppressive properties applicable for the control of the graft-versushost disease. Stem Cells. 2006;24:2582–91.
52. Spaggiari GM, Capobianco A, Abdelrazik H, et al. Mesenchymal stem cells inhibit natural killer-cell proliferation, cytotoxicity, and cytokine production: role of indoleamine 2,3-dioxygenase and prostaglandin E2. Blood. 2008;111:1327–33.
53. Spaggiari GM, Capobianco A, Becchetti S, et al. Mesenchymal stem cell-natural killer cell interactions: evidence that activated NK cells are capable of killing MSCs, whereas MSCs can inhibit IL-2-induced NK-cell proliferation. Blood. 2006;107:1484–90.
54. McMannus PM, Weiss L. Busulfan-induced chronic bone marrow failure: changes in cortical bone, marrow stromal cells, and adherent cell colonies. Blood. 1984;64:1036–41.
55. O'Flaherty E, Sparrow R, Szer J. Bone marrow stromal function from patients after bone marrow transplantation. Bone Marrow Transplant. 1995;15:207–12.
56. Domenech J, Gihana E, Dayan A, et al. Haemopoiesis of transplanted patients with autologous marrows assessed by long-term marrow culture. Br J Haematol. 1994;88:488–96.
57. Uhlman DL, Verfaillie C, Jones RB, Luikart SD. BCNU treatment of marrow stromal monolayers reversibly alters haematopoiesis. Br J Haematol. 1991;78:3304–9.
58. Lazarus HM, et al. Cotransplantation of HLA-identical sibling culture-expanded mesenchymal stem cells and hematopoietic stem cells in hematologic malignancy patients. Biol Blood Marrow Transpl. 2005;11:389–98.
59. Bernardo ME, et al. Co-infusion of ex vivo-expanded, parental MSCs prevents life-threatening acute GVHD, but does not reduce the risk of graft failure in pediatric patients undergoing allogeneic umbilical cord blood transplantation. Bone Marrow Transpl. 2011;46:200–7.
60. Klyushnenkova E, Mosca JD, McIntosh KR. Human mesenchymal stem cells suppress allogeneic T cell responses in vitro: implications for allogeneic transplantation. Blood. 1998;92(suppl 1, part 1):642a.
61. McIntosh KR, Klyushnenkova E, Shustova V, Moseley A, Deans RJ. Suppression of alloreactive T cell responses by human mesenchymal stem cells involves CD8+ cells. Blood. 1999;94(suppl 1, part 1):133a.
62. Bartholomew A, Sturgeon C, Siatskas M, et al. Mesenchymal stem cells suppress lymphocyte proliferation in vitro and prolong skin graft survival in vivo. Exp Hematol. 2002;30:42–8.
63. Djouad F, et al. Mesenchymal stem cells inhibit the differentiation of dendritic cells through an interleukin-6-dependent mechanism. Stem Cells. 2007;25:2025–32.
64. Bai L, et al. Hepatocyte growth factor mediates mesenchymal stem cell-induced recovery in multiple sclerosis models. Nat Neurosci. 2012;15:862–70.
65. Ortiz LA, et al. Interleukin 1 receptor antagonist mediates the antiinflammatory and antifibrotic effect of mesenchymal stem cells during lung injury. Proc Natl Acad Sci USA. 2007;104:11002–7.
66. Sato K, et al. Nitric oxide plays a critical role in suppression of T-cell proliferation by mesenchymal stem cells. Blood. 2007;109:228–34.
67. Kim J, Hematti P. Mesenchymal stem cell-educated macrophages: a novel type of alternatively activated macrophages. Exp Hematol. 2009;37:1445–53.
68. Lee RH, et al. Intravenous hMSCs improve myocardial infarction in mice because cells embolized in lung are activated to secrete the anti-inflammatory protein TSG-6. Cell Stem Cell. 2009;5:54–63.
69. Liu Y, et al. MSCs inhibit bone marrow-derived DC maturation and function through the release of TSG-6. Biochem Biophys Res Commun. 2014;450:1409–15.

70. Nasef A, et al. Identification of IL-10 and TGF-beta transcripts involved in the inhibition of T-lymphocyte proliferation during cell contact with human mesenchymal stem cells. Gene Expr. 2007;13:217–26.
71. Gao S, Mao F, Zhang B, et al. Mouse bone marrow-derived mesenchymal stem cells induce macrophage M2 polarization through the nuclear factor-κB and signal transducer and activator of transcription 3 pathways. Exp Biol Med. 2014;239:366–75.
72. Chen P-M, Liu K-J, Hsu P-J, et al. Induction of immunomodulatory monocytes by human mesenchymal stem cell-derived hepatocyte growth factor through ERK1/2. J Leukoc Biol. 2014;96:295–303.
73. Corcione A, Benvenuto F, Ferretti E, et al. Human mesenchymal stem cells modulate B-cell functions. Blood. 2006;107:367–72.
74. Di Nicola M, et al. Human bone marrow stromal cells suppress T-lymphocyte proliferation induced by cellular or nonspecific mitogenic stimuli. Blood. 2002;99:3838–43.
75. Klyushnenkova E, et al. T cell responses to allogeneic human mesenchymal stem cells: immunogenicity, tolerance, and suppression. J Biomed Sci. 2005;12:47–57.
76. LaPrade RF, Geeslin AG, Murray IR, Musahl V, Zlotnicki JP, Petrigliano F, Mann BJ. Biologic treatments for sports injuries II think tank-current concepts, future research, and barriers to advancement, part 1: biologics overview, ligament injury, tendinopathy. Am J Sports Med. 2016;44(12):3270–83. https://doi.org/10.1177/0363546516634674.
77. Atesok K, Fu FH, Sekiya I, Stolzing A, Ochi M, Rodeo SA. Stem cells in degenerative orthopaedic pathologies: effects of aging on therapeutic potential. Knee Surg Sports Traumatol Arthrosc. 2017;25(2):626–36. https://doi.org/10.1007/s00167-015-3763-9.
78. Gobbi A, Chaurasia S, Karnatzikos G, Nakamura N. Matrix-induced autologous chondrocyte implantation versus multipotent stem cells for the treatment of large patellofemoral chondral lesions: a nonrandomized prospective trial. Cartilage. 2015;6(2):82–97.
79. Gobbi A, Karnatzikos G, Sankineani SR. One-step surgery with multipotent stem cells for the treatment of large full-thickness chondral defects of the knee. Am J Sports Med. 2014;42(3):648–57.
80. Gobbi A, Scotti C, Karnatzikos G, Mudhigere A, Castro M, Peretti GM. One-step surgery with multipotent stem cells and hyaluronanbased scaffold for the treatment of full-thickness chondral defects of the knee in patients older than 45 years. Knee Surg Sports Traumatol Arthrosc. 2017;25(8):2494–501.
81. Gobbi A, Whyte GP. One-stage cartilage repair using a hyaluronic acid–based scaffold with activated bone marrow–derived mesenchymal stem cells compared with microfracture: five-year follow-up. Am J Sports Med. 2016;44(11):2846–54.
82. Gobbi A, Whyte GP. Long-term clinical outcomes of one-stage cartilage repair in the knee with hyaluronic acid-based scaffold embedded with mesenchymal stem cells sourced from bone marrow aspirate concentrate. Am J Sports Med. 2019;47(7):1621–8. https://doi.org/10.1177/0363546519845362.
83. Kim YS, Kwon OR, Choi YJ, Suh DS, Heo DB, Koh YG. Comparative matched-pair analysis of the injection versus implantation of mesenchymal stem cells for knee osteoarthritis. Am J Sports Med. 2015;43(11):2738–46. https://doi.org/10.1177/0363546515599632.
84. de Girolamo L, Kon E, Filardo G, Marmotti AG, Soler F, Peretti GM, Vannini F, Madry H, Chubinskaya S. Regenerative approaches for the treatment of early OA. Knee Surg Sports Traumatol Arthrosc. 2016;24(6):1826–35. https://doi.org/10.1007/s00167-016-4125-y.
85. Nakamura N, Miyama T, Engebretsen L, Yoshikawa H, Shino K. Cell-based therapy in articular cartilage lesions of the knee. Arthroscopy. 2009;25(5):531–52. https://doi.org/10.1016/j.arthro.2009.02.007.
86. Koga H, Engebretsen L, Brinchmann JE, Muneta T, Sekiya I. Mesenchymal stem cell-based therapy for cartilage repair: a review. Knee Surg Sports Traumatol Arthrosc. 2009;17(11):1289–97. https://doi.org/10.1007/s00167-009-0782-4.
87. Filardo G, Madry H, Jelic M, Roffi A, Cucchiarini M, Kon E. Mesenchymal stem cells for the treatment of cartilage lesions: from preclinical findings to clinical application in orthopaedics. Knee Surg Sports Traumatol Arthrosc. 2013;21(8):1717–29. https://doi.org/10.1007/s00167-012-2329-3.
88. Onishi K, Jones DL, Riester SM, Lewallen EA, Lewallen DG, Sellon JL, Dietz AB, Qu W, van Wijnen AJ, Smith J. Human adipose-derived mesenchymal stromal/stem cells remain viable and metabolically active following needle passage. PM R. 2016;8(9):844–54. https://doi.org/10.1016/j.pmrj.2016.01.010.
89. Salem HK, Thiemermann C. Mesenchymal stromal cells: current understanding and clinical status. Stem Cells (Dayton, Ohio). 2010;28(3):585–96. https://doi.org/10.1002/stem.269.
90. Koh YG, Jo SB, Kwon OR, Suh DS, Lee SW, Park SH, Choi YJ. Mesenchymal stem cell injections improve symptoms of knee osteoarthritis. Arthroscopy. 2013;29(4):748–55. https://doi.org/10.1016/j.arthro.2012.11.017.
91. Koh YG, Choi YJ, Kwon SK, Kim YS, Yeo JE. Clinical results and second-look arthroscopic findings after treatment with adipose-derived stem cells for knee osteoarthritis. Knee Surg Sports Traumatol Arthrosc. 2015;23(5):1308–16. https://doi.org/10.1007/s00167-013-2807-2.
92. Freitag J, Bates D, Wickham J, Shah K, Huguenin L, Tenen A, Paterson K, Boyd R. Adipose-derived mesenchymal stem cell therapy in the treatment of knee osteoarthritis: a randomized controlled trial. Regen Med. 2019;14(3):213–30. https://doi.org/10.2217/rme-2018-0161.

93. Gobbi A, Dallo I, Rogers C, Striano RD, Mautner K, Bowers R, Rozak M, Bilbool N, Murrell WD. Two-year clinical outcomes of autologous microfragmented adipose tissue in elderly patients with knee osteoarthritis: a multi-centric, international study. Int Orthop. 2021;45(5):1179–88. https://doi.org/10.1007/s00264-021-04947-0.
94. Gobbi A, Dallo I, D'Ambrosi R. Autologous microfragmented adipose tissue and leukocyte-poor platelet-rich plasma combined with hyaluronic acid show comparable clinical outcomes for symptomatic early knee osteoarthritis over a two-year follow-up period: a prospective randomized clinical trial. Eur J Orthop Surg Traumatol. 2023;33(5):1895–904. https://doi.org/10.1007/s00590-022-03356-2.
95. Sciarretta FV, Ascani C, Sodano L, et al. One-stage cartilage repair using the autologous matrix-induced chondrogenesis combined with simultaneous use of autologous adipose tissue graft and adipose tissue mesenchymal cells technique: clinical results and magnetic resonance imaging evaluation at five-year follow-up. Int Orthop (SICOT). 2024;48:267–77. https://doi.org/10.1007/s00264-023-05921-8.
96. Acosta M, Gobbi A, Bizzoco L. Novel P.A.S.T.A. Technique Combining PRP and MSCs from SVF Shows Improved Outcomes at Short-/Long-Term Follow-Up for Osteoarthritis and Chondropathy. J Orthop Muscular Syst. 2025. (IN PRESS)
97. Gobbi A, Whyte GP. Long-term outcomes of primary repair of the anterior cruciate ligament combined with biologic healing augmentation to treat incomplete tears. Am J Sports Med. 2018;46(14):3368–77. https://doi.org/10.1177/0363546518805740.
98. Kanaya A, Deie M, Adachi N, Nishimori M, Yanada S, Ochi M. Intra-articular injection of mesenchymal stromal cells in partially torn anterior cruciate ligaments in a rat model. Arthroscopy. 2007;23(6):610–7. https://doi.org/10.1016/j.arthro.2007.01.013.
99. Figueroa D, Espinosa M, Calvo R, Scheu M, Vaisman A, Gallegos M, Conget P. Anterior cruciate ligament regeneration using mesenchymal stem cells and collagen type I scaffold in a rabbit model. Knee Surg Sports Traumatol Arthrosc. 2014;22(5):1196–202. https://doi.org/10.1007/s00167-013-2471-6.
100. Gobbi A, Whyte GP. Emerging Orthobiologic approaches to ligament injury. In: Gobbi A, Espregueira-Mendes J, Lane J, Karahan M, editors. Bio-orthopaedics. Berlin, Heidelberg: Springer; 2017. p. 313–24. https://doi.org/10.1007/978-3-662-54181-4_24.
101. Gobbi A, Whyte GP. Biological augmentation in acute ACL repair. Bio-orthopaedics. 2017 325–335. Springer, Berlin, Heidelberg. https://doi.org/10.1007/978-3-662-54181-4_24.
102. Sekiya I, Koga H, Otabe K, Nakagawa Y, Katano H, Ozeki N, Mizuno M, Horie M, Kohno Y, Katagiri K, Watanabe N, Muneta T. Additional use of synovial mesenchymal stem cell transplantation following surgical repair of a complex degenerative tear of the medial meniscus of the knee: a case report. Cell Transplant. 2019;28(11):1445–54. https://doi.org/10.1177/0963689719863793.
103. Ding G, Du J, Hu X, Ao Y. Mesenchymal stem cells from different sources in meniscus repair and regeneration. Front Bioeng Biotechnol. 2022;10:796367. https://doi.org/10.3389/fbioe.2022.796367.
104. Rhim HC, Jeon OH, Han SB, Bae JH, Suh DW, Jang KM. Mesenchymal stem cells for enhancing biological healing after meniscal injuries. World J Stem Cells. 2021;13(8):1005–29. https://doi.org/10.4252/wjsc.v13.i8.1005.
105. Oshita T, Tobita M, Tajima S, Mizuno H. Adipose-derived stem cells improve collagenase-induced tendinopathy in a rat model. Am J Sports Med. 2016;44(8):1983–9. https://doi.org/10.1177/0363546516640750.
106. Sevivas N, Teixeira FG, Portugal R, Direito-Santos B, Espregueira-Mendes J, Oliveira FJ, Silva RF, Sousa N, Sow WT, Nguyen LTH, Ng KW, Salgado AJ. Mesenchymal stem cell secretome improves tendon cell viability in vitro and tendon-bone healing in vivo when a tissue engineering strategy is used in a rat model of chronic massive rotator cuff tear. Am J Sports Med. 2018;46(2):449–59. https://doi.org/10.1177/0363546517735850.
107. Goldberg AJ, Masci L, O'Donnell P, Green R, Brooking D, Bassett P, Lowdell MW, Smith RKW. Autologous bone marrow derived mesenchymal stem cells are safe for the treatment of Achilles tendinopathy. Sci Rep. 2024;14(1):11421. https://doi.org/10.1038/s41598-024-61399-3.
108. Wakitani S, Mera H, Nakamura N, Gobbi A. Review of Caplan (1991) on cell-based therapeutic technology using mesenchymal stem cells. J ISAKOS. joint disorders & orthopaedic sports medicine. 2024;9(3):426–30. https://doi.org/10.1016/j.jisako.2023.08.010.

Evidence-Based Use of Regenerative Medicine

4

Ignacio Dallo, Pedro Bernáldez, José Fábio Lana, Peter A. Everts, Ansar Mahmood, and Alberto Gobbi

4.1 Introduction

Regenerative medicine (RM) is now a key component in modern orthopedics and sports medicine. There is growing interest in using biological treatments incorporating tissue engineering strategies: cells, scaffolds, and signaling molecules [1, 2]. Furthermore, patients are currently seeking "stem cells" or regenerative medicine treatments in response to unmet treatment needs and marketing efforts often outpacing clinical evidence and regulatory control [3–5]. With the advent of injectable therapies using endogenous growth factors and cells inserted directly into the tissue to facilitate healing, decrease inflammation potentially, and subsequently provoke an analgesic effect after an injury or illness, injections have the advantage of being "minimally invasive" with a relatively low risk of complications. Biological regenerative approaches commonly include platelet-rich plasma (PRP), bone marrow aspirate or bone marrow concentrate (BMA/BMC), and adipose tissue. These injectable treatments may contribute to a regenerative microenvironment with the potential to improve healing rates and function in patients with musculoskeletal problems; however, to date, only symptomatic improvements have been reported, and structure-modifying treatments are lacking.

The application of regenerative medicine can facilitate the healing mechanism of tissues with limited healing potential and vascularity, such as

I. Dallo (✉)
Department of Orthopaedic Surgery and Sports Medicine, Unit of Biological Therapies and MSK Interventionism, Sport Me Medical Center, Seville, Spain

Centro Universitario Max Plank, OrthoRegen, Indaiatuba São Paulo, São Paulo, Brazil

P. Bernáldez
Sport Me Medical Center, Department of Orthopaedic Surgery and Sports Medicine, Unit of Biological Therapies and MSK Interventionism, Seville, Spain

J. F. Lana
Sport Me Medical Center, Department of Orthopaedic Surgery and Sports Medicine, Unit of Biological Therapies and MSK Interventionism, Seville, Spain

Department of Orthopaedics, The Bone and Cartilage Institute, Indaiatuba, SP, Brazil

P. A. Everts
Centro Universitario Max Plank, OrthoRegen, Indaiatuba São Paulo, São Paulo, Brazil

Gulf Coast Biologics, Scientific and Research Department, Fort Myers, FL, USA
e-mail: peter@gulfcoastbiologics.com

A. Mahmood
Trauma and Orthopaedics, Queen Elizabeth Hospital, Birmingham, UK
e-mail: ansar@doctors.org.uk

A. Gobbi
O.A.S.I. Bioresearch Foundation Gobbi N.P.O., Milan, Italy
e-mail: gobbi@cartilagedoctor.it

A. Gobbi et al. (eds.), *Regenerative Medicine in Sports and Orthopaedics*,
https://doi.org/10.1007/978-3-031-84693-9_4

tendons, cartilage, meniscus, and ligaments. However, to understand and advance regenerative medicine, it is essential to understand the potential and limitations of the different components of tissue engineering approaches. The term "stem cell" has been overused based on the consensus of the expert's opinion [6]. Stem and progenitor cells contribute to homeostasis, remodeling, and repair of tissues. Therefore, it is recommended that minimally manipulated cell products and tissue-derived culture-expanded cells be referred to as "cell therapy" to better represent the nature of these treatments.

Basic science research has proved that some cell therapy approaches (e.g., bone marrow-derived mesenchymal stem cells MSCs) may downregulate inflammation and produce an analgesic effect. Nonetheless, conclusive clinical evidence is still missing despite studies showing promising clinical results for cartilage injuries and ligament tears [7]. As such, there is a continued need for high-quality basic science and clinical investigation into the safety and efficacy of biological therapies, including cell-based therapies. It is recommended that physicians and institutions offering biologic therapies establish patient registries for surveillance and quality assessments. Several clinical trials are currently being performed evaluating these non-invasive therapies despite limited understanding of the underlying pathologic basis of the disease and without a complete characterization of their components. Additional studies are needed to identify optimal formulations and to define the ideal dose and timing of orthobiologics for various orthopedic conditions [8].

4.2 Regenerative Medicine Treatments

The regenerative medicine nomenclature can be confusing and, in many cases, even misleading, depending on the product being discussed. Lack of consensus on how to describe orthobiologics included in regenerative medicine treatments has led to challenges in interpreting the literature and comparing one biologic product to another. Bio-orthopedics is the modern branch of orthopedics that includes all regenerative medicine treatments aimed to enhance the biological response of connective tissues to optimize the repair process and improve clinical outcomes [9]. Accurate nomenclature and reporting of currently available biological treatments will facilitate advancement in the field. Different tissue sources have been used to obtain orthobiologic therapies, including blood, bone marrow, and fat, among others. For example, platelet-rich plasma (PRP) is a blood-derived product where the blood is centrifuged to allow density separation of its components [10, 11]. There are multiple ways of preparing PRP: manual and commercial. Depending on the protocol used, the final preparation will vary [12]. This autologous product can then be used within the same patient to treat various conditions. Several classification systems have been published to further describe the various PRP products based on the concentrations of the specific contents of the final product and the methods used to obtain them [10]. The most recent classification systems generally include an absolute number for platelet concentration, leukocytes or white blood cells and neutrophils, red blood cell presence, and if activation of platelets occurs synthetically by exogenous agents or naturally [13]. Some versions have additional details about the preparation of the product and how it was applied to the patient [14]. Overall, various terminologies describing PRP and methods of classifying the numerous types of products exist, but no nomenclature standardization for PRP has been established.

Bone marrow aspirate concentrate (BMAC) is obtained by the centrifugation of bone marrow aspirate (BMA). This process concentrates the mononucleated cells and increases the ratio of stem and progenitor cells. Often, BMAC is referred to as a "stem cell" therapy. However, it is essential to mention that only 0.01–0.0001% of the heterogeneous nucleated cells in BMAC are stem or progenitor cells [15]. The composition describes the cellular makeup of the aspirate obtained from an average individual, which has been reported to be mainly neutrophils and erythroblasts in addition to other cell types, and a low

Table 4.1 Summary of growth factors in bone marrow aspirate concentrate (BMAC) [1]

Growth factor/ cytokine	Principle action	Signaling pathway
TGF β1, TGF β2, TGF β3	Chondrocyte proliferation + differentiation	SMAD-2 and SMAD-3
BMP-2	Chondrocyte proliferation, matrix synthesis, and hypertrophy	SMAD-1, SMAD-5, SMAD-8, TAK-1
BMP-7	Increased ECM production	Mitogen-activated kinases (JNK, P38, ERK1/2)
IL-8	Inflammatory response; MSC homing to the site of injury; increased VEGF production; chondrocyte hypertrophy	Mitogen-activated kinase; P38
VEGF	Promotes angiogenesis to sub-chondral bone and supports cartilage growth	HIF-1, Runx2
PDGF	Wound healing, collagen synthesis, angiogenesis, suppression of IL-1β, enhanced BMP signaling	ERK 1/2, down-regulation of NF-kB signaling
IGF-1	Increased synthetic and metabolic activity: Increased collagen and proteoglycan synthesis; chondrogenic differentiation	PI-3 K, ERK 1/2
FGF-2	Chondrogenic differentiation, MSC homing	ERK 1/2, STAT1/P21
FGF-18	Chondrogenic differentiation, enhanced BMP signaling	
L-1/IL-1β	Inflammatory response: Cell migration/recruitment to the site of injury	Mitogen-activated kinases (JNK, P38, ERK1/2)

JNK C-Jun N-terminal kinase, *ERK* extracellular signal-related kinases, *TAK-1* TGF-β activating kinase 1 (TAK-1), *STAT1* signal transducer and activator of transcription-1, *PI-3 K* phosphoinositide 3-Kinase, *Runx2* Runt-domain transcription factor family-2, *HIF-1* hypoxia-inducible factor-1, *NK-kB* nuclear factor kappa beta

percentage of stem and progenitor cells (Tables 4.1 and 4.2) [16, 17] [18].

The International Society for Cellular Therapy proposes minimal criteria to define human mesenchymal stem cells (MSC). First, MSCs must be plastic-adherent when maintained in standard culture conditions. Second, MSCs must express CD105, CD73, and CD90, and lack expression of CD45, CD34, CD14 or CD11b, CD79alpha or CD19, and HLA-DR surface molecules [19]. Third, MSCs must differentiate into osteoblasts, adipocytes, and chondroblasts in vitro. Recently, as per Arnold Caplan, it has become more appropriate to call MSCs medicinal signaling cells, as these cells in vivo respond to injury or disease by secreting bioactive factors that have immunomodulatory effects, providing promising therapeutic options [20].

The DOSES tool for describing cell therapies must be utilized by clinicians, researchers, regulators, and industry professionals to improve transparency and to allow clinicians and patients to understand the characteristics of current and future cell preparations [21–23]. Adipose-derived therapies have recently received more attention in the field due to their ease of use and the basic science finding that they have one of the highest concentrations of stem and progenitor cells per nucleated cell among all potential sources [24]. However, not all adipose-derived therapies are equal since some only employ mechanical fat fragmentation, while others employ enzymatic digestion and culture expansion techniques. Micro-fragmented fat (MF) is a term used to describe the minimally manipulated product of the mechanical breakdown of adipose tissue into tiny particles without requiring additives or ex vivo expansion. This process creates a product with stem and progenitor cells and claims to maintain the vascular niche and extracellular matrix. Differently, stromal vascular fraction (SVF) refers to the product obtained from the enzymatic digestion of adipose tissue with collagenase and then centrifugation to remove adipocytes and free fat, producing a heterogeneous mixture of cells that includes stem and progenitor

Table 4.2 Cellular characterization of bone marrow aspirate (BMA) and bone marrow aspirate concentrate (BMAC) [1]

	n	Median (Range)
Pre-spin measures	24	
Viability, %	25	97.8 (75.2–99.4)
MNCs, %	25	38.5 (26.0–57.5)
Total MNCs/uL	25	6100 (1950–27,000)
HSCs, %	25	3.2 (0.04–21.0)
MSCS, %	25	0.03 (0.00–0.60)
Total MNCS × MSCS, %	25	198 (0–2673)
WBCs, 1000/μL	25	13.0 (3.9–62.8)
RBCs, Mil/μL	25	3.33 (0.17–4.44)
HCTs, %	25	32.0 (1.6–38.2)
Platelet, 1000/μL	25	95 (7–399)
Post-spin measures	22	
Viability, %	23	97.0 (85.4–99.6)
MNCs, %	23	56.2 (25.8–87.9)
Total MNCs/μL	23	16,000 (2900–210,000)
HSCs, %	23	4.4 (1.2–14.0)
MSCS, %	23	0.05 (0.0–09)
Total MNCS × MSCS, %	23	688 (8.7–28,980)
WBCs, 1000/μL	23	31.4 (5.6–97.2)
RBCs, Mil/μL	23	0.96 (0.63–3.65)
HCTs, %	23	8.5 (3.5–1515)
Platelet, 1000/μL	23	422 (52–1515)
Total HSCs injected	23	4,620,000 (174,000-
Total MSCs injected	23	130,200,000)
		34,400 (435–1,449,000)

HCT hematocrit, *HSC* hematopoietic stem cell, *Mil* million, *MNC* mononuclear cell, *MSC* mesenchymal stem cell, *n* number of patient samples analyzed, *RBC* red blood cell, *WBC* white blood cell

cells [25]. Overall, these methods result in products that have been referred to as adipose-derived stem cells (ASCs) or adipose-derived mesenchymal stem cells (AMSCs), and they are currently being studied for their efficacy in treatments of various orthopedic and sports medicine conditions.

4.3 Basic Sciences of Regenerative Medicine

4.3.1 Platelet-Rich Plasma (PRP)

Platelet-rich plasma (PRP) is a blood derivative produced by centrifugation or filtration of whole blood to concentrate or isolate platelets at a higher level (usually 2–5 times higher) than normal plasma levels [26]. The reason for this variability in platelet concentration is explained by the different methods available to produce PRP (e.g., different centrifugation protocols and the development of several commercial kits) [27]. This contributed to confusion in the literature, as several products may differ in terms of effects and indications, making it difficult to compare studies and understand the actual effectiveness of PRP in the various applications and study conditions.

Regardless of the production methods and the concentration of platelets, PRP is characterized by a high concentration of a large composition of growth factors (FCs) and cytokines involved in both the healing process and immunoregulation. These biological mediators are released by exocytosis of the granules α platelets and by membrane disruption that can occur during the handling of this blood derivative after its activation with an initial burst followed by an additional sustained release. Different methods of activating PRP could influence the release curve, but calcium chloride is currently the most common. In particular, 5% calcium chloride appears to be effective in obtaining a high concentration of biological mediators in PRP. However, some products are also promoted without the use of an activator, depending on the spontaneous activation induced by contact with collagen in situ once PRP is injected [28]. The high amount of growth factors released are responsible for the multiple effects of PRP, such as the anabolic effect and the enhancement of cell migration, proliferation and differentiation, the neoformation of vessels, the modulation of inflammation and immunoregulation, and the synthesis of extracellular matrix proteins [29, 30]. In particular, among the most studied FCs secreted by platelets are vascular endothelial growth factor (VEGF), fibroblast growth factor (FGF), platelet-derived growth factor (PDGF), insulin-like growth factor-1 (IGF-1), interleukin-10 (IL-10), transforming growth factor-β1 (TGF-β1), epidermal growth factor (EGF), and hepatocyte growth factor (HGF) [31]. In Table 4.3, an overview is given of the most important PRP granular-based platelet growth factors and biomolecules and their effects on tissue repair processes [32].

The effects of these FCs have been extensively analyzed through in vitro and in vivo studies [33]: TGF-β1 has been shown to enhance matrix deposition, induce chondrogenic differentiation and cell proliferation, and promote the expression of type II collagen (hyaline cartilage) and proteoglycan expression in contrast to the effect of IL-1. In addition, it induces migration to the site of injury, proliferation, and chondrogenic differentiation of bone marrow stromal cells (BMSCs). VEGF is a signaling protein involved in angiogenesis and other angiogenic factors. FGF stimulates fibroblast proliferation and contributes to angiogenesis. PDGF stimulates cell proliferation and angiogenesis and has a mitogenic role for fibroblasts. IGF-1 enhances matrix deposition and promotes cell proliferation and fibroblast differentiation.

On the other hand, it is a mediator in the growth and repair of skeletal muscle. Recent evidence has identified IGF-1 as a possible inhibitor of apoptosis, regulating the expression of the PDCD 5 protein [34]. EGF stimulates the proliferation and migration of mesenchymal and epithelial cells. HGF stimulates angiogenesis and proliferation of endothelial cells, has an antifibrotic role, and increases anti-inflammatory cytokines. IL-10 is an anti-inflammatory cytokine.

In addition to the biological logic, current evidence supports the role of PRP in modulating the inflammatory intra-articular environment in degenerative diseases: after an initial pro-inflammatory action, with the stimulation of synoviocytes to release metalloproteinases and cytokines, a next phase of modulation and reduction of the inflammatory response has been demonstrated, with a decrease in pro-inflammatory cytokines, that contrasts the chemotaxis of monocyte-like cells [35]. In addition, the preclinical literature also showed an analgesic effect of PRP, possibly by increasing cannabinoid receptors [36]. Based on this evidence, it is essential to underline that PRP may not directly promote tissue regeneration but can act through its different bioactive molecules that affect tissue homeostasis, slowing down catabolic and degenerative processes and offering a benefit in symptom relief and functional improvement. Thanks to these biological effects, different clinical applications have been found in PRP over the years, from dentistry, maxillofacial surgery, plastic surgery, and dermatology to even orthopedic surgery with different evidence of effectiveness. The confusion regarding the efficacy of PRP derives mainly from the various methods and protocols for obtaining and preparing it, with varying characteristics in terms of platelets, leukocytes, red blood cells, and fibrin content, which can cause different biological and mechanical characteristics. For these reasons, in 2009, Dohan Ehrenfest et al. divided PRP into four categories according to fibrin architecture and leukocyte content: pure platelet-rich plasma (P-PRP), platelet-leukocyte-rich plasma (L-PRP), pure platelet-rich fibrin (P-PRF), platelet-leukocyte rich fibrin (L-PRF) [37]. Each category presents some main specificities, and within the same category, each product has individuality. For example, several studies evaluated P-PRP behaviors, revealing that P-PRF

Table 4.3 Origen and function of growth factors and cytokines. Modified from Everts et al. [2]

PGF and cytokines	Cell sources	Contribution to tissue repair
PDGF (AA-BB-AB)	Platelets, endothelial cells, macrophages, smooth muscle cells	Mitogenic for mesenchymal cells and osteoblasts; stimulates chemotaxis and mitogenesis in fibroblast/glial/smooth muscle cells; regulates collagenase secretion and collagen synthesis; stimulates macrophage and neutrophil chemotaxis
TGF (α–β)	Macrophages, T lymphocytes, keratinocytes	Stimulates undifferentiated mesenchymal cell proliferation; regulates endothelial, fibroblastic, and osteoblastic mitogenesis; regulates collagen synthesis and collagenase secretion; regulates mitogenic effects of other growth factors; stimulates endothelial chemotaxis and angiogenesis; inhibits macrophage and lymphocyte proliferation
VEGF	Platelets, macrophages, keratinocytes, endothelial cells	Increases angiogenesis and vessel permeability; stimulates mitogenesis for endothelial cells
EGF	Platelets, macrophages, monocytes	Proliferation of keratinocytes, fibroblasts; stimulates mitogenesis for endothelial cells
(a-b)-FGF	Platelets, macrophages, mesenchymal cells, chondrocytes, osteoblasts	Promotes growth and differentiation of chondrocytes and osteoblasts; mitogenic for mesenchymal cells, chondrocytes, and osteoblasts
CTGF	Platelets, fibroblasts	Promotes angiogenesis, cartilage regeneration, fibrosis, and platelet adhesion
IGF-1	Platelets, plasma, epithelial cells, endothelial cells, fibroblasts, osteoblasts, bone matrix	Chemotactic for fibroblasts and stimulates protein synthesis. Enhances bone formation by proliferation and differentiation of osteoblasts
HGF	Platelets, mesenchymal cells	Regulates cell growth and motility in epithelial/endothelial cells, supporting epithelial repair and neovascularization during wound healing
KGF	Fibroblasts, mesenchymal cells	Regulates epithelial migration and proliferation
Ang-1	Platelets, neutrophils	Induces angiogenesis stimulating migration and proliferation of endothelial cells. Supports and stabilizes blood vessel development via the recruitment of pericyte
PF4	Platelets	Calls leucocytes and regulates their activation. Microbiocidal activities
SDF-1α	Platelets, endothelial cells, fibroblasts	Calls CD34+ cells; induces their homing, proliferation, and differentiation into endothelial progenitor cells stimulating angiogenesis. Calls mesenchymal stem cells and leucocytes
TNF	Macrophages, mast cells, T lymphocytes	Regulates monocyte migration, fibroblast proliferation, macrophage activation, angiogenesis

PDGF platelet-derived growth factors, *TGF* transforming growth factor, *VEGF* vascular endothelial growth factor, *EGF* epidermal growth factor, *FGF* fibroblast growth factor, *CTCG* connective tissue growth factor, *IGF* insulin-like growth factor, *HGF* hepatocyte growth factor, *KGF* keratinocyte growth factor, Ang-1

remains solid and intact and continues to release a large amount of FC produced by the cell population for a longer time.

Conversely, P-PRP, on the other hand, releases most of the FC in the first few hours [37, 38]. Another parameter of this classification is the presence/absence of leukocytes. Some authors argue that the presence of leukocytes may provide stimulation of the inflammatory process, while other authors argue that the presence of leukocytes [39] in PRP can increase FC and the production of anti-nociceptive mediators [40, 41], and they also provide antimicrobial activity [42]. In general, leukocytes have a crucial role in regulating inflammatory and healing processes, but this remains one of the most critical points of discussion about PRP today. Regardless of these differences, different types of PRP have been applied in clinical practice, showing promising results as minimally invasive treatments for knee osteoarthritis, which appear to provide short-term symptomatic benefits [43].

However, although the general biological rationale for the use of PRP is known, and despite some promising clinical findings, more studies are still needed to understand better the different biological and clinical effects of various formulations, as well as the best ways to exploit the properties of PRP to provide better clinical practice and to identify the best indications for PRP use.

4.3.2 Cellular Therapies

The International Society of Cell Therapy proposes minimum criteria for defining human MSCs. First, MSCs must adhere to plastic when kept under standard growing conditions. Second, they must express CD105, CD73, and CD90, and lack expression of surface molecules CD45, CD34, CD14 or CD11b, CD79alpha or CD19, and HLA-DR [19]. Third, MSCs should be differentiated into osteoblasts, adipocytes, and chondroblasts in vitro. According to Arnold Caplan, it has become more appropriate to call MSCs medicinal signaling cells since these cells respond in vivo to injury by secreting bioactive factors that have an immunomodulatory effect, providing promising therapeutic options [20]. The application of MSCs could facilitate the healing mechanism of tissues with limited potential for repair and vascularization, such as articular cartilage, menisci, and anterior cruciate ligament (ACL). The term stem cell has been overused according to the consensus of expert opinion [6]. It is recommended that the use of minimally manipulated cell products and culture-expanded tissue-derived cells be referred to as cell therapy, and the nature of these treatments should be clearly understood. Basic scientific research confirmed the proof-of-concept that MSC therapy regulates inflammation and produces an analgesic effect. Few studies show promising long-term clinical results for joint injuries. As such, there is an ongoing need for high-quality basic scientific and clinical research on the safety and efficacy of cell-based therapies. The DOSES tool for describing cell therapies should be used by clinicians, researchers, regulators, and industry professionals to improve transparency and allow clinicians and patients to understand the characteristics of current and future cell preparations [21]. Several therapeutic protocols have been developed involving the use of MSCs for clinical applications and have demonstrated the potential to improve regenerative processes for many conditions. These cells can be isolated from muscle, periosteum, synovial membrane, bone marrow, and adipose tissue. Bone marrow aspirate concentrate (BMAC) is classified through the US Food and Drug Administration (FDA) as a 361 product and is therefore not subject to premarket review and approval. The European Union's (EU) regulatory base is similar to the US system, and processes such as centrifugation are considered minimal manipulation. BMAC has progenitor cells and growth factors with repairing, self-directing, and trophic properties that cause them to migrate to areas of damage. Once at the site of injury, MSCs release cytokines and growth factors that induce paracrine and autocrine activity, aiding in the healing process and modulation of inflammation [44, 45]. The paracrine effects of MSCs are responsible for angiogenesis by the release of vascular endothelial growth factor (VEGF) and hepatocyte growth factor (HGF), as well as immunosuppression of lymphocytes and peripheral blood mononuclear cells by the release of prostaglandin E2, leukemia inhibitory factor, and kynurenine [46]. MCS can be isolated from the bone marrow (BM), periosteum, trabecular bone, adipose tissue, synovial membrane, skeletal muscle, and teeth. Recently, Cassano et al. found an increased concentration of interleukin-1 receptor antagonist (IL-1RA), which, combined with the other components, may provide anti-inflammatory and immunomodulatory effects [47]. Adipose tissue-derived cell therapy, also known as adipose stromal vascular fraction (ASC) or microfragmented autologous adipose tissue, has recently gained popularity as a minimally manipulated product. Adipose tissue, which is usually structured with constant vascularization, has been increasingly recognized as a reliable source of these cells. Also, compared to the BMAC, it has the advantage of obtaining a source with a higher amount of MSCs.

4.4 Clinical Applications

Studies have shown that extra-articular tissues heal following a predefined sequence of events. At first, there is an extravasation of blood and the formation of a primary fibrin-platelet clot, filling the space between the ends of the injured tissue. This process creates a scaffold for the cells to migrate and remodel the primary clot. In the following steps, a scar is formed that fills the defect. Conversely, an injury to intra-articular tissues (cartilage, ligaments, and menisci) activates synoviocytes and increases the expression of plasminogen activator, which is excreted in the intra-articular environment. As a result, the plasminogen present in the synovial fluid is converted into plasmin, which degrades fibrin. Therefore, fibrin in the intra-articular space cannot form a stable clot. The early loss of this temporary scaffold is considered one of the main reasons why the tissues within the joints do not heal quickly [48].

4.4.1 Articular Cartilage Injuries

The pathological process of articular cartilage is highly prevalent in the general population. The mechanisms by which this process occurs are the aggressions caused by a single joint trauma or when the cartilage is subjected to excessive or repetitive pressure loads.

The avascularity of the articular cartilage, the low presence of chondrocytes, and the inability to migrate to the injured site lead to slow healing and, in many cases, a permanent defect of the cartilage [49]. Under physiological conditions, the application of mechanical force to the joint produces an increase in the synthesis of collagen and proteoglycans, especially in the superficial area [50].

4.4.1.1 Platelet-Rich Plasma (PRP)

By understanding the biological processes of tissue repair in advance, the use of biological therapies based on signaling proteins could be advisable and useful. It is well known that platelets play an important role in hemostasis, inflammation, and proliferation for tissue remodeling and healing. This inherent healing potential has led some authors to use plasma growth factors for the treatment of cartilage lesions.

Platelet-rich plasma (PRP) has been shown to have a strong positive effect on chondrocyte proliferation in vitro, observed in a monolayer or three-dimensional environment. It stimulates the synthesis of proteoglycans and type II collagen and could also induce the expression of chondrocyte differentiation proteins such as aggrecan and Sox9 [51]. PRP also led to the decrease of multiple inflammatory effects mediated by IL-1 beta in human osteoarthritic chondrocytes, including inhibition of NF-κB activation [52]. It should also be noted that some PRP formulations could be pro-inflammatory, and concentrated leukocytes increased levels of catabolic and pro-inflammatory signaling molecules.

Two studies have been published in full-thickness knee cartilage lesions treated with an autologously fabricated matrix of hyaline cartilage chips combined with clot-like platelet-poor mixed plasma and liquid-stage growth factor-rich plasma (PRGF). They reported good clinical and magnetic resonance imaging (MRI) outcomes with a return to high-level football [53]. This technique seems to be an excellent alternative for cartilage lesions, as it can be performed on very active young individuals as a single surgery, is inexpensive, has no intolerance or rejection potential, and has demonstrated histological and immunohistochemical features, very similar to healthy articular cartilage in animal studies [54, 55].

4.4.1.2 Cellular Therapies

Gobbi et al. [56] in a prospective 10-year follow-up study concluded that repair of full-thickness cartilage injury in the knee with a concentrated bone marrow-embedded hyaluronic acid scaffold (HA-BMAC) provides good clinical outcomes at long-term follow-up in the treatment of small to large, isolated, or multiple lesions, and injuries to one or two compartments, as well as in cases of treatment of associated injuries. This

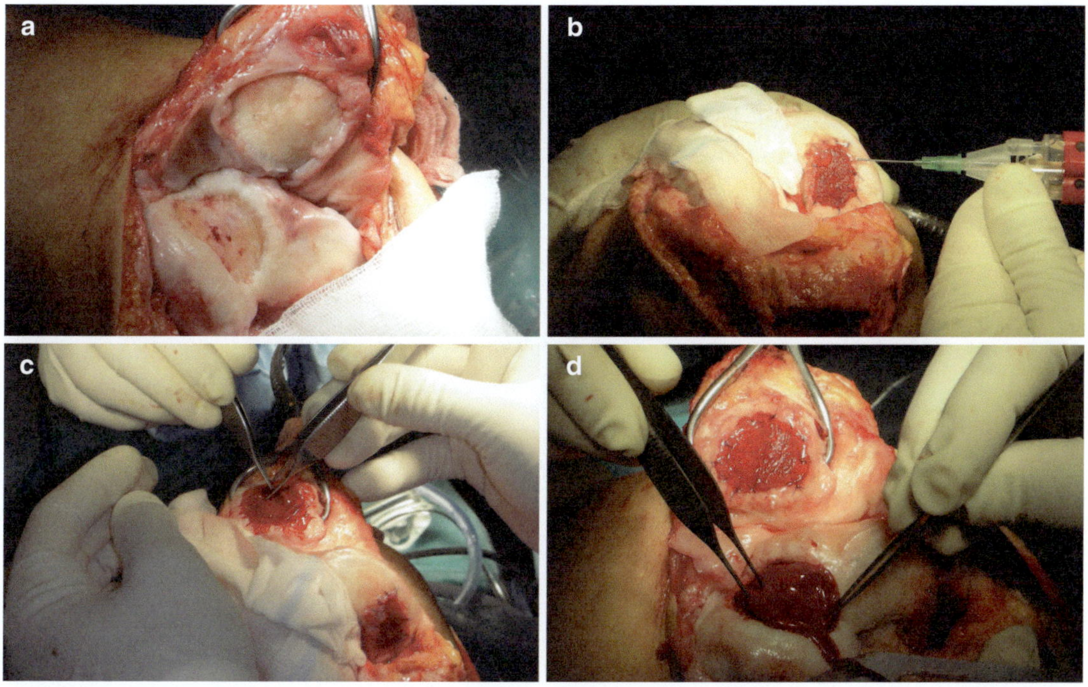

Fig. 4.1 Image showing the one-step technique using a concentrated bone marrow-embedded hyaluronic acid scaffold (HA-BMAC) for the treatment of full-thickness lesions of the articular cartilage of the knee. (**a**) Preparation of chondral defects in the patella and trochlea in a right knee. (**b**) Application of fibrin glue to secure the HA-BMAC graft within an internal femoral condyle lesion. (**c**) Application of the graft in the patellar defect. (**d**) Application of the graft for a chondral defect in the trochlea

one-stage cartilage repair technique could play a prominent role in the treatment of chondral defects as it is a simple, safe, and accessible technique [57] (Fig. 4.1). While good results can be expected among patients over 45, results may be comparatively more successful in younger patients. Based on their experience quantifying colony-forming units (CFUs) in 25 patients, the authors found no correlation between clinical outcomes and the number of CFUs. An intriguing explanation for these results may come from Caplan's recently proposed new view of MSCs as "medicinal signaling cells." According to this concept, MSCs, rather than participating in tissue formation, function as "pharmacies" regulated at the site in vivo by releasing trophic and immunomodulatory factors and are activated by local lesions [58].

In a comparative study of 46 patients with full-thickness chondral defects in the femoral condyle treated with a synthetic osteochondral graft (TruFit) alone, augmented with PRP or BMAC, Aaron Krych et al. [59] showed improvements in cartilage maturation with greater filling and mean T2 values closer to those of superficial native hyaline cartilage at 12 months in patients treated with BMAC-augmented TruFit. The authors conclude that further study will be needed to determine whether this translates into better clinical outcomes.

4.4.2 Anterior Cruciate Ligament (ACL) Injuries

Unlike other knee ligaments, such as the posterior cruciate ligament (PCL) or medial collateral ligament (MCL), the ACL possesses a limited intrinsic ability for spontaneous healing after injury. This is why the gold standard technique for the treatment of ACL injuries is arthroscopic reconstruction. However, some published case series have demonstrated spontaneous healing after ACL injury [60]. Due to these findings and the increased knowledge of the basic sciences of the ACL and the advent of biological therapies, there are currently some strategies to try to help improve the healing process of the ACL and preserve the ligament tissue or improve the surgical reconstruction of the ACL.

4.4.3 Partial ACL Injuries

4.4.3.1 Platelet-Rich Plasma (PRP)

In 19 professional football players with partial ACL rupture, platelet-derived growth factors (PRGF-Endoret) were applied with a spinal needle to the proximal origin of the healthy fascicle and to the medial portion of the fascicle in an amount of approximately 4 cc [61]. At the end of the surgery, another injection of PRGF-Endoret (6 cc) was administered into the joint space. Fifteen patients returned to physical activity at an average of 16.2 weeks (a re-break at the seventh month), while three patients returned at 12.3 weeks. Post-operative magnetic resonance imaging was performed on all patients to observe the rest of the anterior cruciate ligament, which had complete ligamentization 1 year after surgery and a good anatomical disposition.

4.4.3.2 Cellular Therapies

Injection of autologous bone marrow concentrate and PRP under fluoroscopic or arthroscopic guidance (Fig. 4.2) have reported good short-term clinical results with magnetic resonance imaging showing ACL healing with partial lesions (Fig. 4.3) [62, 63]. Gobbi et al. [64] concluded that primary repair with suturing of the injured band to the healthy ACL band combined with PRP and BMAC to treat select cases of knee instability secondary to incomplete injury demonstrated good results at 10 years of follow-up, with high rates of restoration of knee stability and return to pre-injury sports activities in a series of 50 patients. The ambulatory or surgical application of this technique in a tissue with low regenerative capacity is a promising tool in these pathologies [65].

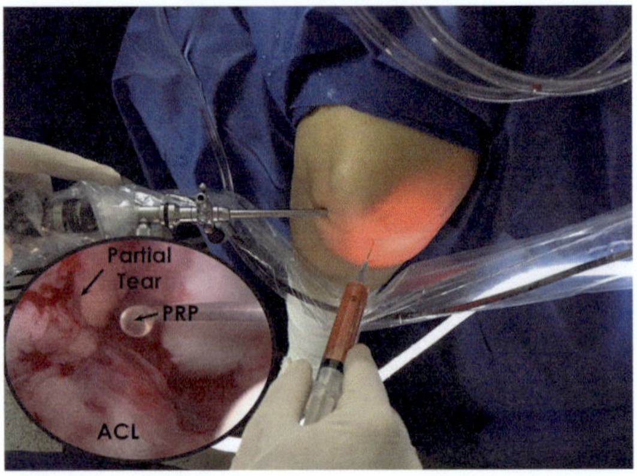

Fig. 4.2 Image showing arthroscopic and surgeon view of a platelet-rich plasma (PRP) infiltration in a partial anterior cruciate ligament (ACL) injury

ACL Reconstruction

Regarding the use of PRP in ACL reconstruction, the translation of the findings achieved in basic research and clinical practice has generated variable but promising results [66]. This variability of the results could be due to the early stage of the use of PRP in this type of surgery, whose protocols are not optimized or standardized, and its application is very heterogeneous. In some cases, the choice of PRP may be key as the use of leukocyte-rich plasma could hinder anti-inflammatory action [67], so the results when used in this surgery may not be as expected [68]. In addition, the way PRP is applied is also an important factor in getting the right result. When applying PRP in ACL reconstruction, both the tunnels and the graft must be injected (Fig. 4.4). This favors the ligamentization of the tendons used as grafts, an anti-apoptotic effect and better bone integration [69]. Lack of application in any of these components could lead to a decrease in the effectiveness of the technique [70–72]. The authors of several papers observed improvements in aspects such as bone tunnel [73], joint stability [74], and return to the previous level of activity with surgery in patients with ACL injuries [75]. Further clinical studies are needed to optimize the application protocol used in ligament injuries.

4.4.4 Meniscal Injuries

Meniscus injury that occurs in close proximity to the perimeniscal capillary plexus (red-red zone) is amenable to repair. Unfortunately, lesions in

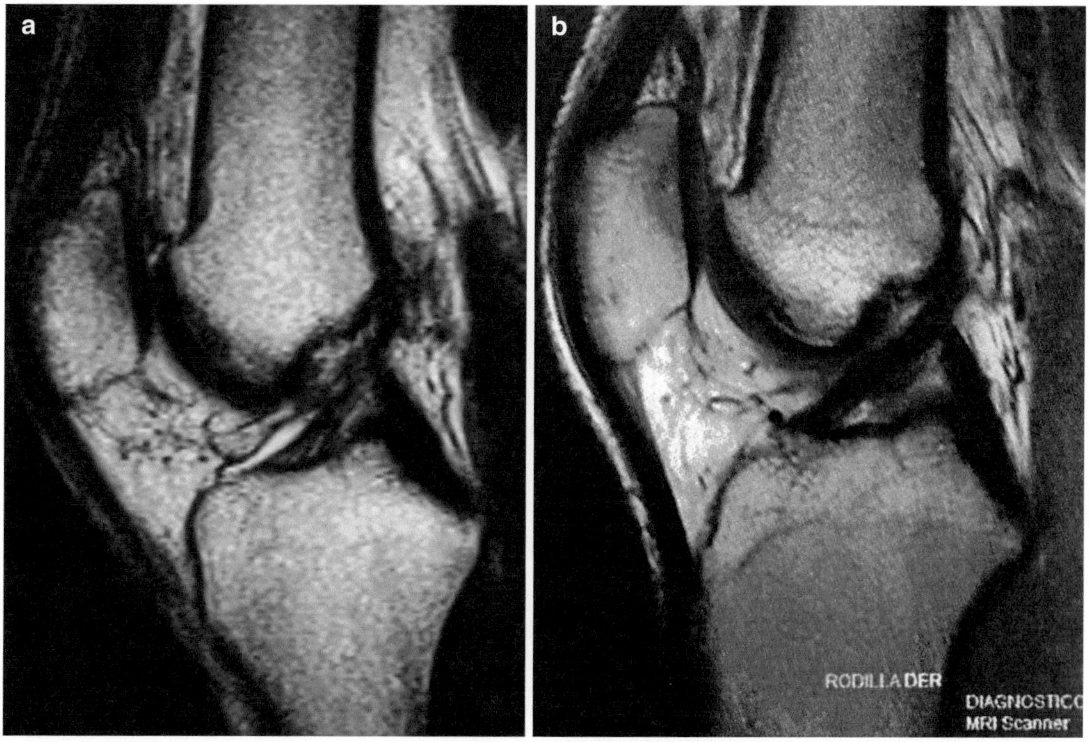

Fig. 4.3 (**a**) Magnetic resonance imaging (MRI), sagittal view, of the knee with hypertensive and heterogeneous ACL signal, compatible with a partial injury. (**b**) MRI of the same patient at 8 months post-injection of BMAC and intraligamentary PRP with hypointense and homogeneous and isointense signal with the PCL, compatible with ACL healing

the avascular region (white-white zone) are less likely to heal. These patients often undergo partial meniscectomy to preserve the undamaged part of the meniscus, relieve pain, and regain range of motion in the locked knee. Disadvantages of treatment include increased load on knee cartilage and erosion of the joint surface leading to the development of osteoarthritis and usually total knee replacement in the near future [44, 76]. The progression encouraged orthopedic surgeons to develop new options for the treatment of lesions located in the red-white or white-white zone, starting with simple options such as scaling or abrasion of the lesion site and synovial membrane [77]. Over the years, many orthobiological techniques have been demonstrated to improve the healing of meniscal injuries in clinical trials, including vascular access channels, microfractures, fibrin clot placement, platelet-rich plasma (PRP), concentrated bone marrow, or adipose or synovial tissue cells [78].

4.4.4.1 Vascular Access Channels and Synovial Abrasion

These simple methods were developed in the 1980s. The first evidence was provided by Arnoczky et al. [79] and Gao [80] in a model of canine meniscal injury. Later, Fox et al. [81] analyzed the efficacy of trepanation in incomplete meniscal lesions with good to excellent results in 90% of patients. Synovial abrasion was influential in the clinical setting for peripheral and small lesions and can induce meniscus healing without surgical fixation [82, 83].

4.4.4.2 Fibrin Clot

Exogenous fibrin clots are an initial augmentation technique in meniscal repair. Good results were demonstrated in preclinical and clinical studies [84]. Recently, newly developed ultrafiltration technologies methods of normally discarded platelet-poor plasma, can be employed, eliminating in a fast and efficient manner plasma

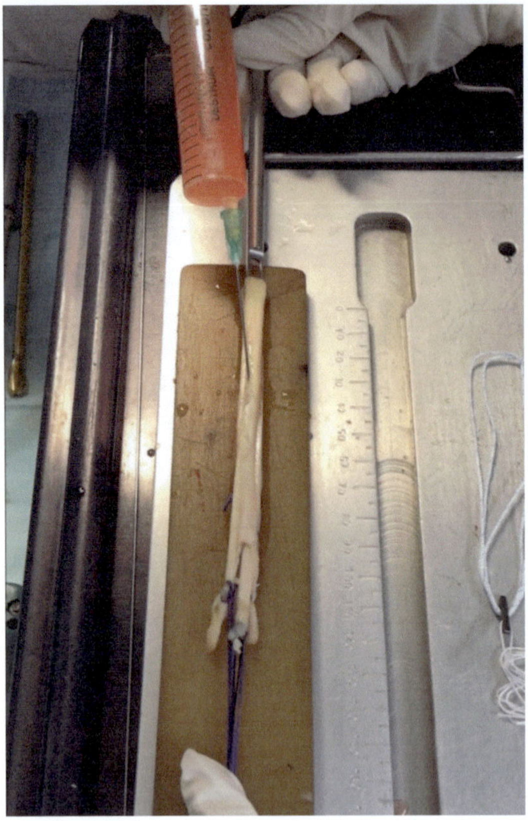

Fig. 4.4 Photograph during platelet-rich plasma (PRP) infiltration of the autologous semitendinosus and internal rectus graft as a biological augmentation of anterior cruciate ligament (ACL) reconstruction in a complete injury

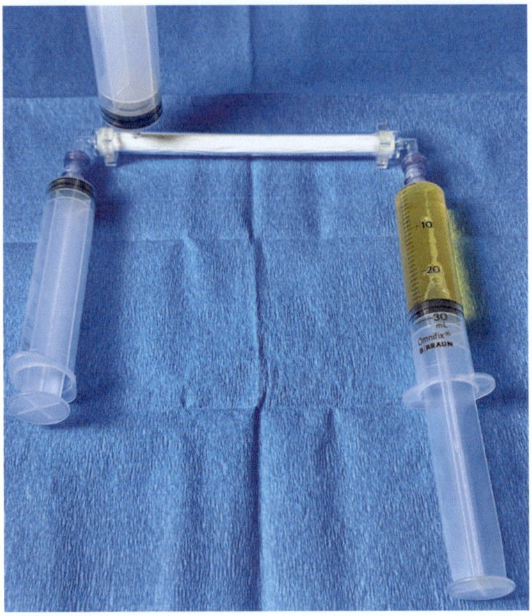

Fig. 4.5 CORE™ ultrafiltration to eliminate plasma water from PPP, to produce a protein-rich PPP concentrate which can be used as a stand-alone biological preparation to inject a high concentration of alpha-2-macroglobulin, to capture deleterious protease. Eventually, the protein-rich plasma can be mixed with PRP to produce a PR-PRP with dual functional activity, protease scavenging and immunomodulation with tissue repair

water, and plasma proteins with a molecular mass (weight) less than the pore size of the fibers. A viscous protein concentrate of functional proteins, like alpha-2-macroglobulin, fibrinogen, and albumin, is created.

After the activation of the concentrated fibrinogen, the fibrin scaffold, in particular alpha-2-macroglobulin, will capture catabolic joint-destroying proteases, facilitating an anti-inflammatory environment in order to prevent cartilage destruction.

Additionally, the concentrated PPP can be mixed with PRP before application to create a protein-rich PRP (PR-PRP) bioformulation to combine the effects of anti-protease activities with tissue immunomodulation and repair. The PR-PRP matrix retains and facilitates interactions between invading resident cells, like macrophages, fibroblasts, and mesenchymal stem cells (MSCs), and the embedded concentrated PRP cells and molecules. The administered PR-PRP biologic will undergo fibrinolysis, leading to a sustained release of concentrated cells and molecules that have been retained in the matrix (Fig. 4.5).

4.4.4.3 Platelet-Rich Plasma (PRP)

Platelet-rich plasma clots also stimulate DNA synthesis, extracellular matrix (ECM) synthesis, and biglycan and decorin mRNA expression [44, 48].

Several clinical trials have been conducted with PRP as an adjuvant to meniscus surgical repair (Fig. 4.6). In a study by Pujol et al. [85], 51 young patients underwent open meniscal repair. Seventeen of them received an additional injection of PRP during the procedure. After 1 year, an MRI was performed and five cases with no signal increase within the repaired meniscus (chronic

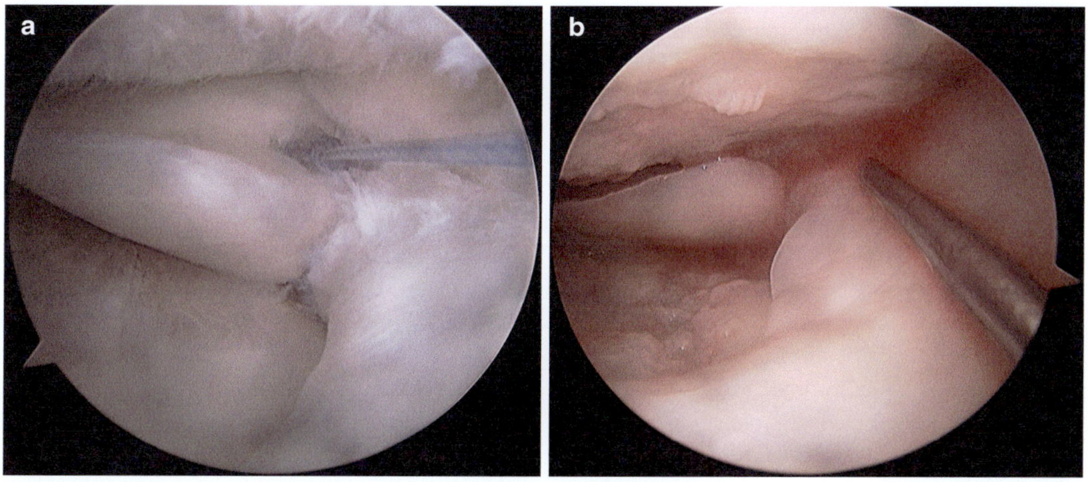

Fig. 4.6 (**a**) Arthroscopic image from the anterolateral portal of the left knee showing the meniscal suture all inside in a radial lesion in zones 2–3 of the internal meniscus body. (**b**) Arthroscopic image with dry knee, showing PRP infiltration at the site of the meniscal suture

tear) were observed in the PRP injection group in contrast to no cases as described above in the non-PRP group. After a minimum of 2 years of follow-up, better scores were reported for the PRP group on the KOOS and IKDC scales. A study by Kaminski et al. [86] showed a significant improvement in the meniscal healing rate of complete vertical meniscal tears after arthroscopy. The improvement recorded was from 47% (without PRP) to 85% of the repaired menisci treated with PRP arthroscopy. Similar to previous studies, a significant improvement in PROM scores was observed. Griffin et al. [87] performed a retrospective analysis, with a second surgery and the Lysholm score as the main outcomes analyzed, with a minimum of 2 years of follow-up. The authors did not show any benefit of increasing PRP in the treatment of meniscal lesions. A nonrandomized study by Kemmochi et al. [88] included 22 patients with meniscal lesions. Seventeen underwent meniscus repair with platelet-rich plasma and fibrin. Five patients were assigned to a control group with the menisci simply sutured. After 6 months of follow-up, a comparison of pre-and postoperative MRIs and functional assessments (Tegner, Lysholm, and IKDC) was performed. There were no significant differences between the two groups in terms of PROMs, but both showed a gradual improvement in functional assessment. MRI scans also did not indicate significant improvements. Similarly, a recent study of arthroscopic repair of lateral disc meniscus injury demonstrated no significant differences in pain relief, functional improvement, or failure rate between groups at medium-term follow-up [89]. However, all of these studies were small, with the risk of being underpowered in all outcomes. In addition, the type of injury and location differed for each study.

Everhart et al. [90] conducted a study in 550 patients undergoing meniscal repair with or without PRP administration and ACL reconstruction. Results after 3 years showed that meniscal repair with PRP (14% failure) has better outcomes than the same procedure performed without PRP (17% and 20% for isolated meniscal repairs), but no difference was recorded when surgery was performed with concomitant ACL reconstruction.

Some studies have also been done on chronic meniscal injuries. Blanke et al. [91] revealed that percutaneous PRP injections were able to relieve pain and halt the progression of meniscal injuries on MRI over a 6-month period. A recent report by Betancourt et al. [92] described a favorable outcome of PRP treatment in a patient with grade

3 internal meniscus injury during a 30-month follow-up. Another study by Strümper et al. [93] demonstrated that intra-articular injection of PRP was an effective option in knee pain associated with meniscal injuries. The study showed significantly improved function in 83% of patients. In addition, concomitant injection of PRP reduced the need for future arthroscopy during the 6-month observation period. No progression of meniscal lesions was observed on MRI, as measured by the Boston Leeds Osteoarthritis Knee Score. Kaminski et al. [76] conducted a prospective, randomized, double-blind, parallel-group, placebo-controlled study of 72 patients who underwent meniscal trephination with or without PRP injection. Failures controlled by second examination MRI/arthroscopy were rarer in the group with concomitant injection of PRP (70–48%), as was the need for future surgery in the PRP-treated group. In both groups, the KOOS subscale, the IKDC score, and the WOMAC score improved, but the values of VAS and KOOS symptoms exceeding the MCID (Minimally Important Clinical Difference) were significantly elevated in the PRP group.

4.4.4.4 Microfractures: Bone Marrow Stimulation

The healing potential is described during meniscal repair with concomitant ACL reconstruction, where drill tunnels in the femur and tibia also release bone marrow with progenitor cells and growth factors [94, 95]. A potential mechanism leading to regeneration and repair of the injured meniscus may involve differentiation of MSCs into desired tissue, production of growth factors, and fusion with meniscus cells [46].

The clinical trials of Dean et al. [94] have provided high-quality evidence for the efficacy of bone marrow stimulation or microfractures in increasing the rate of meniscal healing after repair. The failure rate was similar between meniscal repair with microfractures (12.9%) and meniscal repair plus ACL reconstruction (7.8%). Kaminski et al. [96] conducted a prospective, randomized, double-blind, parallel-group, placebo-controlled study to support the hypothesis that OM stimulation improves meniscus healing as assessed by arthroscopic control. Twenty patients underwent meniscus repair alone, while another 20 received surgery plus microfractures in the intercondylar sulcus; 100% of menisci in the microfracture group healed compared to 76% in the control group. In both groups, patients improved pain scores (VAS and KOOS) with no significant differences. Functional outcomes (IKDC, WOMAC, KOOS) were significantly better in the microfracture group.

Chronic Meniscal Injuries

Clinical trials have shown that arthroscopy is as effective as a placebo in treating chronic meniscal injuries [97, 98].

Non-surgical treatment (physiotherapy, exercise, anti-inflammatory pharmacotherapy, intra-articular injections) and X-rays (PA + side view + Schuss) for at least 3 months is the first-line strategy in degenerative lesions. If treatment fails and the patient develops symptoms, the diagnosis should be verified by magnetic resonance imaging, and non-urgent arthroscopic meniscal repair may be considered. An exception would be advanced structural osteoarthritis when arthroscopic meniscal surgery is not indicated unless mechanical symptoms are present. In such a case, early treatment of arthritis is recommended [99–101].

Degenerative meniscal injuries remain a mystery. They tend to be part of degenerative knee disease and no definitive treatment is established. We believe that these patients may benefit from percutaneous procedures, which carry very little risk compared to surgery. PRP, in particular, is already fairly well documented for playing a positive role in the treatment of knee osteoarthritis.

4.4.4.5 Cellular Therapies

Clinical studies using cell therapies for meniscal repair are currently limited. Vangsness et al. [102], in a Level 1 study of evidence, found statistically significant meniscus growth on MRI at 12 months post-injection, as well as improvements in clinical and functional outcomes using expanded allogenic bone marrow mesenchymal cells injected percutaneously into the knees. A clinical study by Pak et al. [103] reported repair of a grade II meniscal lesion after percutaneous injection of autologous adipose tissue (ASC) cells along with PRP, hyaluronic acid, and CaCl2. Using mesenchymal stem cells to stimulate meniscal tissue regeneration appears to be a promising approach to restoring as much meniscal tissue as possible. However, these regenerative technologies still need to be optimized.

4.4.5 Osteoarthritis

4.4.5.1 Platelet-Rich Plasma

The literature presents numerous studies, including recent RCTs, systematic reviews, and meta-analyses, where PRP significantly improves osteoarthritic patients, especially in pain relief and functional gains [1–6]. Studies reveal the positive attributes of intra-articular PRP, which have more excellent success rates, efficacy, and safety, but mostly in cases of knee osteoarthritis (KOA). One recent systematic review [2] revealed that PRP injection therapy can safely and effectively enhance functional activity in patients with osteoarthritis and provide significant analgesic effects for individuals with KOA, temporomandibular joint osteoarthritis (TMJOA), and ankle osteoarthritis (AOA). However, it did not significantly alleviate pain symptoms in patients with hip osteoarthritis (HOA). This may logically be attributed to the fact that the presence of leukocytes can trigger flare-ups in joints and tissues already suffering from damage and inflammation.

PRP has shown superior effects compared to other interventions such as hyaluronic acid, corticosteroid, and saline placebo. PRP injection therapy can lead to enhanced symptom relief, reduced rates of reintervention, and more frequent attainment of the minimum clinically significant difference (MCID) for pain improvement compared to alternative nonoperative treatment options [3, 4, 7].

Another recent systematic review indicated that a higher platelet dose may lead to better clinical outcomes for platelet-rich plasma treating KOA [6]. In this investigation, 29 studies were analyzed, encompassing 31 arms that utilized PRP as a treatment method. Of these, 28 arms demonstrated statistically significant positive outcomes at 6 months compared to the control group. The mean platelet dose in the 28 arms with positive outcomes was 5500 ± 474 × 10^6. In contrast, the three arms without positive differences had a mean platelet dose of 2302 ± 437 × 10^6 ($P < 0.01$). Additionally, 18 studies reported 12-month outcomes, with 16 of these showing positive results. The positive studies had an average platelet dose of 5464 ± 511 × 10^6, while those without statistically significant differences had an average platelet dose of 2253 ± 753 × 10^6 ($P < 0.05$). Regarding the frequency of injections, one study found that administering three intra-articular PRP injections, spaced one week apart, appears to be more effective than a single injection for the symptomatic management of knee OA [3].

Although PRP shows varying degrees of effectiveness across different joints, its role in enhancing the quality of life for osteoarthritic patients is undeniable. Further research with more robust sample sizes, more extended follow-up periods, refinement in PRP application protocols, and comparison with additional non-operative management alternatives will likely expand its therapeutic potential and solidify its place in regenerative medicine.

4.4.5.2 Cellular Therapies

Bone Marrow Aspirate

Several studies have examined using the MC BMA described here, although they all have relatively small sample sizes. One recent study of 17 knee OA patients reported a significant increase in all 5 subscales of the KOOS 12 weeks post-procedure, as well as a decrease in VAS scores [104]. In another study that involved intra-articular knee MC BMA in ten patients, there was more than a 50% reduction in pain 12 weeks post-procedure [105]. These studies followed patients for a relatively limited timeframe, though, 12 weeks versus the 24 weeks reported here. Finally, a study following 11 patients (15 knees) with mild knee OA found that MC BMA injections significantly improved KOOS-JR scores and Numerical Rating Scale (NRS) pain scores up to 12 months post-injection [106]. Toward this end, a large-scale randomized placebo-controlled study is warranted to determine if the findings described here using MC BMA injections are reproducible in a controlled setting. Future studies should take a closer look at BMA cell analysis to verify the yields of MSCs in the MC aspirate and measure if levels of MSCs correlate with patient outcomes, as seen in previous studies. Previous studies have found that MSC levels in BMCs correlate with good outcomes in lumbar disc pain [107], bone unions [108], and rotator cuff repairs [109]. Nonetheless, a study examining this relationship using this novel BMA aspiration technique to treat OA is needed. Additionally, radiographic imaging may help understand if any structural changes occur in the osteoarthritic joint following a BMA intra-articular injection and whether these changes correlate with positive patient-reported outcomes. Dallo et al., in a prospective study including 15 patients with BMLs of the knee, concluded that biological subchondral bone augmentation by the osteo-core plasty technique is an effective and safe procedure (Figs. 4.7 and 4.8) which significantly reduced pain and better joint function at 1 year. In addition, the MRI showed resolution of the BMLs at 6 and 12 months of follow-up, regardless of age and BMI [110].

4.4.5.3 Bone Marrow Concentrate

Multiple studies have shown improvement in patients with knee osteoarthritis using BMCs or BMAs [104, 111–115]. However, many of these studies have used BMCs with other biologics such as PRP, PPP, or adipose tissue [111, 112, 115–117] (Fig. 4.9).

Lino's technique utilizes a close aspiration system composed of a flexible cannula and vac-

Fig. 4.7 Osteo-core-plasty surgical instruments

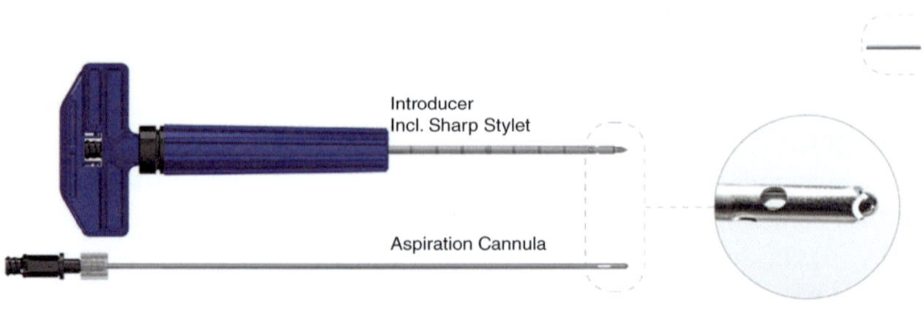

Fig. 4.8 Osteo-core-plasty surgical instruments. Closed aspiration cannula tip to prevent aspiration of excess blood from the entry channel

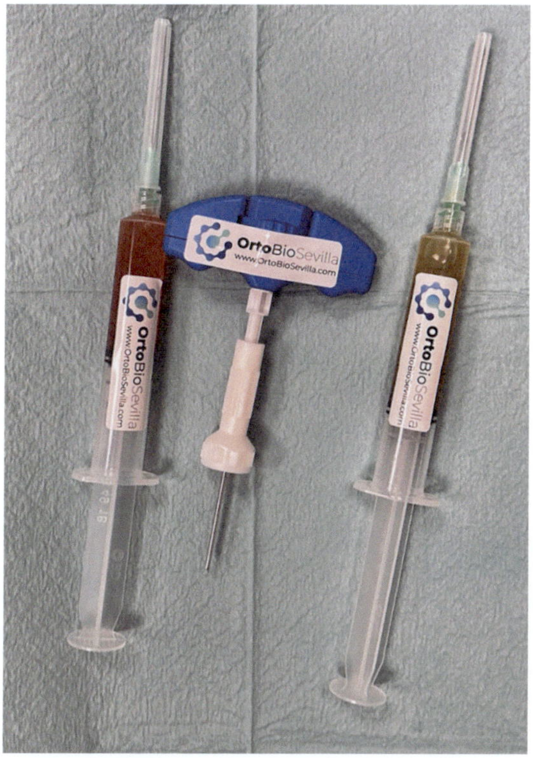

Fig. 4.9 BMA/BMC ready to treat a patient

uum tubes; it is a safe and reproducible technique used in SportMe Medical Center—OrtoBioSevilla with good results in patients with knee OA (Fig. 4.10).

In one study using BMC alone, the treatment was equivalent to PRP in significantly improving WOMAC and IKDC scores over 12 months in patients with KL grades 1–3 OA [113]. Likewise, Kim et al. reported improvement in patient outcomes after an autologous BMC injection, as measured with the VAS, KOOS, and Lysholm Knee Questionnaire [114]; in contrast, Centeno et al. reported modest improvements in pain (NPS) and functional (IKDC) scores following BMC intra-articular injections [118]. While these studies generally agree with the data presented here, the use of different BMCs in these studies makes direct comparisons difficult.

Hernigou et al., in an RCT in 120 knees, compared subchondral bone infiltrations with intra-articular injection of BMC in bilateral knee osteoarthritis. They concluded that implantation of MSCs in the subchondral bone was more effective in postponing TKA than the intra-articular injection of the same dose in the contralateral knee with the same osteoarthritis grade. In a new pilot study of a combined subchondral and intraarticular BMC injective treatment, Kon et al. showed an overall positive outcome in patients with symptomatic knee OA associated with subchondral bone pathology. They presented a reduction of bone marrow edema in MRI at 12 months follow-up [119].

In one study with a placebo-control group, the knee OA patients who received a BMC injection showed significant improvement. However, the improvement was statistically indistinguishable from the improvements in the placebo-controlled group (saline infusion) [112]. Another controlled study found that BMC-treated patients displayed some function improvements compared to an exercise therapy-treated group. However, the study only compared the two groups for 3 months, and PRP was also used in the BMC treatment group, which also received PRP [117]. These results highlight the need for additional studies using either placebo-injected controls or control groups receiving hyaluronic acid injections or other conservative treatments.

Finally, a long-term outcome study is warranted to precisely define the timeframe for statistically significant improvement and determine if the pain eventually returns to baseline.

4.4.6 Tendon Healing

4.4.6.1 PRP

Tendons, the dense connective tissues that transmit the force of muscle contraction to bones, are often subject to injuries due to direct trauma in sports or accidents. Traditional healing after tendon repair is often suboptimal due to the formation of fibrovascular scar tissues with low mechanical properties. PRP therapy has been applied to various types of tendinopathies, including rotator cuff tears, lateral epicondylitis (tennis elbow), Achilles tendinopathy, and patellar tendinopathy [120, 121]. The biological rationale behind using PRP is to provide a

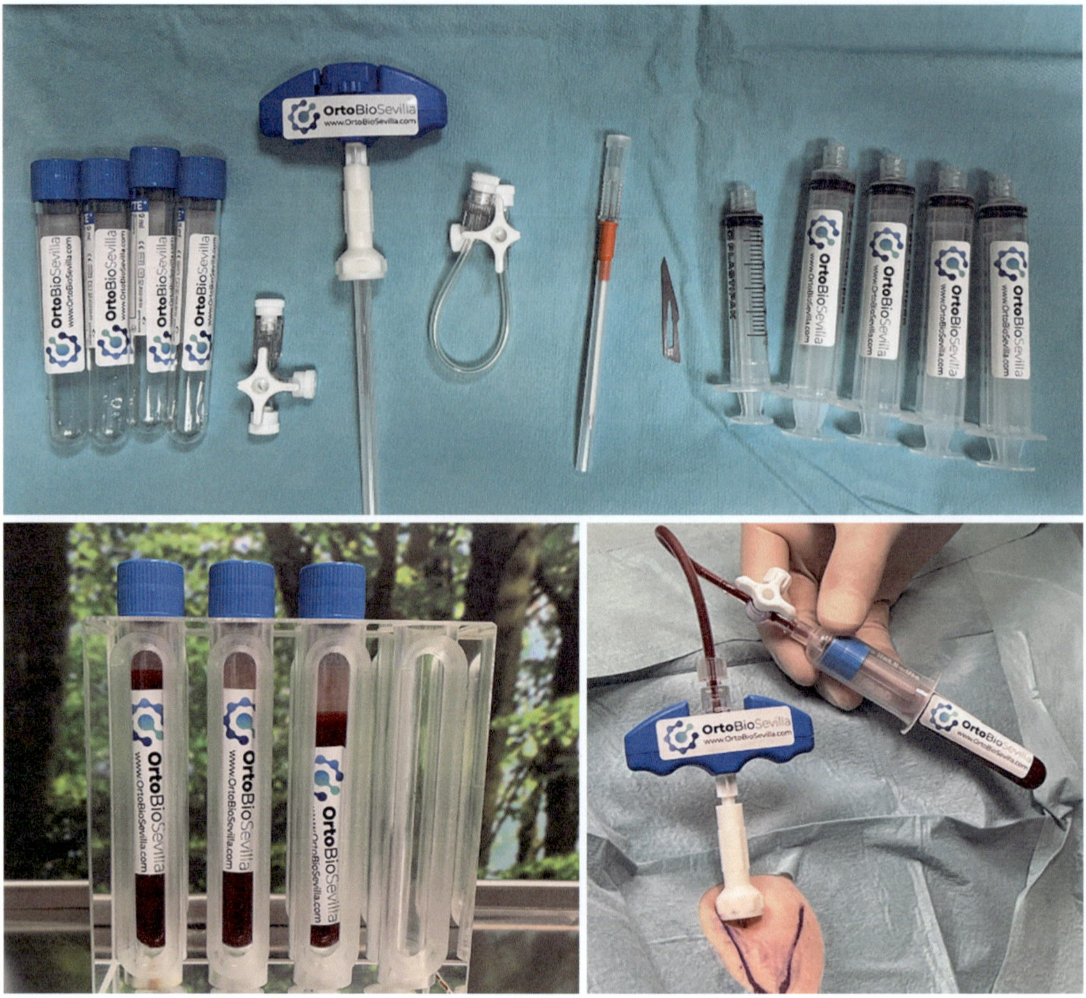

Fig. 4.10 Instruments needed to obtain BMA/BMC using a closed system with tubes

supraphysiological concentration of growth factors to the injury site, promoting tendon healing and potentially accelerating recovery. In vivo evidence for PRP use in tendon injury may be challenging as many factors could affect the results, including the type of lesion, patients' characteristics, or type of PRP [122].

4.4.7 Muscle Healing

4.4.7.1 Platelet-Rich Plasma

Muscle injuries are common in both athletic and general populations, and the healing process is often slow and incomplete. PRP, with its rich growth factors, has been explored as a potential treatment to enhance muscle repair. However, the efficacy of PRP in muscle healing remains inconclusive, necessitating further research. In a cohort study analyzing the return to play of National Football League players with grade 2 hamstring injuries treated with LP-PRP compared to controls, players who received PRP were able to return to play faster with a 1-game overall difference.

References

1. Piuzzi NS, Dominici M, Long M, Pascual-Garrido C, Rodeo S, Huard J, et al. Proceedings of the signature series symposium "cellular therapies for orthopaedics and musculoskeletal disease proven and unproven therapies-promise, facts and fan-

tasy," international society for cellular therapies, Montreal, Canada, may 2, 2018. Cytotherapy. 2018;20(11):1381–400.
2. Piuzzi NS, Khlopas A, Sodhi N, Oak S, Sultan AA, Chughtai M, et al. Cellular therapies in orthopedics: where are we? Surg Technol Int. 2017;31:359–64.
3. Ng MK, Mont MA, Piuzzi NS. Analysis of readability, quality, and content of online information available for "stem cell" injections for knee osteoarthritis. J Arthroplast. 2020;35(3):647–51.e2.
4. Piuzzi NS, Ng M, Chughtai M, Khlopas A, Ng K, Mont MA, et al. The stem-cell market for the treatment of knee osteoarthritis: a patient perspective. J Knee Surg. 2018;31(6):551–6.
5. Ramkumar PN, Navarro SM, Haeberle HS, Chughtai M, Demetriades C, Piuzzi NS, et al. Cellular therapy injections in today's orthopedic market: a social media analysis. Cytotherapy. 2017;19(12):1392–9.
6. Chu CR, Rodeo S, Bhutani N, Goodrich LR, Huard J, Irrgang J, et al. Optimizing clinical use of biologics in Orthopaedic surgery: consensus recommendations from the 2018 AAOS/NIH U-13 conference. J Am Acad Orthop Surg. 2019;27(2):e50–63.
7. Chahla J, Piuzzi NS, Mitchell JJ, Dean CS, Pascual-Garrido C, LaPrade RF, et al. Intra-articular cellular therapy for osteoarthritis and focal cartilage defects of the knee: a systematic review of the literature and study quality analysis. J Bone Joint Surg Am. 2016;98(18):1511–21.
8. Dallo Ignacio VK, Irlandini E. State of the art for injections in orthopaedics. ISAKOS Newsletter; 2020.
9. Alberto Gobbi JE-M, Lane J, Karahan M, Bio-Orthopaedic. A new approach, vol. XIX. Berlin: Springer-Verlag; 2017. p. 696.
10. Rossi LA, Murray IR, Chu CR, Muschler GF, Rodeo SA, Piuzzi NS. Classification systems for platelet-rich plasma. Bone Joint J. 2019;101-b(8):891–6.
11. Piuzzi NS, Chughtai M, Khlopas A, Harwin SF, Miniaci A, Mont MA, et al. Platelet-rich plasma for the treatment of knee osteoarthritis: a review. J Knee Surg. 2017;30(7):627–33.
12. Franco D, Franco T, Schettino AM, Filho JMT, Vendramin FS. Protocol for obtaining platelet-rich plasma (PRP), platelet-poor plasma (PPP), and thrombin for autologous use. Aesth Plast Surg. 2012;36(5):1254–9.
13. Mautner K, Malanga GA, Smith J, Shiple B, Ibrahim V, Sampson S, et al. A call for a standard classification system for future biologic research: the rationale for new PRP nomenclature. PM&R. 2015;7(4):S53–S9.
14. Lana J, Purita J, Paulus C, Huber SC, Rodrigues BL, Rodrigues AA, et al. Contributions for classification of platelet rich plasma—proposal of a new classification: MARSPILL. Regen Med. 2017;12(5):565–74.
15. Piuzzi NS, Khlopas A, Newman JM, Ng M, Roche M, Husni ME, et al. Bone marrow cellular therapies: novel therapy for knee osteoarthritis. J Knee Surg. 2018;31(1):22–6.
16. Piuzzi NS, Mantripragada VP, Kwee E, Sumski A, Selvam S, Boehm C, et al. Bone marrow-derived cellular therapies in Orthopaedics: part II: recommendations for reporting the quality of bone marrow-derived cell populations. JBJS Rev. 2018;6(11):e5.
17. Piuzzi NS, Mantripragada VP, Sumski A, Selvam S, Boehm C, Muschler GF. Bone marrow-derived cellular therapies in orthopaedics: part I: recommendations for bone marrow aspiration technique and safety. JBJS Rev. 2018;6(11):e4.
18. Dallo I, Frank RM, Bradsell H, Piuzzi NS, Gobbi A. Overview of orthobiologics and joint function. In: Gobbi A, Lane JG, Longo UG, Dallo I, editors. Joint function preservation: a focus on the osteochondral unit. Cham: Springer; 2022. p. 21–31.
19. Dominici M, Le Blanc K, Mueller I, Slaper-Cortenbach I, Marini F, Krause D, et al. Minimal criteria for defining multipotent mesenchymal stromal cells. The International Society for Cellular Therapy position statement. Cytotherapy. 2006;8(4):315–7.
20. Caplan AI. Mesenchymal stem cells: time to change the name! Stem Cells Transl Med. 2017;6(6):1445–51.
21. Murray IR, Chahla J, Safran MR, Krych AJ, Saris DBF, Caplan AI, et al. International expert consensus on a cell therapy communication tool: DOSES. J Bone Joint Surg Am. 2019;101(10):904–11.
22. Chahla J, Cinque ME, Piuzzi NS, Mannava S, Geeslin AG, Murray IR, et al. A call for standardization in platelet-rich plasma preparation protocols and composition reporting: a systematic review of the clinical Orthopaedic literature. J Bone Joint Surg Am. 2017;99(20):1769–79.
23. Piuzzi NS, Hussain ZB, Chahla J, Cinque ME, Moatshe G, Mantripragada VP, et al. Variability in the preparation, reporting, and use of bone marrow aspirate concentrate in musculoskeletal disorders: a systematic review of the clinical Orthopaedic literature. J Bone Joint Surg Am. 2018;100(6):517–25.
24. Mantripragada VP, Bova WA, Piuzzi NS, Boehm C, Obuchowski NA, Midura RJ, et al. Native-osteoarthritic joint resident stem and progenitor cells for cartilage cell-based therapies: a quantitative comparison with respect to concentration and biological performance. Am J Sports Med. 2019;47(14):3521–30.
25. Kubrova E, D'Souza RS, Hunt CL, Wang Q, Van Wijnen AJ, Qu W. Injectable biologics. Am J Phys Med Rehabil. 2020;99:1.
26. Nguyen RT, Borg-Stein J, McInnis K. Applications of platelet-rich plasma in musculoskeletal and sports medicine: an evidence-based approach. PM R. 2011;3(3):226–50.
27. Zhu Y, Yuan M, Meng HY, Wang AY, Guo QY, Wang Y, et al. Basic science and clinical application of platelet-rich plasma for cartilage defects and osteoarthritis: a review. Osteoarthr Cartil. 2013;21(11):1627–37.
28. Cavallo C, Roffi A, Grigolo B, Mariani E, Pratelli L, Merli G, et al. Platelet-rich plasma: the choice

of activation method affects the release of bioactive molecules. Biomed Res Int. 2016;2016:6591717.
29. Boakye LA, Ross KA, Pinski JM, Smyth NA, Haleem AM, Hannon CP, et al. Platelet-rich plasma increases transforming growth factor-beta1 expression at graft-host interface following autologous osteochondral transplantation in a rabbit model. World J Orthop. 2015;6(11):961–9.
30. Wu CC, Chen WH, Zao B, Lai PL, Lin TC, Lo HY, et al. Regenerative potentials of platelet-rich plasma enhanced by collagen in retrieving pro-inflammatory cytokine-inhibited chondrogenesis. Biomaterials. 2011;32(25):5847–54.
31. Fice MP, Miller JC, Christian R, Hannon CP, Smyth N, Murawski CD, et al. The role of platelet-rich plasma in cartilage pathology: an updated systematic review of the basic science evidence. Arthroscopy. 2019;35(3):961–76.e3.
32. Everts P, Onishi K, Jayaram P, Lana JF, Mautner K. Platelet-rich plasma: new performance understandings and therapeutic considerations in 2020. Int J Mol Sci. 2020;21(20):7794.
33. Re'em T, Kaminer-Israeli Y, Ruvinov E, Cohen S. Chondrogenesis of hMSC in affinity-bound TGF-beta scaffolds. Biomaterials. 2012;33(3):751–61.
34. Yin Z, Yang X, Jiang Y, Xing L, Xu Y, Lu Y, et al. Platelet-rich plasma combined with agarose as a bioactive scaffold to enhance cartilage repair: an in vitro study. J Biomater Appl. 2014;28(7):1039–50.
35. Pereira RC, Scaranari M, Benelli R, Strada P, Reis RL, Cancedda R, et al. Dual effect of platelet lysate on human articular cartilage: a maintenance of chondrogenic potential and a transient proinflammatory activity followed by an inflammation resolution. Tissue Eng Part A. 2013;19(11–12):1476–88.
36. Lee HR, Park KM, Joung YK, Park KD, Do SH. Platelet-rich plasma loaded hydrogel scaffold enhances chondrogenic differentiation and maturation with up-regulation of CB1 and CB2. J Control Release. 2012;159(3):332–7.
37. Dohan Ehrenfest DM, Rasmusson L, Albrektsson T. Classification of platelet concentrates: from pure platelet-rich plasma (P-PRP) to leucocyte-and platelet-rich fibrin (L-PRF). Trends Biotechnol. 2009;27(3):158–67.
38. Dohan Ehrenfest DM, Bielecki T, Jimbo R, Barbe G, Del Corso M, Inchingolo F, et al. Do the fibrin architecture and leukocyte content influence the growth factor release of platelet concentrates? An evidence-based answer comparing a pure platelet-rich plasma (P-PRP) gel and a leukocyte-and platelet-rich fibrin (L-PRF). Curr Pharm Biotechnol. 2012;13(7):1145–52.
39. Mishra A, Harmon K, Woodall J, Vieira A. Sports medicine applications of platelet rich plasma. Curr Pharm Biotechnol. 2012;13(7):1185–95.
40. Bielecki T, Dohan Ehrenfest DM, Everts PA, Wiczkowski A. The role of leukocytes from L-PRP/L-PRF in wound healing and immune defense: new perspectives. Curr Pharm Biotechnol. 2012;13(7):1153–62.
41. Cavallo C, Filardo G, Mariani E, Kon E, Marcacci M, Pereira Ruiz MT, et al. Comparison of platelet-rich plasma formulations for cartilage healing: an in vitro study. J Bone Joint Surg Am. 2014;96(5):423–9.
42. Moojen DJ, Everts PA, Schure RM, Overdevest EP, van Zundert A, Knape JT, et al. Antimicrobial activity of platelet-leukocyte gel against Staphylococcus aureus. J Orthop Res. 2008;26(3):404–10.
43. Di Martino A, Di Matteo B, Papio T, Tentoni F, Selleri F, Cenacchi A, et al. Platelet-rich plasma versus hyaluronic acid injections for the treatment of knee osteoarthritis: results at 5 years of a double-blind, randomized controlled trial. Am J Sports Med. 2019;47(2):347–54.
44. Longo UG, Campi S, Romeo G, Spiezia F, Maffulli N, Denaro V. Biological strategies to enhance healing of the avascular area of the meniscus. Stem Cells Int. 2012;2012:528359.
45. Koch M, Hammer S, Fuellerer J, Lang S, Pfeifer CG, Pattappa G, et al. Bone marrow aspirate concentrate for the treatment of avascular meniscus tears in a one-step procedure-evaluation of an in vivo model. Int J Mol Sci. 2019;20(5):1120.
46. Pak J, Lee JH, Lee SH. Regenerative repair of damaged meniscus with autologous adipose tissue-derived stem cells. Biomed Res Int. 2014;2014:436029.
47. Cassano JM, Kennedy JG, Ross KA, Fraser EJ, Goodale MB, Fortier LA. Bone marrow concentrate and platelet-rich plasma differ in cell distribution and interleukin 1 receptor antagonist protein concentration. Knee Surg Sports Traumatol Arthroscopy. 2018;26(1):333–42.
48. Hutchinson ID, Rodeo SA, Perrone GS, Murray MM. Can platelet-rich plasma enhance anterior cruciate ligament and meniscal repair? J Knee Surg. 2015;28(1):19–28.
49. Snyder TN, Madhavan K, Intrator M, Dregalla RC, Park D. A fibrin/hyaluronic acid hydrogel for the delivery of mesenchymal stem cells and potential for articular cartilage repair. J Biol Eng. 2014;8:10.
50. McNulty AL, Guilak F. Mechanobiology of the meniscus. J Biomech. 2015;48(8):1469–78.
51. Infante A, Rubio-Azpeitia E, Sanchez P, Alberdi R, Rodriguez CI, Andia I. Platelet rich plasma and culture configuration affect the matrix forming phenotype of bone marrow stromal cells. Tissue Eng Regen Med. 2017;14(5):567–77.
52. van Buul GM, Koevoet WL, Kops N, Bos PK, Verhaar JA, Weinans H, et al. Platelet-rich plasma releasate inhibits inflammatory processes in osteoarthritic chondrocytes. Am J Sports Med. 2011;39(11):2362–70.
53. Cugat R, Alentorn-Geli E, Steinbacher G, Alvarez-Diaz P, Cusco X, Seijas R, et al. Treatment of knee osteochondral lesions using a novel clot of autologous plasma rich in growth factors mixed with healthy

hyaline cartilage chips and intra-articular injection of PRGF. Case Rep Orthop. 2017;2017:8284548.
54. Cugat R, Alentorn-Geli E, Navarro J, Cusco X, Steinbacher G, Seijas R, et al. A novel autologous-made matrix using hyaline cartilage chips and platelet-rich growth factors for the treatment of full-thickness cartilage or osteochondral defects: preliminary results. J Orthop Surg (Hong Kong). 2020;28(1):2309499019887547.
55. Dominguez Perez JM, Fernandez-Sarmiento JA, Aguilar Garcia D, Granados Machuca MDM, Morgaz Rodriguez J, Navarrete Calvo R, et al. Cartilage regeneration using a novel autologous growth factors-based matrix for full-thickness defects in sheep. Knee Surg Sports Traumatol Arthrosc. 2019;27(3):950–61.
56. Gobbi A, Whyte GP. Long-term clinical outcomes of one-stage cartilage repair in the knee with hyaluronic acid-based scaffold embedded with mesenchymal stem cells sourced from bone marrow aspirate concentrate. Am J Sports Med. 2019;47(7):1621–8.
57. Gobbi A, Dallo I, Kumar V. Editorial commentary: biological cartilage repair technique-an "effective, accessible, and safe" surgical solution for an old difficult biological problem. Arthroscopy. 2020;36(3):859–61.
58. Caplan AI, Correa D. The MSC: an injury drugstore. Cell Stem Cell. 2011;9(1):11–5.
59. Krych AJ, Nawabi DH, Farshad-Amacker NA, Jones KJ, Maak TG, Potter HG, et al. Bone marrow concentrate improves early cartilage phase maturation of a scaffold plug in the knee: a comparative magnetic resonance imaging analysis to platelet-rich plasma and control. Am J Sports Med. 2016;44(1):91–8.
60. Costa-Paz M, Ayerza MA, Tanoira I, Astoul J, Muscolo DL. Spontaneous healing in complete ACL ruptures: a clinical and MRI study. Clin Orthop Relat Res. 2012;470(4):979–85.
61. Seijas R, Ares O, Cusco X, Alvarez P, Steinbacher G, Cugat R. Partial anterior cruciate ligament tears treated with intraligamentary plasma rich in growth factors. World J Orthop. 2014;5(3):373–8.
62. Centeno CJ, Pitts J, Al-Sayegh H, Freeman MD. Anterior cruciate ligament tears treated with percutaneous injection of autologous bone marrow nucleated cells: a case series. J Pain Res. 2015;8:437–47.
63. Dallo I. Aumentación Biológica Intra-operatoria en Lesiones Parciales del LCA. Serie de Casos, Técnica y Resultados: Revista de la Asociación Argentina de Traumatologia del Deporte; 2017.
64. Gobbi A, Whyte GP. Long-term outcomes of primary repair of the anterior cruciate ligament combined with biologic healing augmentation to treat incomplete tears. Am J Sports Med. 2018;46(14):3368–77.
65. Dallo I, Chahla J, Mitchell JJ, Pascual-Garrido C, Feagin JA, LaPrade RF. Biologic approaches for the treatment of partial tears of the anterior cruciate ligament: a current concepts review. Orthop J Sports Med. 2017;5(1):2325967116681724.
66. Figueroa D, Figueroa F, Calvo R, Vaisman A, Ahumada X, Arellano S. Platelet-rich plasma use in anterior cruciate ligament surgery: systematic review of the literature. Arthroscopy. 2015;31(5):981–8.
67. Komatsu DE, King L, Gurevich M, Kahn B, Paci JM. The In vivo impact of leukocyte injections on normal rat achilles tendons: potential detriment to tendon morphology, cellularity, and vascularity. Am J Orthop (Belle Mead, NJ). 2018;47(10):1.
68. Magnussen RA, Flanigan DC, Pedroza AD, Heinlein KA, Kaeding CC. Platelet rich plasma use in allograft ACL reconstructions: two-year clinical results of a MOON cohort study. Knee. 2013;20(4):277–80.
69. Sanchez M, Anitua E, Azofra J, Prado R, Muruzabal F, Andia I. Ligamentization of tendon grafts treated with an endogenous preparation rich in growth factors: gross morphology and histology. Arthroscopy. 2010;26(4):470–80.
70. Mirzatolooei F, Alamdari MT, Khalkhali HR. The impact of platelet-rich plasma on the prevention of tunnel widening in anterior cruciate ligament reconstruction using quadrupled autologous hamstring tendon: a randomised clinical trial. Bone Joint J. 2013;95-b(1):65–9.
71. Vogrin M, Rupreht M, Crnjac A, Dinevski D, Krajnc Z, Recnik G. The effect of platelet-derived growth factors on knee stability after anterior cruciate ligament reconstruction: a prospective randomized clinical study. Wien Klin Wochenschr. 2010;122(Suppl 2):91–5.
72. Vadala A, Iorio R, De Carli A, Ferretti M, Paravani D, Caperna L, et al. Platelet-rich plasma: does it help reduce tunnel widening after ACL reconstruction? Knee Surg Sports Traumatol Arthroscopy. 2013;21(4):824–9.
73. Rupreht M, Vogrin M, Hussein M. MRI evaluation of tibial tunnel wall cortical bone formation after platelet-rich plasma applied during anterior cruciate ligament reconstruction. Radiol Oncol. 2013;47(2):119–24.
74. Marui T, Niyibizi C, Georgescu HI, Cao M, Kavalkovich KW, Levine RE, et al. Effect of growth factors on matrix synthesis by ligament fibroblasts. J Orthop Res. 1997;15(1):18–23.
75. Di Matteo B, Loibl M, Andriolo L, Filardo G, Zellner J, Koch M, et al. Biologic agents for anterior cruciate ligament healing: a systematic review. World J Orthop. 2016;7(9):592–603.
76. Kaminski R, Maksymowicz-Wleklik M, Kulinski K, Kozar-Kaminska K, Dabrowska-Thing A, Pomianowski S. Short-term outcomes of percutaneous trephination with a platelet rich plasma Intrameniscal injection for the repair of degenerative meniscal lesions. A prospective, randomized, double-blind, parallel-group, placebo-controlled study. Int J Mol Sci. 2019;20(4):856.
77. Uchio Y, Ochi M, Adachi N, Kawasaki K, Iwasa J. Results of rasping of meniscal tears with and without anterior cruciate ligament injury as evalu-

ated by second-look arthroscopy. Arthroscopy. 2003;19(5):463–9.
78. Chirichella PS, Jow S, Iacono S, Wey HE, Malanga GA. Treatment of knee meniscus pathology: rehabilitation, surgery, and orthobiologics. PM R. 2019;11(3):292–308.
79. Arnoczky SP, Warren RF. The microvasculature of the meniscus and its response to injury. An experimental study in the dog. Am J Sports Med. 1983;11(3):131–41.
80. Gao JZ. Experimental study on healing of old tear in the avascular portion of menisci in dogs. Zhonghua Wai Ke Za Zhi. 1990;28(12):726–9, 82.
81. Fox JM, Rintz KG, Ferkel RD. Trephination of incomplete meniscal tears. Arthroscopy. 1993;9(4):451–5.
82. Henning CE, Lynch MA, Clark JR. Vascularity for healing of meniscus repairs. Arthroscopy. 1987;3(1):13–8.
83. Shelbourne KD, Gray T. Meniscus tears that can be left in situ, with or without trephination or synovial abrasion to stimulate healing. Sports Med Arthrosc Rev. 2012;20(2):62–7.
84. Chahla J, Kennedy NI, Geeslin AG, Moatshe G, Cinque ME, DePhillipo NN, et al. Meniscal repair with fibrin clot augmentation. Arthrosc Tech. 2017;6(6):e2065–e9.
85. Pujol N, Tardy N, Boisrenoult P, Beaufils P. Long-term outcomes of all-inside meniscal repair. Knee Surg Sports Traumatol Arthrosc. 2015;23(1):219–24.
86. Kaminski R, Kulinski K, Kozar-Kaminska K, Wielgus M, Langner M, Wasko MK, et al. A prospective, randomized, double-blind, parallel-group, placebo-controlled study evaluating meniscal healing, clinical outcomes, and safety in patients undergoing meniscal repair of unstable, complete vertical meniscal tears (bucket handle) augmented with platelet-rich plasma. Biomed Res Int. 2018;2018:9315815.
87. Griffin JW, Hadeed MM, Werner BC, Diduch DR, Carson EW, Miller MD. Platelet-rich plasma in meniscal repair: does augmentation improve surgical outcomes? Clin Orthop Relat Res. 2015;473(5):1665–72.
88. Kemmochi M, Sasaki S, Takahashi M, Nishimura T, Aizawa C, Kikuchi J. The use of platelet-rich fibrin with platelet-rich plasma support meniscal repair surgery. J Orthop. 2018;15(2):711–20.
89. Dai WL, Zhang H, Lin ZM, Shi ZJ, Wang J. Efficacy of platelet-rich plasma in arthroscopic repair for discoid lateral meniscus tears. BMC Musculoskelet Disord. 2019;20(1):113.
90. Everhart JS, Cavendish PA, Eikenberry A, Magnussen RA, Kaeding CC, Flanigan DC. Platelet-rich plasma reduces failure risk for isolated meniscal repairs but provides no benefit for meniscal repairs with anterior cruciate ligament reconstruction. Am J Sports Med. 2019;47(8):1789–96.
91. Blanke F, Vavken P, Haenle M, von Wehren L, Pagenstert G, Majewski M. Percutaneous injections of platelet rich plasma for treatment of intrasubstance meniscal lesions. Muscles Ligaments Tendons J. 2015;5(3):162–6.
92. Betancourt JP, Murrell WD. Leukocyte-poor platelet-rich plasma to treat degenerative meniscal tear: a case report. J Clin Orthop Trauma. 2016;7(Suppl 1):106–9.
93. Strumper R. Intra-articular injections of autologous conditioned serum to treat pain from meniscal lesions. Sports Med Int Open. 2017;1(6):E200–e5.
94. Dean CS, Chahla J, Matheny LM, Mitchell JJ, LaPrade RF. Outcomes after biologically augmented isolated meniscal repair with marrow venting are comparable with those after meniscal repair with concomitant anterior cruciate ligament reconstruction. Am J Sports Med. 2017;45(6):1341–8.
95. Driscoll MD, Robin BN, Horie M, Hubert ZT, Sampson HW, Jupiter DC, et al. Marrow stimulation improves meniscal healing at early endpoints in a rabbit meniscal injury model. Arthroscopy. 2013;29(1):113–21.
96. Kaminski R, Kulinski K, Kozar-Kaminska K, Wasko MK, Langner M, Pomianowski S. Repair augmentation of unstable, complete vertical meniscal tears with bone marrow venting procedure: a prospective, randomized, double-blind, parallel-group, placebo-controlled study. Arthroscopy. 2019;35(5): 1500–8.e1.
97. Katz JN, Brophy RH, Chaisson CE, de Chaves L, Cole BJ, Dahm DL, et al. Surgery versus physical therapy for a meniscal tear and osteoarthritis. N Engl J Med. 2013;368(18):1675–84.
98. Sihvonen R, Paavola M, Malmivaara A, Itala A, Joukainen A, Nurmi H, et al. Arthroscopic partial meniscectomy versus sham surgery for a degenerative meniscal tear. N Engl J Med. 2013;369(26):2515–24.
99. Beaufils P, Becker R, Kopf S, Englund M, Verdonk R, Ollivier M, et al. Surgical management of degenerative meniscus lesions: the 2016 ESSKA meniscus consensus. Knee Surg Sports Traumatol Arthrosc. 2017;25(2):335–46.
100. Abram SGF, Beard DJ, Price AJ. Arthroscopic meniscal surgery: a national society treatment guideline and consensus statement. Bone Joint J. 2019;101-b(6):652–9.
101. Doral MN, Bilge O, Huri G, Turhan E, Verdonk R. Modern treatment of meniscal tears. EFORT Open Rev. 2018;3(5):260–8.
102. Vangsness CT Jr, Farr J 2nd, Boyd J, Dellaero DT, Mills CR, LeRoux-Williams M. Adult human mesenchymal stem cells delivered via intra-articular injection to the knee following partial medial meniscectomy: a randomized, double-blind, controlled study. J Bone Joint Surg Am. 2014;96(2):90–8.
103. Pak J, Chang JJ, Lee JH, Lee SH. Safety reporting on implantation of autologous adipose tissue-derived stem cells with platelet-rich plasma into human articular joints. BMC Musculoskelet Disord. 2013;14:337.

104. Varady NH, Cate G, Barghi A, Jobe N, Yakin D, Ylanan RC, et al. Positive early clinical outcomes of bone marrow aspirate concentrate for osteoarthritis using a novel fenestrated trocar. Knee. 2020;27(5):1627–34.
105. Colberg RE, Jurado Vélez JA, Walsh KP, Fleisig G. Short-term outcomes after pure bone marrow aspirate injection for severe knee osteoarthritis: a case series. Regen Med. 2020;15(7):1851–9.
106. Wells K, Klein M, Hurwitz N, Santiago K, Cheng J, Abutalib Z, et al. Cellular and clinical analyses of autologous bone marrow aspirate Injectate for knee osteoarthritis: a pilot study. PM&R. 2020;13(4):387.
107. Pettine KA, Murphy MB, Suzuki RK, Sand TT. Percutaneous injection of autologous bone marrow concentrate cells significantly reduces lumbar discogenic pain through 12 months. Stem Cells. 2015;33:146–56.
108. Hernigou P, Poignard A, Beaujean F, Rouard H. Percutaneous autologous bone-marrow grafting for nonunions. J Bone Joint Surg. 2005;86:1430–7.
109. Hernigou P, Flouzat Lachaniette CH, Delambre J, Zilber S, Duffiet P, Chevallier N, et al. Biologic augmentation of rotator cuff repair with mesenchymal stem cells during arthroscopy improves healing and prevents further tears: a case-controlled study. Int Orthop. 2014;38(9):1811–8.
110. Dallo I, D'Ambrosi R, Szwedowski D, Mobasheri A, Gobbi A. Minimally invasive cell-based therapy for symptomatic bone marrow lesions of the knee: a prospective clinical study at 1 year. Stem Cells Dev. 2022;31:488.
111. Centeno C, Pitts J, Al-Sayegh H, Freeman M. Efficacy of autologous bone marrow concentrate for knee osteoarthritis with and without adipose graft. Biomed Res Int. 2014;2014:370621.
112. Shapiro SA, Kazmerchak SE, Heckman MG, Zubair AC, O'Connor MI. A prospective, single-blind, placebo-controlled trial of bone marrow aspirate concentrate for knee osteoarthritis. Am J Sports Med. 2017;45(1):82–90.
113. Anz AW, Hubbard R, Rendos NK, Everts PA, Andrews JR, Hackel JG. Bone marrow aspirate concentrate is equivalent to platelet-rich plasma for the treatment of knee osteoarthritis at 1 year: a prospective, randomized trial. Orthop J Sports Med. 2020;8(2):2325967119900958.
114. Kim JD, Lee GW, Jung GH, Kim CK, Kim T, Park JH, et al. Clinical outcome of autologous bone marrow aspirates concentrate (BMAC) injection in degenerative arthritis of the knee. Eur J Orthop Surg Traumatol. 2014;24(8):1505–11.
115. Sampson S, Smith J, Vincent H, Aufiero D, Zall M, Botto-van-Bemden A. Intra-articular bone marrow concentrate injection protocol: short-term efficacy in osteoarthritis. Regen Med. 2016;11(6):511–20.
116. Prodromos C, Finkle S. Autologous biologic treatment with fat, bone marrow aspirate and platelet rich plasma is an effective alternative to Total knee arthroplasty for patients with moderate knee arthrosis. Medicines (Basel). 2020;7(6):37.
117. Centeno C, Sheinkop M, Dodson E, Stemper I, Williams C, Hyzy M, et al. A specific protocol of autologous bone marrow concentrate and platelet products versus exercise therapy for symptomatic knee osteoarthritis: a randomized controlled trial with 2 year follow-up. J Transl Med. 2018;16(1):355.
118. Centeno CJ, Al-Sayegh H, Bashir J, Goodyear S, Freeman MD. A dose response analysis of a specific bone marrow concentrate treatment protocol for knee osteoarthritis. BMC Musculoskelet Disord. 2015;16:258.
119. Kon E, Boffa A, Andriolo L, Di Martino A, Di Matteo B, Magarelli N, et al. Subchondral and intra-articular injections of bone marrow concentrate are a safe and effective treatment for knee osteoarthritis: a prospective, multi-center pilot study. Knee Surg Sports Traumatol Arthrosc. 2021;29:4232.
120. Randelli P, Arrigoni P, Ragone V, Aliprandi A, Cabitza P. Platelet rich plasma in arthroscopic rotator cuff repair: a prospective RCT study, 2-year follow-up. J Shoulder Elb Surg. 2011;20(4):518–28.
121. Gosens T, Peerbooms JC, van Laar W, den Oudsten BL. Ongoing positive effect of platelet-rich plasma versus corticosteroid injection in lateral epicondylitis: a double-blind randomized controlled trial with 2-year follow-up. Am J Sports Med. 2011;39(6):1200–8.
122. Bradley JP, Lawyer TJ, Ruef S, Towers JD, Arner JW. Platelet-rich plasma shortens return to play in National Football League Players with Acute Hamstring Injuries. Orthop J Sports Med. 2020;8(4):2325967120911731.

Biological Products in Regenerative Medicine

5

Mikel Sánchez, Ramón Cugat, Isabel Andía, João Espregueira-Mendes, and Diego Delgado

5.1 Introduction: Biological Products in Regenerative Sports Medicine

Regenerative sports medicine focuses on fast-tracking recovery from injury and mitigating overuse injury in athletes, emphasizing new non-invasive treatments beyond corticosteroids. This shift toward cellular-focused therapies marks the rise of orthobiology, leveraging stem cell biology to enhance healing and tissue regeneration. It emphasizes the vital link between cellular health and tissue functionality, exploring complex cellular interactions facilitated by signaling molecules [1]. Blood-derived products and cell-based therapies are rich in signaling molecules that are intended to restore tissue homeostasis. They are explored for their potential in enhancing healing and tissue repair.

5.1.1 Exploring Cellular Therapies: From Minimal Processing to Advanced Preparations

Mesenchymal stem/stromal cells (MSCs) play a vital role in the cell-based therapies (Fig. 5.1), with the term "MSC" encompassing various interpretations like "Mesenchymal Stromal Cells" or "Mesenchymal Stem Cells." The International

M. Sánchez (✉) · D. Delgado
Arthroscopic Surgery Unit, Hospital Vithas Vitoria, Vitoria-Gasteiz, Spain

Advanced Biological Therapy Unit, Hospital Vithas Vitoria, Vitoria-Gasteiz, Spain
e-mail: mikel.sanchez@ucatrauma.com

R. Cugat
Instituto Cugat, Hospital Quironsalud Barcelona, Barcelona, Spain

Fundación García Cugat, Barcelona, Spain

Mutualidad de Futbolistas Españoles - Delegación Catalana, Barcelona, Spain

I. Andía
Clínica Espregueira—FIFA Medical Centre of Excellence, Porto, Portugal

Regenerative Therapies, Biobizkaia Health Research Institute, Cruces University Hospital, Barakaldo, Bizkaia, Spain

J. Espregueira-Mendes
Dom Henrique Research Centre, Porto, Portugal

School of Medicine, University of Minho, Braga, Portugal

ICVS/3B's–PT Government Associate Laboratory, Braga/Guimarães, Portugal

3B's Research Group—Biomaterials, Biodegradables and Biomimetics, University of Minho, Headquarters of the European Institute of Excellence on Tissue Engineering and Regenerative Medicine, Barco, Guimarães, Portugal

A. Gobbi et al. (eds.), *Regenerative Medicine in Sports and Orthopaedics*,
https://doi.org/10.1007/978-3-031-84693-9_5

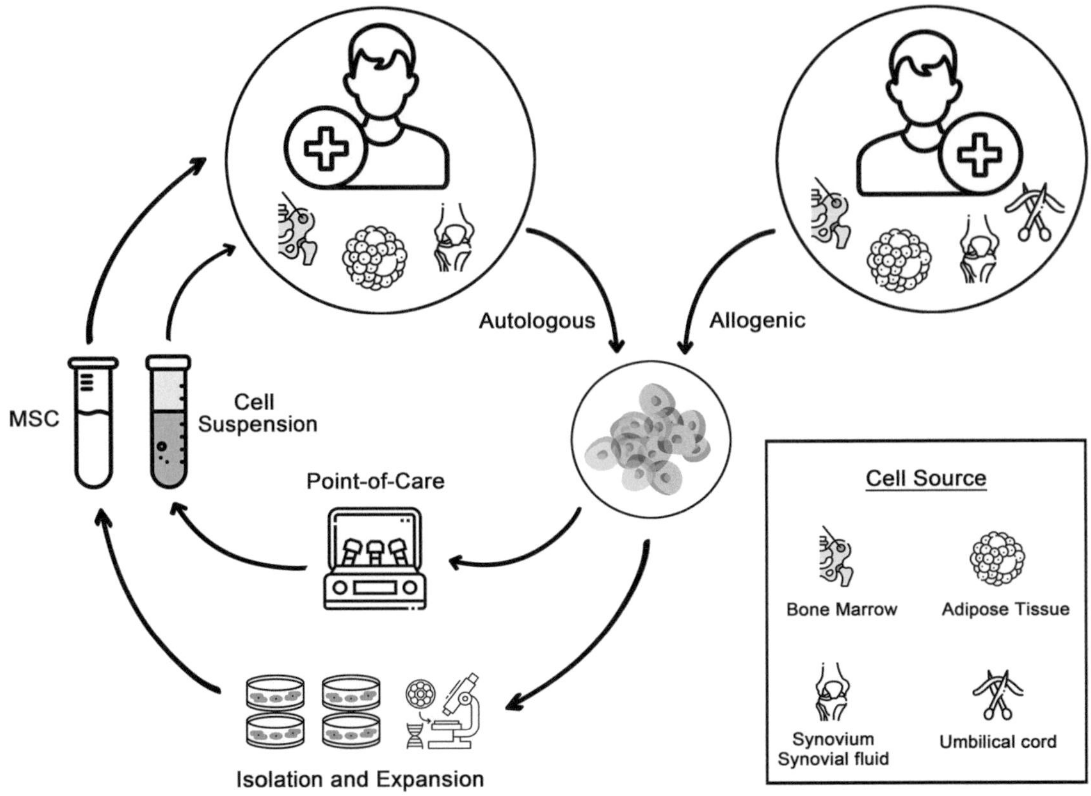

Fig. 5.1 Heterogeneity of cell-based therapies. Cell products can be of both autologous and allogeneic origin, obtained from a variety of cell sources. This, in addition to processing with point-of-care systems or by isolation and expansion of MSC in the laboratory, results in very different products. Resources from https://www.flaticon.es/ were used

Society for Cell & Gene Therapy (ISCT) outlines criteria for the MSC phenotype, including plastic adherence, specific surface markers, and trilineage differentiation capabilities. However, the complexity arises from the existence of different subsets within MSCs and variability in surface markers. The ISCT MSC committee introduces new recommendations, distinguishing between "mesenchymal stem cells" and "mesenchymal stromal cells" based on their respective attributes [2]. To account for disparities among MSCs from different sources, the acronym "MSC" is supplemented with tissue source annotation following ISCT guidelines.

On the one hand, implantable "point-of-care" products primarily derive from adipose tissue and bone marrow, containing various cell phenotypes and a supportive molecular environment. Bone marrow concentrate (BMC) is renowned for its chondrogenic potential [3], while stromal vascular fraction (SVF) encompasses a wide array of cell types. Adipose-derived MSCs, constituting a fraction of total nucleated cells, offer versatility but vary based on preparation protocols [4]. Mechanical breakdown of adipose tissue is preferred due to regulatory constraints on enzymatic manipulation [5].

On the other hand, expanded autologous MSCs, essential for advanced therapy medical products (ATMPs), undergo stringent evaluation for quality and safety [6]. Manufacturing entails adherence to good manufacturing practice (GMP) protocols and potency assays, contributing to therapy costs. In vitro expansion, typically lasting 2–3 weeks, necessitates optimization to maintain safety and functionality. Alternatives to fetal bovine serum (FBS), such as chemically fortified medium formulations and platelet lysates, mitigate risks associated with xenogeneic components [7]. Monitoring cell characteristics throughout expansion ensures quality control and regulatory compliance. MSCs are recognized for their ability to respond to environmental cues and modulate immune and local cells through diverse cytokine outputs. They release extracellular vesicles containing genetic material like miRNAs, regulating processes such as inflammation and angiogenesis [8]. "Cell priming" protocols enhance MSC resilience through exposure to soluble factors in vitro, improving their therapeutic potential. The MSC secretome, comprising various trophic factors, is being explored for clinical applications, with conditioned media formulations offering advantages in manufacturing and storage [9].

Finally, the use of allogeneic MSC therapies poses significant regulatory hurdles in biomanufacturing and biobanking processes. MSCs are isolated from healthy donors, and stringent regulatory compliance involves establishing master and working cell banks, validating manufacturing stages, and ensuring cell stability and functionality. Despite these challenges, allogeneic MSCs offer advantages as readily available "off-the-shelf" products. While adverse immune events in humans are rare, theoretical concerns exist regarding potential immunogenic responses [10].

5.1.2 Controlling Cell Microenvironment: The Role of Blood-Derived Products

Blood-derived products, specifically platelet-rich plasma (PRP), play a key role in orthopedic sports medicine. However, fundamental questions regarding PRP's mechanisms of action and optimal formulation and application protocols remain unanswered [11]. The molecular characterization of PRP is complex and variable, making it challenging to establish quality criteria [12].

Understanding platelet biology and its historical milestones is essential for grasping the potential of PRP in tissue regeneration. Platelets originate from bone marrow megakaryocytes and play crucial roles beyond hemostasis such as contributing to regenerative processes through the release of growth factors and cytokines stored in alpha granules. Furthermore, each platelet presents 5–8 dense granules, which contain small molecules [13]. Platelet extracellular vesicles (pEVs) have emerged as a focus of interest, containing a plethora of bioactive molecules involved in tissue repair [14]. Upon activation, platelets release their granule content, including cytokines, chemokines, and growth factors, which contribute to tissue repair processes.

PRP's application in musculoskeletal therapy is grounded in hypotheses regarding its anti-inflammatory, immune-modulatory, and anabolic effects. It is recognized as safe and capable of reducing pain and improving function in select patients. Further research is needed to unravel PRP signaling within specific tissue contexts such as tendons, muscles, ligaments, and joints to enhance treatment efficacy [15]. Growth factors and cytokines are pivotal in PRP discussions, encompassing a broad range of secreted molecules influencing cell growth, division, and differentiation. These soluble proteins facilitate cell communication by diffusing between cells, exerting localized effects crucial for immediate cell-to-cell communication and response. These molecules play a pivotal role in orchestrating immune responses and inflammatory processes (Fig. 5.2). In addition, the molecular network of PRP causes the action of certain molecules to condition the levels and effects of others [16].

In addition to the biological action triggered by the molecules released by PRP, its activation also leads to the formation of a fibrin matrix, enabling slow release of molecular cargo stored in platelet granules [17]. Thus, PRP can serve dual roles in regenerative medicine: as a primer for host tissue prior to cell implantation and as a delivery vehicle for cell implantation. The integration of PRP with cell therapies provides a controlled microenvironment and a fibrin-formed cytokine delivery system [18]. It confines molecules and cells to the target site, fostering a molecular healing milieu with chemotactic cytokines and growth factors to modulate inflamma-

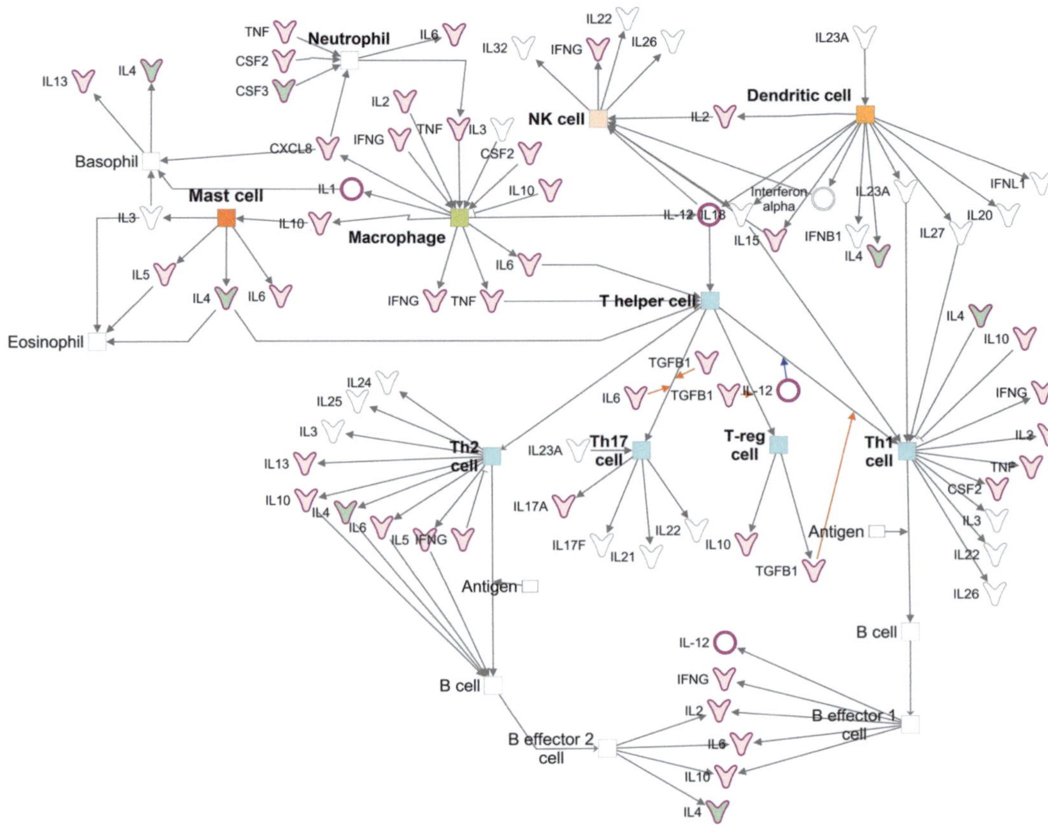

Fig. 5.2 Role of cytokines in mediating the communication between immune cells. In color, the molecules upregulated in PRP compared to platelet-poor plasma (PPP) [16] (This figure was created using Ingenuity Pathway Analyses software, IPA®, QIAGEN)

tion and angiogenesis. Research is ongoing to elucidate the interactions between PRP and MSCs. Recent findings suggest that PRP promotes stemness and prolongs the survival of transplanted cells [19]. Moreover, PRP modulates the secretory function of cells, with varying effects depending on the PRP formulation and cell phenotype [20].

5.2 Clinical Uses for Orthobiologic Products

The use of biological treatments in sports medicine can significantly improve the recovery of the athlete, enhancing tissue repair, improving symptomatology, and shortening the recovery period. The tissues involved in these injuries are highly specific but have a low self-healing capacity due to their composition. Therefore, the biological signals provided by these treatments could enhance tissue healing and the final clinical outcome.

Assessing the quality of orthobiologic treatments poses challenges due to variations among products and individuals. Determining the optimal clinical indications for orthobiological products is crucial, as treatment effectiveness relies on both product quality and application procedure. Informed by clinical experience, detailed protocols for PRP application are outlined to enhance sports injury management. The follow-

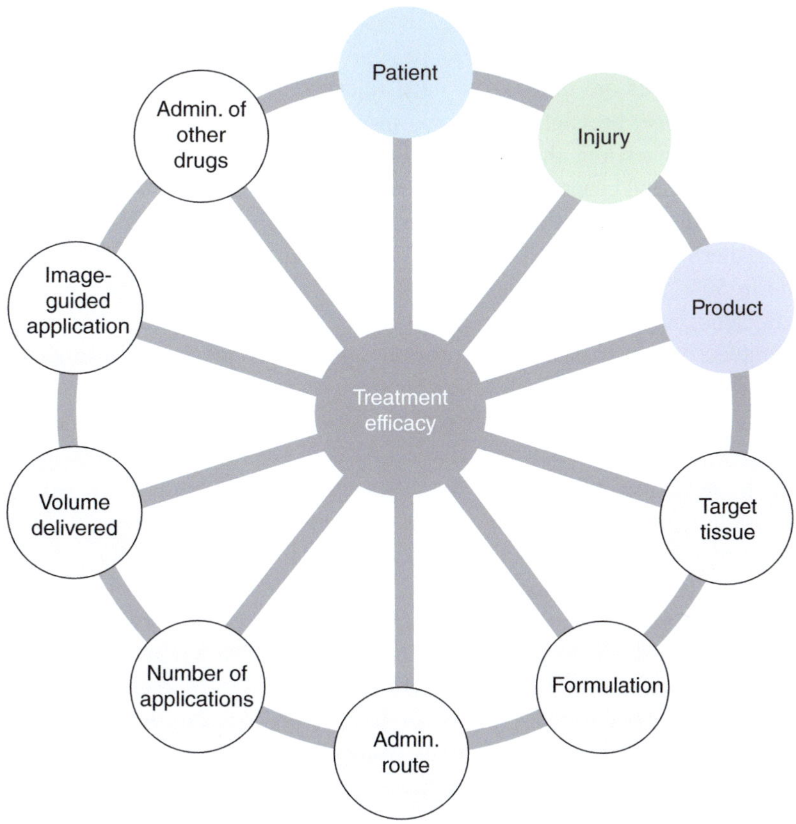

Fig. 5.3 Factors influencing the effectiveness of orthobiological treatments. In addition to the patient, the pathology, and the product, the application protocol also has a great influence on the efficacy of the treatment, encompassing many factors

ing is an overview of the wide variety of treatments and protocols that are applied to the different musculoskeletal tissues (Fig. 5.3).

5.2.1 Ligament Injuries

Ligament injuries are one of the most widespread injuries in sports practice. These injuries range from an elongation of the ligament without rupture (Grade I), through partial rupture (Grade II) to complete rupture (Grade III). A typical example of milder grades are injuries often occurring in the ankle [21], while injuries in the anterior cruciate ligament of the knee are a clearer example of a severe grade [22].

The use of PRP in the treatment of ligament injuries has been broadly studied [23]. Some authors used the PRP to decrease the intensity and the frequency of donor-site pain after ACL reconstruction [24]. For those purposes, most of the studies used coagulated PRP in forms of clots or matrixes that were then placed at the donor site. Other protocols also included intraarticular infiltration at the graft and at the femoral and tibial tunnels as well as graft immersion in PRP [25]. Some research has been done also regarding the role of PRP in the treatment of lateral ankle sprains. Protocols described include single [26] and multiple PRP [27] infiltration as a part of a conservative treatment or postoperatively.

Stem cell therapies have been suggested for use as surgical enhancement in both ligament tears and during ligament reconstruction, aiming to facilitate graft maturation and integration, with most of the protocols having imaging guidance. Several articles report direct injection of bone marrow stem cells into the ACL [28, 29], the use of autograft enhanced with adipose-derived regenerative stem cells injection under direct dry-arthroscopy [30], and single intra-articular injection of combined allogenic precursor mesenchymal stem cells (MPC) with hyaluronic acid (HA) under ultrasound guidance [31].

5.2.2 Tendon Injuries

Tendon injuries are among the most relevant in sports practice, highlighting notably problems in the Achilles tendon, the patellar tendon, or the rotator cuff, and generating a great deal of interest in both the scientific and medical community. Together with ligaments, tendon injuries represent up to 50% of musculoskeletal injuries [32]. The etiopathogenesis of tendon disorders is diverse and complex, which makes the therapeutic approach difficult. Thus, among the different injuries we can find tendon ruptures and tendinopathies caused by overuse that degenerates this tissue and can be accompanied by inflammatory processes and pain (tendinitis), or degeneration processes without inflammation (tendinosis). This sustained deterioration may also lead to tendon rupture [33].

Several administration protocols for PRP are described according to the treatment of the tendon injury. However, whether in the Achilles tendon [34, 35], rotator cuff [36, 37], epicondylitis [38], or patellar tendon [39], application protocols are affected by a great deal of variability. Both surgically and conservatively, this variability includes key factors such as the volume used, the number of injections, image guidance, use of local anesthetics, type of PRP, or PRP activation.

MSC therapy holds promise for enhancing tendon healing, but optimal protocols vary widely. From direct injections to guided interventions, the diverse approaches highlight the seek for precision and efficacy in addressing tendon pathologies. Several authors have advocated performing an injection in the tendon and bone after a surgical rotator cuff repair [40] while others reported a fluoroscopy or ultrasound-guided injection of bone marrow stem cells into the affected region to accurately identify the injury area [41].

5.2.3 Meniscal Tears

Meniscal injuries are common in orthopedic and sports medicine, with an incidence of around 61 per 100,000 population [42]. The type of injury may condition subsequent treatment, and there have been several attempts to standardize a classification system. One of the most recent is that proposed by the International Society of Arthroscopy, Knee Surgery, and Orthopaedic Sports Medicine (ISAKOS) that is based on eight arthroscopic findings [43].

Most orthobiologic applications in meniscus injuries are focused on their use as augmentation of meniscal repair surgeries. The application of biological products is conducted during the surgical process of meniscal suturing, either by injection or by applying fibrin clots [44]. In addition to biological augmentation in meniscus surgery, the application of these treatments as a conservative technique has also been explored. The application of injections in these pathologies has been both intra-articular [45] and intrameniscal [46], achieving promising results in both cases.

Regarding cell-based products, literature describes the use of this therapy as augmentation in surgery, intraarticular, intrasubstance, and scaffolding. Protocols include bone marrow-derived MSC implanted in collagen matrices [47], synovial MSC placed onto the repaired meniscus [48], post-partial meniscectomy one single intra-articular allogenic MSC injection [49], or BMC injected into the meniscus repair site [50].

5.2.4 Muscle Injuries

Muscle injuries are one of the most prevalent injuries in sports. For many years there have been several attempts to classify these types of injuries to achieve a more uniform and standard nomenclature of muscle injuries [51]. The latest classifications in this respect are the Munich classification [52], the British Athletic classification [53], and MLG-R system [54], which classify muscle injuries on the basis of clinical and magnetic resonance imaging findings.

Regardless of their severity, muscle lesions can be a paradigmatic example of acute injuries, as they tend to be diagnosed and treated mostly within a short time after their occurrence. Unlike the application in other tissues, the use of orthobiological treatments has not been as widely used and it is mostly limited to PRP. However, there are several PRP treatment protocols described in the literature [55]. Most of the studies describe a single PRP infiltration of about 2–6 mL [56], but some of them also report posterior successive injections [57]. Regarding the infiltration method, most of the studies described ultrasound-guided procedures [58] while some authors also describe infiltration under direct palpation [56, 57].

5.2.5 Bone Fractures

Bone fractures in sports medicine can be divided into two main groups: acute fractures and stress fractures. As for the former, they occur when the mechanical force or impact to which the bone is subjected exceeds the strength of the bone; they involve long recovery periods prior to return to sporting activity [59]. By contrast, stress fractures are generated when the bone is subjected to repetitive impacts or loads, even if they are not as strong as those that produce acute fractures [60].

Most of the studies described PRP application in combination with bone grafts such as iliac crest autograft [61] and including homologous bone, xenografts, and synthetic bone composites [62]. Other studies described single [63] or double [64] intraoperative PRP infiltrations directly into the fracture as well as the use of PRP to obtain autologous thrombin [65].

Cell-based therapies have been employed to enhance approaches for treating bone defects, aiming to enhance bone graft incorporation and, in certain instances, substitute for it. For non-union injuries, different approaches have been used, applying to the non-union area BMC [66], SVF [67], or osteogenic 3D adipose-derived grafts [68].

5.2.6 Osteochondral Injuries

Chondral and osteochondral lesions are a type of injury that damages joint tissue and whose impact can extend beyond the chondral tissue affecting also the subchondral bone. One of the joints most affected by these injuries is the knee, where they are present in up to 60% of knees undergoing arthroscopic evaluation [69]. The talus is another area where the incidence of these injuries is very high reaching, often associated with sprains [70]. These types of injuries can be due to acute trauma or be the result of other pathological processes such as osteoarthritis, osteonecrosis, or osteochondritis dissecans [71].

The manner in which PRP is delivered will be defined mostly by the treatment indication by the physician [72]. The conservative application of PRP in these pathologies is based on intra-articular or intra-osseous infiltrations in more severe cases [11]. When the application is during surgery, procedures include both infiltrations [73] or the use of fibrin matrices such as particulated autologous chondral—platelet-rich plasma matrix implantation (PACI) procedure in patients with large chondral injuries [74].

Cell-based products derived from MSC or SVF may encompass intraarticular injections, direct delivery into the cartilage defect, or incorporation of cells within a supportive matrix to enhance tissue regeneration and structural integrity [75, 76]. Moreover, various cell-based matrix has been widely described, consisting of autologous chondrocyte implantation (ACI), matrix-induced autologous chondrocyte implantation (MACI), and autologous matrix-induced chondrogenesis in knee and ankle [77–79].

5.3 Conclusion and Future Perspective

Orthobiological treatments represent a great opportunity in the field of sports medicine to achieve optimal recovery results while shortening recovery time. These therapies have advantages related to the presence of bio-signals that could stimulate and enhance the repair of tissues that have a very limited self-regenerative capacity. In addition, their versatility allows them to be used both conservatively and to augment surgical interventions.

However, despite their enormous therapeutic potential, the current lack of knowledge and understanding of these therapies prevents the best possible results from being obtained. As we have seen throughout this chapter, the great variability that exists between these treatments, even between products of the same nature, as well as between the different forms of application, is an enormous impediment to the medical and scientific progress of these therapies. These factors, as well as pathology and patient variability, sometimes lead to contradictory results even when the same product is used.

Therefore, studies carried out in the field of orthobiologics must consider and report as much detailed information as possible about the patient, the pathology, the product, and the application protocol. This is the only way to advance our knowledge of these treatments, which will lead to greater efficacy and better patient response.

References

1. Andia I, Maffulli N. Biological therapies in regenerative sports medicine. Sports Med. 2017;47(5):807–28.
2. Viswanathan S, Shi Y, Galipeau J, Krampera M, Leblanc K, Martin I, et al. Mesenchymal stem versus stromal cells: International Society for Cell & gene therapy (ISCT®) mesenchymal stromal cell committee position statement on nomenclature. Cytotherapy. 2019;21(10):1019–24.
3. Dulic O, Lalic I, Kecojevic V, Gavrilovic G, Abazovic D, Miskulin M, et al. Do knee injection portals affect clinical results of bone marrow aspirate concentrate injection in the treatment of osteoarthritis? A prospective randomized controlled study. Regen Med. 2020;15(8):1987–2000.
4. Andia I, Maffulli N, Burgos-Alonso N. Stromal vascular fraction technologies and clinical applications. Expert Opin Biol Ther. 2019;19(12):1289–305.
5. Senesi L, De Francesco F, Farinelli L, Manzotti S, Gagliardi G, Papalia GF, et al. Mechanical and enzymatic procedures to isolate the stromal vascular fraction from adipose tissue: preliminary results. Front Cell Dev Biol. 2019;7:88.
6. Kouroupis D, Bowles AC, Greif DN, Leñero C, Best TM, Kaplan LD, et al. Regulatory-compliant conditions during cell product manufacturing enhance in vitro immunomodulatory properties of infrapatellar fat pad-derived mesenchymal stem/stromal cells. Cytotherapy. 2020;22(11):677–89.
7. Burnouf T, Barro L, Nebie O, Wu YW, Goubran H, Knutson F, et al. Viral safety of human platelet lysate for cell therapy and regenerative medicine: moving forward, yes, but without forgetting the past. Transfus Apher Sci. 2019;58(6):102674.
8. Waterman RS, Tomchuck SL, Henkle SL, Betancourt AM. A new mesenchymal stem cell (MSC) paradigm: polarization into a pro-inflammatory MSC1 or an immunosuppressive MSC2 phenotype. PLoS One. 2010;5(4):e10088.
9. Arrigoni C, D'Arrigo D, Rossella V, Candrian C, Albertini V, Moretti M. Umbilical cord MSCs and their secretome in the therapy of Arthritic diseases: a research and industrial perspective. Cells. 2020;9(6):1343.
10. Gupta PK, Chullikana A, Rengasamy M, Shetty N, Pandey V, Agarwal V, et al. Efficacy and safety of adult human bone marrow-derived, cultured, pooled, allogeneic mesenchymal stromal cells (Stempeucel®): preclinical and clinical trial in osteoarthritis of the knee joint. Arthritis Res Ther. 2016;18(1):301.
11. Sánchez M, Jorquera C, de Dicastillo LL, Fiz N, Knörr J, Beitia M, et al. Real-world evidence to assess the effectiveness of platelet-rich plasma in the treatment of knee degenerative pathology: a prospective observational study. Ther Adv Musculoskelet Dis. 2022;14:1759720X221100304.
12. Kon E, Di Matteo B, Delgado D, Cole BJ, Dorotei A, Dragoo JL, et al. Platelet-rich plasma for the treatment of knee osteoarthritis: an expert opinion and proposal for a novel classification and coding system. Expert Opin Biol Ther. 2020;20:1447–60.
13. Nurden AT. The biology of the platelet with special reference to inflammation, wound healing and immunity. Front Biosci (Landmark Ed). 2018;23(4):726–51.
14. Saumell-Esnaola M, Delgado D, García Del Caño G, Beitia M, Sallés J, González-Burguera I, et al. Isolation of platelet-derived exosomes from human platelet-rich plasma: biochemical and morphological characterization. Int J Mol Sci. 2022;23(5):2861.
15. Andia I, Maffulli N. A contemporary view of platelet-rich plasma therapies: moving toward refined clinical protocols and precise indications. Regen Med. 2018;13(6):717–28.
16. Del Amo C, Perez-Valle A, Atilano L, Andia I. Unraveling the signaling secretome of platelet-rich plasma: towards a better understanding of its therapeutic potential in knee osteoarthritis. J Clin Med. 2022;11(3):473.
17. Anitua E, Prado R, Azkargorta M, Rodriguez-Suárez E, Iloro I, Casado-Vela J, et al. High-throughput proteomic characterization of plasma rich in growth factors (PRGF-Endoret)-derived fibrin clot interactome. J Tissue Eng Regen Med. 2015;9(11):E1–12.
18. Garate A, Sánchez P, Delgado D, Bilbao AM, Muiños-López E, Granero-Moltó F, et al. Autologous bioscaffolds based on different concentrations of

platelet rich plasma and synovial fluid as a vehicle for mesenchymal stem cells. J Biomed Mater Res. 2018;106(2):377–85.

19. Mellado-López M, Griffeth RJ, Meseguer-Ripolles J, Cugat R, García M, Moreno-Manzano V. Plasma rich in growth factors induces cell proliferation, migration, differentiation, and cell survival of adipose-derived stem cells. Stem Cells Int. 2017;2017:5946527.
20. Tobita M, Tajima S, Mizuno H. Adipose tissue-derived mesenchymal stem cells and platelet-rich plasma: stem cell transplantation methods that enhance stemness. Stem Cell Res Ther. 2015;6:215.
21. Dias S, Lewis TL, Alkhalfan Y, Ahluwalia R, Ray R. Current concepts in the surgical management of chronic ankle lateral ligament instability. J Orthop. 2022;33:87–94.
22. Sanders TL, Maradit Kremers H, Bryan AJ, Larson DR, Dahm DL, Levy BA, et al. Incidence of anterior cruciate ligament tears and reconstruction: a 21-year population-based study. Am J Sports Med. 2016;44(6):1502–7.
23. McRobb J, Kamil KH, Ahmed I, Dhaif F, Metcalfe A. Influence of platelet-rich plasma (PRP) analogues on healing and clinical outcomes following anterior cruciate ligament (ACL) reconstructive surgery: a systematic review. Eur J Orthop Surg Traumatol Orthop Traumatol. 2023;33(2):225–53.
24. Seijas R, Rius M, Ares O, García-Balletbó M, Serra I, Cugat R. Healing of donor site in bone-tendon-bone ACL reconstruction accelerated with plasma rich in growth factors: a randomized clinical trial. Knee Surg Sports Traumatol Arthrosc. 2015;23(4):991–7.
25. Sánchez M, Delgado D, Sánchez P, Fiz N, Azofra J, Orive G, et al. Platelet rich plasma and knee surgery. Biomed Res Int. 2014;2014:890630.
26. Blanco-Rivera J, Elizondo-Rodríguez J, Simental-Mendía M, Vilchez-Cavazos F, Peña-Martínez VM, Acosta-Olivo C. Treatment of lateral ankle sprain with platelet-rich plasma: a randomized clinical study. Foot Ankle Surg. 2020;26(7):750–4.
27. Sabaghzadeh A, Zarei Kurdkandi H, Ebrahimpour A, Biglari F, Jafari KM. Efficacy of platelet-rich plasma for chronic lateral ankle instability after modified Broström-Gould surgery: a randomized, single-blinded, prospective controlled trial. Foot Ankle Orthop. 2023;8(2):24730114231168630.
28. Centeno C, Markle J, Dodson E, Stemper I, Williams C, Hyzy M, et al. Symptomatic anterior cruciate ligament tears treated with percutaneous injection of autologous bone marrow concentrate and platelet products: a non-controlled registry study. J Transl Med. 2018;16(1):246.
29. Gobbi A, Whyte GP. Long-term outcomes of primary repair of the anterior cruciate ligament combined with biologic healing augmentation to treat incomplete tears. Am J Sports Med. 2018;46(14):3368–77. https://doi.org/10.1177/0363546518805740.
30. Alentorn-Geli E, Seijas R, Martínez-De la Torre A, Cuscó X, Steinbacher G, Álvarez-Díaz P, et al. Effects of autologous adipose-derived regenerative stem cells administered at the time of anterior cruciate ligament reconstruction on knee function and graft healing. J Orthop Surg Hong Kong. 2019;27(3):2309499019867580.
31. Wang Y, Shimmin A, Ghosh P, Marks P, Linklater J, Connell D, et al. Safety, tolerability, clinical, and joint structural outcomes of a single intra-articular injection of allogeneic mesenchymal precursor cells in patients following anterior cruciate ligament reconstruction: a controlled double-blind randomised trial. Arthritis Res Ther. 2017;19(1):180.
32. James R, Kesturu G, Balian G, Chhabra AB. Tendon: biology, biomechanics, repair, growth factors, and evolving treatment options. J Hand Surg. 2008;33(1):102–12.
33. Wu F, Nerlich M, Docheva D. Tendon injuries: basic science and new repair proposals. EFORT Open Rev. 2017;2(7):332–42.
34. Boksh K, Elbashir M, Thomas O, Divall P, Mangwani J. Platelet-rich plasma in acute Achilles tendon ruptures: a systematic review and meta-analysis. Foot Edinb. 2022;53:101923.
35. Liu CJ, Yu KL, Bai JB, Tian DH, Liu GL. Platelet-rich plasma injection for the treatment of chronic Achilles tendinopathy: a meta-analysis. Medicine (Baltimore). 2019;98(16):e15278.
36. Peng Y, Du L, Yang B, Fan D, Jia S, Zheng C. Efficacy of platelet-rich plasma and platelet-rich fibrin in arthroscopic rotator cuff repair: a systematic review and meta-analysis. PM R. 2023;15(12):1643–53.
37. Pang L, Xu Y, Li T, Li Y, Zhu J, Tang X. Platelet-rich plasma injection can be a viable alternative to corticosteroid injection for conservative treatment of rotator cuff disease: a meta-analysis of randomized controlled trials. Arthrosc J Arthrosc Relat Surg. 2023;39(2):402–421.e1.
38. Xu Y, Li T, Wang L, Yao L, Li J, Tang X. Platelet-rich plasma has better results for long-term functional improvement and pain relief for lateral epicondylitis: a systematic review and meta-analysis of randomized controlled trials. Am J Sports Med. 2024;52:2646.
39. Barman A, Sinha MK, Sahoo J, Jena D, Patel V, Patel S, et al. Platelet-rich plasma injection in the treatment of patellar tendinopathy: a systematic review and meta-analysis. Knee Surg Relat Res. 2022;34(1):22.
40. Hernigou P, Flouzat Lachaniette CH, Delambre J, Zilber S, Duffiet P, Chevallier N, et al. Biologic augmentation of rotator cuff repair with mesenchymal stem cells during arthroscopy improves healing and prevents further tears: a case-controlled study. Int Orthop. 2014;38(9):1811–8.
41. Centeno CJ, Al-Sayegh H, Bashir J, Goodyear S, Freeman MD. A prospective multi-site registry study of a specific protocol of autologous bone marrow concentrate for the treatment of shoulder rotator cuff tears and osteoarthritis. J Pain Res. 2015;8:269–76.
42. Luvsannyam E, Jain MS, Leitao AR, Maikawa N, Leitao AE. Meniscus tear: pathology, incidence, and management. Cureus. 2022;14(5):e25121.

43. Sayegh ET, Matzkin E. Classifications in Brief: the International Society of Arthroscopy, Knee Surgery, and Orthopaedic Sports Medicine Classification of Meniscal Tears. Clin Orthop. 2022;480(1):39–44.
44. Blough CL, Bobba CM, DiBartola AC, Everhart JS, Magnussen RA, Kaeding C, et al. Biologic augmentation during meniscal repair. J Knee Surg. 2021;36:498.
45. Alessio-Mazzola M, Felli L, Trentini R, Formica M, Capello AG, Lovisolo S, et al. Efficacy of autologous platelet-rich plasma injections for grade 3 symptomatic degenerative meniscal lesions: a 1-year follow-up prospective study. Sports Health. 2022;14(2):227–36.
46. Sánchez M, Jorquera C, Bilbao AM, García S, Beitia M, Espregueira-Mendes J, et al. High survival rate after the combination of intrameniscal and intraarticular infiltrations of platelet-rich plasma as conservative treatment for meniscal lesions. Knee Surg Sports Traumatol Arthrosc. 2023;31:4246.
47. Whitehouse MR, Howells NR, Parry MC, Austin E, Kafienah W, Brady K, et al. Repair of torn avascular meniscal cartilage using undifferentiated autologous mesenchymal stem cells: from in vitro optimization to a first-in-human study. Stem Cells Transl Med. 2017;6(4):1237–48.
48. Sekiya I, Koga H, Otabe K, Nakagawa Y, Katano H, Ozeki N, et al. Additional use of synovial mesenchymal stem cell transplantation following surgical repair of a complex degenerative tear of the medial meniscus of the knee: a case report. Cell Transplant. 2019;28(11):1445–54.
49. Vangsness CT, Farr J, Boyd J, Dellaero DT, Mills CR, LeRoux-Williams M. Adult human mesenchymal stem cells delivered via intra-articular injection to the knee following partial medial meniscectomy: a randomized, double-blind, controlled study. J Bone Joint Surg Am. 2014;96(2):90–8.
50. Dancy ME, Marigi EM, Krych AJ, Werner BC, Camp CL. Impact of biologic augmentation on revision surgery rates after meniscus repair: a matched-cohort analysis of 3420 patients. Orthop J Sports Med. 2023;11(8):23259671231186990.
51. Grassi A, Quaglia A, Canata GL, Zaffagnini S. An update on the grading of muscle injuries: a narrative review from clinical to comprehensive systems. Joints. 2016;4(1):39–46.
52. Mueller-Wohlfahrt HW, Haensel L, Mithoefer K, Ekstrand J, English B, McNally S, et al. Terminology and classification of muscle injuries in sport: the Munich consensus statement. Br J Sports Med. 2013;47(6):342–50.
53. Pollock N, James SLJ, Lee JC, Chakraverty R. British athletics muscle injury classification: a new grading system. Br J Sports Med. 2014;48(18):1347–51.
54. Valle X, Alentorn-Geli E, Tol JL, Hamilton B, Garrett WE, Pruna R, et al. Muscle injuries in sports: a new evidence-informed and expert consensus-based classification with clinical application. Sports Med. 2017;47(7):1241–53.
55. Seow D, Shimozono Y, Tengku Yusof TNB, Yasui Y, Massey A, Kennedy JG. Platelet-rich plasma injection for the treatment of hamstring injuries: a systematic review and meta-analysis with Best-worst case analysis. Am J Sports Med. 2021;49(2):529–37.
56. Hamilton B, Tol JL, Almusa E, Boukarroum S, Eirale C, Farooq A, et al. Platelet-rich plasma does not enhance return to play in hamstring injuries: a randomised controlled trial. Br J Sports Med. 2015;49(14):943–50.
57. Bradley JP, Lawyer TJ, Ruef S, Towers JD, Arner JW. Platelet-rich plasma shortens return to play in National Football League Players with Acute Hamstring Injuries. Orthop J Sports Med. 2020;8(4):2325967120911731.
58. Guillodo Y, Madouas G, Simon T, Dauphin HL, Saraux A. Platelet-rich plasma (PRP) treatment of sports-related severe acute hamstring injuries. Muscles Ligaments Tendons J. 2016;5:284.
59. Robertson GAJ, Wood AM. Return to sport after Tibial shaft fractures: a systematic review. Sports Health. 2016;8(4):324–30.
60. Moreira CA, Bilezikian JP. Stress fractures: concepts and therapeutics. J Clin Endocrinol Metab. 2017;102(2):525–34.
61. Acosta-Olivo C, Garza-Borjon A, Simental-Mendia M, Vilchez-Cavazos F, Tamez-Mata Y, Peña-Martinez V. Delayed union of humeral shaft fractures: comparison of autograft with and without platelet-rich plasma treatment: a randomized, single blinded clinical trial. Arch Orthop Trauma Surg. 2017;137(9):1247–52.
62. Calori GM, Tagliabue L, Gala L, d'Imporzano M, Peretti G, Albisetti W. Application of rhBMP-7 and platelet-rich plasma in the treatment of long bone non-unions: a prospective randomised clinical study on 120 patients. Injury. 2008;39(12):1391–402.
63. Griffin XL, Achten J, Parsons N, Costa ML. Platelet-rich therapy in the treatment of patients with hip fractures: a single centre, parallel group, participant-blinded, randomised controlled trial. BMJ Open. 2013;3(6):e002583.
64. Samuel G, Menon J, Thimmaiah S, Behera G. Role of isolated percutaneous autologous platelet concentrate in delayed union of long bones. Eur J Orthop Surg Traumatol Orthop Traumatol. 2018;28(5):985–90.
65. Singh R, Rohilla R, Gawande J, Kumar SP. To evaluate the role of platelet-rich plasma in healing of acute diaphyseal fractures of the femur. Chin J Traumatol. 2017;20(1):39–44.
66. Hernigou P, Guissou I, Homma Y, Poignard A, Chevallier N, Rouard H, et al. Percutaneous injection of bone marrow mesenchymal stem cells for ankle non-unions decreases complications in patients with diabetes. Int Orthop. 2015;39(8):1639–43.
67. Khalpey Z, Marsh KM, Ferng A, Riaz IB, Hemphill C, Johnson K, et al. First in man: sternal reconstruction with autologous stem cells. ASAIO J. 2015;61(5):e31–2.

68. Dufrane D, Docquier PL, Delloye C, Poirel HA, André W, Aouassar N. Scaffold-free three-dimensional graft from autologous adipose-derived stem cells for large bone defect reconstruction: clinical proof of concept. Medicine (Baltimore). 2015;94(50):e2220.
69. Merkely G, Ackermann J, Lattermann C. Articular cartilage defects: incidence, diagnosis, and natural history. Oper Tech Sports Med. 2018;26(3):156–61.
70. Shimozono Y, Yasui Y, Ross AW, Kennedy JG. Osteochondral lesions of the talus in the athlete: up to date review. Curr Rev Musculoskelet Med. 2017;10(1):131–40.
71. Gorbachova T, Melenevsky Y, Cohen M, Cerniglia BW. Osteochondral lesions of the knee: differentiating the Most common entities at MRI. Radiogr Rev Publ Radiol Soc N Am Inc. 2018;38(5):1478–95.
72. Laver L, Filardo G, Sanchez M, Magalon J, Tischer T, Abat F, et al. The use of injectable orthobiologics for knee osteoarthritis: a European ESSKA-ORBIT consensus. Part 1-Blood-derived products (platelet-rich plasma). Knee Surg Sports Traumatol Arthrosc. 2024;32:783.
73. Danieli MV, Guerreiro JPF, Vimercati TA, Mendes PHF, Miyazaki PRTK, Cataneo DC. Platelet-rich plasma versus hyaluronic acid for knee chondral injuries in young patients. Rev Bras Ortop. 2021;56(5):634–40.
74. Cugat R, Samitier G, Vinagre G, Sava M, Alentorn-Geli E, García-Balletbó M, et al. Particulated autologous chondral-platelet-rich plasma matrix implantation (PACI) for treatment of full-thickness cartilage osteochondral defects. Arthrosc Tech. 2021;10(2):e539–44.
75. Gobbi A, Whyte GP. One-stage cartilage repair using a hyaluronic acid-based scaffold with activated bone marrow-derived mesenchymal stem cells compared with microfracture: five-year follow-up. Am J Sports Med. 2016;44(11):2846–54.
76. Jo CH, Chai JW, Jeong EC, Oh S, Shin JS, Shim H, et al. Intra-articular injection of mesenchymal stem cells for the treatment of osteoarthritis of the knee: a 2-year follow-up study. Am J Sports Med. 2017;45(12):2774–83.
77. Brittberg M, Recker D, Ilgenfritz J, Saris DBF, SUMMIT Extension Study Group. Matrix-applied characterized autologous cultured chondrocytes versus microfracture: five-year follow-up of a prospective randomized trial. Am J Sports Med. 2018;46(6):1343–51.
78. Kon E, Filardo G, Brittberg M, Busacca M, Condello V, Engebretsen L, et al. A multilayer biomaterial for osteochondral regeneration shows superiority vs microfractures for the treatment of osteochondral lesions in a multicentre randomized trial at 2 years. Knee Surg Sports Traumatol Arthrosc. 2018;26(9):2704–15.
79. Niemeyer P, Schubert T, Grebe M, Hoburg A. Treatment costs of matrix-associated autologous chondrocyte implantation compared with microfracture: results of a matched-pair claims data analysis on the treatment of cartilage knee defects in Germany. Orthop J Sports Med Diciembre De. 2019;7(12):2325967119886583.

Regulation and Regenerative Treatments

6

Leonard Tiger Onsen, Jason Grieshober, Ralph Gambardella, and Bert Mandelbaum

6.1 Introduction

Regenerative medicine refers to a broad field that focuses on the process of replacing, engineering, or regenerating human cells, tissues, or organs with the goal of restoring normal function [1]. This field has sparked significant interest in orthopedics and sports medicine to aid in treating various pathologies that lack significant innate healing potential such as cartilage, meniscus, intra-articular ligaments, or even fractures in some cases. In turn, orthopedics and sports medicine have become regenerative medicine leaders both in research and clinical practice by using certain applications now for over three decades [2]. Various regenerative treatment options in orthopedics include cell-based therapies, scaffolds, tissue engineering, and morphogenic stimuli. Although a complete review of the products within these grouping is beyond the scope of this chapter, we will review stem cell therapy to better understand the regulation of these products at various levels both within the United States and abroad.

L. T. Onsen · R. Gambardella · B. Mandelbaum (✉)
Cedars Sinai Kerlan Jobe institute,
Los Angeles, CA, USA

J. Grieshober
Platinum Orthopaedics, Anaheim, CA, USA

6.2 Stem Cell Therapy

Stem cells are defined as cells that can self-renew, multiply, and differentiate into a variety of functional cells in response to stimuli or specific conditions [3]. Stem cells can be further classified as being pluripotent or multipotent. Pluripotent cells have the potential to differentiate into any type of cell, whereas multipotent cells can differentiate into all cell types within a given lineage [4]. Adult stem cells such as mesenchymal stem cells (MSCs) are considered multipotent and have less safety or ethical concerns compared to pluripotent cells. This has led MSCs to being a focus of stem cell therapy research.

The International Society of Cellular Therapy (ISCT) issued a consensus statement detailing the minimum criteria to define MSCs. Minimum criteria state that cells must be plastic adherent, express certain cell surface antigens (CD105, CD73, and CD90) but not others (CD45, CD34, CD14, CD11b, CD79α, CD19, or human leukocyte antigen–antigen D related), and have the capacity to differentiate into osteoblasts, adipocytes, and chondroblasts in vitro [5]. MSCs are believed to be present within nearly every tissue in the body with known sources including bone marrow aspirate, adipose tissue, synovial tissue, periosteal tissue, peripheral blood, and perivascular cells [6, 7]. These properties allow MSCs the potential to aid in tissue regeneration, promoting

A. Gobbi et al. (eds.), *Regenerative Medicine in Sports and Orthopaedics*,
https://doi.org/10.1007/978-3-031-84693-9_6

angiogenesis, reducing inflammation, and aiding in tissue healing by recruiting cellular messengers [8].

Pluripotent stem cells possess the greatest potential for uses given their ability to differentiate into any cell type. However, with this potential comes the greater ethical concerns and reduced opportunities for research. These cells can be produced from embryonic sources, somatic cell nuclear transfer, or induction of adult stem cells into induced pluripotent stem cells (iPSCs) [4].

Regulation of regenerative medicine products is diverse overall and specific to individual countries. The purpose of this chapter is to review regulation of these products in the United States and in other countries. In doing this we will review regulatory steps from product development to commercialization. We will also cover alternative pathways to marketplace approval and review reimbursement considerations for these products.

6.2.1 Regulation of Regenerative Medicine in the United States

The Food and Drug Administration (FDA) is responsible for the regulation of regenerative medicine treatments in the United States. Divisions of the FDA responsible for the regulation of medical products include the Center for Drug Evaluation and Research (CDER), Center for Devices and Radiological Health (CDRH), and Center for Biologics Evaluation and Research (CBER) [9]. CDER is responsible for regulation of therapeutic drugs and biologics. CDRH focuses on regulating devices and products that emit radiation. CBER has the responsibility of regulating biologic products and devices.

The broad spectrum of regenerative medicine products allows for debate over their classification as a drug, medical device, biologic agent, or a combination of any of the three [10]. This makes it difficult to cleanly fit regenerative medicine within the jurisdiction of any specific FDA division. Debate can occur over classification and what division will handle the regulation of the product. In these situations, the FDA Office of Combination Products will review the product and assign a division to oversee it based on its primary mode of action [9].

6.3 FDA Review Process

As expected, the process of product development from conception to widespread public availability is time and labor intensive. As products are developed, they must first undergo a review of local institution review board (IRB) and FDA. Preclinical studies with animals are the next step to determine the safety and pharmacology of the product [10]. Following these evaluations, developers must complete an investigational device exemption application for devices or an investigational new drug application for drugs and biologic product before starting clinical trials [9]. Phase 1 trials assess the drug's safety and dose range in a small sample of healthy individuals. Phase 2 trials study the drug in a larger group of people with a particular pathology. Efficacy and short-term adverse reactions are evaluated at this time. For phase 3 a large multicenter double-blinded randomized controlled trial is performed with an even larger sample size and over a longer period of time. This phase provides much of the data for product labeling and allows for a thorough risk-benefit analysis [10]. After completion of trials, official applications for marketing must be completed through the FDA. Premarket applications (PMA) are completed for devices, new drug applications (NDA) are required for drugs, and biologic license application (BLA) is needed for those classified as biologics. Though specific applications may differ by product, the steps and amount of regulation remain similar. Only after completion of these steps can a product be marketed and made available to the public for use. The overall process can be time-consuming and expensive leading to partnerships with industry or private investors [11]. Finally, post-approval studies may still be required by the FDA despite completion of the above steps.

6.4 Three-Tiered Approach

A three-tiered approach to the regulation of all human cells, tissues, or cellular and tissue-based products (HCT/Ps) intended for human use was set forth by the US FDA in 1997 via Title 21, Part 1271 of the Code of Federal Regulations [8]. Products are stratified into categories based on risk. Category 1 is considered low risk and not regulated as a HCT/P [4]. This category includes whole blood, blood-derived products, bone marrow, human organs for transplants, and extracted human products.

Category 2 includes lower risk HCT/Ps that require minimal supervision. These products are regulated by section 361 of the Public Health Service Act that only requires products be manufactured under good tissue practices transmission of communicable disease. Commonly referenced as 361 products, they do not require premarket studies or approval [4]. Within this group includes biologics that are minimally manipulated, used in a homologous fashion, not combined with other cells or chemicals, and do not have systemic effects [4].

Category 3 are higher risk products and thus require more stringent regulation and approval processes. Regulation of these products falls under section 351 of the Public Health Service Act. Referred to as 351 s, they must complete all the premarketing studies and approval process described in the FDA review process section as well as the steps outlined for the lower risk categories. Additionally, manufacturing specifications are determined on an individual basis during pre-clinical process by the FDA. This category includes biologic products that are more than minimally manipulated, used in non-homologous manner, combined with other substances or cells, have systemic effects, or are used in an allogeneic manner [4].

As mentioned in the category descriptions, the concepts of "minimal manipulation" and "more than minimal manipulation" have been established within this approach. Minimal manipulation is described as only slight modification with maintenance of core properties [12]. Examples of this include cell extraction from tissue, centrifugation, cutting, grinding, antibiotic treatment, gamma irradiation, and cryopreservation [13].

"More than minimal manipulation" is defined as modifications that alter the biologic characteristics of cells or tissue [13]. Activation, encapsulation, genetic modification, or expansion of cells or tissues fall under this classification [13]. Enzyme degradation, chemical additions, or biologic element additions are further examples [12]. The importance of this concept cannot be overstated as those that do not fall under "minimal manipulation" must undergo the time and financially taxing FDA premarket approval process [14].

6.5 Alternative Pathways

Mechanisms to accelerate the FDA approval process do exist in the United States. Some of these pathways include the "Fast Track" program, "Accelerated Approval" program, "Priority Review" program, "breakthrough therapy" designation, and "compassionate use" exemption. At the state level, "right to try" laws have been attempted for biologic products [12].

The Fast Track program expedites the review and approval process as well as allows for rolling reviews where application portions can be reviewed prior to complete submission. This consideration can be applied to products that treat a serious condition and have clinical or nonclinical data showing potential to address an unmet medical need [15]. Accelerated Approval program works to make the final approval process quicker. This is accomplished by setting more easily achievable surrogate clinical endpoints as opposed to focusing on survivability or irreversible morbidity [15]. Priority Review program speeds up the process by taking action on application within 6 months as opposed to the standard 10-month timeline [15].

Breakthrough therapy designation helps to accelerate the review process for particularly inventive products. Qualifying criteria for this designation include preliminary data in place showing a substantial improvement in a clinically significant endpoint compared to available products [12]. To date, this designation has not been applied to a stem cell therapy and instead has primarily been used for monoclonal antibodies and traditional pharmaceutical products.

Compassionate use also referred to as "expanded access" provides a pathway for product use in terminally ill or those with immediate life-threatening conditions. Criteria physician and product manufacturer must agree on a usage and then the FDA must authorize the use. This allows products to move directly from phase 1 trials to clinical use [12]. Traditionally, this pathway has less stringent approval standards and thus a much higher approval rate compared to other pathways.

At the state government level, "right to try" laws have been created as an alternative pathway for approval. Many times, the laws passed in this vein have directly opposed the in-place federal regulations. Like "compassionate use," these laws require patients to be gravely ill and products to have demonstrated safety in phase 1 trials. Colorado was the first [16] state to pass a right to try law in 2014 and a total of 41 states had passed a similar law by 2018 [12]. Overall alternative pathways aid in improving product access and speed of approval process, but come with ethical debates for both industry and regulators.

6.6 Worldwide Regulation of Regenerative Medicine

Regulation of regenerative medicine products around the world is best generalized as variable- and country-specific in nature. Many countries have systems in place similar to the United States and European Union (EU), though much diversity still exists. On the extreme end, some countries are largely deregulated for these products. This allows for therapies to be offered that are not available in other countries. Such availability can lead to medical tourism and offer an economic advantage to these areas [14]. However, without regulation there may be less reliability in terms of product safety and efficacy. Issues with product safety and efficacy pose the risk of damaging the field's perception and inhibit future progress.

6.7 Europe

Regulatory infrastructure of the European Union (EU) is well organized and mirrors the US system in many facets. The European Medical Agency (EMA) serves as the core regulatory body for regenerative medicine in the EU. Regenerative medicine products fall under the grouping advanced therapy medicinal products (ATMPs). This grouping was defined under the regulation EC No. 1394/2007 and became effective in 2008 [17, 18].

Under the EMA, the process of product development from conception to widespread public use closely resembles that of the US FDA. Preclinical trials to determine product safety are performed in accordance with International Conference on Harmonisation of Technical Requirements for Registration of Pharmaceuticals for Human Use (ICH) guidelines [10]. Clinical trial authorization (CTA) next must be obtained at the national level prior to beginning clinical trials [19]. EU clinical trials phases 1-3 are similar to the US. The Committee on Advanced Therapies (CAT) monitors the quality of trials that establish the safety and efficacy of ATMPs. CAT reviews both the product and its trials before making a recommendation to the Committee for Medicinal Products for Human Use (CHMP) of the EMA for final approval [20]. Upon clinical trial completion, marketing authorization application (MAA) must be completed prior to widespread public use and marketing [10].

Alternative pathways to work around the system described above exist. One such pathway is termed "Hospital Exemption". In this exemption, a physician within a member state hospital can administer an ATMP in a nonroutine manner [20]. This is viewed similarly to off-label use of products in the United States that fall outside of FDA regulation [17]. Another alternative pathway is known as conditional marketing authorization. This path works similar to the Fast Track program by accelerating a product if it fulfills an unmet need when the benefit to public health with immediate availability outweighs the risks of additional data needed [21, 22].

Despite many similarities, the EU system has several differences from the United States. In the

United States, there is a distinction between federal (NIH) and private funding, whereas there is no distinction in the EU system [20]. Under the Bush Administration, federal funding restrictions were placed on human embryonic stem cell research [20]. This placed the United States at a disadvantage for research in the field; however, these restrictions were later loosened under the Obama Administration. On the other hand, the EU was able to continue research without any funding restrictions.

6.8 Australia

Australia's regulatory system is also quite like the United States'. The Therapeutic Goods Administration (TGA) serves as the regulatory body for regenerative medicine products in Australia. A three-tiered approach is employed in this system as well [17]. First category is biologics that are regulated as therapeutic goods. These products include blood, plasma, vaccines, and hematopoietic progenitor cells. This group undergoes similar trials to pharmaceuticals in the United States [17]. Second category is biologics regulated as biologic goods. Members of this group include stem cells, combined cell and tissue products, and genetically modified and expanded cellular products. Classification for this group is based on risk stratification according to the Australian Regulatory Guidelines for Biologicals (ARBG) [17]. Classified from 1 to 4 with classes 3 and 4 being higher risk products with more than minimal manipulation and nonhomologous use. Those classes require additional monitoring over safety and efficacy. Final category includes human organs, reproductive tissue for reproductive therapy, hematopoietic progenitor cells, and cellular and tissue products collected from a patient for single treatment under the supervision of a medical professional. Products in this category are excluded from regulation as biologics [17].

6.9 Canada

The regulatory system of Canada also shares many similarities to the US system. Health Canada regulates regenerative medicine products under the Canadian Food and Drugs Act. Products must undergo preclinical studies, clinical trials, and marketing approval. Regenerative medicine products are subject to divisions 1A, 2, 4, 5, and 8 of the Food and Drugs Act requirements [23]. After passing clinical trials, a notice of compliance (NOC) is issued and then marketing can begin. Manufacturing sites also require establishment license [23].

In a similar trend, Canada employs a risk-based three-tiered system for regulation. Low-risk products are not under formal regulation. To qualify for this category, products must be autologous, homologous, minimally manipulated, and not have systemic effects. The next risk group falls under the Safety of Human Cells, Tissues, and Organs for Transplantation (CTO) regulations. This group includes allogenic, minimally manipulated, and homologous use products. Examples include established products such as bone marrow, peripheral blood, cord blood, or pancreatic islet cells. These products do not require an NOC or pre-marketing approval, but do require marketing site registration. Finally, the higher risk group undergoes stricter premarketing approval under division 5 of the Food and Drugs Act. Characteristics of this product group include more than minimally manipulated cells, xenogeneic cells, nonhomologous use, and cells with systemic effects. Also similar to America is the time-consuming and expensive nature of this process with some estimates of hundreds of millions of dollars to take a new product to marketing approval [23].

6.10 Russia

On the opposite end of the spectrum, stem cells are largely unregulated in Russia [24]. The Ministry of Health is responsible for regulating these products in Russia. Although the administration of stem cells requires a license, there are no regulations on the safety, efficacy, or type of cells used [24]. The Academy of Medical Sciences is tasked with institutional research monitoring but does not have jurisdiction over the private sector [24]. Thus, private clinics can provide stem cell therapy without much impediment.

Although lack of regulation can promote growth and research, it also comes with risk. Adverse events have been reported with unproven stem cell therapies in Russia [24]. In fact, this was such of an issue that the Ministry of Health set up an expert council to investigate the problem in 2005. Many clinics were found violating safety standards. Many clinics were shut down, but in 2007 an estimated 500 clinics were still providing unlicensed stem cell therapy in Moscow alone [24].

6.11 India

Though more tightly regulated than Russia, India has been known to offer stem cell therapies that do not meet international standards [24]. The Indian Council of Medical Research (ICMR) funds and oversees research on these products. Overall structure of India's regulatory system is confusing and contradictory. Guidelines to deal with stem cell research were set forth by the Department of Biotechnology (DBT) in 2001. However, just 1 year later, the ICMR established its own conflicting set of guidelines on the topic. Eventually, the ICMR and DBT teamed up to form the National Guidelines for Stem Cell Research and Therapy. This in turn led to a National Apex Committee to oversee all stem cell use and Institutional Committees to oversee all stem cell research in India. Despite this, noncompliance is thought to be common and the handling of violations is still unclear [24]. With the possibility of economic benefit of a stem cell tourism location, India has committed significant financial resources to stem cell research [25]. As their system continues to evolve, the ICMR has invited opinions from the public and medical community to be incorporated into policy.

6.12 Mexico

The Federal Commission for the Protection Against Sanitary Risk (COFEPRIS) serves as the governing body over regenerative medicine in Mexico [14]. However, limited resources and lack of standardized guidelines have led to concerns over COFEPRIS lack of authority. Mexico's General Health Law prohibits falsely advertising proposed benefits of medical treatments, but this law does not apply to stem cell treatments [14]. Furthermore, only a handful of private company stem cell treatments have underwent clinical trials with COFEPRIS [14]. Regulatory lines were further blurred in 2012 when new law allowed commercialization of organs, tissues, and cells not previously permitted [14]. These reasons have led Mexico to become a medical tourism destination particularly for patients from America and Canada seeking stem cell therapies not available in their home country. As always lack of regulation comes with risk and without assurances of safety or efficacy, and it is prudent for countries to strive for an environment with more practical supervision.

6.13 China

China has embraced stem cell research and is the second most published country in the world on the topic [26]. Previous lax regulation led China to become a common location of stem tourism. However, since 2009 China has been more thorough in evaluation of safety and efficacy for stem cell therapies. Stem cell research now follows the guidelines established by the Ministry of Health and the Ministry of Science and Technology [26]. The National Stem Cell Research and Coordination committee monitors stem cell research [14]. All stem cell research must be registered. Overall, despite early hands-off approach to regulation, China now appears to be taking a more active role in ensuring the safety of stem cell products.

6.14 Japan

Japan's regulatory system more closely resembles that of the United States and EU. Regenerative medicine was officially classified in this system in 2013 under the Pharmaceuticals and Medical Devices (PMD) Act [27]. Under this act, the

Pharmaceuticals and Medical Devices Agency (PMDA) is responsible for product review. PMDA reviews are finally evaluated by the Ministry of Health, Labor, and Welfare (MHLW) which then makes a final approval decision [14]. With Japan's growing involvement in stem cell research the PMDA created the Office of Cellular and Tissue-Based Products the handle the growing number of clinical trials [14]. These trials go through similar phases to the United States. Important to note is the distinction between "clinical research" and "clinical trials" in Japan. Clinical research denotes product development for academic or private medical practice purposes and is governed by the Act on the Safety of Regenerative Medicine (ASRM) [27]. Clinical trials denote studies on product development for marketing purposes [27].

2013 saw the Regenerative Medicine Promotion Law passed in Japan to help streamline the process for these products to reach public use [27]. This law allows for marketing authorization applications (MAA) to be submitted after early trials if safety and efficacy has been demonstrated [17]. If approved, products can reach the public with a time-limited conditional approval. Conditional approval lasts for 7 years after which full approval must be applied for under the MHLW [27]. The Regenerative Medicine Promotion Law is a creative solution to the rigors of marketing approval process. It allows patients earlier access to therapies while still monitoring safety and efficacy. It also helps manufacturers stay active in the market while a product progresses through the approval process. This law promotes an exciting opportunity in the world of regenerative medicine and with success offers the possibility to be adapted in other parts of the world.

6.15 The REGROW Act

In 2016 bill S. 2689 "Reliable and Effective Growth of Regenerative Health Options that Improve Wellness" or REGROW Act was introduced [28]. The goal of this bill is to optimize the regulatory process and accelerate the timeline for regenerative medicine products to make it to public availability. Under this bill, there will not be restrictions or modifications to sections 351 or 361 of the Public Service Health Act. Instead, the REGROW Act adds an amendment to section 351 stating a program will be created to allow conditional approval of a product prior to phase 3 clinical trials [28]. A conditional approval period of 5 years is allowed where a product can be introduced into the marketplace [28].

The REGROW Act offers a similar solution to the Regenerative Medicine Promotion Law in Japan. Both allow products to reach the marketplace prior to completing phase 3 trials but while still collecting data on safety and efficacy. Marketplace introduction helps offset costs associated with product development and clinical trials. This bill has the potential to promote growth and advances in regenerative medicine while still promoting a safe regulatory environment. To date, this bill has not yet been passed [28].

6.16 Reimbursement

Reimbursement provides further insight into a country's regulatory environment and financial status. Regulatory systems where products demonstrate safety and efficacy better allow products to be reimbursed by payors. Clinical factors also play a role in a product's reimbursement. Some examples include the clinical setting of administration and added benefit of a product over standard treatment [21, 22]. Other factors favoring reimbursement involve products that treat pathology with high disease burden and those that target smaller populations [29]. Next, we will briefly review the reimbursement environment of the United States and EU.

6.17 Reimbursement in the United States

The US health insurance environment is inherently complex with many hundreds of different plans available from both public and private payors. Public options include Medicaid, Medicare, and the Veterans system. Private payors are commercial companies that sell policies to individuals, employers, and unions. Although the United States has the largest pharmaceutical market, it does not regulate pricing [30]. Payors must negotiate prices and reimbursement with manufacturers. Public companies commonly get discounts from manufacturers [30]. Private companies determine pricing based on product market share [30].

Regenerative medicine products are still new in the grand scheme of things and thus their reimbursement is ever-changing. Reimbursement process involves initial FDA approval and companies understanding the how the products fit within the market. However, even with FDA approval, there is no guarantee a product will be covered by insurance. For coverage determination, payors further evaluate product safety and efficacy. Oftentimes, this leads to even greater scrutiny than the initial FDA process. Ultimate coverage decisions are largely based on clinical evidence with the more the better. Medicare tends to be quicker than commercial payors in determining coverage and largely bases decisions on FDA review results. Commercial coverage decisions can take up to years. Even then products can be deemed "investigational" and will not receive coverage [31].

If a product is covered by an insurance, reimbursement must next be determined. Reimbursement is influenced by the type of payor, clinical setting of product administration, added value of the product, and cost of the product [31]. An appropriate billing code is also required for reimbursement to take place. In some cases, existing codes do not accurately describe a regenerative medicine product and thus a company must create or modify an existing code. This process can add significant time to the process and in some cases years [31].

Variability in regenerative medicine reimbursement is demonstrated in a few examples. Platelet-rich plasma (PRP) and bone marrow aspirate concentrate (BMAC) are examples of products that do not require FDA approval. Despite this, most payors consider their use "investigational" and do not reimburse for them. Therefore, patients are responsible for paying for these products. Another example is matrix-induced autologous chondrocyte implantation (MACI). This product is FDA approved and has its own designated CPT code. Unlike PRP and BMAC, MACI procedures are commonly covered by insurances.

Overall, the reimbursement system in the United States adds another challenge for regenerative medicine products. Therefore, it is imperative for manufacturers to consider and plan for reimbursement aspects from the early stages of product development. This will allow companies to stay afloat and further their products after finally passing the FDA approval process. Proper planning is essential to the success of products and future advances in the field of regenerative medicine.

6.18 Reimbursement in the EU

Regenerative medicine products in the EU follow the same reimbursement processes as standard pharmaceuticals. Reimbursement is determined by value-based, competitor-based, or cost-based assessments. Value-based assessment reviews a product's added value over current standard treatments. Competitor-based assessment is common in the EU and focuses on reference group pricing for similar products. Cost-based assessment is not common in the EU [29]. The centralized status of countries also plays a role. Centralized countries like France or Germany decide reimbursement at the national level and are more uniform in nature. Such set reimbursements are likely included in hospital formularies. Decentralized countries such as Italy or Spain make decisions at the regional level. In these countries, a product ceiling price is set nationally, but specific prices and general implementation of products are determined regionally.

In the United Kingdom, the National Institute for Health and Care Excellence (NICE) takes a value-based approach and performs cost-effectiveness analysis using quality-adjusted life years (QALYs). The Department of Health uses this information to make final decisions on product pricing and reimbursement. Products are more likely to be approved if they are cost-effective and under 30,000 pounds [29].

As in the United States, reimbursement must be considered at early stages of product development in the EU. There is a greater emphasis on value-based assessments in this system. Manufacturers should focus on maximizing their product benefits over standard treatments. A result of this is many products targeting areas of unmet medical need to demonstrate benefits more easily in a value-based model.

6.19 Conclusion

Regenerative medicine is an exciting field that offers significant potential. These products may one day offer solutions to many orthopedic problems as well as pathologies across all medical specialties. The potential of this field makes it natural for patients to seek out these products whether domestically or via medical tourism. However, it is critical that patients be well informed on the evidence regarding product safety and efficacy. Unfortunately, resources with this information can be challenging to find at times. Regardless informed patients and proper product use is critical to patient care, the continued research of these products, and overall public reputation of the field.

Many countries have standardized risk-based approaches to the regulation of regenerative medicine products [32]. Countries with limited, lax, or even no regulation do still exist. Less regulated settings allow certain treatments to be offered that are unavailable in other parts of the world. This can provide economic gain for countries via medical tourism. Again these scenarios re-emphasize the importance of patients informing themselves on treatment before use to avoid disappointing or potentially harmful results. Ineffective and unsafe use of regenerative medicine products pose a risk to the field worldwide reputation of the field and could impede future advances [17].

The United States has a rigorous approval process and regulatory system for regenerative medicine products [32]. This creates a difficult environment for economic development and innovation in the field. The approval process is extremely costly. Average cost to take a product from inception through phase 4 trials was found to be 60–80 million dollars [33]. Other reports have estimated total cost as high as 160 million dollars [33]. In addition to cost, there is also significant time investment required. Average time for clinical trial completion is reported as 7.5 years [33]. These challenges pose a problem for smaller companies to have a chance in this space.

Efforts have been made to help better facilitate product approval and commercialization. Alternative pathways in the United States focus on speeding up the approval process so that products can reach the market quickly. For each pathway, specific requirements must still be met such as meeting specific clinical need or demonstrating substantial improvement. In Japan, regulatory changes have allowed products to reach the marketplace after proving safety and efficacy in early phase trials. The REGROW Act is an attempt at a similar change in the US system but has yet to be passed. Optimizing product approval helps to offset some of the time and cost burden the standard process places on developers. These efforts demonstrate an investment in the growth of regenerative medicine field.

Although they present challenges, regulatory systems are essential to regenerative medicine product development. Despite time and cost requirements, the ultimate goal of regulatory systems is to promote a safe environment for product development and patient safety. A risk-based approach is most common among regulatory systems. In this approach, extent of regulations correlates with amount of risk associated with a product. Even with development of alternative pathways, safety should remain the primary focus of regulation and not be sacrificed. As we continue to learn more within the field of regenerative medicine, policy will have to be adapted to not lag behind. Ultimately between policy makers and developers is critical for proper regulation as well as continued progress in the field of regenerative medicine.

References

1. Mason C, Dunnill P. A brief definition of regenerative medicine. Regen Med. 2008;3(1):1–5.
2. Brittberg M, Lindahl A, Nilsson A, Ohlsson C, Isaksson O, Peterson L. Treatment of deep cartilage defects in the knee with autologous chondrocyte transplantation. N Engl J Med. 1994;331(14):889–95.
3. Weissman IL, Anderson DJ, Gage F. Stem and progenitor cells: origins, phenotypes, lineage commitments, and transdifferentiations. Annu Rev Cell Dev Biol. 2001;17:387–403.
4. Chirba MA, Sweetapple B, Hannon CP, Anderson JA. FDA regulation of adult stem cell therapies as used in sports medicine. J Knee Surg. 2015;28(01):055–62.
5. Dominici M, Le Blanc K, Mueller I, Slaper-Cortenbach I, Marini F, Krause D, et al. Minimal criteria for defining multipotent mesenchymal stromal cells. The International Society for Cellular Therapy position statement. Cytotherapy. 2006;8(4):315–7.
6. LaPrade RF, Geeslin AG, Murray IR, Musahl V, Zlotnicki JP, Petrigliano F, et al. Biologic treatments for sports injuries II think tank-current concepts, future research, and barriers to advancement, part 1: biologics overview, ligament injury, Tendinopathy. Am J Sports Med. 2016;44(12):3270–83.
7. Mosna F, Sensebé L, Krampera M. Human bone marrow and adipose tissue mesenchymal stem cells: a user's guide. Stem Cells Dev. 2010;19(10):1449–70.
8. Anz AW, Hackel JG, Nilssen EC, Andrews JR. Application of biologics in the treatment of the rotator cuff, meniscus, cartilage, and osteoarthritis. J Am Acad Orthop Surg. 2014;22(2):68–79.
9. Levine DW, Mondano L, Halpin M. FDA regulatory pathways for knee cartilage repair products. Sports Med Arthrosc Rev. 2008;16(4):202–7.
10. Kingham R, Klasa G, Carver K. Key regulatory guidelines for the development of biologics in the United States and Europe. 1st ed. New York: Wiley; 2013. p. 75–109.
11. Bara JJ, Herrmann M, Evans CH, Miclau T, Ratcliffe A, Richards RG. Improving translation success of cell-based therapies in orthopaedics. J Orthop Res. 2016;34(1):17–21.
12. Knoepfler PS. From bench to FDA to bedside: US regulatory trends for new stem cell therapies. Adv Drug Deliv Rev. 2015;82-83:192–6.
13. Administration FaD. Proposed approach to regulation of cellular and tissue-based products; 1997. p. 9721–2.
14. Board on Health Sciences P, Board on Life S, Division on E, Life S, Institute of M, National Academy of S. Stem cell therapies: opportunities for ensuring the quality and safety of clinical offerings: summary of a joint workshop. Washington, DC: National Academies Press; 2014. Copyright 2014 by the National Academy of Sciences. All rights reserved.
15. Seoane-Vazquez E, Rodriguez-Monguio R, Powers JH 3rd. Analysis of US Food and Drug Administration new drug and biologic approvals, regulatory pathways, and review times, 1980-2022. Sci Rep. 2024;14(1):3325.
16. Agarwal R, Saltz LB. Understanding the right to try act. Clin Cancer Res. 2020;26(2):340–3.
17. Lysaght T, Kerridge IH, Sipp D, Porter G, Capps BJ. Ethical and regulatory challenges with autologous adult stem cells: a comparative review of international regulations. J Bioeth Inq. 2017;14(2):261–73.
18. Dwenger A, Strassburger J, Schwerdtfeger W. Regulation (EC) no. 1394/2007 on advanced therapy medicinal products: incorporation into national law. Bundesgesundheitsblatt Gesundheitsforschung Gesundheitsschutz. 2010;53(1):14–9.
19. George B. Regulations and guidelines governing stem cell based products: clinical considerations. Perspect Clin Res. 2011;2(3):94–9.
20. G D. Stem cell therapy regulations: the US vs. EU Regulatory Rapporteur. 2009;6:4–6.
21. Mount NM, Ward SJ, Kefalas P, Hyllner J. Cell-based therapy technology classifications and translational challenges. Philos Trans R Soc B: Biol Sci. 2015;370(1680):20150017.
22. Feigal EG, Tsokas K, Viswanathan S, Zhang J, Priest C, Pearce J, et al. Proceedings: international regulatory considerations on development pathways for cell therapies. Stem Cells Transl Med. 2014;3(8):879–87.
23. Ridgway A, Agbanyo F, Wang J, Rosu-Myles M. Regulatory oversight of cell and gene therapy products in Canada. In: Galli MC, Serabian M, editors. Regulatory aspects of gene therapy and cell therapy products: a global perspective. Cham: Springer; 2015. p. 49–71.
24. Cohen CB, Cohen PJ. International stem cell tourism and the need for effective regulation. Part I: stem cell tourism in Russia and India: clinical research, innovative treatment, or unproven hype? Kennedy Inst Ethics J. 2010;20(1):27–49.
25. Salter B, Cooper M, Dickins A, Cardo V. Stem cell science in India: emerging economies and the politics of globalization. Regen Med. 2007;2(1):75–89.
26. McMahon DS, Thorsteinsdóttir H, Singer PA, Daar AS. Cultivating regenerative medicine innovation in China. Regen Med. 2010;5(1):35–44.
27. Maeda D, Yamaguchi T, Ishizuka T, Hirata M, Takekita K, Sato D. Regulatory frameworks for gene and Cell therapies in Japan. In: Galli MC, Serabian M, editors. Regulatory aspects of gene therapy and cell therapy products: a global perspective. Cham: Springer International Publishing; 2015. p. 147–62.
28. Congress t. S.2689: govtrack.us; 2016 [.
29. Jørgensen J, Kefalas P. Reimbursement of licensed cell and gene therapies across the major European healthcare markets. J Market Access Health Policy. 2015;3(1):29321.

30. Danzon PM. Pricing and reimbursement of biopharmaceuticals and medical devices in the USA. In: Culyer AJ, editor. Encyclopedia of health economics. San Diego: Elsevier; 2014. p. 127–35.
31. T H. The challenge of orthobiologics: nativagating the coverage and reimbursement landscape in a complex martketplace. Orthopedic Design and Technology; 2008.
32. Vonk LA, de Windt TS, Slaper-Cortenbach ICM, Saris DBF. Autologous, allogeneic, induced pluripotent stem cell or a combination stem cell therapy? Where are we headed in cartilage repair and why: a concise review. Stem Cell Res Ther. 2015;6(1):94.
33. Sertkaya ABA, Berlind A, Eryaud J. Examination of clinical trial costs and barriers for drug development. Washington, DC: US Department of Health and Human Services; 2014.

7 Biological Factors in Maintaining Joint Health

Noah Knezic, Jacob Singer, Dane Lind,
Ashley Groesbeck, Meredith Luhmann,
Luis Lopez, Sara Robinson, and Johnny Huard

7.1 Introduction

A joint is a complex structure where two or more bones meet and allow for movement. Joints consist of many distinct parts that work together to maintain a healthy, well-functioning system. This chapter will take a closer look at the structure of a healthy synovial joint, common pathologies, and biological factors to consider as potential therapeutic options. Some of the components of a synovial joint include articular cartilage (AC), synovial membrane (SM), synovial fluid (SF), joint capsule, ligaments, tendons, and muscles (Fig. 7.1). AC plays a crucial role in providing structural support and protection from wear and tear. AC is maintained by chondrocyte cells, which excrete an extracellular matrix (ECM) composed primarily of collagen II and numerous other proteins, proteoglycans, and glycosaminoglycans [1]. These building blocks provide AC with the structure to cushion the joint and reduce friction from movement. The joint capsule is a tough fibrous layer that provides structure while allowing for mobility and encloses the joint cavity [1]. The SM lines the capsule and is responsible for producing SF, exchanging nutrients and waste, and monitoring and controlling the milieu within the joint [2]. SF helps maintain the environment surrounding the cartilage by providing chondrocytes within the avascular cartilage with nutrients and removing waste products [3, 4]. Further, SF provides lubrication [4]. Ligaments play a key role in providing structure and stability to the joint, joining bone to bone in a flexible manner [5]. Muscles control joint movement, anchored to the bone with tendons [6]. For the joint to operate in its intended fashion, each of these components must function normally. The complexity of a healthy joint is further increased by the complexity of each of the individual tissues within the joint. Each of these individual tissues has its own slew of potential injuries and considerations for repair. Pathologies of the joint often involve more than one of these tissues, and each must be considered when determining the best treatment protocol [7].

Noah Knezic and Jacob Singer have contributed equally to this work.

N. Knezic · J. Singer · D. Lind · A. Groesbeck
M. Luhmann · L. Lopez · S. Robinson · J. Huard (✉)
Linda and Mitch Hart Regenerative and Personalized Medicine, Steadman Philippon Research Institute, Vail, CO, USA
e-mail: jhuard@sprivail.org

A. Gobbi et al. (eds.), *Regenerative Medicine in Sports and Orthopaedics*,
https://doi.org/10.1007/978-3-031-84693-9_7

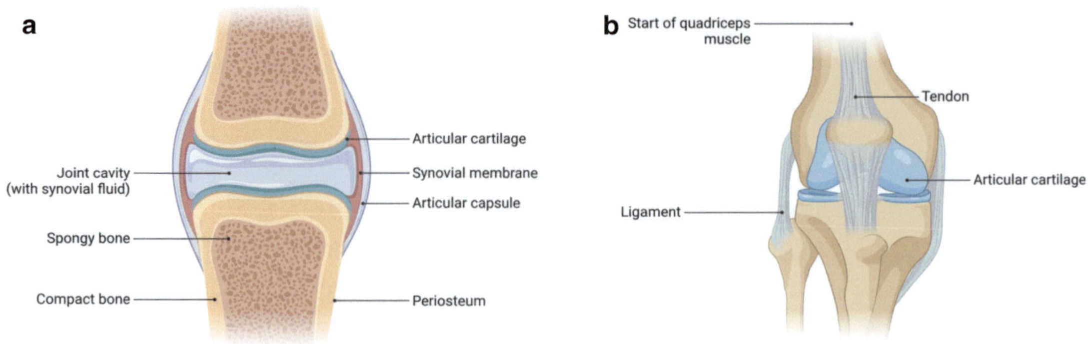

Fig. 7.1 Structure of a healthy synovial joint. (**a**) Synovial joint structures, including capsule, synovial membrane, synovial fluid, articular cartilage, and subchondral bone. (**b**) Soft tissue structure of a knee joint, including muscle, tendon, ligament, and articular cartilage. (Created with Biorender)

7.2 Common Pathologies

7.2.1 Osteoarthritis

One of the most common pathologies thought of when discussing joint health is osteoarthritis (OA). OA is a joint condition characterized by chronic pain, inflammation, and motor dysfunction that impacts quality of life [8]. Over 500 million people worldwide live with OA, and most of these individuals are over 55 years [9]. As a whole joint disease, OA consists of damage to the AC, SM thickening, subchondral bone sclerosis, osteophyte formation, and structural changes in the ligaments and muscles [8] (Fig. 7.2). Risk factors include aging, obesity, previous joint damage, and high-intensity activities [8]. Currently, there is no known cure or intervention for OA that leads to new cartilage growth or prevents the disease from progressing further.

The pathology of OA is multifactorial and influenced by inflammation, metabolism, and biochemical machinery [8]. Changes in cytokine levels, matrix-degrading enzymes, and reactive oxygen species (ROS) affect the joint environment and joint function [10]. Cartilage ECM proteins like collagen II and aggrecan degrade due to increased catabolic matrix metalloproteinases (MMPs) and aggrecanases [11]. Macrophage-secreted interleukin-1ß (IL-1ß) triggers inflammatory mediator production, MMPs, aggrecanases, and ROS, which further damages the cartilage [12]. IL-1ß activates nuclear factor-kappa B (NF-κB), inhibiting collagen II and increasing MMPs [12]. Tumor necrosis factor-alpha is secreted and leads to pathways initiating an inflammatory response, such as interleukin-6 secretion (IL-6) [13]. IL-6 increases inflammatory responses and impacts the subchondral bone by promoting osteoclast formation and bone resorption [13]. The Wnt/ß-catenin signaling pathway, involved in chondrocyte homeostasis, can lead to cartilage degradation, osteophyte formation, and reduced ECM synthesis [14]. Excess ROS causes oxidative stress, leading to protein alterations, DNA damage, and genomic instability [15].

Post-traumatic osteoarthritis (PTOA) can develop following an injury to the joint and can account for as much as 12% of all OA cases [16, 17]. Intra-articular fractures, meniscal tears, ligamentous injuries, and subchondral injuries can lead to PTOA [16, 18]. PTOA, unlike OA, develops due to an acute injury and affects younger patients [16]. The rapid onset of PTOA is due to immediate cartilage damage causing chondrocyte apoptosis from mitochondrial biogenesis imbalance, inflammation, and cartilage loss [18, 19]. Mitochondrial dysfunction triggers calcium ion influx, IL-6 upregulation, cytochrome C leakage, and ROS increases, all of which induce apoptosis [19]. Current surgical therapies aim to repair the injury and prevent or slow the onset and progression of PTOA post-injury [18].

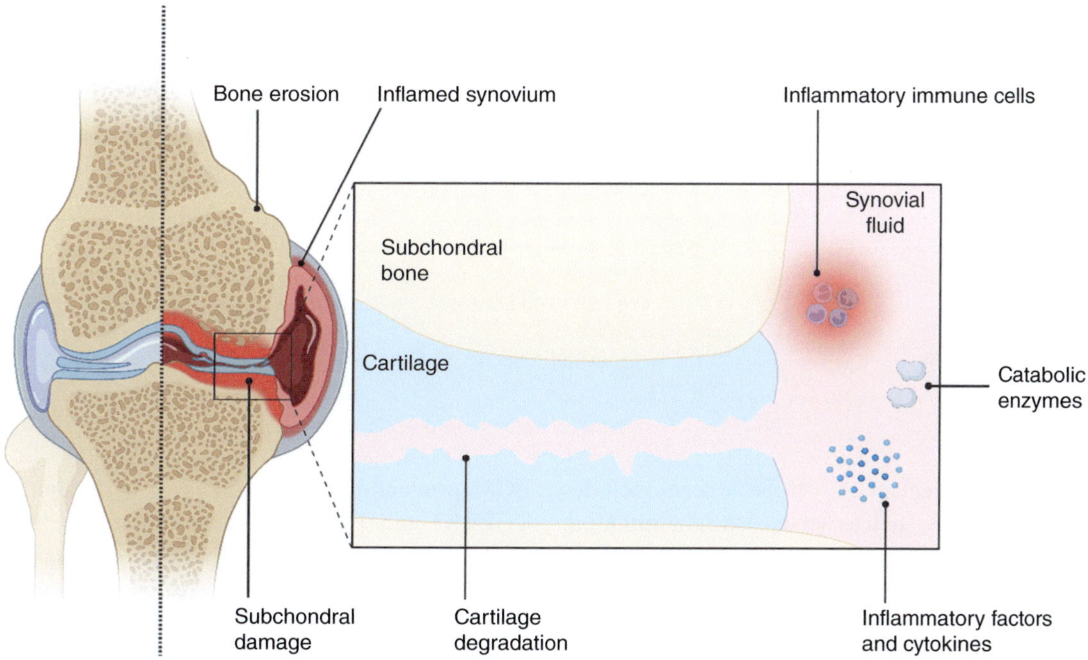

Fig. 7.2 Pathological changes in osteoarthritis. (Created with Biorender)

7.2.2 Acute Cartilage Injuries

Acute cartilage injuries are often caused by trauma or high-impact activities, including direct impacts, twisting motions, or fractures that extend into the joint space [20]. These injuries lead to significant pain and inflammation, causing impaired mobility and altered biomechanics [21]. These injuries can result in instability or mechanical symptoms of locking or catching due to loose cartilage fragments or defects to the cartilage surface that compromise the smooth protective layer [21, 22]. Timely treatment is necessary to restore joint function and limit long-term complications.

Treatments of acute cartilage injuries aim to alleviate symptoms, restore joint function, and prevent further deterioration. The conservative approach aims to lower pain and inflammation with rest, followed by physical therapy to improve strength and maintain range of motion [23]. When this approach fails, surgical methods, including arthroscopy, microfracture, or osteochondral grafts, are implemented [24]. Decisions for treatment are made based on the size of the defect, the level of discomfort to the patient, and the involvement of the subchondral surface [24].

7.2.3 Soft Tissue Injuries

Many of the soft tissues in the joint are susceptible to injury. As discussed previously, soft tissue plays a key role in maintaining stability and mobility in the joint [25, 26]. Therefore, injury to soft tissue can significantly impact the function of the joint. Tears can happen as an acute injury or chronically in the form of numerous small tears, as seen in many rotator cuff tears. Muscle, tendon, and ligament tears significantly impact joint health by reducing the stability and functionality of the joint [27, 28]. Often, tears lead to alternative loading through altered biomechanics; this can lead to further damage to other tissues and an increased risk of another injury [29, 30]. The trauma-induced inflammation paired with altered loading and poor stability can lead to more degenerative conditions such as OA if left untreated [30].

Table 7.1 Common synovial joint pathologies

Pathology	Key changes	Impact to function
Osteoarthritis	Cartilage degradation/thinning, synovial thickening, subchondral sclerosis	Chronic pain, inflammation, reduced mobility
Post-traumatic osteoarthritis	Caused by acute injury, chondrocyte apoptosis, cartilage degradation/thinning, synovial thickening, subchondral sclerosis	Rapid onset, joint instability, pain, inflammation, reduced mobility, younger patients affected
Acute cartilage injury	Caused by direct impact or twisting motion can lead to noticeable chondral defect	Pain, inflammation, reduced mobility, often active patients
Soft tissue injury	Tears in ligaments, tendons, or muscles	Reduced stability and functionality, pain, inflammation

The management of soft tissue tears includes several options, such as acute care, rehabilitation, and surgical intervention. Many noninvasive treatments focus on rest and the management of pain and inflammation [31, 32]. Rehabilitation plays a vital role in restoring function post-injury, with emphasis placed on recovering strength, flexibility, and proprioception [28]. Surgical intervention may be necessary for severe or complete tears to reconstruct the tissue so it can manage the required loads [33] (Table 7.1).

7.2.4 Complications with Repair

Tissue regeneration is sometimes impaired by the dysregulation of biological factors. One common issue is excessive or prolonged inflammation [34]. In the case of dysregulated inflammation, pro-inflammatory cytokines such as tumor necrosis factor-α (TNF-α), interleukin-1β (IL-1β), and interleukin-6 (IL-6) remain at an elevated level leading to chronic inflammation [35, 36]. As previously discussed, elevated levels of these factors can lead to the development of chronic conditions like OA in the joint through the activation of pathways such as NF-κB [13, 36, 37]. Another frequent problem is fibrosis, resulting in the formation of scar tissue. Prolonged inflammation leads to the activation of fibroblasts and increased deposition of ECM components, which often leads to fibrosis [38]. This overproduction of ECM components leads to a stiff, mechanically inferior tissue. Transforming growth factor-β (TGF-β) plays a key role in the regulation of ECM production. TGF-β binds to its receptors TGFBR1 and TGFBR2 on the cell surface and activates Smad signaling pathways through phosphorylation [39]. These activated Smads translocate to the nucleus and regulate the transcription of pro-fibrotic genes. This pathway controls fibroblast proliferation and synthesis of collagen and other ECM proteins. Repair can be impaired by reduced vascularization, especially in tissues like ligaments and cartilage, which already have low vascularization [40, 41]. The vascular endothelial growth factor (VEGF) is a primary driver of angiogenesis [42]. VEGF binds its receptors VEGFR1 and VEGFR2, activating signaling cascades such as PI3k/Akt and MARK pathways, promoting endothelial proliferation, migration, and blood vessel formation [43]. On the other hand, disorganized or excessive angiogenesis can also lead to complications, which can lead to an unstable environment, resulting in increased inflammation through a continuous supply of inflammatory mediators [44–46]. Excessive inflammation, fibrosis, and altered angiogenesis can all decrease the quality of soft tissue repair and increase the time it takes to heal, as well as increase the chance of re-injury.

7.3 Biologic Factors in Maintaining Joint Health

The microenvironment of the joint is critical to joint health, and age-related changes occur in the composition of this microenvironment [47]. The factors that make up this milieu have a direct impact on the tissues that make a healthy joint. If the microenvironment is disrupted through injury or disease, the mechanistic pathways react to the specific injury and disrupt the homeostatic balance of the joint milieu [48]. Once this balance is disrupted, inflammatory pathways are activated to recruit various pro- and anti-inflammatory growth factors and cytokines to the injury site for the healing response [49]. However, the body does not always induce the necessary type of healing response. Additionally, dysregulation of several biological factors can be detrimental to healthy joints and can hinder repair of joint tissues (Table 7.2).

7.3.1 Transforming Growth Factor-β

One such biological factor is TGF-β, which plays a crucial role in the regulation of chondrocyte proliferation. The dysregulation of TGF-β is associated with the development of OA [50]. TGF-β is responsible for the controlled proliferation of chondrocytes and the differentiation of precursor cells, such as stem cells, into cartilage-producing chondrocytes [50, 51]. As previously discussed, TGF-β activates Smad-dependent signaling pathways, which are responsible for controlling many aspects of a cell's life [52]. In this way, TGF-β is responsible for maintaining the structure and function of healthy joint cartilage.

TGF-β is also responsible for stimulating the production of the ECM components, such as collagen and proteoglycans, necessary for the structural integrity and mechanical properties of hyaline AC [53]. TGF-β regulates both anabolic and catabolic pathways allowing it to achieve the delicate balance necessary to maintain cartilage [53]. In response to injury, TGF-β plays a role in the repair of damaged cartilage by recruiting and activating mesenchymal stem cells (MSCs) to the injury site [54]. Through facilitation of this process, TGF-β preserves the integrity of the joint and prevents further joint deterioration.

While TGF-β1 is a necessary factor in a healthy joint, it can have a detrimental effect in a dysregulated state. As previously discussed, TGF-β1 is essential for many cellular processes; as such, underactivity of TGF-β1 would result in dysfunction of the joint. Overactivation of TGF-β1 has also proven to be detrimental, as it can lead to a heightened state of inflammation and

Table 7.2 Biological factors to consider in joint health

Biological factor/cytokine	Role in healthy joint	Pathological impact	Therapeutic target/ intervention
Transforming growth factor-β	Regulates chondrocyte proliferation, ECM production, cartilage maintenance	Overactivation leads to inflammation, fibrosis, joint degradation	TGF-β inhibitors (e.g., fresolimumab), losartan, ligand traps
Tumor necrosis factor-α	Immune surveillance, regulation of inflammation, apoptosis	Chronic inflammation, synovial inflammation, ECM degradation	Anti-TNF-α therapies (e.g., infliximab, adalimumab)
Interleukin-1β	Modulates SF composition, cartilage ECM turnover	Increased MMPs and aggrecanases inhibit collagen synthesis, chronic inflammation	IL-1 receptor antagonists (e.g., anakinra)
Interleukin-6	Cartilage metabolism, bone metabolism	Chronic inflammation, systemic inflammation, increased osteoclastogenesis	IL-6 receptor antagonist (e.g., tocilizumab)
Vascular endothelial growth factor	Promotes angiogenesis, ensures nourishment and waste removal	Excessive VEGF causes inflammation and leaky vessels. Insufficient VEGF impairs healing	Anti-VEGF (e.g., bevacizumab), recombinant VEGF injections
Senescence-associated secretory phenotype	Lower activity in healthy tissue and older patients	Promotes chronic inflammation, altered chondrocyte formation	Senolytic drugs (e.g., Dasatinib, Quercetin), Senomorphics (e.g., metformin, rapamycin

ultimately lead to tissue fibrosis or excessive ECM production [55, 56]. Overexpression of TGF-β1 can be observed after traumatic or chronic injury, causing increased inflammation and further production of TGF-β1 [56, 57]. Once the overactivation begins, it creates a cycle of upregulated TGF-β1 that results in increased joint degradation [56, 58]. If left untreated, the cycle of dysregulated TGF-β can lead to more severe conditions such as OA [55]. In its dysregulated state, TGF-β1 promotes more catabolic processes, which eventually deteriorate the joint surface and result in pain for the patient.

The treatment of TGF-β1 overexpression is critical for the preservation of a healthy joint. This is complicated by the fact that some TGF-β1 is necessary for joint function, so outright inhibition is detrimental. As such, research has focused on titrating the amount of active TGF-β1 in the joint to optimize the microenvironment to promote joint health. Areas of exploration include the use of antibodies like fresolimumab to target TGF-β isoforms, small molecules such as galunisertib to target TGF-βR1, and ligand traps [59–61]. Additional drugs, such as Losartan, have been explored for their potential to reduce the action of TGF-β1 and have shown positive effects during surgery to prevent the overgrowth of fibrotic tissue [62, 63]. Future research aims to use nanofibers as a mechanism to promote a time-controlled release of drugs such as Losartan into the joint, aiming to have an intrinsic effect that can provide an immediate and controlled response [64]. Some concerns with these treatments are their specificity and safety, as systemic inhibition can be problematic. Additionally, medications such as Losartan, normally prescribed for the treatment of high blood pressure, have different effects at different doses. These drugs need to be studied further to understand the mechanistic pathways blocking fibrosis and their long-term impact [65]. Our recent publication suggests that low dose (1 mg/knee) intra-articular injection of losartan improved cartilage regeneration more effectively than high dose (100 mg/knee) injection, which may damage normal articular cartilage [63].

7.3.2 Tumor Necrosis Factor-α

In a healthy joint, TNF-α plays a crucial role in immune surveillance and the regulation of inflammation. TNF-α is a pro-inflammatory cytokine secreted in response to injury or disease and modulates the immune response to favor repair [66]. TNF-α plays a vital role in the recruitment and activation of immune cells like macrophages and neutrophils. Additionally, TNF-α plays a role in apoptosis by helping to clear damaged cells and maintain tissue integrity [67, 68]. TNF-α activates the NF-κB signaling pathway and leads to the transcription of genes responsible for immune response to injury and disease [69].

Elevated levels of TNF-α are closely associated with chronic inflammation and joint degradation [70]. TNF-α promotes the expression of MMPs and other catabolic enzymes responsible for degradation of ECM [71, 72]. TNF-α is associated with the expression of additional pro-inflammatory factors. Overexpression of TNF-α can lead to a perpetuating cycle of inflammation, resulting in a chronic environment that supports joint tissue degradation [73–75]. Left untreated, this chronic inflammation can contribute to the development of OA. Finally, TNF-α is an apoptotic factor, and it can induce apoptosis in chondrocytes at high levels, further degrading the joint space by removing cells responsible for maintaining cartilage homeostasis [75].

Targeting pathological levels of TNF-α has proven effective in managing chronic inflammatory joint diseases, including OA. Anti-TNF-α therapies utilize monoclonal antibodies, such as infliximab and adalimumab, or TNF-α receptors, like etanercept, to bind and reduce the concentration of active TNF-α in the body [76, 77]. These biologics prevent the activation of the NF-κB pathway, therefore reducing the expression of MMPs and other inflammatory factors. This effectively reduces levels of inflammation and prevents further degradation and progression to chronic conditions like OA. One risk of targeting TNF-α is the potential to increase the risk of infection due to its crucial role in immunosurveillance and recruitment [78].

7.3.3 Interleukin-1β

IL-1β is essential in the maintenance of a healthy joint. Like TNF-α, IL-1β regulates immune response [79]. At physiological levels, IL-1β modulates SF composition, which maintains lubrication of the joint and nutrient supply to chondrocytes [80, 81]. Additionally, IL-1β signals for turnover and remodeling of cartilage ECM, which controls the activity of chondrocytes [82, 83].

IL-1β plays a crucial role in the pathogenesis of inflammatory joint diseases such as OA. Elevated levels of IL-1β can lead to higher levels of expression of MMPs and aggrecanases, which are responsible for the breakdown of cartilage components by targeting several vital proteins that make up the ECM [71, 84]. At pathological levels, IL-1β has been shown to significantly inhibit the synthesis of collagen by chondrocytes, which impairs the repair and regeneration of this protective tissue [85]. High levels of IL-1β lead to continuous activation of catabolic pathways and immune responses to cause a perpetual state of chronic inflammation that further degrades the joint.

Dysregulation of IL-1β can be targeted using antagonists for IL-1 receptors. One such antagonist, anakinra, blocks the binding of IL-1β to its receptor, thereby reducing its pro-inflammatory actions [86]. Inhibition of IL-1β prevents the activation of the NF-κB and MAPK pathways and reduces the production of catabolic enzymes, including MMPs and aggrecanases [71].

7.3.4 Interleukin-6

IL-6 is both a pro- and anti-inflammatory cytokine that plays a pivotal role in maintaining tissue homeostasis [87]. Like IL-1β, IL-6 modulates immune response and contributes to the composition of SF [88, 89]. IL-6 supports cartilage metabolism and promotes the synthesis of proteoglycans and collagen [90]. Additionally, IL-6 supports bone metabolism by balancing resorption and formation, which is crucial to joint health and stability [91].

At pathological levels, IL-6 contributes significantly to local and systemic inflammation. IL-6 promotes the differentiation and proliferation of pro-inflammatory T-cells [92]. Additionally, IL-6 has been shown to stimulate the production of acute-phase proteins by the liver to lead to chronic systemic inflammation [93]. Systemic inflammation can have a negative effect on the health of multiple joints by favoring catabolic pathways [34, 94]. Osteoclastogenesis is also promoted by IL-6 through the JAK/STAT pathway, which increases bone resorption and erodes the tissue, leading to conditions like osteoporosis (a loss of bone density and an increase in fragility) [95]. Further, elevated levels of IL-6 contribute to the upregulation of other inflammatory markers, perpetuating a chronic inflammatory cycle, as other dysregulated factors have been shown to do.

Like IL-1β, therapeutics for IL-6 are typically antagonists for IL-6 receptors. The monoclonal antibody tocilizumab blocks IL-6 signaling through competitive binding of the receptor to reduce inflammation and limit further joint degradation [96]. By inhibiting the JAK/STAT pathway, tocilizumab prevents the activation of genes involved in inflammation and immune responses. Treatment targeting the dysregulation of IL-6 leads to a reduction in local and systemic inflammation. Further, it decreases bone resorption by reducing osteoclastogenesis.

7.3.5 Vascular Endothelial Growth Hormone

VEGF is known for its role in the promotion of angiogenesis by ensuring adequate blood supply to the SM, subchondral bone, muscle, and tendons [97]. Vascularization of these tissues is necessary to ensure proper nourishment and waste removal. VEGF also plays a vital role in the repair of the vascular network after injury to any of these tissues [98].

VEGF can be pathological with over-expression or under-expression. Excessive VEGF can lead to the formation of immature, disorganized, leaky blood vessels, which serve to exacerbate inflammation in the tissue [46, 99]. This can also lead to increased blood flow and permeation of inflammatory immune cells and mediators into the joint tissues, which can be observed as increased swelling and pain [46]. Furthermore, too much VEGF in tissue can cause cartilage to turn into bone; without this, cartilaginous joints may have less flexibility and cushioning, along with more pain and inflammation due to increased bone-on-bone contact [100]. Additionally, too little VEGF can be pathological, resulting in a reduced capacity to provide nourishment and remove waste [98]. This can be particularly problematic with acute injury by limiting the healing potential.

Treating VEGF dysregulation involves addressing both under- and overexpression. Anti-VEGF therapies, like bevacizumab, can be used to inhibit the interaction between overexpressed VEGF and its receptors [101]. This can reduce inflammation in the body. Under-expression of VEGF results in poor vascularization of joint tissues and limits the potential for regeneration after injury. Strategies should look at increasing VEGF levels, promoting production, and increasing effectiveness. The most promising treatment to date involves the use of recombinant VEGF as an injectable into the site of injury to promote angiogenesis and improve the healing potential [102].

7.3.6 Senescence

Senescence was classified in the 1960s as a cell state characterized by arrest in the cell cycle with the lack of proliferation [103]. A variety of environmental factors can trigger cellular senescence, but it is mainly associated with aging [104]. One of the most detrimental characteristics of senescence regarding joint health is the promotion of an inflammatory environment through the senescence-associated secretory phenotype (SASP) [105]. SASP is caused by a stress response leading to the secretion of several pro-inflammatory cytokines that can induce a chronic state of inflammation, which, as mentioned, can be detrimental to joint health. Beyond a chronic state of inflammation, senescence can have a more direct impact on joint health. Senescent chondrocytes exhibit altered function, characterized by a reduction in ECM production and increased secretion of catabolic enzymes, which can contribute to cartilage loss seen in conditions like OA [106]. Xu et al. (2017) found that intra-articular injections of senescent cells in mice induced an OA-like state, with cartilage deterioration, increased pain, and inflammation [107].

In recent years, research has focused on the ability to clear senescent cells to reduce the effects of characteristics like SASP. The original emphasis was to eliminate senescent cells contributing to the inflammatory state. For this purpose, senolytics such as Dasatinib, Quercetin, and Fisetin were explored as drugs that can promote apoptosis of senescent cells by disabling the pro-survival networks that senescent cells need [108–111]. Hickson et al tested the combination of Dasatinib and Quercetin; he found a decrease in senescent cells in diabetes and chronic kidney disease by the reduction of well-known senescent markers p16, p21, and senescent associated β-galactosidase [112]. After treatment, the subjects showed improvement in gait function and pain reduction. Further exploration has led to the identification of senomorphics such as Metformin and Rapamycin, which have been shown to reduce SASP and improve cellular homeostasis [113–115]. Finally, the focus has shifted to regenerative therapies by focusing on the goal of rejuvenating senescent cells to contribute to the overall health of the joint [116, 117]. These regenerative therapies often focus on the use of MSCs and their secretomes.

7.4 Orthobiologics and Exosomes

As aging progresses, our bodies become more resistant to our natural mechanistic defenses. Injuries become challenging to rehabilitate, causing lengthy recovery times, which can increase the chances of developing a chronic condition such as OA. A specific injury can activate anti-inflammatory and pro-inflammatory pathways, which, when regenerating and promoting healing, need to be in balance as inflammatory homeostasis is crucial to healing [118]. If one pathway is over- or under-activated, then healing mechanisms cannot reproduce tissue that mimics the native joint state. Surgical interventions can be a resource that can help acute orthopedic injuries; however, maintaining the joint's native tissue and structure is complex, and issues like fibrosis are still a problem. While surgeries are sometimes needed, these procedures can introduce further trauma into the joint. This trauma can be necessary to establish an environment favoring healing and repair [119]. New interventions have been developed to enhance the tissue quality in the joint to aid traditional surgeries; these can range from autologous products to the use of cadaver parts to help create an environment for the native joint to heal.

Orthobiologic use is a rapidly developing subset of therapies using patients' cells to help initiate a healing response [120]. The concept behind orthobiologics is taking a patient sample, concentrating the solution, and then re-introducing it back into the injury site to stimulate the intrinsic healing mechanisms and improve the healing response [121]. Biological products can be used in a clinical setting or in conjunction with a procedure. The most common orthobiologics currently being explored are bone marrow aspirate concentrate (BMAC), platelet-rich plasma (PRP), and extracellular vesicles (EVs) or exosomes [122, 123] (Table 7.3).

PRP is an autologous orthobiologic product that uses the patient's whole blood, commonly drawn via venipuncture, with a small amount of anticoagulant, either acid citrate dextrose solution (ACD-A) or heparin [122]. The whole blood is spun through a centrifuge to separate the blood layers. The buffy coat is then captured and concentrated with plasma to reduce the amount of white blood cells (WBCs) and erythrocytes while increasing the concentration of platelets, mononuclear cells, and bioactive molecules. Platelets are essential in forming clots and initiating inflammation, proliferation, and remodeling [124–126]. In addition, the PRP product contains peripheral blood mononuclear cells (PBMCs), specifically monocytes and lymphocytes, which play a role in tissue regeneration by releasing cytokines that can stimulate tissue repair [127].

BMAC is a unique orthobiologic product as it contains the body's own stem cells, known as bone marrow-derived stem cells (BMSCs), while also containing properties similar to PRP. However, it is drawn from the bone marrow rather than peripheral blood [128]. Bone marrow is a spongy soft tissue found within the larger bones of the body in the medullary cavity. Bone marrow is crucial in fighting disease and injury as it creates erythrocytes, WBCs, and platelets. It is composed of red bone marrow and yellow bone marrow [129]. Red bone marrow is comprised of fibrous tissues that contain hematopoietic cells and is where all erythrocytes and platelets are formed [130]. While lymphocytes begin their formation in the red bone marrow, they fully mature when they reach their destination, such as the thymus or spleen [130]. Yellow bone marrow is home to hematopoietic stem cells, some of which are classified as MSCs, and it is the fattier portion of the marrow with more nutrients than the red bone marrow [131].

BMAC is usually drawn from the iliac crest as bone marrow aspirate (BMA), which contains an added anticoagulant, such as ACD-A or heparin [128]. The BMA is then centrifuged, the blood layers are separated, and the buffy coat is captured and concentrated into an injectable form. The one issue is that BMAC cannot be seen as a "pure" stem cell product as it is only 0.01–0.001% of MSCs [132, 133]. It is hard to distinguish if such a small amount of MSCs is impacting the treatment or if other factors within BMAC are responsible for the regenerative effect. Thus, the benefit to OA patients may rely on paracrine secretion of potent trophic factors from cells within BMAC rather than (or in addition to) the direct differentiation of stem cells into the surrounding tissue [133].

Autologous biologics (BMAC and PRP) have improved pain and function in some clinical trials, but benefits have been inconsistent. Recent findings suggest that the mix of beneficial and detrimental cytokines and other factors modulate the efficacy of these biologics. The joint injection has shown efficacy in reducing knee OA symptoms

Table 7.3 Common orthobiologic treatments under investigation

Therapy	Source	Key components	Mechanism of action
Platelet-rich plasma	Whole blood	Platelets, growth factors, cytokines, mononuclear cells	Activates healing mechanisms through the promotion of inflammation, proliferation, remodeling
Bone marrow aspirate concentrate	Bone marrow	Stem cells (BMSCs), growth factors, cytokines, mononuclear cells	Tissue regeneration, anti-inflammatory
Extracellular vesicles	PRP, BMAC	MiRNA, proteins, lipids	Cell-to-cell communication, gene expression modulation

with minimal adverse effects [134–136]. However, clinical results of PRP and BMAC injections are highly variable across patients, and a single infusion may provide only relatively short-term pain relief. Both PRP and BMAC contain many EVs, which are small vesicles that carry proteins, lipids, and nucleic acids, including mRNA, microRNA (miRNA), and long noncoding RNA. EVs isolated from PRP and BMAC can benefit OA by promoting cartilage repair, extracellular matrix synthesis, and bone remodeling, as well as attenuating inflammation and fibrosis. Consequently, EVs offer potential as a non-cellular therapy for OA.

7.4.1 Exosomes

Orthobiologics present a promising future for restoring joint damage, and EVs are an exciting area of new research. EVs are lipid-bound vesicles categorized as microvesicles (MVs), exosomes, and apoptotic bodies, each with different protein profiles [137]. MVs and apoptotic bodies are generated by outward budding of the cell's plasma membrane, whereas exosomes are formed from the endocytic pathway [138]. These are nanometer-sized lipid molecules of 30-150 nm diameter, which function as mediators facilitating cell-to-cell communication [139]. This communication is delivered as cargo, including proteins, nucleic acids, and lipids that react in diverse ways to stimuli and physiological conditions when transported. EVs serve as a delivery system for both endogenous and exogenous cargo, offering the potential of being a non-cellular therapy for OA.

Many RNAs are packaged in EVs, including mRNA, miRNA, and long noncoding RNA. miRNAs are small noncoding RNAs that can regulate gene expression by binding to the 3′-untranslated regions of target mRNAs. Exosomal miRNAs can affect recipient cells' biological processes and functions by modulating their gene expression [140]. It has been shown that EVs derived from BMAC and PRP have therapeutic potential for treating OA by delivering various biomolecules, such as miRNAs, to the damaged joint tissues [141, 142]. However, OA patients have altered, non-typical miRNA profiles and EVs from OA patients may contain deleterious miRNA associated with OA pathogenesis and progression [143].

PRP has known benefits as it contains TGF-b1, platelet-derived growth factor (PDGF), and VEGF. However, most patients in need of PRP have an injury or disease, which limits the effectiveness of this autologous product as it contains the inflammatory factors and proteins induced by their injury or disease. PRP is a blood product, so it cannot be drawn from a healthy patient and used on a diseased patient, as immunoreactivity and rejection could occur. PRP-derived exosomes provide an approach to produce more reliable and consistent results in PRP therapy and, in the future, could be potentially used as an allogenic therapy on patients if, indeed, EVs are non-immunogenic [144]. PRP-derived exosomes have shown the ability to cause MSCs to proliferate and migrate to specific injury sites [145]. A study compared the difference between activated and inactivated PRP to PRP-derived exosomes, revealing that PRP-derived exosomes proliferate and recruit cells significantly better [132]. In addition, this study highlighted the potential for PRP-derived exosome's ability to decrease the apoptotic rate of OA chondrocytes acting as a carrier for the specific growth factors that activate the Wnt/β-catenin signaling pathway. PRP-derived exosomes provide a more reliable and consistent means of delivering regenerative signals to the injury site. This precision in molecular messaging ensures that the recruitment of endogenous stem cells is not dependent on platelet activation.

BMAC is another orthobiologic product that can be utilized similarly for exosome therapeutics. Bone marrow contains a subset of stem cells of the "hematopoietic niche," which are multipotent [146]. MSCs are part of this niche and are mainly studied for their differentiation capabilities to neural progenitor-like cells, adipocytes, osteocytes, and chondrocytes, making them a prime candidate for therapeutic use [147]. MSCs promote paracrine anti-inflammatory and trophic effects, supporting tissue repair and regeneration. Specifically, bone marrow MSCs can promote a pro-regenerative response through their secretome profile [148]. Recent studies reveal that MSCs' paracrine effect is the driving factor in tissue regeneration and chondrocyte formation, making exosomes an emerging field of study regarding MSCs [149]. Bone marrow-derived exosomes have shown the ability to combat OA by reducing the death of chondrocytes as a protective mechanism through the polarization of M2 macrophages [150]. This mechanism indicates that bone marrow-derived exosomes are bioenergetically reorganizing the microenvironment of the joint space to allow proper healing [151].

7.5 Future Directions

Orthobiologics has provided the blueprint for the future of regenerative medicine; it has shown promise in regenerating native tissues and can aid in the success of surgical intervention. EVs are an area of intense research as they provide an opportunity to direct cellular communication to enhance healing. EV research has provided evidence that EVs contain both enhancing and deleterious miRNA, limiting their therapeutic potential [152]. However, recent research has shown that the production of EVs containing inflammatory miRNA can be inhibited using pharmacologic agents such as glucocorticoids and antisense oligonucleotides (AMOs) that target deviant miRNAs [153]. In addition, it has been revealed that EVs coming from senescent cells have properties in them that can lead to the progression of OA and other diseases [154–156]. Future research could seek to utilize senolytic drugs to help reduce the amount of senescence in the body before the isolation of EVs. Another possibility is to use senolytic therapies in a combinatorial treatment with EVs. Furthermore, using an anti-fibrotic agent such as Losartan could help combat specific miRNA that could lead to fibrosis, driving a more hyaline-like cartilage repair. EVs have a bright future in regenerative medicine.

7.6 Conclusion

Maintaining a healthy joint involves a complex interplay of tissues and is mediated by a further complicated system of biological factors. Each part has its own intricate structures and potential injuries. Synovial joints are complex and require the coordination of AC, SM, SF, ligaments, tendons, and muscles to function correctly. If not addressed, disruption to any of these tissues can lead to chronic conditions such as OA.

Current treatment methods for joint pathologies are often limited in their ability to restore native joint tissue and function. Conservative treatments and surgical interventions often provide some relief but can neglect to address underlying biological dysfunctions. Emerging therapies in the orthobiologic field offer promising ways to coordinate and enhance the body's intrinsic healing mechanisms. Additionally, these therapies can address some of the dysregulations in cytokine and growth factor balance that have been demonstrated to negatively impact repair and increase systemic inflammation. As therapies continue to develop and address existing shortcomings, joint health will only improve.

Future research should focus on the optimization of therapeutic EVs. The potential to develop a non-immunogenic, conditional, biologic treatment has numerous implications in tissue regeneration treatments. Further, these therapies could be used to target specific dysregulations associated with poor repairs and control these signaling mechanisms.

Our group is currently completing multiple placebo-controlled clinical trials administering Fisetin and Losartan for the treatment of OA. However, we have recently identified pain as a key area of study; if we can develop novel analgesics that lack the negative side effects associated with current medications like opioids or NSAIDS, patients may be less likely to need further intervention for their condition. Novel analgesics should be an area of focus in reducing patient burden from OA and keeping patients active and load-bearing.

In conclusion, maintaining a healthy joint requires a comprehensive approach that considers the complex anatomy and biology of the synovial joint. As the field continues to develop, it will be important to integrate existing treatment methods with innovative biological therapies to optimize outcomes. By addressing underlying biologic factors, the body's intrinsic healing processes can be optimized to regenerate damaged tissues.

References

1. Sophia Fox AJ, Bedi A, Rodeo SA. The basic science of articular cartilage: structure, composition, and function. Sports Health. 2009;1(6):461–8.
2. Berumen-Nafarrate E, et al. Synovial tissue and synovial fluid. J Knee Surg. 2002;15(1):46–8.
3. Li Y, et al. Recent advances in understanding the role of cartilage lubrication in osteoarthritis. Molecules. 2021a;26(20):6122.
4. Seidman AJ, Limaiem F. Synovial fluid analysis. In: StatPearls. Treasure Island, FL: StatPearls; 2024.
5. Asahara H, Inui M, Lotz MK. Tendons and ligaments: connecting developmental biology to musculoskeletal disease pathogenesis. J Bone Miner Res. 2017;32(9):1773–82.
6. Thorpe CT, Screen HR. Tendon structure and composition. Adv Exp Med Biol. 2016;920:3–10.
7. Tong L, et al. Current understanding of osteoarthritis pathogenesis and relevant new approaches. Bone Res. 2022;10(1):60.
8. Li Z, Huang Z, Bai L. Cell interplay in osteoarthritis. Front Cell Dev Biol. 2021b;9:720477.
9. Osteoarthritis [Website]. 2023. https://www.who.int/news-room/fact-sheets/detail/osteoarthritis. Accessed 25 Jun 2024.
10. Brandt KD, et al. Yet more evidence that osteoarthritis is not a cartilage disease. Ann Rheum Dis. 2006;65(10):1261–4.
11. Mehana EE, Khafaga AF, El-Blehi SS. The role of matrix metalloproteinases in osteoarthritis pathogenesis: an updated review. Life Sci. 2019;234:116786.
12. Luo P, et al. Effects of immune cells and cytokines on different cells in OA. J Inflamm Res. 2023;16:2329–43.
13. Wojdasiewicz P, Poniatowski LA, Szukiewicz D. The role of inflammatory and anti-inflammatory cytokines in the pathogenesis of osteoarthritis. Mediators Inflamm. 2014;2014:561459.
14. Zhou Y, et al. Wnt/beta-catenin signaling in osteoarthritis and in other forms of arthritis. Curr Rheumatol Rep. 2017;19(9):53.
15. Drevet S, et al. Reactive oxygen species and NADPH oxidase 4 involvement in osteoarthritis. Exp Gerontol. 2018;111:107–17.
16. Haller JM, et al. Posttraumatic osteoarthritis: from basic science to clinical implications. OTA Int. 2023;6(2 Suppl):e232.
17. Brown TD, et al. Posttraumatic osteoarthritis: a first estimate of incidence, prevalence, and burden of disease. J Orthop Trauma. 2006;20(10):739–44. https://doi.org/10.1097/01.bot.0000246468.80635.ef.
18. Kramer WC, Hendricks KJ, Wang J. Pathogenetic mechanisms of posttraumatic osteoarthritis: opportunities for early intervention. Int J Clin Exp Med. 2011;4(4):285–98.
19. Dilley JE, et al. Post-traumatic osteoarthritis: a review of pathogenic mechanisms and novel targets for mitigation. Bone Rep. 2023;18:101658.
20. Buckwalter JA. Articular cartilage injuries. Clin Orthop Relat Res. 2002;402:21–37.
21. Hayes DW Jr, Brower RL, John KJ. Articular cartilage. Anatomy, injury, and repair. Clin Podiatr Med Surg. 2001;18(1):35–53.
22. Moyad TF. Cartilage injuries in the adult knee: evaluation and management. Cartilage. 2011;2(3):226–36.
23. Hanaoka C, Fausett C, Jayabalan P. Nonsurgical management of cartilage defects of the knee: who, when, why, and how? J Knee Surg. 2020;33(11):1078–87.
24. Krych AJ, et al. Cartilage injury in the knee: assessment and treatment options. J Am Acad Orthop Surg. 2020;28(22):914–22.
25. Kaya Keles CS, Ates F. How mechanics of individual muscle-tendon units define knee and ankle joint function in health and cerebral palsy—a narrative review. Front Bioeng Biotechnol. 2023;11:1287385.

26. Woo SL, et al. Biomechanics of knee ligaments: injury, healing, and repair. J Biomech. 2006;39(1):1–20.
27. Mococain P, et al. Biomechanical effect on joint stability of including deltoid ligament repair in an ankle fracture soft tissue injury model with deltoid and syndesmotic disruption. Foot Ankle Int. 2020;41(9):1158–64.
28. Jurecka A, et al. Evaluating the effectiveness of soft tissue therapy in the treatment of disorders and postoperative conditions of the knee joint—a systematic review. J Clin Med. 2021;10(24):5944.
29. Maly MR. Abnormal and cumulative loading in knee osteoarthritis. Curr Opin Rheumatol. 2008;20(5):547–52.
30. Guilak F. Biomechanical factors in osteoarthritis. Best Pract Res Clin Rheumatol. 2011;25(6):815–23.
31. Papaleontiou A, et al. Conservative vs surgical treatment of anterior cruciate ligament rupture: a systematic review. Cureus. 2024;16(3):e56532.
32. Almeida MO, et al. Conservative interventions for treating exercise-related musculotendinous, ligamentous and osseous groin pain. Cochrane Database Syst Rev. 2013;2013(6):CD009565.
33. Al Meklef R, Rein S, Kremer T. Acute and post-acute soft tissue reconstruction. Unfallchirurgie (Heidelb). 2024;127:560–6.
34. Sokolove J, Lepus CM. Role of inflammation in the pathogenesis of osteoarthritis: latest findings and interpretations. Ther Adv Musculoskelet Dis. 2013;5(2):77–94.
35. Schuerwegh AJ, et al. Influence of pro-inflammatory (IL-1 alpha, IL-6, TNF-alpha, IFN-gamma) and anti-inflammatory (IL-4) cytokines on chondrocyte function. Osteoarthr Cartil. 2003;11(9):681–7.
36. Molnar V, et al. Cytokines and chemokines involved in osteoarthritis pathogenesis. Int J Mol Sci. 2021;22(17):9208.
37. Kapoor M, et al. Role of proinflammatory cytokines in the pathophysiology of osteoarthritis. Nat Rev Rheumatol. 2011;7(1):33–42.
38. Mack M. Inflammation and fibrosis. Matrix Biol. 2018;68-69:106–21.
39. Miyazono K. TGF-beta signaling by Smad proteins. Cytokine Growth Factor Rev. 2000;11(1–2):15–22.
40. Fenwick SA, Hazleman BL, Riley GP. The vasculature and its role in the damaged and healing tendon. Arthritis Res. 2002;4(4):252–60.
41. Bray RC, Leonard CA, Salo PT. Vascular physiology and long-term healing of partial ligament tears. J Orthop Res. 2002;20(5):984–9.
42. Hoeben A, et al. Vascular endothelial growth factor and angiogenesis. Pharmacol Rev. 2004;56(4):549–80.
43. Shibuya M. Vascular endothelial growth factor (VEGF) and its receptor (VEGFR) signaling in angiogenesis: a crucial target for anti- and pro-angiogenic therapies. Genes Cancer. 2011;2(12):1097–105.
44. Spilsbury K, et al. Overexpression of vascular endothelial growth factor (VEGF) in the retinal pigment epithelium leads to the development of choroidal neovascularization. Am J Pathol. 2000;157(1):135–44.
45. Elebiyo TC, et al. Reassessing vascular endothelial growth factor (VEGF) in anti-angiogenic cancer therapy. Cancer Treat Res Commun. 2022;32:100620.
46. Liu X, et al. The role of vascular endothelial growth factor in tendon healing. Front Physiol. 2021;12:766080.
47. Rezus E, et al. The link between inflammaging and degenerative joint diseases. Int J Mol Sci. 2019;20(3):614.
48. Lotz M, Loeser RF. Effects of aging on articular cartilage homeostasis. Bone. 2012;51(2):241–8.
49. Bolander J, et al. The synovial environment steers cartilage deterioration and regeneration. Sci Adv. 2023;9(16):eade4645.
50. Kim JG, Rim YA, Ju JH. The role of transforming growth factor beta in joint homeostasis and cartilage regeneration. Tissue Eng Part C Methods. 2022;28(10):570–87.
51. Wang W, Rigueur D, Lyons KM. TGFbeta signaling in cartilage development and maintenance. Birth Defects Res C Embryo Today. 2014;102(1):37–51.
52. Hu HH, et al. New insights into TGF-beta/Smad signaling in tissue fibrosis. Chem Biol Interact. 2018;292:76–83.
53. Thielen NGM, van der Kraan PM, van Caam APM. TGFbeta/BMP signaling pathway in cartilage homeostasis. Cells. 2019;8(9):969.
54. Du X, et al. The role of TGF-beta3 in cartilage development and osteoarthritis. Bone Res. 2023;11(1):2.
55. Bakker AC, et al. Overexpression of active TGF-beta-1 in the murine knee joint: evidence for synovial-layer-dependent chondro-osteophyte formation. Osteoarthr Cartil. 2001;9(2):128–36.
56. van der Kraan PM. Differential role of transforming growth factor-beta in an osteoarthritic or a healthy joint. J Bone Metab. 2018;25(2):65–72.
57. Cheon H, et al. Increased expression of pro-inflammatory cytokines and metalloproteinase-1 by TGF-beta1 in synovial fibroblasts from rheumatoid arthritis and normal individuals. Clin Exp Immunol. 2002;127(3):547–52.
58. Shen J, Li S, Chen D. TGF-beta signaling and the development of osteoarthritis. Bone Res. 2014;2:14002.
59. Morris JC, et al. Phase I study of GC1008 (fresolimumab): a human anti-transforming growth factor-beta (TGFbeta) monoclonal antibody in patients with advanced malignant melanoma or renal cell carcinoma. PLoS One. 2014;9(3):e90353.
60. Herbertz S, et al. Clinical development of galunisertib (LY2157299 monohydrate), a small molecule inhibitor of transforming growth factor-beta signaling pathway. Drug Des Dev Ther. 2015;9:4479–99.
61. Yung LM, et al. A selective transforming growth factor-beta ligand trap attenuates pulmonary hypertension. Am J Respir Crit Care Med. 2016;194(9):1140–51.
62. Utsunomiya H, et al. Biologically regulated marrow stimulation by blocking TGF-beta1 with losartan Oral administration results in hyaline-like cartilage

repair: a rabbit osteochondral defect model. Am J Sports Med. 2020;48(4):974–84.
63. Logan CA, et al. The beneficial effect of an intra-articular injection of losartan on microfracture-mediated cartilage repair is dose dependent. Am J Sports Med. 2021;49(9):2509–21.
64. Yamaura K, et al. Sustained-release losartan from peptide nanofibers promotes chondrogenesis. Front Bioeng Biotechnol. 2023;11:1122456.
65. Nakama GY, et al. Effect of Oral losartan on orthobiologics: implications for platelet-rich plasma and bone marrow concentrate—a rabbit study. Int J Mol Sci. 2020;21(19):7374.
66. Pesce B, et al. TNF-alpha affects signature cytokines of Th1 and Th17 T cell subsets through differential actions on TNFR1 and TNFR2. Int J Mol Sci. 2022;23(16):9306.
67. Rath PC, Aggarwal BB. TNF-induced signaling in apoptosis. J Clin Immunol. 1999;19(6):350–64.
68. Salamone G, et al. Promotion of neutrophil apoptosis by TNF-alpha. J Immunol. 2001;166(5):3476–83.
69. Liu T, et al. NF-kappaB signaling in inflammation. Signal Transduct Target Ther. 2017;2:17023.
70. Nees TA, et al. Inflammation and osteoarthritis-related pain. Schmerz. 2019;33(1):4–12.
71. Burrage PS, Mix KS, Brinckerhoff CE. Matrix metalloproteinases: role in arthritis. Front Biosci. 2006;11:529–43.
72. Fernandes JC, Martel-Pelletier J, Pelletier JP. The role of cytokines in osteoarthritis pathophysiology. Biorheology. 2002;39(1–2):237–46.
73. Popa C, et al. The role of TNF-alpha in chronic inflammatory conditions, intermediary metabolism, and cardiovascular risk. J Lipid Res. 2007;48(4):751–62.
74. Sekut L, et al. Evaluation of the significance of elevated levels of systemic and localized tumor necrosis factor in different animal models of inflammation. J Lab Clin Med. 1994;124(6):813–20.
75. Aizawa T, et al. Induction of apoptosis in chondrocytes by tumor necrosis factor-alpha. J Orthop Res. 2001;19(5):785–96.
76. Chen YF, et al. A systematic review of the effectiveness of adalimumab, etanercept and infliximab for the treatment of rheumatoid arthritis in adults and an economic evaluation of their cost-effectiveness. Health Technol Assess. 2006;10(42):iii–iv, xi–xiii, 1–229.
77. Harty LC, et al. Joint tenderness and swelling in biologic-treated inflammatory arthritis patients—a tricky trade off? Int J Clin Pract. 2012;66(2):128–31.
78. Fernandez-Ruiz M, Aguado JM. Risk of infection associated with anti-TNF-alpha therapy. Expert Rev Anti Infect Ther. 2018;16(12):939–56.
79. Aarreberg LD, et al. Interleukin-1beta induces mtDNA release to activate innate immune signaling via cGAS-STING. Mol Cell. 2019;74(4):801–815e6.
80. Yang J, et al. IL-1beta increases the expression of inflammatory factors in synovial fluid-derived fibroblast-like synoviocytes via activation of the NF-kappaB-mediated ERK-STAT1 signaling pathway. Mol Med Rep. 2019;20(6):4993–5001.
81. Sluzalska KD, et al. Interleukin-1beta affects the phospholipid biosynthesis of fibroblast-like synoviocytes from human osteoarthritic knee joints. Osteoarthr Cartil. 2017;25(11):1890–9.
82. Aida Y, et al. The effect of IL-1beta on the expression of inflammatory cytokines and their receptors in human chondrocytes. Life Sci. 2006;79(8):764–71.
83. van Vulpen LF, et al. IL-1beta, in contrast to TNFalpha, is pivotal in blood-induced cartilage damage and is a potential target for therapy. Blood. 2015;126(19):2239–46.
84. Tabeian H, et al. IL-1beta damages fibrocartilage and upregulates MMP-13 expression in fibrochondrocytes in the condyle of the temporomandibular joint. Int J Mol Sci. 2019;20(9):2260.
85. Goldring MB, et al. Interleukin 1 suppresses expression of cartilage-specific types II and IX collagens and increases types I and III collagens in human chondrocytes. J Clin Invest. 1988;82(6):2026–37.
86. Cohen SB. The use of anakinra, an interleukin-1 receptor antagonist, in the treatment of rheumatoid arthritis. Rheum Dis Clin North Am. 2004;30(2):365–80, vii.
87. Nara H, Watanabe R. Anti-inflammatory effect of muscle-derived Interleukin-6 and its involvement in lipid metabolism. Int J Mol Sci. 2021;22(18):9889.
88. Tanaka T, Narazaki M, Kishimoto T. IL-6 in inflammation, immunity, and disease. Cold Spring Harb Perspect Biol. 2014;6(10):a016295.
89. Randau TM, et al. Interleukin-6 in serum and in synovial fluid enhances the differentiation between periprosthetic joint infection and aseptic loosening. PLoS One. 2014;9(2):e89045.
90. Tsuchida AI, et al. Interleukin-6 is elevated in synovial fluid of patients with focal cartilage defects and stimulates cartilage matrix production in an in vitro regeneration model. Arthritis Res Ther. 2012;14(6):R262.
91. Wang J, et al. IL-6 regulates the bone metabolism and inflammatory microenvironment in aging mice by inhibiting Setd7. Acta Histochem. 2021;123(5):151718.
92. Li B, Jones LL, Geiger TL. IL-6 promotes T cell proliferation and expansion under inflammatory conditions in association with low-level RORgammat expression. J Immunol. 2018;201(10):2934–46.
93. Gabay C. Interleukin-6 and chronic inflammation. Arthritis Res Ther. 2006;8 Suppl 2(Suppl 2):S3.
94. Herrero-Beaumont G, et al. Systemic osteoarthritis: the difficulty of categorically naming a continuous condition. Aging Clin Exp Res. 2024;36(1):45.
95. Harmer D, Falank C, Reagan MR. Interleukin-6 interweaves the bone marrow microenvironment, bone loss, and multiple myeloma. Front Endocrinol (Lausanne). 2018;9:788.
96. Sebba A. Tocilizumab: the first interleukin-6-receptor inhibitor. Am J Health-Syst Pharm. 2008;65(15):1413–8.

97. Melincovici CS, et al. Vascular endothelial growth factor (VEGF)—key factor in normal and pathological angiogenesis. Rom J Morphol Embryol. 2018;59(2):455–67.
98. Johnson KE, Wilgus TA. Vascular endothelial growth factor and angiogenesis in the regulation of cutaneous wound repair. Adv Wound Care (New Rochelle). 2014;3(10):647–61.
99. Niu G, Chen X. Vascular endothelial growth factor as an anti-angiogenic target for cancer therapy. Curr Drug Targets. 2010;11(8):1000–17.
100. Dai J, Rabie AB. VEGF: an essential mediator of both angiogenesis and endochondral ossification. J Dent Res. 2007;86(10):937–50.
101. Gerriets V, Kasi A. Bevacizumab. In: StatPearls. Treasure Island, FL: StatPearls; 2024.
102. Abraham S, Rangaswamy SP, Chinnaiah A. Evaluation of recombinant human vascular endothelial growth factor VEGF121-loaded poly-l-lactide microparticles as a controlled release delivery system. Turk J Biol. 2020;44(1):34–47.
103. Hayflick L. The limited in vitro lifetime of human diploid cell strains. Exp Cell Res. 1965;37:614–36.
104. Rodier F, Campisi J. Four faces of cellular senescence. J Cell Biol. 2011;192(4):547–56.
105. Tchkonia T, Kirkland JL. Aging, cell senescence, and chronic disease: emerging therapeutic strategies. JAMA. 2018;320(13):1319–20.
106. Rim YA, Nam Y, Ju JH. The role of chondrocyte hypertrophy and senescence in osteoarthritis initiation and progression. Int J Mol Sci. 2020;21(7):2358.
107. Xu M, et al. Transplanted senescent cells induce an osteoarthritis-like condition in mice. J Gerontol A Biol Sci Med Sci. 2017;72(6):780–5.
108. Islam MT, et al. Senolytic drugs, dasatinib and quercetin, attenuate adipose tissue inflammation, and ameliorate metabolic function in old age. Aging Cell. 2023;22(2):e13767.
109. Saccon TD, et al. Senolytic combination of Dasatinib and quercetin alleviates intestinal senescence and inflammation and modulates the gut microbiome in aged mice. J Gerontol A Biol Sci Med Sci. 2021;76(11):1895–905.
110. Yousefzadeh MJ, et al. Fisetin is a senotherapeutic that extends health and lifespan. EBioMedicine. 2018;36:18–28.
111. Mullen M, et al. Fisetin attenuates cellular senescence accumulation during culture expansion of human adipose-derived stem cells. Stem Cells. 2023;41(7):698–710.
112. Hickson LJ, et al. Senolytics decrease senescent cells in humans: preliminary report from a clinical trial of Dasatinib plus quercetin in individuals with diabetic kidney disease. EBioMedicine. 2019;47:446–56.
113. Reifsnyder PC, et al. Rapamycin/metformin co-treatment normalizes insulin sensitivity and reduces complications of metabolic syndrome in type 2 diabetic mice. Aging Cell. 2022;21(9):e13666.
114. Acar MB, et al. Proteomic and biological analysis of the effects of metformin Senomorphics on the mesenchymal stromal cells. Front Bioeng Biotechnol. 2021;9:730813.
115. Wang R, et al. Rapamycin inhibits the secretory phenotype of senescent cells by a Nrf2-independent mechanism. Aging Cell. 2017;16(3):564–74.
116. Sikora E. Rejuvenation of senescent cells-the road to postponing human aging and age-related disease? Exp Gerontol. 2013;48(7):661–6.
117. Sekelova T, Danisovic L, Cehakova M. Rejuvenation of senescent mesenchymal stem cells to prevent age-related changes in synovial joints. Cell Transplant. 2023;32:9636897231200065.
118. Peng Z, et al. The regulation of cartilage extracellular matrix homeostasis in joint cartilage degeneration and regeneration. Biomaterials. 2021;268:120555.
119. Vundelinckx B, et al. Surgical indications and technique for anterior cruciate ligament reconstruction combined with lateral extra-articular tenodesis or anterolateral ligament reconstruction. Clin Sports Med. 2017;36(1):135–53. https://doi.org/10.1016/j.csm.2016.08.009.
120. Jang K, Berrigan WA, Mautner K. Regulatory considerations of orthobiologic procedures. Phys Med Rehabil Clin N Am. 2023;34(1):275–83.
121. Allickson JG. Emerging translation of regenerative therapies. Clin Pharmacol Ther. 2017;101(1):28–30.
122. Boivin J, et al. The biological use of platelet-rich plasma in skeletal muscle injury and repair. Am J Sports Med. 2023;51(5):1347–55.
123. Dulic O, et al. Bone marrow aspirate concentrate versus platelet rich plasma or hyaluronic acid for the treatment of knee osteoarthritis. Medicina (Kaunas). 2021;57(11):1193.
124. Fang J, et al. Platelet-rich plasma therapy in the treatment of diseases associated with orthopedic injuries. Tissue Eng Part B Rev. 2020;26(6):571–85.
125. Borrione P, et al. The role of platelet-rich plasma in muscle healing. Curr Sports Med Rep. 2017;16(6):459–63.
126. Graca FA, et al. Platelet-derived chemokines promote skeletal muscle regeneration by guiding neutrophil recruitment to injured muscles. Nat Commun. 2023;14(1):2900.
127. Andia I, Abate M. Platelet-rich plasma in the treatment of skeletal muscle injuries. Expert Opin Biol Ther. 2015;15(7):987–99.
128. Cavallo C, et al. Bone marrow concentrate injections for the treatment of osteoarthritis: evidence from preclinical findings to the clinical application. Int Orthop. 2021;45(2):525–38.
129. Travlos GS. Normal structure, function, and histology of the bone marrow. Toxicol Pathol. 2006;34(5):548–65.
130. Lucas D. Structural organization of the bone marrow and its role in hematopoiesis. Curr Opin Hematol. 2021;28(1):36–42.

131. Gulati GL, Ashton JK, Hyun BH. Structure and function of the bone marrow and hematopoiesis. Hematol Oncol Clin North Am. 1988;2(4):495–511.
132. Keeling LE, et al. Bone marrow aspirate concentrate for the treatment of knee osteoarthritis: a systematic review. Am J Sports Med. 2022;50(8):2315–23.
133. Kim GB, et al. Bone marrow aspirate concentrate: its uses in osteoarthritis. Int J Mol Sci. 2020;21(9):3224.
134. Pintore A, et al. Intra-articular injection of bone marrow aspirate concentrate (BMAC) or adipose-derived stem cells (ADSCs) for knee osteoarthritis: a prospective comparative clinical trial. J Orthop Surg Res. 2023;18(1):350.
135. Fortier LA, et al. Concentrated bone marrow aspirate improves full-thickness cartilage repair compared with microfracture in the equine model. J Bone Joint Surg Am. 2010;92(10):1927–37.
136. Chu CR, et al. Minimally manipulated bone marrow concentrate compared with microfracture treatment of full-thickness chondral defects: a one-year study in an equine model. J Bone Joint Surg Am. 2018;100(2):138–46.
137. Doyle LM, Wang MZ. Overview of extracellular vesicles, their origin, composition, purpose, and methods for exosome isolation and analysis. Cells. 2019;8(7):727.
138. Gurung S, et al. The exosome journey: from biogenesis to uptake and intracellular signalling. Cell Commun Signal. 2021;19(1):47.
139. Jella KK, et al. Exosomes, their biogenesis and role in inter-cellular communication, tumor microenvironment and cancer immunotherapy. Vaccines (Basel). 2018;6(4):69.
140. Nail HM, et al. Exosomal miRNA-mediated intercellular communications and immunomodulatory effects in tumor microenvironments. J Biomed Sci. 2023;30(1):69.
141. Yin B, et al. Harnessing tissue-derived extracellular vesicles for osteoarthritis theranostics. Theranostics. 2022;12(1):207–31.
142. Ragni E, et al. Endogenous controls for the evaluation of osteoarthritis-related miRNAs in extracellular vesicles from bone-marrow-derived mesenchymal stromal cells and the impact of osteoarthritis synovial fluid. Biomol Ther. 2022;12(2):316.
143. Panagopoulos PK, Lambrou GI. The involvement of microRNAs in osteoarthritis and recent developments: a narrative review. Mediterr J Rheumatol. 2018;29(2):67–79.
144. Guo SC, et al. Exosomes derived from platelet-rich plasma promote the re-epithelization of chronic cutaneous wounds via activation of YAP in a diabetic rat model. Theranostics. 2017;7(1):81–96.
145. Anitua E, et al. Advances in platelet rich plasma-derived extracellular vesicles for regenerative medicine: a systematic-narrative review. Int J Mol Sci. 2023;24(17):13043.
146. Morrison SJ, Scadden DT. The bone marrow niche for haematopoietic stem cells. Nature. 2014;505(7483):327–34.
147. Reger RL, Tucker AH, Wolfe MR. Differentiation and characterization of human MSCs. Methods Mol Biol. 2008;449:93–107.
148. Squecco R, et al. Bone marrow-mesenchymal stromal cell secretome as conditioned medium relieves experimental skeletal muscle damage induced by ex vivo eccentric contraction. Int J Mol Sci. 2021;22(7):3645.
149. Kou M, et al. Mesenchymal stem cell-derived extracellular vesicles for immunomodulation and regeneration: a next generation therapeutic tool? Cell Death Dis. 2022;13(7):580.
150. Wu CL, et al. The role of macrophages in osteoarthritis and cartilage repair. Osteoarthr Cartil. 2020;28(5):544–54.
151. Ning G, et al. MicroRNA-92a upholds bmp signaling by targeting noggin3 during pharyngeal cartilage formation. Dev Cell. 2013;24(3):283–95.
152. Dos Santos CC, et al. The MSC-EV-microRNAome: a perspective on therapeutic mechanisms of action in sepsis and ARDS. Cells. 2024;13(2):122.
153. Esau CC. Inhibition of microRNA with antisense oligonucleotides. Methods. 2008;44(1):55–60.
154. Jérémy B, et al. Extracellular vesicles from senescent mesenchymal stromal cells are defective and cannot prevent osteoarthritis. J Nanobiotechnol. 2024;22(1):255.
155. Yin Y, et al. Roles of extracellular vesicles in the aging microenvironment and age-related diseases. J Extracell Vesicles. 2021;10(12):e12154.
156. Herrmann M, et al. Extracellular vesicles in musculoskeletal pathologies and regeneration. Front Bioeng Biotechnol. 2020;8:624096.

Part II

Basic Science

Stem Cells: Different Families

8

Ri Han, Zheng Yang, and James Hoi Po Hui

8.1 Introduction

Stem cells are a unique type of cell characterized by two key properties including the ability to self-renew and differentiate. Self-renewal allows them to produce more cells like themselves, and differentiation enables them to transform into various other cell types with distinct functions. Stem cells are present in nearly all body tissues and play a crucial role in tissue maintenance and repair following injury.

The term "stem cell" originates from the late nineteenth century. Ernst Haeckel (1834–1919) published on *Natürliche Schöpfungsgeschichte* (1868) about unicellular organisms or protozoa, and used "Stammzellen" (stem cells) as the phylogenetic ancestors of multicellular organisms. A few years later, Haeckel explained in another series of lectures on *Anthropogenie*: "The name 'stem cell' seems to me the most simple and appropriate one because all other cells stem from it and because it is in its most literal sense the stem father as well as the stem mother of all the countless generations of cells of which later on the multicellular organism is composed." Significant milestones in defining and demonstrating the properties of stem cells were achieved in the 1960s by scientists such as Ernest A. McCulloch and James E. Till, who provided the first concrete evidence of stem cells in bone marrow. In 1998, James Thomson at the University of Wisconsin-Madison successfully isolated the first human embryonic stem cells, which sparked both excitement and ethical debates. The development of induced pluripotent stem cells (iPSCs) by Shinya Yamanaka in 2006 represented another major advancement. This breakthrough allowed adult cells to be reprogrammed into a pluripotent state, thus bypassing ethical issues associated with embryonic stem cells and opening new avenues for research and therapy.

Today, the field of stem cell research is rapidly expanding and represents one of the most exciting areas in life sciences. Due to their ability to differentiate into distinct tissue types, stem cells hold great potential to revolutionize modern clinical medicine. These undifferentiated, self-renewing cells offer hope to address various

R. Han
Department of Orthopaedic Surgery, Yong Loo Lin School of Medicine, National University of Singapore, Singapore, Singapore
e-mail: ri.han@nus.edu.sg

Z. Yang
NUS Tissue Engineering Program, Life Sciences Institute, National University of Singapore, Singapore, Singapore
e-mail: dosyz@nus.edu.sg

J. Hoi Po Hui (✉)
Department of Orthopaedic Surgery, Yong Loo Lin School of Medicine, National University of Singapore, National University Hospital, Singapore, Singapore
e-mail: doshuij@nus.edu.sg

A. Gobbi et al. (eds.), *Regenerative Medicine in Sports and Orthopaedics*,
https://doi.org/10.1007/978-3-031-84693-9_8

illnesses such as Parkinson's and Alzheimer's diseases, spinal cord injuries, and insulin-dependent diabetes.

The importance of studying stem cell biology is attributed to their wide range of applications. In basic research, stem cells provide a powerful platform to investigate gene function and the physiological processes that occur during development. In biomedical research, they are used to study the pathogenesis of human genetic diseases, identify new diagnostic and prognostic biomarkers, and test improved drugs. However, the true significance of stem cell research is found in their vast clinical potential. Their ability to differentiate into specific cell types can be harnessed in regenerative medicine to treat damaged or diseased tissues through cell replacement therapies. Food and Drug Administration (FDA) has already approved clinical trials using stem cells for treating heart disease. Although some stem cell therapies are in clinical trials, much more basic research is needed before therapies using differentiated stem cell derivatives can be widely applied to humans. The secretome and extracellular vesicles (EVs) derived from stem cells have garnered significant interest in regenerative medicine due to their potential therapeutic benefits. The secretome refers to the array of bioactive molecules secreted by cells, including cytokines, growth factors, and proteins, while EVs are membrane-bound vesicles that carry proteins, lipids, and nucleic acids, playing a crucial role in intercellular communication.

This chapter categorizes the different families of stem cells with their unique characteristics. Starting with the differences between embryonic stem cells (ESCs) and adult stem cells (ASCs), their origins, properties, and roles in tissue maintenance and repair are explained in detail. Meanwhile, another classification based on differentiation capabilities, including totipotent, pluripotent, multipotent, and unipotent stem cells, is explored and discussed. In addition, the development of induced pluripotent stem cells (iPSCs) is highlighted as a method to reprogram adult cells into a pluripotent state. Advances and applications of stem cell therapy in orthopedics are emphasized, showcasing promising results in cartilage, bone, tendon, and nerve repair. Despite their significant potential, the applications of stem cells face challenges such as ethical considerations, differentiation control, and regulatory hurdles. Ongoing research aims to address these issues, paving the way for groundbreaking therapies and innovations in regenerative medicine and orthopedics.

8.2 Categories of Stem Cells

8.2.1 Embryonic Stem Cells (ESCs)

Human embryonic stem cells (hESCs) were successfully grown in the laboratory for the first time in 1998 by James Thompson and colleagues. Under appropriate culture conditions, these cells have demonstrated a remarkable ability to self-renew continuously. In mammals, embryonic stem cells are found in the inner cell mass of the blastocyst, an early stage of the developing embryo lasting from the 4th to the 7th day after fertilization. In their undifferentiated state, ESCs retain the potential to differentiate into cells of all three embryonic tissue layers: the endoderm (inner layer), the ectoderm (outer layer), and the mesoderm (middle layer). ESCs are characterized by their pluripotency, capable to differentiate into nearly all cell types in the body. This remarkable capability makes ESCs a valuable tool for research and potential therapeutic applications.

Over the past two decades, the potential of hESCs to revolutionize the treatment of various diseases has been increasingly recognized [1, 2]. Their capacity to differentiate into diverse cell types offers promising prospects for repairing or replacing damaged tissues, particularly in conditions where current treatments are limited. The top three categories of trials involving hESCs are ophthalmologic conditions, neurologic conditions, and diabetes [3]. For example, the research group led by Douglas A. Melton at the Harvard Stem Cell Institute has developed methods to produce hundreds of millions of functional beta cells from human stem cells, including both embryonic stem cells and induced pluripotent stem cells. One study demonstrated that functional human stem-cell-derived beta cells (SC-β) can be directly generated from human pluripotent stem cells, specifically from the human embryonic stem cell line HUES8 and human-induced pluripotent stem cells. These SC-β cells express markers found in mature beta cells, respond to glucose by fluxing Ca^{2+}, package insulin into secretory granules, and secrete insulin in quantities comparable to adult beta cells in response to multiple glucose challenges. This advancement provides a new platform for in vitro drug discovery and characterization for human diabetes [4]. Rudolf Jaenisch, also well-known as one of the pioneers in the field, focuses on using embryonic stem cells to study the genetic and epigenetic basis of diseases such as Parkinson's, Alzheimer's, autism, and cancer. Research on hESCs also promises to enhance drug discovery and development by providing reproducible, cost-effective tools for toxicity testing of drugs under development and for studying disease mechanisms and pathways [5–7].

8.2.2 Adult (Somatic) Stem Cells (ASCs)

Adult stem cells, also referred to as somatic or tissue-specific stem cells, are undifferentiated cells present in various tissues throughout the body, such as bone marrow, brain, blood, skeletal muscles, skin, and liver. In contrast to embryonic stem cells, adult stem cells are typically restricted to differentiating into specialized cell types of their tissue of origin. Their primary function is to facilitate tissue repair and maintain homeostasis.

The first adult stem cells were identified in bone marrow in 1961 by James Till and Ernest McCulloch as cells capable of giving rise to multilineage hematopoietic colonies in the spleen. These hematopoietic stem cells (HSCs) were first isolated from mouse bone marrow and termed spleen-colony forming units (CFU-S) [8, 9]. In 1982, Alexander Friedenstein discovered mesenchymal stem cells (MSCs) and expanded the understanding of ASCs differentiation [10]. Although both HSCs and MSCs are located in bone marrow, MSCs exhibit a broader differentiation potential including osteoblasts, chondrocytes, myoblasts, adipocytes, and fibroblasts.

Adult stem cells are rare and generally small in number. Research has shown that only 1 in 10,000 cells in bone marrow is a hematopoietic stem cell [11, 12]. Various techniques are employed to isolate adult stem cells according to specific source tissues. These methods include bone marrow aspiration, where a needle extracts stem cells from the bone marrow; peripheral blood collection, which involves drawing blood and isolating the mononuclear cell fraction; umbilical cord blood collection, where stem cells are obtained from the blood in the umbilical cord after childbirth; adipose tissue harvesting through liposuction or surgical removal; dental pulp and papilla dissection, which extracts stem cells from teeth; and synovial fluid aspiration, where human mesenchymal stem cells are obtained through aspiration of synovial fluid from joints.

As described above, ASCs have been isolated from numerous tissues with a remarkable ability to transform into other cell types to repair damaged tissues. For instance, when bone marrow stem cells were injected into the eyes of mice, they integrated with retinal astrocytes and became part of the vascular network [13]. In addition, some in vitro studies have demonstrated that bone marrow stem cells can form cartilage tissue, and the research on rats and mice has shown these cells can develop into mesangial cells to repopulate the kidney's glomerulus [14–16]. Additionally, clinical trials have explored using bone marrow-derived stem cells to treat patients with damaged cardiac tissue [17–19]. The trials have produced promising results and indicated that these cells can regenerate damaged cardiac tissue and improve cardiac performance in humans. In terms of the other ASCs, human cord blood stem cells have been reported to produce functional liver cells [20–22]. Neural stem cells and neural progenitor cells have been studied for their low immunogenicity and immunoprivileged status upon transplantation. This characteristic raises the possibility of using donor neural stem cells to treat degenerative brain conditions [23].

Currently, ASCs have diverse applications in treating diseases and advancing medical research. They are used in regenerative medicine for bone marrow transplants, cardiac regeneration after heart attacks, and repairing spinal cord injuries. Meanwhile, ASCs can also promise in treating diabetes, multiple sclerosis, and osteoarthritis. In research, ASCs are instrumental in modeling diseases and testing new drugs, thereby enhancing the understanding and facilitating the development of new therapies. Additionally, these cells can also serve as effective vehicles for gene therapy applications. Engineered adult stem cells may possess improved functionality, increased proliferative capacity, or enhanced stimulatory capabilities [24–26].

8.2.3 Totipotent, Pluripotent, Multipotent, and Unipotent Stem Cells

Stem cells can be categorized based on their source and developmental stage into embryonic stem cells (ESCs) and adult stem cells (ASCs). Additionally, they are categorized according to their differentiation potential into totipotent, pluripotent, multipotent, and unipotent cells. As developmental potency decreases, unipotent stem cells (USCs) exhibit the most limited capacity to differentiate into various cell types compared to other stem cells. In contrast, totipotent stem cells (TSCs) are the most versatile type of stem cells, and they can differentiate into all cell types required for the development of a complete organism. To provide a concise overview, Table 8.1 presents a summary of the examples, advantages, and disadvantages of each type of stem cell categorized according to their developmental potency.

The stringent definition of totipotent stem cells (TSCs) refers to their unique ability to contribute to both embryonic and extraembryonic lineages in a developing embryo. According to this strict definition, only the zygote (fertilized egg) and the two-cell (2C) stages of developing embryos (in the earliest stages of embryonic development) are genuinely totipotent stem cells. These cells are referred to as in vivo canonical totipotent stem cells, which will subsequently commit to two distinct lineages: the embryonic cell lineage (inner cell mass, ICM) to form the embryo proper and the extraembryonic cell lineage (trophectoderm, TE) to form placental tissue [27, 28]. The primary applications of totipotent stem cells are in developmental biology research, where they provide critical insights into the earliest stages of human development, cell differentiation, and tissue formation. Although not directly used in current regenerative therapies due to ethical and technical challenges, knowledge of totipotent stem cells informs broader stem cell research and the development of new regenerative treatments. Additionally, understanding totipotent stem cells is crucial for reproductive technologies, including cloning and in vitro fertilization.

Pluripotent stem cells (PSCs) have the remarkable ability to differentiate into almost all cell types, excluding extra-embryonic tissues such as the placenta. Following the previous description, a zygote can develop into any of the three germ layers or form a placenta as TSCs. Around 4 days post-fertilization, the blastocyst's inner cell mass becomes PSCs, which can differentiate into cells from all three germ layers: ectoderm (skin and nervous system), mesoderm (muscle, bone, blood), and endoderm (gut, liver, lungs), making them invaluable for research and potential therapeutic applications. Human pluripotent stem cells (hPSCs), including human embryonic stem cells (hESCs) and human-induced pluripotent stem cells (hiPSCs), can self-renew indefinitely in culture while retaining the ability to differentiate into nearly any cell type in the human body [29–31]. Recent technological advancements have provided exciting opportunities to use hPSCs to gain novel insights into human embryonic development. This knowledge will not only enhance our understanding of the underlying causes of human diseases but also guide efforts to develop treatments for conditions such as Parkinson's disease, heart disease, and diabetes by providing the excellent experimental platform for studying human development, screening drugs, and developing cell therapies [32].

Table 8.1 Definitions and characteristics of totipotent, pluripotent, multipotent, and unipotent stem cells

Terminology	Definition and ability	Example of cells	Pros	Cons
Totipotent	A single cell to divide and produce all the differentiated cells in an organism, including both the embryonic and extraembryonic tissues	• Zygote • Two-cell stages of developing embryos	• Easy to isolate and grow	• Ethical concerns
Pluripotent	A single cell to develop into most, but not all, of the body's cell types. They can form all cell types of the embryo and adult except for extraembryonic tissues such as the placenta	• Embryonic stem cells • Induced pluripotent stem cells	• Broad differentiation potential • Easy to isolate and grow • Unlimited self-renewal	• Ethical issues • Immune rejection • Tumor (teratoma) formation
Multipotent	The type of cells can differentiate into a limited range of cell types within a particular tissue or organ, typically within a specific lineage or group of related cell types	• Mesenchymal stem cells • Hematopoietic stem cells • Epithelial stem cells • Neural stem cells	• Less likely of tissue rejection • Less ethical issues	• Hard to isolate • Limited differentiation • Limited proliferative potential
Unipotent	The ability of a single cell to develop into only one specific cell type. They have the lowest differentiation potential	• Muscle stem cells • Germline stem cells • Epidermal stem cells	• Specialized function • Reduced tumor risk	• Limited application for specific cell therapy

Multipotent stem cells (MSCs) are specialized cells capable of differentiating into multiple cell types within a specific tissue or organ lineage, distinguishing them from PSCs. In other words, the differentiation potential of MSCs is restricted to the cell types found in their tissue of origin. These cells are believed to exist in most body tissues and are essential for tissue maintenance by renewing and replacing old or dysfunctional cells throughout the life of a person [33]. Adult stem cells, such as neural stem cells, mesenchymal stem cells, and hematopoietic stem cells, are all multipotent. Hematopoietic stem cells can produce various blood cells, while mesenchymal stem cells are able to differentiate into bone, cartilage, and fat cells. These stem cells are being researched for their therapeutic applications in treating conditions such as osteoarthritis, blood disorders, and neurological injuries due to their ability to regenerate specific tissues and enhance targeted therapies. Progress in this field also includes isolating MSCs from fetal brain tissues, where they differentiate into new nerve cells. In summary, MSCs have been widely tested in late-stage pre-clinical studies and clinical trials as the best option for clinical application. Additionally, another significant benefit of multipotent stem cells is their anti-inflammatory and immunomodulatory properties [34].

Unlike mesenchymal stem cells (MSCs), unipotent stem cells (USCs) have the capacity to differentiate into only one cell type, which makes them with limited competence toward a single cell lineage. An example of USCs is muscle stem cells (satellite cells). They are found in muscle tissue and develop into mature muscle cells, but not into any other cell type [35]. Germline stem cells (responsible for producing sperm) and epidermal stem cells (responsible for producing skin) are also categorized as USCs. Clinically, the transplantation of cultured epithelial autografts (CEA), which contain epidermal stem cells and keratinocytes, has become a well-established treatment for extensive burn wounds. These autografts can be delivered as either cell suspensions or cell sheets, often in conjunction with fibrin matrices to facilitate cell delivery. This method is recognized as a promising stem cell-based therapy and is employed in numerous burn centers worldwide [36, 37].

8.2.4 Induced Pluripotent Stem Cells (iPSCs)

Induced pluripotent stem cells (iPSCs) are adult somatic cells genetically reprogrammed to an embryonic stem cell-like state by introducing specific genes that reset their developmental stage. This reprogramming process involves key factors such as Oct4, Sox2, Klf4, and c-Myc, which restore the cells' pluripotency [38]. Unlike naturally occurring adult stem cells, which are typically multipotent, iPSCs possess a pluripotent state that allows them to differentiate into most somatic cell types, thereby offering broader potential for regenerative medicine and personalized therapies (Fig. 8.1).

Fibroblasts were initially used as a source of induced pluripotent stem cells (iPSCs). Due to the requirement of a biopsy to obtain these cells, further research was conducted to identify more accessible cell sources, such as exploring the potential of using peripheral blood cells, keratinocytes, and renal epithelial cells found in urine for iPSC generation. Another approach to stem cell therapy involves stimulating a patient's endogenous stem cells to divide or differentiate, a process that naturally occurs during skin wound healing. In 2008, it was demonstrated that pancreatic exo-

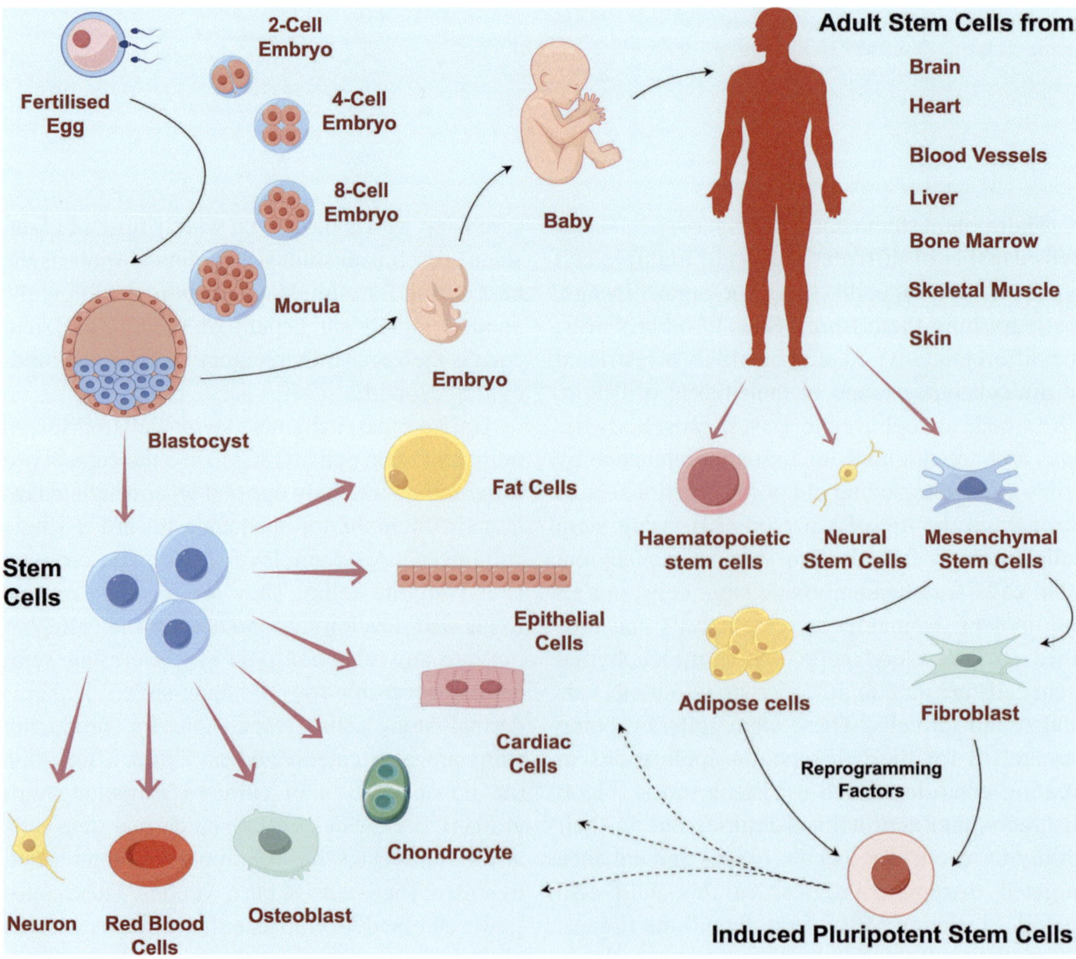

Fig. 8.1 Stem cells: differentiation pathways from embryo to adult stem cells and induced pluripotent stem cells

crine cells could be reprogrammed into functional, insulin-producing β cells [39]. iPSCs offer a novel method to create pluripotent stem cells without the ethical concerns associated with using embryonic stem cells. Since iPSCs are derived from the patient's own cells, they are autologous, thereby minimizing the risk of immune rejection.

The development of iPSC technology also revolutionized in vitro research and holds significant promise for advancing regenerative medicine. To meet the rigorous standards required for clinical applications, clinically grade iPSCs are produced under stringent current Good Manufacturing Practice (cGMP) conditions with extensive quality control to ensure safety, purity, and genetic stability. These cells are designed for use in human clinical trials and therapies, adhering to regulatory standards set by organizations such as the US Food and Drug Administration (FDA) and the European Medicines Agency (EMA). This compliance ensures their suitability for therapeutic applications, including regenerative medicine and cell therapy [40]. The advancement of iPSC technology has created opportunities for standardized in vitro modeling of human development and diseases, drug screening, and cell therapy application [31, 41, 42]. Enhanced understanding of the mechanisms governing iPSC induction has illuminated cell fate decisions, which indeed accelerate the development of efficient iPSC derivation methods and protocols for differentiating iPSCs into adult somatic cells [43]. In addition, modeling human biology with iPSCs and iPSC-derived cells is particularly advantageous due to their human origin and the capability to derive patient-specific iPSCs with disease-relevant genetic backgrounds [42]. iPSC-based cellular models can reveal human-specific phenotypes and molecular mechanisms that may not be evident in animal models [44, 45].

8.3 Applications of Stem Cell Therapy for Orthopedic

Stem cell research plays a critical role in the field of orthopedic regenerative medicine. Current literature provides compelling evidence from animal studies demonstrating significant advancements in bone, tendon, and cartilage repair. Furthermore, early clinical outcomes in bone and cartilage repair have been promising and are well-documented. Stem cell therapy, in particular, has the potential to revolutionize orthopedic practice, offering innovative solutions for the healing of bone defects caused by trauma, tumors, or infections, as well as for cartilage defects, and the regeneration of nerves, tendons, and ligaments (Fig. 8.2).

8.3.1 Cartilage Repair and Regeneration

Articular cartilage is a highly specialized tissue with limited intrinsic capacity for self-repair. The primary objective of any cartilage repair procedure is to restore the tissue's functionality and integrity, enabling it to withstand the wear and tear of daily activities. Beyond autograft transplantation and autologous chondrocyte transplantation, current therapeutic approaches for cartilage defects include the recruitment of mesenchymal stem cells (MSCs). MSCs have demonstrated a beneficial effect in osteoarthritis (OA) due to their role in inhibiting the catabolic activity of matrix metalloproteinases (MMPs) [46]. In another study, Sato et al. demonstrated that guinea pigs with age-related OA treated with MSC-laden hyaluronic injections exhibited superior cartilage regeneration, characterized by increased type II collagen and reduced MMP content [47]. To date, the application of MSCs for cartilage repair has progressed significantly beyond animal studies with clinical trials to investigate their use for osteoarthritis (OA) [48, 49]. Notably, a study by Nejadnik et al. compared the clinical outcomes of patients treated with first-generation autologous chondrocyte implantation (ACI) to those treated with autologous bone marrow-derived mesenchymal stem cells (BMSCs). The findings indicated that BMSCs are as effective as chondrocytes for articular cartilage repair, while requiring one fewer knee surgery, reducing costs, and minimizing donor-site morbidity [50]. To evaluate the long-term effects, Teo et al. conducted a study comparing the clinical outcomes of ACI versus BMSCs at a 10-year follow-up. The results con-

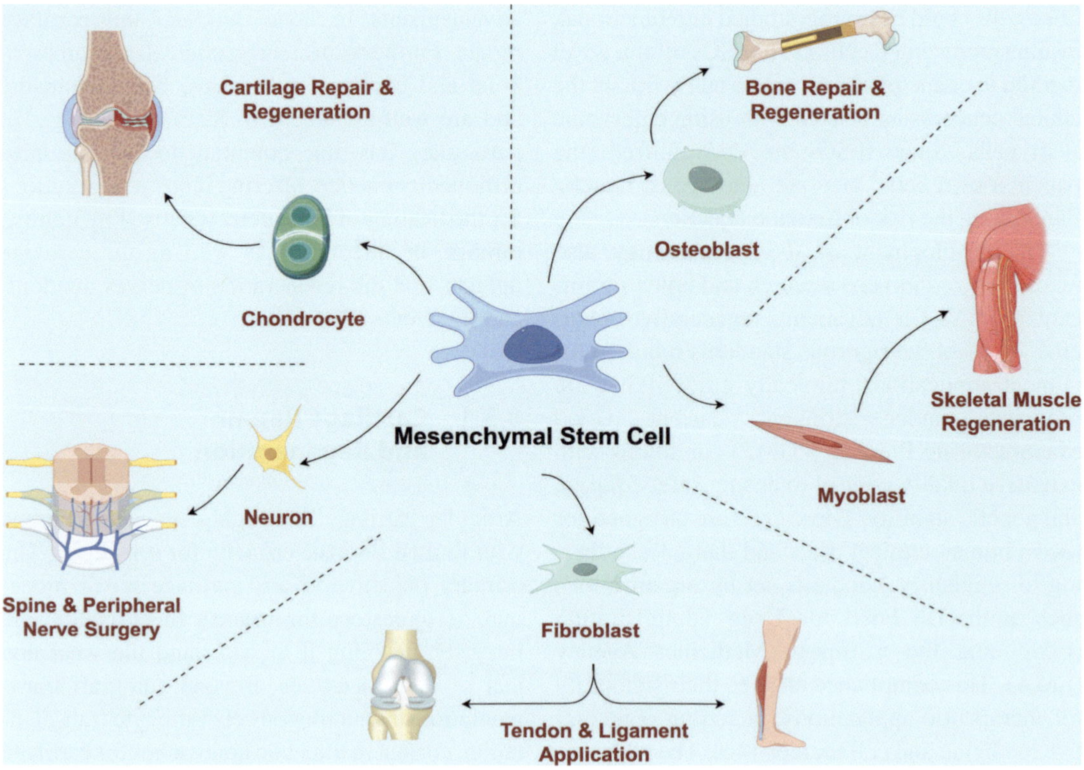

Fig. 8.2 Mesenchymal stem cells: applications in orthopedic treatment and regeneration

cluded that BMSC implantation for the treatment of chondral defects of the knee yields clinical outcomes equivalent to first-generation ACI after 10 years with no increased risk of tumor formation [51]. Gobbi et al. showed that the repair of full-thickness cartilage injury in the knee with a HA-BMAC provides good to excellent clinical outcomes at long-term follow-up in the treatment of small to large lesions [52].

Innovative matrices incorporating mesenchymal stem cells (MSCs) have been increasingly tested in vivo using animal models. Shafiee et al. explored MSC-based cartilage repair in a rabbit model with full-thickness cartilage defects, utilizing poly (vinyl alcohol)/polycaprolactone (PVA/PCL) nanofibers as a scaffold matrix. This study demonstrated the scaffold's ability to support MSC proliferation and chondrogenic differentiation in vitro [53]. In another study addressing focal cartilage defects, Tay et al. employed alginate-embedded MSCs for cartilage repair in a rabbit model, with autologous chondrocytes as the control group [54]. After 6 months post operation, the MSC treatment showed comparable effectiveness to chondrocyte transplantation and achieved significantly better macroscopic scores.

8.3.2 Bone Repair and Regeneration

Large bone defects are typically caused by trauma, infection, or tumors. The current treatment strategies primarily involve autologous bone grafts from the iliac crest, which are considered the gold standard. This includes autologous fibula graft transfer and allogenic bone graft transplantation. However, these methods come with significant limitations. The supply of autologous bone is limited, and its harvesting is associated with high morbidity rates. Additionally, allogenic transplantation carries risks of disease transmission and graft rejection.

Ethical and safety concerns surrounding traditional methods have shifted the focus toward the use of adult stem cells in orthopedic therapy. Specifically, in the context of high tibial osteotomy (HTO), Dallari et al. demonstrated that lyophilized bone chips treated with platelet gels and mesenchymal stem cells (MSCs) resulted in a higher rate of osteointegration in HTO procedures [55]. Furthermore, a study by Wong KL et al. reported that intra-articular injection of cultured MSCs effectively improves both short-term clinical outcomes and MOCART scores in patients undergoing HTO and microfracture for varus knees with cartilage defects [56]. Moreover, regenerative medicine has advanced to the point where cell therapy approaches are being used to treat localized bone defects and systemic skeletal diseases in patients [57]. These innovative therapies offer promising alternatives to traditional methods, potentially overcoming the limitations of bone supply and the risks associated with autologous and allogenic grafts [58].

8.3.3 Spine and Peripheral Nerve Surgery

Spine surgery encompasses a variety of procedures aimed at treating conditions affecting the spinal column. These conditions include degenerative disc disease, spinal stenosis, herniated discs, scoliosis, fractures, infections, and tumors. The primary goals of spine surgery are to alleviate pain, restore function, and improve the patient's quality of life.

Neen et al., in a prospective study, showed that unselected stem cells used with hydroxyapatite (HA) scaffolds had similar healing rates as autologous grafting, thereby preventing donor site morbidity [59]. Similarly, Gan et al. achieved comparable results using β-tricalcium phosphate (β-TCP) [60]. For the treatment of intervertebral disc degeneration, animal experiments have demonstrated increased proteoglycan content and maintenance of disc height with percutaneous stem cell injections [61]. Additionally, an animal study by Tamaki et al. indicated that muscle-derived mesenchymal stem cells (MSCs) successfully aided in the regeneration of a crushed peripheral nerve [62].

8.3.4 Tendon, Ligament Repair and Regeneration

Tendon and ligament injuries are common causes of physical disability, and they present a significant clinical challenge to orthopedic surgeons. Injured tendons typically respond poorly to current treatments, which do not promote tissue regeneration and require long rehabilitation periods. Stem cell-based therapies hold great promise for tendon regeneration due to their high proliferative, synthetic, and immunomodulatory activities, as well as their ability to differentiate into target cell types and undergo genetic modification.

A study by Adams et al. demonstrated that rats with Achilles tendon tears treated with stem cell-bearing sutures exhibited higher failure strength and better histological properties [63]. Similarly, in a rat rotator cuff model, Gulotta et al. applied MSCs in a fibrin sealant, placing it between the tendon and bone before refixation, which showed promising results [64]. Awad et al. explored a gel-based application technique, using a collagen gel with varying concentrations of MSCs fixed to suture material to fill defects in rabbits' patellar tendons. After 12 and 26 weeks, significantly higher maximum stresses and moduli were observed compared to natural repair tissues [65]. In addition, results from Burk et al. suggested that MSCs are effective in treating early-phase tendon disease in horses, with the animals remaining free of re-injuries and returning to training within 12 to 18 months [66]. In clinical trials, a case series by Pascual-Garrido et al. demonstrated that unselected mesenchymal stem cells (MSCs) were used for ultrasound-guided injections in patients with chronic patellar tendinopathy, resulting in good clinical outcomes [67]. BMSCs were also studied as a safe biological treatment with no adverse effects. Patients exhibited significant clinical improvement and notable enhancements in tendon structure as observed by magnetic resonance imaging (MRI) [68].

8.4 Challenges and Limitations in Stem Cell Biology

Although stem cells appear to be an ideal solution for medicine and clinical applications, their use in research and therapy is accompanied by significant challenges and ethical considerations. One of the foremost problems is the ethical concern surrounding the use of embryonic stem cells (hESCs). The destruction of embryos in obtaining hESCs raises moral concerns and leads to strict regulations in many countries. Despite these ethical debates, hESCs remain potentially powerful tools for therapeutic applications in tissue regeneration. To address these ethical considerations, induced pluripotent stem cells (iPSCs) have garnered significant attention. iPSCs are created by reprogramming adult somatic cells to an embryonic stem cell-like state, bypassing the need to use embryos. Furthermore, a recent study by Wruck et al. suggested that using rejuvenated iPSCs can also bypass the tissue- and age-related heterogeneity inherent in native mesenchymal stem cells (MSCs) [69]. This MSC heterogeneity can occur at multiple levels and restrict further research on MSC secretome potency, future clinical application, and commercialization of MSC exosome therapy [70].

However, iPSC therapies pose unique challenges regarding safety, potency, genetic stability, immunogenicity, tumorigenicity, cell reproducibility, and scalability. One major concern limiting the use of iPSCs is their potential for tumorigenicity. During the reprogramming process, there is a risk that the expression of oncogenes may increase, potentially leading to tumor formation [71]. Additionally, the immunogenicity of iPSCs poses a challenge, as the immune system may recognize and reject these reprogrammed cells. These issues underscore the need for continued research to address the safety and efficacy of stem cell-based therapies and ensure that their potential benefits can be realized without compromising patient safety.

8.5 Conclusion

Stem cells are a cornerstone of contemporary biomedical research due to their unique ability to differentiate into various cell types and their self-renewal capabilities. These properties make them invaluable for regenerative medicine, allowing for the potential repair and replacement of damaged tissues and organs. Moreover, stem cells are powerful tools for understanding disease mechanisms, facilitating drug discovery, and developing personalized medical treatments.

In orthopedic surgery, stem cells offer promising potential for repairing and regenerating musculoskeletal tissues such as bone, cartilage, and tendons. They can significantly enhance healing processes to provide new treatments for conditions like osteoarthritis, fractures, and tendon injuries. Studies have shown that mesenchymal stem cells (MSCs) can improve the quality and speed of tissue regeneration. However, challenges remain in ensuring proper cell differentiation and integration, overcoming immune rejection, and maintaining the long-term viability and function of transplanted cells. Additionally, regulatory hurdles and ethical concerns must be addressed to ensure safe and effective clinical applications.

Furthermore, concerns about tumorigenicity, immunogenicity, and cellular heterogeneity pose significant barriers. There is a risk that stem cells, particularly iPSCs, could form tumors if not properly controlled. Immunogenicity issues can lead to rejection of transplanted cells, and heterogeneity among stem cells can result in inconsistent therapeutic outcomes. Despite these challenges, the research and advancements from iPSCs are promising to drive the field forward with the potential to significantly improve patient outcomes in orthopedic surgery.

References

1. Doğan A. Embryonic stem cells in development and regenerative medicine. Adv Exp Med Biol. 2018;1079:1–15.
2. Kolios G, Moodley Y. Introduction to stem cells and regenerative medicine. Respiration. 2013;85(1):3–10.
3. Park SJ, Kim YY, Han JY, Kim SW, Kim H, Ku SY. Advancements in human embryonic stem cell research: clinical applications and ethical issues. Tissue Eng Regen Med. 2024;21(3):379–94.
4. Pagliuca FW, Millman JR, Gürtler M, Segel M, Van Dervort A, Ryu JH, Peterson QP, Greiner D, Melton DA. Generation of functional human pancreatic β cells in vitro. Cell. 2014;159(2):428–39.

5. Lian Q, Chow Y, Esteban MA, Pei D, Tse HF. Future perspective of induced pluripotent stem cells for diagnosis, drug screening and treatment of human diseases. Thromb Haemost. 2010;104(1):39–44.
6. Menendez P, Bueno C, Wang L. Human embryonic stem cells: a journey beyond cell replacement therapies. Cytotherapy. 2006;8(6):530–41.
7. Pouton CW, Haynes JM. Embryonic stem cells as a source of models for drug discovery. Nat Rev Drug Discov. 2007;6(8):605–16.
8. Becker AJ, McCulloch EA, Till JE. Cytological demonstration of the clonal nature of spleen colonies derived from transplanted mouse marrow cells. Nature. 1963;197(4866):452–4.
9. Siminovitch L, McCulloch EA, Till JE. The distribution of colony-forming cells among spleen colonies. J Cell Comp Physiol. 1963;62:327–36.
10. Latsinik NV, Sidorovich S, Fridenshteĭn A. Effect of bone marrow trypsinization on the efficiency of fibroblast colony formation in monolayer cultures. Biull Eksp Biol Med. 1981;92(9):356–8.
11. Laurenti E, Göttgens B. From haematopoietic stem cells to complex differentiation landscapes. Nature. 2018;553(7689):418–26.
12. Till JE, Mc CE. A direct measurement of the radiation sensitivity of normal mouse bone marrow cells. Radiat Res. 1961;14:213–22.
13. Otani A, Kinder K, Ewalt K, Otero FJ, Schimmel P, Friedlander M. Bone marrow-derived stem cells target retinal astrocytes and can promote or inhibit retinal angiogenesis. Nat Med. 2002;8(9):1004–10.
14. Sekiya I, Vuoristo JT, Larson BL, Prockop DJ. In vitro cartilage formation by human adult stem cells from bone marrow stroma defines the sequence of cellular and molecular events during chondrogenesis. Proc Natl Acad Sci U S A. 2002;99(7):4397–402.
15. Ito T, Suzuki A, Imai E, Okabe M, Hori M. Bone marrow is a reservoir of repopulating mesangial cells during glomerular remodeling. J Am Soc Nephrol. 2001;12(12):2625–35.
16. Masuya M, Drake CJ, Fleming PA, Reilly CM, Zeng H, Hill WD, Martin-Studdard A, Hess DC, Ogawa M. Hematopoietic origin of glomerular mesangial cells. Blood. 2003;101(6):2215–8.
17. Perin EC, Dohmann HF, Borojevic R, Silva SA, Sousa AL, Mesquita CT, Rossi MI, Carvalho AC, Dutra HS, Dohmann HJ, Silva GV, Belém L, Vivacqua R, Rangel FO, Esporcatte R, Geng YJ, Vaughn WK, Assad JA, Mesquita ET, Willerson JT. Transendocardial, autologous bone marrow cell transplantation for severe, chronic ischemic heart failure. Circulation. 2003;107(18):2294–302.
18. Stamm C, Westphal B, Kleine HD, Petzsch M, Kittner C, Klinge H, Schümichen C, Nienaber CA, Freund M, Steinhoff G. Autologous bone-marrow stem-cell transplantation for myocardial regeneration. Lancet. 2003;361(9351):45–6.
19. Tse HF, Kwong YL, Chan JK, Lo G, Ho CL, Lau CP. Angiogenesis in ischaemic myocardium by intramyocardial autologous bone marrow mononuclear cell implantation. Lancet. 2003;361(9351):47–9.
20. Ishikawa F, Drake CJ, Yang S, Fleming P, Minamiguchi H, Visconti RP, Crosby CV, Argraves WS, Harada M, Key LL Jr, Livingston AG, Wingard JR, Ogawa M. Transplanted human cord blood cells give rise to hepatocytes in engrafted mice. Ann N Y Acad Sci. 2003;996:174–85.
21. Kakinuma S, Tanaka Y, Chinzei R, Watanabe M, Shimizu-Saito K, Hara Y, Teramoto K, Arii S, Sato C, Takase K, Yasumizu T, Teraoka H. Human umbilical cord blood as a source of transplantable hepatic progenitor cells. Stem Cells. 2003;21(2):217–27.
22. Wang X, Ge S, McNamara G, Hao QL, Crooks GM, Nolta JA. Albumin-expressing hepatocyte-like cells develop in the livers of immune-deficient mice that received transplants of highly purified human hematopoietic stem cells. Blood. 2003;101(10):4201–8.
23. Hori J, Ng TF, Shatos M, Klassen H, Streilein JW, Young MJ. Neural progenitor cells lack immunogenicity and resist destruction as allografts. Stem Cells. 2003;21(4):405–16.
24. Aiuti A, Slavin S, Aker M, Ficara F, Deola S, Mortellaro A, Morecki S, Andolfi G, Tabucchi A, Carlucci F, Marinello E, Cattaneo F, Vai S, Servida P, Miniero R, Roncarolo MG, Bordignon C. Correction of ADA-SCID by stem cell gene therapy combined with nonmyeloablative conditioning. Science. 2002;296(5577):2410–3.
25. Cavazzana-Calvo M, Hacein-Bey S, de Saint Basile G, Gross F, Yvon E, Nusbaum P, Selz F, Hue C, Certain S, Casanova JL, Bousso P, Deist FL, Fischer A. Gene therapy of human severe combined immunodeficiency (SCID)-X1 disease. Science. 2000;288(5466):669–72.
26. Simonsen JL, Rosada C, Serakinci N, Justesen J, Stenderup K, Rattan SI, Jensen TG, Kassem M. Telomerase expression extends the proliferative life-span and maintains the osteogenic potential of human bone marrow stromal cells. Nat Biotechnol. 2002;20(6):592–6.
27. Malik V, Wang J. Pursuing totipotency: authentic totipotent stem cells in culture. Trends Genet. 2022;38(7):632–6.
28. Riveiro AR, Brickman JM. From pluripotency to totipotency: an experimentalist's guide to cellular potency. Development. 2020;147(16):dev189845.
29. Takahashi K, Tanabe K, Ohnuki M, Narita M, Ichisaka T, Tomoda K, Yamanaka S. Induction of pluripotent stem cells from adult human fibroblasts by defined factors. Cell. 2007;131(5):861–72.
30. Thomson JA, Itskovitz-Eldor J, Shapiro SS, Waknitz MA, Swiergiel JJ, Marshall VS, Jones JM. Embryonic stem cell lines derived from human blastocysts. Science. 1998;282(5391):1145–7.
31. Yu J, Vodyanik MA, Smuga-Otto K, Antosiewicz-Bourget J, Frane JL, Tian S, Nie J, Jonsdottir GA, Ruotti V, Stewart R, Slukvin II, Thomson JA. Induced pluripotent stem cell lines derived from human somatic cells. Science. 2007;318(5858):1917–20.
32. Yang S, Cho Y, Jang J. Single cell heterogeneity in human pluripotent stem cells. BMB Rep. 2021;54(10):505–15.

33. Martinez-Morales PL, Revilla A, Ocana I, Gonzalez C, Sainz P, McGuire D, Liste I. Progress in stem cell therapy for major human neurological disorders. Stem Cell Rev Rep. 2013;9(5):685–99.
34. Biehl JK, Russell B. Introduction to stem cell therapy. J Cardiovasc Nurs. 2009;24(2):98–103.
35. Bentzinger CF, Wang YX, Dumont NA, Rudnicki MA. Cellular dynamics in the muscle satellite cell niche. EMBO Rep. 2013;14(12):1062–72.
36. Wood FM, Kolybaba ML, Allen P. The use of cultured epithelial autograft in the treatment of major burn wounds: eleven years of clinical experience. Burns. 2006;32(5):538–44.
37. Chester DL, Balderson DS, Papini RP. A review of keratinocyte delivery to the wound bed. J Burn Care Rehabil. 2004;25(3):266–75.
38. Takahashi K, Yamanaka S. Induction of pluripotent stem cells from mouse embryonic and adult fibroblast cultures by defined factors. Cell. 2006;126(4):663–76.
39. Zhou Q, Brown J, Kanarek A, Rajagopal J, Melton DA. In vivo reprogramming of adult pancreatic exocrine cells to β-cells. Nature. 2008;455(7213):627–32.
40. Rehakova D, Souralova T, Koutna I. Clinical-grade human pluripotent stem cells for cell therapy: characterization strategy. Int J Mol Sci. 2020;21(7):2435.
41. Rowe RG, Daley GQ. Induced pluripotent stem cells in disease modelling and drug discovery. Nat Rev Genet. 2019;20(7):377–88.
42. Shi Y, Inoue H, Wu JC, Yamanaka S. Induced pluripotent stem cell technology: a decade of progress. Nat Rev Drug Discov. 2017;16(2):115–30.
43. Takahashi K, Yamanaka S. A decade of transcription factor-mediated reprogramming to pluripotency. Nat Rev Mol Cell Biol. 2016;17(3):183–93.
44. Breschi A, Gingeras TR, Guigó R. Comparative transcriptomics in human and mouse. Nat Rev Genet. 2017;18(7):425–40.
45. Gharib WH, Robinson-Rechavi M. When orthologs diverge between human and mouse. Brief Bioinform. 2011;12(5):436–41.
46. Lozito TP, Tuan RS. Mesenchymal stem cells inhibit both endogenous and exogenous MMPs via secreted TIMPs. J Cell Physiol. 2011;226(2):385–96.
47. Sato M, Uchida K, Nakajima H, Miyazaki T, Guerrero AR, Watanabe S, Roberts S, Baba H. Direct transplantation of mesenchymal stem cells into the knee joints of Hartley strain guinea pigs with spontaneous osteoarthritis. Arthritis Res Ther. 2012;14(1):R31.
48. Copp G, Robb KP, Viswanathan S. Culture-expanded mesenchymal stromal cell therapy: does it work in knee osteoarthritis? A pathway to clinical success. Cell Mol Immunol. 2023;20(6):626–50.
49. Anil U, Markus DH, Hurley ET, Manjunath AK, Alaia MJ, Campbell KA, Jazrawi LM, Strauss EJ. The efficacy of intra-articular injections in the treatment of knee osteoarthritis: a network meta-analysis of randomized controlled trials. Knee. 2021;32:173–82.
50. Nejadnik H, Hui JH, Feng Choong EP, Tai BC, Lee EH. Autologous bone marrow-derived mesenchymal stem cells versus autologous chondrocyte implantation: an observational cohort study. Am J Sports Med. 2010;38(6):1110–6.
51. Teo AQA, Wong KL, Shen L, Lim JY, Toh WS, Lee EH, Hui JHP. Equivalent 10-year outcomes after implantation of autologous bone marrow-derived mesenchymal stem cells versus autologous chondrocyte implantation for chondral defects of the knee. Am J Sports Med. 2019;47(12):2881–7.
52. Gobbi A, Whyte GP. Long-term clinical outcomes of one-stage cartilage repair in the knee with hyaluronic acid–based scaffold embedded with mesenchymal stem cells sourced from bone marrow aspirate concentrate. Am J Sports Med. 2019;47(7):1621–8.
53. Shafiee A, Soleimani M, Chamheidari GA, Seyedjafari E, Dodel M, Atashi A, Gheisari Y. Electrospun nanofiber-based regeneration of cartilage enhanced by mesenchymal stem cells. J Biomed Mater Res A. 2011;99(3):467–78.
54. Tay LX, Ahmad RE, Dashtdar H, Tay KW, Masjuddin T, Ab-Rahim S, Chong PP, Selvaratnam L, Kamarul T. Treatment outcomes of alginate-embedded allogenic mesenchymal stem cells versus autologous chondrocytes for the repair of focal articular cartilage defects in a rabbit model. Am J Sports Med. 2012;40(1):83–90.
55. Dallari D, Savarino L, Stagni C, Cenni E, Cenacchi A, Fornasari PM, Albisinni U, Rimondi E, Baldini N, Giunti A. Enhanced tibial osteotomy healing with use of bone grafts supplemented with platelet gel or platelet gel and bone marrow stromal cells. J Bone Joint Surg Am. 2007;89(11):2413–20.
56. Wong KL, Lee KB, Tai BC, Law P, Lee EH, Hui JH. Injectable cultured bone marrow-derived mesenchymal stem cells in varus knees with cartilage defects undergoing high tibial osteotomy: a prospective, randomized controlled clinical trial with 2 years' follow-up. Arthroscopy. 2013;29(12):2020–8.
57. Horwitz EM, Gordon PL, Koo WK, Marx JC, Neel MD, McNall RY, Muul L, Hofmann T. Isolated allogeneic bone marrow-derived mesenchymal cells engraft and stimulate growth in children with osteogenesis imperfecta: implications for cell therapy of bone. Proc Natl Acad Sci U S A. 2002;99(13):8932–7.
58. Grayson WL, Bunnell BA, Martin E, Frazier T, Hung BP, Gimble JM. Stromal cells and stem cells in clinical bone regeneration. Nat Rev Endocrinol. 2015;11(3):140–50.
59. Neen D, Noyes D, Shaw M, Gwilym S, Fairlie N, Birch N. Healos and bone marrow aspirate used for lumbar spine fusion: a case controlled study comparing healos with autograft. Spine (Phila Pa 1976). 2006;31(18):E636–40.
60. Gan Y, Dai K, Zhang P, Tang T, Zhu Z, Lu J. The clinical use of enriched bone marrow stem cells combined with porous beta-tricalcium phosphate in posterior spinal fusion. Biomaterials. 2008;29(29):3973–82.
61. Miyamoto T, Muneta T, Tabuchi T, Matsumoto K, Saito H, Tsuji K, Sekiya I. Intradiscal transplantation of synovial mesenchymal stem cells prevents intervertebral disc degeneration through suppression of matrix

metalloproteinase-related genes in nucleus pulposus cells in rabbits. Arthritis Res Ther. 2010;12(6):R206.
62. Tamaki T, Hirata M, Soeda S, Nakajima N, Saito K, Nakazato K, Okada Y, Hashimoto H, Uchiyama Y, Mochida J. Preferential and comprehensive reconstitution of severely damaged sciatic nerve using murine skeletal muscle-derived multipotent stem cells. PLoS One. 2014;9(3):e91257.
63. Adams SB Jr, Thorpe MA, Parks BG, Aghazarian G, Allen E, Schon LC. Stem cell-bearing suture improves Achilles tendon healing in a rat model. Foot Ankle Int. 2014;35(3):293–9.
64. Gulotta LV, Kovacevic D, Ehteshami JR, Dagher E, Packer JD, Rodeo SA. Application of bone marrow-derived mesenchymal stem cells in a rotator cuff repair model. Am J Sports Med. 2009;37(11):2126–33.
65. Awad HA, Boivin GP, Dressler MR, Smith FN, Young RG, Butler DL. Repair of patellar tendon injuries using a cell-collagen composite. J Orthop Res. 2003;21(3):420–31.
66. Burk J, Wittenberg-Voges L, Schubert S, Horstmeier C, Brehm W, Geburek F. Treatment of naturally occurring tendon disease with allogeneic multipotent mesenchymal stromal cells: a randomized, controlled, triple-blinded pilot study in horses. Cells. 2023;12(21):2513.
67. Pascual-Garrido C, Rolón A, Makino A. Treatment of chronic patellar tendinopathy with autologous bone marrow stem cells: a 5-year-followup. Stem Cells Int. 2012;2012:953510.
68. Soler R, Rodas G, Rius-Tarruella J, Alomar X, Balius R, Ruíz-Cotorro Á, Masci L, Maffulli N, Orozco L. Safety and efficacy of bone marrow-derived mesenchymal stem cells for chronic patellar tendinopathy (with gap >3 mm) in patients: 12-month follow-up results of a phase 1/2 clinical trial. Orthop J Sports Med. 2023;11(9):23259671231184400.
69. Wruck W, Graffmann N, Spitzhorn LS, Adjaye J. Human induced pluripotent stem cell-derived mesenchymal stem cells acquire rejuvenation and reduced heterogeneity. Front Cell Dev Biol. 2021;9:717772.
70. Goh D, Yang Y, Lee EH, Hui JHP, Yang Z. Managing the heterogeneity of mesenchymal stem cells for cartilage regenerative therapy: a review. Bioengineering (Basel). 2023;10(3):355.
71. Zhong C, Liu M, Pan X, Zhu H. Tumorigenicity risk of iPSCs in vivo: nip it in the bud. Precis Clin Med. 2022;5(1):pbac004.

9 Growth Factors in Regenerative Medicine

Carola Cavallo and Brunella Grigolo

9.1 Introduction

Regenerative medicine is an emerging field at the forefront of medical science, integrating principles from biology, engineering, and clinical practice to develop innovative therapies to restore, replace, or regenerate damaged cells, tissues, or organs.

The key players in this field are cells, scaffolds, and growth factors (GFs) each of which plays a central role in the regeneration process and works together to facilitate tissue repair and regeneration [1] (Fig. 9.1).

Cells have been identified as the building blocks for regenerative therapies. These include mesenchymal stromal cells (MSCs), which can differentiate into different cell types upon specific stimuli and can be induced to regenerate specific tissues or organs. These unique properties of MSCs offer great potential for treating conditions ranging from neurodegenerative diseases to musculoskeletal disorders [2–4].

Cellular therapies are complemented by scaffolds, which provide a framework for cells to grow into new tissue in an organized fashion, thereby enhancing the efficacy and functionality of regenerative treatments [5–7]. The combination of stem cells and biomaterials stimulates the onset of specific changes in cell properties and fate, influencing processes such as self-renewal, cell adhesion, and differentiation. These scaffolds are typically composed of biocompatible materials (metals, ceramics, or glass-ceramics) and natural/synthetic polymers that mimic the extracellular matrix and provide a conducive environment for cell adhesion, proliferation, and differentiation.

Several evidence obtained from both preclinical and clinical studies demonstrated that the therapeutic ability of MSCs is mainly related to the secretion of biologically active molecules, opening the way for the use of growth factors and cytokines in regenerative medicine [8]. These molecules play a critical role in regulating cell behavior and tissue development, orchestrating various biological processes such as cell proliferation, migration, and differentiation, thereby exerting precise control over the regeneration process. By targeting growth factors, researchers can modulate tissue repair mechanisms to promote optimal healing and regeneration in diseased or injured tissues. In the orthopedic field, traumatic injuries, cartilage degeneration, and osteochondral defects are very common and represent a major burden on the National Health Service and result in reduced quality of life for patients [9]. Current surgical procedures, such as debridement, microfracture, and autologous

C. Cavallo · B. Grigolo (✉)
RAMSES Laboratory, Rizzoli RIT-Research Innovation and Technology Department, Istituto di Ricerca Codivilla Putti, IRCCS Istituto Ortopedico Rizzoli, Bologna, Italy
e-mail: carola.cavallo@ior.it; brunella.grigolo@ior.it

A. Gobbi et al. (eds.), *Regenerative Medicine in Sports and Orthopaedics*,
https://doi.org/10.1007/978-3-031-84693-9_9

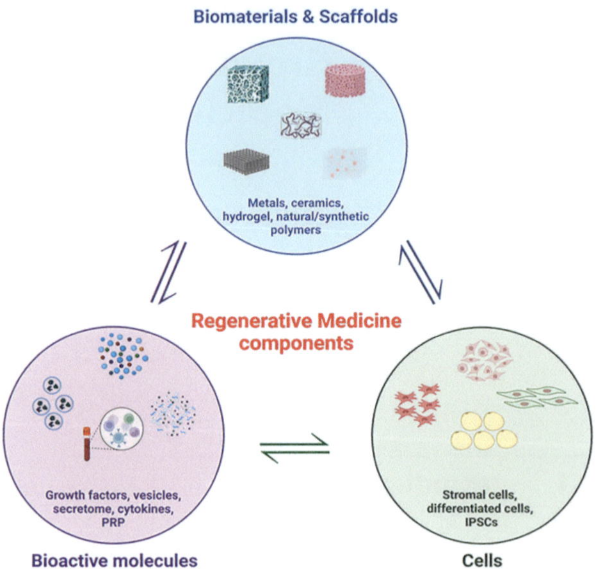

Fig. 9.1 Regenerative medicine key components. (The image was created with BioRender.com)

chondrocyte implantation (ACI), fail to restore damaged tissue and offer only temporary pain relief, and not a long-term clinical solution [10–12]. In this light, regenerative medicine represents an exciting approach to restore, replace, or improve the function of damaged tissue using a multidisciplinary approach, including free cell therapies [13, 14]. The use of trophic factors (chemokines, cytokines, GFs, hormones, and lipid mediators) to treat musculoskeletal diseases could overcome the risks associated with the use of cells, such as the possibility of malignant transformation, transmission of infections, and their immunogenicity. However, despite the promise of using GFs, there are some limitations due to their short half-life, side effects, regulatory and safety concerns, and the potential risk of overstimulation, which could limit their application. Ongoing research and development into the optimization of the delivery and efficacy of growth factors will therefore be essential.

9.2 Growth Factors (GFs)

The term "growth factors" refers to proteins secreted naturally by cells and essential for regenerative processes. These proteins are a large and heterogenous group of peptides signaling molecules. First described effects of growth factors were related to tissue growth, but subsequently it was shown that they act as both hormonal and local paracrine/autocrine regulators of a wide array of biological processes [15].

GFs facilitate and promote the regeneration of new tissues by interacting with specific transmembrane receptors and regulating protein synthesis within cells. Binding to the specific growth factor receptors specifically stimulates cellular signal transduction pathways that trigger events such as cell migration, survival, adhesion, proliferation, growth, and differentiation. Their ability to influence protein synthesis and cellular interactions makes them a key tool in controlling stem cell differentiation [16]. GFs have been classified into several main families, mainly TGF-β, fibroblast growth factors (FGFs), platelet-derived growth factors (PDGFs), and nuclear factor-kappa β (NF-kappa β). The TGF-β superfamily is the largest, comprising major groups of polypeptide morphogens. They are secreted growth factors, which act as multifunctional regulatory peptides involved in a wide range of processes, including cell development, proliferation, and differentiation; wound healing; and carcinogenesis [17]. This family is divided into two subgroups. The first of these subgroups, which includes GFs such as TGF-β, exhibit functions in cell adhesion, growth, cytoskeletal organization, survival, migration, proliferation, differentiation, chemotaxis, and activation of many cells of the immune system during the development of multicellular organisms.

The other group of the TGF-β superfamily is that of BMPs. However, only a part of these (BMP-2, BMP-4, BMP-6, BMP-7, and BMP-9) act significantly on bone formation. BMP-1 is not part of the TGF-β family but belongs to the metalloproteinase family.

9.2.1 TGF-β Superfamily

The TGF-β superfamily represents a large family of cytokines pleiotropic and structurally related to each other which counts in vertebrates approximately 40 distinct members: they, generally, come divided into two branches, TGF-β/Activins/Nodals and BMPs, and are capable of activating one of the two intracellular signaling cascades [17]. These cytokines play an important role in the regulation of the main physiological functions of cells such as proliferation, development, differentiation, migration, cell adhesion, expression of extracellular matrix (ECM) proteins, apoptosis, and repair of wounds. Alterations in production and signal transduction of some members of the TGF-β superfamily are important pathogenetic factors implicated in developmental anomalies and in various human diseases including cancer, fibrosis, and autoimmune diseases. Five different isoforms of the growth factor TGF-β are known. In the mammals only three of these isoforms, TGF-β1, TGF-β2, and TGF-β3, are expressed, each encoded by a specific gene and expressed both in a tissue-specific manner that is regulated during development. TGF-β1 is expressed by endothelial, hematopoietic, and connective tissue cells; TGF-β2 by neuronal and epithelial cells, while TGF-β3 is expressed by mesenchymal cells [18]. The three isoforms have highly conserved sequences with a homology of 70–80%. TGF-β1 is the most abundant and universally expressed isoform and is also the one that is examined in most studies or used to carry out experimental treatments with the growth factor exogenous. TGF-β1 is secreted into the extracellular matrix as a protein complex latent, bound to proteins associated with the latency state and one of the four isoforms of latent TGF-β1 binding proteins. The activation of TGF-β1, which is necessary for its biological activity, takes place through mechanisms that are still poorly understood and involves processing proteolysis of factor-associated proteins and release of the same factor.

9.2.2 Fibroblast Growth Factors (FGFs)

Fibroblast growth factors (FGFs) represent a large family of polypeptides with two main characteristics: a high affinity toward heparin and HLGAGs (heparan-like glycosaminoglycans) and a central region of 140 aa, highly homologous among the various members of this family [19].

FGFs are cell signaling proteins produced by macrophages; they are involved in a wide variety of processes, most notably as crucial elements for normal development in animal cells. Any irregularities in their function lead to a range of developmental defects. These growth factors typically act as systemic or locally circulating molecules of extracellular origin that activate cell surface receptors. FGFs are critical during normal development of both vertebrates and invertebrates and any irregularities in their function lead to a range of developmental defects. FGFs are important players in wound healing, and in particular, FGF1 and FGF2 stimulate angiogenesis and the proliferation of fibroblasts that give rise to granulation tissue, which fills up a wound space/cavity early in the wound-healing process [20].

9.2.3 Platelet-Derived Growth Factor (PDGF)

Platelet-derived growth factor (PDGF) constitutes a family of dimeric isoforms, acting on connective tissue cells and certain other cell types [21]. PDGF was originally discovered as a constituent of platelets, which are released into serum in conjunction with blood coagulation. Although the α-granules of platelets are a major storage site for PDGF, it is also produced by many other cell types.

In the human genome, there are four genes encoding PDGF isoforms, PDGF-A, -B, -C, and -D. The four dimeric isoforms, PDGF-AA, -BB, -CC, and -DD, bind with different specificities to two related tyrosine kinase receptors, forming aa, ab, or bb receptor dimers.

PDGF stimulates the growth of its target cells, but also affects chemotaxis, i.e., directed cell movement, and cell shape through reorganization of the actin filament system. PDGF also affects the differentiation of specific cell types and promotes cell survival.

Through these effects, PDGF has important functions in certain organs during embryonic development, as well as in the adult in the stimulation of wound healing and in the maintenance of connective tissue homeostasis.

9.2.4 Nuclear Factor-Kappa B (*NFK-B*)

NFk-B is a ubiquitous transcription factor that, by regulating the expression of multiple inflammatory and immune genes, plays a critical role in host defense and in chronic inflammatory diseases [22]. It is a heterodimer, present in the cytoplasm in an inactive form complexed to an inhibitory protein—I kappa B. Many extracellular stimuli, including viruses, oxidants, inflammatory cytokines, and immune stimuli, activate NFk-B. Once activated, it binds to recognition elements in the promoter regions of inflammatory and immune genes, such as proinflammatory cytokines, chemokines, inflammatory enzymes, and adhesion molecules.

NFk-B signaling regulates osteoclast and osteoblast formation and is activated in osteoclast precursors in response to the essential osteoclastogenic cytokine, receptor activator of NFk-B ligand (RANKL), which can also control osteoblast formation through RANK-RANKL reverse signaling in osteoblast precursors.

9.2.5 Bone Morphogenetic Proteins (BMPs)

BMP-2 like other bone morphogenetic proteins plays an important role in the development of bone and cartilage [23]. It is involved in the hedgehog pathway, TGF beta signaling pathway, and in cytokine-cytokine receptor interaction. It is also involved in cardiac cell differentiation and epithelial-to-mesenchymal transition. Like many other proteins from the BMP family, BMP-2 has been demonstrated to potently induce osteoblast differentiation in a variety of cell types [24]. BMP-2 may be involved in white adipogenesis and may have metabolic effects.

Preclinical proof-of-concept, feasibility, and efficacy studies in animals resulted in the accumulation of data that served as backbone trials with recombinant human bone morphogenetic protein-2 (rhBMP-2) for orthopedic application [25]. First introduced in the United States in 2002 as approved by the Food and Drug Administration (FDA) for single-level anterior lumbar interbody fusion within a specific threaded titanium tapered cage [26].

As an adjuvant to allograft bone or as a replacement for harvested autograft, it appeared to improve fusion rates after spinal arthrodesis while reducing the donor-site morbidity previously associated with such procedures. BMP-2 was then approved for tibial non-union as an alternative to autograft and for oral maxillofacial reconstructions in 2007. However, an exponential increase in adverse events with the use of rhBMP-2 was observed [27]. Systematic reviews through investigations from the FDA examined the safety and efficacy of BMP-2. These reviews concluded that there was evidence of reporting bias in the initial industry-sponsored publications and no clear advantages of BMP-2 use as compared with iliac crest bone graft [28]. Moreover, adverse events associated with BMP-2 were not only frequent but were also occasionally catastrophic, especially in anterior cervical spine fusions [29]. These included cervical and soft tissue swelling, airway compromise, and need for reoperation. These new insights prompted the FDA to issue a Public Health Notification about BMP-2 use in 2008 [30].

9.3 Therapeutic Use of GFs for Musculoskeletal Pathologies

Musculoskeletal tissue pathologies affect million people worldwide with a great impact on the quality of life. Owing to population aging, the World Health Organization estimates a noticeable increase in the percentage of patients in 2050, consequently rising costs for the healthcare system.

Tissue engineering has been shown to represent and address the shortcomings of existing clinical approaches. It represents a fascinating strategy to regenerate damaged tissues due to trauma or pathologies. The selection of appropriate cell population, suitable scaffolds, and right stimulation plays a definitive role in the success of the treatment. It has been demonstrated that applying GFs could facilitate tissue regeneration by influencing tissue environment promoting cell migration proliferation and differentiation.

Several GFs have been approved by the US Food and Drug Administration (FDA) to facilitate tissue regeneration. Among them, as reported above, BMP-2 is the only FDA-approved GF for adjuvant musculoskeletal tissue regeneration and has elicited broad investigation for use in tissue engineering applications [31, 32]. With annual sales of over $750 million, INFUSE® Bone Graft (Medtronic) is widely used by orthopedic surgeons to stimulate bone formation for spinal fusion and promote fracture repair in patients with a high risk for non-union. The BMP-2-loaded collagen carrier of the INFUSE® system can be used alone or in combination with other treatment techniques (e.g., bone marrow aspirate) to increase bone regeneration and lower failure rates in spinal fusion and tibial fracture repair [33, 34]. However, high dosages of BMP-2 can result in significant complications including inflammation, heterotopic ossification, hematoma, and myelopathy that require additional revision procedures [35].

However, despite promising results in preclinical models, the use of GFs has achieved limited success in translation to the clinic. This is because some problems have not been resolved such as the high cost of their synthesis using recombinant technologies, the choice of the right dosages, and a temporal control to avoid detrimental side effects.

9.4 Platelet-Rich Plasma (PRP)

Regarding the use of GFs in the regenerative medicine field, platelet-rich plasma (PRP) has gained a pivotal role in clinical practice thanks to its reparative potential [36–38]. Numerous pre-clinical and clinical studies of PRP have yielded encouraging results in several areas of medicine, particularly in the orthopedic field as an infiltrative treatment for cartilage degeneration, osteoarthritis (OA), and tendinopathies. PRP is an autologous peripheral blood derivative with a platelet concentration above the baseline and contains platelet-related growth factors capable of exerting biological effects [39]. It contains alpha-granules rich in GFs, including: platelet-derived growth factors (PDGF); TGF: transforming growth factor α and β (TGF-α/ β); vascular endothelial growth factor (VEGF); epidermal growth factor (EGF); fibroblast growth factor (FGF); connective tissue growth factor (CTCG); insulin-like growth factor (IGF); hepatocyte growth factor (HGF); stromal cell-derived factor (SDF-1α), each of them capable to promote angiogenesis, cell migration and proliferation, protein synthesis, cartilage regeneration, and bone formation [40–45]. However, despite the widespread use of PRP in the treatment of several musculoskeletal diseases, its properties, compositions, and effects remain to be elucidated [46]. The main issues with PRP are related to the lack of consensus on its standardization and inadequate reporting of bioformulations in clinical applications, which leads to inconsistencies in reported outcomes [47].

9.5 MSC Secretome

Recent studies have shown that the therapeutic potential of MSCs for tissue repair and regenerative medicine lies in their paracrine effects, thanks to the large number of bioactive factors they produce, which play an important role in the regulation of key biological processes [48]. This set of secreted factors/molecules, referred to as the "secretome," includes not only GFs, cytokines, and chemokines, but also soluble proteins, nucleic acids, lipids, and extracellular vesicles, all of which are involved in tissue repair and regeneration through the regulation of cell-to-cell signaling [49].

The ability of the secretome to play a crucial role in crosstalk between cells and neighboring tissues, influencing their biological function, has attracted the interest of researchers who could use it in cell-free therapies. The use of the secretome could help to overcome some of the problems associated with the transplantation of living and proliferating cells, such as the risk of tumorigenicity, immunogenicity, and transmission of infection [50]. However, despite the encouraging evidence of the beneficial effects of the secretome in musculoskeletal disease, its therapeutic use should be subject to a regulatory requirement for manufacturing and quality control to establish the safety and efficacy profile of this product [51].

9.6 Extracellular Vesicles

The paracrine effects of MSCs (and other cell types) are not only due to the production of GFs but also due to the secretion of extracellular vesicles (EVs), including exosomes, which contain bioactive components such as nucleic acids, proteins, and lipids [52]. They play a central role in cell-cell communication by transferring their contents, including proteins, lipids, and nucleic acids, and modulating the function or fate of target cells [53]. Comparing the use of EVs with stem cells for therapeutic purposes, EVs have shown some advantages such as easy storage requirements, the possibility of allowing allogeneic transplantation (thanks to the lack of MHC I and MHC II antigens), and the reduction of the risk of iatrogenic tumor growth [54]. Several studies have reported that EVs are effective in promoting cartilage repair and regeneration [55] and in attenuating the inflammatory/catabolic environment due to the onset of osteoarthritis [56]. However, EVs released from damaged or inflamed tissue can carry an unwanted message, reprogramming the cell phenotype [57]. Therefore, careful assessment of the environment of the cells from which EVs originate, control of their release, and standardization of EV products must be considered to ensure their safety and make EV therapy effective and efficient.

References

1. Giannoudis PV, Einhorn TA, Schmidmaier G, Marsh D. The diamond concept-open questions. Injury. 2008;39(Suppl 2):S5–8. https://doi.org/10.1016/S0020-1383(08)70010-X.
2. Wakitani S, Mera H, Nakamura N, Gobbi A. Review of Caplan (1991) on cell-based therapeutic technology using mesenchymal stem cells. J ISAKOS. 2024;9(3):426–30. https://doi.org/10.1016/j.jisako.2023.08.010.
3. Rohban R, Pieber TR. Mesenchymal stem and progenitor cells in regeneration: tissue specificity and regenerative potential. Stem Cells Int. 2017;2017:5173732. https://doi.org/10.1155/2017/5173732.
4. Gobbi A, Whyte GP. Emerging orthobiologic approaches to ligament injury. In: Gobbi A, Espregueira-Mendes J, Lane J, Karahan M, editors. Bio-orthopaedics. Berlin, Heidelberg: Springer; 2017. https://doi.org/10.1007/978-3-662-54181-4_24.
5. Adhikari B, Stager MA, Krebs MD. Cell-instructive biomaterials in tissue engineering and regenerative medicine. J Biomed Mater Res A. 2003;11:660–81. https://doi.org/10.1002/jbm.a.37510.
6. Gobbi A, Whyte GP. Long-term clinical outcomes of one-stage cartilage repair in the knee with hyaluronic acid-based scaffold embedded with mesenchymal stem cells sourced from bone marrow aspirate concentrate. Am J Sports Med. 2019;47(7):1621–8. https://doi.org/10.1177/0363546519845362.
7. Whyte GP, Herman K, Gobbi A. One-step cell-based cartilage repair in the knee using hyaluronic acid-based scaffold embedded with mesenchymal stem cells sourced from bone marrow aspirate concentrate (HA-BMAC). In: Brittberg M, Slynarski K, editors. Lower extremity joint preservation. Cham: Springer; 2021. https://doi.org/10.1007/978-3-030-57382-9_5.
8. Fu Y, Karbaat L, Wu L, Leijten J, Both SK, Karperien M. Trophic effects of mesenchymal stem cells in tissue regeneration. Tissue Eng Part B. 2017;6:515–28. https://doi.org/10.1089/ten.TEB.2016.0365.
9. Aigner T, Rose J, Martin J, Buckwalter J. Aging theories of primary osteoarthritis: from epidemiology to molecular biology.

Rejuvenation Res. 2004;7(2):134–45. https://doi.org/10.1089/1549168041552964.
10. Mobasheri A, Kalamegam G, Musumeci G, Batt ME. Chondrocyte and mesenchymal stem cell-based therapies for cartilage repair in osteoarthritis and related orthopaedic conditions. Maturitas. 2014;78(3):188–98. https://doi.org/10.1016/j.maturitas.2014.04.017.
11. Favreau H, Pijnenburg L, Seitlinger J, Fioretti F, Keller L, Scipioni D, et al. Osteochondral repair combining therapeutics implant with mesenchymal stem cells spheroids. Nanomedicine. 2020;29:102253. https://doi.org/10.1016/j.nano.2020.102253.
12. Gobbi A, Whyte GP. One-stage cartilage repair using a hyaluronic acid-based scaffold with activated bone marrow-derived mesenchymal stem cells compared with microfracture: five-year follow-up. Am J Sports Med. 2016;44(11):2846–54. https://doi.org/10.1177/0363546516656179.
13. Amsar RM, Wijaya CH, Ana ID, Hidajah AC, Notobroto HB, Kencana Wungu TD, et al. Extracellular vesicles: a promising cell-free therapy for cartilage repair. Future Sci OA. 2021;8(2):FSO774. https://doi.org/10.2144/fsoa-2021-0096.
14. Dallo I, Frank RM, Bradsell H, Piuzzi NS, Gobbi A. Overview of Orthobiologics and joint function. In: Gobbi A, Lane JG, Longo UG, Dallo I, editors. Joint function preservation. Cham: Springer; 2022. https://doi.org/10.1007/978-3-030-82958-2_2.
15. Lloyd RV. Growth factors. Endocr Pathol. 1997;8:121–7. https://doi.org/10.1007/BF02739940.
16. Lee K, Silva EA, Mooney DJ. Growth factor delivery-based tissue engineering: general approaches and a review of recent developments. J R Soc Interface. 2011;8(55):153–70. https://doi.org/10.1098/rsif.2010.0223.
17. Chen G, Deng C, Li YP. TGF-beta and BMP signaling in osteoblast differentiation and bone formation. Int J Biol Sci. 2012;8:272–88. https://doi.org/10.7150/ijbs.2929.
18. Massagué J, Blain SW, Lo RS. TGFbeta signaling in growth control, cancer, and heritable disorders. Cell. 2000;13(103):295–309. https://doi.org/10.1016/s0092-8674(00)00121-5.
19. Burgess WH, Maciag T. The heparin-binding (fibroblast) growth factor family of proteins. Annu Rev Biochem. 1989;58:575–606. https://doi.org/10.1146/annurev.bi.58.070189.003043.
20. Ornitz DM, Itoh N. Fibroblast growth factors. Genome Biol. 2001;2(3):REVIEWS3005. https://doi.org/10.1186/gb-2001-2-3-reviews3005.
21. Ross R. Platelet-derived growth factor. Annu Rev Med. 1987;38:71–9. https://doi.org/10.1146/annurev.me.38.020187.000443.
22. Barnes PJ. Nuclear factor-kappa B. Int J Biochem Cell Biol. 1997;29(6):867–70. https://doi.org/10.1016/s1357-2725(96)00159-8.
23. Chen D, Zhao M, Mundy GR. Bone morphogenetic proteins. Growth Factors. 2004;22:233–41. https://doi.org/10.1080/08977190412331279890.
24. Ingwersen LC, Frank M, Naujokat H, Loger K, Bader R, Jonitz-Heincke A. BMP-2 long-term stimulation of human pre-osteoblasts induces osteogenic differentiation and promotes transdifferentiation and bone remodeling processes. Int J Mol Sci. 2022;12(23):3077. https://doi.org/10.3390/ijms23063077.
25. Khan SN, Lane JM. The use of recombinant human bone morphogenetic protein-2 (rhBMP-2) in orthopaedic applications. Expert Opin Biol Ther. 2004;4:741–8. https://doi.org/10.1517/14712598.4.5.741.
26. Gautschi OP, Frey SP, Zellweger R. Bone morphogenetic proteins in clinical applications. ANZ J Surg. 2007;77:626–31. https://doi.org/10.1111/j.1445-2197.2007.04175.x.
27. Burkus JK, Gornet MF, Dickman CA, Zdeblick TA. Anterior lumbar interbody fusion using rhBMP-2 with tapered interbody cages. J Spinal Disord Tech. 2002;15:337–49. https://doi.org/10.1097/00024720-200210000-00001.
28. James AW, LaChaud G, Shen J, Asatrian G, Nguyen V, Zhang X, et al. A review of the clinical side effects of bone morphogenetic protein-2. Tissue Eng Part B. 2016;22:284–97. https://doi.org/10.1089/ten.TEB.2015.0357.
29. Simmonds MC, Brown JV, Heirs MK, Higgins JP, Mannion RJ, Rodgers MA, et al. Safety and effectiveness of recombinant human bone morphogenetic protein-2 for spinal fusion: a meta-analysis of individual-participant data. Ann Intern Med. 2013;18(158):877–89. https://doi.org/10.7326/0003-4819-158-12-201306180-00005.
30. Center for Devices and Radiological Health Public Health Notifications (Medical devices)-FDA Public Health Notification: life-threatening complications associated with recombinant human bone morphogenetic protein in cervical spine fusion. wwwfda.gov/MedicalDevices/Safety/AlertsandNotices/PublicHealthNotifications/ucm062000.htm.
31. Kim DH, Rhim R, Li L, Martha J, Swaim BH, Banco RJ, et al. Prospective study of iliac crest bone graft harvest site pain and morbidity. Spine J. 2009;9:886–92. https://doi.org/10.1016/j.spinee.2009.05.006.
32. Copuroglu C, Calori GM, Giannoudis PV. Fracture non-union: who is at risk? Injury. 2013;44:1379–82. https://doi.org/10.1016/j.injury.2013.08.003.
33. Burkus JK, Gornet MF, Schuler TC, Kleeman TJ, Zdeblick TA. Six-year outcomes of anterior lumbar interbody arthrodesis with use of interbody fusion cages and recombinant human bone morphogenetic protein-2. J Bone Joint Surg Am. 2009;91:1181–9. https://doi.org/10.2106/JBJS.G.01485.
34. Starr AJ. Recombinant human bone morphogenetic protein-2 for treatment of open tibial fractures. J Bone Joint Surg Am. 2003;85:2049–50. https://doi.org/10.2106/00004623-200310000-00027.
35. Epstein NE. Complications due to the use of BMP/INFUSE in spine surgery: the evidence continues to mount. Surg Neurol Int. 2013;4(Suppl 5):S343–52. https://doi.org/10.4103/2152-7806.114813.

36. Cecerska-Heryć E, Goszka M, Serwin N, Roszak M, Grygorcewicz B, Heryć R, et al. Applications of the regenerative capacity of platelets in modern medicine. Cytokine Growth Factor Rev. 2022;64:84–94. https://doi.org/10.1016/j.cytogfr.2021.11.003.
37. Jacob G, Nakamura N, Parker D, Lane J, Gobbi A. The use of ortho-biologics in sports medicine: an ISAKOS survey report. In: ISAKOS newsletter 2024—current perspectives on arthroscopy, knee surgery & orthopaedic sports medicine, vol. 1; 2024. p. 12–5.
38. Hurley ET, Sherman SL, Stokes DJ, Rodeo SA, Shapiro SA, Mautner K, Buford DA, Dragoo JL, Mandelbaum BR, Zaslav KR, Cole BJ, Frank RM, Members of the Biologics Association. Experts achieve consensus on a majority of statements regarding platelet-rich plasma treatments for treatment of musculoskeletal pathology. Arthroscopy. 2024;40(2):470–477.e1. https://doi.org/10.1016/j.arthro.2023.08.020.
39. Filardo G, Di Matteo B, Kon E, Merli G, Marcacci M. Platelet-rich plasma in tendon-related disorders: results and indications. Knee Surg Sports Traumatol Arthrosc. 2018;26:1984–99. https://doi.org/10.1007/s00167-016-4261-4.
40. Lubkowska A, Dolegowska B, Banfi G. Growth factor content in PRP and their applicability in medicine. J Biol Regul Homeost Agents. 2012;26:3S–22S.
41. Giusti I, D'Ascenzo S, Macchiarelli G, Dolo V. In vitro evidence supporting applications of platelet derivatives in regenerative medicine. Blood Transfus. 2020;18:117–29. https://doi.org/10.2450/2019.0164-19.
42. Marmotti A, Rossi R, Castoldi F, Roveda E, Michielon G, Peretti GM. PRP and articular cartilage: a clinical update. Biomed Res Int. 2015;2015:542502. https://doi.org/10.1155/2015/542502.
43. Cavallo C, Filardo G, Mariani E, Kon E, Marcacci M, Pereira Ruiz MT, et al. Comparison of platelet-rich plasma formulations for cartilage healing: an in vitro study. J Bone Joint Surg Am. 2014;96(5):423–9. https://doi.org/10.2106/JBJS.M.00726.
44. Xie Y, Chen M, Chen Y, Xu Y, Sun Y, Liang J, et al. Effects of PRP and LyPRP on osteogenic differentiation of MSCs. J Biomed Mater Res A. 2020;108:116–26. https://doi.org/10.1002/jbm.a.36797.
45. Xu HH, Li SM, Fang L, Xia CJ, Zhang P, Xu R, et al. Platelet-rich plasma promotes bone formation, restrains adipogenesis and accelerates vascularization to relieve steroids-induced osteonecrosis of the femoral head. Platelets. 2021;3(32):950–9. https://doi.org/10.1080/09537104.2020.1810221.
46. Beitzel K, Allen D, Apostolakos J, Russell R, McCarthy M, Gallo G, et al. US definitions, current use, and FDA stance on use of platelet-rich plasma in sports medicine. J Knee Surg. 2015;28:29–34. https://doi.org/10.1055/s-0034-1390030.
47. Everts P, Onishi K, Jayaram P, Lana JF, Mautner K. Platelet-rich plasma: new performance understandings and therapeutic considerations in 2020. Int J Mol Sci. 2020;21:7794. https://doi.org/10.3390/ijms21207794.
48. Eleuteri S, Fierabracci A. Insights into the secretome of mesenchymal stem cells and its potential applications. Int J Mol Sci. 2015;20:4597. https://doi.org/10.3390/ijms20184597.
49. Madrigal M, Rao KS, Riordan NH. A review of therapeutic effects of mesenchymal stem cell secretions and induction of secretory modification by different culture methods. J Transl Med. 2014;12:260. https://doi.org/10.1186/s12967-014-0260-8.
50. Vizoso FJ, Eiro N, Cid S, Schneider J, Perez-Fernandez R. Mesenchymal stem cell secretome: toward cell-free therapeutic strategies in regenerative medicine. Int J Mol Sci. 2017;25(18):1852. https://doi.org/10.3390/ijms18091852.
51. Miclau K, Hambright WS, Huard J, Stoddart MJ, Bahney CS. Cellular expansion of MSCs: Shifting the regenerative potential. Aging Cell. 2023;22:e13759. https://doi.org/10.1111/acel.13759.
52. D'Arrigo D, Roffi A, Cucchiarini M, Moretti M, Candrian C, Filardo G. Secretome and extracellular vesicles as new biological therapies for knee osteoarthritis: a systematic review. J Clin Med. 2019;8(11):1867. https://doi.org/10.3390/jcm8111867.
53. Lo Cicero A, Stahl PD, Raposo G. Extracellular vesicles shuffling intercellular messages: for good or for bad. Curr Opin Cell Biol. 2015;35:69–77. https://doi.org/10.1016/j.ceb.2015.04.013.
54. Abraham A, Krasnodembskaya A. Mesenchymal stem cell-derived extracellular vesicles for the treatment of acute respiratory distress syndrome. Stem Cells Transl Med. 2020;9(1):28–38. https://doi.org/10.1002/sctm.19-0205.
55. Liao Q, Li BJ, Li Y, Xiao Y, Zeng H, Liu JM, et al. Low-intensity pulsed ultrasound promotes osteoarthritic cartilage regeneration by BMSC-derived exosomes via modulating the NF-kappaB signaling pathway. Int Immunopharmacol. 2021;97:107824. https://doi.org/10.1016/j.intimp.2021.107824.
56. Cavallo C, Merli G, Borzì RM, Zini N, D'Adamo S, Guescini M, et al. Small extracellular vesicles from adipose derived stromal cells significantly attenuate in vitro the NF-κB dependent inflammatory/catabolic environment of osteoarthritis. Sci Rep. 2021;11(1):1053. https://doi.org/10.1038/s41598-020-80032-7.
57. Cavallo C, Merli G, Zini N, D'Adamo S, Cattini L, Guescini M, et al. Small extracellular vesicles from inflamed adipose derived stromal cells enhance the NF-κB-dependent inflammatory/catabolic environment of osteoarthritis. Stem Cells Int. 2022;2022:9376338. https://doi.org/10.1155/2022/9376338.

10 Multipotent Adult Progenitor Cells

Pooja Swami, Henintsoa Fanjaniaina Andriamifidy, Apratim Maity, and Daniel A. Grande

10.1 Introduction

The classification of stem cells is primarily based on their potency, which includes totipotent, pluripotent, multipotent, and unipotent designations. Totipotent stem cells, such as those from the zygote, have the ability to differentiate into all cell types, including both embryonic and extra-embryonic tissues. Pluripotent stem cells, like those found in the inner cell mass of the blastocyst, can form nearly all cell types but are limited to the three germ layers: ectoderm, mesoderm, and endoderm. Multipotent stem cells, exemplified by hematopoietic and mesenchymal stem cells, can produce a limited range of cell types within a specific lineage. Unipotent stem cells, although capable of self-renewal, are restricted to producing a single cell type. These varying levels of potency are crucial for the dynamic processes of growth, development, and healing throughout an organism's life.

Adult stem cells, typically recognized for their multipotent nature, tend to be restricted in the types of cells that they can differentiate into. Since the 1990s, research has revealed that adult stem cells can exhibit a broader differentiation potential under specific conditions [1]. Several research groups have successfully isolated these types of stem cells from non-embryonic sources, such as bone marrow, amniotic fluid, and umbilical cord blood [1]. Notable among these cell types are multipotent adult progenitor cells (MAPCs), very small embryonic stem cell-like stem cells (VSELs), marrow-isolated adult multilineage inducible stem cells (MIAMI), and multipotent adult stem cells (MASCs).

This chapter will primarily focus on MAPCs and their potential applications in regenerative medicine and orthopedics. We will begin by exploring the history of MAPCs, followed by a detailed characterization of these cells. Subsequently, we will examine their practical applications in medical treatments. Finally, we will address both the ethical considerations surrounding MAPC research and prospective future directions within this promising field.

P. Swami · H. F. Andriamifidy
Orthopaedic Research Laboratory, The Feinstein Institute for Medical Research, Northwell Health, Manhasset, NY, USA
e-mail: Pswami@northwell.edu; HFanjaniai@northwell.edu

A. Maity
Donald and Barbara Zucker School of Medicine at Hofstra/Northwell, Hempstead, NY, USA
e-mail: AMaity@northwell.edu

D. A. Grande (✉)
Orthopaedic Research Laboratory, The Feinstein Institute for Medical Research, Northwell Health, Manhasset, NY, USA

Donald and Barbara Zucker School of Medicine at Hofstra/Northwell, Hempstead, NY, USA
e-mail: dgrande@northwell.edu

A. Gobbi et al. (eds.), *Regenerative Medicine in Sports and Orthopaedics*,
https://doi.org/10.1007/978-3-031-84693-9_10

10.1.1 What Are MAPCs?

MAPCs were first discovered by Catherine Verfaillie's research group in 2002 at the University of Minnesota Medical School's Department of Neurosurgery. During their work with human bone marrow mesenchymal stem cells, they identified a rare cell type capable of extensive expansion [1, 2]. Named "multipotent adult progenitor cells," these cells demonstrated unique differentiation potential, not only forming mesenchymal lineage cells such as those of bone, cartilage, and fat, but also differentiating into endothelial cells, hematopoietic stem cells, hepatic stem cells, and neural stem cells. Furthermore, MAPCs isolated from rodent bone marrow exhibited pluripotent characteristics, differentiating into cells from all three germ layers: ectoderm, mesoderm, and endoderm. When injected into an early-stage blastocyst, MAPCs integrated into most developing tissues in the organism and differentiated into various cell types in response to signals from neighboring tissues [1]. MAPC lineage, their differentiation potential, relevant features, and applications are highlighted in Fig. 10.1.

The researchers isolated these cells from human, murine, and rat bone marrow. The main difference between the isolation protocols of human and murine MAPCs was the requirement of leukemia inhibitory factor (LIF) in the expansion of murine MAPCs (mMAPC) [1, 2]. The mMAPCs were successfully cultured for more than 120 population doublings, indicating their ability to proliferate extensively without losing their undifferentiated state. Even after extensive culture, the morphology and phenotype of mMAPCs remained stable. Similarly, MAPCs

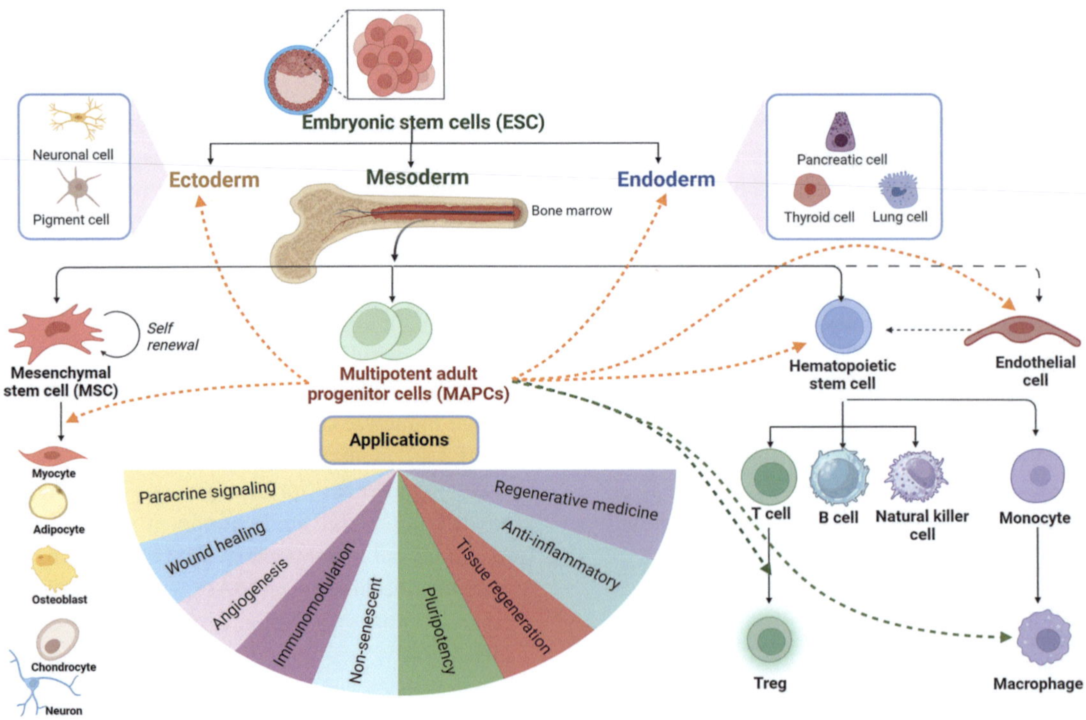

Fig. 10.1 Differentiation potential of MAPCs: Embryonic stem cells (ESCs) are able to differentiate into cells of ectodermal (pigment, neuronal cells), endodermal (pancreatic cells, lung cells), and mesodermal (hematopoietic and MSC endothelial progenitor cells) lineage. MAPCs are found in bone marrow. Orange dashed lines refer to different germ layers and cell types that MAPCs have the potential to differentiate into. Green dashed line refers to cells that MAPCs are able to influence through paracrine signaling-Treg differentiation and macrophage polarization. The pie chart represents various applications and biologically relevant features of MAPCs

isolated from rat bone marrow (rMAPCs) were successfully cultured for more than 100 population doublings, requiring LIF, epidermal growth factor (EGF), and platelet-derived growth factor (PDGF)-BB for successful culture. Human MAPCs (hMAPCs), in contrast, do not require LIF but need fetal calf serum (FCS), EGF, and PDGF-BB for expansion, achieving over 70 population doublings. hMAPCs exhibit extensive immunomodulatory properties both *in vitro* and *in vivo*, and possess significant therapeutic potential, making them a major research interest in regenerative medicine. When grafted into ischemia models, both hMAPCs and rodent MAPCs significantly enhanced angiogenesis and the proliferation of endogenous stem cells, leading to improved functions in the central nervous system (CNS), cardiac, and skeletal muscle tissues [1].

10.2 Distinctive Features of MAPCs: A Comparative Analysis with MSCs and ESCs

MAPCs are non-hematopoietic stem cells that were initially found to co-purify with mesenchymal stem cells (MSCs). MAPCs exhibit multipotency [3], immunomodulatory properties [4], proangiogenic capabilities [5], and self-renewal capacity like MSCs. In addition, they share the three key criteria defining MSCs provided by the International Society for Cell Therapy (ISCT): (1) plastic adherence under standard culture conditions; (2) positive expression of CD105, CD90, and CD73, and negative expression of MHC-II, CD11b, CD14, CD34, CD45, and CD31; and (3) in-vitro tri-lineage differentiation capacity into mesodermal derivatives like osteocytes, chondrocytes, and adipocytes. However, MAPCs have broader plasticity, differentiating into cell types of neuronal, ectodermal, and endodermal origins [6–10]. This difference has led to speculation that they are functionally and phenotypically distinct from MSCs.

Human MAPCs (hMAPCs) phenotypically express pluripotent markers, such as octamer-binding transcription factor 4 (OCT-4) and homeobox protein NANOG, both at the gene and protein levels [10]. In contrast, rodent MAPCs express only OCT-4 [1]. These transcription factors are crucial for maintaining pluripotency in stem cells, particularly embryonic stem cells (ESCs), suggesting that MAPCs are more biologically primitive than classical adult MSCs. Recent studies have shown that dynamic changes in epigenetic regulation within stem cells dictate their plasticity and pluripotency [11]. Gene ontology studies of ESCs revealed enrichment of genes related to epigenetic repression of differentiation genes through DNA methylation and histone modifications [12]. In contrast, MAPCs repress differentiation genes primarily through histone modifications, rather than DNA methylation. This difference explains MAPCs' wider multi-lineage potential compared to classical MSCs but also suggests that they are a distinct stem cell population capable of maintaining an undifferentiated state under specific epigenetic conditions. Interestingly, principal component analysis of gene expression signatures in rodent MAPCs and ESCs showed a clear separation between them. Genes related to primitive mesendodermal differentiation during embryonic development, such as Laminin A1, Laminin A4, Laminin C1, and Collagen IV, as well as Pitx2, were uniquely expressed in rodent MAPCs [13]. Among ESC-associated transcripts, only six were highly expressed in MAPCs, including Spalt Family Member (Sall4), a key modulator of OCT-4 activation [14]. Understanding how the global epigenomic landscape of MAPCs facilitates dynamic transcription changes and gene activation governing their lineage commitment remains to be elucidated. Such work is pertinent in defining the distinctive genes involved in maintaining MAPCs' pluripotent nature. Nevertheless, the abovementioned similarities with ESCs demonstrate that MAPCs are stem cells that retain aspects of transcription programs found in ESCs.

Like adult MSCs, MAPCs display extensive self-renewal capacity. However, MAPCs surpass MSCs in their ability to proliferate, achieving over 60–120 cell doublings without undergoing senescence or experiencing reduced telomerase activity. Optimal culture conditions for MAPCs require lower oxygen levels (5%), low serum (2%), and a combination of growth factors, such as human platelet-derived growth factor (PDGF) and human epidermal growth factor (EGF) [3]. For rodent and murine MAPCs, the culture media is supplemented with the cytokine leukemia inhibitory factor (LIF), which helps preserve their self-renewing state [15]. LIF is known to inhibit lineage specification and maintain the "naïve" state of embryonic stem cells (ESCs) during in-vitro culture by initiating the STAT3 signaling pathway through its cytokine receptor GP130 [16]. This pathway perpetually maintains ESCs in a pluripotent state, highlighting a shared feature between MAPCs and ESCs.

The in-vitro conditions necessary for MAPCs have sparked debate about whether they are a distinct stem cell population, or a product of cell programming induced by culture conditions. The requirement for low oxygen levels suggests a potential physiological relevance to their in-vivo niche and functionality. Hypoxia induces the activation of hypoxia-dependent genes, such as vascular endothelial growth factor (VEGF) and fibroblast growth factors (bFGF, FGF2), as well as the transcription factor hypoxia-inducible factor (HIF). Under hypoxic conditions, the α-subunit of HIF stabilizes and forms an active heterodimer with the β-subunit, leading to the transcription of genes associated with cell differentiation, metabolism, and angiogenesis [17]. A prolonged exposure to hypoxia in ESCs has been shown to upregulate hypoxia-related genes, such as the c-MYC transcription factor, which controls their undifferentiated phenotype, and stage-specific embryonic antigen 3 (SSEA-3), another pluripotency marker [18]. Suppression of HIF-2α results in the downregulation of OCT4, SOX2, and NANOG, core regulators of pluripotency in ESCs [19, 20]. These findings suggest a link between hypoxia and pluripotency gene mediators, supporting the idea that MAPCs may share molecular mechanisms with ESCs under specific conditions. Nevertheless, it remains inconclusive whether MAPCs acquire ESC-like characteristics through environmental cues or are intrinsically primitive stem cells derived from ESCs and maintained in adult tissues. Acquiring a pluripotent state through genetic reprogramming was demonstrated through exogenous expression of specific pluripotency genes (Oct3/4, Sox2, Klf4, and c-Myc), which led to the discovery of induced pluripotent stem cells (iPSCs). However, despite genetic reprogramming, iPSCs exhibit molecular and functional heterogeneity in contrast to ESCs. Notably, iPSCs can undergo senescence and retain an epigenomic memory of their somatic origin [21]. Therefore, identifying key differences in the global transcriptional and epigenomic profiles between iPSCs and MAPCs could further clarify the unique nature of MAPCs. Despite these debates, MAPCs possess distinctive characteristics that underscore their unique nature within the realm of stem cells.

10.3 Characterization of MAPCs: Insights from Phenotypic, Transcriptomic, Functional Analysis

MAPCs are distinguishable by their morphology, being smaller in size and triangular in shape [10]. Phenotypically, they can be characterized by their surface markers, notably CD34−, CD44 low, CD45−, CD117 (cKit)−, Class I–HLA−, HLA-DR−, CD90+, CD13+, CD105+, and CD73+ [22]. Transcriptome analysis of MAPCs show overexpression of genes that are functionally involved in biological pathways related to vascular integrity and vessel formation, such as angiopoietin-like 1 and 4 (ANGPTL-1 and 4) and other proangiogenic factors [23]. The same study showed that this transcriptomic profile is consistent with their observed cellular behavior. MAPCs demonstrate a greater commitment to endothelial differentiation and vascular tube formation in in-vitro assays compared to MSCs. *In vivo*, only MAPCs, and not MSCs, induced the formation of functional human CD34+ vessels in subcutaneous implantation models in nude mice.

Interestingly, MAPCs do not express CD140a and CD140b, platelet-derived growth factor receptors (PDGFR-β) that are constitutively expressed in MSCs and known to modulate their angiogenic potency. It is important to note that

Table 10.1 List of adult human stem cell populations and their characteristics

Stem cell type	Morphology	Surface markers	Differentiation potential
Embryonic stem cells (ESCs)	Round, large nucleus to cytoplasm ratio	SSEA-3, SSEA-4, Tra-1-60, Oct4	Pluripotent (all three germ layers)
Adult stem cells (e.g., mesenchymal stem cells)	Spindle-shaped	CD73, CD90, CD105	Multipotent (multiple cell types within a lineage)
Induced pluripotent stem cells (iPSCs)	Similar to ESCs	Similar to ESCs	Pluripotent (all three germ layers)
Multipotent adult progenitor stem cells (MAPCs)	Small, triangular shape	CD73, CD90, CD105, Oct4	Pluripotent (all three germ layers)

in-vitro culture of adult stem cells can refer to a heterogeneous cell population. Thus, assessing the differentiation of engrafted MSCs and MAPCs post-implantation cannot fully determine their intrinsic functional status at the individual cellular level. Nevertheless, MAPCs seem to intrinsically harbor a robust progeny of endothelial precursors through activation of the Notch signaling pathway both *in vitro* and *in vivo* [24]. The activation of Notch in these cells upregulates Hey-2, a target gene essential for the formation of blood vessels during embryonic development [25]. These findings highlight specific functional properties of MAPCs, particularly a potentially distinct role in tissue repair and regeneration (Table 10.1).

10.4 Applications of MAPCs

10.4.1 Role in the Immune System

10.4.1.1 Immunomodulation

MAPCs have immunomodulatory properties that reduce inflammation by regulating immune system functions. Tregs or regulatory T cells are either thymus-derived (tTregs) or peripherally derived from naïve T cells (pTregs) and are crucial for maintaining tolerance to self-antigens and suppressing excessive immune response. Valentin-Torres et al. have demonstrated *in vitro*, that upon co-culture of peripheral blood mononuclear cells (PBMCs) with MAPCs, MAPCs release transforming growth factor (TGFb), stimulating the CD4+ T-cells to differentiate into Tregs. These Tregs further modulate the immune response through the secretion of anti-inflammatory factors, such as IL-10 and TGFb, suppressing the cytokine cascade produced by effector T-cells. Tregs also inhibit maturation of antigen presenting cells, preventing them from initiating an immune response, thus taking charge of the inflammatory environment [26]. Tregs are being explored for therapy for autoimmune diseases.

The immunomodulatory potential of MAPCs has been derived through assessment of components released by them under various local stimuli. Secretome analysis of MAPCs cultured *in vitro* have revealed the presence of immune system-related proteins, such as pentraxins and lactotransferrins, as well as immunomodulatory proteins, such as soluble intercellular adhesion molecule 1 (sICAM1), semaphorins, and galectins [27]. Pentraxins identify foreign antigens, triggering an immune response. They play a crucial role in activating classical cascade of complement leading to lysis and removal of pathogens. Lactotransferrins are known to have antimicrobial and anti-inflammatory properties. The proteins provide immediate protection against invading pathogens, while adaptive immune response is being mounted. These proteins contribute to early innate immune response by recognizing and opsonizing pathogens, activating complement, modulating inflammation, and exerting direct antimicrobial effects [28, 29].

10.4.1.2 Non-Immunogenic Properties

In-depth characterization of MAPCs has also revealed low and negative expression of major histocompatibility complex class I and II, respectively [30]. As a result of this, MAPCs are unable to trigger allogeneic T-cells in mixed lymphocyte reactions (MLR), inhibiting subsequent immune response cascade and rejection. This is significantly relevant in graft-vs-host-disease (GVHD) resulting from stem cell transplantation [31].

In an *in vitro* wound healing model, MAPCs have been shown to decrease infiltration of neutrophils, pro-inflammatory macrophages at the injury site, while promoting an anti-inflammatory, pro-resolution phenotype in macrophages. This helps balance the inflammatory response and create a more favorable environment for tissue repair [32].

10.4.2 Role in Tissue Repair

10.4.2.1 Paracrine Signaling

MAPCs secrete a wide range of bioactive factors like growth factors, cytokines, and other molecules that can activate and modulate resident cells to promote tissue repair processes. The paracrine signaling is considered a primary mechanism rather than direct differentiation and engraftment of MAPCs.

Factors secreted by MAPCs include fibroblast growth factor 2 (FGF-2), platelet-derived growth factor (PDGF), transforming growth factor (TGF-b), epidermal growth factor (EGF), keratinocyte growth factor (KGF), and hepatocyte growth factor (HGF), which play a significant role in would healing. These factors enhance the migration and proliferation of fibroblasts and keratinocytes at the site of injury.

Fibroblasts contribute to wound contraction, by formation of granulation tissue and deposition of collagen during tissue repair process. Keratinocytes facilitate re-epithelialization and wound closure. They migrate across the wound bed to form neo-epidermis [33]. The coordinated activities of these cells, driven by MAPC-secreted factors, ensure effective wound healing and tissue regeneration.

10.4.2.2 Promotion of Angiogenesis

Paracrine secretions of MAPCs also contain factors such as VEGF, FGF-2, and IL-8. These factors stimulate the migration and proliferation of endothelial cells and can enhance the formation of new blood vessels. They also secrete factors such as stromal cell-derived factor 1 (SDF-1/CXCL12), which are known to assist in the recruitment of endothelial progenitor cells (EPCs) from the bone marrow or circulation to the sites of tissue injury [34]. MAPCs, like any other stem cells, also secrete exosomes and microvesicles encapsulating pro-angiogenic factors. These vesicles may be taken up by local endothelial cells to upregulate angiogenesis.

MAPCs can accelerate wound healing by simultaneously contributing to modulation of inflammation and promotion of angiogenesis, making it a promising approach to treat chronic wounds.

10.4.2.3 ECM Production

In addition to stimulation of fibroblasts, MAPCs are capable of augmenting the production of collagen I and III, which are important extracellular matrix (ECM) components. These proteins are critical for structure and function of ECM, which provides a scaffold for cell attachment and tissue organization during repair. They also secrete matrix metalloproteinases (MMPs) and their inhibitors known as tissue inhibitors of metalloproteinases (TIMPs). These enzymes play a crucial role in regulating ECM remodeling and degradation, ensuring a balanced ECM turnover. This balance is essential for proper tissue repair as it prevents excessive breakdown or accumulation of ECM components [32]. MAPCs have also demonstrated the ability to differentiate into chondrocyte-like cells upon stimulation *in vitro*, expressing collagen II, a crucial component of ECM [35]. Thus, MAPCs play a crucial role in ECM production and regulation, making them valuable for tissue repair and regenerative medicine.

10.4.3 Pre-Clinical Studies

10.4.3.1 Traumatic Brain Injury (TBI)

MAPCs have demonstrated neuroprotective effects in pre-clinical models of TBI. In one study, administration of MAPCs after TBI within a specific time window led to improved functional recovery and reduced neuroinflammation in rats. They also found that MAPCs could modulate the inflammatory response and preserve microvascular permeability after TBI. In another spinal cord injury model in rats, injection of MAPCs resulted in suppression of MMP-9 released from macrophages, inducing proinflammatory macrophages (M1) to shift to anti-inflammatory macrophages (M2). A control in the inflammation brought about therapeutic changes, such as promotion of sensory neurite outgrowth and increased axon growth capacity [36, 37].

10.4.3.2 Wound Healing

Preclinical studies have investigated the use of MAPC-conditioned medium (MAPC-CM) for accelerating would healing. *In vitro* experiments showed that MAPC-CM improved migration, proliferation, and functionality of skin cells like keratinocytes, fibroblasts, and endothelial cells. In a murine excisional wound model, intradermal injection of MAPC-CM led to faster wound closure, increased re-epithelialization, and enhanced angiogenesis [32]. Collagen deposition was also found to be augmented, which is particularly important during wound contraction and tissue remodeling. The enhanced ECM stability supports mechanical integrity in the healing tissue, facilitating effective wound closure and long-term resilience.

10.4.3.3 Myocardial Infarction

In preclinical studies of myocardial infarction in rodents, MAPC administration has been shown to decrease left ventricular injury and have a favorable cytokine profile compared to other cellular therapies. The results also demonstrated differentiation of MAPCs into functional endothelial cells both *in vitro* and *in vivo*, stimulating superior angiogenesis [38].

10.4.3.4 Bone Repair

Various research groups have demonstrated a higher differentiation potential and stability of MAPCs, compared to MSCs. This was further supported by a group that first compared the protein release profiles of MAPCs and MSCs *in vitro*, confirming greater osteogenic and angiogenic profiles exhibited by the MAPCs. The results were further validated in a rat fibular defect model. Both cell types were mixed with demineralized bone matrix prior to implantation, and it was found that the rats with the MAPCs demonstrated a significantly higher vasculature and augmented bone healing process [39].

10.5 Application in Regenerative Medicine and Orthopedics

Characteristics defining MAPCs, their diverse differentiation and expansion potential, their mechanisms of action, and their applications have been discussed in this book chapter. Highlighting their potential specifically in musculoskeletal system, MAPCs are capable of differentiating into cells of cartilage, bone, and muscle, making them highly valuable in orthopedic regenerative medicine. There is tremendous scope for MAPCs to be utilized to fulfill or ameliorate the gaps in orthopedic regenerative medicine research. While great strides have been made in harnessing the potential of orthobiologics, such as platelet-rich plasma (PRP), bone marrow aspirate concentrate (BMAC), and various stem cells [40–46], there remains significant scope for improvement. Challenges, such as risk of immune rejection, difficulties in harvesting BM-MSCs, limited cell expansion potential, and restricted differentiation capabilities, draw attention to superior potential of MAPCs to overcome these limitations.

MAPCs exhibit several advantages that make them particularly suited for orthopedic applications. They have low immunogenicity, reducing the risk of immune rejection when using stem cell implantation as a therapy for conditions, such as osteoarthritis or rheumatoid arthritis. Additionally, MAPCs possess immunomodulatory properties that can mitigate inflammatory responses, further promoting a healing environment in orthopedic treatments. Their high proliferation rates without loss of differentiation potential are particularly beneficial for treating large defects or injuries in the musculoskeletal system. Their ability to differentiate into multiple lineages pertinent to the musculoskeletal system, including osteoblasts, chondrocytes, and myocytes, adds to their versatility in repairing and regenerating bone, cartilage, and muscle tissue. Furthermore, MAPCs contribute to tissue repair not only through direct differentiation but also by secreting bioactive molecules that stimulate endogenous repair mechanisms, enhancing angiogenesis, reducing apoptosis, and modulating the local immune response, which are critical factors in successful orthopedic therapies.

10.6 Clinical Trials

Athersys, a biotechnology company, has successfully isolated and developed a product from human MAPCs called MultiStem. MultiStem is designed for a wide range of therapeutic applications and has been the focus of numerous clinical trials. Presently, these trials target conditions, such as acute respiratory distress syndrome (ARDS), COVID-induced ARDS, leukemia, ischemic stroke, heart attacks, acute myocardial infarction, stroke treatment, and trauma-induced multiple organ failure [47–50].

In patients with acute ischemic stroke, administration of MAPCs was found to be safe and well tolerated. However, no significant improvement in neurological outcomes was observed [51]. Conversely, in a trial involving patients with acute myocardial infarction, MultiStem treatment was safe and resulted in improved ejection fraction—the percentage of blood that is pumped out of the left ventricle with each heartbeat—and stroke volume—the amount of blood pumped out of the left ventricle—for 4 months post-operation, suggesting the potential of MAPC delivery in patients with acute coronary syndrome [48]. A clinical trial tested out the effects of MAPC administration in patients immediately post-hematopoietic cell transplantation to prevent GVHD. MAPCs were well tolerated with successful engraftment and no significant toxicity. The MAPCs were administered at various doses, and the highest dose group showed a lower incidence of acute GVHD through their anti-inflammatory and immunomodulatory effects. Results suggest MAPCs have potential as a novel GVHD prophylactic therapy, warranting further investigation [50].

Overall, the outcomes of these trials have been either neutral or positive, indicating a promising future for MAPCs in various therapeutic applications.

10.7 Controversy

Despite their promising potential, the discovery of MAPCs has not been without controversy. Figures in the original 2002 paper showing MAPC surface proteins contained several errors, including duplications [52]. Furthermore, an investigation by the University of Minnesota revealed that the Verfaillie group used incorrect procedures in the 2002 paper, calling the validity of the data into question, but a Nature peer-review panel analyzed the data and concluded that the paper's conclusions remained valid. Additionally, while some studies have claimed that MAPCs can differentiate into a wide variety of cell types, other researchers have struggled to replicate these findings, leading to debates regarding the true nature and potential of MAPCs. The significant overlap in phenotypic markers, expression profiles, and functional characteristics with other stem cells, such as MSCs and HSCs, has been the source of debate regarding the true identity of MAPCs. Nevertheless, the discovery of MAPCs presents exciting possibilities for regenerative medicine, despite their unique identity and capabilities remaining the subject of scientific scrutiny and debate.

10.8 Future Directions

Future directions in MAPCs encompass a multifaceted approach aimed at advancing our understanding of these unique stem cells and harnessing their therapeutic potential. Efforts should focus on unraveling the molecular mechanisms governing MAPC behavior and differentiation, including the exploration of epigenetic regulation underlying signaling pathways. Additionally, there is a need to improve characterization techniques through the development of advanced phenotypic and transcriptomic profiling methods, as well as standardization of protocols to ensure consistency across studies. In terms of therapeutic applications, further investigation into MAPCs immunomodulatory properties holds promise for treating autoimmune diseases and inflammatory disorders. Research in tissue engineering and organ regeneration should explore their interaction with biomaterials and scaffolds. Translation into clinical practice will require conducting extensive trials to assess safety and efficacy, optimizing delivery methods, and addressing ethical and regulatory considerations. Technological advances, such as CRISPR-Cas9 gene editing, will also play a crucial role in enhancing the therapeutic potential and safety profile of MAPCs. Overall, these future directions will aim to accelerate the development and translation of MAPC-based therapies, ultimately benefiting patients in need of regenerative treatments.

References

1. Sohni A, Verfaillie CM. Multipotent adult progenitor cells. Best Pract Res Clin Haematol. 2011;24(1):3–11.
2. Jiang Y, Jahagirdar BN, Reinhardt RL, Schwartz RE, Keene CD, Ortiz-Gonzalez XR, et al. Pluripotency of mesenchymal stem cells derived from adult marrow. Nature. 2002;418(6893):41–9.
3. Jiang Y, Vaessen B, Lenvik T, Blackstad M, Reyes M, Verfaillie CM. Multipotent progenitor cells can be isolated from postnatal murine bone marrow, muscle, and brain. Exp Hematol. 2002;30(8):896–904.
4. Jacobs SA, Pinxteren J, Roobrouck VD, Luyckx A, van't Hof W, Deans R, et al. Human multipotent adult progenitor cells are nonimmunogenic and exert potent immunomodulatory effects on alloreactive T-cell responses. Cell Transplant. 2013;22(10):1915–28.
5. Ryu JC, Davidson BP, Xie A, Qi Y, Zha D, Belcik JT, et al. Molecular imaging of the paracrine proangiogenic effects of progenitor cell therapy in limb ischemia. Circulation. 2013;127(6):710–9.
6. Jiang Y, Henderson D, Blackstad M, Chen A, Miller RF, Verfaillie CM. Neuroectodermal differentiation from mouse multipotent adult progenitor cells. Proc Natl Acad Sci U S A. 2003;100 Suppl 1(Suppl 1):11854–60.
7. Lehman N, Cutrone R, Raber A, Perry R, Van't Hof W, Deans R, et al. Development of a surrogate angiogenic potency assay for clinical-grade stem cell production. Cytotherapy. 2012;14(8):994–1004.
8. Luttun A, Ross JJ, Verfaillie C, Aranguren XL, Prósper F. Differentiation of multipotent adult progenitor cells into functional endothelial and smooth muscle cells. Curr Protoc Immunol. 2006;Chapter 22:Unit 22F.9.
9. Schwartz RE, Reyes M, Koodie L, Jiang Y, Blackstad M, Lund T, et al. Multipotent adult progenitor cells from bone marrow differentiate into functional hepatocyte-like cells. J Clin Invest. 2002;109(10):1291–302.
10. Singh SP, Tripathy NK, Nityanand S. Comparison of phenotypic markers and neural differentiation potential of multipotent adult progenitor cells and mesenchymal stem cells. World J Stem Cells. 2013;5(2):53–60.
11. Kraushaar DC, Zhao K. The Epigenomics of embryonic stem cell differentiation. Int J Biol Sci. 2013;9(10):1134–44.
12. Aranda P, Agirre X, Ballestar E, Andreu EJ, Román-Gómez J, Prieto I, et al. Epigenetic signatures associated with different levels of differentiation potential in human stem cells. PLoS One. 2009;4(11):e7809.
13. Ulloa-Montoya F, Kidder BL, Pauwelyn KA, Chase LG, Luttun A, Crabbe A, et al. Comparative transcriptome analysis of embryonic and adult stem cells with extended and limited differentiation capacity. Genome Biol. 2007;8(8):R163.
14. Zhang J, Tam WL, Tong GQ, Wu Q, Chan HY, Soh BS, et al. Sall4 modulates embryonic stem cell pluripotency and early embryonic development by the transcriptional regulation of Pou5f1. Nat Cell Biol. 2006;8(10):1114–23.
15. Breyer A, Estharabadi N, Oki M, Ulloa F, Nelson-Holte M, Lien L, et al. Multipotent adult progenitor cell isolation and culture procedures. Exp Hematol. 2006;34(11):1596–601.
16. Romito A, Cobellis G. Pluripotent stem cells: current understanding and future directions. Stem Cells Int. 2016;2016:9451492.
17. Loboda A, Jozkowicz A, Dulak J. HIF-1 and HIF-2 transcription factors—similar but not identical. Mol Cells. 2010;29(5):435–42.
18. Närvä E, Pursiheimo JP, Laiho A, Rahkonen N, Emani MR, Viitala M, et al. Continuous hypoxic culturing of human embryonic stem cells enhances SSEA-3 and MYC levels. PLoS One. 2013;8(11):e78847.
19. Petruzzelli R, Christensen DR, Parry KL, Sanchez-Elsner T, Houghton FD. HIF-2α regulates NANOG expression in human embryonic stem cells following hypoxia and reoxygenation through the interaction with an Oct-sox cis regulatory element. PLoS One. 2014;9(10):e108309.
20. Zhang S, Zhao L, Wang J, Chen N, Yan J, Pan X. HIF-2α and Oct4 have synergistic effects on survival and myocardial repair of very small embryonic-like mesenchymal stem cells in infarcted hearts. Cell Death Dis. 2017;8(1):e2548.

21. Narsinh KH, Plews J, Wu JC. Comparison of human induced pluripotent and embryonic stem cells: fraternal or identical twins? Mol Ther. 2011;19(4):635–8.
22. Reyes M, Lund T, Lenvik T, Aguiar D, Koodie L, Verfaillie CM. Purification and ex vivo expansion of postnatal human marrow mesodermal progenitor cells. Blood. 2001;98(9):2615–25.
23. Roobrouck VD, Clavel C, Jacobs SA, Ulloa-Montoya F, Crippa S, Sohni A, et al. Differentiation potential of human postnatal mesenchymal stem cells, mesoangioblasts, and multipotent adult progenitor cells reflected in their transcriptome and partially influenced by the culture conditions. Stem Cells. 2011;29(5):871–82.
24. Aranguren XL, McCue JD, Hendrickx B, Zhu XH, Du F, Chen E, et al. Multipotent adult progenitor cells sustain function of ischemic limbs in mice. J Clin Invest. 2008;118(2):505–14.
25. Fischer A, Schumacher N, Maier M, Sendtner M, Gessler M. The notch target genes Hey1 and Hey2 are required for embryonic vascular development. Genes Dev. 2004;18(8):901–11.
26. Valentin-Torres A, Day C, Taggart JM, Williams N, Stubblefield SR, Roobrouck VD, et al. Multipotent adult progenitor cells induce regulatory T cells and promote their suppressive phenotype via TGFβ and monocyte-dependent mechanisms. Sci Rep. 2021;11(1):13549.
27. Burrows GG, van't Hof W, Newell LF, Reddy A, Wilmarth PA, David LL, et al. Dissection of the human multipotent adult progenitor cell secretome by proteomic analysis. Stem Cells Transl Med. 2013;2(10):745–57.
28. Du Clos TW. Pentraxins: structure, function, and role in inflammation. ISRN Inflamm. 2013;2013:379040.
29. Kowalczyk P, Kaczyńska K, Kleczkowska P, Bukowska-Ośko I, Kramkowski K, Sulejczak D. The Lactoferrin phenomenon—A miracle molecule. Molecules. 2022;27(9):2941.
30. Kovacsovics-Bankowski M, Streeter PR, Hof WJV, Deans RJ, Maziarz RT. Multipotent adult progenitor cells (MAPC) are non-immunogenic and display immunosuppressive properties on activated T cells. Biol Blood Marrow Transplant. 2006;12(2):106–7.
31. Metheny L, Eid S, Wuttisarnwattana P, Auletta JJ, Liu C, Van Dervort A, et al. Human multipotent adult progenitor cells effectively reduce graft-vs-host disease while preserving graft-vs-leukemia activity. Stem Cells. 2021;39(11):1506–19.
32. Ahangar P, Mills SJ, Smith LE, Strudwick XL, Ting AE, Vaes B, et al. Human multipotent adult progenitor cell-conditioned medium improves wound healing through modulating inflammation and angiogenesis in mice. Stem Cell Res Ther. 2020;11(1):299.
33. Gurtner GC, Werner S, Barrandon Y, Longaker MT. Wound repair and regeneration. Nature. 2008;453(7193):314–21.
34. Ho TK, Shiwen X, Abraham D, Tsui J, Baker D. Stromal-cell-derived factor-1 (SDF-1)/CXCL12 as potential target of therapeutic angiogenesis in critical leg ischaemia. Cardiol Res Pract. 2012;2012:143209.
35. Yu L, Weng Y, Shui X, Fang W, Zhang E, Pan J. Multipotent adult progenitor cells from bone marrow differentiate into chondrocyte-like cells. J Arthroplasty. 2015;30(7):1273–6.
36. Busch SA, Hamilton JA, Horn KP, Cuascut FX, Cutrone R, Lehman N, et al. Multipotent adult progenitor cells prevent macrophage-mediated axonal dieback and promote regrowth after spinal cord injury. J Neurosci. 2011;31(3):944–53.
37. Bedi SS, Aertker BM, Liao GP, Caplan HW, Bhattarai D, Mandy F, et al. Therapeutic time window of multipotent adult progenitor therapy after traumatic brain injury. J Neuroinflammation. 2018;15(1):84.
38. Pelacho B, Aranguren XL, Mazo M, Abizanda G, Gavira JJ, Clavel C, et al. Plasticity and cardiovascular applications of multipotent adult progenitor cells. Nat Rev Cardiol. 2007;4(S1):S15–20.
39. LoGuidice A, Houlihan A, Deans R. Multipotent adult progenitor cells on an allograft scaffold facilitate the bone repair process. J Tissue Eng. 2016;7:2041731416656l4.
40. Gobbi A, Dallo I, D'Ambrosi R. Autologous microfragmented adipose tissue and leukocyte-poor platelet-rich plasma combined with hyaluronic acid show comparable clinical outcomes for symptomatic early knee osteoarthritis over a two-year follow-up period: a prospective randomized clinical trial. Eur J Orthopaed Surg Traumatol. 2023;33(5):1895–904. https://doi.org/10.1007/s00590-022-03356-2.
41. Dallo I, Morales M, Gobbi A. Platelets and adipose stroma combined for the treatment of the arthritic knee. Arthrosc Tech. 2021;10(11):e2407–14. https://doi.org/10.1016/j.eats.2021.07.021.
42. Whyte GP, Herman K, Gobbi A. One-step cell-based cartilage repair in the knee using hyaluronic acid-based scaffold embedded with mesenchymal stem cells sourced from bone marrow aspirate concentrate (HA-BMAC). In: Brittberg M, Slynarski K, editors. Lower extremity joint preservation. Cham: Springer; 2021. https://doi.org/10.1007/978-3-030-57382-9_5.
43. Gobbi A, Whyte GP. Long-term clinical outcomes of one-stage cartilage repair in the knee with hyaluronic acid-based scaffold embedded with mesenchymal stem cells sourced from bone marrow aspirate concentrate. Am J Sports Med. 2019;47(7):1621–8. https://doi.org/10.1177/0363546519845362.
44. Rogers C, Gobbi A. The optimization of natural healing. In: Gobbi A, Espregueira-Mendes J, Lane J, Karahan M, editors. Bio-orthopaedics. Berlin, Heidelberg: Springer; 2017. https://doi.org/10.1007/978-3-662-54181-4_1.
45. Gobbi A, Whyte GP. Emerging orthobiologic approaches to ligament injury. In: Gobbi A, Espregueira-Mendes J, Lane J, Karahan M, editors. Bio-orthopaedics. Berlin, Heidelberg: Springer; 2017. https://doi.org/10.1007/978-3-662-54181-4_24.
46. Gobbi A, de Girolamo L, Whyte GP, Sciarretta FV. Clinical applications of adipose tissue-derived stem cells. In: Gobbi A, Espregueira-Mendes J, Lane J, Karahan M, editors. Bio-orthopaedics.

Berlin, Heidelberg: Springer; 2017. https://doi.org/10.1007/978-3-662-54181-4_44.

47. Hess DC, Sila CA, Furlan AJ, Wechsler LR, Switzer JA, Mays RW. A double-blind placebo-controlled clinical evaluation of MultiStem for the treatment of ischemic stroke. Int J Stroke. 2014;9(3):381–6.
48. Penn MS, Ellis S, Gandhi S, Greenbaum A, Hodes Z, Mendelsohn FO, et al. Adventitial delivery of an allogeneic bone marrow–derived adherent stem cell in acute myocardial infarction: phase I clinical study. Circ Res. 2012;110(2):304–11.
49. Houkin K, Osanai T, Uchiyama S, Minematsu K, Taguchi A, Maruichi K, et al. Allogeneic stem cell therapy for acute ischemic stroke: the phase 2/3 TREASURE randomized clinical trial. JAMA Neurol. 2024;81(2):154–62.
50. Maziarz RT, Devos T, Bachier CR, Goldstein SC, Leis JF, Devine SM, et al. Single and multiple dose MultiStem (multipotent adult progenitor cell) therapy prophylaxis of acute graft-versus-host disease in myeloablative allogeneic hematopoietic cell transplantation: a phase 1 trial. Biol Blood Marrow Transplant. 2015;21(4):720–8.
51. Hess DC, Wechsler LR, Clark WM, Savitz SI, Ford GA, Chiu D, et al. Safety and efficacy of multipotent adult progenitor cells in acute ischaemic stroke (MASTERS): a randomised, double-blind, placebo-controlled, phase 2 trial. Lancet Neurol. 2017;16(5):360–8.
52. Check E. Stem-cell paper corrected. Nature. 2007;447(7146):763.

Preclinical Evidence of MSC-Derived Exosomes

11

Lucienne A. Vonk and Rachel M. Frank

11.1 Introduction

While culture-expanded mesenchymal stromal/stem cells (MSCs) have found their way into clinical trials for various musculoskeletal pathologies [1], MSC-derived extracellular vesicles are emerging as their potential cell-free alternative. The evidence for engraftment and direct participation in tissue regeneration after MSC treatment is limited [2, 3], but their therapeutic effects through paracrine signaling are well established [4, 5]. Part of the paracrine signaling is through the form of extracellular vesicles (EVs). Extracellular vesicles comprise exosomes, microvesicles, and small apoptotic bodies [6]. Exosomes are small (30–150 nm) EVs that are formed in multivesicular endosomes and secreted upon the fusion of these structures with the cell membrane. Microvesicles (100–1000 nm) are directly budded of the cell membrane and apoptotic bodies (50–5000 nm) are released by apoptotic cells. All of them can fall within the size range of small EVs (up to 200 nm) that are believed to have more therapeutic effect than the larger EVs. As there are currently no markers to discriminate between them, it is recommended by the International Society for Extracellular Vesicles (ISEV) to use the term extracellular vesicle [7], although the term exosomes is sometimes used in a popular way to describe all small EVs.

Cells secrete EVs as part of their intercellular communication mechanism, and they can carry a variety of bioactive molecules (DNA fragments, various types of RNA among which also microRNAs, cytokines, lipids, enzymes, and other proteins) that can be taken up by other cells, and they have ligands on their membrane for receptor-mediated signaling. As EVs retain certain properties of their donor cells, there is a special interest in MSC-derived EVs for regenerative medicine [6]. Similar to the donor MSCs, MSC-EVs can have anti-inflammatory and immunomodulatory capacities that are helpful in creating a regenerative microenvironment, and they can orchestrate other regenerative processes such as regulating cell proliferation and survival, autophagy, and promote vascularization and extracellular matrix production. These properties also make them attractive as an off-the-shelf biological treatment suitable for various applications in sports medicine and orthopedics.

L. A. Vonk (✉)
Xintela AB, Lund, Sweden

Department of Orthopaedics, University Medical Center Utrecht, Utrecht University, Utrecht, The Netherlands
e-mail: L.A.Vonk-3@umcutrecht.nl

R. M. Frank
Department of Orthopedics, University of Colorado School of Medicine, Aurora, CO, USA
e-mail: Rachel.Frank@cuanschutz.edu

A. Gobbi et al. (eds.), *Regenerative Medicine in Sports and Orthopaedics*,
https://doi.org/10.1007/978-3-031-84693-9_11

11.2 Cartilage and Osteoarthritis

Several studies have shown the potential of MSC-derived EVs as a novel therapeutic option for cartilage damage and OA. In vitro, it has been shown that MSC-derived EVs can directly affect chondrocytes and synoviocytes as well as exert anti-inflammatory and immunomodulatory capacities. Chondrocytes can internalize EVs, at least partly mediated by the hyaluronic acid receptor CD44 [8, 9]. MSC-EVs can increase the deposition of hyaline cartilage matrix molecules and the expression of chondrogenic genes [9–12] and decrease the expression of catabolic and hypertrophic markers and proteolytic enzymes [8, 10–13]. In addition, MSC-EVs stimulated proliferation and autophagy, increased cell survival, and inhibited apoptosis of chondrocytes under inflammatory culture conditions [8–12]. Besides chondrocytes, synoviocytes are able to internalize EVs in a hyaluronic acid– CD44 mediated process [14]. MSC-derived EVs were able to decrease proliferation and facilitate apoptosis of inflammatory synoviocytes and reduce the expression of pro-inflammatory cytokines and chemokines [14–17]. Both for chondrocytes and synoviocytes, it has been suggested that the anti-inflammatory and anti-catabolic capacities of MSC-derived EVs are a result of inhibition of nuclear translocation of p65 subunit of nuclear factor kappa-light-chain-enhancer of activated B cells (NF-κB) [8, 16].

An overview of in vivo studies with EVs for cartilage repair and osteoarthritis treatment is provided in Table 11.1. A few in vivo studies have focused on using MSC-derived EVs for the treatment of (osteo)chondral defects. Most of them promoted healing and improved the quality of the repair tissue [9, 18–21].

One study augmented bone marrow stimulation with MSC-derived EVs in minipigs. Although the addition of MSC-derived EVs resulted in enhanced subchondral bone healing, osseous ingrowth impaired optimal cartilage repair resulting in more fibrocartilage repair tissue [22].

When it comes to OA, a variety of experimental models have been used. EVs from bone marrow MSCs, adipose MSCs, synovium MSCs, and induced pluripotent cell-derived MSCs have shown to protect against OA in collagenase-induced OA in mice [12, 23–25].

Injection of MSC-EVs also alleviated OA in the surgical destabilization of the medial meniscus (DMM)-induced OA model in mice [10, 26–28].

Besides mice, both chemically and surgically induced OA models in rats have been used. Injection of amniotic fluid stem cell-derived EVs in monoiodoacetate (MIA)-induced OA rats resulted in a pain tolerance level comparable to healthy rats and improved histological joint scores [29]. Another study showed the capacity of both rabbit bone marrow MSC-derived EVs and EVs derived from cocultures of rabbit articular chondrocytes and MSCs, especially with higher ratios of chondrocytes, to improve OA as assessed by radiography and histology [30]. Injection of bone marrow MSC-derived EVs resulted in pain relief and improved articular cartilage histology in a sodium iodoacetate-induced OA rat model [31].

Using an anterior cruciate ligament transection (ACLT) model, therapeutic effects of human umbilical cord-derived MSC EVs [32, 33], human Wharton's jelly MSC-derived EVs [34], and human hypoxia cultured adipose MSC-derived EVs [35] have been shown. These treatments alleviated cartilage damage, improved subchondral bone repair, and suppressed inflammation. The study by Cao et al. traced the EVs and that a small part of the (non-modified) EVs were retained in the joint, while most were traced back in the kidneys, suggesting renal clearance. With injection of infrapatellar MSC-derived EVs in the rat ACLT model, it was noted that it alleviated pain, inflammation, and osteophyte formation, and it could delay but not fully suppress cartilage degradation and OA progression [36].

Liang et al. showed that injection of synovial fluid MSC-derived EVs could alleviate both OA changes in joints, as well as change systemic levels of serum cytokines in an ACLT-DMM rat model [37]. Two independent studies reported prevention of OA by injecting EVs derived from miR-140 overexpressing MSCs

Table 11.1 Overview of extracellular vesicles (EV) source, experimental model, results, and molecular mechanism of in vivo studies for cartilage repair and osteoarthritis treatment

Source of EVs	Experimental model	Results	Molecular mechanism	Ref
H BM MSCs	Rabbit articular cartilage defect	Improved repair of cartilage defects	ns	[9]
H IFP MSCs	DMM mice	Amelioration of OA severity	miRNA-100-5p-regulated inhibition of mTOR-autophagy	[10]
M BM MSCs	Collagenase-induced OA in mice	Protection against joint damage	ns	[12]
H IFP MSCs	Acute synovitis/IFP fibrosis in rats	Reduction of proliferation and inflammatory mediators, macrophage polarization to M2 phenotype	ns	[15]
H ESC MSCs	Osteochondral defects in rabbit knees	Improved cartilage repair and mechanical properties	ns	[18]
H BM MSCs	Chondral defects in rabbits	Improved healing	ns	[19]
H ESC MSCs	Osteochondral defects in rats	Improved repair of the defects	ns	[20]
H ESC MSCs	Osteochondral defects in micropigs	Improved functional cartilage and subchondral bone repair and biomechanical properties	ns	[21]
H ESC MSCs	Chondral defects in minipigs, combined with BM stimulation	Improved subchondral bone healing, osseous ingrowth that impaired cartilage repair	ns	[22]
M BM MSCs	Delayed-type hypersensitivity and collagen-induced arthritis in mice	Suppression of inflammation	ns	[23]
H AMSCs	Collagenase-induced OA in mice	Protection against joint damage	ns	[24]
H iPS and S MSCs	Collagenase-induced OA in mice	Both attenuated OA, but iPS-MSC-EVs had a stronger effect	ns	[25]
H ESC MSCs	DMM mice	Protection against cartilage destruction	ns	[26]
H S MSCs	DMM mice	Prevention of OA development	Let-7b involvement	[27]
M chondrogenic progenitor cells	DMM mice	Attenuation of OA, stronger effect by EVs from the MRL/MpJ-derived cells	miRNA 221-3p mediated proliferation and migration	[28]
Amniotic fluid SCs	Monoiodoacatate induced OA in rats	Enhanced pain tolerance and improved histology	Positive correlation with TGFβ content	[29]
Rabbit BM MSC in coculture with chondrocytes	Monoiodoacatate induced OA in rats	Recovery of OA, especially by EVs from cocultures with higher chondrocyte to MSC ratio	ns	[30]
Rat BM MSCs	Sodium idoacetate-induced OA in rats	Alleviation of pain and improved ECM synthesis	ns	[31]
H UC MSCs	ACLT in rats	Protection against cartilage damage	PI3K-AKT and Rap1 signalling	[32]
H UC MSCs	ACLT in rats	Protection against OA changes	P53 signalling	[33]
H WJ MSCs	ACLT in rats	Promotion of chondrogenesis	Let-7e-5p mediated regulation of STAT3 and IGF1R	[34]

(continued)

Table 11.1 (continued)

Source of EVs	Experimental model	Results	Molecular mechanism	Ref
H A MSCs	ACLT in rats	Suppression of OA progression	ns	[35]
H IFP MSCs	ACLT in mice	Alleviation of pain, suppression of cartilage degeneration, osteophyte formation, and synovial inflammation	ns	[36]
Rat SF MSCs	ACLT and DMM in rats	Less cartilage damage and catabolic enzyme expression, lower pro-inflammatory cytokines in serum	ns	[37]
H U MSCs	ACLT and DMM in rats	Enhancement of cartilage regeneration and subchondral bone remodelling	ns	[38]
H miR-140-5p overexpressing S MSCs	ACLT and DMM in rats	Prevention of OA	ns	[39]
H BM MSCs	High-fat diet combined with grooving cartilage surface in rats	No significant effects	ns	[40]

A adipose, *ACLT* anterior cruciate ligament transection, *BM* bone marrow, *DMM* destabilization of the medial meniscus, *ESC* embryonic stem cell, *H* human, *IFP* infrapatellar fat pad, *iPS* induced pluripotent stem cell, *M* mouse, *MSC* mesenchymal stromal/stem cell, *ns* not specified, *OA* osteoarthritis, *Ref* reference, *S* synovium, *SC* stem cells, *SF* synovial fluid, *U* urine, *UC* umbilical cord, *WJ* Wharton's jelly

(urine and synovial MSCs) in the ACLT-DMM rat model [38, 39].

One in vivo rat study used a metabolic high-fat diet combined with articular cartilage groove mild OA model. In this study, no significant therapeutic effect of bone marrow MSC-derived EVs was shown, while the injection with the parent MSCs aggravated OA [40].

11.3 Ligaments and Tendon

EVs have been exploited for ligament and tendon repair. Uptake of EVs by tenocytes and tendon stem cells has been shown, resulting in increased proliferative, migratory, anti-inflammatory, and anti-apoptotic effects in vitro [41–47]. EVs derived from inflammatory-licensed mouse adipose MSCs promoted proliferation and type I collagen production of tendon cells in vitro [48, 49]. When injected in a mouse Achilles tendon injury model, less inflammation was observed leading to less scar formation, faster functional recovery, and lower incidence of gaps and rupture. Similar results were reported using rat Achilles tendon injury and rupture models [41, 45–47, 50, 51], but only when the EVs were derived from relative early passage and non-senescent MSCs [52]. MSC-EVs have also been shown to be effective in reducing fibrosis and adhesion [53, 54] and tendinopathy [55–57] in rat Achilles tendon models.

MSC-EVs have also been shown to be beneficial in tendon-to-bone integration after ligament reconstruction. MSC-EVs in a hydrogel delivered in a mouse tendon-bone reconstruction model decreased inflammation, increased cell proliferation, and promoted the formation of fibrocartilage, resulting in improved biomechanical properties [58]. In rat anterior cruciate ligament reconstruction (ACLR) models, MSC-EVs stimulated osseous ingrowth into a tendon graft, prevented peri-tunnel bone loss, increased fibrocartilage at the tendon-bone tunnel interface, and accelerated tendon-bone healing and intra-articular graft remodelling [44, 59–61].

To show therapeutic effects of MSC-EVs for the treatment of rotator cuffs, both small and larger animal models have been used. In a murine rotator cuff injury model, MSC-EVs remained present at the injury site for at least 1 week after injection, attracting autologous MSCs [62]. Furthermore, MSC-EVs treatments resulted in better tissue structure and biomechanical properties [42, 63]. Injection of human adipose MSCs-EVs in a rat model prevented muscle atrophy and improved muscle regeneration and biomechanical properties in torn rotator cuff muscles [64]. Co-treatment of adipose MSC-EVs with glucocorticoids showed a synergistic anti-inflammatory effect in a chronic rotator cuff injury model in rats, while the EVs showed a protective effect against the negative effects of glucocorticoids such as cell senescence, apoptosis, transcription of reactive oxygen species, and proteolytic enzymes [65]. In rotator cuff reconstruction models, MSC-EVs improved the tendon-bone interface, decreased fatty tissue infiltration, and improved the biomechanical properties [66–70].

In an ovine model of acute infraspinatus tendon injury, it was shown that application of a type I collagen scaffold with human umbilical cord MSC-derived EVs improved both tendon and tendon-to-bone repair, dampened inflammation, and attenuated osteophyte formation [71].

In rat quadriceps tendinopathy models, treatment with iPSC-MSC derived EVs alleviated pain and inflammation, improved histology, and increased proliferation of tenocytes [72–74].

Therapeutic effects of bone marrow MSC-EVs, tendon stem cell-EVs, and adipose MSC-EVs were demonstrated in rat patellar defect models. Less inflammation and improved structure and biomechanical properties of the repaired tendons, as well as increased tendon-related gene expression and induced proliferation of local tendon stem cells, were shown [41, 43, 75]. More details on in vivo studies with EVs for tendon and ligament repair is provided in Table 11.2.

Table 11.2 Overview of extracellular vesicles (EV) source, experimental model, results, and molecular mechanism of in vivo studies for tendon and ligament repair

Source of EVs	Experimental model	Results	Molecular mechanism	Ref
Rat T SCs	Rat Achilles tendon injury	Promotion of high-quality healing of tendon	ns	[41]
H A MSCs	Rat rotator cuff injury	Promotion of rotator cuff repair	ns	[42]
Rat T SCs	Patellar tendon defect in rats	Facilitation of tendon repair	miR-144-3p-regulated proliferation and migration	[43]
Rat BM MSCs	Rat ACL reconstruction	Accelerated grafted tendon-bone tunnel integration, especially by hypoxic cultured MSC-EVs	ns	[44]
Rabbit A MSCs	Rabbit Achilles tendon repair	Improved tendon healing	ns	[45]
Rat BM MSCs	Rat Achilles tendon injury	Acceleration of the remodelling stage and better restoration of tendon architecture	ns	[46]
H UC MSCs	Rat Achilles tendon injury	Reduction of tendon injury	miR-27b-3p-mediated suppression of ARHGAP5, resulting in RhoA activation, leading to increased cell proliferation and invasion of primary injured tenocytes	[47]
M A MSCs, inflammatory primed	Mouse Achilles tendon injury and repair	Enhancement of tendon repair by attenuating inflammation and promoting intrinsic healing	miR-147-3p-mediated blocking of toll-like receptor 4/NF-κB signalling, thereby suppressing M1 macrophage activation	[48]

(continued)

Table 11.2 (continued)

Source of EVs	Experimental model	Results	Molecular mechanism	Ref
M A MSCs, control and inflammatory primed	Mouse Achilles tendon injury and repair	Reduction of post-repair tendon gap formation and rupture and facilitated collagen formation at the injury site. Attenuation of the early tendon inflammatory response	Blocking macrophage NF-κB activity	[49]
H A MSCs, control and bioglass activated	Rat Achilles tendon rupture and repair	Promotion of tenogenesis, reduction of morphological changes, improved biomechanical properties. Positive correlation between M2/M1 macrophage ratio	ns	[50]
H UC MSCs	Rat Achilles tendon injury	Beneficial effects on tendon regeneration	Involvement of TGF-β1 and mTOR signalling	[51]
H BM MSCs	Mouse Achilles tendon rupture	Promotion of Achilles tendon healing	ns	[52]
H UC MSCs	Rat Achilles tendon injury adhesion	Relief of tendon adhesion	miR-21a-3p mediated suppression of p65 activation	[53]
H UC MSCs	Rat Achilles tendon injury	Inhibition of adhesion	Provoking ER stress in fibroblasts	[54]
T SCs	Rat Achilles tendon tendinopathy	Suppression of inflammation and facilitation of healing	ns	[55]
Rat T SCs	Collagenase-induced Achilles tendon tendinopathy in rats	Improved tendon injury healing	ns	[56]
Rat A MSCs	Collagenase-induced Achilles tendon tendinopathy in rats	Beneficial effects on tendon repair	ns	[57]
M BM MSCs	Mouse tendon-bone reconstruction	Promotion of fibrocartilage formation by increasing M2 macrophage polarization, leading to improved biomechanical properties	ns	[58]
H BM MSCs, either or not magnetically actuated	Rat anterior cruciate ligament reconstruction	Prevention of peri-tunnel bone loss, more osseous ingrowth, increases fibrocartilage at tendon-bone interface and higher maximum load to failure, especially by magnetically actuated MSCs	miR-21-5p and SMAD7 signaling	[59]
Rat BM MSCs	Rat anterior cruciate ligament reconstruction	Promotion of macrophage polarization from M1 to M2, reduction of early inflammation and promotion of ACL healing	miR-23a-3p induced macrophage polarization	[60]
Rat IFP MSCs	Rat anterior cruciate ligament reconstruction	Acceleration of tendon-bone healing and graft remodeling, immunomodulation by macrophage polarization to M2	ns	[61]
M A and BM MSCs	Mice rotator cuff injury	Acceleration of bone-tendon injury healing, promotion of osteogenesis and chondrogenesis	ns	[62]
M BM MSCs	Mice chronic overuse tendinopathy of supraspinatus tendon	Promotion of macrophage polarization from M1 to M2, decreased pathological changes and higher maximum failure load	ns	[63]

(continued)

Table 11.2 (continued)

Source of EVs	Experimental model	Results	Molecular mechanism	Ref
H A MSCs	Rat massive rotator cuff tear	Decrease in muscle atrophy and degeneration, improvement in muscle regeneration and biomechanical properties	ns	[64]
A MSCs concomitant with glucocorticoids (GC)	Rat rotator cuff injury	Anti-inflammation, overriding detrimental effects of GC on rotator cuff tendons	ns	[65]
Rat BM MSCs, young and aged	Rat rotator cuff tear	Accelerated healing, increased ECM remodeling, osteogenesis, angiogenesis and stemness of T SCs	ns	[66]
Rat BM MSCs	Rat rotator cuff reconstruction	Promotion of tendon-bone healing by stimulating angiogenesis and inhibiting inflammation	ns	[67]
H A MSCs	Rabbit rotator cuff tear	Inhibition of fatty infiltration, promotion of tendon-bone healing and improvement of biomechanical properties	ns	[68]
A MSCs	Rabbit partial thickness rotator cuff tear	Improved healing and biomechanical properties	ns	[69]
Rabbit BM MSCs	Rabbit chronic rotator cuff tear	Improved repair and biomechanical properties, decreased degeneration	ns	[70]
H UC MSCs	Sheep acute infraspinatus tendon injury	Promotion of tendon and enthesis healing	ns	[71]
H iPS MSCs	Rat quadriceps tendon tendinopathy	Relieve of pain through inhibition of mast cell activation	HIF-1 signaling	[72]
H iPS MSCs	Rat quadriceps tendon tendinopathy	Promotion of macrophage polarization from M1 to M2, pain mitigation	DUSP2 and DUSP3 mediated regulation of p38 MAPK pathway	[73]
H iPS MSCs	Rat quadriceps tendon tendinopathy	Relieve of pain, increase tenocyte proliferation, decrease in pro-inflammatory cytokines	ns	[74]
Rat BM MSCs	Rat patellar defect model	Beneficial effects in tendon regeneration, improved biomechanical properties, promoted proliferation of T SCs	ns	[75]

A adipose, *ACL* anterior cruciate ligament, *BM* bone marrow, *H* human, *IFP* infrapatellar fat pad, *iPS* induced pluripotent stem cell, *M* mouse, *MSC* mesenchymal stromal/stem cell, *ns* not specified, *Ref* reference, *SC* stem cells, *T* tendon, *UC* umbilical cord

11.4 Bone and Osteoporosis and Osteosarcoma

An overview of in vivo studies with EVs for bone repair and treatment of osteoporosis is provided in Table 11.3. For the treatment of large bone defects, EVs have been loaded in several kinds of bioactive scaffolds and implants. Several studies showed enhanced bone regeneration after implantation of MSC-derived EV-loaded hydrogels in rat calvarial critical sized defects [76–79]. The promotion of bone regeneration was also observed upon the implantation of MSC-EV–loaded ceramics [80–82] and polymers [83, 84] in rat critical sized calvarial defects, in a rat femoral drilling [85] model, and in a rat distal femur defect [86]. Regarding metal implants, an induced osteogenic effect has been demonstrated after the implantation of titanium alloy scaffolds with MSC-EVs in rat radial defects [87].

Table 11.3 Overview of extracellular vesicles (EV) source, experimental model, results, and molecular mechanism of in vivo studies for bone repair and osteoporosis

Source of EVs	Experimental model	Results	Molecular mechanism	Ref
H BM MSCs	Rat calvarial bone defect	More bone formation	miR-196a	[76]
H MSCs overexpressing BMP2	Rat calvarial bone defect	Promotion of bone repair	ns	[77]
H A MSCs overexpressing miR-375	Rat calvarial bone defect	Enhanced bone regeneration	miR-375 mediated downregulation of IGFBP3	[78]
H UC MSCs	Rat calvarial bone defect	Enhanced bone regeneration	ns	[79]
Rat BM and A MSCs	Rat cranial bone defect	Enhanced bone regeneration	miRNA/Acvr2b/Acvr1/Bmpr2/Smad1/5/9 phosphorylation	[80]
H iPS MSCs	Rat calvarial bone defect	Enhanced osteogenesis	PI3K/Akt signaling	[81]
H iPS MSCs	Calvarial bone defects in ovariectomized rats	Enhanced bone regeneration by enhancing angiogenesis and osteogenesis	ns	[82]
H A MSCs	Mouse calvarial bone defect	Enhanced bone regeneration	ns	[83]
H DP SC	Mouse calvarial bone defect	Accelerated bone healing	ns	[84]
Rat BM MSCs	Rat femoral drilling	Enhanced bone formation and osseointegration, promotion of macrophage polarization from M1 to M2	NF-κB modulation of macrophages	[85]
Rat BM MSCs	Rat distal femur defect	Accelerated bone regeneration and promotion of angiogenesis	miR-146a mediated inhibition of Smad4 and NF2	[86]
H MSCs, osteogenically pre-differentiated	Rat radial bone defect	Bone regeneration and stimulation of osteogenic differentiation	miRNA mediated activation of PI3K/Akt and MAPK signalling	[87]
WJ MSCs	Ovariectomized mice	Increase in osteoblastic activity and improved bone structural parameters	miRNA regulated PI3K/Akt signalling	[89]
M BM MSCs	Ovariectomized mice	Increased bone density, improved bone damage recovery, decreased bone resorption markers and number of osteoclasts	miR-27a /DKK2/Wnt/β-catenin signaling pathway	[90]
A MSCs	Ovariectomized mice	Attenuation of bone loss, inhibition of osteoclast differentiation and promoted and increased BM MSC migration	Osteoprotegerin, miR-21-5p and let-7b-5p mediated inhibition of osteoclast differentiation	[91]
M MSCs	Ovariectomized rats	Protection against osteoporosis and bone targeting when coupled to alendronate	ns	[92]
H BM MSCs	Ovariectomized rats	Promotion of osteogenesis	miR-186 regulation of Hippo signaling	[93]

(continued)

Table 11.3 (continued)

Source of EVs	Experimental model	Results	Molecular mechanism	Ref
Rat A MSCs	Diabetic osteoporosis in rats	Reduction of bone resorption and decreased inflammation	Suppression of NLRP3 inflammasome activation in osteoclasts	[94]
Rat A MSCs	Glucocorticoid-induced osteoporosis in rats	Restoration of bone mass and reduction of oxidative stress	Regulation of Nrf2/HO-1	[95]
H UC MSCs	Disuse osteoporosis in rats	Amelioration of osteoporotic changes	miR-1263/Mob1/Hippo signalling	[96]

A adipose, *BM* bone marrow, *BMP* bone morphogenetic protein, *DP* dental pulp, *H* human, *iPS* induced pluripotent stem cell, *M* mouse, *MSC* mesenchymal stromal/stem cell, *ns* not specified, *SC* stem cells, *UC* umbilical cord, *WJ* Wharton's jelly

Moreover, MSC-EVs have been suggested as a possible treatment for osteoporosis. In vitro, MSC derived-EVs can stimulate the proliferation and bone-forming activities of osteoblasts. In addition, they can stimulate osteogenic differentiation of MSCs, which could lead to a healthier bone metabolism in case of pathology [88]. In vivo therapeutic efficacy of MSC-EVs has been shown in ovariectomized mice [89–91] and rats [92, 93], diabetic osteoporotic rats [94], glucocorticoid-induced osteoporotic rats [95], and disuse osteoporotic rats [96]. Although many positive findings have been reported on the use of MSC-derived EVs for the treatment of osteoporosis, it must be noted that these were found with EVs from non-osteoporotic parent MSCs. EVs from bone marrow MSCs of bone-losing post-menopausal mice carry a profile that could lead to an aggravation of osteoporosis [97].

A few studies have focused on the effects of MSC-EVs on osteosarcoma. Most of these studies suggested that MSC-EVs can promote osteosarcoma cell invasion, migration, metastasis, and protect against their apoptosis [98–101].

11.5 Mechanisms of Action

MSC-EVs stimulate cell proliferation, differentiation, and matrix deposition. This has been shown for chondrocytes [8–10, 12], tenocytes, and tendon stem cells [41–47], and osteogenically differentiating MSCs and osteoblasts [76, 78, 88]. Many factors have been suggested to play a role in this, such as the transfer of growth factors, microRNA-mediated effects, and mitochondrial transfer [6, 102].

Besides the regenerative capacities, they have anti-inflammatory and immunomodulatory abilities. Many studies have reported a decrease in the expression of pro-inflammatory cytokines and an increased expression of anti-inflammatory cytokines [8, 12, 16, 35, 41, 48, 103].

One of the main suggested immunomodulatory roles of MSC-EVs is their ability to switch macrophages from the pro-inflammatory M1 subtype to the anti-inflammatory M2 subtype. The pro-inflammatory M1 subtype is responsible for the production of pro-inflammatory cytokines that aggravate tissue degradation and worsen pain, whereas an increase in the anti-inflammatory M2 subtype is associated with symptom relief and can protect against tissue degradation, synovitis, and osteophyte formation. M1 macrophages that are present in the synovial tissue play a pivotal role in the progression of and pain related to OA [104]. An increased presence of M1 macrophages is also associated with a fibrotic response and tenocyte apoptosis [41, 105]. Especially in the tendon-related studies, regulating macrophage polarization has been connected to beneficial outcomes [50, 58, 60, 63, 67, 73].

11.6 Conclusions

The majority of the studies investigating therapeutic effects of MSC-EVs for orthopedic and sports medicine applications have found promising results. Most of the available data has been generated in vitro and in small animal studies. So far, two studies have reported on biodistribution after injection of MSC-EVs, but no dedicated safety-related biodistribution or tumorigenesis data have been published yet. Only a handful of studies have been performed in larger animals, such as minipigs and sheep. Similar to the development of cell therapies with expanded cells, GMP manufacturing and process development, including quality control and product release assays for identity, purity, and potency, need to catch up with the exponential growth of early preclinical studies. As with most biological products, time will tell us if these promising results in experimental models can be translated to the clinical situation where these pathologies will be present for a longer period before MSC-EV treatment will be sought.

References

1. Copp G, Robb KP, Viswanathan S. Culture-expanded mesenchymal stromal cell therapy: does it work in knee osteoarthritis? A pathway to clinical success. Cell Mol Immunol. 2023;20:626–50.
2. Andersen C, Jacobsen S, Uvebrant K, Griffin JF, Vonk LA, Walters M, et al. Integrin α10β1-selected mesenchymal stem cells reduce pain and cartilage degradation and increase immunomodulation in an equine osteoarthritis model. Cartilage. 2023. Advance online publication. https://doi:10.1177/19476035231209402.
3. Andersen C, Uvebrant K, Mori Y, Aarsvold S, Jacobsen S, Berg LC, et al. Human integrin α10β1-selected mesenchymal stem cells home to cartilage defects in the rabbit knee and assume a chondrocyte-like phenotype. Stem Cell Res Ther. 2022;13(1):206.
4. Caplan AI, Dennis JE. Mesenchymal stem cells as trophic mediators. J Cell Biochem. 2006;98(5):1076–84.
5. de Windt TS, Vonk LA, Slaper-Cortenbach ICM, Nizak R, van Rijen MHP, Saris DBF. Allogeneic MSCs and recycled autologous chondrons mixed in a one-stage cartilage cell transplantion: a first-in-man trial in 35 patients. Stem Cells. 2017;35(8):1984–93.
6. Varderidou-Minasian S, Lorenowicz MJ. Mesenchymal stromal/stem cell-derived extracellular vesicles in tissue repair: challenges and opportunities. Theranostics. 2020;10(13):5979–97.
7. Witwer KW, Théry C. Extracellular vesicles or exosomes? On primacy, precision, and popularity influencing a choice of nomenclature. J Extracell Vesicles. 2019;8(1):1648167. https://doi.org/10.1080/20013078.2019.1648167.
8. Vonk LA, van Dooremalen SFJ, Liv N, Klumperman J, Coffer PJ, Saris DBF, et al. Mesenchymal stromal/stem cell-derived extracellular vesicles promote human cartilage regeneration in vitro. Theranostics. 2018;8(4):906–20.
9. Xiang C, Yang K, Liang Z, Wan Y, Cheng Y, Ma D, et al. Sphingosine-1-phosphate mediates the therapeutic effects of bone marrow mesenchymal stem cell-derived microvesicles on articular cartilage defect. Transl Res. 2018;193:42–53.
10. Wu J, Kuang L, Chen C, Yang J, Zeng WN, Li T, et al. miR-100-5p-abundant exosomes derived from infrapatellar fat pad MSCs protect articular cartilage and ameliorate gait abnormalities via inhibition of mTOR in osteoarthritis. Biomaterials. 2019;206:87–100.
11. Liu Y, Lin L, Zou R, Wen C, Wang Z, Lin F. MSC-derived exosomes promote proliferation and inhibit apoptosis of chondrocytes via lncRNA-KLF3-AS1/miR-206/GIT1 axis in osteoarthritis. Cell Cycle. 2018;17(21–22):2411–22.
12. Cosenza S, Ruiz M, Toupet K, Jorgensen C, Noël D. Mesenchymal stem cells derived exosomes and microparticles protect cartilage and bone from degradation in osteoarthritis. Sci Rep. 2017;7(1):1–12.
13. Tofiño-Vian M, Guillén MI, Pérez Del Caz MD, Silvestre A, Alcaraz MJ. Microvesicles from human adipose tissue-derived mesenchymal stem cells as a new protective strategy in osteoarthritic chondrocytes. Cell Physiol Biochem. 2018;47(1):11–25.
14. Ragni E, Perucca Orfei C, De Luca P, Lugano G, Viganò M, Colombini A, et al. Interaction with hyaluronan matrix and miRNA cargo as contributors for in vitro potential of mesenchymal stem cell-derived extracellular vesicles in a model of human osteoarthritic synoviocytes. Stem Cell Res Ther. 2019;10(1):109.
15. Kouroupis D, Kaplan LD, Best TM. Human infrapatellar fat pad mesenchymal stem cells show immunomodulatory exosomal signatures. Sci Rep. 2022;12(1):3609.
16. Cavallo C, Merli G, Borzì RM, Zini N, D'Adamo S, Guescini M, et al. Small extracellular vesicles from adipose derived stromal cells significantly attenuate in vitro the NF-κB dependent inflammatory/catabolic environment of osteoarthritis. Sci Rep. 2021;11(1):1053.
17. Zhang Y, Tu B, Sha Q, Qian J. Bone marrow mesenchymal stem cells-derived exosomes suppress miRNA-5189-3p to increase fibroblast-like synoviocyte apoptosis via the BATF2/JAK2/STAT3 signaling pathway. Bioengineered. 2022;13(3):6767–80.

18. Wong KL, Zhang S, Wang M, Ren X, Afizah H, Lai RC, et al. Intra-articular injections of mesenchymal stem cell exosomes and hyaluronic acid improve structural and mechanical properties of repaired cartilage in a rabbit model. Arthroscopy. 2020;36(8):2215–2228.e2.
19. Yang H, Cong M, Huang W, Chen J, Zhang M, Gu X, et al. The effect of human bone marrow mesenchymal stem cell-derived exosomes on cartilage repair in rabbits. Stem Cells Int. 2022;2022:1.
20. Zhang S, Chu WC, Lai RC, Lim SK, Hui JHP, Toh WS. Exosomes derived from human embryonic mesenchymal stem cells promote osteochondral regeneration. Osteoarthr Cartil. 2016;24(12):2135–40.
21. Zhang S, Wong KL, Ren X, Teo KYW, Afizah H, Choo ABH, et al. Mesenchymal stem cell exosomes promote functional osteochondral repair in a clinically relevant porcine model. Am J Sports Med. 2022;50(3):788–800.
22. Hede KTC, Christensen BB, Olesen ML, Thomsen JS, Foldager CB, Toh WS, et al. Mesenchymal stem cell extracellular vesicles as adjuvant to bone marrow stimulation in chondral defect repair in a Minipig model. Cartilage. 2021;13(2_suppl):254S–66S.
23. Cosenza S, Toupet K, Maumus M, Luz-Crawford P, Blanc-Brude O, Jorgensen C, et al. Mesenchymal stem cells-derived exosomes are more immunosuppressive than microparticles in inflammatory arthritis. Theranostics. 2018;8(5):1399–410.
24. Hanai H, Hart DA, Jacob G, Shimomura K, Ando W, Yoshioka Y, et al. Small extracellular vesicles derived from human adipose-derived mesenchymal stromal cells cultured in a new chemically-defined contaminate-free media exhibit enhanced biological and therapeutic effects on human chondrocytes in vitro and in a mouse osteoarthritis model. J Extracell Vesicles. 2023;12(7):e12337.
25. Zhu Y, Wang Y, Zhao B, Niu X, Hu B, Li Q, et al. Comparison of exosomes secreted by induced pluripotent stem cell-derived mesenchymal stem cells and synovial membrane-derived mesenchymal stem cells for the treatment of osteoarthritis. Stem Cell Res Ther. 2017;8(1):64.
26. Wang Y, Yu D, Liu Z, Zhou F, Dai J, Wu B, et al. Exosomes from embryonic mesenchymal stem cells alleviate osteoarthritis through balancing synthesis and degradation of cartilage extracellular matrix. Stem Cell Res Ther. 2017;8(1):189.
27. Duan A, Shen K, Li B, Li C, Zhou H, Kong R, et al. Extracellular vesicles derived from LPS-preconditioned human synovial mesenchymal stem cells inhibit extracellular matrix degradation and prevent osteoarthritis of the knee in a mouse model. Stem Cell Res Ther. 2021;12(1):427.
28. Wang R, Jiang W, Zhang L, Xie S, Zhang S, Yuan S, et al. Intra-articular delivery of extracellular vesicles secreted by chondrogenic progenitor cells from MRL/MpJ superhealer mice enhances articular cartilage repair in a mouse injury model. Stem Cell Res Ther. 2020;11(1):93.
29. Zavatti M, Beretti F, Casciaro F, Bertucci E, Maraldi T. Comparison of the therapeutic effect of amniotic fluid stem cells and their exosomes on monoiodoacetate-induced animal model of osteoarthritis. Biofactors. 2020;46(1):106–17.
30. Hosseinzadeh M, Kamali A, Baghaban Eslaminejad M, Hosseini S. Higher ratios of chondrocyte to mesenchymal stem cells elevate the therapeutic effects of extracellular vesicles harvested from chondrocyte/mesenchymal stem cell co-culture on osteoarthritis in a rat model. Cell Tissue Res. 2023;394(1):145–62.
31. He L, He T, Xing J, Zhou Q, Fan L, Liu C, et al. Bone marrow mesenchymal stem cell-derived exosomes protect cartilage damage and relieve knee osteoarthritis pain in a rat model of osteoarthritis. Stem Cell Res Ther. 2020;11(1):276.
32. Tang S, Chen P, Zhang H, Weng H, Fang Z, Chen C, et al. Comparison of curative effect of human umbilical cord-derived mesenchymal stem cells and their small extracellular vesicles in treating osteoarthritis. Int J Nanomedicine. 2021;16:8185–202.
33. Cao H, Chen M, Cui X, Liu Y, Liu Y, Deng S, et al. Cell-free osteoarthritis treatment with sustained-release of chondrocyte-targeting exosomes from umbilical cord-derived mesenchymal stem cells to rejuvenate aging chondrocytes. ACS Nano. 2023;17(14):13358–76.
34. Chen P, Tang S, Gao H, Zhang H, Chen C, Fang Z, et al. Wharton's jelly mesenchymal stem cell-derived small extracellular vesicles as natural nanoparticles to attenuate cartilage injury via microRNA regulation. Int J Pharm. 2022;623:121952.
35. Chang LH, Wu SC, Chen CH, Chen JW, Huang WC, Wu CW, et al. Exosomes derived from hypoxia-cultured human adipose stem cells alleviate articular chondrocyte inflammaging and post-traumatic osteoarthritis progression. Int J Mol Sci. 2023;24(17):13414.
36. Liu Q, Wu J, Wang H, Jia Z, Li G. Human infrapatellar fat pad mesenchymal stem cell–derived extracellular vesicles purified by anion exchange chromatography suppress osteoarthritis progression in a mouse model. Clin Orthop Relat Res. 2024;482:1246.
37. Liang H, Li D, Neufeld EV, Sayegh MJ, Kiridly A, Palacios P, et al. Extracellular vesicles from synovial fluid-derived mesenchymal stem cells confer chondroprotective effects on in vitro and in vivo osteoarthritic chondrocytes. J Cartilage Joint Preserv. 2023;3(4):100146.
38. Liu Y, Zeng Y, Si HB, Tang L, Xie HQ, Shen B. Exosomes derived from human urine–derived stem cells overexpressing miR-140-5p alleviate knee osteoarthritis through downregulation of VEGFA in a rat model. Am J Sports Med. 2022;50(4):1088–105.
39. Tao SC, Yuan T, Zhang YL, Yin WJ, Guo SC, Zhang CQ. Exosomes derived from miR-140-5p-overexpressing human synovial mesenchymal stem cells enhance cartilage tissue regeneration and

prevent osteoarthritis of the knee in a rat model. Theranostics. 2017;7(1):180–95.
40. Warmink K, Rios JL, Varderidou-Minasian S, Torres-Torrillas M, van Valkengoed DR, Versteeg S, et al. Mesenchymal stem/stromal cells-derived extracellular vesicles as a potentially more beneficial therapeutic strategy than MSC-based treatment in a mild metabolic osteoarthritis model. Stem Cell Res Ther. 2023;14(1):137.
41. Zhang M, Liu H, Cui Q, Han P, Yang S, Shi M, et al. Tendon stem cell-derived exosomes regulate inflammation and promote the high-quality healing of injured tendon. Stem Cell Res Ther. 2020;11(1):402.
42. Fu G, Lu L, Pan Z, Fan A, Yin F. Adipose-derived stem cell exosomes facilitate rotator cuff repair by mediating tendon-derived stem cells. Regen Med. 2021;16(4):359–72.
43. Song K, Jiang T, Pan P, Yao Y, Jiang Q. Exosomes from tendon derived stem cells promote tendon repair through miR-144-3p-regulated tenocyte proliferation and migration. Stem Cell Res Ther. 2022;13(1):80.
44. Zhang T, Yan S, Song Y, Chen C, Xu D, Lu B, et al. Exosomes secreted by hypoxia-stimulated bone-marrow mesenchymal stem cells promote grafted tendon-bone tunnel healing in rat anterior cruciate ligament reconstruction model. J Orthop Transl. 2022;36:152–63.
45. Chen SH, Chen ZY, Lin YH, Chen SH, Chou PY, Kao HK, et al. Extracellular vesicles of adipose-derived stem cells promote the healing of traumatized achilles tendons. Int J Mol Sci. 2021;22(22):12373.
46. Gissi C, Radeghieri A, Lamorgese Passeri CA, Gallorini M, Calciano L, Oliva F, et al. Extracellular vesicles from rat-bone-marrow mesenchymal stromal/stem cells improve tendon repair in rat Achilles tendon injury model in dose-dependent manner: a pilot study. PLoS One. 2020;15(3):e0229914.
47. Han Q, Wang S, Chen D, Gan D, Wang T. Exosomes derived from human umbilical cord mesenchymal stem cells reduce tendon injuries via the miR-27b-3p/ ARHGAP5/RhoA signaling pathway. Acta Biochim Biophys Sin. 2022;54(2):232–42.
48. Shen H, Lane RA. Extracellular vesicles from primed adipose-derived stem cells enhance Achilles tendon repair by reducing inflammation and promoting intrinsic healing. Stem Cells. 2023;41(6):617–27.
49. Shen H, Yoneda S, Abu-Amer Y, Guilak F, Gelberman RH. Stem cell-derived extracellular vesicles attenuate the early inflammatory response after tendon injury and repair. J Orthop Res. 2020;38(1):117–27.
50. Xu H, Zhu Y, Hsiao AWT, Xu J, Tong W, Chang L, et al. Bioactive glass-elicited stem cell-derived extracellular vesicles regulate M2 macrophage polarization and angiogenesis to improve tendon regeneration and functional recovery. Biomaterials. 2023;294:121998.
51. Yao Z, Li J, Xiong H, Cui H, Ning J, Wang S, et al. MicroRNA engineered umbilical cord stem cell-derived exosomes direct tendon regeneration by mTOR signaling. J Nanobiotechnol. 2021;19(1):169.
52. Hayashi Y, Yimiti D, Sanada Y, Ding C, Omoto T, Ogura T, et al. The therapeutic capacity of bone marrow MSC-derived extracellular vesicles in Achilles tendon healing is passage-dependent and indicated by specific glycans. FEBS Lett. 2022;596(8):1047–58.
53. Yao Z, Li J, Wang X, Peng S, Ning J, Qian Y, et al. Microrna-21-3p engineered umbilical cord stem cell-derived exosomes inhibit tendon adhesion. J Inflamm Res. 2020;13:303–16.
54. Li J, Yao Z, Xiong H, Cui H, Wang X, Zheng W, et al. Extracellular vesicles from hydroxycamptothecin primed umbilical cord stem cells enhance anti-adhesion potential for treatment of tendon injury. Stem Cell Res Ther. 2020;11(1):500.
55. Liu A, Wang Q, Zhao Z, Wu R, Wang M, Li J, et al. Nitric oxide Nanomotor driving exosomes-loaded microneedles for Achilles tendinopathy healing. ACS Nano. 2021;15(8):13339–50.
56. Wang Y, He G, Guo Y, Tang H, Shi Y, Bian X, et al. Exosomes from tendon stem cells promote injury tendon healing through balancing synthesis and degradation of the tendon extracellular matrix. J Cell Mol Med. 2019;23(8):5475–85.
57. Xu T, Lin Y, Yu X, Jiang G, Wang J, Xu K, et al. Comparative effects of exosomes and Ectosomes isolated from adipose-derived mesenchymal stem cells on Achilles tendinopathy in a rat model. Am J Sports Med. 2022;50(10):2740–52.
58. Shi Y, Kang X, Wang Y, Bian X, He G, Zhou M, et al. Exosomes derived from bone marrow stromal cells (BMSCs) enhance tendon-bone healing by regulating macrophage polarization. Med Sci Monit. 2020;26:e923328.
59. Wu XD, Kang L, Tian J, Wu Y, Huang Y, Liu J, et al. Exosomes derived from magnetically actuated bone mesenchymal stem cells promote tendon-bone healing through the miR-21-5p/SMAD7 pathway. Mater Today Bio. 2022;15:100319.
60. Li Z, Li Q, Tong K, Zhu J, Wang H, Chen B, et al. BMSC-derived exosomes promote tendon-bone healing after anterior cruciate ligament reconstruction by regulating M1/M2 macrophage polarization in rats. Stem Cell Res Ther. 2022;13(1):295.
61. Xu J, Ye Z, Han K, Zheng T, Zhang T, Dong S, et al. Infrapatellar fat pad mesenchymal stromal cell–derived exosomes accelerate tendon-bone healing and intra-articular graft remodeling after anterior cruciate ligament reconstruction. Am J Sports Med. 2022;50(3):662–73.
62. Tan X, Xiao H, Yan A, Li M, Wang L. Effect of exosomes from bone marrow–derived mesenchymal stromal cells and adipose-derived stromal cells on bone-tendon healing in a murine rotator cuff injury model. Orthop J Sports Med. 2024;12(1):23259671231210304.
63. Wang C, Zhang Y, Zhang G, Yu W, He Y. Adipose stem cell–derived exosomes ameliorate chronic rotator cuff tendinopathy by regulating macrophage polarization: from a mouse model to a study in human tissue. Am J Sports Med. 2021;49(9):2321–31.

64. Wang C, Song W, Chen B, Liu X, He Y. Exosomes isolated from adipose-derived stem cells: a new cell-free approach to prevent the muscle degeneration associated with torn rotator cuffs. Am J Sports Med. 2019;47(13):3247–55.
65. Zhang X, Li A, Han K, Zhang H, Huangfu X, Huang J, et al. Anti-inflammatory and tendon-protective effects of adipose stem cell-derived exosomes with concomitant use of glucocorticoids. Stem Cells Int. 2022;2022:1455226.
66. Zhang J, Cai Z, Feng F, Peng Y, Cui Y, Xu Y. Age-different BMSCs-derived exosomes accelerate tendon-bone interface healing in rotator cuff tears model. Gene. 2024;895:148002.
67. Huang Y, He B, Wang L, Yuan B, Shu H, Zhang F, et al. Bone marrow mesenchymal stem cell-derived exosomes promote rotator cuff tendon-bone healing by promoting angiogenesis and regulating M1 macrophages in rats. Stem Cell Res Ther. 2020;11(1):496.
68. Wang C, Hu Q, Song W, Yu W, He Y. Adipose stem cell–derived exosomes decrease fatty infiltration and enhance rotator cuff healing in a rabbit model of chronic tears. Am J Sports Med. 2020;48(6):1456–64.
69. Wang C, Tan J, Zhang Y, Chen D, He Y. In situ-forming fibrin gel encapsulation of MSC-exosomes for partial-thickness rotator cuff tears in a rabbit model: effectiveness shown in preventing tear progression and promoting healing. J Bone Joint Surg. 2022;104(16):1492–502.
70. Zhang X, Han Z, Han K, Zhang H, Huang J, Huangfu X, et al. Loading mesenchymal stem cell–derived exosomes into a traditionally designed rotator cuff patch: a potential strategy to enhance the repair of chronic rotator cuff tear associated with degenerative changes. Am J Sports Med. 2022;50(8):2234–46.
71. Jenner F, Wagner A, Gerner I, Ludewig E, Trujanovic R, Rohde E, et al. Evaluation of the potential of umbilical cord mesenchymal stromal cell–derived small extracellular vesicles to improve rotator cuff healing: a Pilot Ovine Study. Am J Sports Med. 2023;51(2):331–42.
72. Gao R, Ye T, Zhu Z, Li Q, Zhang J, Yuan J, et al. Small extracellular vesicles from iPSC-derived mesenchymal stem cells ameliorate tendinopathy pain by inhibiting mast cell activation. Nanomedicine. 2022;17(8):513–29.
73. Ye T, Chen Z, Zhang J, Luo L, Gao R, Gong L, et al. Large extracellular vesicles secreted by human iPSC-derived MSCs ameliorate tendinopathy via regulating macrophage heterogeneity. Bioact Mater. 2023;21:194–208.
74. Zhu Z, Gao R, Ye T, Feng K, Zhang J, Chen Y, et al. The therapeutic effect of iMSC-derived small extracellular vesicles on tendinopathy related pain through alleviating inflammation: an in vivo and in vitro study. J Inflamm Res. 2022;15:1421–36.
75. Yu H, Cheng J, Shi W, Ren B, Zhao F, Shi Y, et al. Bone marrow mesenchymal stem cell-derived exosomes promote tendon regeneration by facilitating the proliferation and migration of endogenous tendon stem/progenitor cells. Acta Biomater. 2020;106:328–41.
76. Qin Y, Wang L, Gao Z, Chen G, Zhang C. Bone marrow stromal/stem cell-derived extracellular vesicles regulate osteoblast activity and differentiation in vitro and promote bone regeneration in vivo. Sci Rep. 2016;6:21961.
77. Huang CC, Kang M, Shirazi S, Lu Y, Cooper LF, Gajendrareddy P, et al. 3D encapsulation and tethering of functionally engineered extracellular vesicles to hydrogels. Acta Biomater. 2021;126:199–210.
78. Chen S, Tang Y, Liu Y, Zhang P, Lv L, Zhang X, et al. Exosomes derived from miR-375-overexpressing human adipose mesenchymal stem cells promote bone regeneration. Cell Prolif. 2019;52(5):e12669.
79. Yang S, Zhu B, Yin P, Zhao L, Wang Y, Fu Z, et al. Integration of human umbilical cord mesenchymal stem cells-derived exosomes with hydroxyapatite-embedded hyaluronic acid-alginate hydrogel for bone regeneration. ACS Biomater Sci Eng. 2020;6(3):1590–602.
80. Liu A, Lin D, Zhao H, Chen L, Cai B, Lin K, et al. Optimized BMSC-derived osteoinductive exosomes immobilized in hierarchical scaffold via lyophilization for bone repair through Bmpr2/Acvr2b competitive receptor-activated Smad pathway. Biomaterials. 2021;272:120718.
81. Zhang J, Liu X, Li H, Chen C, Hu B, Niu X, et al. Exosomes/tricalcium phosphate combination scaffolds can enhance bone regeneration by activating the PI3K/Akt signaling pathway. Stem Cell Res Ther. 2016;7(1):136.
82. Qi X, Zhang J, Yuan H, Xu Z, Li Q, Niu X, et al. Exosomes secreted by human-induced pluripotent stem cell-derived mesenchymal stem cells repair critical-sized bone defects through enhanced angiogenesis and osteogenesis in osteoporotic rats. Int J Biol Sci. 2016;12(7):836–49.
83. Li W, Liu Y, Zhang P, Tang Y, Zhou M, Jiang W, et al. Tissue-engineered bone immobilized with human adipose stem cells-derived exosomes promotes bone regeneration. ACS Appl Mater Interfaces. 2018;10(6):5240–54.
84. Swanson WB, Zhang Z, Xiu K, Gong T, Eberle M, Wang Z, et al. Scaffolds with controlled release of pro-mineralization exosomes to promote craniofacial bone healing without cell transplantation. Acta Biomater. 2020;118:215–32.
85. Fan L, Guan P, Xiao C, Wen H, Wang Q, Liu C, et al. Exosome-functionalized polyetheretherketone-based implant with immunomodulatory property for enhancing osseointegration. Bioact Mater. 2021;6(9):2754–66.
86. Liu L, Yu F, Li L, Zhou L, Zhou T, Xu Y, et al. Bone marrow stromal cells stimulated by strontium-substituted calcium silicate ceramics: release of exosomal miR-146a regulates osteogenesis and angiogenesis. Acta Biomater. 2021;119:444–57.

87. Zhai M, Zhu Y, Yang M, Mao C. Human mesenchymal stem cell derived exosomes enhance cell-free bone regeneration by altering their miRNAs profiles. Adv Sci. 2020;7(19):2001334.
88. An Q, Wu D, Ma Y, Zhou B, Liu Q. Suppression of Evi1 promotes the osteogenic differentiation and inhibits the adipogenic differentiation of bone marrow-derived mesenchymal stem cells in vitro. Int J Mol Med. 2015;36(6):1615–22.
89. Lu CH, Chen YA, Ke CC, Chiu SJ, Jeng FS, Chen CC, et al. Multiplexed molecular imaging strategy integrated with rna sequencing in the assessment of the therapeutic effect of Wharton's jelly mesenchymal stem cell-derived extracellular vesicles for osteoporosis. Int J Nanomedicine. 2021;16:7813–30.
90. Wang Y, Zhou X, Wang D. Mesenchymal stem cell–derived extracellular vesicles inhibit osteoporosis via microRNA-27a-induced inhibition of DKK2-mediated Wnt/β-Catenin pathway. Inflammation. 2022;45(2):780–99.
91. Lee KS, Lee J, Kim HK, Yeom SH, Woo CH, Jung YJ, et al. Extracellular vesicles from adipose tissue-derived stem cells alleviate osteoporosis through osteoprotegerin and miR-21-5p. J Extracell Vesicles. 2021;10(12):e12152.
92. Wang Y, Yao J, Cai L, Liu T, Wang X, Zhang Y, et al. Bone-targeted extracellular vesicles from mesenchymal stem cells for osteoporosis therapy. Int J Nanomedicine. 2020;15:7967–77.
93. Li L, Zhou X, Zhang JT, Feng LA, Zhang C, Han JC, et al. Exosomal miR-186 derived from BMSCs promote osteogenesis through hippo signaling pathway in postmenopausal osteoporosis. J Orthop Surg Res. 2021;16(1):23.
94. Zhang L, Wang Q, Su H, Cheng J. Exosomes from adipose derived mesenchymal stem cells alleviate diabetic osteoporosis in rats through suppressing NLRP3 inflammasome activation in osteoclasts. J Biosci Bioeng. 2021;131(6):671–8.
95. Yao XW, Liu ZY, Ma NF, Jiang WK, Zhou Z, Chen B, et al. Exosomes from adipose-derived stem cells alleviate dexamethasone-induced bone loss by regulating the Nrf2/HO-1 axis. Oxid Med Cell Longev. 2023;2023:3602962.
96. Yang BC, Kuang MJ, Kang JY, Zhao J, Ma JX, Ma XL. Human umbilical cord mesenchymal stem cell-derived exosomes act via the miR-1263/Mob1/hippo signaling pathway to prevent apoptosis in disuse osteoporosis. Biochem Biophys Res Commun. 2020;524(4):883–9.
97. Wang X, Li X, Li J, Zhai L, Liu D, Abdurahman A, et al. Mechanical loading stimulates bone angiogenesis through enhancing type H vessel formation and downregulating exosomal miR-214-3p from bone marrow-derived mesenchymal stem cells. FASEB J. 2021;35(1):e21150.
98. Li F, Chen X, Shang C, Ying Q, Zhou X, Zhu R, et al. Bone marrow mesenchymal stem cells-derived extracellular vesicles promote proliferation, invasion and migration of osteosarcoma cells via the lncrna malatl/mir-l43/nrsn2/wnt/p-catenin axis. Onco Targets Ther. 2021;14:737–49.
99. Vallabhaneni KC, Hassler MY, Abraham A, Whitt J, Mo YY, Atfi A, et al. Mesenchymal stem/stromal cells under stress increase osteosarcoma migration and apoptosis resistance via extracellular vesicle mediated communication. PLoS One. 2016;11:e0166027.
100. Wang Y, Chu Y, Li K, Zhang G, Guo Z, Wu X, et al. Exosomes secreted by adipose-derived mesenchymal stem cells Foster metastasis and osteosarcoma proliferation by increasing COLGALT2 expression. Front Cell Dev Biol. 2020;8:353.
101. He H, Ding M, Li T, Zhao W, Zhang L, Yin P, et al. Bone mesenchymal stem cell-derived extracellular vesicles containing NORAD promote osteosarcoma by miR-30c-5p. Lab Invest. 2022;102(8):826–37.
102. Korpershoek JV, Rikkers M, Wallis F, Dijkstra K, Saris DBF, Vonk LA. Mitochondrial transport between chondrocytes and mesenchymal stromal cells. Berlin: International Cartilage Regeneration and Joint Preservation Society; 2022.
103. Casado JG, Blázquez R, Vela FJ, Álvarez V, Tarazona R, Sánchez-Margallo FM. Mesenchymal stem cell-derived exosomes: immunomodulatory evaluation in an antigen-induced synovitis porcine model. Front Vet Sci. 2017;4:39.
104. Griffin TM, Scanzello CR. Innate inflammation and synovial macrophages in osteoarthritis pathophysiology. Clin Exp Rheumatol. 2019;37(Suppl 120):57–63.
105. Galipeau J. Macrophages at the nexus of mesenchymal stromal cell potency: the emerging role of chemokine cooperativity. Stem Cells. 2021;39:1145.

Introduction to the Concept of the Secretome in Orthopaedics

12

Antonio Marmotti, Martina Coco, Laura Mangiavini, Laura de Girolamo, Enrico Bellato, Enrico Ragni, Daniela Taverna, and Filippo Castoldi

12.1 Introduction to the Concept of Secretome

> You can change someone's life in three minutes with the right song (Springsteen B, the Guardian, October 30th, 2016).

Do you remember the music of our lives? Everyone remembers some songs and music that deeply moved the heart in our lives, lyrics and notes that talked to our soul easing our pains, giving hope and strength and making us see our present time somewhat differently, healing our inner part and our mind in some of the most important moments of our lives… Music has done that since the beginning of human history. Music has been always present, and it has been capable to inspire and change something in our inner part. From the primitive sound of joy and war to the complex structure of classical music as Beethoven's symphonies, to the "social music" of Miles Davis, the fusion sound of Herbie Hancock, the jazz solos of Coltrane, the rock songs of U2 and many more, to the Springsteen's poems in songs, music makes people feel different and do something different in our lives. Secretome is that; secretome is the music of the mesenchymal stem cells (MSC) and, at large, of every cell, a music made by growth factors, transcription factors, proteins enzymes, and miRNAs that can "talk" deeply to the surrounding cells of a specific microenvironment. So, by interacting with the inner part of the cells as cytoplasm and nucleus, the secretome promotes protein and gene expression and the behavior of the cells may change. "Secretome music" makes cells do something and, thus, changes the course of the healing

A. Marmotti (✉) · E. Bellato
Department of Orthopaedics and Traumatology, University of Turin, Torino, Italy

Department of Molecular Biotechnology and Health Sciences, Molecular Biotechnology Center "Guido Tarone", University of Torino, Torino, Italy
e-mail: enrico.bellato@unito.it

M. Coco · D. Taverna
Department of Molecular Biotechnology and Health Sciences, Molecular Biotechnology Center "Guido Tarone", University of Torino, Torino, Italy
e-mail: martina.coco@unito.it; daniela.taverna@unito.it

L. Mangiavini
Department of Biomedical Sciences for Health, University of Milan, Milan, Italy

IRCCS Istituto Ortopedico Galeazzi, Milan, Italy
e-mail: laura.mangiavini@unimi.it

L. de Girolamo · E. Ragni
Laboratorio di Biotecnologie Applicate all'Ortopedia, IRCCS Istituto Ortopedico Galeazzi, Milan, Italy
e-mail: laura.degirolamo@grupposandonato.it; enrico.ragni@grupposandonato.it

F. Castoldi
Department of Orthopaedics and Traumatology, University of Turin, Torino, Italy
e-mail: filippo.castoldi@unito.it

A. Gobbi et al. (eds.), *Regenerative Medicine in Sports and Orthopaedics*,
https://doi.org/10.1007/978-3-031-84693-9_12

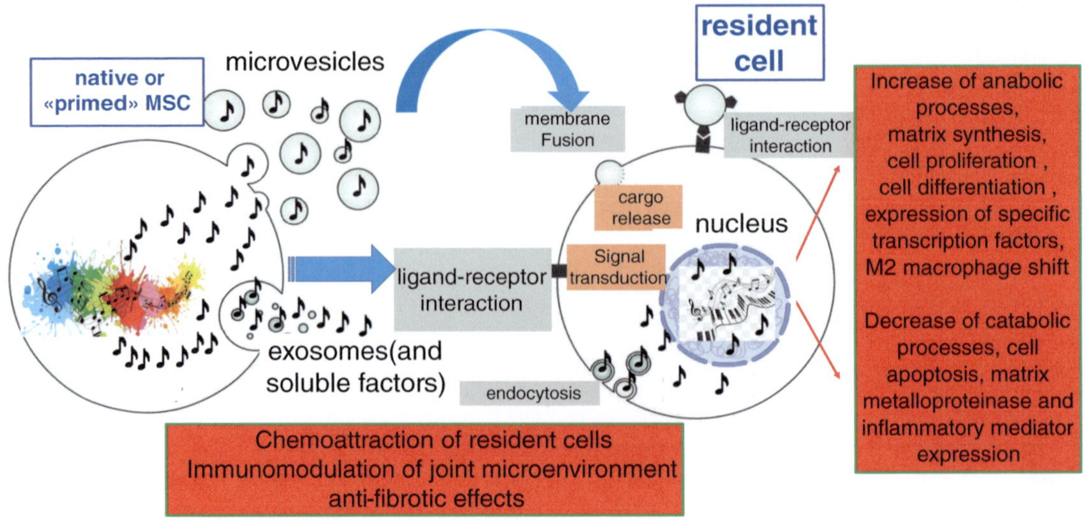

Fig. 12.1 The "secretome music"; MSC (native MSC or MSC primed in vitro with different biochemical stimuli to increase the expression of specific factors, proteins or miRNAs) release the EVs (exosome, macrovesicles, soluble factors) close to resident cells at a lesion site or at a repair site; the EVs are internalized by resident cells through endocytosis, membrane fusion or ligand-receptor interaction; the "cargo" of EVs interact with the intracellular environment of the target cells promoting anabolic processes, matrix synthesis, cell proliferation and cell differentiation, macrophage shift from M1 to M2 by the expression of specific transcription factors, and decreasing catabolic processes, cell apoptosis, matrix metalloproteinase and inflammatory mediator expression. Chemoattraction of resident cells, immunomodulation of the lesion site or the joint microenvironment, anti-fibrotic effects are also linked to secretome release at the lesion site

(see Fig. 12.1). This is the fascinating power and potential of cell secretome. Indeed, in regenerative medicine, tissue destruction is the cause of loss of function in chronic degenerative diseases, and at present, multiple evidence suggests the tissues can be "rejuvenated" if treated with secretomes from MSC. This has been observed in multiple clinical pictures than orthopaedic diseases, such as type 1 diabetes, macular degeneration, chronic kidney disease, ischemic stroke, Alzheimer's, multiple sclerosis, sepsis, hepatitis, chronic liver disease, and skin disease, in which MSC-derived secretome can modulate the immune system, stimulate cell proliferation, promote angiogenesis, prevent apoptosis, suppress oxidative stress, and promote the macrophage shift between M1 and M2.

The secretome is mainly composed of soluble factors, which are small molecules released by cells under certain biological stimuli, and extracellular vesicles (EVs), which can be further subdivided into exosomes, microvesicles, and apoptotic bodies [1]. All these elements are responsible for the paracrine "musical cross-talking" between MSC and surrounding cells through receptor activations (i.e., the growth factors) or cytoplasm interaction (i.e., miRNAs, which interact directly with the intracellular structures, changing protein expression and nuclear factor activation).

Apoptotic bodies have a size of 50–4000 nm, and they are produced by apoptotic cells in the final stage of apoptosis; they are the less important component of the great group of EVs. They contain membrane components (such as phosphatidylserine), nuclear material, and cellular organelles.

Microvesicles shed from the membrane of healthy cells and have a size from 100 to 1000 nm; they are not released from the cell via endocytosis, so they lack endocytosis-related proteins. Exosomes are the smallest EVs, having a diameter between 30 and 150 nm, and are formed during late endosome membrane inward invagination and the formation of multiple vesicular bodies (MVBs). Exosomes are formed inside MVBs and secreted to the extracellular environment via endocytosis by the MVB membrane fusing with the cell membrane. Exosomes can be further classified into three types based on their size: large exosomes (exo-L, size between 90 and 120 nm), small exosomes (exo-S, size between 60 and 80 nm), and exomers (35 nm). In literature, exosomes are one of the most common components of secretomes to be analyzed and used in preclinical and in vitro studies, due to the possibility of separation by methods such as ultracentrifugation, size-based techniques, density gradient centrifugation, immunoaffinity purification, precipitation, and microfluidics-based isolation techniques. Exosome identification is based on morphological features, particle size, and signature external proteins, such as CD9, CD63, CD81, and HSP90 [1].

In a "paradigm of cure," advantages of exosome and secretome administration are multiple when compared to MSC direct administration [2]: (i) secretomes do not imply the transfer of cells containing immunogenic molecules as well as mutated or damaged DNA, and thus they do not have the risk of genetic instability or immunoreaction after allogeneic use; (ii) being nanosized, secretomes can easily enter and move within any organ, whereas cells are larger and have more difficulties in migrating to the site of injury through adhesion protein mechanism through capillaries circulation; (iii) secretomes do have homing molecules on their surface and can also migrate to different parts of the body; (iv) having surface biochemical properties similar to those of their derived MSC, secretomes may avoid major phagocytosis; (v) secretomes may be easier preserved because EVs are stored at −20 °C, while stem cells must be maintained in liquid nitrogen to retain viability.

As for its counterpart (the MSC direct administration), some critical issues may be summarized in a possible future clinical application of secretome. Indeed, more studies are needed to identify the best method for large-scale production and isolation, for quantification and characterization of EVs and their cargo, the pharmacokinetics, targeting, and transfer mechanisms of EVs to the target sites, and the optimal clinical dosage upon the foreseeable repeated administration needed to complete the tissue healing and rejuvenation.

12.2 Early Evidences of Secretome ("The First Recordings")

The story of the secretome is not recent in clinical literature. Trams et al. proposed the term "exosome" in 1981 as "exfoliation of membrane ecto-enzymes in the form of micro-vesicles" [3, 4], even if some brilliant intuitions of the existence of extracellular vesicles were present in the hematological studies of Chargaff and West in 1946 about blood clotting, in which they observed "a variety of minute breakdown products of blood corpuscles," in the studies of Peter Wolf in 1967 [5], who described clearly a "material in minute particulate form, originating from platelets, but distinguishable from intact platelets" and in the work of Crawford [6] in 1971, who described the corpuscles as containing lipid and carried cargo including ATP and contractile proteins. Subsequently, the coming of electron microscopy allowed to describe structures consistent with the sub-micron size of EVs by several authors as Nunez et al. [7] in 1974, who first observed the release of the vesicles contained within the cell "into the luminal space." From the 1980s, more specific research on EVs led to a better understanding of their form and function [8] and to the demonstration that the exosomes are enzymatically active by Rose Johnstone in 1989 [9].

Subsequently, from 2000, evidences in literature appeared about: a) the possible release of cytokines via EVs, b) the key role of EVs derived from immune cells in the function of the immune system, c) the immunosuppressive role of MSC-derived EVs, d) the potential of tumor-derived EVs as anti-tumor therapy, e) the effects of EVs in protecting animal models from disease, f) the functional transfer of nucleic acids, and g) the fact that even plant cells can use EVs as a means of communication [8].

In the fields of orthopaedic, one of the first pieces of evidence that MSCs can release "something" has been shown by Schaper et al. in 2004 [10], when they observed in a mouse model of hindlimb that bone marrow-derived cells from littermates, injected to improve the angiogenesis, do not promote vascular growth by incorporating into vessel walls, but they functioned as supporting cells in a position around the growing collateral arteries. This reminds a very important intuition of the "father" of MSC Arnold Caplan who, around the same period, wrote that "MSCs are naturally found as perivascular cells, summarily referred to as pericytes, which are released at sites of injury, where they secrete large quantities of bioactive factors that are both immunomodulatory and trophic" [11]. This observation of Caplan about the behavior of pericytes-MSCs and their ability to secrete immunomodulatory and trophic bioactive factors may be considered the ideal "bridge" between the previous studies of EVs and the modern concept of MSC role for tissue repair; indeed, Caplan and Correa [12] later in 2011 wrote that resident pericytes-MSCs may be considered as "injury drugstores," and that during a local injury or at a surgical site, resident pericytes-MSCs leave by chemotaxis their perivascular location and, in an inflammatory setting, they become activated, and they create a regenerative and antifibrotic microenvironment by secreting bioactive molecules and regulating the local immune response through "secretome." Thus, the primary form of healing mechanism of MSCs was understood as a repair dependent not only on the "MSC differentiation into something," but, more importantly, and more efficiently, on "what MSCs tell to do to the resident cells at a repair site." The secretome was finally recognized as the primary form of healing mechanism of MSCs.

A critical aspect of the use of secretome appeared in the same period, by the work of van Osch's group in 2012 [13]. The authors obtained a peculiar secretome derived from MSCs grown in presence of TNF-alpha and IFN gamma. Then, they treated synovial explants observing decreased gene expression of interleukin-1 beta, matrix metalloproteinase (MMP)1 and MMP13, and cartilage explant, observing the upregulation of IL-1 receptor antagonist and reduction in reduced NO production in cartilage explants, compared to controls that were not exposed to primed MSC-derived secretome. This introduced a new concept in the form of "priming the MSCs": MSCs undergo several stimuli during cell culture, allowing for obtaining a more active and specific secretome. The principle of "priming the MSCs" was observed in a clinical setting during the "ADIPOA 1" study of Perks in 2016, in which autologous MSCs from adipose tissue were injected in the knees of patients to treat knee osteoarthritis. In this study, a better anabolic effect was obtained when the autologous MSCs from adipose tissue were injected into the joints with a more marked inflammation. This demonstrates the priming effect of an inflammatory joint microenvironment on the behavior of autologous MSCs. Indeed, inflammation "primed" the injected MSCs and induced the cells to produce a more anabolic secretome comparted to that secreted in the joints with less inflammation [14]. These observations are widely present in literature and the concept of selecting specific culture conditions to obtain a more specific and efficient secretome cargo may be considered as the "next-generation" paracrine therapy.

12.3 "In Vitro" Work ("In Studio Soundtrack")

Following these observations, several in vitro studies [15] exploited the regenerative and differentiative potential of the secretome derived from different sources, such as adipose-derived-MSCs, bone marrow-MSCs, and even umbilical cord stromal MSCs (aka Wharton's jelly derived MSCs) [16] in chondrocyte cultures. The results were consistent with the anti-senescence potential, the anti-inflammatory and chondroprotective effects, the increase of matrix synthesis and cell proliferation, and the inhibition of cell apoptosis. In these studies, most of the extracellular vesicles were internalized after 30 min from their administration, reaching a peak of uptake after 12–18 h. This is consistent with the very quick action of secretome and miRNA; indeed, miRNA may survive only 200 h in a pericellular microenvironment.

Furthermore, the in vitro studies involving MSC preconditioning to improve secretome cargo allowed for identifying the most relevant culture conditions and regenerative factors involved in the secretome-dive repair processes. Redondo-Castro et al. [17] in 2017 showed the efficacy of "priming" MSCs with IL-1 to obtain a greater anti-inflammatory and pro-trophic potential. In the same period, Lo Sicco et al. [18] discovered that secretome derived from MSC grown in hypoxic condition has a stronger anti-inflammatory effect than that from normoxic MSCs in a mouse model of cardiotoxin-induced muscle injury. They also showed the pivotal role of the secretome in promoting the macrophage shift from M1 to M2, and M2 polarization has been recently confirmed in the study of Luo et al. in 2021 [19] as one of the essential secretome mechanisms to promote muscle regeneration and reduce inflammation and fibrosis in muscle contusion. On the contrary, MSC secretomes from adipose stem cells in normoxic conditions are more efficient than that from ASCs cultured under hypoxic settings in promoting survival and decreasing the inflammatory response in rats after a generated sepsis syndrome, in a study by Chang et al. [20]. In cultures with TGF-beta 3 primed MSCs, Ruiz et al. discovered a key regulator of the anabolic action of EVs, named TGFβ-induced gene product-h3 (TGFBI) [21]. Ragni et al. [22–24] recently identified some of the most relevant miRNAs in EVs of adipose-derived MSCs for chondrogenic development as miR-26a-5p, miR-22-5p, and miR-29a-5p. Very recently, in the field of tendon repair, An et al. [25] have demonstrated that miR-337-3p is a very important tenogenic factor protecting tenocytes from apoptosis by targeting caspase-3, which is a member of the cysteine-aspartic acid protease (caspase) family whose sequential activation plays a central role in the execution-phase of cell apoptosis. They demonstrated this observation in an in vitro study in which exosomes were derived from tendon stem cells overexpressing microRNA(miR)-377-3p.

These works are only a few of the studies in the literature that are focused on the field of "secretome functionalization." Indeed, improving secretome repair by MSC priming is more promising strategy for identifying the main factors involved in the exosome-driven healing process and for recognizing the most effective conditions for targeting specific pathological settings with an "enhanced secretome." At present, the in vitro strategies for improving secretome cargo may be summarized as (i) enriching native exosomes with specific miRNA (i.e., utilizing in vitro MSCs transfection with specific miRNA or by directly incorporating specific miRNA MSCs-EVs through biochemical systems as electroporation); (ii) enriching native exosomes with specific circular RNAs (circRNAs) that "work" as miRNA but are more stable in cell microenvironment; (iii) adding chondrogenic molecules in native EVs (i.e., adding the chondrogenic small molecule named kartogenin in MSCs culture, allowing for obtaining kartogenin-enriched exosomes) [26].

12.4 Preclinical Studies ("Secretome Live Music")

Along with the in vitro studies, the literature is rich in preclinical models of secretome-driven tissue regeneration, and the field of cartilage repair is certainly one of the most promising settings. From the group of Montpellier, in 2017 Stella Cosenza et al. observed cartilage regeneration in a rat model of collagenase-induced OA by one intra-articular injection of exosome from murine BM-MSCs [27]. Their work, first presented at the ICRS meeting in 2016, represents one of the first evidence of the preclinical efficacy of exosomes in promoting cartilage repair, and a most fascinating aspect of this study was that the results were similar to those obtained by intra-articular MSCs injection. Contemporarily, Zhu et al. [28] in a rat model observed a similar repair with 3 weekly intra-articular injections of exosomes derived from iPS-derived MSCs (induced pluripotent stem cell-derived MSCs) and synovial membrane-derived MSCs, showing new promising sources of the secretome. Wu et al. in 2019 [29] demonstrated cartilage repair in a mouse model of OA obtained by destabilization of the medial meniscus, using an intra-articular injection of exosomes from the human infrapatellar fat pad. In this study, these authors showed that one of the most relevant mechanisms was linked to miR100-5p and the miR100-5p inhibition of the mTOR-autophagy pathway. In Italy, Zavatti et al. [30] observed the anti-osteoarthritic potential of two intra-articular injections of exosomes derived from amniotic stem cells in a mouse model of OA induced by intra-articular injection of monosodium iodoacetate. Again, migration and polarization of M2 macrophages during cartilage repair were identified as a key mechanism induced by exosomes for the secretome-driven cartilage repair. Zhang in Singapore in 2016 [31] treated osteochondral defects in trochlear grooves of adult rats with 12 weekly intra-articular injections of exosomes derived from human embryonic MSC observing a strong chondrogenic potential, and even a "secretome scaffold" was first conceived by Liu et al. in 2017 [32], who treated rabbit patellar groove defect with iPSC-MSCs. Exosomes vehiculated in a hydrogel glue able to preserve in-vitro exosomes at >90% for 1 month. Encapsulating EVs in a suitable biomaterial, thus creating a "bioactive" tissue patch to be applied to cover focal cartilage defects, is an appealing application of secretome to treat chondral and osteochondral lesions. Indeed, the biomaterial allows for preventing rapid clearance and obtaining a prolonged release of EVs from the scaffold matrix. This model has been focused in several recent preclinical studies [26]. Human umbilical cord MSCs-exosomes were loaded in a methacrylate/nano clay hydrogel by Hu et al. [33] and implanted into knee cartilage defect of rats, obtaining positive in vivo results. Besides, this allowed the authors for identifying miR-23a-3p as one of the most relevant chondrogenic miRNAs in EVs. Zhang et al. [34] produced an injectable adhesive hydrogel of alginate–dopamine, chondroitin sulfate, and regenerated silk fibroin as a vehicle for bone marrow MSCs-exosomes. These authors treated trochlear osteochondral defects of rats, achieving cartilage regeneration and observing a massive and relevant local MSCs recruitment from the tissues surrounding the defects, thus showing a high chemotactic activity of their construct. Acellular cartilage extracellular matrix (ACECM) has been also studied as an exosome carrier in a preclinical model; Jiang et al. [35] loaded the ACECM with human Wharton's jelly-MSCs exosomes and treated trochlear osteochondral defects of rabbits, reaching favorable results in term of cartilage repair and immunomodulation of joint inflammation by macrophage polarization toward the M2 phenotype. These authors identified miR26a and miR199a as key elements in the regenerative potential of Wharton's jelly-MSCs exosomes. These are only examples of the different studies [26] that have been performed following the concept of "exosome scaffold."

Furthermore, the frontier of "exosome functionalization has also been investigated in preclinical models. Tao et al. in 2017 [36] established a rat model of osteoarthritis by medial collateral ligament and anterior cruciate ligament transection and treated the animals with four intra-articular injections of exosomes derived from synovial MSCs, previously transfected by lentivirus containing miR-140-5p. These miR-140-5p-enriched exosomes slowed the progression of early OA and prevented the severe damage to knee articular cartilage in the OA model caused by instability of the knee joint. The authors identified miR-140-5p as a key factor in targeting and inhibiting the molecule named RalA, a specific enzyme of the hydrolase family, that downregulates the expression of SOX9 (one of the most famous chondrogenic transcription factors) and aggrecan (also known as cartilage-specific proteoglycan core protein or chondroitin sulfate proteoglycan 1, a basic component of cartilage matrix). Their in vivo observations paralleled a previous in vitro study of Karlsen [37] et al. that, in 2014, demonstrated the close relationship between miR-140, RalA inhibition, and translational enhancement of SOX9 and aggrecan to promote chondrogenic differentiation of human mesenchymal stem cells. This shows that in vitro studies often precede preclinical evidence in the fascinating evolution of research. In a setting similar to Tao's work, Mao et al. [38] transfected human bone marrow MSCs with a cationic-lipid transfection reagent (non-viral transfections) carrying miR-92a-3p, a strong chondrogenic miRNA. In a preclinical model of collagenase-induced OA model in mice, the authors performed three intra-articular injections of MSCs-miR-92a-3p-exosomes obtaining promising histological results and showing the potential of miR-92a in suppressing WNT5A, a protein belonging to the Wnt signaling pathway that is involved in cartilage formation. The authors observed that WNT5A suppression led to the upregulation of aggrecan, COL2A1 (a gene for collagen II) and SOX9 and the downregulation of COL10A1 (a gene for collagen I), RUNX2 (a transcription factor for bone development) and MMP13 (a matrix metalloproteinase). This complex mechanism is based on the role of WNT5A, which is activated in the early stage of chondrogenesis but, if expressed in the late stage, it promotes MMP13 expression, leading to the loss of aggrecan, collagen 2, and several cartilage matrix proteins as COMP (cartilage oligomeric matrix protein). Thus, inhibiting WNT5A may suppress cartilage degeneration, and recently, a Wnt inhibitor M04690 has even been developed as a disease-modifying OA drug (DMOAD). In their fascinating study, Mao et al. showed that exosomal miR-92a-3p has a function similar to that of a Wnt inhibitor. Taken together, these observations show the superior role of "functionalized secretome." Indeed, pure natural EVs and exosomes have some critical drawbacks as the short time of activity when injected into the joint microenvironment, the low yield, the low targeting capacity, and a reduced function compared to functionalized "engineered" EVs obtained from cells subjected to specific differentiation processes by gene transfection or growth media exposure during culture. In this context, a new horizon of research is also represented by the surface modification of EVs to improve their target ability, adding targeting ligands onto the surface of the exosomes, to achieve precise delivery of their cargo to lesion sites [39]. Further studies will certainly exploit these critical points in the field of cartilage repair.

But even in the setting of tendon regeneration, recent studies have shown the promising potential of the secretome. The first observations of Sevivas et al. in [40, 41] showed the potential of human MSCs secretome to increase human tendon cells viability and density at 7 days in vitro. Furthermore, the authors conducted an in vivo preclinical study on a chronic rat cuff tear, and they described an improvement in the histological and biomechanical properties of the cuff treated by human MSCs secretome as well as less fatty degeneration and less atrophy of the rotator cuff muscles. Several recent shreds of evidence demonstrated also the positive effect of in vivo secretome application. Chen et al. [42] showed in 2021 the potential of human bone marrow-derived MSC secretome in enhancing tendon-bone healing in a rat rotator cuff model, and they observed, even in this tendon repair setting, the importance of macrophage polarization and M2 phenotype promotion, asserting that the benefit of MSC secretome relied basically on its immune-regulative property. Jenner et al. [43], in 2023, demonstrated the great potential of human umbilical cord MSC-derived secretome loaded onto a type 1 collagen scaffold (a local single dose application of 2×10^{10} hUC-MSC-EVs loaded onto a type 1 collagen sponge) in an ovine model of acute infraspinatus tendon injury. They observed an improved rotator cuff healing and tendon-to-bone insertion at 6 weeks. Fredianto et al. [44, 45] showed the positive effect of secretome derived from hypoxia-preconditioned mesenchymal stem cell in promoting the healing process of tendon-to-bone interface in a rat acute rotator cuff tear model. In their study, these authors observed at 2 and 8weeks an increased expression of b-FGF, tenomodulin (biomarker for tendon proliferation), Runt-related transcription factor 2 (RUNX2, that has a pivotal role in osteogenesis and enthesis regulation controlling calcium deposition and osteoblast-specific protein expression), and hypoxi-ainducible factor (HIF)-1a.2 and 8 weeks. This agrees with the previous observations of Yu et al. [46] who demonstrated in 2016 the positive role of HIF-1a in promoting the expression of scleraxis. In December 2023, Zhang et al. [47] demonstrated in a rat model that extracellular vesicles from tendon stem cells may facilitate the healing of ruptured Achilles tendon sutured with

4–0 Vicryl absorbable sutures. They conceived a gelatin scaffold that allowed for slow release of exosome cargo, and they applied topically the construct on the injured tendon. The result was amazing: the TSC-sEVs loaded with gelatin methacryloyl (GelMA) hydrogel promoted tenogenesis and hampered osteogenesis at the injured site already at 2 weeks, and most notably at 4 and 8 weeks, and they observed also that one of the key factors in their model was the miR-145-3p that may be considered actually a tenogenic miRNA. Thus, a hypothetic possible cell-free therapeutic approach for tendon lesions is conceivable! Following the same proof of concepts, Dou et al. [48] have published in July 2024 their work showing that, in a similar preclinical model of rat, Achilles tendon transection sutured with the Kessler technique. The authors treated the injured site with 0.15 mL of a gel made of polichitosan mixed with human umbilical vein endothelial cells-derived exosomes. Following the same proof of concepts, Dou et al. have published in July 2024 a study with a similar preclinical model of rat, in which Achilles tendon was transected and subsequently repaired with the Kessler suture technique. The authors treated the injured site with 0.15 mL of a gel made of polichitosan mixed with exosomes derived from human umbilical vein endothelial cells. They described at 4 weeks superior histological properties, decreased tendon cell apoptosis, decreased inflammation and enhanced tendon regeneration in the treated group. Moreover, the animals who received the exosomes demonstrated, at 4 weeks, superior tensile strength and greater load-tofailure ratio, compared to those of the control group [49]. They did not observe any significant difference between exosomes from adipose or bone marrow stem cells. Furthermore, even unrepairable massive rotator cuff tears have been studied as a target for secretome therapy: Zhang et al. [50] have set up a rabbit model of massive rotator cuff tear by section of the supraspinatus tendon, and after 12 weeks, they bridged the retracted tendon with a polycaprolactone-based scaffold loaded with rabbit tendon stem cell-derived exosomes and type 1 collagen. The treated group exhibited at 16 weeks greater tendinous tissue regeneration and better load and stress to failure and stiffness than those of the untreated control group, and these results were similar to those obtained by treating the tendinous gap by autogenous fascia lata bridging. A promising possible solution for unrepairable massive rotator cuff tears may thus be hypothesized with the help of secretome boosting effects. But what about the patient's age in the effect of secretome? This has been the main concern of a recent study by Zhang et al. [51] who set up a rat model of supraspinatus tendon rupture, sectioned and sutured with a "figure of 8" 6–0 suture. The lesion site was treated by immediate injection of a chitosan-nano-hydroxyapatite-collagen-β-glycerophosphate hydrogel, loaded with exosome from young and adult bone marrow MSC, at the junction of supraspinatus tendon and humerus. They demonstrated a significantly better effect of the "young" exosome toward the growth of Sharpey-like fibers, the formation of "tendon-non-calcified fibrocartilage-calcified cartilage-bone" structures at the lesion site, and the expression of tendon-related genes as collage, aggrecan, scleraxis, and decorin. So, for a future application of these brilliant results, we do not have to forget that age matters in the choice of exosome source.

12.5 From PRP to PRP-Secretome Sinergy ("Secretome Jam Session with PRP")

In a fantastic landscape like this, what is left for the "oldies goldies" orthobiologics as platelet-rich plasma (PRP)? "Secretome wave" has also involved this field, suggesting a possible solution to overcome the main drawbacks of autologous PRP as patient's specific variations and differences in PRP's extraction methods. What about a "next-generation PRP" made of allogeneic PRP exosomes, obtained by selected donors with precise and reproducible extraction procedures? This has been the idea of some recent studies that may revolutionize our present knowledge about the clinical use of PRP. Indeed, Xu et al. [52] have shown promising anabolic properties of exosomes derived from human apheresis platelets in an in vitro and preclinical mouse model of osteoarthritis. They demonstrated that PRP exosomes can "dramatically promote" cartilage regeneration and attenuate tibial subchondral bone remodeling in a way much more effective than that observed in previous studies involving the use of PRP. Han et al. [53] studied a rabbit model of rotator cuff tear in which the supraspinatus tendon was sectioned and, after 1 week, sutured in situ, and at the tendon-bone interface, the authors injected rabbit PRP exosome suspended in PBS before wound closure. They obtained at 9 weeks a better repair tissue than that of the control group, showing that PRP-derived exosomes can promote tendon-bone interface healing. And the contents of collagen type II, SOX-9, and tissue inhibitors of metalloproteinases-1 were also increased in the treated group. Li et al. [54] obtained similar improvement in the post-injury healing process with a rabbit model of infraspinatus section and repair treated at the bone-tendon interface with a human PRP exosome-loaded fibrin glue scaffold. The relevance of their study is multiple: the exosome source is human, validating the positive effect of human donors in this specific pathological picture; the scaffold used is very simple and already used at present in human surgery, suggesting an easiness of applicability of this model in a future possible clinical setting, and the result analyses were very accurate and detailed, including histological assessment and multi-modality imaging examinations, as ultrasound and micro-CT. Chen et al. [55] confirmed these concepts in a model of supraspinatus tendon section and suture (by a modified Mason–Allen stitch) treated with a weekly ultrasound-guided subacromial injection of 20 μL ($2 \times 10E10$ particles) of PL-exosomes in PBS. At 8 weeks after the operation, they observed a more compact extracellular matrix in the treated group, with increased alignment of fibers, resembling a healthy tendon composition. But the most original aspect of this study is that it revealed new key elements for tendon regeneration; indeed, PRP exosomes were able to rejuvenate tendon cells by decreasing ferroptosis, an iron-dependent form of programmed cell death, and this effect was obtained by PRP-exosome-driven expression of nuclear factor erythroid 2-related factor 2 (Nrf2), a transcription factor pivotal for regulating antioxidant stress and inflammatory responses, and ferroptosis, and the activation of AMP-activated protein kinase, also involved in ferroptosis mechanisms. Thus, following the stream of secretome therapy, the biology of tendon regeneration is more and more unraveled, and new possible clinical therapeutic scenarios may be conceived for the future of orthopaedic tissue regeneration.

12.6 Present Clinical Trials ("Secretome on Stage Around the World")

But, at the end of this brief journey in the secretome world, an unquestionable doubt arises: Are all these in vitro and preclinical studies still too far from a "human experience"?

Maybe not, and the clinical landscape at present may be surprising. In 2018, in a different setting than that of orthopaedic, a phase 1 and 2 trial started, assessing the potential of allogenic MSCs-derived engineered exosomes (enriched by miR-124) in patients with acute ischemic stroke, to improve the disability of patients with intravenous administration of the engineered exosomes 1 month after the ischemic attack. And, in the world of orthopaedics, in July 2024, eight clinical trials were enlisted (with the keywords joint, secretome, and exosome) to assess the value of secretome in preventing osteoarthritis progression and related

Table 12.1 Eight Clinical Human Trials with the Keywords Joint, Secretome, and Exosome are described (last accessed: July 2024); these are the current Clinical Trials involving Secretome for the Treatment of Osteoarthritis and Related Conditions

		Brief summary	Study start (estimated)	Phase
Intra-articular injection of MSC-derived exosomes in knee osteoarthritis (ExoOA-1)	ClinicalTrials.gov ID NCT05060107 Sponsor Francisco Espinoza Universidad de los Andes, Chile (Responsible Party)	Safety of exosomes (sEVs) from allogeneic mesenchymal stromal cells delivered by an intra-articular injection in the knee of patients with mild to moderate symptomatic osteoarthritis. The sEVs will be produced in a GMP-facility: 10 patients in this phase 1 trial; follow-up 12 months	10/2021	Phase 1
Mesenchymal stem cell (MSC) therapy and conditioned medium therapy for osteoarthritis (OA)	ClinicalTrials.gov ID NCT04314661 Sponsor PT. Prodia Stem Cell Indonesia	Compare the efficacy of umbilical cord mesenchymal stem cell and secretome between arthroscopy and without arthroscopy intervention in OA patients. 4 arms: arthroscopy + booster, arthroscopy + pre-conditioning, non-arthroscopy + booster, non-arthroscopy + pre-conditioning, follow-up 6 months	8/2020	Phase 1 Phase 2
Effectiveness of PRP, conditioned medium UC-MSC secretome and hyaluronic acid for the treatment of knee osteoarthritis	ClinicalTrials.gov ID NCT05579665 Sponsor Universitas Sriwijaya	Compare the efficacy of platelet-rich plasma, conditioned medium from umbilical cord mesenchymal stem cell (MSC) culture secretome and hyaluronic acid for the treatment of knee osteoarthritis; 5 times injection 3 cc platelet-rich plasma or 2 cc hyaluronic acid or 2 cc conditioned medium from umbilical cord MSC secretome at intervals 1 week Clinical data in the form of subjective outcome measures will be collected pre-treatment and 6 months	10/2022	Phase 1 Phase 2
Safety and tolerability of IMMUNA(IMM01-STEM) in patients with muscle atrophy related to knee osteoarthritis	ClinicalTrials.gov ID NCT05211986 Sponsor Immunis, Inc.	Safety and tolerability of IMMUNA (IMM01-STEM), a secretome product derived from partially differentiated pluripotent stem cells that contain regenerative molecules, in participants with muscle atrophy related to knee osteoarthritis (KOA) Up to 18 participants will receive twice weekly intramuscular administration of IMMUNA (IMM01-STEM) for 4 weeks in up to 3 dose cohorts: Cohort A, IMMUNA 225 μg; Cohort B, IMMUNA 450 μg; and Cohort C, IMMUNA 900 μg. After their last injection, participants will enter a safety follow-up (SFU) period, with clinic visits 3 days after their last injection and then monthly for 3 months	9/2022	Phase 1 Phase 2

(continued)

Table 12.1 (continued)

		Brief summary	Study start (estimated)	Phase
Clinical efficacy of exosome in degenerative meniscal injury (KNEEXO)	ClinicalTrials.gov ID NCT05261360 Sponsor Eskisehir Osmangazi University	Comparison of intra-articular administration of synovial fluid-derived autologous mesenchymal stem cell-derived exosomes with synovial fluid-derived mesenchymal stem cells on the same patient; patients with stage 0, 1, or 2 gonarthrosis according to Kellgren–Lawrence gonarthrosis staging; the control group does not receive treatment. SF-MSC (synovial fluid-derived mesenchymal stem cells) will be applied intraarticularly to the experimental group's right knee. SF-MSC-EX (synovial fluid-derived mesenchymal stem cell-derived exosomes) will be applied intraarticularly to the experimental group's left knee, follow-up 12 months	3/2022	Phase 2
Mesenchymal stem cell-derived exosomes in osteoarthritis patients	ClinicalTrials.gov ID NCT06466850 Sponsor Isfahan University of Medical Sciences	Patients with moderate to severe knee osteoarthritis, Kellgren–Lawrence grade 2,3; 45–75 years, intra-articular injection of allogenic mesenchymal stem cell-derived exosomes at day 1 and day 90	1/2024	Not applicable
Phase 1b clinical trial to evaluate PEP and EUFLEXXA for knee osteoarthritis (KOA)	ClinicalTrials.gov ID NCT06463132 Sponsor Rion Inc.	Single intra-articular injections of PEP (purified exosome product) reconstituted with 0.9% normal saline at a low dose (one vial PEP) and high dose (two vials PEP), with and without EUFLEXXA, for the treatment of Kellgren–Lawrence Grade 2–3 in knee osteoarthritis, 18–90 years	11/2024	Phase 1
Intra-articular injection of UC-MSC exosome in knee osteoarthritis (EXO-OA01)	ClinicalTrials.gov ID NCT06431152 Sponsor Universidad de los Andes, Chile	Patients 30–70 years with symptomatic Kellgren II–III knee OA, dose escalation pilot study in which three cohorts of subjects with OA will receive increasing doses of umbilical cord MSC-sEV administered as a single intra-articular (IA) injection	6/2024	Early phase 1

conditions (see Table 12.1). Thus, in a few years, along with the new preclinical and in vitro achievements, their results will be able to start a new ready-to-use, cell-free therapeutic alternative to cell-based MSC therapies, based on repeated injections of secretome (or on direct secretome application by scaffolds) to obtain cartilage, tendon and muscle repair, and reduction of osteoarthritis.

So, an ultimate question arises for each of us who will always love and follow orthopaedic basic science: "Should we choose cells, PRP, or secretome to make a difference in our future therapies?"

Acknowledgments We thank Fondazione Carlo Enrico Pilone Brunet e.t.s. for the support.

References

1. Ni Z, Zhou S, Li S, Kuang L, Chen H, Luo X, Ouyang J, He M, Du X, Chen L. Exosomes: roles and therapeutic potential in osteoarthritis. Bone Res. 2020;8:25. https://doi.org/10.1038/s41413-020-0100-9.
2. Mahindran E, Wan Kamarul Zaman WS, Ahmad Amin Noordin KB, Tan Y-F, Nordin F. Mesenchymal stem cell-derived extracellular vesicles: hype or hope for skeletal muscle anti-frailty. Int J Mol Sci. 2023;24:7833. https://doi.org/10.3390/ijms24097833.
3. MacDonald E, Salem N. Dr. Eberhard Trams—the man who coined the name "exosomes"—a prescient but largely forgotten pioneer. J Extracell Vesicles. 2023;12:e12370. https://doi.org/10.1002/jev2.12370.
4. Trams EG, Lauter CJ, Salem N, Heine U. Exfoliation of membrane ecto-enzymes in the form of microvesicles. Biochim Biophys Acta. 1981;645:63–70. https://doi.org/10.1016/0005-2736(81)90512-5.
5. Wolf P. The nature and significance of platelet products in human plasma. Br J Haematol. 1967;13:269–88. https://doi.org/10.1111/j.1365-2141.1967.tb08741.x.
6. Crawford N. The presence of contractile proteins in platelet microparticles isolated from human and animal platelet-free plasma. Br J Haematol. 1971;21:53–69. https://doi.org/10.1111/j.1365-2141.1971.tb03416.x.
7. Nunez EA, Wallis J, Gershon MD. Secretory processes in follicular cells of the bat thyroid. 3. The occurrence of extracellular vesicles and colloid droplets during arousal from hibernation. Am J Anat. 1974;141:179–201. https://doi.org/10.1002/aja.1001410203.
8. Couch Y, Buzàs EI, Di Vizio D, Gho YS, Harrison P, Hill AF, Lötvall J, Raposo G, Stahl PD, Théry C, et al. A brief history of nearly EV-erything—the rise and rise of extracellular vesicles. J Extracell Vesicles. 2021;10:e12144. https://doi.org/10.1002/jev2.12144.
9. Johnstone RM, Bianchini A, Teng K. Reticulocyte maturation and exosome release: transferrin receptor containing exosomes shows multiple plasma membrane functions. Blood. 1989;74:1844–51.
10. Ziegelhoeffer T, Fernandez B, Kostin S, Heil M, Voswinckel R, Helisch A, Schaper W. Bone marrow-derived cells do not incorporate into the adult growing vasculature. Circ Res. 2004;94:230–8. https://doi.org/10.1161/01.RES.0000110419.50982.1C.
11. Caplan AI. Why are MSCs therapeutic? New data: new insight. J Pathol. 2009;217:318–24. https://doi.org/10.1002/path.2469.
12. Caplan AI, Correa D. The MSC: an injury drugstore. Cell Stem Cell. 2011;9:11–5. https://doi.org/10.1016/j.stem.2011.06.008.
13. van Buul GM, Villafuertes E, Bos PK, Waarsing JH, Kops N, Narcisi R, Weinans H, Verhaar JN, Bernsen MR, van Osch GJVM. Mesenchymal stem cells secrete factors that inhibit inflammatory processes in short-term osteoarthritic synovium and cartilage explant culture. Osteoarthr Cartil. 2012;20:1186–96. https://doi.org/10.1016/j.joca.2012.06.003.
14. Pers Y-M, Rackwitz L, Ferreira R, Pullig O, Delfour C, Barry F, Sensebe L, Casteilla L, Fleury S, Bourin P, et al. Adipose mesenchymal stromal cell-based therapy for severe osteoarthritis of the knee: a phase I dose-escalation trial. Stem Cells Transl Med. 2016;5:847–56. https://doi.org/10.5966/sctm.2015-0245.
15. D'Arrigo D, Roffi A, Cucchiarini M, Moretti M, Candrian C, Filardo G. Secretome and extracellular vesicles as new biological therapies for knee osteoarthritis: a systematic review. J Clin Med. 2019;8:E1867. https://doi.org/10.3390/jcm8111867.
16. Hassan Famian M, Montazer Saheb S, Montaseri A. Conditioned medium of Wharton's jelly derived stem cells can enhance the cartilage specific genes expression by chondrocytes in monolayer and mass culture systems. Adv Pharm Bull. 2017;7:123–30. https://doi.org/10.15171/apb.2017.016.
17. Redondo-Castro E, Cunningham C, Miller J, Martuscelli L, Aoulad-Ali S, Rothwell NJ, Kielty CM, Allan SM, Pinteaux E. Interleukin-1 primes human mesenchymal stem cells towards an anti-inflammatory and pro-trophic phenotype in vitro. Stem Cell Res Ther. 2017;8:79. https://doi.org/10.1186/s13287-017-0531-4.
18. Lo Sicco C, Reverberi D, Balbi C, Ulivi V, Principi E, Pascucci L, Becherini P, Bosco MC, Varesio L, Franzin C, et al. Mesenchymal stem cell-derived extracellular vesicles as mediators of anti-inflammatory effects: endorsement of macrophage polarization. Stem Cells Transl Med. 2017;6:1018–28. https://doi.org/10.1002/sctm.16-0363.
19. Luo Z, Lin J, Sun Y, Wang C, Chen J. Bone marrow stromal cell-derived exosomes promote muscle healing following contusion through macrophage polarization. Stem Cells Dev. 2021;30:135–48. https://doi.org/10.1089/scd.2020.0167.
20. Chang C-L, Sung P-H, Chen K-H, Shao P-L, Yang C-C, Cheng B-C, Lin K-C, Chen C-H, Chai H-T, Chang H-W, et al. Adipose-derived mesenchymal stem cell-derived exosomes alleviate overwhelming systemic inflammatory reaction and organ damage and improve outcome in rat sepsis syndrome. Am J Transl Res. 2018;10:1053–70.
21. Ruiz M, Toupet K, Maumus M, Rozier P, Jorgensen C, Noël D. TGFBI secreted by mesenchymal stromal cells ameliorates osteoarthritis and is detected in extracellular vesicles. Biomaterials. 2020;226:119544. https://doi.org/10.1016/j.biomaterials.2019.119544.
22. Ragni E, De Luca P, Marmotti A, de Girolamo L. miR-26a-5p is a stable reference gene for miRNA studies in chondrocytes from developing human cartilage. Cells. 2019;8:E631. https://doi.org/10.3390/cells8060631.
23. Ragni E, Perucca Orfei C, De Luca P, Colombini A, Viganò M, Lugano G, Bollati V, de Girolamo L. Identification of miRNA reference genes in extracellular vesicles from adipose derived mesenchymal stem cells for studying osteoarthritis. Int J

Mol Sci. 2019;20:E1108. https://doi.org/10.3390/ijms20051108.
24. Ragni E, Perucca Orfei C, De Luca P, Viganò M, Colombini A, Lugano G, Bollati V, de Girolamo L. miR-22-5p and miR-29a-5p are reliable reference genes for analyzing extracellular vesicle-associated miRNAs in adipose-derived mesenchymal stem cells and are stable under inflammatory priming mimicking osteoarthritis condition. Stem Cell Rev Rep. 2019;15:743–54. https://doi.org/10.1007/s12015-019-09899-y.
25. An Q, Zhou Z, Xu C, Xiao Q. Exosomes derived from mir-337-3p over-expressing tendon stem cells protect against apoptosis of tenocytes via targeting caspase3. BMC Musculoskelet Disord. 2024;25:561. https://doi.org/10.1186/s12891-024-07691-9.
26. Wu R, Li H, Sun C, Liu J, Chen D, Yu H, Huang Z, Lin S, Chen Y, Zheng Q. Exosome-based strategy for degenerative disease in orthopedics: recent progress and perspectives. J Orthop Translat. 2022;36:8–17. https://doi.org/10.1016/j.jot.2022.05.009.
27. Cosenza S, Ruiz M, Toupet K, Jorgensen C, Noël D. Mesenchymal stem cells derived exosomes and microparticles protect cartilage and bone from degradation in osteoarthritis. Sci Rep. 2017;7:16214. https://doi.org/10.1038/s41598-017-15376-8.
28. Zhu Y, Wang Y, Zhao B, Niu X, Hu B, Li Q, Zhang J, Ding J, Chen Y, Wang Y. Comparison of exosomes secreted by induced pluripotent stem cell-derived mesenchymal stem cells and synovial membrane-derived mesenchymal stem cells for the treatment of osteoarthritis. Stem Cell Res Ther. 2017;8:64. https://doi.org/10.1186/s13287-017-0510-9.
29. Wu J, Kuang L, Chen C, Yang J, Zeng W-N, Li T, Chen H, Huang S, Fu Z, Li J, et al. miR-100-5p-abundant exosomes derived from infrapatellar fat pad MSCs protect articular cartilage and ameliorate gait abnormalities via inhibition of mTOR in osteoarthritis. Biomaterials. 2019;206:87–100. https://doi.org/10.1016/j.biomaterials.2019.03.022.
30. Zavatti M, Beretti F, Casciaro F, Bertucci E, Maraldi T. Comparison of the therapeutic effect of amniotic fluid stem cells and their exosomes on monoiodoacetate-induced animal model of osteoarthritis. Biofactors. 2020;46:106–17. https://doi.org/10.1002/biof.1576.
31. Zhang S, Chu WC, Lai RC, Lim SK, Hui JHP, Toh WS. Exosomes derived from human embryonic mesenchymal stem cells promote osteochondral regeneration. Osteoarthr Cartil. 2016;24:2135–40. https://doi.org/10.1016/j.joca.2016.06.022.
32. Liu X, Yang Y, Li Y, Niu X, Zhao B, Wang Y, Bao C, Xie Z, Lin Q, Zhu L. Integration of stem cell-derived exosomes with in situ hydrogel glue as a promising tissue patch for articular cartilage regeneration. Nanoscale. 2017;9:4430–8. https://doi.org/10.1039/c7nr00352h.
33. Hu H, Dong L, Bu Z, Shen Y, Luo J, Zhang H, Zhao S, Lv F, Liu Z. miR-23a-3p-abundant small extracellular vesicles released from Gelma/nanoclay hydrogel for cartilage regeneration. J Extracell Vesicles. 2020;9:1778883. https://doi.org/10.1080/20013078.2020.1778883.
34. Zhang F-X, Liu P, Ding W, Meng Q-B, Su D-H, Zhang Q-C, Lian R-X, Yu B-Q, Zhao M-D, Dong J, et al. Injectable mussel-inspired highly adhesive hydrogel with exosomes for endogenous cell recruitment and cartilage defect regeneration. Biomaterials. 2021;278:121169. https://doi.org/10.1016/j.biomaterials.2021.121169.
35. Jiang S, Tian G, Yang Z, Gao X, Wang F, Li J, Tian Z, Huang B, Wei F, Sang X, et al. Enhancement of acellular cartilage matrix scaffold by Wharton's jelly mesenchymal stem cell-derived exosomes to promote osteochondral regeneration. Bioact Mater. 2021;6:2711–28. https://doi.org/10.1016/j.bioactmat.2021.01.031.
36. Tao S-C, Yuan T, Zhang Y-L, Yin W-J, Guo S-C, Zhang C-Q. Exosomes derived from miR-140-5p-overexpressing human synovial mesenchymal stem cells enhance cartilage tissue regeneration and prevent osteoarthritis of the knee in a rat model. Theranostics. 2017;7:180–95. https://doi.org/10.7150/thno.17133.
37. Karlsen TA, Jakobsen RB, Mikkelsen TS, Brinchmann JE. microRNA-140 targets RALA and regulates chondrogenic differentiation of human mesenchymal stem cells by translational enhancement of SOX9 and ACAN. Stem Cells Dev. 2014;23:290–304. https://doi.org/10.1089/scd.2013.0209.
38. Mao G, Zhang Z, Hu S, Zhang Z, Chang Z, Huang Z, Liao W, Kang Y. Exosomes derived from miR-92a-3p-overexpressing human mesenchymal stem cells enhance chondrogenesis and suppress cartilage degradation via targeting WNT5A. Stem Cell Res Ther. 2018;9:247. https://doi.org/10.1186/s13287-018-1004-0.
39. Cheng J, Sun Y, Ma Y, Ao Y, Hu X, Meng Q. Engineering of MSC-derived exosomes: a promising cell-free therapy for osteoarthritis. Membranes (Basel). 2022;12:739. https://doi.org/10.3390/membranes12080739.
40. Sevivas N, Teixeira FG, Portugal R, Direito-Santos B, Espregueira-Mendes J, Oliveira FJ, Silva RF, Sousa N, Sow WT, Nguyen LTH, et al. Mesenchymal stem cell secretome improves tendon cell viability in vitro and tendon-bone healing in vivo when a tissue engineering strategy is used in a rat model of chronic massive rotator cuff tear. Am J Sports Med. 2018;46:449–59. https://doi.org/10.1177/0363546517735850.
41. Sevivas N, Teixeira FG, Portugal R, Araújo L, Carriço LF, Ferreira N, Vieira da Silva M, Espregueira-Mendes J, Anjo S, Manadas B, et al. Mesenchymal stem cell secretome: a potential tool for the prevention of muscle degenerative changes associated with chronic rotator cuff tears. Am J Sports Med. 2017;45:179–88. https://doi.org/10.1177/0363546516657827.
42. Chen W, Sun Y, Gu X, Cai J, Liu X, Zhang X, Chen J, Hao Y, Chen S. Conditioned medium of human bone marrow-derived stem cells promotes tendon-bone healing of the rotator cuff in a rat model. Biomaterials.

2021;271:120714. https://doi.org/10.1016/j.biomaterials.2021.120714.

43. Jenner F, Wagner A, Gerner I, Ludewig E, Trujanovic R, Rohde E, von Rechenberg B, Gimona M, Traweger A. Evaluation of the potential of umbilical cord mesenchymal stromal cell-derived small extracellular vesicles to improve rotator cuff healing: a pilot ovine study. Am J Sports Med. 2023;51:331–42. https://doi.org/10.1177/03635465221145958.
44. Fredianto M, Herman H, Ismiarto YD, Purba A, Putra A, Hidayah N. Combination effect of rotator cuff repair with secretome-hypoxia MSCs ameliorates TNMD, RUNX2, and healing histology score in rotator cuff tear rats. Arch Bone Jt Surg. 2023;11:617–24. https://doi.org/10.22038/ABJS.2023.67933.3218.
45. Fredianto M, Herman H, Dias Ismiarto Y, Putra A, Alif I, Amalina ND, Ariq Nazar M. Secretome of hypoxia-preconditioned mesenchymal stem cells enhance the expression of HIF-1a and bFGF in a rotator cuff tear model. Med Glas (Zenica). 2023;20. https://doi.org/10.17392/1605-23.
46. Yu Y, Zhou Y, Cheng T, Lu X, Yu K, Zhou Y, Hong J, Chen Y. Hypoxia enhances tenocyte differentiation of adipose-derived mesenchymal stem cells by inducing hypoxia-inducible factor-1α in a co-culture system. Cell Prolif. 2016;49:173–84. https://doi.org/10.1111/cpr.12250.
47. Zhang T, Wu Y, Li X, Zhang A, Liu H, Shi M, Zhang Z, Lu W, Guo Y, Tang X, et al. Small extracellular vesicles derived from tendon stem cells promote the healing of injured Achilles tendons by regulating miR-145-3p. Acta Biomater. 2023;172:280–96. https://doi.org/10.1016/j.actbio.2023.10.004.
48. Dou Y, Zhai H, Li H, Xing H, Zhu C, Xuan Z. Endothelial cells-derived exosomes-based hydrogel improved tendinous repair via anti-inflammatory and tissue regeneration-promoting properties. J Nanobiotechnol. 2024;22:401. https://doi.org/10.1186/s12951-024-02607-0.
49. Tan X, Xiao H, Yan A, Li M, Wang L. Effect of exosomes from bone marrow-derived mesenchymal stromal cells and adipose-derived stromal cells on bone-tendon healing in a murine rotator cuff injury model. Orthop J Sports Med. 2024;12:23259671231210304. https://doi.org/10.1177/23259671231210304.
50. Zhang X, Wu Y, Han K, Fang Z, Cho E, Hu Y, Huangfu X, Zhao J. 3-Dimensional bioprinting of a tendon stem cell-derived exosomes loaded scaffold to bridge the unrepairable massive rotator cuff tear. Am J Sports Med. 2024;52(9):2358–71. https://doi.org/10.1177/03635465241255918.
51. Zhang J, Cai Z, Feng F, Peng Y, Cui Y, Xu Y. Age-different BMSCs-derived exosomes accelerate tendon-bone interface healing in rotator cuff tears model. Gene. 2024;895:148002. https://doi.org/10.1016/j.gene.2023.148002.
52. Xu C, Mi Z, Dong Z, Chen X, Ji G, Kang H, Li K, Zhao B, Wang F. Platelet-derived exosomes alleviate knee osteoarthritis by attenuating cartilage degeneration and subchondral bone loss. Am J Sports Med. 2023;51:2975–85. https://doi.org/10.1177/03635465231188122.
53. Han L, Hu N, Wang C, Ye Z, Wang T, Lan F. Platelet-rich plasma-derived exosomes promote rotator cuff tendon-bone healing. Injury. 2024;55:111212. https://doi.org/10.1016/j.injury.2023.111212.
54. Li M, Shi L, Chen X, Yi D, Ding Y, Chen J, Xing G, Chen S, Wang L, Zhang Y, et al. In-situ gelation of fibrin gel encapsulating platelet-rich plasma-derived exosomes promotes rotator cuff healing. Commun Biol. 2024;7:205. https://doi.org/10.1038/s42003-024-05882-7.
55. Chen D, Tang Q, Song W, He Y. Platelet-derived exosomes alleviate tendon stem/progenitor cell senescence and ferroptosis by regulating AMPK/Nrf2/GPX4 signaling and improve tendon-bone junction regeneration in rats. J Orthop Surg Res. 2024;19:382. https://doi.org/10.1186/s13018-024-04869-8.

Micro RNA in Orthopaedics 13

George Jacob, Kazunori Shimomura,
Hiroto Hanai, and Norimasa Nakamura

13.1 Introduction

MicroRNAs (miRNAs) are principal molecules responsible for regulating tissue homeostasis by silencing RNA and post-transcriptional regulation of gene suppression [1]. MiRNAs are small non-coding ribonucleic acid (RNA) molecules around 19–24 nucleotides in length [2]. These molecules play pivotal roles in cellular processes starting from embryonic development to tissue repair and regeneration [3]. This makes them of particular interest in regenerative domains of medicine. First identified in nematodes [4, 5] and then in other organisms, miRNAs are mainly found within cells; however, extracellular miRNAs also exist in extracellular matrices (ECM) and various biological fluids [6, 7]. In humans, 2588 mature miRNAs have been identified and listed [8].

With miRNAs being so heavily involved in all biological processes, research has focused on harnessing their natural function of signalling and communication to halt disease progression and initiate reparative processes to enhance management of pathologies on a genetic level. With respect to orthopaedics, most published research remains pre-clinical and aimed at understanding what effects these molecules have on each cellular process. In this chapter, we aim to discuss the current available literature reporting on use of miRNA in orthopaedics.

13.2 Definition/Characteristics

MiRNAs are non-coding RNAs with an average length of 22 nucleotides [9]. They originate from deoxyribonucleic acid (DNA) sequences and undergo processing to become a mature miRNA. These miRNAs function by suppressing

G. Jacob
Department of Orthopaedics, VPS Lakeshore Hospital, Kochi, Kerala, India

Avant Orthopaedics, Cochin, Kerala, India

K. Shimomura (✉)
Department of Rehabilitation, Kansai University of Welfare Sciences, Osaka, Japan

Department of Orthopaedic Surgery, Osaka University Graduate School of Medicine, Osaka, Japan
e-mail: kazunori-shimomura@umin.net

H. Hanai
Department of Orthopaedic Surgery, Osaka University Graduate School of Medicine, Osaka, Japan

N. Nakamura
Department of Orthopaedic Surgery, Osaka University Graduate School of Medicine, Osaka, Japan

Institute for Medical Science in Sports, Osaka Health Science University, Osaka, Japan

Global Center for Medical Engineering and Informatics, Osaka University, Osaka, Japan
e-mail: norimasa.nakamura@ohsu.ac.jp

A. Gobbi et al. (eds.), *Regenerative Medicine in Sports and Orthopaedics*,
https://doi.org/10.1007/978-3-031-84693-9_13

expression of genes by interacting with a 3′ untranslated region (UTR); however, other regions of interaction have been reported as well [9]. Overall, miRNAs are deployed through various cellular compartments to control rates of transcription and translation [10]. Additionally, under certain conditions they have been shown to activate gene expression.

13.3 Mechanisms of Action

MiRNAs have several mechanisms of actions that result in the regulation of cellular functions and overall tissue homeostasis and repair [11] (Fig. 13.1). The process starts with the genesis of the microRNA, which is transcribed from a DNA sequence into a primary miRNA (pri-miRNA) by RNA polymerase II. The pri-miRNA is then processed within the cell nucleus to form precursor miRNA (pre-miRNA). These are then moved into the cell cytoplasm. Next, the pre-miRNA undergoes maturation by the enzyme Dicer cleaving its stem-loop structure into short double-stranded RNA molecules with a length of 22 nucleotides. At this point, it consists of one mature miRNA strand and its complementary sequence. The mature miRNA is then incorporated into the RNA-induced silencing complex (RISC) and the complementary sequence degraded. Next is the process of target recognition; here the mature miRNA guides the complex to target mRNAs through base pairing interactions. This is done by binding to complementary sequences on the 3′UTR as mentioned earlier. Once bound to the mRNA, the miRNA regulates gene expression through mRNA degradation and translation inhibition [9]. Where there is imperfection in the target mRNA, the miRNA can inhibit translation and signal a premature dissociation, decreasing protein synthesis [10].

Activity of miRNA can be regulated at various of the above-mentioned levels, including transcriptional control of the miRNA genes, pri-miRNA processing, loading of the mature miRNAs, and stability of miRNA-target mRNA interactions [9]. Finally, competition for shared target mRNAs also influence miRNA activity [12].

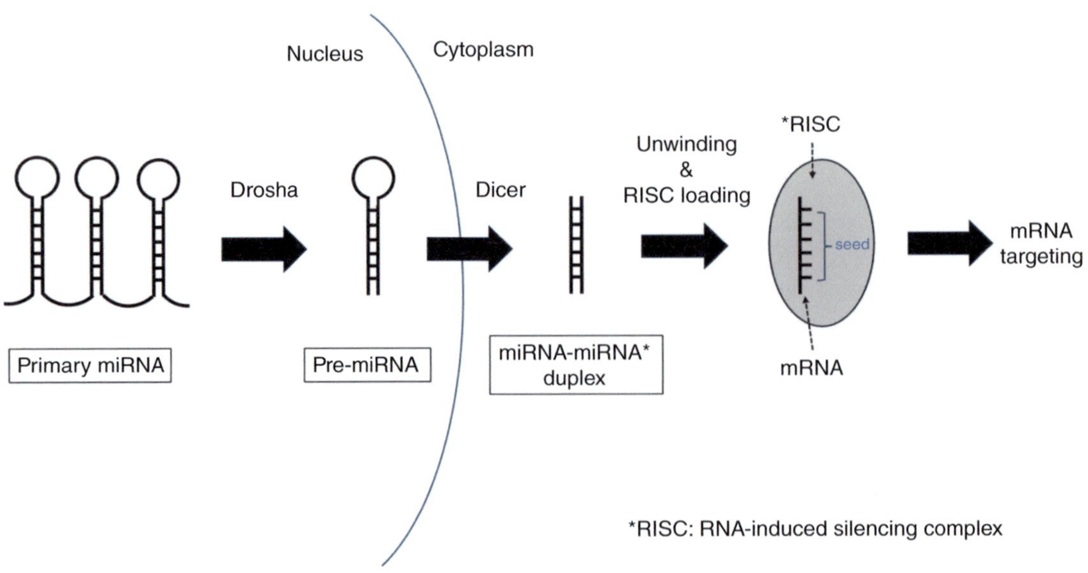

Fig. 13.1 Representation of miRNA structure, biogenesis and mechanism of action. The miRNA has a 5′ end and a 3′ end. The sequence consists of about 21–25 nucleotides. The primary miRNA forms a hairpin structure with a short region of complementarity. The nucleotides in the top strands pair with those on the bottom strand. The seed region in the center is crucial for target mRNA recognition and binding

13.4 Extraction Methods

To study and use miRNA for therapeutic purposes, it must first be extracted from tissue samples and biological fluids. Owing to the small size of miRNA and high levels of protein contamination, it has remained quite a challenge to isolate and accurately quantify miRNA particularly from complex samples, such as plasma [13]. Over time, various methods have been utilized to extract miRNA, and each has its own advantages and drawbacks. The method chosen usually depends on the source, available resources, and desired use and analyses. Table 13.1 is a summary of the various methods of extraction.

13.5 Delivery Systems

Delivery methods for miRNA can broadly be divided into viral and non-viral methods. Viral methods include the use of retroviruses, lentiviruses, and adenoviruses [14]. These viruses are genetically modified in a specified genomic area to prevent their replication to make them safe for use. The advantage with this method is high transfection and a constant high level of expression [15]. They do, however, come with a limitation on the size of DNA that can be transferred and possible toxic, immunogenic reaction. Non-viral methods negate these problems but have a lower transfection efficiency [16]. Table 13.2

Table 13.1 Available extraction methods

Phenol-chloroform extraction	Phenol-chloroform is used to separate the miRNA from other cellular components and then precipitation of the miRNA
Silica-based column purification	miRNA is bound to silica columns, and then contaminants are washed away. This is a simple and reproducible technique
Trizol (or TRI reagent) method	Trizol is used to denature the proteins and extract RNA and miRNA. After which, chloroform is used to separate the phases and precipitate the miRNA with isopropanol
miRNA magnetic bead-based purification	Magnetic beads with miRNA-specific oligonucleotides to capture miRNA are used. This is a rapid and efficient method for hi-throughout application
Acid-phenol method	Acidified phenol is used to extract the miRNA. Good method to extract smaller miRNA
Commercial kits	Standardized protocols may be offered as commercial kits optimized for various ample types. These tend to be more expensive
Microfluidic-based extraction	Microfluidic devices provide rapid and automated extraction. Good in settings for diagnostics and research
Enzymatic methods	Enzymes such as proteinase K or lysosome are used for digestion of proteins and cell components. RNA is then purified

Table 13.2 Available delivery methods

Viral vector	Efficient for transfer of miRNA to cells but have significant safety concerns. Being potential immunogenicity and insertional mutagenesis
Non-viral vectors	Lipid-based delivery: Lipid nanoparticles encapsulating miRNAs are safe but have a low transfection rate. Polymer-based delivery: Polymers like polyethyleneimine (PEI) or poly(lactic-co-glycolic acid) (PLGA) can be engineered into a nanoparticle for cell delivery. Their properties can be modified for efficiency and have low immunogenicity Peptide-based delivery: These are protein-based carriers conjugated to miRNAs to enhance cellular uptake, which offer targeted delivery and low cytotoxicity
Exosome-based delivery	Exosomes are extracellular vesicles secreted by cells and can transport across cell membranes. They have a high biocompatibility, traverse biological barriers with ease, and can have a targeted delivery
Electroporation	Electric pulses are used to cause cell membrane disruption allowing miRNAs to enter the cells. It is a simple and efficient method but can induce cell stress and damage
Physical methods	Ultrasound-mediated delivery or magnetic field-assisted delivery enhance miRNA uptake into cells. This is a non-invasive and localized delivery option
Cellular carriers	Stem cells can be engineered to express miRNAs. This method offers the benefit of cell-based homing and sustained release of miRNA

summarizes various delivery methods used for transfer of miRNA.

13.6 Applications of miRNA in Orthopaedics

13.6.1 Bone

MiRNA plays a significant job in bone metabolism by regulating both osteoblast and osteoclasts differentiation effectively controlling bone formation and resorption [17]. Their defined roles in bone metabolism can, therefore, be harnessed to treat bone-related disorders. MiRNAs signal bone formation by targeting regulators of bone morphogenic protein (BMP), Wnt/β-catenin, and transforming growth factor beta (TGF-β). Regulators that function to reduce osteoblast differentiation are miR-29, miR-34, and miR-138 [18]. These function by inhibiting osteogenic transcription factors (Runx2 and Osterix). MiR-2861, miR-133, and miR-204 promote bone formation and mineralization by targeting inhibitors of osteogenesis (DKK1 and sclerostin) [19].

Bone resorption is mediated by coordinated osteoclast differentiation, which is imperative in maintaining calcium homeostasis [20]. Receptor activation of nuclear factor-kappa B ligand (RANKL) and osteoprotegerin (OPG) result in the expression of osteoclastogenic factors. MiR-21, miR-34, and miR-155 are responsible for this pathway by targeting negative regulators of RANKL signalling [19–21]. On the other hand, miR-223, miR-124, and miR-125b inhibit osteoclast formation by targeting transcription factors essential for osteoclast differentiation [22].

Over expression and imbalanced of these miRNAs can result in adverse processes such as inflammation, uncontrolled cell proliferation, and angiogenesis. An effective strategy to treating disorders such as osteoporosis, osteoarthritis, and bone metastasis potentially lies in modulating these miRNAs [23]. Additionally, miRNA expression may be used as a tool to diagnoses pathologies and bone tumors, where their upregulated expression can raise suspicion to an underlying disease process [24]. Further research into miRNA in bone metabolism will help develop both a diagnostic advantage and targeted therapeutic therapy.

13.6.2 Cartilage

The various specialized biomechanical properties of cartilage are maintained by cells known as chondrocytes, which produce and maintain the surrounding ECM [25]. The ECM provides structure and supports the over cartilage tissue, which is essential for efficient joint function [26]. Chondrocyte metabolism is controlled by miRNA expression which in turn affects ECM production, content, and maintenance [27]. This directly affects the overall strength and integrity of the joint cartilage. The production of aggrecan and collagen type II, which are the main constituents of the chondral ECM, is regulated by miR-140 [28]. Imbalance of this gene expression is linked to cartilage degradation and osteoarthritis of the joint as shown in animal models. Inhibition of miR-27a has shown increased production of insulin-like growth factor binding protein-5 (Igfbp5) and matrix metalloproteinase 13 (mmp13), which both significantly contribute to progression of osteoarthritis [29]. Pre-clinical studies have proven various other miRNAs to contribute to the regulation of different gene expressions that can promote or inhibit chondrogenesis and regulate ECM production. Some therapeutic strategies have already aimed to modulate these mechanisms and slow down the progression of cartilage degradation. One such example is the use of interleukin-1 antagonists to reduce expression of interleukin-1β which would downregulate the expression of miR0146a expression, which stimulates chondrocyte apoptosis [30]. In this way, such miRNA therapies can also be used as a treatment to reduce inflammation and chondral damage after a sports injury or even as an adjuvant to a cartilage resurfacing procedure.

The further understanding of miRNA can additionally be used to diagnose and determine the prognosis of cartilage-related disorders early aiming to manage them early on in disease progression rather than once the process has advanced. Additionally, it would allow for serial monitoring of the joint health after an injury or treatment interventions.

13.6.3 Meniscus

Meniscus injuries are commonly associated with sports injuries but also well documented in osteoarthritic knee joints [31, 32]. They may be a part of a mechanical trauma but also thought to be due to the negative biochemical mechanisms and the upregulation of adversely impacting genetic factors. MiRNA could be used to shut down negative mechanisms that increase intra-articular inflammation and adverse chemical events.

In the setting of a meniscus tear, the use of miRNA has a few potential uses. Like the intended use of miRNA in osteoarthritis, reducing inflammatory mediators improves the harsh joint environment, which effectively tackles the negative cartilage and meniscus factors [33]. Intra-articular joint environments are avascular and the meniscal tissues even more so [34, 35]. The use of miRNAs could promote angiogenesis and improved the delivery of natural healing factors to the tear site, allowing for enhanced repair and healing. Intra-articular injection of miRNA-210 has shown some potential benefit in vivo [36]. Using miRNA for diagnosis of a meniscal tear remains limited as further characterization of miRNA expression in synovial fluid is necessary.

13.6.4 Ligament

Anterior Cruciate ligament (ACL) injuries are a common sporting injury [37] is well researched condition, and the post-traumatic joint degeneration is thought to be a result of unregulated biochemical changes within the joint [38]. Delayed or lack of ACL healing is attributed to the avascular joint environment. Much research has focused on modulating this chemical environment to reduce negative changes to the surround structures and accelerate the injured ACL healing [39, 40]. Downregulation of miR-210 has been noted after ACL injury and so supplementing the joint with miR-210 could induce a healing response by upregulating vascular endothelial growth factor (VEGF) and fibroblast growth factor 2 (FGF2) within the knee joint. A small animal study by Shoji et al. was able to demonstrate better ACL healing using this method [41]. MiR-29 and miR-133 have been shown to regulate ECM remodelling, especially regarding collagen synthesis [42]. Reduced collagen production would have a negative impact on ligament healing and integrity. In the setting of diagnostic value and an ACL injury, miRNA expression would help in determining injury severity and prognostic outcome and monitoring the responses in the joint environment to administered treatment and surgery.

13.6.5 Tendon

Tendon injuries are a common occurrence in the athletic population and can be quite difficult to treat in severe cases. With the lack of focused treatments for tendon injuries being available, it is thought that regulating gene expression may improve healing and control the reparative processes. Various miRNAs have been documented to be involved in tendon healing and tenogenic differentiation [43].

MiRNAs have been shown to control tendinocyte apoptosis and proliferation, which are specific to miR-499 and miRNA-205-5p [43]. This could be beneficial when treating injuries such as rotator cuff tears to promote healing of the tendon onto the proximal humerus. In the process of tendon healing, adhesion has continually been a challenge, and it has been known to be governed by the expression of various genes, namely transforming growth factor β, Smad3, and p65 [44]. Using miRNA therapies, tendon adhesion can be reduced by modulating expression of these specified genes. As mentioned earlier, miRNA controls ECM production and tissue homeostasis. In most tendon injuries, the rate of remodelling is high, which results in decreased mechanical strength predisposing the tendon to reinjury [45]. The following miRNAs have been proven to be

Table 13.3 Summary of miRNA in each tissue

Tissue	Type of miRNA	Effects
Bone	miR-21, miR-29, miR-30, miR-34, miR-124, miR-125b, miR-133, miR-138, miR-155, miR-204, miR-223, miR-2861	Control bone formation and resorption by regulating osteoblast and osteoclast function
Cartilage	miR-9, miR-22, miR-140, miR-146a, miR-199, miR-210, miR-455, miR-675	Regulate chondrocyte development, maintenance of cartilage ECM, and influence pathologic processes associated with cartilage
Meniscus	miR-140, miR-146a	Involved in meniscus health and pathology. Similar to chondrocyte mediating miRNAs
Ligament	miR-29, miR-133a, miR-140, miR-146a, miR-155, miR-210	Regulate synthesis of elastin and collagen. Mediate differentiation of MSCs to ligament fibroblasts
Tendon	miR-29, miR-133a, miR-140, miR-146a, miR-208a, miR-210	Essential in tendon cell proliferation, differentiation, matrix synthesis, and remodeling
Nerve	miR-9, miR-21, miR-124, miR-132, miR-133b, miR-137, miR-155	Involved in processes of neuronal differentiation, axon guidance, synaptic plasticity, and nerve regeneration

essential to tendon matrix production, miR-145-5p, miR-151a-3p, miR-382-5p, miR-199a-5p, miR-21-5p, miR-125a-5p, and miR-498 [43]. Modifying these miRNAs can support improved tendon healing and ECM production. Lastly, miRNAs control tenogenic differentiation of tendon stem cells and mesenchymal stem cells. Using miRNA, tendon healing can be supported, and in the case of a tendon repair, it can be used to augment the repair to result in a superior repair outcome.

13.6.6 Nerve

It is well known that miRNAs have a role in neural cell proliferation and differentiation [46]. They have been linked to stimulating cell apoptosis, inflammation, neurogenesis, and angiogenesis. MiRNAs, such as miR-124 and miR-132, are known play a significant role in neuronal development and synaptic plasticity. Additionally, miRNA roles in neurodegenerative disorders have been studied, and much research has focused on utilizing them for early detection and possible treatment [47].

With respect to orthopaedics, miRNAs may have a significant role to play in spine cord injuries, nerve injuries, and neuropathic pain. After a nerve injury cell apoptosis and degeneration occur if this is not controlled it can be detrimental to final recovery [48]. MiR-195-5p inhibition has been shown to have close relationship with X-linked inhibitor of apoptosis protein (XIAP) and inhibition of miR-192-5p demonstrated upregulation of XIAP which decreased neural apoptosis and improve nerve recovery in a rat model [49]. This application of miRNA can be used in spinal cord and nerve injuries to promote healing and reduce the deleterious factors that inhibit neural regeneration (Table 13.3).

13.7 Future Perspective

With further understanding and research into miRNA mechanisms and functions, they can be effectively used to modulate body function. The challenge remains first in studying and mapping each miRNA with biological effects and functions, which is well underway, and then its delivery into the tissue. As discussed earlier in this chapter, delivery methods have their pros and cons, so finding them most effective method to deliver a high transfection rate with minimum side effects remains the goal. Additionally, it is known that contaminant-free isolation of miRNAs is not an easy task. Addressing these difficulties will be the way forward in making miRNA therapies a successful treatment option in all fields of medicine.

13.8 Conclusion

MiRNAs appear to be the next frontier in diagnostic and targeted therapeutic medicine. With the ability to modulate all biological tissue processes within the body, allowing early diagnosis of a pathologic process, halt a disease process, or enhance tissue healing. Further research and translation into clinical trials will truly elucidate the clinical benefit of miRNA.

References

1. Guarnieri DJ, DiLeone RJ. MicroRNAs: a new class of gene regulators. Ann Med. 2008;40(3):197–208.
2. Bartel DP. MicroRNAs: genomics, biogenesis, mechanism, and function. Cell. 2004;116(2):281–97.
3. Li MA, He L. microRNAs as novel regulators of stem cell pluripotency and somatic cell reprogramming. BioEssays. 2012;34(8):670–80.
4. Reinhart BJ, Slack FJ, Basson M, Pasquinelli AE, Bettinger JC, Rougvie AE, Horvitz HR, Ruvkun G. The 21-nucleotide let-7 RNA regulates developmental timing in Caenorhabditis elegans. Nature. 2000;403(6772):901–6.
5. Lee RC, Feinbaum RL, Ambros V. The C. elegans heterochronic gene lin-4 encodes small RNAs with antisense complementarity to lin-14. Cell. 1993;75(5):843–54.
6. Mitchell PS, Parkin RK, Kroh EM, Fritz BR, Wyman SK, Pogosova-Agadjanyan EL, Peterson A, Noteboom J, O'Briant KC, Allen A, Lin DW. Circulating microRNAs as stable blood-based markers for cancer detection. Proc Natl Acad Sci U S A. 2008;105(30):10513–8.
7. Ko SY, Lee W, Naora H. Harnessing microRNA-enriched extracellular vesicles for liquid biopsy. Front Mol Biosci. 2024;11:1356780.
8. Kozomara A, Griffiths-Jones S. miRBase: annotating high confidence microRNAs using deep sequencing data. Nucleic Acids Res. 2014;42(D1):D68–73.
9. O'Brien J, Hayder H, Zayed Y, Peng C. Overview of microRNA biogenesis, mechanisms of actions, and circulation. Front Endocrinol. 2018;9:388354.
10. Oliveto S, Mancino M, Manfrini N, Biffo S. Role of microRNAs in translation regulation and cancer. World J Biol Chem. 2017;8(1):45.
11. Sen CK, Ghatak S. miRNA control of tissue repair and regeneration. Am J Pathol. 2015;185(10):2629–40.
12. Bosia C, Sgrò F, Conti L, Baldassi C, Brusa D, Cavallo F, Cunto FD, Turco E, Pagnani A, Zecchina R. RNAs competing for microRNAs mutually influence their fluctuations in a highly non-linear microRNA-dependent manner in single cells. Genome Biol. 2017;18:1–4.
13. Siddika T, Heinemann IU. Bringing MicroRNAs to light: methods for MicroRNA quantification and visualization in live cells. Front Bioeng Biotechnol. 2021;8:619583.
14. Nayerossadat N, Maedeh T, Ali PA. Viral and nonviral delivery systems for gene delivery. Adv Biomed Res. 2012;1(1):27.
15. Yang N. An overview of viral and nonviral delivery systems for microRNA. Int J Pharm Investig. 2015;5(4):179.
16. Kelly EJ, Russell SJ. MicroRNAs and the regulation of vector tropism. Mol Ther. 2009;17(3):409–16.
17. Weivoda MM, Lee SK, Monroe DG. miRNAs in osteoclast biology. Bone. 2021;143:115757.
18. Garcia J, Delany AM. MicroRNAs regulating TGFβ and BMP signaling in the osteoblast lineage. Bone. 2021;143:115791.
19. Fang S, Deng Y, Gu P, Fan X. MicroRNAs regulate bone development and regeneration. Int J Mol Sci. 2015;16(4):8227–53.
20. Sun Y, Li J, Xie X, Gu F, Sui Z, Zhang K, Yu T. Recent advances in osteoclast biological behavior. Front Cell Dev Biol. 2021;9:788680.
21. Kim K, Kim JH, Kim I, Lee J, Seong S, Park YW, Kim N. MicroRNA-26a regulates RANKL-induced osteoclast formation. Mol Cells. 2015;38(1):75–80.
22. Inoue K, Ng C, Xia Y, Zhao B. Regulation of osteoclastogenesis and bone resorption by miRNAs. Front Cell Dev Biol. 2021;9:651161.
23. Sharma AR, Lee YH, Lee SS. Recent advancements of miRNAs in the treatment of bone diseases and their delivery potential. Curr Res Pharmacol Drug Discov. 2023;4:100150.
24. Sharma PC, Gupta A. MicroRNAs: potential biomarkers for diagnosis and prognosis of different cancers. Transl Cancer Res. 2020;9(9):5798.
25. Chen C, Tambe DT, Deng L, Yang L. Biomechanical properties and mechanobiology of the articular chondrocyte. Am J Phys Cell Phys. 2013;305(12):C1202–8.
26. Jacob G, Shimomura K, Hart DA, Fujie H, Nakamura N. Mechanical and biologic properties of articular cartilage repair biomaterials. Orthop Biomech Sports Med. 2021:57–71.
27. Ping Li Y, Chun Wei X, Cui Li P, Wei Chen C, Hu Wang X, Jiao Q, Ming Wang D, Yuan Wei F, Zhong Zhang J, Wei L. The role of miRNAs in cartilage homeostasis. Curr Genomics. 2015;16(6):393–404.
28. Li SH, Wu QF. MicroRNAs target on cartilage extracellular matrix degradation of knee osteoarthritis. Eur Rev Med Pharmacol Sci. 2021;25(3):1185–97.
29. Duan L, Liang Y, Xu X, Xiao Y, Wang D. Recent progress on the role of miR-140 in cartilage matrix remodelling and its implications for osteoarthritis treatment. Arthritis Res Ther. 2020;22(1):194.
30. Dinarello CA. Reduction of inflammation by decreasing production of interleukin-1 or by specific receptor antagonism. Int J Tissue React. 1992;14(2):65–75.
31. Jacob G, Shimomura K, Hart DA, Nakamura N. Meniscus injury and early osteoarthritis. In: Early

osteoarthritis: state-of-the-art approaches to diagnosis, treatment and controversies; 2022. p. 259–69.
32. Jacob G, Shimomura K, Krych AJ, Nakamura N. The meniscus tear: a review of stem cell therapies. Cells. 2019;9(1):92.
33. Panagopoulos PK, Lambrou GI. The involvement of microRNAs in osteoarthritis and recent developments: a narrative review. Mediterr J Rheumatol. 2018;29(2):67–79.
34. Fox AJ, Bedi A, Rodeo SA. The basic science of human knee menisci: structure, composition, and function. Sports Health. 2012;4(4):340–51.
35. Arnoczky SP, Warren RF. Microvasculature of the human meniscus. Am J Sports Med. 1982;10(2): 90–5.
36. Kawanishi Y, Nakasa T, Shoji T, Hamanishi M, Shimizu R, Kamei N, Usman MA, Ochi M. Intra-articular injection of synthetic microRNA-210 accelerates avascular meniscal healing in rat medial meniscal injured model. Arthritis Res Ther. 2014;16:1–2.
37. Montalvo AM, Schneider DK, Webster KE, Yut L, Galloway MT, Heidt RS Jr, Kaeding CC, Kremcheck TE, Magnussen RA, Parikh SN, Stanfield DT. Anterior cruciate ligament injury risk in sport: a systematic review and meta-analysis of injury incidence by sex and sport classification. J Athl Train. 2019;54(5):472–82.
38. Wang LJ, Zeng N, Yan ZP, Li JT, Ni GX. Post-traumatic osteoarthritis following ACL injury. Arthritis Res Ther. 2020;22:1–8.
39. Li H, Chen C, Chen S. Posttraumatic knee osteoarthritis following anterior cruciate ligament injury: potential biochemical mediators of degenerative alteration and specific biochemical markers. Biomed Rep. 2015;3(2):147–51.
40. Higuchi H, Shirakura K, Kimura M, Terauchi M, Shinozaki T, Watanabe H, Takagishi K. Changes in biochemical parameters after anterior cruciate ligament injury. Int Orthop. 2006;30:43–7.
41. Shoji T, Nakasa T, Yamasaki K, Kodama A, Miyaki S, Niimoto T, Okuhara A, Kamei N, Adachi N, Ochi M. The effect of intra-articular injection of microRNA-210 on ligament healing in a rat model. Am J Sports Med. 2012;40(11):2470–8.
42. Suh EJ, Remillard MY, Legesse-Miller A, Johnson EL, Lemons JM, Chapman TR, Forman JJ, Kojima M, Silberman ES, Coller HA. A microRNA network regulates proliferative timing and extracellular matrix synthesis during cellular quiescence in fibroblasts. Genome Biol. 2012;13:1–4.
43. Ding L, Wang M, Qin S, Xu L. The roles of MicroRNAs in tendon healing and regeneration. Front Cell Dev Biol. 2021;9:687117.
44. Jiang K, Chun G, Wang Z, Du Q, Wang A, Xiong Y. Effect of transforming growth factor-β3 on the expression of Smad3 and Smad7 in tenocytes. Mol Med Rep. 2016;13(4):3567–73.
45. Li ZJ, Yang QQ, Zhou YL. Basic research on tendon repair: strategies, evaluation, and development. Front Med. 2021;8:664909.
46. Stappert L, Roese-Koerner B, Brüstle O. The role of microRNAs in human neural stem cells, neuronal differentiation and subtype specification. Cell Tissue Res. 2015;359:47–64.
47. Rashidi SK, Kalirad A, Rafie S, Behzad E, Dezfouli MA. The role of microRNAs in neurobiology and pathophysiology of the hippocampus. Front Mol Neurosci. 2023;16:1226413.
48. Menorca RM, Fussell TS, Elfar JC. Peripheral nerve trauma: mechanisms of injury and recovery. Hand Clin. 2013;29(3):317.
49. Liu X, Cui X, Guan G, Dong Y, Zhang Z. microRNA-192-5p is involved in nerve repair in rats with peripheral nerve injury by regulating XIAP. Cell Cycle. 2020;19(3):326–38.

14 Gene Therapy in Regenerative Medicine

Saseendar Samundeeswari,
Girinivasan Chellamuthu,
Gokulesh Devannagoundanur Gurumurthy,
Saseendar Shanmugasundaram,
and Ignacio Dallo

14.1 Introduction to Gene Therapy

Gene therapy is a method of altering the genome by introducing the engineered genetic material into the target cells. It is done for therapeutic purposes by expression of specific cellular characteristics. The primary aim is to incorporate the gene into the cell and promote its expression. The American Society of Gene and Cell Therapy (ASGCT), the European Medicines Agency (EMA), and the US Food and Drug Administration (FDA) define gene therapy as modification to gene expression that is accomplished through the replacement or correction of disease-causing genes, the inactivation of target genes, or the insertion of new or modified genes using genetically modified viruses, microorganisms, or cells as a vector or delivery system of genetic sequence or gene [1, 2] (Fig. 14.1).

Gene addition and gene editing are the types of gene therapy. Gene editing can in turn be divided into gene silencing and gene correction [3]. As implied by their names, gene addition inserts new functional genes into the nucleus, which may or may not incorporate itself into the host cell DNA. Gene silencing is the inhibition of host cell gene expression by exogenous genetic material delivered into the cell. Gene editing can be understood with an example of the CRISPR-Cas9 system, which consists of two components. The first is a guide RNA that identifies the target DNA sequence and then comes the Cas nuclease that cleaves the desired fragment of DNA making repair possible. Other methods of gene editing under research are adeno-associated virus-based homologous recombination, transcription activator-like effector nuclease (TALEN), and zinc finger nucleases (ZFN) [4].

S. Samundeeswari (✉) · S. Shanmugasundaram
Department of Orthopaedics, Sri Lakshmi Narayana Institute of Medical Sciences, Puducherry, India

G. Chellamuthu
Faculty of Research, Department of Biotechnology, Karpagam Academy of Higher Education, Coimbatore, India

Orthopaedic Research Group, Coimbatore, Tamil Nadu, India

G. D. Gurumurthy
Orthopaedic Research Group, Coimbatore, Tamil Nadu, India

Rajiv Gandhi Government General Hospital and Madras Medical College, Chennai, India

I. Dallo
Department of Orthopaedic Surgery and Sports Medicine, Unit of Biological Therapies and MSK Interventionism, Sport Me Medical Center, Seville, Spain

A. Gobbi et al. (eds.), *Regenerative Medicine in Sports and Orthopaedics*,
https://doi.org/10.1007/978-3-031-84693-9_14

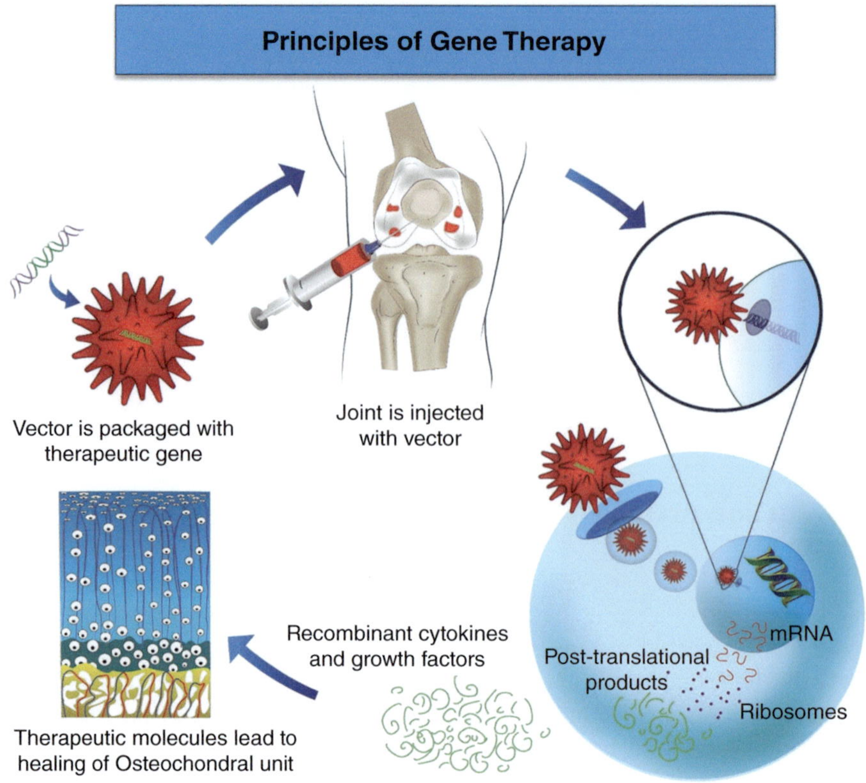

Fig. 14.1 Illustration of the vector principle of gene therapy. (Modified and reproduced with written permission from *Saseendar S, Shetty A, Kim SJ, Kumar M (2020). Gene Therapy in Osteoarthritis. In: Dhillon M (ed). Orthobiologics. Salubris Publishing, India*)

14.2 Gene Therapy Models

There are two major avenues in planning gene therapy strategies for regenerative medicine. One is direct delivery of genetic material in vivo using viral or non-viral vectors; other is to create desired cells in vitro later to be scaffolded into the host (Fig. 14.2). Both approaches require genetic material to be delivered and integrated effectively into the cell. The absence of a non-toxic efficient gene delivery system is the major roadblock to be bypassed given the availability of numerous therapeutic genes.

Maintenance of transgene expression and integration and its consequences are important aspects to be considered. The vector must be able to evade immune surveillance for a successful transfection of the carried genetic material. The vector must cross biological barriers only to reach the targeted cells and no other cells. This is referred to as cell targeting. Viral vectors being derived from wild-type parent viruses necessitate a thorough understanding of their replication and host interaction. Viral vectors should be non-pathogenic to human beings [5].

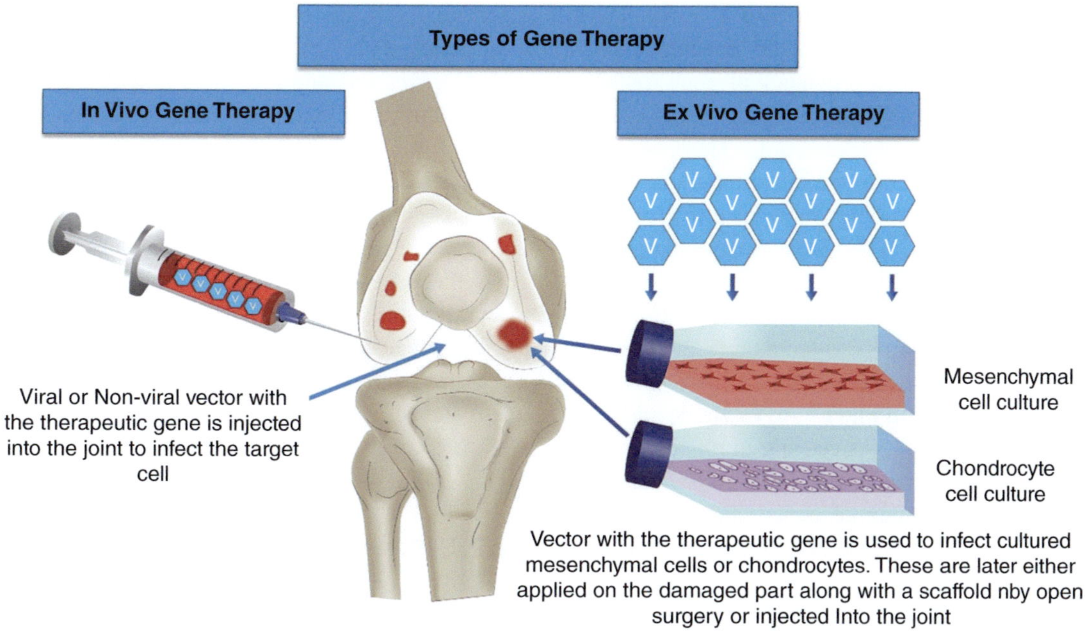

Fig. 14.2 Illustration of in vivo and ex vivo types of gene therapy. (Modified and reproduced with written permission from *Saseendar S, Shetty A, Kim SJ, Kumar M (2020). Gene Therapy in Osteoarthritis. In: Dhillon M (ed). Orthobiologics. Salubris Publishing, India*)

14.2.1 Viral Models

Currently, replication-deficient viral constructs are the most commonly used vectors in gene therapy. They have an inherently high efficiency of gene transfection and can be targeted locally or systemically. This is because of their natural ability to transduce their genes into host cells for replication during their life cycle, which they have acquired over millions of years of evolution [6].

However, there are several resolutions to be made before the clinical application of viral vectors. Immunogenicity and toxic reactions are major drawbacks that limit multiple injections of viral carriers. This occurs initially due to innate immune response via Toll-like receptors resulting in non-specific global stimulation and pro-inflammatory cytokine production. Later on, antigen presenting cell contribute to the initiation of adaptive immunity resulting in activated CD4+ and CD8+ T cells resulting in antiviral antibodies and cytotoxicity for elimination of the vector virus [7]. Carcinogenesis due to clonal dominance can be caused by retroviral vector-mediated insertion in or near proto-oncogenes or other signalling genes. This has been demonstrated in both normal and malignant haematopoiesis in both murine models and human studies. This is an adversity to be addressed. DNA packaging capacity of a viral vector is limited, unlike non-viral systems. A broad range of tropism is the next issue to be looked into. For example, adeno and adeno-associated viruses do not infect CNS naturally but do so when injected into the brain. Hence, systemic delivery of a vector may lead to undesired uptake by many cell types in multiple organs. This seems to be a liability rather than a benefit. Also, even leakage of local delivery can cause systemic effects. Transduction targeting must be stressed to overcome this hurdle [8].

Adenoviruses (ADV), retroviruses (γ retroviruses and lentiviruses), poxviruses, herpes simplex virus, and adeno-associated viruses (AAV) have been used in many gene therapy trials of which AAV, ADV, and Retroviridae appeared promising and were extensively investigated and used.

14.2.1.1 Adenoviral Vectors

It was the first therapeutically developed DNA virus due to its high efficiency of gene transduction. The main challenge as stated earlier was the immunogenicity of Adcapsid. This has been addressed by ensuring the safest route of administration and dosage. They have been the most successful in vivo gene delivery system, being a more common vector in many gene therapy trials. Many issues relating to immunogenicity, packaging capacity, helper virus contamination, and viral longevity can be resolved with high-capacity adenoviral vectors (HC-AdV). HC-AdVs are devoid of all viral coding sequences causing immune reduction that prevents tissue damage and also prolongs viral lifespan. Also, minimal contamination of helper virus is needed for the HC-AdV system [3, 6, 9].

14.2.1.2 Adeno-Associated Viral Vectors

They are non-enveloped ss-DNA viruses of the Parvoviridae family that rely on helper viruses for replication. The AAV genome has two genes namely *rep* (replicase) and *cap (*structural proteins) flanked by two short inverted terminal repeats between which a gene of interest is inserted. AVV has low immunogenicity but a high capacity to infect both quiescent and actively dividing cells. It also has site-specific integration, making it a very good viral vector candidate. The major setbacks are the large-scale production and cost. The multilevel interaction of protein capsid, DNA, and protein product with the host immune system and the time-consuming process of conversion of DNA from single to double-stranded structure are other obstacles. Despite these barriers, AAV are used exhaustively in preclinical trials of Parkinson's disease, giant cell axonal neuropathy, achromatopsia, Pompe disease, etc., and AAV9 has been shown to cross the blood-brain barrier and transfect neurons [10].

14.2.1.3 Retroviral Vectors

Retroviruses are RNA viruses that undergo reverse transcription within the host cell with a special reverse transcriptase enzyme to form DNA that later on integrates itself into the host cell genome. Then the host cell machinery is used for the expression of proteins encoded by the viral genes. The viral proteins and enzymes are coded by *gag*, *pol*, and *env* genes, and the genome has two long terminal repeats at 5′ and 3′ ends to act as promoters for these genes. Deletion of all three open reading frames (*gag*, *pol*, *env*) permits the packaging of 8 Kbp DNA. Murine leukaemia virus is the retrovirus most often employed in gene therapy. The limitations are inability to infect quiescent cells, which make most cells in the human body non-targetable by retroviruses. They also result in oncogenesis due to insertional oncogene activation and tumor suppressor inactivation [11, 12].

14.2.2 Non-viral Models

Apart from viruses, synthetic polymers, lipids, peptides, and DNA can be used as chemical vectors. Cellular uptake of these vectors is through endocytic pathways, which can be clathrin-mediated, caveolae-mediated, phagocytosis, or macropinocytosis. After internalization by endosome, the genetic material can be either destroyed by exposure to acidic lysosomes or expelled outside the cell by recycling endosomes. Hence, a proton sponge effect is created by chemical modifications to make the genetic material survive in acidic endosomes. Non-viral carriers are being studied extensively because they are less likely to induce immune reactions and have low toxicity. Easy production and modifications can be done on a large scale. Large molecular-size genetic material can be introduced into host cells by non-viral delivery systems. The major disadvantage is the efficacy of transfection, which is not as high as viral vectors, and also genetic integration is impossible [13, 14].

Various methods have been explored in this domain. Application of calcium phosphate-plasmid DNA co-precipitate to cells can lead to transfection. Lipoplexes are produced when DNA and cationic lipids combine. These lipoplexes have positively charged hydrophilic heads that bind to the negatively charged phosphate groups of nucleic acids. A linker connects the hydrophobic tail to the positively charged hydrophilic heads. Likely, this interaction between the positively charged lipoplexes and the negatively charged cell membrane molecules facilitates their cellular absorption. Cationic polymers like polyethylene imines, poly-L-lysine, chitosan, and dendrimers have been promising. Polyethyleneimine has shown high transfection efficiency and versatility apart from having low molecular weight. Polylysine was found to be immunogenic, and many others were found to be non-biodegradable and toxic. One major success was the dextran-spermine-based polycations that have shown exceptional effectiveness among other polycations. Modest amounts of reporter gene expression have been obtained in mouse lungs with a dextran-spermine/plasmid DNA gene delivery system. Further modification of physio-chemical properties can increase the efficiency of transfection. Transposons are naturally occurring discrete segments of DNA that can move and replicate within the genome. Considerable success has been achieved in transposing mammalian cells with several transposons. These include Tc1/mariner-like Frog Prince, cabbage looper moth PiggyBac, and medaka fish Tol2 [13, 15].

14.3 Regenerative Medicine Approaches Based on Gene Therapy

Regenerative medicine is an upcoming interdisciplinary branch of medicine combining biosciences and engineering principles. The broader aim is to rejuvenate or replace tissues impaired by trauma, degeneration, or other disease processes [16]. This can promisingly treat chronic illnesses and acute insults. However, there are several challenges with regenerative therapy. Targeted delivery of biomolecules or stem cells to the affected region can be difficult. The half-life of most regenerative modalities is limited. These modalities aim to accelerate the natural healing response. In injury and disease, the healing pathway can be affected making the regenerative modalities less effective [17]. Gene therapy broadly involves the transfer of genetic material into an individual's cells to achieve the desired therapeutic response [1]. Gene therapy is mainly being explored in the treatment of various genetic diseases. Combining gene therapy with regenerative medicine would increase the scope of this combination therapy in treating a variety of genetic, degenerative, and traumatic pathologies with synergistic benefits. Gene therapy increases the biomolecule availability for longer periods by modifying the native cells.

14.4 Applications of Gene Therapy in Musculoskeletal Regeneration

14.4.1 Bone Regeneration

Various molecules like recombinant bone morphogenic protein (BMP) have been tried in bone regeneration [18]. Factors like low bioavailability, the need for repeated doses, and high cost have limited the use of these molecules. With the integration of gene therapy, stem cells have been integrated with BMP genes using viral and non-viral vectors ex vivo and tried in treating bone defects. Animal studies have been encouraging [19–22]. Commonly BMP–2 cDNA has been used for this purpose. Viral vectors used include the adenovirus, lentivirus, and retrovirus. Lentivirus and retrovirus invoke less immune response when compared to the adenovirus. Park et al. [21] compared the use of adenovirus and non-viral vector–liposome for gene therapy on bone marrow stromal cells (BMSC) for bone regeneration. They concluded that though adenovirus is more effective, liposomes are better considering the ease of use, safety, and large amount of DNA carried. Other than BMSC, mesenchymal stem cells (MSC) from adipose tissue and muscle-derived stem cells have also been used for bone regeneration based on gene therapy. Cyclooxygenase 2 (COX 2) gene has been explored to accelerate bone healing in addition to the BMP 2 gene. COX 2 is said to promote angiogenesis and new bone formation. Rundle et al., in rat fracture models, showed good expression of COX 2 with bony union at 21 days post-therapy [23].

Non-viral methods of transfection are constantly being explored for their obvious advantages. Apart from liposomes, newer techniques are being explored to directly transfer the naked DNA into stem cells. These include electroporation and sonoporation. In electroporation, high-voltage currents are being used to make the cells more permeable for the transfection of DNA materials. Sheyn et al. [24] used electroporation to modify the adipose tissue-derived stem cells (ATSC) to over-express BMP 6. They showed that these stem cells can be used to treat vertebral defects. In sonoporation, ultrasound is used for the transfection of plasmid DNA [25]. Another mode of transfection of naked DNA is by using gene-activated matric (GAM). GAM contains biodegradable scaffolds with plasmid genes [26].

14.4.2 Meniscal Regeneration

Despite recent advances in knee arthroscopy, meniscal healing remains challenging, particularly in cases with complex meniscal tears. Repair techniques are not standardized. Even with best repair techniques, some patterns of meniscal tears like radial tears show poor healing response. Regenerative medicine products, such as platelet rich fibrin clot and bone marrow aspirate concentrate, have been applied with promising outcomes. Combing gene therapy would enhance healing response in a more controlled manner. Cucchiarini et al. [27] in their review on advances in combining gene therapy with tissue engineering to enhance meniscal healing and describe some candidate genes and vectors. Increase in healing has been noted with the transfer of IGF 1 and TGF beta genes to meniscal and bone marrow cells.

14.4.3 Cartilage Regeneration and Management of Arthritis

Cartilage regeneration/restoration after injury has been a challenge to clinicians despite a wide range of treatment modalities available [28]. Gene therapy has been tried in combination with regenerative modalities to improve cartilage regeneration. Traditional regenerative modalities like platelet-rich plasma provide a milieu of growth factors enhancing cartilage healing. The duration of action is limited, and these modalities rely on the host cells' healing potential. When gene therapy is added, the availability of such growth factors increases substantially. Genetically modified bone marrow stromal cells have been tried in cartilage regeneration. Overexpression of growth factors like transforming growth factor beta, insulin-like growth factor, BMP-2, and Indian Hedgehog gene have been tried [29–31]. However, animal studies on the dose of gene therapy required with the size of the defect are lacking. The long-term survival of regenerated cartilage should also be evaluated.

In the treatment of arthritis including rheumatoid arthritis, research is quite promising. There were early challenges in the use of gene therapy in arthritis. They include high cost and the need for two surgeries. Most common modalities of gene therapy use ex vivo means of transfection. This requires extraction of autologous cells during the first procedure, their transfection and re-injection during the second procedure. The inflammatory reactions while using vectors during in vivo transfection made in vivo transfection difficult. However, over the years, these challenges have been addressed. To bring down the cost and number of procedures, allogenic transfected cells are being tried. For example, Invossa, a product approved for gene therapy in arthritis, uses allogenic chondrocytes. The allogeneic chondrocytes are engineered to overexpress transforming growth factor beta. In vivo transfection has been made possible using adeno-associated viruses (AAV). These vectors seem to be safe and less immunogenic [32, 33]. Clinical trials using gene therapy for arthritis use retrovirus-based models for ex vivo transfection and AAV for in vivo transfection.

Collagen and hyaluronic acid (HA)-based hydrogels are employed in tissue engineering-based cartilage repair modalities [34]. These hydrogels are capable of binding to chondrocytes and preserving the mature differentiated chondrocytes. Hence these hydrogel-based scaffolds are being exploited in regenerative medicine-based gene therapy [35]. Other scaffolds with better mechanical stability and biological compatibility are being tried in cartilage repair. They include 3D woven poly e-caprolactone (PCL) and poly L-lactic acid (PLLA).

14.4.4 Intervertebral Disc Regeneration

In disc degeneration, the nucleus pulposus cells are deranged and fail to maintain the natural anabolic and catabolic balance in the disc leading to a spectrum of degenerative disc diseases. Various gene therapy modalities have targeted the nucleus pulposus cells to genetically modify them and reverse degeneration. Viral vectors like AAV serotype 2 and non-viral transfection mechanisms like ultrasound-mediated transfection and polyplex micelles-mediated transfection have been tried. Feng et al. showed that heme oxygenase 1 (HO1) plasmid DNA transfected using polyplex micelles was capable of reversing disc degeneration [36–38]. HO1 caused an increase in type 2 collagen and aggrecans and reduced the matrix metalloproteinase and COX2. The glycosaminoglycan content increased, and the inflammatory response was reduced following gene therapy. Sudo et al. studied the effect of caspase 3 in treating intervertebral disc degeneration and concluded that the knockdown of caspase 3 can improve disc regeneration [39].

14.4.5 Muscle Regeneration

Muscle injuries are common in sports. The degree of these injuries might vary from simple sprains to full thickness tears. The major challenge in dealing with these muscle tears is that the natural healing process of muscle is associated with significant fibrosis. Hence the gene therapy must address the fibrosis along with muscle regeneration. IGF-1 has myogenic property. Molecules like suramin, relaxin, INF-gamma, and decorin have anti-fibrotic properties. Hence, IGF-1 and these anti-fibrotic agents have been tried for muscle regeneration [40, 41]. Li et al. showed that decorin, a leucin-rich proteoglycan, acts by inhibiting the transforming growth factor beta. This prevents fibrosis. Myoblasts expressing decorin had increased capacity to differentiate into myotubes. This led to enhanced expression of certain myogenic genes. Thus decorin, in addition to preventing fibrosis, improves muscle regeneration [40].

14.4.6 Tendon and Ligament Regeneration

Tendons and ligaments have poor healing potential after injury. They often require surgical intervention. Healing after surgical intervention is often complicated by fibrosis. Gene therapy has been tried in tendon regeneration mainly in three models: Achilles tendon, digital flexor tendon, and patellar tendon. Increased availability of growth factors, such as vascular endothelial growth factor (VEGF), basic fibroblast growth factor (bFGF), and platelet derived growth factor B (PDGF B), have been shown to increase healing [42]. Wang et al. showed that adeno-associated virus type 2 (AAV 2) is the most efficient vector in gene therapy for tendon healing. Basic fibroblast growth factor showed increased expression over 3 weeks after gene therapy and enhanced tendon healing [43]. Tang et al. showed that bFGF or VEGF gene therapy using AAV 2 increased the type 1 collagen production and enhanced strength of healed tendons [44].

14.5 Conclusion

Gene therapy in combination with regenerative medicine seems to be an attractive approach to enhance biological healing in many musculoskeletal conditions. The preclinical studies are promising. However, clinical studies are limited to certain conditions. There are some gaps to be addressed, which include best candidate gene or best combination of genes to be transduced, ideal vectors, and most adapted cell source for transfer. Requirements for medical regulatory boards should be considered to adapt the combined regenerative and gene therapy solutions into clinical practice.

References

1. Rafi MA. Gene and stem cell therapy: alone or in combination? BioImpacts BI. 2011;1(4):213–8.
2. Center for Biologics Evaluation and Research (CBER). What is gene therapy? FDA [Internet]. 2020 [cited 2024 May 26]. https://www.fda.gov/vaccines--blood-biologics/cellular-gene-therapy-products/what-gene-therapy.
3. Raper SE, Chirmule N, Lee FS, Wivel NA, Bagg A, Gao G, et al. Fatal systemic inflammatory response syndrome in a ornithine transcarbamylase deficient patient following adenoviral gene transfer. Mol Genet Metab. 2003;80(1–2):148–58.
4. Different approaches | ASGCT—American Society of Gene & Cell Therapy | [Internet]. [cited 2024 May 26]. https://patienteducation.asgct.org/gene-therapy-101/gene-therapy-approaches.
5. Bouard D, Alazard-Dany D, Cosset FL. Viral vectors: from virology to transgene expression. Br J Pharmacol. 2009;157(2):153–65.
6. Crystal RG. Adenovirus: the first effective in vivo gene delivery vector. Hum Gene Ther. 2014;25(1):3–11.
7. Bessis N, GarciaCozar FJ, Boissier MC. Immune responses to gene therapy vectors: influence on vector function and effector mechanisms. Gene Ther. 2004;11(Suppl 1):S10–7.
8. Baum C, Kustikova O, Modlich U, Li Z, Fehse B. Mutagenesis and oncogenesis by chromosomal insertion of gene transfer vectors. Hum Gene Ther. 2006;17(3):253–63.
9. Marshall E. Gene therapy death prompts review of adenovirus vector. Science. 1999;286(5448):2244–5.
10. Wang D, Tai PWL, Gao G. Adeno-associated virus vector as a platform for gene therapy delivery. Nat Rev Drug Discov. 2019;18(5):358–78.
11. Urban JH, Merten CA. Retroviral display in gene therapy, protein engineering, and vaccine development. ACS Chem Biol. 2011;6(1):61–74.
12. Yamashita M, Emerman M. Retroviral infection of non-dividing cells: old and new perspectives. Virology. 2006;344(1):88–93.
13. Ramamoorth M, Narvekar A. Non viral vectors in gene therapy—an overview. J Clin Diagn Res JCDR. 2015;9(1):GE01–6.
14. Hosseinkhani H, Tabata Y. Self assembly of DNA nanoparticles with polycations for the delivery of genetic materials into cells. J Nanosci Nanotechnol. 2006;6(8):2320–8.
15. Yeo WWY, Hosseinkhani H, Rahman SA, Rosli R, Domb AJ, Abdullah S. Safety profile of dextran-spermine gene delivery vector in mouse lungs. J Nanosci Nanotechnol. 2014;14(5):3328–36.
16. Mao AS, Mooney DJ. Regenerative medicine: current therapies and future directions. Proc Natl Acad Sci U S A. 2015;112(47):14452–9.
17. Kami D, Gojo S. Tuning cell fate: from insights to vertebrate regeneration. Organogenesis. 2014;10(2):231–40.
18. Hustedt JW, Blizzard DJ. The controversy surrounding bone morphogenetic proteins in the spine: a review of current research. Yale J Biol Med. 2014;87(4):549–61.
19. Turgeman G, Pittman DD, Müller R, Kurkalli BG, Zhou S, Pelled G, et al. Engineered human mesenchymal stem cells: a novel platform for skeletal cell mediated gene therapy. J Gene Med. 2001;3(3):240–51.
20. Gates CB, Karthikeyan T, Fu F, Huard J. Regenerative medicine for the musculoskeletal system based on muscle-derived stem cells. J Am Acad Orthop Surg. 2008;16(2):68–76.
21. Park J, Ries J, Gelse K, Kloss F, von der Mark K, Wiltfang J, et al. Bone regeneration in critical size defects by cell-mediated BMP-2 gene transfer: a comparison of adenoviral vectors and liposomes. Gene Ther. 2003;10(13):1089–98.
22. Lieberman JR, Daluiski A, Stevenson S, Wu L, McAllister P, Lee YP, et al. The effect of regional gene therapy with bone morphogenetic protein-2-producing bone-marrow cells on the repair of segmental femoral defects in rats. J Bone Joint Surg Am. 1999;81(7):905–17.
23. Rundle CH, Strong DD, Chen ST, Linkhart TA, Sheng MHC, Wergedal JE, et al. Retroviral-based gene therapy with cyclooxygenase-2 promotes the union of bony callus tissues and accelerates fracture healing in the rat. J Gene Med. 2008;10(3):229–41.
24. Sheyn D, Kallai I, Tawackoli W, Yakubovich DC, Oh A, Su S, et al. Gene-modified adult stem cells regenerate vertebral bone defect in a rat model. Mol Pharm. 2011;8(5):1592–601.
25. Bez M, Sheyn D, Tawackoli W, Avalos P, Shapiro G, Giaconi JC, et al. In situ bone tissue engineering via ultrasound-mediated gene delivery to endogenous progenitor cells in mini-pigs. Sci Transl Med. 2017;9(390):eaal3128.
26. Fang J, Zhu YY, Smiley E, Bonadio J, Rouleau JP, Goldstein SA, et al. Stimulation of new bone formation by direct transfer of osteogenic plasmid genes. Proc Natl Acad Sci U S A. 1996;93(12):5753–8.
27. Cucchiarini M, McNulty AL, Mauck RL, Setton LA, Guilak F, Madry H. Advances in combining gene therapy with cell and tissue engineering-based approaches to enhance healing of the meniscus. Osteoarthr Cartil. 2016;24(8):1330–9.
28. Muthu S, Viswanathan VK, Chellamuthu G, Thabrez M. Clinical effectiveness of various treatments for cartilage defects compared with microfracture: a network meta-analysis of randomized controlled trials. J Cartil Jt Preserv. 2023;4:100163.
29. Sieker JT, Kunz M, Weißenberger M, Gilbert F, Frey S, Rudert M, et al. Direct bone morphogenetic protein 2 and Indian hedgehog gene transfer for articu-

lar cartilage repair using bone marrow coagulates. Osteoarthr Cartil. 2015;23(3):433–42.
30. Madry H, Kaul G, Cucchiarini M, Stein U, Zurakowski D, Remberger K, et al. Enhanced repair of articular cartilage defects in vivo by transplanted chondrocytes overexpressing insulin-like growth factor I (IGF-I). Gene Ther. 2005;12(15):1171–9.
31. Ivkovic A, Pascher A, Hudetz D, Maticic D, Jelic M, Dickinson S, et al. Articular cartilage repair by genetically modified bone marrow aspirate in sheep. Gene Ther. 2010;17(6):779–89.
32. Evans CH, Ghivizzani SC, Robbins PD. Arthritis gene therapy is becoming a reality. Nat Rev Rheumatol. 2018;14(7):381–2.
33. Evans CH, Ghivizzani SC, Robbins PD. Gene delivery to joints by intra-articular injection. Hum Gene Ther. 2018;29(1):2–14.
34. Vega SL, Kwon MY, Burdick JA. Recent advances in hydrogels for cartilage tissue engineering. Eur Cell Mater. 2017;33:59–75.
35. Girotto D, Urbani S, Brun P, Renier D, Barbucci R, Abatangelo G. Tissue-specific gene expression in chondrocytes grown on three-dimensional hyaluronic acid scaffolds. Biomaterials. 2003;24(19):3265–75.
36. Antoniou J, Steffen T, Nelson F, Winterbottom N, Hollander AP, Poole RA, et al. The human lumbar intervertebral disc: evidence for changes in the biosynthesis and denaturation of the extracellular matrix with growth, maturation, ageing, and degeneration. J Clin Invest. 1996;98(4):996–1003.
37. Leckie SK, Bechara BP, Hartman RA, Sowa GA, Woods BI, Coelho JP, et al. Injection of AAV2-BMP2 and AAV2-TIMP1 into the nucleus pulposus slows the course of intervertebral disc degeneration in an in vivo rabbit model. Spine J. 2012;12(1):7–20.
38. Nishida K, Doita M, Takada T, Kakutani KI, Miyamoto H, Shimomura T, et al. Sustained transgene expression in intervertebral disc cells in vivo mediated by microbubble-enhanced ultrasound gene therapy. Spine. 2006;31(13):1415.
39. Caspase 3 as a therapeutic target for regulation of intervertebral disc degeneration in rabbits. [cited 2024 May 26]; https://onlinelibrary.wiley.com/doi/10.1002/art.30251.
40. Li Y, Li J, Zhu J, Sun B, Branca M, Tang Y, et al. Decorin gene transfer promotes muscle cell differentiation and muscle regeneration. Mol Ther J Am Soc Gene Ther. 2007;15(9):1616–22.
41. Schertzer JD, Lynch GS. Comparative evaluation of IGF-I gene transfer and IGF-I protein administration for enhancing skeletal muscle regeneration after injury. Gene Ther. 2006;13(23):1657–64.
42. Mehrzad R, Wang XT, Zhou YL, Wu YF, Liu PY. Gene therapy in tendon repair and regeneration. In: Giatsidis G, editor. Gene therapy in reconstructive and regenerative surgery. Cham: Springer; 2018. p. 83–9. [cited 2024 Jun 23]. https://doi.org/10.1007/978-3-319-78957-6_6.
43. Wang XT, Liu PY, Tang JB, Mizukami H, Xin KQ, Ozawa K, et al. Tendon healing in vitro: adeno-associated virus-2 effectively transduces intrasynovial tenocytes with persistent expression of the transgene, but other serotypes do not. Plast Reconstr Surg. 2007;119(1):227–34.
44. Tang JB, Wu YF, Cao Y, Chen CH, Zhou YL, Avanessian B, et al. Basic FGF or VEGF gene therapy corrects insufficiency in the intrinsic healing capacity of tendons. Sci Rep. 2016;6:20643.

15 Aragonite Off-the-Shelf Implant for Articular Cartilage in Early Osteoarthritis

Alberto Bulgarelli, Pietro Conte,
Giuseppe Anzillotti, and Elizaveta Kon

15.1 Introduction

Articular cartilage defects of the knee have always been a significant challenge for surgeons due to their limited spontaneous healing potential and propensity to worsen over time. Mature cartilage is, by definition, avascular and has a slow cellular turnover. The relationship between these lesions and the underlying bone further complicates things, creating a multifaceted problem that demands precise intervention to relieve pain and restore joint function [1]. If left untreated, chondral and osteochondral lesions slowly but steadily progress to osteoarthritis (OA).

The biomechanical relationship between the articular cartilage and subchondral bone is well established. In a healthy joint, hyaline cartilage maintains a near-frictionless articulating surface while the subchondral bone distributes joint forces. Nonetheless, after a trauma, damage to the subchondral bone may alter this equilibrium, leading to force distribution changes and cartilage degeneration [2]. Several studies have also explored the concept of biochemical crosstalk between cartilage and bone. The subchondral bone normally supplies cartilage with nutrients. However, during early OA, subchondral bone angiogenesis has been shown to lead to cartilage degradation and innervation. Walsh et al. [3] explored the relationship between new microvasculature and nerve growth in osteochondral tissue and found elevated levels of vascular endothelial growth factor (VEGF) and nerve growth factor (NGF) in the subchondral bone, extending into the noncalcified cartilage [2].

Several techniques have been developed over the years to address articular cartilage defects. However, most traditional regenerative techniques (e.g., autologous chondrocyte implantation) were expensive, usually involved two steps, and had lengthy postoperative care. More importantly, though, they did not provide an optimal solution for osteochondral lesions since not addressing the complexity of the cartilage-bone interface. This is why, with the advent of tissue bioengineering, we have witnessed a shift towards developing innovative cell-free treatment strategies that target chondral and osteochondral lesions as a whole [4]. Recognizing the integral role of the subchondral bone in lesion pathogenesis, researchers have focused on devising scaffolds that facilitate regeneration across

A. Bulgarelli · P. Conte · G. Anzillotti · E. Kon (✉)
IRCCS Humanitas Research Hospital, Milan, Italy

Department of Biomedical Sciences, Humanitas University, Milan, Italy
e-mail: elisaveta.kon@humanitas.it

A. Gobbi et al. (eds.), *Regenerative Medicine in Sports and Orthopaedics*,
https://doi.org/10.1007/978-3-031-84693-9_15

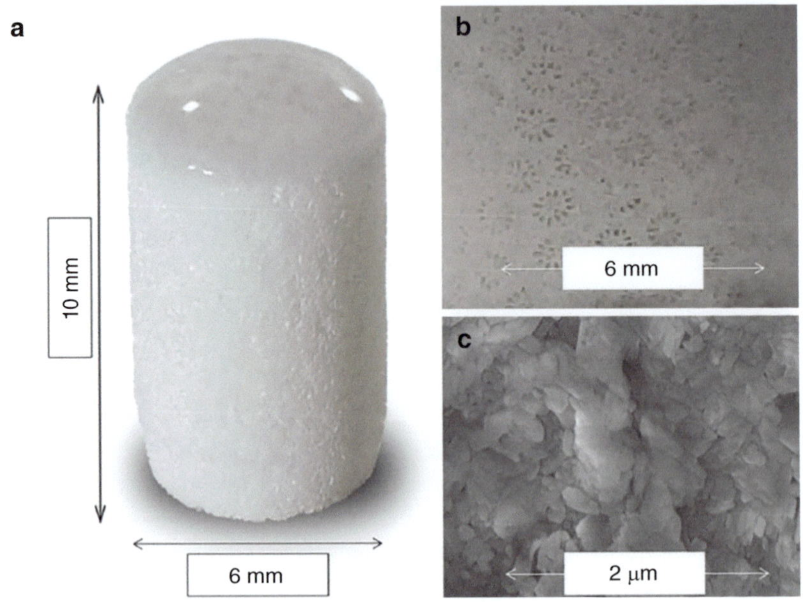

Fig. 15.1 Micrograph of the Agili-C™ (CartiHeal, Smith & Nephew) scaffold. (**a**) Interconnected natural inorganic calcium carbonate (aragonite) derived from purified, inorganic, coral exoskeleton. (**b**) and (**c**) are close-ups of the bone phase that reveal its porous microstructure [8]

both cartilage and bone regions. This approach aims to restore the osteochondral unit, offering a more comprehensive solution to joint surface defects [5]. On top of this, scaffolds are easy to handle, are not associated with donor site morbidity, and do not depend on the viability of grafts but consist of an off-the-shelf solution. Moreover, they can be used in one-step procedures, avoiding issues related to cell manipulation and culture, while reducing costs [6].

Among these innovative therapies, one has been gaining particular attention. *Agili-C® (CartiHeal, Smith & Nephew)* is an aragonite-based (inorganic calcium carbonate) scaffold approved for clinical use in 2022. It is biocompatible, porous, and resorbable and mimics the microarchitecture of bone. It is a bi-phasic scaffold (Fig. 15.1), consisting of interconnected natural inorganic calcium carbonate (aragonite) derived from purified, inorganic, coral exoskeleton [7, 8].

This chapter aims to illustrate the scientific evidence surrounding *Agili-C™*, exploring its composition, scientific rationale, clinical indications, advantages, and long-term efficacy.

15.2 Cartiheal Microstructure and the Bioengineering Idea Behind It

The optimal scaffold should support bone repair in the subchondral area while facilitating the migration of mesenchymal stem cells into the superficial cartilaginous layers. It should also enable cartilage repair at the articular surface and promote the integration of the neocartilage with the surrounding cartilage.

Coral exoskeletons (aragonite) resemble human bone, sharing 3D structure and internal composition (calcium carbonate). They have osteoconductive and osteogenetic properties and, with time, are fully reabsorbed [8–10].

The Agili-C™ scaffold (CartiHeal, Smith & Nephew) is a porous, interconnected calcium carbonate (crystalline aragonite) implant derived from a purified, inorganic coral exoskeleton. It is biocompatible, resorbable, and has a two-part structure (Fig. 15.1).

The lower segment, the subchondral phase, is made of inorganic aragonite and has large pores (100–200 μm) that allow new blood vessels to grow into it. Osteoclasts and osteoblasts degrade and reconstitute this part into new subchondral bone.

The upper segment, the chondral phase, is mechanically processed to create a grid of small channels (2 mm deep), as shown in Fig. 15.1. This configuration facilitates adhesion, differentiation, and proliferation of bone marrow and synovial mesenchymal stem cells to chondrocytes, thereby fostering articular cartilage formation [5]. Finally, the implant design is tapered and angled at 2° to improve press-fit implantation and enhance patient outcomes [11].

Several preclinical and clinical studies have clarified and confirmed the rationale for the Agili-C scaffold design and safety profile.

Kon et al. were the first to show how mechanical modification with drilled channels and biological augmentation with HA enhanced the scaffold's regenerative potential [8]. They chose goats as their animal model and created a total of 28 osteochondral lesions. Implants were placed in two locations: the medial femoral condyle (MFC) and the lateral femoral condyle (LFC). In each site, a single device was inserted into the central part of the condyle in a weight-bearing area. Animals were randomized into four groups: group 1 received implants with drilled channels in the cartilage phase, group 2 received implants with drilled channels in the bone phase, group 3 received unmodified implants, and Group 4 received no implants. Half of the implants within each group had HA impregnated in the cartilage phase.

Each joint underwent macroscopic and microscopic evaluation. The macroscopic evaluation was blindly assessed using three scaling systems (Fontier [9], Niederauer [12], ICRS). For each site, five central serial sections were prepared for microscopic analysis. Three sections were stained with safranin-haematoxylin-eosin for inflammation analysis, Masson trichrome for general morphology, and Safranin-O/Fast green for proteoglycan content. The remaining two sections were used for immunohistochemistry to determine collagen type I and collagen type II.

Group 1 implants impregnated with HA outperformed all other tested groups in all evaluation methods. Group 1 implants were replaced by newly formed hyaline cartilage and subchondral bone. The repair tissue had a smooth contour, integrated well with adjacent cartilage, and showed characteristics of hyaline cartilage, with high levels of proteoglycans and collagen type II and no collagen type I.

The early potential for bone and cartilage formation at 6 months was confirmed in a later study on the same animal model at the 12-month follow-up [7]. In this work, the Agili-C scaffold (Ar-HA test group) was compared with a defect in which a blood clot extruding from the bone marrow is allowed to form. Each operated joint underwent ultrasound imaging, X-Ray, micro-CT, MRI, histology, and immunohistochemistry evaluation. At the 12-month mark, only the Ar-HA test group displayed the formation of a hyaline cartilage repair tissue, with little to no signs of neovascularisation and inflammation. The hyaline cartilage features were confirmed by the presence of proteoglycans and collagen type II, as well as the absence of collagen type I. The repair tissue integrated well with the adjacent native cartilage and basal tissues at all sites. A reconstructed subchondral bone plate with a clear tidemark was observed in all specimens except one.

Early human trials [11] by Kon et al. confirmed the findings of preclinical studies. Overall, a total of 97 patients without severe osteoarthritis (OA) were included and treated: 21 patients with a newly optimized tapered implants were compared with 76 patients previously with classical cylindrical-shaped implants. The main inclusion criteria for both groups were as follows: evidence of a chondral-osteochondral lesion of the femoral condyles and trochlea, young age (<50 years old), and

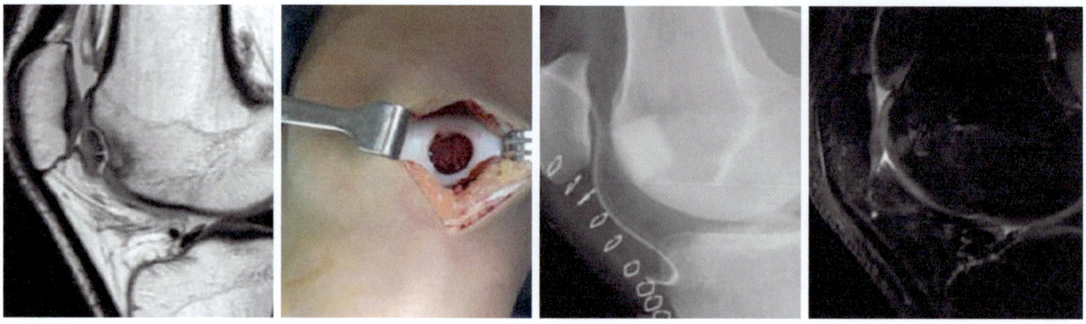

Fig. 15.2 20-year-old patient treated for a significant osteochondral defect in the trochlea. After 12 months of follow-up, his IKDC score improved from 60 to 86, and his overall KOOS increased from 46 to 90 [11]

Kellgren-Lawrence grade less than 3. Clinical outcomes were compared using the IKDC subjective score, the Lysholm score, and all five KOOS subscales, administered preoperatively and at 6 and 12 months post-surgery. Nineteen patients in the tapered implants group also had imaging evaluations at the 12-month follow-up. A statistically significant improvement in all clinical scores and in MRI findings was documented in both groups. In particular, tapered scaffolds maintained the clinical improvement seen in cylindrical implants while reducing the risk of revision surgery. Indeed, for the Agili-C scaffold to be effective, it must remain stable postoperatively and during the subsequent phases of weight-bearing and physical activity. Stability ensures proper graft fixation, which is critical for successful integration with the host bone, maintenance of articular surface congruency, and vascular formation. The tapered design makes the implant more self-centered, avoiding possible damage during plug introduction while also maintaining a better press fit (Fig. 15.2).

15.3 Clinical Indications

A variety of scaffolds have been approved for clinical use in recent years, sometimes making it quite tricky for orthopedic surgeons to determine the correct indication. To address this issue and fill in the gap in the literature, a panel of experts belonging to the ICRS (International Cartilage Regenerative & Joint Preservation Society) published in 2021 the "Scaffold Consensus" [13]. Using the RAND/UCLA method (RAM),[1] they redefined the appropriateness of chondral and osteochondral scaffold surgery, starting from 5 clinical variables and working on 72 clinical scenarios.

The panel's conclusions can be summarised in three levels of appropriateness (Table 15.1).

[1] The RAND/UCLA method is used to develop guidelines and clinical criteria by achieving expert consensus in situations where the scientific literature is limited or controversial. It aims to systematically incorporate expert opinions while minimizing the influence of social pressure or panel hierarchy, making the guideline development process more transparent and rigorous.

Table 15.1 Main indications and controindications for osteochondral scaffold surgery based on the icrs "Scaffold Consensus" (2021)

Contraindications	Relative indications	Absolute indications
Osteonecrosis	Age >50	Age 18–50
Associated lesions and/or malalignment[a]	Degenerative	Trauma
OA grade IV	OA grade II–III	OA grade 0–I

[a] Associated lesions and malalignment can be contextually treated as long as they lead to "restitutio ad integrum" of the joint (es. meniscal transplant or suture)

Certainly, representing every indication is not easy, and this was not even the 2021 "Scaffold Consensus" goal. Consider lesion size, for example. While scaffold therapy is generally recommended for non-OA knees independently of lesion size (even if more expensive than other options, i.e., microfractures), their use is typically not recommended in OA knees, except for extensive lesions in young patients. However, it is vital to still reach a conclusion, as expert consensus on scenarios where scaffolds are appropriate or inappropriate supports decision-making in daily cartilage surgery practice. Based on the clinical results reported in section 5, Agili-C is now indicated for use in cartilage and osteochondral defects in traumatic and osteoarthritic joints. Specifically, FDA indicated its use for the treatment of an International Cartilage Repair Society grade III or above knee-joint surface lesion(s), with a total treatable area of 1–7 cm^2, without severe osteoarthritis (Kellgren-Lawrence grade 0–3).

15.4 Surgical Technique and Postoperative Care

The lesion is directly exposed through a mini-arthrotomy. A cannulated aligner is pointed at the lesion, and a K wire is passed through so that it goes perpendicular to the lesion floor and right through its center. At this point, the lesion is drilled away so that all it's left is a hole with healthy cartilage around it. Before proceeding, ensure there are no violations near the notch. The implant should be surrounded by at least 3 mm of bone. A manual reamer and a cartilage cutter may then be used to prepare the floor and walls of the lesion. Finally, the implant is manually inserted and pushed to the bottom of the lesion until it is 2 mm below the cartilage surface (Fig. 15.3).

The lesion's margins must be covered with healthy cartilage and stable. If, during surgery, the margins (or "shoulders") are unsuitable to host the scaffold, the procedure cannot proceed as planned, as the rate of failure of the implant is unacceptable. It is essential always to analyze the preoperative MRI to minimize the risk of changing the indication during surgery.

Postoperative care and rehabilitation for osteochondral scaffold surgery include 4 weeks of toe-touch weight bearing and 2 weeks of partial weight bearing. Continuous passive motion (CPM), active-assisted ROM exercises, and quadriceps isometric strengthening are advised during this period. Hydrotherapy is allowed and highly recommended after suture removal [11] (Fig. 15.3).

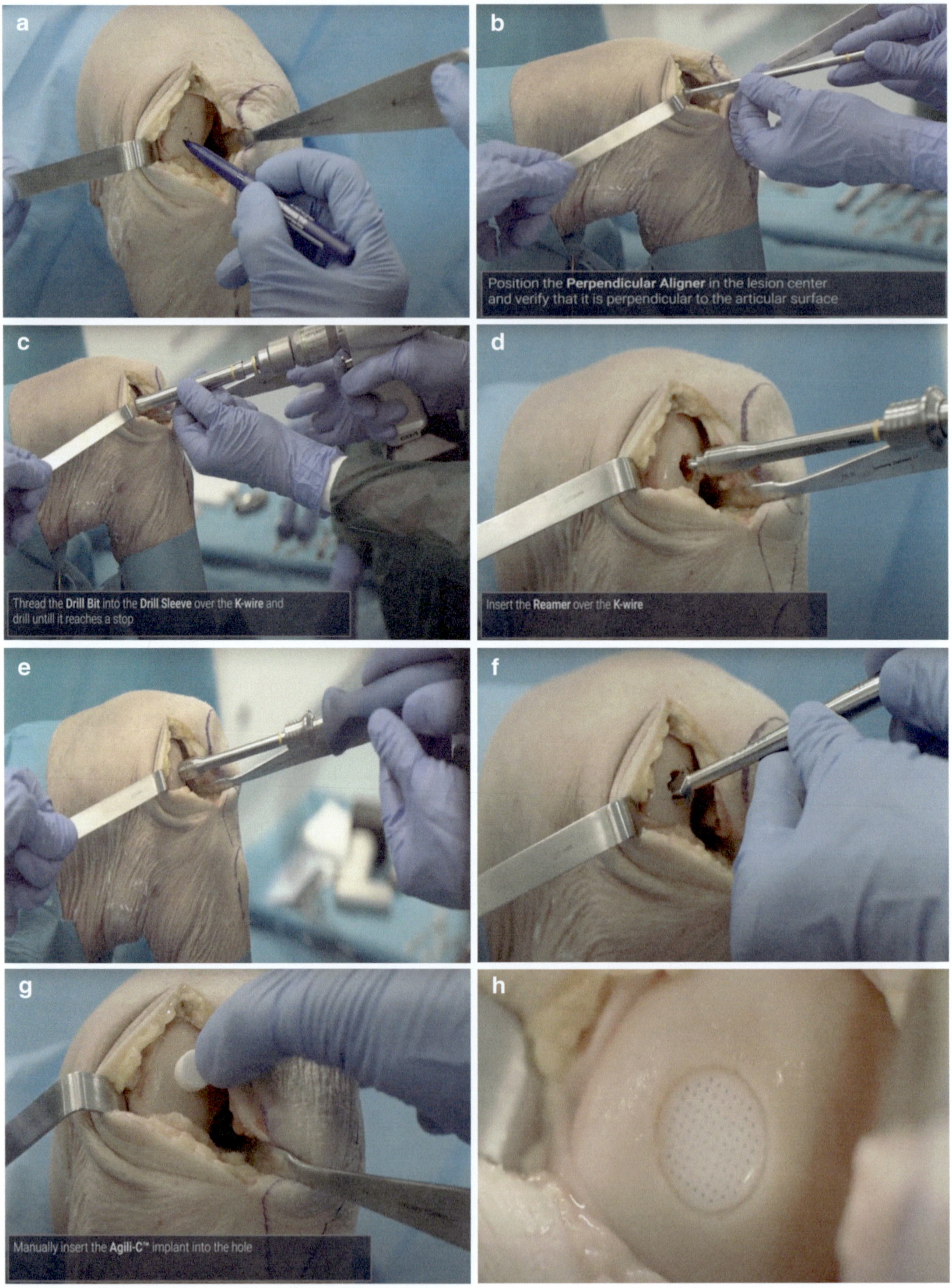

Fig. 15.3 Surgical technique: C® (**a**) Lesions exposure through mini-arthrotomy; (**b**) Positioning of the cannulated aligner and K wire placement; (**c**) Placement of the Drill Sleeve and drilling of the lesion with the Drill Bit; (**d**) Reaming of the bottom of the lesion; (**e**) Positioning of the Shaper; (**f**) Trimming of the peripheral cartilage using the Cartilage Cutter; (**g**) Implant insertion; (**h**) Final result. Watch the full surgical technique at this address https://www.youtube.com/watch?si=7DZpWLx2v2UiB5BQ&v=WgcyyxI-Xho&feature=youtu.be [14]

15.5 Follow-Up and Comparison with Other Scaffolds

The mid-term data on Agili-C® (CartiHeal) are largely derived from three clinical studies that followed the initial human clinical trial from Kon et al. [11] previously described.

The first study is a single follow-up case series [15] published by Van *Genechten* et al. in 2021 focused on clinical response, safety outcomes, and MRI imaging results up to 3 years post-surgery. Thirteen subjects with knee joint surface lesions (JSL) were successfully treated using one or two aragonite-based scaffolds, depending on lesion size. Patients experienced significant improvements in pain and overall function, both shortly after the surgery and in the long term. While they did not fully regain their pre-injury activity levels, significant improvement was still observed at 18 and 24 months following the procedure. No serious adverse events (AEs) were reported during surgery or during the 36-month follow-up period. Potential device-related AEs included joint arthralgia (50%) and joint effusion (33%). No cases of implant loosening, failure, or infection were observed throughout the study. Bony integration and remodeling of the scaffold were completed in 66–100% of cases for 6 implants (40%) at 12 months and 11 implants (78.7%) at the final follow-up. MRI frequently revealed a typical pit-like healing pattern, with peripheral ingrowth of newly formed subchondral bone, subsequent scaffold biodegradation, and the formation of a cartilaginous layer.

Later that year, *Kon* et al. published the results of a 2-year multicenter prospective study [16] in the *American Journal of Sports Medicine* (AJSM). It was the first multicenter trial investigating the utilization of an osteochondral scaffold in knees affected by osteoarthritis. A total of 86 patients, with a mean age of 37.4 ± 10.0 years, were recruited from 8 medical centers. All patients had joint surface lesions (JSLs) in the context of mild to moderate OA (Kellgren II and III) and a mean defect size of 3.0 ± 1.7 cm^2. The inclusion criteria were as follows: radiographic evidence of mild-to-moderate knee OA (Kellgren-Lawrence grade 2 or 3); up to three treatable chondral/osteochondral defects (International Cartilage Repair Society grades 3 and 4) on the femoral condyles or trochlea; a total defect size ≤ 7 cm^2; and no concurrent knee instability, severe axial malalignment, or systemic arthropathy. All patients were assessed before surgery and at 6, 12, 18, and 24 months. Clinical examinations and patient questionnaires were used to assess symptoms, physical condition, and knee function. All patients underwent concomitantly MRI evaluation. Histological evaluation was performed on a single patient who later underwent knee replacement. The primary endpoint was the change in KOOS value from baseline to the 24-month follow-up. Failure was defined as the removal of the scaffold for any reason during follow-up.

The main finding of the study was that Agili-C® may provide significant clinical improvement in patients with JSLs. KOOS and IKDC subjective scores significantly improved after 6 months and continued to increase further during the subsequent follow-up visits at 12, 18, and 24 months. Main AEs included swelling and pain (15 patients). During the 2-year follow-up, 8 patients (9.3%) underwent implant removal and were classified as failures. In each instance, there was a recurrence of severe knee pain accompanied by swelling and limited motion. Reasons for implant removal varied: two cases were due to procedure-related infections, five cases involved partial loosening of the scaffold due to lack of integration, and one case experienced progression of osteoarthritis in the patellofemoral compartment, leading to total knee replacement 14 months after scaffold implantation. On MRI, a notable increase in the area of the defect filled with regenerated cartilage was evident. Improvement in defect fill was noticeable as early as 6 months post-implantation, persisting through the 12 and 18-month follow-ups and culminating in its peak after 24 months.

Finally, in 2023, Altschuler et al. published in the AJSM the first randomized clinical trial [17] comparing the efficacy of aragonite-based scaffolds with arthroscopic debridement/microfractures for the treatment of knee chondral and osteochondral lesions. It was the largest study to date on Agili-C® and clearly demonstrates the superiority of aragonite-based scaffolds over the most common surgical standard of care.

A total of 251 patients were recruited from 26 medical centers. Patients had to be between 21 and 75 years old, with a maximum of three joint surface lesions graded 3a or higher according to the ICRS on the femoral condyles or trochlea. The total treatable area had to range from 1 to 7 cm^2. Patients were randomly assigned to receive either the aragonite-based implant or undergo debridement/microfracture treatment, with a 2:1 ratio. Patients with focal defects and Kellgren-Lawrence (KL) scores of 0 or 1 were randomized against similar patients treated with microfracture. Patients with mild-to-moderate OA (KL 2 or 3) were randomized against those treated with arthroscopic debridement. Assessments were conducted at 6, 12, 18, and 24 months, focusing on the overall KOOS as the primary endpoint, along with the KOOS subscales, percentage of responders, and IKDC subjective score as secondary endpoints. Patients underwent MRI evaluations at 12 and 24 months to assess defect fill grade. Failure was defined as any secondary invasive intervention in the treated joint (e.g., open, mini-open surgical, or arthroscopic procedures, as well as any intraarticular injection), regardless if related or unrelated to the original treatment.

The implant group exhibited significantly superior results in both the primary and all secondary endpoints during each follow-up assessment. At the 2-year mark, the magnitude of improvement in the implant group was double that of the control group in terms of mean KOOS enhancement. The same applied to the responder rate and imaging outcomes. At 24 months, 88.5% of patients treated with Agili-C® achieved >75% defect fill, compared to just 30.9% of those treated with the standard surgical care. The failure rate was 7.2% in the implant group compared to 21.4% in the control group. The most common adverse event was transient knee pain following surgery, experienced by 15.0% of the scaffold group compared to 39.3% of the control group. Moreover, it is important to underline how this was also true for patients who received two scaffolds. Indeed, this study illustrates how multiple scaffolds may be used to treat patients presenting either big or various lesions.

The results of these three studies can be compared to the outcomes of other scaffolds. In 2021, Boffa et al. published their systematic review and meta-analysis [6] on multi-layer cell-free scaffolds in the *Journal of Experimental Orthopedics*. They focused on articles reporting results of the International Knee Documentation Committee (IKDC) and Tegner scores. The scores were analyzed as improvement from baseline to 1, 2, and ≥3 years of follow-up. They included 34 studies (1022 patients), centered around 3 scaffolds: TruFit® (11 studies; 9 case series, 2 retrospective comparative techniques), MaioRegen (20 studies; 19 case series and one RCT), and Agili-C™ (3 studies; 2 case series and one retrospective study).

The main finding of this meta-analysis was that multi-layer cell-free scaffolds can offer clinical benefits for treating knee osteochondral lesions. Quantitative analysis revealed statistically significant improvements in the IKDC subjective score and the Tegner score at each follow-up compared to baseline.

TruFit® [18] was the first scaffold introduced into clinical practice. This bilayer scaffold is composed of semi-porous polylactic (PLGA), polyglycolic acid (PGA), and calcium sulfate biopolymer. Initially designed to backfill graft donor sites during OAT procedures, TruFit® was later predominantly used as a one-step osteochondral treatment. However, it has been withdrawn from the market due to poor outcomes, high failure rates, and unfavorable histological outcomes. Shivji et al. conducted a long-term follow-up study (121 months) on this scaffold. They reported no statistically significant improvements in any scores from baseline. MRI evaluations revealed incomplete or absent plug incorporation and persistent chondral loss.

Maioregen [19] is a nanostructured biomimetic scaffold that targets the calcified tide mark with a triphasic design (superficial part 100% type I collagen; middle 60% C-I and 40% HA; deep 30% C-I 70% HA). It is also resorbable and osteoconductive, aiding in the regeneration of cartilage, as confirmed by histological analysis.

Significant clinical improvement has been reported in almost all studies, with further improvement in clinical scales observed up to 5 years after surgery. However, it is important to note that while outcomes are comparable, the experience with Maioregen is, at the moment, superior given the higher number of published trials.

15.6 Conclusion

Agili-C® (CartiHeal, Smith & Nephew) is an innovative aragonite-based (inorganic calcium carbonate) scaffold that was approved for clinical use from FDA in 2022. It is an off-the-shelf, cell-free, and cost-effective implant indicated for the treatment of an ICRS grade III or above knee-joint surface lesion(s), with a total treatable area of 1–7 cm^2, without severe osteoarthritis (Kellgren-Lawrence grade 0–3). This implant can effectively treat single or multiple lesions and has been demonstrated as a reliable option in several clinical and preclinical studies. Its main indications are post-traumatic etiology, young age, and minimal to no osteoarthritis.

References

1. Pape D, Filardo G, Kon E, van Dijk CN, Madry H. Disease-specific clinical problems associated with the subchondral bone. Knee Surg Sports Traumatol Arthrosc. 2010;18(4):448–62. https://pubmed.ncbi.nlm.nih.gov/20151111/. [cited 2024 May 16].
2. Lepage SIM, Robson N, Gilmore H, Davis O, Hooper A, St John S, et al. Beyond cartilage repair: the role of the osteochondral unit in joint health and disease. Tissue Eng Part B Rev. 2019;25(2):114. [cited 2024 May 22]. Available from: /pmc/articles/PMC6486663/.
3. Walsh DA, Mcwilliams DF, Turley MJ, Dixon MR, Fransès RE, Mapp PI, et al. Angiogenesis and nerve growth factor at the osteochondral junction in rheumatoid arthritis and osteoarthritis. Rheumatology (Oxford). 2010;49(10):1852–61. https://pubmed.ncbi.nlm.nih.gov/20581375/. [cited 2024 May 22].
4. Jeon JE, Vaquette C, Klein TJ, Hutmacher DW. Perspectives in multiphasic osteochondral tissue engineering. Anat Rec (Hoboken). 2014;297(1):26–35. https://pubmed.ncbi.nlm.nih.gov/24293311/. [cited 2024 May 16].
5. Matta C, Szűcs-Somogyi C, Kon E, Robinson D, Neufeld T, Altschuler N, et al. Osteogenic differentiation of human bone marrow-derived mesenchymal stem cells is enhanced by an aragonite scaffold. Differentiation. 2019;107:24–34.
6. Boffa A, Solaro L, Poggi A, Andriolo L, Reale D, Di Martino A. Multi-layer cell-free scaffolds for osteochondral defects of the knee: a systematic review and meta-analysis of clinical evidence. J Exp Orthop. 2021;8(1):56.
7. Kon E, Filardo G, Shani J, Altschuler N, Levy A, Zaslav K, et al. Osteochondral regeneration with a novel aragonite-hyaluronate biphasic scaffold: up to 12-month follow-up study in a goat model. J Orthop Surg Res. 2015;10(1):81.
8. Kon E, Filardo G, Robinson D, Eisman JA, Levy A, Zaslav K, et al. Osteochondral regeneration using a novel aragonite-hyaluronate bi-phasic scaffold in a goat model. Knee Surg Sports Traumatol Arthrosc. 2014;22(6):1452–64.
9. Kreklau B, Sittinger M, Mensing MB, Voigt C, Berger G, Burmester GR, et al. Tissue engineering of biphasic joint cartilage transplants. Biomaterials. 1999;20(18):1743–9. https://pubmed.ncbi.nlm.nih.gov/10503975/. [cited 2024 May 16].
10. Shahgaldi BF. Coral graft restoration of osteochondral defects. Biomaterials. 1998;19(1–3):205–13.
11. Kon E, Robinson D, Verdonk P, Drobnic M, Patrascu M, Dulic O, et al. A novel aragonite-based scaffold for osteochondral regeneration: early experience on human implants and technical developments. Injury. 2016;47:S27.
12. Guillemin G, Meunier A, Dallant P, Christel P, Pouliquen J-C, Sedel L. Comparison of coral resorption and bone apposition with two natural corals of different porosities. J Biomed Mater Res. 1989;23(7):765–79. https://pubmed.ncbi.nlm.nih.gov/2738087/. [cited 2024 May 16].
13. Filardo G, Andriolo L, Angele P, Berruto M, Brittberg M, Condello V, et al. Scaffolds for knee chondral and osteochondral defects: indications for different clinical scenarios. A consensus statement. Cartilage. 2021;13(1_suppl):1036S–46S.
14. Agili-C™—The future of joint repair. https://www.cartiheal.com/agili-c/. [cited 2024 May 29].
15. Van Genechten W, Vuylsteke K, Struijk C, Swinnen L, Verdonk P. Joint surface lesions in the knee treated with an acellular aragonite-based scaffold: a 3-year follow-up case series. Cartilage. 2021;13(1_suppl):1217S–27S.
16. Kon E, Di Matteo B, Verdonk P, Drobnic M, Dulic O, Gavrilovic G, et al. Aragonite-based scaffold for the treatment of joint surface lesions in mild to moderate osteoarthritic knees: results of a 2-year multicenter prospective study. Am J Sports Med. 2021;49(3):588–98.
17. Altschuler N, Zaslav KR, Di Matteo B, Sherman SL, Gomoll AH, Hacker SA, et al. Aragonite-based scaffold versus microfracture and debridement for the treatment of knee chondral and osteochondral lesions:

results of a multicenter randomized controlled trial. Am J Sports Med. 2023;51(4):957–67.
18. Bedi A, Foo LF, Williams RJ, Potter HG. The maturation of synthetic scaffolds for osteochondral donor sites of the knee: an MRI and T2-mapping analysis. Cartilage. 2010;1(1):20–8. https://pubmed.ncbi.nlm.nih.gov/26069533/. [cited 2024 May 29].
19. D'Ambrosi R, Valli F, De Luca P, Ursino N, Usuelli FG. MaioRegen osteochondral substitute for the treatment of knee defects: a systematic review of the literature. J Clin Med. 2019;8(6):783. https://pubmed.ncbi.nlm.nih.gov/31159439/. [cited 2024 May 17].

Growth and Maturation of Articular Cartilage

16

Pedro Guillén García, Guillén-Vicente Isabel, Guillén-Vicente Marta, Rodríguez-Iñigo Elena, López-Alcorocho Juan Manuel, Tomas Fernandez Jaen, and Steve Abelow

16.1 Introduction: My View on Cartilage

I have observed cartilage in all its stages, from the embryonic period to old age, passing through adolescence, its prime. I have seen how it changes color from bright white in youth to dirty straw-yellow in old age as a result of joint wear. I have also noticed how it varies in hardness from soft and almost displaceable on the bone in the fetal stage to hard, stony, and thin in old age.

I have seen clean and beautiful cartilage in adolescence and in other cases covered with crystals on the surface as in gout or throughout its thickness as in chondrocalcinosis. Through my studies of the human body in its different stages—embryonic, fetal, adolescent, adult, and elderly—I have found that time causes evident biological deterioration. This change turns a structure full of hope into one marked by frustration and disappointment.

The growth of articular cartilage is a quantitative process that can be measured and occurs continuously during the embryonic and fetal phases, and after birth in childhood and until puberty (16–17 years) [1].

The main cause of deterioration and death in humans is time. All tissues lose their functionality over time, and this deterioration is accentuated when they suffer trauma or disease.

16.2 Structure and Characteristics of Articular Cartilage

Articular cartilage, or hyaline cartilage, is an avascular, alymphatic, and aneural tissue that covers the ends of bones in joints, supporting deforming and compressive forces thanks to its elasticity. Its repair capacity is limited, as Hunter observed in 1743. Chondrocytes produce the extracellular matrix, becoming embedded in it and not establishing contact with each other. These chondrocytes detect changes in the matrix and perform continuous turnover of its macromolecules, essential for maintaining hyaline articular cartilage.

Cartilage is a dense connective tissue specialized to meet its demanding functions, with an abundant extracellular matrix (ECM) and few cells, the chondrocytes, which are metabolically very active mesenchymal origin cells whose sole and essential function is the formation, maintenance, and repair of the ECM in which they are imprisoned or buried [2].

P. G. García (✉) · G.-V. Isabel · G.-V. Marta
R.-I. Elena · L.-A. J. Manuel · T. F. Jaen · S. Abelow
Department of Traumatology and Research Unit, Clinica CEMTRO, Madrid, Spain
e-mail: isabel.guillen@clinicacemtro.com; marta.guillen@clinicacemtro.com; jm.lopez@amplicel.com; tomas.fernandez@clinicacemtro.com

A. Gobbi et al. (eds.), *Regenerative Medicine in Sports and Orthopaedics*,
https://doi.org/10.1007/978-3-031-84693-9_16

The ECM is responsible for the homeostasis and nutrition of chondrocytes. The most abundant component is water, representing 70–80% of its wet weight, and it is not homogeneously distributed, representing 65% in the deep zone and 80% in the superficial layers. Collagen is another fundamental component in a proportion of 10–20% of its weight, with type II collagen making up 95% of it. The third component of articular cartilage is proteoglycans, in a proportion of 10–15% of the wet weight. All these components, as well as minor ones like non-collagenous proteins, lipids, or carbohydrates, are produced by chondrocytes and secreted into the ECM. Unlike other types of cartilage, such as elastic cartilage or fibrocartilage, 95% of the collagen in articular cartilage is type II and does not contain type I collagen [2–4].

In hyaline cartilage, we establish four layers or zones that, from the articular surface to the subchondral bone, are:

- *Superficial or sliding layer*, occupying 10–20% of its thickness, with a size of 40 μm, with a few flattened chondrocytes and a high water and type II collagen ECM. It is the thinnest zone, and some authors divide it into two layers: the most superficial without chondrocytes that sometimes detaches, and below, the second layer that already contains some chondrocytes and is resistant to shear.
- *Transitional or middle layer*, thicker than the superficial one at 500 μm and less than the deep layer. The chondrocytes are spherical, with an ECM with more proteoglycans and a lower percentage of water and collagen. Collagen fibers are arranged tangentially and better resist compressive forces.
- *Radial or deep layer*, the widest at 1000 μm, with larger and more abundant collagen fibers, with more numerous chondrocytes aligned in columns. It is rich in proteoglycans and poor in water. Collagen fibers are vertical, giving the cartilage high compression tolerance.
- *Tide-mark layer*, very thin at only 5 μm, basophilic, with tangentially oriented fibers.
- *Calcified layer*, 300 μm in size, with fewer and smaller cells than those in the radial zone. In this layer, hydroxyapatite crystals can be seen anchoring or fixing to the subchondral bone. Beneath this is spongy tissue with rich vascularization, which cartilage lacks.

This histological structure of articular cartilage described (Fig. 16.1) and its composition are configured to perform its two essential functions: reducing friction between the bone ends in the joint (reducing friction forces) and load distribution. The coefficient of friction in humans is 0.002–0.004. Factors that decrease the coefficient are the formation of the synovial fluid film, elastic deformation of the articular cartilage, synovial

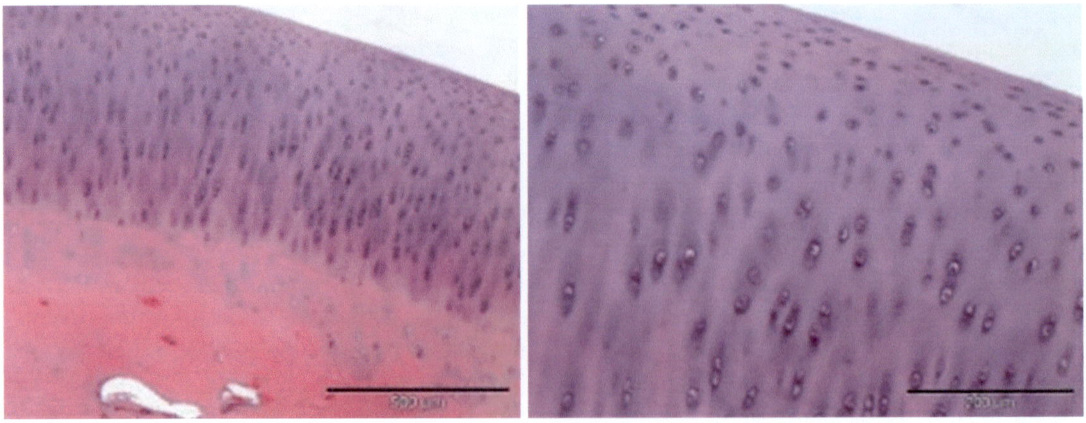

Fig. 16.1 Histological structure of articular hyaline cartilage. The arrangement of chondrocytes is observed: more flattened on the surface and more spherical and stacked in the deeper layers

membrane, and fluid in the articular cartilage. Conversely, the factor that increases the friction coefficient is cartilage fibrillation. Although it may seem exaggerated, the fact is that between two surfaces of healthy cartilage, there is less friction than between two pieces of ice [3].

16.3 Cartilage Nutrition

The nutrition of articular cartilage is still a topic of debate. Cartilage researchers ask what the nutrition of hyaline cartilage is and where it comes from. Given its location at the ends of the joint bones and its contact with the synovial medium, its nutrition can only come from the subchondral bone where the cartilage is anchored or the synovial fluid (SF) in the joints. According to different authors, we know that the superficial layers of articular cartilage are nourished by SF and the deep layers by the vascularization of the subchondral bone. This theory has many more followers than detractors [5].

In the 1970s, when we began studying SF and the anatomopathological study of the synovial membrane, we observed several gaps in the previous theory. The first was the frequent failures in osteochondritis dissecans (OCD) after fixing it to its bed with Kirschner needles, screws, or biodegradable nails. We only obtained good results in unstable OCD when we made the bed bleed abundantly and the bone content was abundant or in acute osteochondral fractures of the patella or femoral condyles after fixing them to their bed.

In 1996, the period when we began culturing autologous chondrocytes from biopsies of cartilage taken from healthy areas, we found that when the biopsy was taken from a free osteochondral fragment (free OS) or from the fragment of OCD, we could not get the chondrocytes to multiply in culture. Indeed, although cartilage from a free body thrives well in the synovial environment, when we cultured it, we observed that the cells did not divide. In contrast, they did cultivate and divide normally when the cartilage came from a healthy area.

This work on the viability of free OS was published in the journal *Cartilage* in the year 2014, demonstrating that the synovial environment only keeps chondrocytes "alive," but they lose the significant biochemical capacities necessary for division, which the synovial environment does not provide. In contrast, chondrocytes from healthy areas in a culture medium divide easily, reaching millions of cells [6].

Based on the above, we theorized that perhaps only the superficial layer of articular cartilage (the chondrocyte) can be nourished by the synovial medium but that these cells require the biochemical messages of the blood from the subchondral bone to maintain their very demanding functions. The rest of the articular cartilage thickness, representing 90% of its thickness, is nourished by the blood from the subchondral bone, maintaining the demanding conditions required by this tissue under pressure, deformation, and shear [3].

Furthermore, considering that many authors divide the superficial layer of hyaline cartilage into two zones, a more superficial layer with flattened, more mature cells that could survive and be nourished by the synovial medium, and a deeper layer with more spherical cells that would not have sufficient nutrition from the SF and would degenerate, as happens when subchondral bone nutrition is lacking in cases of subchondral bone osteonecrosis (ON) and OCD.

16.4 Embryonic Development of Articular Cartilage

As described in the previous section, articular cartilage is a specialized tissue that covers the surfaces of bones in synovial joints, providing a smooth and lubricated surface for joint movement. Its development and maturation are complex processes that begin in the womb and continue into adulthood. Understanding these processes is crucial for treating various joint diseases and developing regenerative therapies. This chapter will explore in detail the stages of embryonic development of articular cartilage and the factors that influence its maturation. As Aristotle said in his book on embryology: "He who sees

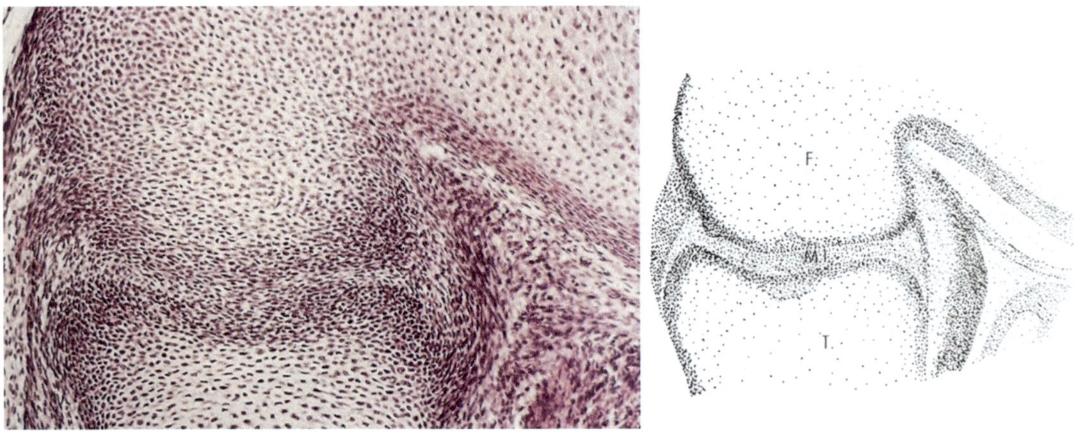

Fig. 16.2 12 mm human embryo, stage 17 of O'Rahilly, showing the precartilaginous anlagen of the femur and tibia joined and continuous by undifferentiated mesenchyme. Drawing of the anatomical preparation to the right of the figure. *F* femur, *T* tibia, *MI* undifferentiated mesenchyme

things from their beginning will have the best and most perfect understanding of them."

I have studied articular cartilage for many years, across all ages. In this 12 mm human embryo at stage 17 of O'Rahilly (Fig. 16.2), we see the precartilaginous rudiments of the femur and tibia connected and continuous through undifferentiated mesenchyme. Also, in the embryonic period, as shown in Fig. 16.3, a histological section of a 23 mm human embryo at stage 21 of O'Rahilly, we see the first rudiments of the precartilaginous blastema of the femur, tibia, and fibula [1].

Several stages can be distinguished during the development of cartilage, which we will describe below.

16.4.1 Condensation of the Articular Blastema

The development of articular cartilage begins with the condensation of mesenchymal cells. This process starts around the fourth week of gestation in humans, when mesenchymal cells begin to cluster and differentiate into chondroblasts. This cellular clustering is called the articular blastema, and it is the precursor of articular cartilage. During this phase, mesenchymal cells start expressing specific transcription factors that guide them in differentiating into mature chondrocytes [1–3].

The condensation of mesenchymal cells in the region that will form the joint is induced by morphogenetic signals, such as fibroblast growth factor 2 (FGF2) and transforming growth factor-beta (TGF-β), which are crucial for the initial formation of cartilage [5].

16.4.2 Chondrification

Chondrification is the process during which chondroblasts become chondrocytes, the mature cells of cartilage. Chondroblasts secrete an extracellular matrix rich in type II collagen and proteoglycans, which provide the cartilage with its biomechanical properties and resistance to compression.

The process of chondrification also involves the accumulation of proteoglycans, especially aggrecan, which bind to collagen to form a three-dimensional network that helps maintain the integrity of the cartilage.

16.4.3 Interzone Phase

The interzone phase is characterized by the appearance of an intermediate layer between the

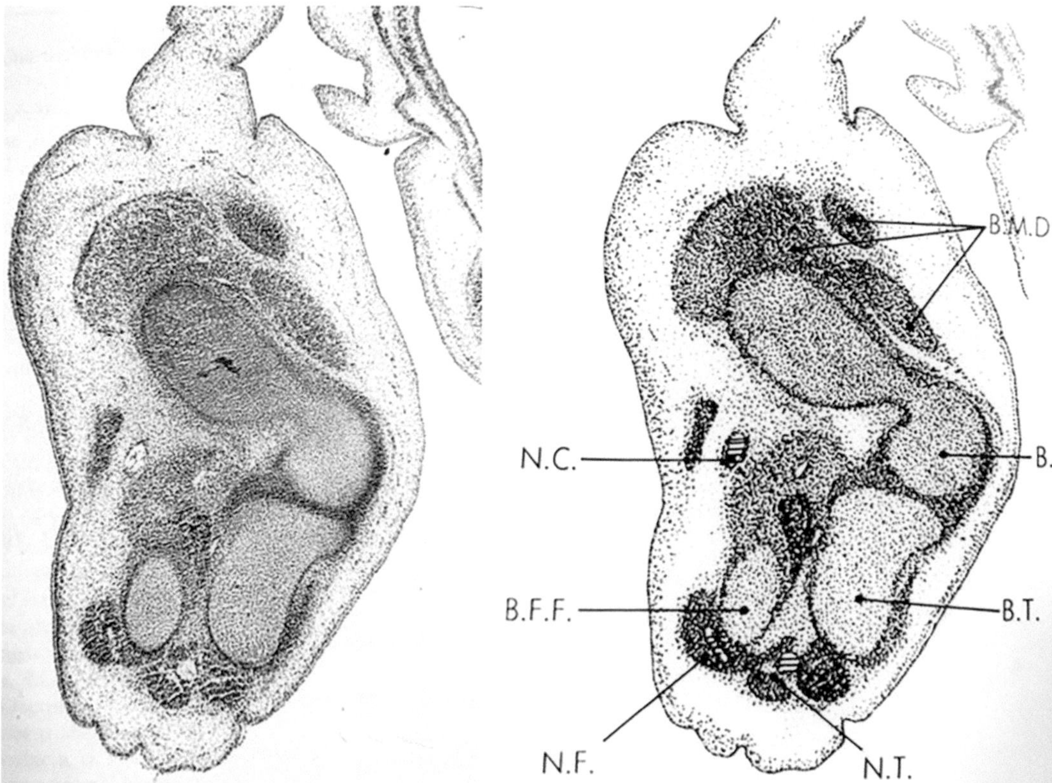

Fig. 16.3 Histological section of a 23 mm human embryo, stage 21 of O'Rahilly, showing the first outlines of the precartilaginous blastema of the femur, tibia, and fibula. Drawing of the anatomical preparation to the right of the figure. *B.M.D* dorsal premuscular blastema of the thigh, *B.F* femoral precartilaginous blastema, *B.T* tibial precartilaginous blastema, *B.F.F* fibular precartilaginous blastema, *N.C* sciatic nerve, *N.T* tibial nerve, *N.F* fibular nerve

developing bone epiphyses (Fig. 16.4a). This phase is critical for the formation of the joint cavity. The organization of the cells into a structure that will give rise to the future joint cavity is also observed (Fig. 16.4b). The formation of the interzone is regulated by signaling through the retinoic acid receptor (RAR) and the epidermal growth factor receptor (EGFR), which coordinate the development and formation of the joint cavity.

16.4.4 Cavitation

Cavitation is a key process in the formation of synovial joints. As cells in the intermediate zone of the interzone begin to undergo apoptosis simultaneously, a space forms that will become the joint cavity (Fig. 16.5). The regulation of this process is mediated by various signaling pathways, including the matrix metalloproteinases (MMPs) pathway and ME proteins, which coordinate the remodeling and formation of the cavity.

16.4.5 Fetal Remodeling

During the embryonic and fetal period until birth, the articular cartilage continues to grow and mature. Chondrocytes continue to produce and organize the extracellular matrix in response to increasing mechanical loads. This process ensures that the cartilage is functional and capa-

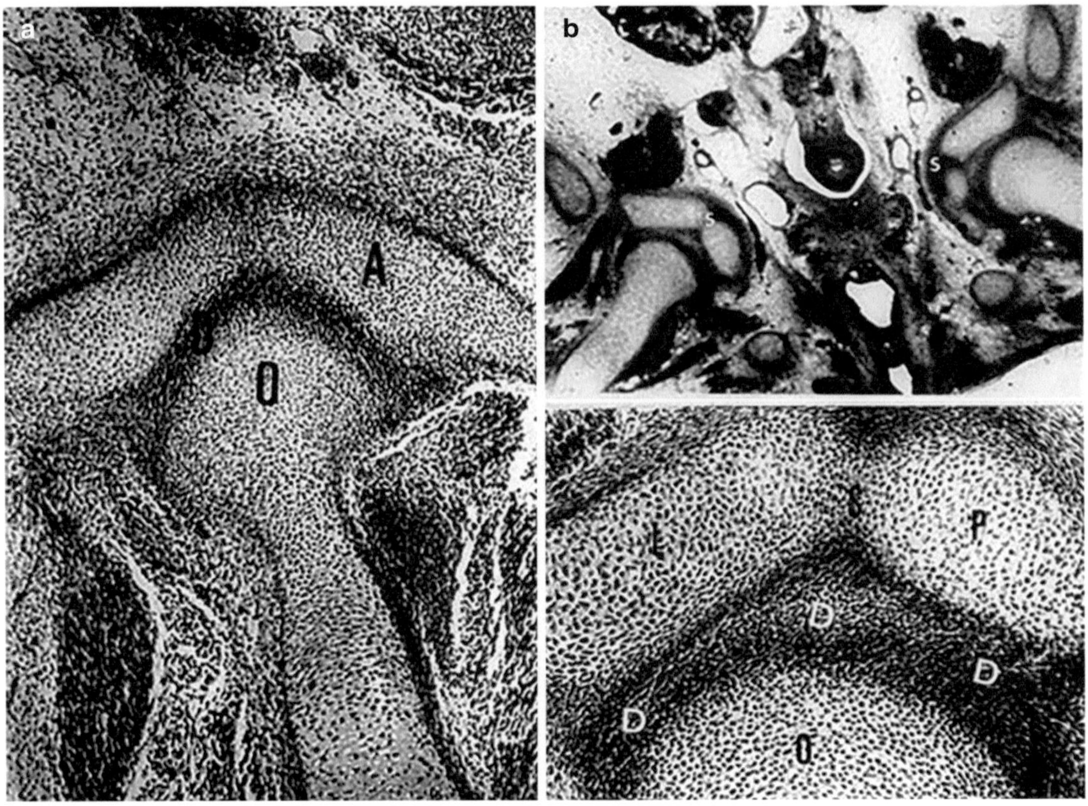

Fig. 16.4 Interzone phase showing an intermediate layer between the developing bones (**a**) that will later form the joint cavity (**b**). *O* femoral head, *A* acetabulum

ble of withstanding mechanical loads at birth and during postnatal growth. Figure 16.6 shows a sagittal section of a 55 mm fetus where the cartilage surrounding the internal femoral condyle is visible. It also shows how the superior tibioperoneal joint cavity communicates with the knee joint. Figures 16.7, 16.8, 16.9, and 16.10 show these interesting dissections of fetal knees, where we studied the growth and maturation of cartilage tissue and menisci. These preparations were very difficult and required patience and a lot of time [1, 3].

In the case of knee cartilage development, some authors hold phylogenetic criteria, arguing that the knee joint is the result of adaptation processes and static and functional changes that the lower limbs undergo upon contact with the ground. This includes the establishment of highly specialized dynamic support levers, the organization of intermediate cartilages, and the reinforcement of capsuloligamentous structures. Other authors consider additional factors such as movement, which conditions the dynamic and structural integration of neural, muscular, and chemoenzymatic dependencies that together determine and regulate joint formation. It has been shown that although initial movement provides cavitation, in the absence of premuscular masses, cavitation disorganizes and disappears.

At these stages of development, the cartilage is pearly white. In Fig. 16.11a, the right knee of a full-term fetus is shown, with menisci and white cartilage. After removing the menisci (Fig. 16.11b), we see the white cartilage, and for the first time, we show the meniscal impression on the internal tibial plateau, favored by knee varus in the maternal womb and the significant mobility enjoyed by the lower limbs. In Fig. 16.12, we see the meniscal impression of an adult knee, and after removing the menisci in the internal compartment, the tibial plateau has grown and become eburnated due to sustained load [3].

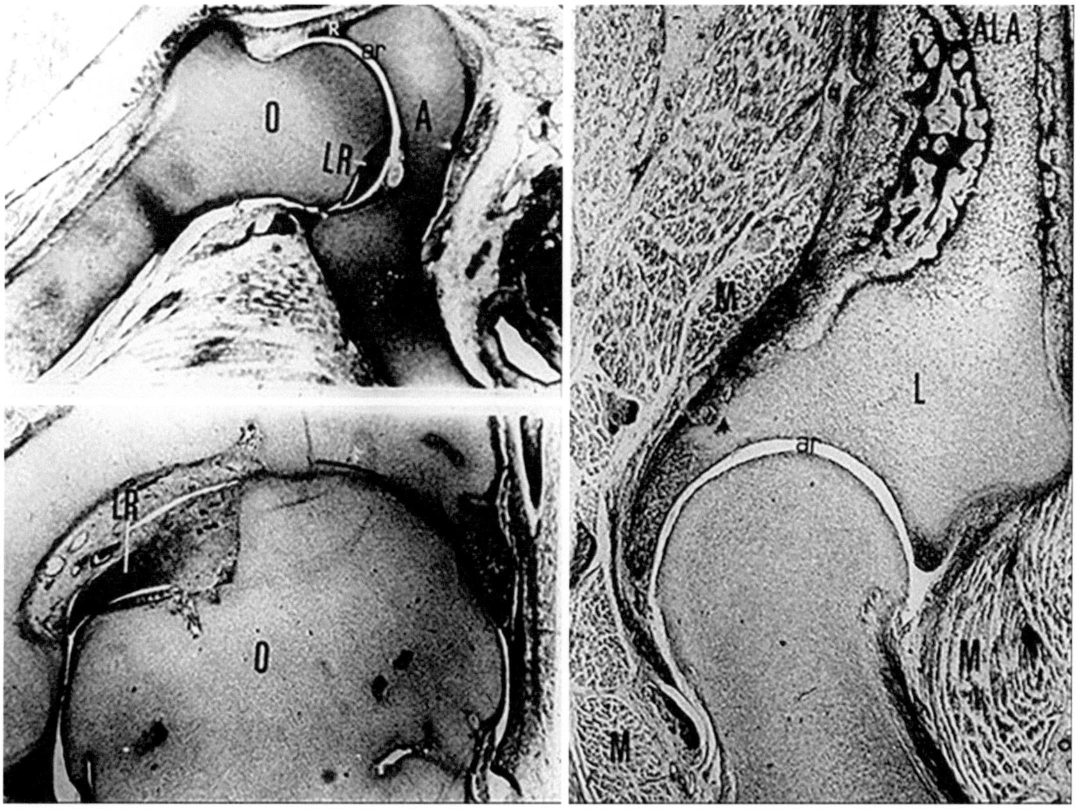

Fig. 16.5 The cells of the intermediate zone of the interzone undergo apoptosis to form the joint cavity. Cell death is life for the joint

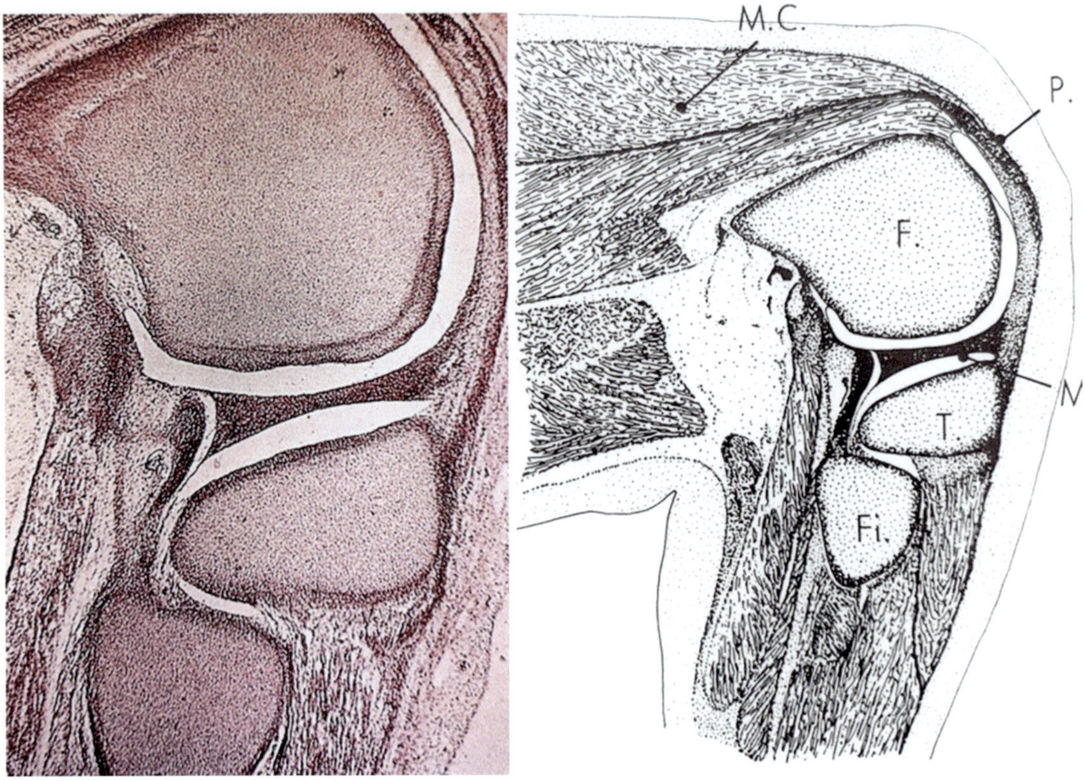

Fig. 16.6 Sagittal section of a 55 mm fetus showing cartilage surrounding the internal femoral condyle and the communication of the superior tibiofibular joint cavity with the knee joint. Drawing of the anatomical preparation to the right of the figure. *M.C* quadriceps muscle, *P* patella, *F* femur, *M* meniscus, *T* tibia, *Fi* fibula

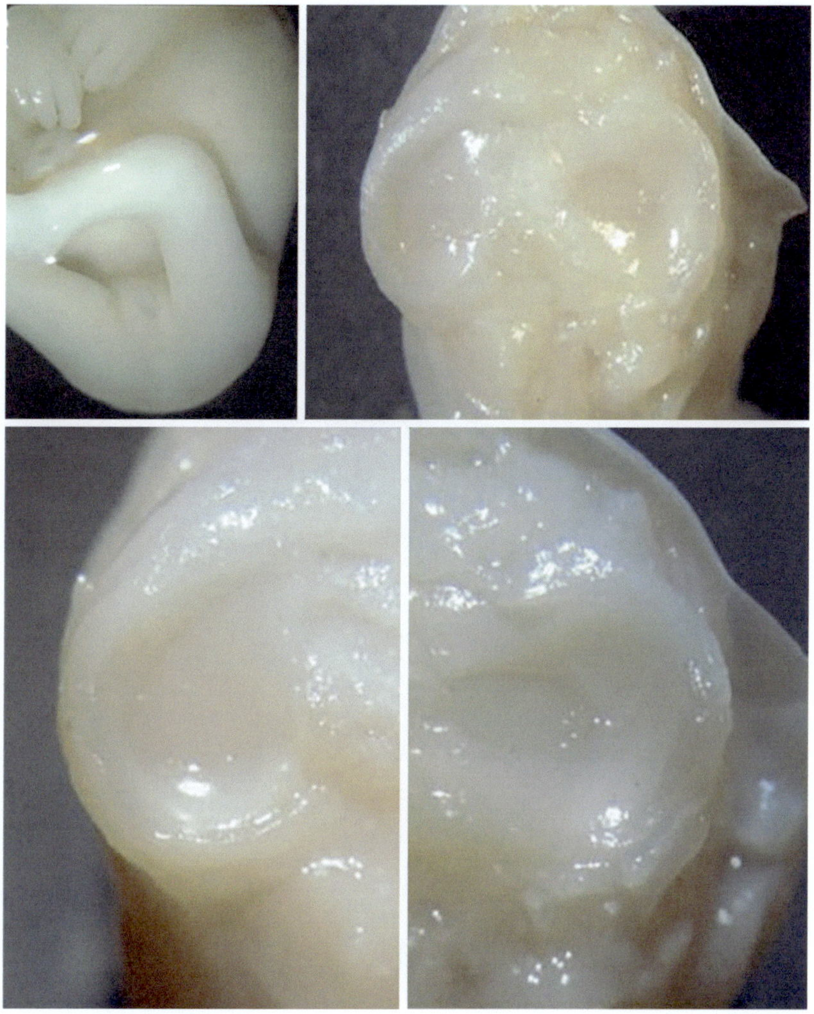

Fig. 16.7 12-week-old, 75 mm fetus showing cartilage and menisci that are not yet configured

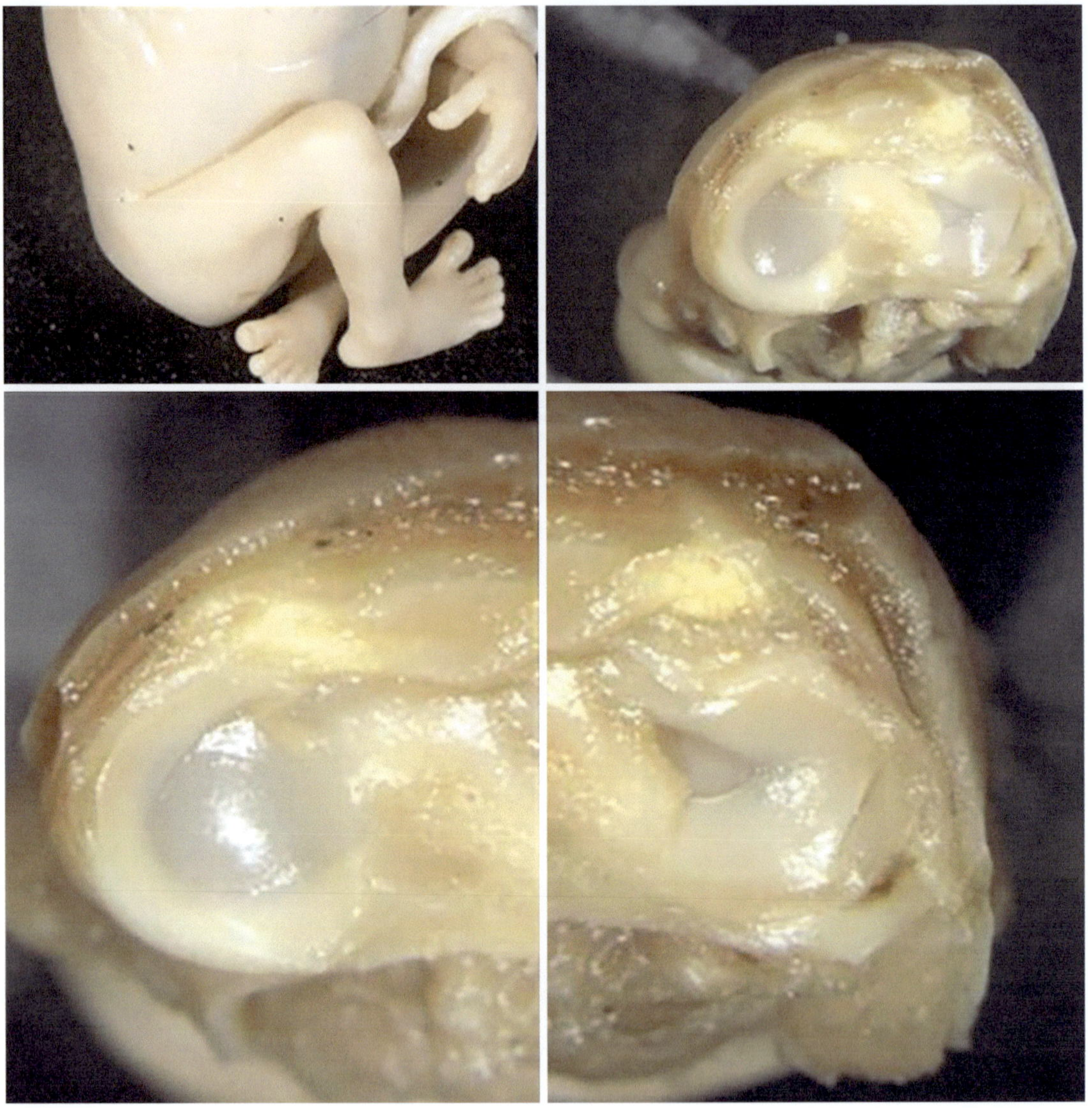

Fig. 16.8 19-week-old, 170 mm fetus with menisci and white cartilage showing greater maturation

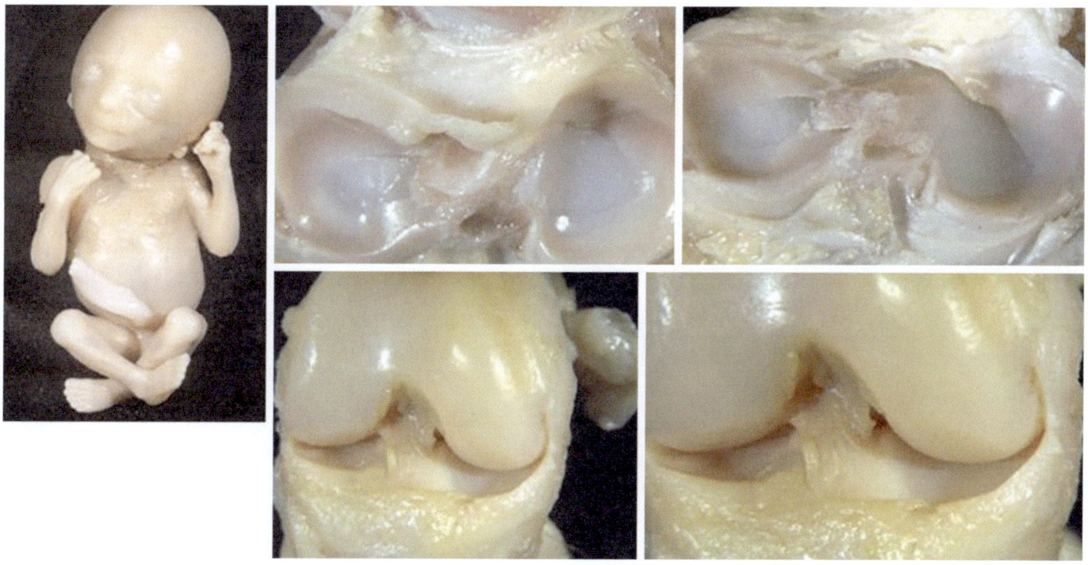

Fig. 16.9 22-week-old, 205 mm fetus with white articular cartilage covering the bone ends of the knee joint, showing more advanced maturation

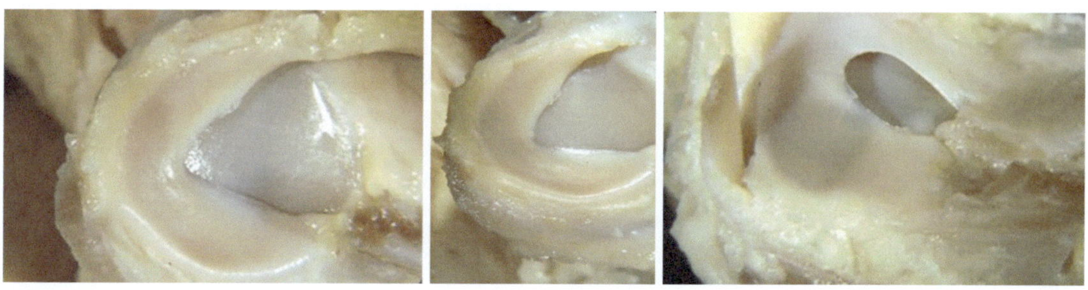

Fig. 16.10 White cartilage of the tibial plateau with fully configured menisci

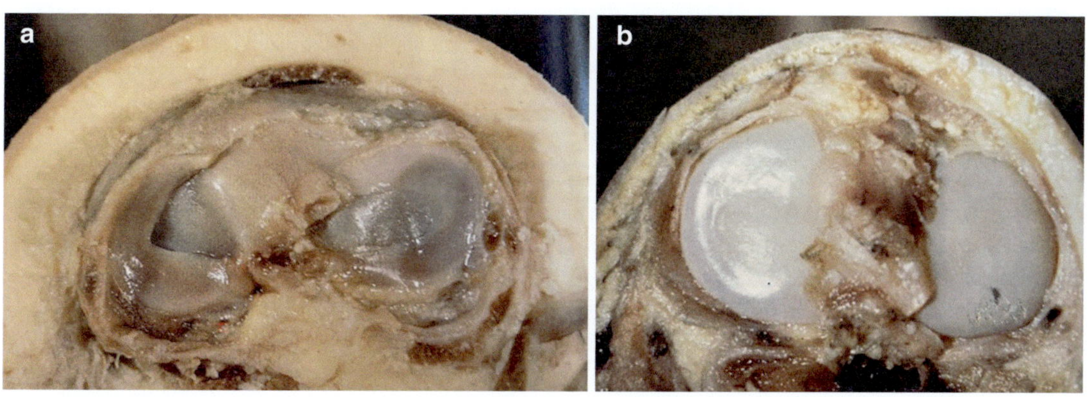

Fig. 16.11 Right knee of a full-term fetus, cranial view, with menisci and white cartilage (**a**). After removing the menisci, white cartilage is seen, and for the first time, the "meniscal impression" (impressio meniscale) on the internal tibial plateau favored by knee varus in the womb and the great mobility enjoyed by the joints is shown (**b**)

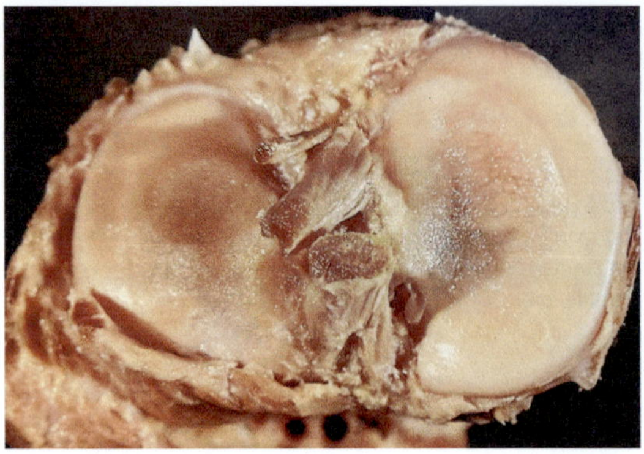

Fig. 16.12 Meniscal impression' (impressio meniscale) of an adult knee, showing that after the removal of the menisci, the internal compartment of the tibial plateau has grown and become ivory-like due to sustained loading

16.5 Factors Influencing the Development of Articular Cartilage

16.5.1 Genetic Signaling

16.5.1.1 Sox9

Sox9 is a crucial transcription factor for chondrogenesis. Its expression is critical for the differentiation of mesenchymal cells into chondrocytes. Sox9 regulates the transcription of essential genes for the production of cartilage matrix, such as Col2a1 and aggrecan. Mutations in the Sox9 gene can cause cartilage development disorders, such as campomelic dysplasia (bone and cartilage abnormalities).

16.5.1.2 Col2a1 and Aggrecan

The Col2a1 gene encodes type II collagen, the main collagen present in articular cartilage. Aggrecan, a proteoglycan, is essential for cartilage hydration and elasticity. The interaction between these components is vital for forming a functional cartilage matrix. Alterations in the expression of Col2a1 or aggrecan can result in defects in cartilage structure and a predisposition to joint diseases [5, 7].

16.5.2 Growth Factors

16.5.2.1 TGF-β

Transforming growth factor-beta (TGF-β) is crucial for regulating chondrocyte proliferation and differentiation. TGF-β acts through the Smad signaling pathway to regulate the expression of cartilage genes and extracellular matrix production. Dysfunction in TGF-β signaling can lead to alterations in cartilage structure and diseases such as osteoarthritis [4].

16.5.2.2 IGF

Insulin-like growth factor (IGF) promotes protein synthesis and cellular proliferation in cartilage. IGF-1, in particular, plays a significant role in regulating cartilage growth during development. IGF-1 deficiency is associated with dwarfism and cartilage development problems.

16.5.2.3 FGF

Fibroblast growth factors (FGF) are important for chondrocyte differentiation and cartilage matrix formation. FGF-2, for example, stimulates chondrocyte proliferation and contributes to angiogenesis in cartilage. Proper regulation of FGF is essential for the formation and maintenance of articular cartilage.

16.5.3 Mechanical Influence

Mechanical stimulation is fundamental for the development and maintenance of articular cartilage. The load and movement within the joint stimulate the production of matrix by chondrocytes and the organization of collagen fibers. Lack of movement or immobilization can lead to cartilage degeneration, underscoring the importance of mechanical stimulus in its maintenance. Studies have shown that prolonged immobilization in animal models results in the loss of cartilage integrity and alterations in its structure. Movement influences joint health.

16.5.4 Hormonal Influence

16.5.4.1 Growth Hormone (GH)

Growth hormone (GH) plays a crucial role in the growth and development of articular cartilage. GH stimulates the proliferation of chondrocytes and the synthesis of extracellular matrix through the production of IGF-1. GH deficiency can result in poor cartilage growth and skeletal development problems.

16.5.4.2 Thyroid Hormone (TH)

Thyroid hormone (TH) is essential for regulating cartilage metabolism and growth. TH regulates protein synthesis and chondrocyte differentiation. TH deficiency during development can lead to abnormalities in cartilage growth and bone formation.

16.5.5 Nutritional Factors

Nutrition is fundamental for the proper development of articular cartilage. Nutrients such as calcium, phosphorus, and vitamins D and K are essential for the synthesis and maintenance of the extracellular matrix. Deficiency in these nutrients can negatively affect cartilage development and increase the risk of diseases.

16.6 Postnatal Maturation of Articular Cartilage

The postnatal maturation of articular cartilage is a complex process that follows the initial formation of cartilage during embryonic development. This process is fundamental for the adaptation of cartilage to biomechanical and functional demands as the individual grows. Postnatal maturation involves structural and biochemical changes in the cartilage that are critical for maintaining its integrity and functionality throughout life.

16.6.1 Structural and Biochemical Changes

16.6.1.1 Organization of the Extracellular Matrix

The extracellular matrix (ECM) of articular cartilage is primarily composed of type II collagen, proteoglycans, and water. Throughout postnatal maturation, significant changes occur in the composition and organization of this matrix. During the early years of life, articular cartilage is highly hydrated, allowing for rapid growth and cellular expansion. As maturation progresses, the concentration of proteoglycans and type II collagen increases, and the matrix organizes into layers with a more structured arrangement of collagen fibers [7, 8].

16.6.1.2 Changes in Chondrocyte Proliferation and Differentiation

During childhood and adolescence, chondrocyte proliferation and matrix synthesis are highly active. The proliferative zone of the cartilage expands to allow for bone and joint growth. However, with maturation, the rate of chondrocyte proliferation decreases, and the rate of differentiation into mature chondrocytes increases. This change in cellular activity correlates with the transition from a more flexible cartilage in childhood to a more resilient and structured one in adulthood.

16.6.2 Influence of Mechanical Load

16.6.2.1 Adaptation to Load

Articular cartilage is exposed to mechanical loads during movement and physical activities. Mechanical load is a determining factor in the maturation and maintenance of cartilage. Regular load application stimulates matrix synthesis by chondrocytes and maintains ECM integrity. Articular cartilage adapts its structure and com-

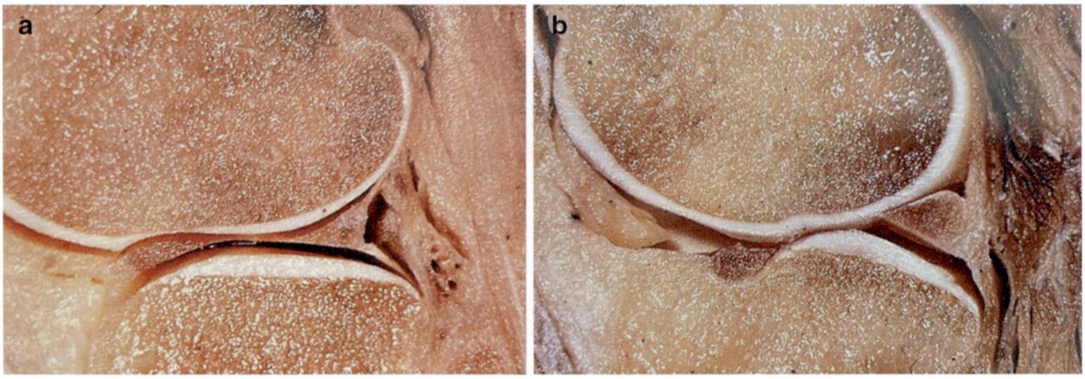

Fig. 16.13 Section of an extended adult knee showing the thinner external femoral condyle as it is a mobility compartment (**a**) compared to its internal compartment (**b**) where the cartilage is thicker as it is a stability and load-bearing compartment

position in response to mechanical loads, with increased production of type II collagen and proteoglycans to withstand the applied forces.

In Fig. 16.13, we observe the thinner thickness of the external femoral condyle, which is a compartment of mobility (Fig. 16.13a). In contrast, a section of the same adult knee in extension, but from its internal compartment (Fig. 16.13b), shows a greater cartilage thickness due to its role as a compartment of stability and load. Thus, the cartilage, the white tissue, varies in thickness from one compartment to another within the same knee [1, 3].

16.6.2.2 Effects of Immobilization

Prolonged immobilization, such as occurs with certain injuries or post-surgery, can have adverse effects on cartilage health. The lack of mechanical load and movement can lead to cartilage atrophy, with decreased matrix synthesis and increased degradation. Studies have shown that immobilization results in a loss of collagen and proteoglycan density and a reduction in the cartilage's impact-absorbing capacity.

16.6.3 Changes in Adulthood

16.6.3.1 Cartilage Maintenance

In its mature state, chondrocytes cease cell division and remain in this state for life [8]. When articular cartilage is adult, it appears white due to the lack of blood supply, while the femoral condyles and tibial plates are orange due to their rich vascularization (Fig. 16.14). Thus, the color of our tissues is influenced by blood supply: red for high blood supply and white for poor blood supply (Fig. 16.15). Over the years, cartilage takes on a light brown color from joint use, but if washed well, it appears white. We have found that during the development of articular cartilage, the cell density is high, and chondrocytes exhibit maximum metabolic activity. Once development is complete, chondrocytes no longer divide but continue to maintain the production of extracellular matrix, thus ensuring the remodeling of the

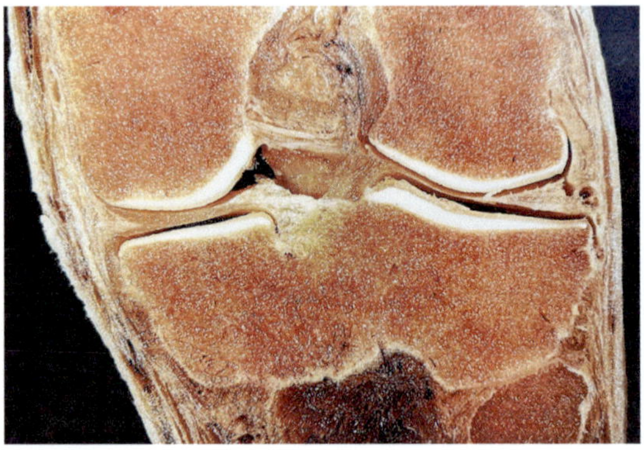

Fig. 16.14 Adult articular cartilage, left knee, appearing white due to lack of blood supply, and femoral condyles and tibial plates appearing orange due to their rich vascularization

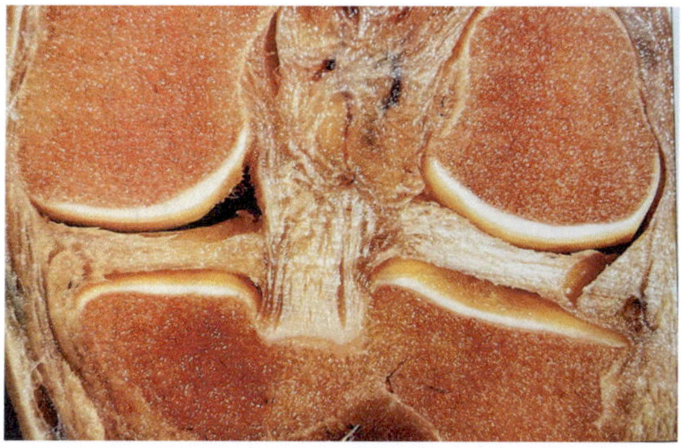

Fig. 16.15 Articular cartilage of the bone ends appearing white due to lack of blood supply, and femoral condyles and tibial plates appearing orange due to their rich vascularization. Same knee as in Fig. 16.14, in a more posterior section

articular surface. However, they are not capable of filling cartilage defects caused by trauma or other reasons.

The biology of articular cartilage depends on the chondrocytes contained within it, which play the essential role of maintaining a stable extracellular matrix composition. The nutrition of the superficial part of articular cartilage comes from synovial fluid, and during tissue development and maturation, it is primarily through diffusion from the subchondral bone. Articular cartilage has low metabolic activity due to its low cellularity. The chondron is the anatomical, cytogenetic, and mechanical unit of cartilage [3, 5, 7].

In adulthood, articular cartilage maintains its structure through a balance between the synthesis and degradation of the extracellular matrix. However, with age, this balance tends to become disrupted, with decreased matrix production and increased activity of matrix metalloproteinases (MMPs), which degrade collagen and proteoglycans. These changes contribute to cartilage aging and conditions such as osteoarthritis [9].

Most long bones initially form in the embryo as cartilage, serving as a template later replaced by bone (endochondral ossification) [2]. The chondrocyte is the first cell of the musculoskeletal system, alongside muscle. Since we know mature chondrocytes do not divide and cannot heal once degenerated or injured, we should recommend avoiding joint traumas and high demands on our joints. The best way to extend the life of the joint chondrocyte is to avoid doing anything that would shorten it.

16.6.4 Osteoarthritis and Degeneration

Osteoarthritis (OA) is a degenerative disease affecting articular cartilage, characterized by the progressive loss of the extracellular matrix and chondrocyte dysfunction. In OA, cartilage degeneration is accompanied by alterations in matrix structure, with increased MMP activity and reduced levels of type II collagen and proteoglycans. Additionally, chronic inflammation in the joint contributes to disease progression [9].

16.6.5 Genetic and Environmental Factors

16.6.5.1 Genetic Influence

Genetic factors play a significant role in cartilage maturation and susceptibility to joint diseases. Genetic variants in genes such as COL2A1, AGC1, and MMP13 can influence cartilage health and predisposition to conditions like osteoarthritis [4].

16.6.5.2 Environmental Impact

Environmental factors, such as physical activity level, body weight, and diet, also affect the health of articular cartilage. Obesity, for example, is associated with increased load on joints, which can accelerate cartilage degeneration. A diet rich in anti-inflammatory nutrients can help maintain cartilage health and prevent joint diseases.

16.7 Hyaline Cartilage: A Nerveless Tissue

The oldest known disease in humans, or rather in a hominid of the human evolutionary line, has been detected in bone. These remains belong to a Java Man, originally named *Pithecanthropus erectus* by its discoverer Eugène Dubois, and later reclassified in light of new discoveries as *Homo erectus*. These fossil remains, at least 500,000 years old, show a lesion that is possibly an exostosis, or a tumor, on the inner face of the upper third of the femur, which, in shape and characteristics, is almost identical to the femur of a twenty-first-century human.

As mentioned earlier, articular hyaline cartilage is aneural. The joint capsule receives proprioceptive endings that inform about the position of the limb; it also receives nerve branches that travel through the adventitial membrane, controlling vasomotor function, and are responsible for the blood supply to the capsule, especially the synovial membrane. Lastly, it has pain-sensitive endings that react to the distension of the capsule. To demonstrate that articular hyaline cartilage is not painful due to the lack of innervation, Dr. Scott F. Dye conducted a self-experiment. He performed arthroscopy on both knees without anesthesia, though the entry portals were anesthetized, but not the joint itself [10].

During the experiment, he and his team probed and pressed on all knee structures: synovial membrane, patellar cartilage, condyles and tibial plates, menisci and ligaments, and Hoffa's fat pad, which were not anesthetized. They demonstrated that pressing up to half a kilogram on the cartilage of the patellar ridge did not cause pain, even in cases of chondromalacia, as was the case in this self-experiment. Palpation of the synovial membrane and fat pad caused intense pain even without pressure.

This self-experimentation proves that cartilage, being aneural, does not cause pain on palpation. In the late twentieth century, there was doubt about whether cartilage caused pain when injured, particularly in patellar chondromalacia. A doctor, convinced that injured cartilage does not intrinsically trigger pain, undertook the task of demonstrating this on his own knees. Believing in his theory, the doctor understood that he had to perform the experiment on himself to validate his hypothesis.

Articular cartilage, impacted by over 200 million steps in a man's lifetime of up to 75 years (excluding athletes whose numbers are significantly higher), endures astronomical loads. In the absence of disease or accident, cartilage tolerates this perfectly, indicating its unparalleled efficiency despite being avascular, having almost no repair capacity, and being only a few millimeters thick. During each step, synovial fluid penetrates the cartilage and then exits into the synovial environment through diffusion. However, this diffusion is not sufficient to nourish it since 80% of synovial fluid is water, leaving very little for nutrients. Nevertheless, it is clear that movement, steps, or load are necessary to lubricate the cartilage and possibly nourish its superficial layer.

We have outlined the growth and maturation of the chondrocyte, that is, its anatomy from the embryonic and fetal stages to adulthood. This is a very important detail, as anatomy is to physiology what geography is to history; it describes the context of action and the place where the disease occurs. Understanding the histology of the chondrocyte is not only crucial for knowing its physiology and pathology but also serves as a fundamental basis for clinical practice. In our upcoming research, which is close and eagerly anticipated, we aim to address the repair of damaged cells.

16.8 Conclusion

In conclusion, to determine how to address a damaged tissue, we recommend studying it at all stages of its growth and anatomical development. Given that articular cartilage is an avascular, aneural, and alymphatic tissue, which conditions its lack of regenerative capacity, and due to the high prevalence of its pathology after the age of 50, we recommend studying and researching the "Synovial Environment." This encompasses more than just cartilage, including the synovial membrane, synovial fluid, and subchondral bone.

We believe that the cell is a medication and that once damaged, we must investigate how to restore it to its normal state. The degenerated (arthritic) chondrocyte represents a therapeutic opportunity, and a deep understanding of the processes of development and deterioration is essential for applying effective regenerative therapies that rejuvenate it.

Regenerative medicine, with its significant advancements, has only dipped the index finger into the fountain of youth.

Articular cartilage injuries and their treatment pose a significant challenge in orthopedic and trauma surgery, whether acute or chronic. Advanced therapies, including tissue engineering, cell therapy, and gene therapy, are showing promising potential in finding a solution to the prevalent condition of osteoarthritis.

I conclude by expressing my support for the study of anatomical biology and research, which are the driving forces of progress. And with a prediction supported by great biotechnological advances: "to heal without stopping the engine of man."

References

1. Jiménez-Collado J, Guillén-García P, Sobrado-Pérez J. The knee: morphogenesis, applied anatomy, approach routes. Madrid: Ed. MAPFRE S.A.; 1994.
2. Orts-Llorca. Anatomía humana. Tomo Segundo. Madrid: Ed. Científico Médica; 1948.
3. Guillen P. The chondrocyte, a therapeutic opportunity in orthopedic and traumatology surgery. Madrid: Ed. Clínica CEMTRO; 2013.
4. Decker RS. Articular cartilage and joint development from embryogenesis to adulthood. Semin Cell Dev Biol. 2017;62:50–6. https://doi.org/10.1016/j.semcdb.2016.10.005.
5. Chijimatsu R, Saito T. Mechanisms of synovial joint and articular cartilage development. Cell Mol Life Sci. 2019;76:3939–52. https://doi.org/10.1007/s00018-019-03191-5.
6. Guillén-García P, Rodríguez-Iñigo E, Guillén-Vicente I, Guillén-Vicente M, Fernández-Jaén T, Concejero V, Val D, Maestro A, Abelow S, López-Alcorocho JM. Viability of pathologic cartilage fragments as a source for autologous chondrocyte cultures. Cartilage. 2016;7:149–56. https://doi.org/10.1177/1947603515621998.
7. Decker RS, Koyama E, Pacifici M. Genesis and morphogenesis of limb synovial joints and articular cartilage. Matrix Biol. 2014;39:5–10. https://doi.org/10.1016/j.matbio.2014.08.006.
8. Archer CW, Francis-West P. The chondrocyte. Int J Biochem Cell Biol. 2003;35:401–4. https://doi.org/10.1016/s1357-2725(02)00301-1.
9. Decker RS, Koyama E, Pacifici M. Articular cartilage: structural and developmental intricacies and questions. Curr Osteoporos Rep. 2015;13:407–14. https://doi.org/10.1007/s11914-015-0290-z.
10. Dye SF, Vaupel GL, Dye CC. Conscious neurosensory mapping of the internal structures of the human knee without intraarticular anesthesia. Am J Sports Med. 1998;26:773–7.

Part III

Clinical Application

17 Advances in Meniscal Tissue Engineering

Tomoyuki Nakasa, Kazunori Shimomura, Shuhei Otsuki, and Nobuo Adachi

17.1 Introduction

Although meniscus injuries are very common in daily clinical practice, there have been no perfect methods to treat them. Partial meniscectomy or meniscus repair is commonly performed to improve the symptoms, but its therapeutic effects are often limited [1]. Meniscus repair has the advantage of preserving the meniscus. However, it has a higher reoperation rate than partial meniscectomy [2]. Meniscectomy, the removal of the damaged part of the meniscus, can improve symptoms such as pain or locking but it has the potential to cause osteoarthritis due to the elevation of the peak contact stress on the articular cartilage and subchondral bone [3]. For large injuries of the meniscus, the ideal treatment is to replace the injured meniscus. Therefore, meniscal allograft transplantation has been conducted to alleviate symptoms and delay osteoarthritis. Although several studies demonstrated the low failure rate of meniscal allograft transplantation, the rate of complications such as tears, removal, and infections is high between 30% and 46% [4]. Graft availability and sizing in this procedure are difficult [5, 6]. In addition, the number of countries where it can be available is limited. Since meniscus allograft transplantation cannot be a common injured meniscus treatment, meniscal scaffolds have been developed. Three meniscal scaffolds, CMI-Menaflex (Stryker, Greenwood Village, CO), Actifit (Orteq Sports Medicine Ltd., London, UK), and Nusurface (Active Implants, Memphis, TN), have been clinically used. However, previous studies revealed that these meniscus scaffolds have a high failure rate and lack of clinical data [7].

With the advancement of new technologies, new scaffolds to repair or reconstruct the injured meniscus are being developed using tissue engineering. Tissue engineering can create scaffolds that mimic the original native tissue and are gradually replaced to form functional tissue through the enhancement of regenerative potential to promote the growth of tissue into the scaffolds.

T. Nakasa
Department of Artificial Joints and Biomaterials, Graduate School of Biomedical and Health Sciences, Hiroshima University, Hiroshima, Japan

K. Shimomura
Department of Rehabilitation, Kansai University of Welfare Sciences, Osaka, Japan

Department of Orthopaedic Surgery, Osaka University Graduate School of Medicine, Osaka, Japan
e-mail: kazunori-shimomura@umin.net

S. Otsuki
Department of Orthopedic Surgery, Osaka Medical and Pharmaceutical University, Osaka, Japan
e-mail: shuhei.otsuki@ompu.ac.jp

N. Adachi (✉)
Department of Orthopaedic Surgery, Graduate School of Biomedical and Health Sciences, Hiroshima University, Hiroshima, Japan
e-mail: nadachi@hiroshima-u.ac.jp

A. Gobbi et al. (eds.), *Regenerative Medicine in Sports and Orthopaedics*,
https://doi.org/10.1007/978-3-031-84693-9_17

Although several tissues such as skin and cartilage created by tissue engineering have been clinically used and obtained good clinical outcomes, there are still several problems to be solved with the meniscus scaffolds due to their unique tissue structure as follows.

17.1.1 Mechanical Properties

The meniscus stabilizes the knee joint and coordinates knee motion with load transmission and shock absorption [8, 9]. In addition, it has important roles in lubricating the joint and proprioception, and these functions ensure the maintenance of normal knee kinematics and even prevent cartilage degeneration [10–12]. For the meniscus scaffold to reproduce these functions, the tissue engineering meniscus scaffold must have sufficient mechanical properties to support axial impact, rotational forces, and shear forces in the knee joint. On the other hand, the tissue engineering scaffold should be gradually replaced by the host tissues by providing the microenvironment for cell migration, proliferation, and matrix synthesis. The meniscus scaffold should have suitable porosity that allows cells/tissues to easily infiltrate into the materials, good biocompatibility, ideal biodegradability, and high-strength capacity to hold weight [13]. Although there is still no consensus regarding the ideal materials for a scaffold of meniscal regeneration, scaffolds for tissue engineering of the meniscus are categorized into four classes: synthetic polymers such as polyurethane, poly-caprolactone (PCL), and polylactic acid (PLGA), hydrogels, natural matrix including collagen or hyaluronan or tissue-derived materials such as decellularized ECM [14]. These materials have been advanced to achieve higher mechanical properties and bioactivity.

17.1.2 Morphology

The meniscus has a unique semilunar shape, with different shapes on the medial and lateral meniscus. The medial meniscus is C-shaped approximately 40–45 mm in length and 27 mm in width. The lateral meniscus is O-shaped and shorter than the medial meniscus, with 32–35 mm in length. The medial meniscus covers 51–74% of the medial joint surface, and the lateral meniscus covers 75–93% of the joint surface [15, 16]. The meniscus is constituted by the hydrophilic extracellular matrix (ECM) containing water (72%), collagen (22%), and glycosaminoglycans (GAGs) (0.8%) with only a few cells. The fiber and fascicles in the meniscus exhibit various arrangements depending on the location [17]. The inner region of the meniscus is composed of small and irregular radial collagen fibers, and the outer region is composed of organized interweaved collagen fibers [18]. In addition, blood vessels and nerves exist in the outer third of the synovial margin of the meniscus, which makes restoring and regeneration of the meniscus difficult. Meniscus scaffolds are required to have these gradient bionic properties in addition to unique shape and size of the medial and lateral meniscus, and sufficient mechanical strength [19]. Recently, emerging technologies of 3D bioprinting of gradient meniscus scaffolds for meniscus repair and regeneration are expected to address these morphological problems [20, 21].

17.1.3 Cells

Cells in the meniscus show high heterogeneity including fibroblast-like, chondrocyte-like cells, and intermediate cells with both characteristics [22, 23]. The cells in the outer region are mainly fibroblast-like cells, surrounded by abundant collagen type I [24]. In the inner region, chondrocyte-like cells mainly appeared, with 40% collagen type I and 60% of collagen type II, and abundant GAGs [14]. This different cell distribution may result from the discrepancy in the force to different areas of the meniscus. Fibroblast-like cells with rich collagen type I in the outer region are adapted to circumferential tension and chondrocyte-like cells with abundant collagen type II and GAGs have advantages to compressive forces [25]. The presence of various cells including vessels and nerves in the outer region also makes the meniscus regeneration challenging. For meniscus regeneration, various cells such as meniscal cells, chondrocytes, and mesenchymal stem cells (MSCs) derived from bone marrow, fat tissue, or synovium have been used in animal studies [26–30]. It is reported that the implantation of minced meniscus could be a potential cell source for meniscus regeneration through meniscus cell migration and proliferation in the scaffold, which is not required for cell isolation and culture [31]. For tissue engineering meniscus, MSCs are widely utilized due to their pluripotency and immunomodulatory effects. However, there is no consensus on the ideal cell source because of the lack of evidence to show which cells are the most effective due to various combinations of different models and cell sources.

As mentioned above, there have been many problems to be solved to achieve the ideal tissue engineering meniscus scaffold. Still, various basic research efforts contributed to advancing the development of better tissue engineering meniscus, and some of them have been developed for clinical trials.

17.2 Experimental Studies and Currently Available Regenerative Therapies

Once the meniscus is injured, the natural healing response will unlikely occur due to its limited healing capacity. Recent experimental studies proposed several therapeutic options to enhance the repair of meniscus as well as reconstruct the resected meniscus and exploit meniscal substitutes. The details are described below.

17.2.1 Biologics

To enhance the repair potential, some biologic augmentation have been attempted to promote the healing of torn meniscal sites in either experimental or clinical practice, and these include platelet-rich plasma (PRP) [32] as well as bone marrow aspirate concentrate [33], and MSCs [34], which has been proven to improve the meniscal healing rate. PRP is considered to aid with meniscal healing by increasing the local concentration of growth factors and cytokines involved in the healing process [35]. However, there have been still limited evidences in the clinical practice because of the heterogeneity of PRP protocols, short-term follow-up, and lack of comparative studies [36]. MSC injections from different sources, including bone marrow, synovium and adipose, have been employed in various animal models to evaluate their effects on tissue regeneration and healing [37, 38]. It has still not been determined which cell source is superior for MSC treatments, but each has shown promising results. In addition, allogeneic cells have also shown promising results, which would have its own benefits for manufacturing, cost, and development of single-stage treatment strategies.

17.2.2 Biomaterial Augmentation

Historically, a sheet-like tissue augmentation was utilized for complex meniscal tear. Henning et al. reported the use of a fascia sheath to cover the meniscal tear in combination with a fibrin clot improved the healing rates, but the repair of tears in the middle one-third of the lateral meniscus still showed a high failure rate [39]. To further enhance the repair of such incurable meniscal tears, Shimomura et al. developed biomimetic-aligned nanofibrous scaffolds produced by electrospinning and tested the feasibility of meniscal radial tear [40]. They demonstrated that an MSC-seeded nanofibrous scaffold, wrapping the meniscal tear, restored the meniscal hoop structure and function, and prevented the progression to cartilage degeneration, using a rabbit clinically relevant model [41]. Thus, such a technique using biomaterial augmentation will be expected as a potential therapeutic method.

17.2.3 Meniscal Substitutes

Several biomaterials have been developed and assessed for efficacy in addressing meniscal deficiency, either in vitro or in vivo. Biomaterials for the meniscus substitute could be categorized into four broad classes: synthetic polymers (e.g., polyurethane (PU), polycaprolactone (PCL), polylactic acid (PLA), polyglycolic acid (PGA), polylactic co-glycolic acid (PLGA)), hydrogels, natural matrix (e.g. collagen, hyaluronan), or tissue-derived materials (e.g., decellularized extracellular matrix), as mentioned above [42]. Among them, there have been two bioabsorbable implants available for clinical practice: one is a collagen meniscus implant (CMI®) and the other, a polyurethane polymeric implant (Actifit®) [43]. Recent long-term follow-up studies reported that CMI® provided significant pain relief and functional improvement with safety and a low rate of implant failure [44]. Similarly, Actifit® showed safety and effectiveness with improved clinical outcomes with middle-term follow-up [45]. On the other hand, several studies reported negative outcomes based on MRI results with these two implants, despite the general observation of improved clinical scores [46, 47]. These implants were partially or totally resorbed during follow-up, and mostly showed hyperintensity and/or extrusion, accompanied by subchondral bone edema. Therefore, these clinical studies suggest there would likely still be room for improvement, regarding the development of such implants serving as meniscal substitutes. As an alternative to this procedure, total meniscus replacement using a synthetic meniscus has been assessed for such patients, although the shape of the implant is anatomically different. Recently, a synthetic polyethylene-reinforced polycarbonate urethane (PCU) meniscus implant (NUsurface®) has been developed, and preliminary data showed significant improvements in pain and function scores with the implant over nonsurgical therapy and a similar adverse event rate [48].

As a new trend, anatomy-based meniscal regeneration has been tested. To date, many biomaterials resemble the shape of native meniscus, but with a porous structure inside [43], although the meniscus has unique collagen fiber arrangements [49]. Recently, a fiber-reinforced scaffold, composed of a type I collagen and hyaluronic acid sponge reinforced with a unique pattern of continuous tyrosine-derived biodegradable polymer fiber, has been developed and tested in large animal studies [50]. The fiber-reinforced scaffolds were implanted into sheep knee joints after a total meniscectomy, and it exhibited the formation of a functional neomeniscus tissue, with the potential to prevent joint degeneration at 2 years postoperatively. Thus, such fiber-reinforced scaffolds might provide the appropriate structural properties to take over the load-bearing role that is required of the meniscus, and provide long-term chondro-protective effects.

17.2.4 Tissue Engineering Approaches

Recently, tissue engineering approaches have gained increasing attention as potential regenerative therapies in the field of musculoskeletal medicine. These approaches are still primarily in the preclinical phases of development but likely will progress to the clinical application stage in near future. Tissue engineering is defined as the application of the principles of biology and engineering to the development of functional substitutes for damaged tissue and usually utilizes a combination of cells, scaffolds, and growth factors [51]. Regarding the selection of cell sources, meniscal cells, chondrocytes, and MSCs derived from bone marrow, fat tissue or synovium have been used [37]. However, there is still no consensus regarding the best cell resource for meniscal regeneration due to the lack of comparative studies being performed. Several scaffolds could be utilized for meniscal tissue engineering, including synthetic polymers, hydrogels, natural matrix, or tissue-derived materials, as mentioned above. For the ideal selection of a scaffold for meniscal regeneration, it is important to know the material properties of each of the potential scaffolds in order to pick the most appropriate for the environment for which they will be used. In general, synthetic polymers are easy to handle and have high mechanical properties, while natural materials retain higher biocompatibility, which might be advantageous for tissue healing and remodeling [42]. Similar to cell sources, there is still no consensus regarding the ideal materials for a scaffold to be used for meniscal regeneration. As listed in Table 17.1, we have outlined the recent preclinical large animal studies of cell-based meniscal tissue engineering, which have been published since 2010. There have been some promising studies reported, and thus, some of these approaches may be expected to lead to the initiation of clinical trials in the near future.

As a unique MSC-based therapy, Moriguchi et al. developed a scaffold-free tissue-engineered construct (TEC) consisting of allogeneic synovial MSCs and extracellular matrix (ECM) synthesized by the cells and tested their feasibility for meniscal repair using a porcine meniscal defect model [30]. All TEC implanted defects were consistently repaired by a fibro-cartilaginous tissue with good tissue integration to the adjacent host meniscal tissue, while the controls remained empty or partially filled. Moreover, TEC treatment significantly reduced the size and severity of chondral lesions on the tibial plateau. Thus, this study concluded that the TEC could be a promising stem cell-based implant to repair meniscal lesions with preventive effects from meniscal body degeneration and the development of osteoarthritis.

As another tissue engineering approach based on the meniscal anatomy, the applicability of decellularized meniscal ECM (dmECM) has been investigated by specifically comparing region-dependent effects of the dmECMs on 3D constructs seeded with human bone marrow MSCs in an experimental study [52, 53]. Such studies have shown that the inner dmECM (avascular zone) enhances the fibrocartilaginous differentiation of MSCs, while the outer dmECM (vascular zone) promotes a more fibroblastic phenotype, supporting the feasibility to engineer a meniscus-like tissue that mimics the anatomy and biochemistry of the native meniscus by using region-specific dmECM preparations.

Additionally, 3D bioprinting technique has been recently applied for meniscal tissue engineering. Since 3D printing was first conceptualized and invented by Charles Hull in the early 1980s, such technologies have garnered attention in the field of regenerative medicine [54]. Moreover, the recent advancements in medical imaging, such as CT and MRI, can provide the exact 3D-reconstructed images for creating 3D objects via printing, making patient-specific scaffolds for meniscal tissue engineering approaches [37]. Most studies trend to reproduce the meniscal fiber arrangement using a 3D printer [55], and such design of the scaffold should potentially improve the clinical outcomes. 3D printing is an innovative and promising technique for regenerative medicine, and it will enable the provision of "made-to-order" medicine according to the needs of the individual condition by the fabrication of size-matching implants.

Table 17.1 Summary of meniscal tissue engineering in preclinical large animal studies

Authors	Animal	Experimental model	Cell/growth factor	Scaffold	Follow-up period	Outcome measure	Outcomes	PMID
Kon E, et al. [56]	Sheep	Total meniscectomy	Chondrocyte	Hyaluronic acid/polycaprolactone	12 months	Gross, histology	Cell-seeded scaffold showed higher gross & histological score and better fibrocartilaginous tissue formation	22500654
Gu Y, et al. [57]	Dog	2.5 mm width defect	Myoblast	PLGA	12 weeks	Gross, histology, immunohistochemistry	Good defect filling & integration, fibrocartilaginous repair	22969884
Moriguchi Y, et al. [30]	Pig	φ4 mm defect	Synovial MSC	Scaffold-free	6 months	Gross, histology	Fibrocartilaginous repair, chondroprotective effect	23261221
Zhu WH, et al. [58]	Dog	2 mm width defect	Myoblast transfected w/ hCDMP-2	PLA/PGA	12 weeks	Gross, histology, immunohistochemistry, collagen I & II, GAG quantification	Meniscal fibrocartilage healing was promoted. Collagen I and II were expressed at the red-red and the white-white zones of the repaired tissue	24626772
Desando G, et al. [59]	Sheep	Meniscectomy	BAC or bone marrow MSC	Hyaluronan-based scaffold	12 weeks	Gross, histology, immunohistochemistry, microcomputed tomography	BMC-HA showed a greater repair in inhibiting OA progression compared to MSC-HA, leading to a reduction of inflammation	27151837
Nakagawa Y, et al. [55]	Sheep	Subtotal meniscectomy	CTGF & TGF-β3	PCL	12 months	Gross, histology, MRI	Growth factors did not increase inflammation and fibrosis. Articular cartilage degeneration and meniscal extrusion was observed in most animals	31386550

BMC bone marrow concentrate, *CTGF* connective tissue growth factor, *GAG* glycosaminoglycan, *HA* hyaluronic acid, *hCDMP-2* human cartilage-derived morphogenetic protein-2, *MRI* magnetic resonance imaging, *MSC* mesenchymal stem cell, *OA* osteoarthritis, *PCL* poly-ε-caprolactone, *PGA* polyglycolic acid, *PLA* polylactic acid, *PLGA* poly(lactic-co-glycolide acid), *TGF- β3* transforming growth factor β3

17.3 New Trend for Meniscal Regeneration: Development of Novel Scaffold and Its Clinical Outcomes

The three elements necessary for regenerative medicine are considered in terms of scaffolds, growth factors, and cell and tissue engineering [51]. The meniscus contributes to shock absorption and stability in the knee joint. The outer margin forms a tough collagen hoop structure, capable of withstanding tensile loads >100 MPa, and the central portion functions to distribute loading from the femur with a radial fiber structure. When considering tissue regeneration in such a biomechanically demanding environment, scaffolds are considered to play the most important role in meniscus regeneration.

Herein, we describe our novel bioabsorbable meniscal scaffold, fabricated with polyglycolic acid (PGA) and coated with polylactic acid/caprolactone (P(LA/CL), used after partial meniscectomy for degenerative and irreparable meniscal tears. As the arthroscopic implementation technique is influenced by the stiffness and strength of the scaffold implant employed, we provide a detailed description of our implantation technique, including a description of potential pitfalls to consider in order to improve clinical outcomes.

17.3.1 Developmental Background

With reference to the results of previous meniscus scaffolds, the meniscal scaffold should have excellent initial strength and not become deformed after being maintained in the implantation environment. Candidate materials were already used clinically, and PGA, P(LA/CL) and poly L-lactic acid (PLLA) were considered as novel meniscal scaffold materials. After histological findings of meniscus regeneration after implantation, the most ideal tissue regeneration was confirmed with PGA coated with P(LA/CL) [55, 56].

Next, we proceeded with a study of mechanical strength of meniscal scaffold materials. The mechanical strength of the human meniscus has been reported to be 10 MPa in the radial direction and 110 MPa in the circumferential direction. We have developed a novel scaffold with these values as our targets. As a result of biomechanical analysis, the compressive strength in the axial direction of meniscal scaffold was elevated when the PGA density was increased, and this was able to achieve the required strength in the radial direction. However, the initial tensile strength was limited to approximately 20 MPa that was not achieved in terms of circumferential strength. These results indicated that the indication of scaffold implantation was considered to preserve the hoop function of the meniscus at implanted menisci. Based on these results, the morphology of the P(LA/CL)-coated PGA meniscal scaffold was completed (Fig. 17.1a).

In the next step, the meniscal scaffold was implanted in porcine menisci and observed until 48 weeks. Regarding the histological findings, autologous tissue migrated from the outer surface surrounding the implanted scaffold at postoperative 4–8 weeks, and the scaffold was almost completely replaced by autologous tissue as observed after Safranin O staining, and the gross appearance was similar to that of the surrounding normal meniscus 24 weeks after implantation (Fig. 17.1b–d). Blood examination and Magnetic resonance image (MRI) findings also revealed no evidence of prolonged inflammatory reactions.

17.3.2 Inclusion and Exclusion Criteria in the Exploratory Clinical Study

Based on successful results in the preclinical study, the exploratory clinical study was initiated. The patients were enrolled as follows. Implantation of a PGA scaffold is indicated after partial meniscectomy for a degenerative meniscal tear that cannot be repaired by suturing in individuals. The ages of this indication were between 18 and 60. The peripheral rim of the meniscus must be preserved, and the size of the defect must be <1/3 of the length of the circumference, as a large defect area might induce implanted scaffold instability. Lastly, the hip-knee-ankle angle should be within

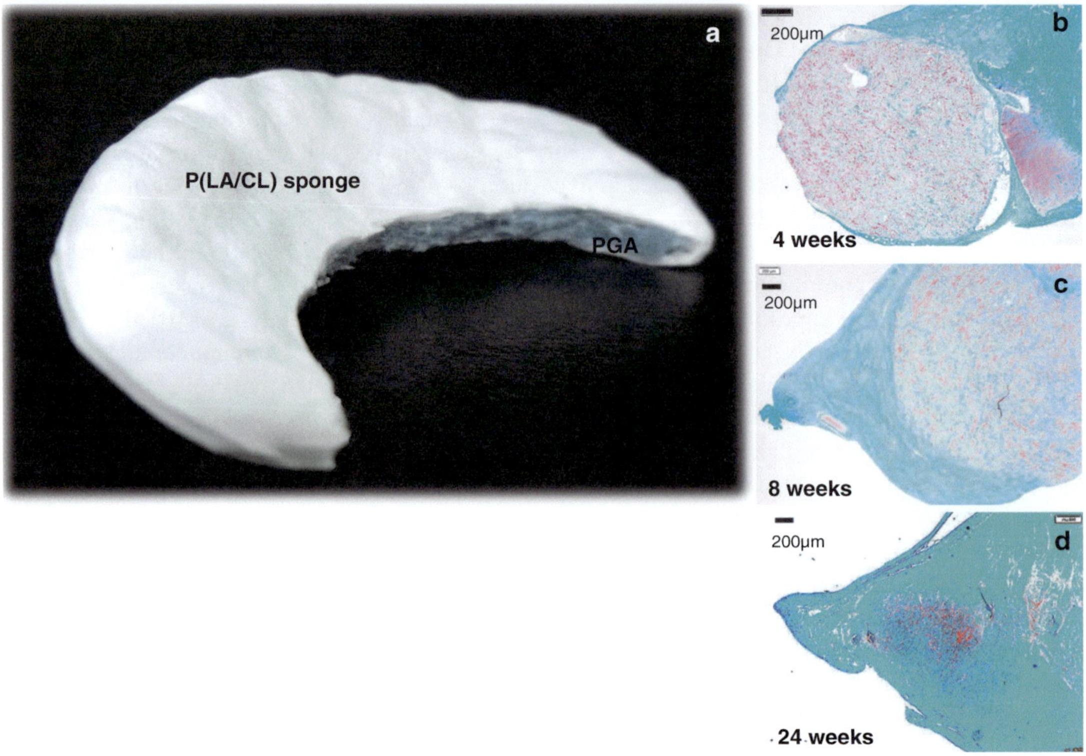

Fig. 17.1 (**a**) Novel meniscal scaffold consisting of PGA coated with P(LA/CL). (B-D) Histological findings with Safranin O staining after scaffold implantation in porcine menisci. Meniscal scaffold was covered with new tissue (**b**, **c**) and completely regenerated at 24 weeks after implantation (**d**)

±3° to ensure normal coronal knee alignment. Preoperative exclusion criteria included predisposition to allergic symptoms, such as bronchial asthma and urticaria; obesity, defined as a body mass index (BMI) ≧30 kg/m^2; disruption of the cruciate ligaments resulting in knee joint instability; and surgery involving the target knee joint, such as ligament repair and osteotomy, within the previous 6 months. The intraoperative exclusion criteria were cartilage damage >1 cm^2 and an International Cartilage Repair Society (ICRS) classification grade ≧III [61].

17.3.3 Surgical Technique

The meniscal scaffold is trimmed to be 1–2 mm larger than the size of the defect area measured to ensure sufficient contact between the scaffold and the native meniscus. To control the position of the scaffold during the arthroscopic procedure, a rein suture is set at the center of the scaffold. The synovium is removed via the anterior portal, allowing for implantation of the scaffold. The prepared meniscus scaffold is appropriately located on the meniscal defect using a grasper. The rein suture is then pulled to avoid peripheral displacement of the scaffold during inside-out suture fixation of the scaffold to the native meniscus. The inside-out suture technique is performed using a double-armed meniscus needle. The scaffold is first sutured to the native posterior horn and then to the central area of the meniscus, as recommended because of the relative lower flexibility of the posterior horn compared to the rest of the meniscus. As a result of this differential stiffness, gapping between the scaffold and the native menisci can easily occur during knee motion.

17.3.4 Postoperative Rehabilitation

Patients completed a postoperative program of rehabilitation as per the recommendations for meniscal scaffolds. Passive range of motion through the full range of the knee joint is initiated on postoperative day 3, using a continuous passive motion device. Partial and full weight bearing are permitted at 4 and 8 weeks after surgery, respectively.

17.3.5 Outcomes

In the current clinical trial, 4 male and 2 female participants with a mean age of 41.0 ± 10.2 years (range: 30–59 years) were included. The median BMI was 23.0 ± 3.3 kg/m^2 (range: 19.4–27.5 kg/m^2). Examinations of white blood cell counts and C-reactive protein levels were performed to evaluate inflammation after implantation, and WBC abundance was within the normal range, while CRP had increased in 50% of patients at 2 months postoperatively; after that, it returned to the normal range. All patients exhibited improved clinical outcomes with Lysholm scores, IKDC, KOOS, and visual analogue scale (VAS). Notably, VAS was significantly decreased at 12 months after implantation [62].

Except for KOOS and VAS, clinical outcome measures improved within 6 months, and all outcomes improved at 12 months post-implantation. Range of knee motion recovered to preoperative levels at 12 months after implantation. In addition, the Tegner activity score improved in three cases and remained unchanged in the other three cases. No osteoarthritic progression was found during the follow-up period based on X-ray examinations. The implant interface, morphology, and bone bruising were not altered in the postoperative MRI findings. However, the scaffold signal changed from iso- to low-intensity (Fig. 17.2).

Second-look arthroscopy was performed 52 weeks after initial surgery and revealed improved scores relative to those at initial arthroscopy (Fig. 17.3). Five participants demonstrated good or excellent meniscal repair. However, repair in one patient was reduced in size, and another demonstrated cheese wiring at suture fixation.

17.3.6 Discussions and Summaries

The experience of this clinical trial revealed a variety of issues. Arthroscopic examination 1 year after transplantation showed scattered images of scaffold resorption and suture failure. The PGA scaffold we have developed begins to replace autologous tissue 8–12 weeks after implantation. This was possibly one of the reasons for the imbalance because the balance of mechanical strength at the suture area had been altered from that of initial fixation between non-absorbable threads for fixation and absorbable scaffold.

According to the experimental data, suture strength exhibited no decrease after 2 weeks of hydrolysis; however, a decrease in strength was observed after 4 weeks, raising concerns regarding suture failure. We will continue to conduct basic studies on factors that may affect suture method, spacing, and tension, with the aim of improving clinical outcomes.

In biomechanical tests, the implantation of scaffolds 20% larger than the circumferential and vertical defect sizes results in greater contact stress than the intact meniscus. In mechanical evaluation associated with the decomposition behavior, the strength decreased after 4 weeks of hydrolysis. Meanwhile, in the biomechanical test considering hydrolysis, contact stress and area equivalent to intact were obtained after 4 weeks of hydrolysis (Fig. 17.4) [63].

Rehabilitation is also an important consideration, and we believe that early ROM exercise and the angle of knee flexion allowed with load-bearing should be progressed carefully. In terms of indication of age, the patients were <60 years of age in this clinical trial; however, it is expected that younger cases may have higher cellular activity and thereby achieve earlier autologous tissue replacement and structural reconstruction. From the viewpoint of load-bearing, we believe meniscus scaffold implantation combined with

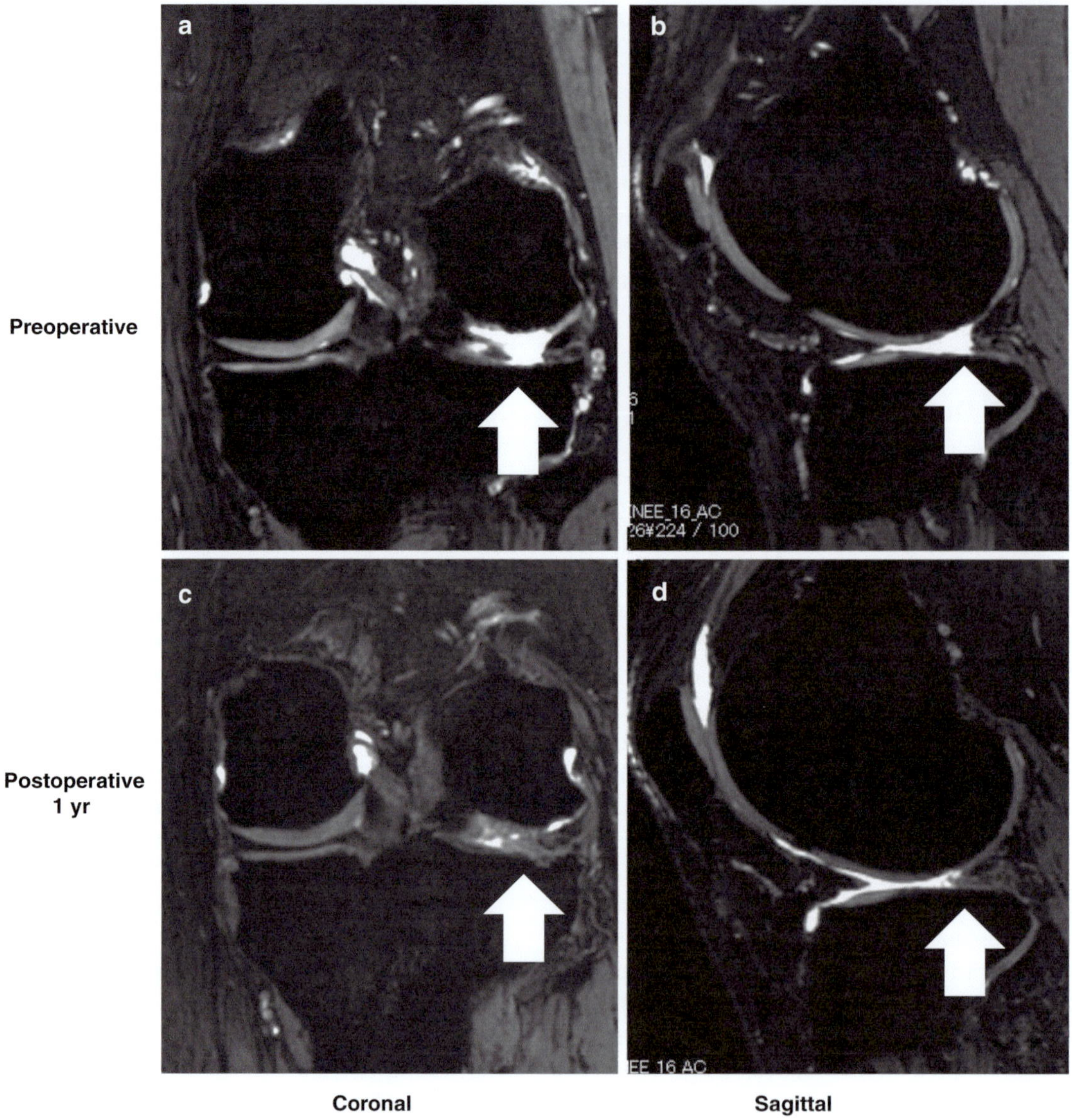

Fig. 17.2 MRI of pre- and post-operative meniscus implantation. Meniscal defects were detected from preoperative MRI (**a**, **c**). Soft tissue was occupied at defect area by postoperative year 1 (**b**, **d**)

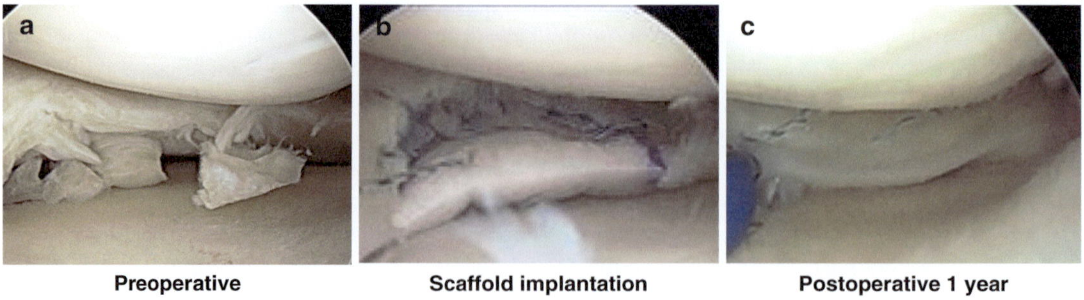

Fig. 17.3 Arthroscopic findings at preoperative (**a**), scaffold implantation (**b**), and postoperative 1 year (**c**)

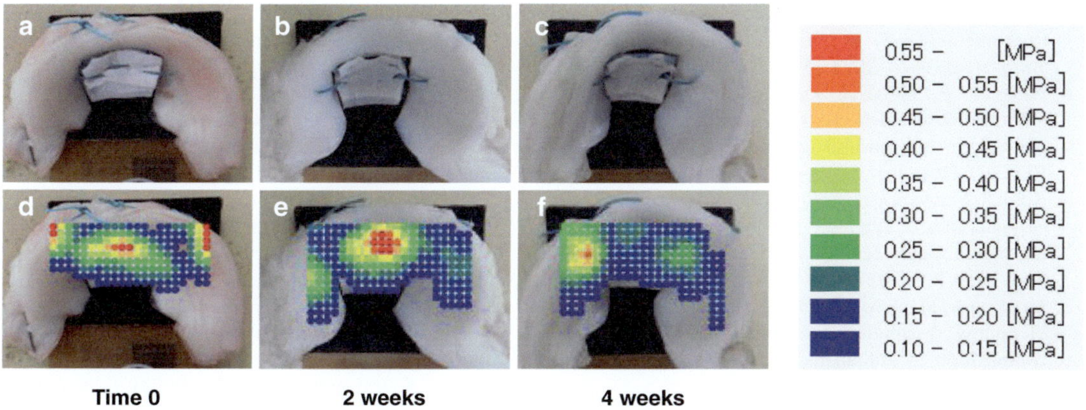

Fig. 17.4 Biomechanical analysis after implantation in the defective menisci. Meniscal scaffold was implanted (**a**, **c**), and contact stress and area were equivalent to intact after 4 weeks of hydrolysis (**b**, **c**, **e**, **f**)

osteotomy may lead to superior results, even in cases involving minor correction of alignment deformities.

In summary, novel therapies for otherwise irreparable meniscus injury are still being investigated and developed worldwide. One of the reasons for this issue is the high demand for scaffolds for repairing meniscus defects, which has not led to increase the use of leading products. Although partial meniscal replacement using PGA covered with P(LA/CL) is safe and achieves significant and encouraging improvement in terms of clinical results, continued research is needed to improve clinical results while addressing various issues.

17.4 Future Perspectives and Conclusions

With recent advancements in biomaterial- and cell-based technologies in tissue engineering, we may have new therapeutic options for addressing meniscal injuries in clinical practice. On the other hand, there are still several potential problems to be solved, considering complete meniscus regeneration remains challenging with currently available techniques. Firstly, the selection and design of biomaterials with sufficient mechanical strength and long-term durability for the optimal repair and remodeling of menisci as well as the prevention of the progression of cartilage degeneration in the long term have not been fully elucidated. Secondly, current approaches indicate that meniscal function could not be fully restored effectively, based on observing absorption of the implanted biomaterials, meniscal extrusion, and the progression of joint space narrowing after biomaterial implantation. Therefore, further studies will be needed and should be conducted in a methodologically rigorous fashion to elucidate such potential problems due to the lack of evidence regarding these issues.

References

1. Abram SGF, Hopewell S, Monk AP, Bayliss LE, Beard DJ, Price AJ. Arthroscopic partial meniscectomy for meniscal tears of the knee: a systematic review and meta-analysis. Br J Sports Med. 2020;54(11):652–63.
2. Schweizer C, Hanreich C, Tscholl PM, Ristl R, Apprich S, Windhager R, et al. Nineteen percent of meniscus repairs are being revised and failures frequently occur after the second postoperative year: a systematic review and meta-anaylsis with a minimum follow-up of 5 years. Knee Surg Sports Traumatol Arthrosc. 2022;30(7):2267–76.
3. Rodeo SA, Monbi F, Dehghani B, Maher S. Biological and mechanical predictors of meniscus function: basic science to clinical transplantation. J Orthop Res. 2020;38:937–45.
4. Dangerlmajer S, Familiari F, Simonetta R, Kaymakoglu M, Huri G. Meniscal transplants and scaffolds: a systematic review of the literature. Knee Surg Relat Res. 2017;29:3–10.
5. Dienst M, Greis PE, Ellis BJ, Bachus KN, Burks RT. Effects of lateral meniscal allograft sizing on

contact mechanics of the lateral tibial plateau: an experimental study in human cadaveric knee joints. Am J Sports Med. 2007;35:34–42.
6. Figueroa F, Figueroa D, Calvo R, Vaisman A, Espregueira-Mendes J. Meniscus allograft transplantation: indications, techniques and outcomes. EFFORT Open Rev. 2019;4:115–20.
7. Kohli S, Schwenck J, Barlow I. Failure rates and clinical outcomes of synthetic meniscal implants following partial meniscectomy: a systematic review. Knee Surg Relat Res. 2022;34:27.
8. Morejon A, Mantero AM, Best TM, et al. Mechanisms of energy dissipation and relationship with tissue composition in human meniscus. Osteoarthr Cartil. 2022;30:605–12.
9. Hart DA, Nakamura N, Shrive NG. Perspective: challenges presented for regeneration of heterogeneous musculoskeletal tissues that normally develop in unique biomechanical environments. Front Bioeng Bioechnol. 2021;9:760273.
10. Syed S, Zaki MN, Lakshmanan J, et al. Knee meniscal retears after repair: a systematic review comparing diagnostic imaging modalities. Libyan J Med. 2022;17:2030024.
11. Gee SM, Posner M. Meniscus anatomy and basic science. Sports Med Arthrosc Rev. 2021;29:e18–23.
12. Canciani B, Millar VR, Pallaoro M, et al. Testing hypoxia in pig meniscal culture: biological role of the vascular-related factors in the differentiation and viability of neonatal meniscus. Int J Mol Sci. 2021;22:12465.
13. Stocco E, Porzionato A, De Rose E, Barbon S, De Caro R, Macchi V. Meniscus regeneration by 3D printing technologies: current advances and future perspectives. J Tissue Eng. 2022;13:20417314211065860.
14. Makris EA, Hadidi P, Athanasiou KA. The knee meniscus: structure-function, pathology, current repair techniques, and prospects for regeneration. Biomaterials. 2011;32(30):7411–31.
15. McDermott ID, Sharifi F, Bull AM, et al. An anatomical study of meniscal allograft sizing. Knee Surg Sports Traumatol Arthrosc. 2004;12:130–5.
16. Shaffer B, Kennedy S, Klimkiewicz J, et al. Preoperative sizing of meniscal allografts in meniscus transplantation. Am J Sports Med. 2000;28:524–33.
17. Twomey-Kozak J, Jayasuriya CT. Meniscus repair and regeneration: a systematic review from a basic and translational science perspective. Clin Sports Med. 2020;39(1):125–63.
18. Andrews SHJ, Adesida AB, Abusara Z, Shrive NG. Current concepts on structure-function relationships in the menisci. Connet Tissue Res. 2017;58(3–4):271–81.
19. Veronesi F, Di Matteo B, Vitale ND, Filardo G, Visani A, Kon E, et al. Biosynthetic scaffolds for partial meniscal loss: a systematic review from animal models to clinical practice. Bioact Mat. 2021;6(11):3782–800.
20. Lv H, Deng G, Lai J, Yu Y, Chen F, Yao J. Advances in 3D bioprinting of biomimetic and engineered meniscal grafts. Macromol Biosci. 2023;23:2300199.
21. Du MZ, Dou Y, Ai LY, Su T, Zhang Z, Chen YR, Jiang D. Meniscus heterogeneity and 3D-printed strategies for engineering anisotropic meniscus. Int J Biorpint. 2023;9(3):359–78.
22. Fox AJ, Bedi A, Rodeo SA. The basic science of human knee menisci: structure, composition, and function. Sports Health. 2012;4:340–51.
23. Li H, Wang X, Liu J, et al. Nanofiber configuration affects the biological performance of decellularized meniscus extracellular matrix incorporated electrospun scaffolds. Biomed Mater. 2021;16:065013.
24. Zhang X, Aoyama T, Ito A, Tajino J, Nagai M, Yamaguchi S, Iijima H, Kuroki H. Regional comparisons of porcine menisci. J Orthop Res. 2014;32:1602–11.
25. Le Graverand MP, Ou Y, Schield-Yee T, et al. The cells of the rabbit meniscus: their arrangement, interrelationship, morphological variations and cytoarchitecture. J Anat. 2001;198:525–35.
26. Kang SW, Son SM, Lee JS, et al. Regeneration of whole meniscus using meniscal cells and polymer scaffolds in a rabbit total meniscetomy model. J Biomed Mater Res A. 2006;77(4):659–71.
27. Kon E, Chiari C, Marcacci M, et al. Tissue engineering for total meniscal substitution: animal study in sheep model. Tissue Eng. 2008;14(6):1067–80.
28. Yamasaki T, Deie M, Shinomiua R, Yasunaga Y, Yanada S, Ochi M. Transplantation of meniscus regenerated tissue engineering with a scaffold derived from a rat meniscus and mesenchymal stromal cells derived from rat bone marrow. Artif Organs. 2008;32(7):519–24.
29. Moutos FT, Guilak F. Functional properties of cell-seeded three-dimensionally woven polu (epsilon-caprolactone) scaffolds for cartilage tissue engineering. Tissue Eng. 2010;16(4):1291–301.
30. Moriguchi Y, Tateishi K, Ando W, et al. Repair of meniscal lesions using a scaffold-free tissue-engineered construct derived from allogenic synovial MSCs in a miniature swine model. Biomaterials. 2013;34(9):2185–93.
31. Matsubara N, Nakasa T, Ishikawa M, Tamura T, Adachi N. Autologous meniscus fragments embedded in atelocollagen gel enhance meniscus repair in a rabbit model. Bone Joint Res. 2021;10(4):269–76.
32. Griffin JW, Hadeed MM, Werner BC, Diduch DR, Carson EW, Miller MD. Platelet-rich plasma in meniscal repair: does augmentation improve surgical outcomes? Clin Orthop Relat Res. 2015;473(5):1665–72.
33. Koch M, Hammer S, Fuellerer J, Lang S, Pfeifer CG, Pattappa G, Weber J, Loibl M, Nerlich M, Angele P, Zellner J. Bone marrow aspirate concentrate for the treatment of avascular meniscus tears in a one-step procedure-evaluation of an in vitro model. Int J Mol Sci. 2019;20(5):1120.

34. Hatsushika D, Muneta T, Nakamura T, Horie M, Koga H, Nakagawa Y, Tsuji K, Hishikawa S, Kobayashi E, Sekiya I. Repetitive allogeneic intraarticular injections of synovial mesenchymal stem cells promote meniscus regeneration in a porcine massive meniscus defect model. Osteoarthr Cartil. 2014;22(7):941–50.
35. Qi Y, Tang R, Shi Z, Feng G, Zhang W. Wnt5a/Platelet-rich plasma synergistically inhibits IL-1β-induced inflammatory activity through NF-κB signaling pathway and prevents cartilage damage and promotes meniscus regeneration. J Tissue Eng Regen Med. 2021;15(7):612–24.
36. Gopinatth V, Batra AK, Chahla J, Smith MV, Matava MJ, Brophy RH, Knapik DM. Degenerative meniscus tears treated nonoperatively with platelet-rich plasma yield variable clinical and imaging outcomes: a systemic review. Arthrosc Sports Med Rehabil. 2024;6(2):100916.
37. Shimomura K, Hamamoto S, Hart DA, Yoshikawa H, Nakamura N. Meniscal repair and regeneration: current strategies and future perspectives. J Clin Orthop Trauma. 2018;9(3):247–53.
38. Jacob G, Shimomura K, Krych A, Nakamura N. The meniscus tear: a review of stem cell therapies. Cells. 2019;9(1):92.
39. Henning CE, Yearour KM, Vequist SW, Stallbaumer RJ, Decker KA. Use of the fascia sheath coverage and exogeneous fibrin clot in the treatment if complex meniscal treas. Am J Sports Med. 1991;19(6):626–31.
40. Shimomura K, Bean AC, Lin H, Nakamura N, Tuan RS. In vivo repair of meniscal radial tear using aligned electrospun nanofibrous scaffold. Tissue Eng Part A. 2015;21(13–14):2066–75.
41. Shimomura K, Rothrauff BB, Hart DA, Hamamoto S, Kobayashi M, Yoshikawa H, Tuan RSD, Nakamura N. Enhanced repair of meniscal hoop structure injuries using an aligned electrospun nanofibrous scaffold combined with a mesenchymal stem cell-derived tissue engineered construct. Biomaterials. 2019;192:346–54.
42. Makris EA, Hadidi P, Athanasiou KA. The knee meniscus: structure-function, pathophysiology, current repair techniques, and prospects for regeneration. Biomaterials. 2011;32(30):7411–31.
43. Vrancken ACT, Buma P, van Tienen TG. Synthetic meniscus replacement: a review. Int Orthop. 2013;37(2):291–9.
44. Bulgheroni E, Grassi A, Bulgheroni P, Mucciolo GMM, Zaffagnini S, Marcacci M. Long-term outcomes of medial CMI implant versus partial medial meniscectomy in patients with concomitant ACL reconstruction. Knee Surg Sports Traumatol Arthrosc. 2015;23(11):3221–7.
45. Leroy A, Beaufils P, Faivre B, Steltzlen C, Boisrenoult P, Pujol N. Actifit polyurethane meniscal scaffold: MRI and functional outcomes after a minimum follow-up of 5 years. Orthop Traumatol Surg Res. 2017;103(4):609–14.
46. de Chanterac A, Garetier M, Gunepin FX, Baynat C, des Déserts MD, Perchoc A, Letissier H, Andro C. ACTIFIT™: 38% full resorption at 8 years in a cohort of 18 cases. Orthop Traumatol Surg Res. 2023;109(6):103616.
47. Hirschmann MT, Keller L, Hirschmann A, Schenk L, Berbig R, Lüthi U, Amsler F, Friederich NF, Arnold MP. One-year clinical and MR imaging outcome after partial meniscal replacement in stabilized knees using a collagen meniscus implant. Knee Surg Sports Traumatol Arthrosc. 2013;21(3):740–7.
48. McKeon BP, Zasalv KR, Alfred RH, Alley RM, Edelson RH, Gersoff WK, Greenleaf JE, Kaeding CC. Preliminary results from a US clinical trial of a novel synthetic polymer meniscal implant. Orthop J Sports Med. 2020;8(9):2325967120952414.
49. Petersen W, Tillmann B. Collagenous fibril texture of the human knee joint menici. Anat Embryol. 1998;197(4):317–24.
50. Patel JM, Ghodbane SA, Brzezinski A, Gatt CJ Jr, Dunn MG. Tissue-engineered total meniscus replacement with a fiber-reinforced scaffold in a 2-year ovine model. Am J Sports Med. 2018;46(8):1844–56.
51. Langer R, Vacanti JP. Tissue engineering. Science. 1993;260(5110):920–6.
52. Shimomura K, et al. Region-specific effect of the decellularized meniscus extracellular matrix on mesenchymal stem cell-based meniscus tissue engineering. Am J Sports Med. 2017;45(3):604–11. PMID: 27895039.
53. Rothrauff BB, Shimomura K, Gottardi R, Alexander PG, Tuan RS. Anatomical region-dependent enhancement of 3-dimensional chondrogenic differentiation of human mesenchymal stem cells by soluble meniscus extracellular matrix. Acta Biomater. 2017;49:140–51.
54. Tack P, Victor J, Gemmel P, Annemans L. 3D-printing techniques in a medial setting: a systematic literature review. Biomed Eng Online. 2016;15(1):115.
55. Nakagawa Y, Fortier LA, Mao JJ, Lee CH, Goodale MB, Koff MF, Uppstrom TJ, Croen B, Wada S, Carballo CB, Potter HG, Rodeo SA. Long-term evaluation of meniscal tissue formation in 3-dimensional-printed scaffolds with sequential release of connective tissue growth factor and TGF-β3 in an ovine model. Am J Sports Med. 2019;47(11):2596–607.
56. Kon E, et al. Tissue engineering for total meniscal substitution: animal study in sheep model--results at 12 months. Tissue Eng Part A. 2012;18(15-16):1573–82. https://doi.org/10.1089/ten.TEA.2011.0572.
57. Gu Y, et al. Repair of meniscal defect using an induced myoblast-loaded polyglycolic acid mesh in a canine model. Exp Ther Med. 2012;3(2):293–298. https://doi.org/10.3892/etm.2011.403.
58. Zhu WH, et al. Effects of canine myoblasts expressing human cartilage-derived morphogenetic protein-2 on the repair of meniscal fibrocartilage injury. Mol Med Rep. 2014;9(5):1767–72. https://doi.org/10.3892/mmr.2014.2047.

59. Desando G, et al. Autologous Bone Marrow Concentrate in a Sheep Model of Osteoarthritis: New Perspectives for Cartilage and Meniscus Repair. Tissue Eng Part C Methods. 2016;22(6):608–19. https://doi.org/10.1089/ten.TEC.2016.0033.
60. Murakami T, Otsuki S, Nakagawa K, Okamoto Y, Inoue T, Sakamoto Y, Sato H, Neo M. Establishment of novel meniscal scaffold structures using polyglycolic and poly-l-lactic acids. J Biomater Appl. 2017;32(2):150–61.
61. Otsuki S, Ikeda K, Tanaka K, Okamoto Y, Sezaki S, Neo M. Implantation of novel meniscus scaffold for irreparable meniscal tear. Arthrosc Tech. 2022;11(5):e775–9.
62. Otsuki S, Sezaki S, Okamoto Y, Ishitani T, Wakama H, Neo M. Safety and efficacy of a novel polyglycolic acid meniscal scaffold for irreparable meniscal tear. Cartilage. 2023;15(2):110–9.
63. Sezaki S, Otsuki S, Ikeda K, Ishitani T, Okamoto Y, Wakama H, Matsuyama J, Nakamura K, Neo M. Biomechanical assessment of a novel meniscal scaffold compared to partial meniscectomy: a study on porcine meniscal injury. J Biomed Mater Res B Appl Biomater. 2022;111(4):895–902.

Orthobiologics in Meniscus Pathology

18

Konrad Slynarski and John G. Lane

18.1 Introduction

Meniscal tears are a frequent orthopedic injury [1–4]. They often result in significant pain and mechanical issues, and if untreated, they can lead to the development of osteoarthritis [5, 6]. Over time, the approach to treating these injuries has progressed from total meniscectomy to partial meniscectomy and, finally, to meniscal repair and substitution with scaffolds or allografts [6]. Preserving as much of the meniscus as possible is preferable because losing any part of it can considerably increase stress on the articular cartilage compared to when it is intact [7, 8]. Nonetheless, the meniscus generally has a limited capacity for self-repair due to its poor blood supply, leading researchers to explore deeper into the biology of meniscal healing and seek ways to enhance this process [9–11]. Several elements affect the inherent healing ability of the meniscus, with the most critical factors including regional variations in blood supply, changes in the composition and density of the matrix, inadequate cellular density, and various soluble and mechanical influences. Healing is a complex process, involving cells that migrate into the damaged area, proliferate, and remodel the extracellular matrix. They also synthesize new matrix proteins, a crucial step in closing the wound. However, this process is not possible without a robust vascular supply. This supply delivers regenerative cells, nutrients, and growth factors, all of which are essential for effective repair. The observed region-specific healing in clinical settings is primarily due to variations in vascularity. Limited blood flow can impede the formation of a fibrin clot, which is necessary for cell infiltration and matrix remodeling in the inner, non-vascular zones of the tissue. In adults, the perimeniscal capillary plexus's blood vessels do not extend beyond the outer third of the meniscus (known as the red-red zone), resulting in a central avascular area (white-white zone) that exhibits poor healing potential [12–15]. Recent advancements in orthopedics have continually aimed at improving surgical or non-surgical treatment outcomes and enhancing tissue regeneration. For very long time, a key area of interest has been the role of growth factors in healing processes, particularly in complex problems as cartilage repair, meniscus repair and ligament repair [16]. Those attempts were developing in two pathways; interventional—nonsurgical treatments and also to improve soft tissue healing in surgical procedures. Several studies have demonstrated the positive impact of various growth factors on meniscal cell activity. These include platelet-derived growth factor, fibroblast growth factors (basic and regular), transforming

K. Slynarski
Słynarski Knee Clinic, Warsaw, Poland

J. G. Lane (✉)
Musculoskeletal and Joint Research Foundation, San Diego, CA, USA
e-mail: jglane@san.rr.com

A. Gobbi et al. (eds.), *Regenerative Medicine in Sports and Orthopaedics*,
https://doi.org/10.1007/978-3-031-84693-9_18

growth factor-β, insulin-like growth factor, bone morphogenetic protein, hepatocyte growth factor, and vascular endothelial growth factor. Conversely, other cytokines like interleukin-1, tumor necrosis factor-α, and matrix metalloproteinases have been shown to hinder meniscal healing in laboratory settings.

18.2 Basics of Orthobiologics

Orthobiologics treatment refers to using biological substances to help musculoskeletal injuries heal more quickly and efficiently. These substances are usually derived from cells and factors that are naturally found in the human body. Orthobiologics aims to improve the healing of broken bones, injured muscles, tendons, cartilage, and ligaments and possibly reduce the need for surgery or improve healing after a surgical procedure [17, 18]. Common types of orthobiologics include autologous substances such as growth factors, autogenic mesenchymal stem cells, and platelet-rich plasma (PRP) or allogenic factors, such as allogeneic mesenchymal stem cells, cord blood, and related tissues or amniotic-derived products. These therapies enhance the body's natural healing processes, potentially leading to faster recovery and a return to function. Growth factors are proteins secreted by cells that play crucial roles in regulating functions like cell growth and differentiation, including the repair of tissues. Research has shown that several growth factors positively influence meniscus cells in tissue samples (explants) or when cultured in isolation [19–21]. Specifically, PDGF-AB, EGF, TGF-b, and b-FGF significantly stimulate meniscus cell proliferation in vitro. These factors also enhance collagen synthesis, vital for directing the cells toward a more favorable phenotype for tissue repair and regeneration [19–23]. The discovery of these growth factors has paved the way for new treatments that involve delivering factors directly to the meniscus to promote better healing.

18.3 Orthobiologics in Meniscus Repair and Regeneration

18.3.1 Growth Factors

Long-term research indicates that repairing the menisci during anterior cruciate ligament (ACL) reconstruction improves clinical outcomes [24]. When a "repairable" meniscal tear is addressed concurrently with ACL surgery, the success rate at the 2-year follow-up increases up to 90% [25]. Additionally, performing meniscal repair alongside ACL reconstruction reduces the risk of needing further surgery by 7% absolutely and 42% relatively over 2 years, compared to performing the meniscal repair alone [26]. PDGF release during ACL reconstruction is significantly higher than in meniscectomy patients, potentially explaining the higher meniscal healing rate in ACL reconstruction patients [27]. In the study confirming this observation, published by Girolamo, two distinct groups of patients: one group consisted of 20 consecutive patients who underwent arthroscopic partial medial meniscectomy (APM), and the other included 20 consecutive patients who received endoscopic single-bundle anterior cruciate ligament reconstruction (ACLR) with hamstring grafts. Peripheral venous blood and joint fluid samples were analyzed using ELISA to measure concentrations of growth factors, such as bFGF, TGF-b, and PDGF. Peripheral blood concentrations of TGF-b, bFGF, and PDGF were similar between the two groups and within normal ranges, except for elevated post-operative levels of TGF-b and bFGF, which were significantly above the typical blood range. However, concentrations of these growth factors in joint fluid did not show significant differences between groups for TGF-b and bFGF. Notably, PDGF levels were significantly higher in the ACLR group compared to the APM group (2022.8 ± 1242.3 vs. 392.2 ± 327.6 pg/ml, respectively; $p < 0.001$). PDGF, predominantly produced by platelets and macrophages, endothelial cells, and fibroblasts during initial inflammatory responses, is among the earliest growth factors expressed following tissue damage. It plays a crucial role in the early stages of the healing process by stimulating the production of other growth factors, including insulin growth factor 1 (IGF-1). PDGF is essential for promoting the healing process, as it encourages fibroblast proliferation and differentiation, collagen formation, and angiogenesis. Several pieces of evidence show that PDGF can enhance matrix construction in meniscus cells and fibroblasts [28–30]. Platelet-rich plasma (PRP) has been utilized in medical treatments since the 1950s, initially in managing maxillofacial and dermatological conditions. More recently, its use has expanded significantly in orthopedics. Preparing PRP involves drawing peripheral blood from the patient and subjecting it to a two-stage centrifugation process. The first stage separates plasma and platelets from red and white blood cells, while the second stage further concentrates the platelets. The resultant PRP is then injected into the knee joint. While some debate exists regarding the benefits of platelet-poor plasma, its inclusion varies across different formulations.

PRP is a rich source of platelets, white blood cells, and proteins. Neutrophils and monocytes in PRP trigger a localized inflammatory response that can aid tissue healing. The therapeutic potential of PRP lies in the growth factors released by platelets, such as platelet-derived growth factor (PDGF) and vascular endothelial growth factor (VEGF). These factors are known to facilitate various stages of tissue repair, including chemotaxis, angiogenesis, cell proliferation, and matrix formation. However, the presence of transforming growth factor (TGF-β1) in PRP can have inhibitory effects, leading to unpredictable outcomes in tissue healing processes. This review underscores the promising applications and complexities of using PRP in orthopedic treatments, with a specific focus on knee disorders. Activating the platelets releases growth factors stored in their alpha and dense granules. These growth factors initiate a healing cascade that promotes cellular chemotaxis, angiogenesis, collagen matrix synthesis, and cell proliferation, offering new hope for effective meniscal repair. Intra-articular autologous conditioned serum injections may be an effective treatment option for knee pain associated with meniscal lesions, potentially avoiding surgery [31]. Strumper et al. aimed to evaluate the effects of intra-articular injections of autologous conditioned serum on patients experiencing knee pain due to meniscal defects. The serum, separated from the patients' blood using a specialized centrifuge device, was administered weekly to 47 outpatients with various knee meniscus injuries, most of which were related to traumatic incidents. The participants, aged between 21 and 79, underwent an average of 5.2 treatments. The results, assessed retrospectively, showed notable improvements. By the end of the 6-month observation period, 83% of the patients avoided the need for surgery. There was a significant increase in the Oxford Knee Score, indicating improved knee function, from 29.1 to 44.3. Additionally, MRI scans assessed by the Boston Leeds Osteoarthritis Knee Score revealed significant structural improvements in the knee. These findings suggest that intra-articular autologous conditioned serum injections could be a viable treatment option for alleviating knee pain associated with meniscal lesions. This study's limitation was its primary focus on clinical scores, leaving it undetermined whether improvements in pain and function were directly linked to meniscal healing. The meniscus's healing process depends on the proliferation of meniscal fibrochondrocytes, which can come from intrinsic sources at the injury site or extrinsic sources such as the blood supply or synovium. In a study by Kasemkijwattana et al., the impact of concentrations of nine growth factors was explored on the synthesis of collagen and non-collagen structures and the proliferation of these cells. The study identified potential agents such as epidermal growth factor (EGF), transforming growth factor alpha (TGFa), basic fibroblast growth factor (bFGF), and platelet-derived growth factor AB (PDGF-AB) to enhance the process of meniscal healing. However, the direct administration of these growth factors may be ineffective as they have short biological half-lives and are quickly cleared from the body. The study also investigated the potential of gene therapy and tissue engineering to deliver these growth factors effectively. The findings suggest that ex vivo gene transfer techniques using direct or myoblast-mediated methods can lead to high-level and sustained expression of these growth factors in the damaged meniscus.

18.3.2 Cells

The meniscus lacks blood vessels and cells, so mobilizing healing cells is a crucial first step in repair enhancement. Different types of cells, including bone marrow and adipose-derived MSCs, synovial stem cells, blood vessel-derived stem cells, and chondrocytes, have been locally delivered in preclinical meniscal suture repair models [32–35]. Synovial stem cells have shown promise in enhancing cellular proliferation and wound closure when injected into meniscal defects. Through injection, growth factors, or scaffold-based delivery methods, cells can be directly applied to the injury site during meniscal repair. Meniscal progenitor cells and fibrochondrocytes have the potential for robust matrix production, although their migration through the dense extracellular matrix (ECM) remains challenging. Techniques to improve cell migration include localized collagenase release, growth factor or serum delivery, and nuclear softening [36–39]. Supplementary cells, whether exogenously added or endogenously recruited, are crucial for healing meniscal lesions and guiding behaviors like vascularization, matrix deposition, and inflammatory relief. Cell-based therapies, especially those using mesenchymal stem cells (MSCs) as medicinal signaling cells and tissue engineering strategies, have great potential for improving meniscal repair, especially in the avascular regions [40–44]. These therapies harness the strong signaling abilities of MSCs to enhance tissue healing and integration. By using advanced scaffolding materials and growth factors, tissue engineering strategies are essential for creating a conducive environment for cell proliferation and differentiation [45, 46]. This combination accelerates the repair process and aims to restore the natural structure and function of the meniscus, providing a promising solution for patients with meniscal injuries. One of the main advantages of using MSCs as medicinal signaling cells, as opposed to platelet-rich plasma (PRP), is their unmatched ability to coordinate a complex healing response. While PRP is well-known for its high concentration of growth factors that can aid in healing, it primarily enhances the body's natural repair mechanisms without the specialized signaling functions of MSCs. MSCs release many bioactive molecules, including anti-inflammatory cytokines, growth factors, and extracellular vesicles. These molecules modulate the local immune response, reduce inflammation, and promote tissue repair and regeneration, offering a robust and comprehensive healing solution.

Moreover, the paracrine effect of MSCs can create a more favorable microenvironment for endogenous repair cells, enhancing their recruitment, proliferation, and activity at the injury site. This effect differs from PRP, which, despite its benefits, has a relatively short-lived impact and lacks the sustained, multifaceted signaling provided by MSCs.

Furthermore, tissue engineering allows for the precise placement of cells within the defect site, optimizing the spatial arrangement and mechanical properties necessary for effective meniscal repair. This level of control and customization is not achievable with PRP alone. By creating a synergistic interaction between MSCs and engineered scaffolds, these advanced therapies can significantly improve the outcomes for meniscal injuries. Importantly, ongoing research and clinical trials are actively refining these approaches, potentially leading to more effective and accessible treatments in the near future.

A notable clinical study investigated using the patient's own mesenchymal stem cells to repair avascular meniscal tears [47]. Autologous MSCs were isolated from an iliac crest bone marrow biopsy, expanded and seeded into the collagen scaffold. The resulting human-MSC/collagen-scaffold implant was placed into the meniscal tear prior to repair with vertical mattress sutures and the patients were followed for 2 years. Three out of five patients experienced significant reductions in pain and improvements in knee function. Two required subsequent meniscectomy due to retear or nonhealing of the meniscal tear at approximately 15 months after implantation. No other adverse events occurred. The use of the patient's own MSCs minimized the risk of immune rejection and adverse reactions, confirming the safety of this approach.

A potential method involves delivering cells through a simple injection. For instance, an animal study demonstrated the effectiveness of this approach in enhancing meniscal regeneration using mesenchymal stem/stromal cells (MSCs). In the study, researchers conducted a hemi-meniscectomy procedure in rats. They injected either rat-MSCs (rMSCs) or human-MSCs (hMSCs) into the right knee joint after the surgery and administered PBS into the left knee for comparison. The results were evaluated through macroscopic and histological assessments at 2, 4, and 8 weeks. Changes in gene transcription were analyzed using species-specific microarrays and real-time RT-PCRs. Despite a decline in hMSC numbers over time, both hMSCs and rMSCs equally promoted meniscal regeneration. hMSCs boosted the expression of rat type II collagen (rat-Col II) and hindered the progression of osteoarthritis. A small portion of hMSCs exhibited high levels of genes, such as Indian hedgehog (Ihh), parathyroid hormone-like hormone (PTHLH), and bone morphogenetic protein 2 (BMP2), which stimulated rat-Col II expression. It was found that hedgehog signaling was crucial for this process, as inhibiting it reduced rat-Col II expression, while activating it increased rat-Col II expression even without hMSCs. Even though the numbers of injected hMSCs decreased rapidly, they were activated within the joint to express Ihh, PTHLH, and BMP2, contributing to meniscal regeneration. Hedgehog signaling played a pivotal role in enhancing rat-Col II expression, with additional factors from hMSCs aiding in the repair process.

18.3.3 Increasing the Blood Supply

The difference in healing between the outer and inner meniscus is mainly due to variations in regional blood supply. This underscores the importance of vascularization in meniscal repair. Enhancing blood flow to the inner meniscus through radial perforations has shown slight improvements, but further investigation is urgently needed. Bioabsorbable conduits and mobilization of blood-vessel-derived stem cells (CD34+, CD146+) have enhanced avascular tear repair in models [48]. Several proangiogenic factors, such as angiogenin, and endothelial growth factors, have increased neovascularization and promoted cellular proliferation, tissue formation, and defect bridging [49]. New blood vessel formation, or angiogenesis, is critical for healing avascular injury sites, and its role in meniscal repair cannot be overstated.

18.4 Emerging Trends in Orthobiologic Research for Meniscus Pathology

Genomic and proteomic analyses could significantly enhance our understanding of the etiopathogenesis and progression of meniscal diseases, yet such studies are currently sparse in the literature. To better characterize the meniscus pathology, the researchers designed a study aimed to bridge this gap by using microarray and proteomic analyses to compare normal and pathologic meniscal tissues. Meniscal tissue samples were collected from five patient groups ($n = 3$ per group), including patients undergoing meniscectomy with or without articular cartilage pathology, patients undergoing total knee arthroplasty with varying degrees of osteoarthritis, and aged-normal controls from organ donors. Tissue sections were taken from the white/white and white/red zones of the posterior medial menisci. Microarray analysis identified 157 genes with at least a 1.5-fold change in expression levels between pathologic and control menisci. These genes were categorized into four subgroups: synthesis, vascularity, degradation and antidegradation, and signaling pathways. Proteomics analysis, conducted using mass spectrometry, identified 173 proteins of interest. These proteins were classified into six categories: extracellular matrix (ECM) proteins, vascularity-associated proteins, degradation and antidegradation proteins, cytoskeleton proteins, glycolysis pathway proteins, and signaling proteins. This study provided novel molecular and biochemical information for continued investigation of etiology, disease mechanisms, and diagnostic and therapeutic strategies related to meniscal disorders.

Advances in 3D printing technology are enabling the creation of personalized meniscal scaffolds and implants that better match patient-specific anatomy and improve the mechanical and biological outcomes of meniscal repair. Recently, researchers created a 3D model of bioengineered medial meniscus tissue using MRI scans from a human volunteer [50]. They processed the Digital Imaging and Communications in Medicine (DICOM) data from these scans with specialized software to produce a detailed STL model of the meniscus structure. Using a microvalve-based inkjet printhead from the 3D Discovery printing tool, they printed the model. To construct the meniscus, primary mesenchymal stem cells (MSCs) isolated from bone marrow were embedded in a collagen-based bio-ink. The researchers then conducted a LIVE/DEAD assay on the printed cell-laden constructs to assess cell distribution and viability. This study successfully produced a human cell-laden collagen meniscus using 3D bioprinting, highlighting the potential of this technology to create anatomically shaped, patient-specific constructs with viable cells on a biocompatible material. The preliminary findings indicate that this custom-made, cell-laden, collagen-based meniscus could be foundational for future advancements in tissue-engineered implants designed to replace damaged menisci.

18.5 Conclusions

Meniscal tears are a common orthopedic injury that can cause significant pain and mechanical issues and, if left untreated, may lead to osteoarthritis. The evolution of treatment options from total meniscectomy to partial meniscectomy, meniscal repair, and the use of scaffolds or allografts emphasizes the importance of preserving the meniscus to reduce stress on the articular cartilage. However, the meniscus has limited self-repair capability due to its poor blood supply, which requires further exploration to enhance its healing.

The healing process of the meniscus relies on a strong vascular supply and involves cell migration, proliferation, and remodeling of the extracellular matrix. Variations in blood flow significantly affect healing, with the central avascular area (white-white zone) showing poor healing potential. Orthobiologics, which utilize biological substances to promote musculoskeletal healing, show promising advancements in this field. These substances include autologous growth factors, mesenchymal stem cells (MSCs), and platelet-rich plasma (PRP), aiming to improve tissue repair and potentially reduce the need for surgery. Inflammation in the synovial environment can hinder regeneration if not properly managed. This underscores the necessity of inflammation control in meniscal repair. Inflammatory cytokines, upregulated in osteoarthritic conditions, can lead to mechanically inferior tissue prone to retear. In vitro studies show that interleukin-1 (IL-1) decreases repair strength, cell migration, and tissue formation, but inhibiting IL-1 enhances repair [51, 52]. Inhibition of other proinflammatory cytokines, such as tumor necrosis factor α and matrix metalloprotease, can improve healing [53]. Controlling inflammation is vital for maximizing repair success, especially in degenerative or inflammatory joint conditions. The importance of this aspect in the overall success of the procedure cannot be overstated, and it should be a focus of future research in meniscal repair.

Clinical evidence supports the effectiveness of meniscal repair during ACL reconstruction, which improves success rates and reduces the need for further surgery, partly due to higher PDGF levels aiding healing. While PRP promotes healing through a rich concentration of growth factors, its effects can be variable due to the presence of TGF-β1. Intra-articular autologous conditioned serum injections have shown promise in reducing knee pain and potentially avoiding surgery.

Cell-based therapies, particularly those using MSCs, offer significant potential in enhancing meniscal repair, especially in avascular regions. MSCs' ability to coordinate complex healing responses and promote tissue regeneration makes them a robust solution for meniscal injuries. Using advanced scaffolds and growth factors, tissue engineering further optimizes cell proliferation and differentiation. Clinical studies have demonstrated the safety and efficacy of using autologous MSCs in a collagen scaffold, with notable improvements in knee function and pain reduction.

Emerging trends, such as genomic and proteomic analyses, can deepen our understanding of meniscal diseases. Advances in 3D printing technology have enabled the creation of personalized meniscal scaffolds and implants, which improve repair outcomes. The successful creation of a 3D-printed, cell-laden collagen meniscus indicates potential for future tissue-engineered implants.

In conclusion, the complexity of meniscal healing necessitates innovative approaches, and the advancements in orthobiologics and technology offer promising solutions for enhancing meniscal repair and regeneration.

References

1. Masini BD, et al. Epidemiology of isolated meniscus tears in young athletes. Orthop J Sports Med. 2015;3:2325967115S00107.
2. Keene GCR, Bickerstaff D, Rae PJ, Paterson RS. The natural history of meniscal tears in anterior cruciate ligament insufficiency. Am J Sports Med. 1993;21:672–9.
3. McDermott ID, Amis AA. The consequences of meniscectomy. Bone Jt J. 2006;88:1549–56.
4. Rao AJ, et al. The meniscus-deficient knee. Orthop J Sports Med. 2015;3:2325967115611386.
5. Englund M, Roemer FW, Hayashi D, Crema MD, Guermazi A. Meniscus pathology, osteoarthritis and the treatment controversy. Nat Rev Rheumatol. 2012;8:412–9.
6. Drobnič M, et al. Treatment options for the symptomatic post-meniscectomy knee. Knee Surg Sports Traumatol Arthrosc. 2019;27:1817–24.
7. McCann L, Ingham E, Jin Z, Fisher J. Influence of the meniscus on friction and degradation of cartilage in the natural knee joint. Osteoarthr Cartil. 2009;17:995–1000.
8. Peña E, Calvo B, Martínez MA, Palanca D, Doblaré M. Finite element analysis of the effect of meniscal tears and meniscectomies on human knee biomechanics. Clin Biomech. 2005;20:498–507.
9. Zadeh LG, Chevrier A, Farr J, Rodeo S, Buschmann M. Augmentation techniques for meniscus repair. J Knee Surg. 2017;31:099–116.
10. Mesiha M, et al. Pathologic characteristics of the torn human meniscus. Am J Sports Med. 2007;35:103–12.
11. Stone RG, Frewin PR, Gonzales S. Long-term assessment of arthroscopic meniscus repair: A two- to six-year follow-up study. Arthrosc J Arthrosc Relat Surg. 1990;6:73–8.
12. Arnoczky SP, Warren RF. Microvasculature of the human meniscus. Am J Sports Med. 1982;10:90–5.
13. Day B, Mackenzie WG, Shim SS, Leung G. The vascular and nerve supply of the human meniscus. Arthrosc J Arthrosc Relat Surg. 1985;1:58–62.
14. Arnoczky SP, Warren RF. The microvasculature of the meniscus and its response to injury. Am J Sports Med. 1983;11:131–41.
15. Michel PA, et al. Age-related changes in the microvascular density of the human meniscus. Am J Sports Med. 2021;49:3544–50.
16. Forriol F. Growth factors in cartilage and meniscus repair. Injury. 2009;40:S12–6.
17. Everts PA, Mazzola T, Mautner K, Randelli PS, Podesta L. Modifying orthobiological PRP therapies are imperative for the advancement of treatment outcomes in musculoskeletal pathologies. Biomedicine. 2022;10:2933.
18. Sezgin EA, Atik OŞ. Are orthobiologics the next chapter in clinical orthopedics? A literature review. Jt Dis Relat Surg. 2018;29:110–6.
19. Bhargava MM, Hidaka C, Hannafin JA, Doty S, Warren RF. Effects of hepatocyte growth factor and platelet-derived growth factor on the repair of meniscal defects in vitro. In Vitro Cell Dev Biol Anim. 2005;41:305–10.
20. Kasemkijwattana C, et al. The use of growth factors, gene therapy and tissue engineering to improve meniscal healing. Mater Sci Eng C. 2000;13:19–28.
21. Pangborn CA, Athanasiou KA. Effects of growth factors on meniscal fibrochondrocytes. Tissue Eng. 2005;11:1141–8.
22. Collier S, Ghosh P. Effects of transforming growth factor beta on proteoglycan synthesis by cell and explant cultures derived from the knee joint meniscus. Osteoarthr Cartil. 1995;3:127–38.

23. Makris EA, Hadidi P, Athanasiou KA. The knee meniscus: structure-function, pathophysiology, current repair techniques, and prospects for regeneration. Biomaterials. 2011;32:1–21.
24. Shelbourne KD, Carr DR. Meniscal repair compared with meniscectomy for bucket-handle medial meniscal tears in anterior cruciate ligament-reconstructed knees. Am J Sports Med. 2003;31:718–23.
25. Toman CV, et al. Success of meniscal repair at anterior cruciate ligament reconstruction. Am J Sports Med. 2009;37:1111–5.
26. Wasserstein D, et al. A matched-cohort population study of reoperation after meniscal repair with and without concomitant anterior cruciate ligament reconstruction. Am J Sports Med. 2013;41:349–55.
27. de Girolamo L, et al. Why menisci show higher healing rate when repaired during ACL reconstruction? Growth factors release can be the explanation. Knee Surg Sports Traumatol Arthrosc. 2015;23:90–6.
28. Zaleskas JM, et al. Growth factor regulation of smooth muscle actin expression and contraction of human articular chondrocytes and meniscal cells in a collagen–GAG matrix. Exp Cell Res. 2001;270:21–31.
29. Rhee S, Grinnell F. P21-activated kinase 1: convergence point in PDGF- and LPA-stimulated collagen matrix contraction by human fibroblasts. J Cell Biol. 2006;172:423–32.
30. Huey DJ, Athanasiou KA. Maturational growth of self-assembled, functional menisci as a result of TGF-β1 and enzymatic chondroitinase-ABC stimulation. Biomaterials. 2011;32:2052–8.
31. Strümper R. Intra-articular injections of autologous conditioned serum to treat pain from meniscal lesions. Sports Med Int Open. 2017;1:E200–5.
32. Nakagawa Y, et al. Synovial mesenchymal stem cells promote healing after meniscal repair in microminipigs. Osteoarthr Cartil. 2015;23:1007–17.
33. Horie M, et al. Implantation of allogenic synovial stem cells promotes meniscal regeneration in a rabbit meniscal defect model. J Bone Jt Surg. 2012;94:701–12.
34. Osawa A, et al. The use of blood vessel–derived stem cells for meniscal regeneration and repair. Med Sci Sports Exerc. 2013;45:813–23.
35. Weinand C, et al. An allogenic cell–based implant for meniscal lesions. Am J Sports Med. 2006;34:1779–89.
36. Heo S-J, et al. Nuclear softening expedites interstitial cell migration in fibrous networks and dense connective tissues. Sci Adv. 2020;6:eaax5083.
37. Freymann U, Endres M, Goldmann U, Sittinger M, Kaps C. Toward scaffold-based meniscus repair: effect of human serum, hyaluronic acid and TGF-ß3 on cell recruitment and re-differentiation. Osteoarthr Cartil. 2013;21:773–81.
38. Lee KI, Olmer M, Baek J, D'Lima DD, Lotz MK. Platelet-derived growth factor-coated decellularized meniscus scaffold for integrative healing of meniscus tears. Acta Biomater. 2018;76:126–34.
39. Qu F, Holloway JL, Esterhai JL, Burdick JA, Mauck RL. Programmed biomolecule delivery to enable and direct cell migration for connective tissue repair. Nat Commun. 2017;8:1780.
40. Caplan AI. Adult mesenchymal stem cells for tissue engineering versus regenerative medicine. J Cell Physiol. 2007;213:341–7.
41. Caplan AI, Dennis JE. Mesenchymal stem cells as trophic mediators. J Cell Biochem. 2006;98:1076 1084.
42. Caplan A. Why are MSCs therapeutic? New data: new insight. J Pathol. 2009;217:318–24.
43. Caplan AI, Correa D. The MSC: an injury drugstore. Cell Stem Cell. 2011;9:11 15.
44. Dimarino AM, Caplan AI, Bonfield TL. Mesenchymal stem cells in tissue repair. Front Immunol. 2013;4:201.
45. McCorry MC, Puetzer JL, Bonassar LJ. Characterization of mesenchymal stem cells and fibrochondrocytes in three-dimensional co-culture: analysis of cell shape, matrix production, and mechanical performance. Stem Cell Res Ther. 2016;7:39.
46. Cui X, Hasegawa A, Lotz M, D'Lima D. Structured three-dimensional co-culture of mesenchymal stem cells with meniscus cells promotes meniscal phenotype without hypertrophy. Biotechnol Bioeng. 2012;109:2369–80.
47. Whitehouse MR, et al. Repair of torn avascular meniscal cartilage using undifferentiated autologous mesenchymal stem cells: from in vitro optimization to a first-in-human study. Stem Cells Transl Med. 2017;6:1237–48.
48. Cook JL, Fox DB. A novel bioabsorbable conduit augments healing of avascular meniscal tears in a dog model. Am J Sports Med. 2007;35:1877–87.
49. Williams LB, Adesida AB. Angiogenic approaches to meniscal healing. Injury. 2018;49:467–72.
50. Filardo G, et al. Patient-specific meniscus prototype based on 3D bioprinting of human cell-laden scaffold. Bone Jt Res. 2019;8:101–6.
51. Wilusz RE, Weinberg JB, Guilak F, McNulty AL. Inhibition of integrative repair of the meniscus following acute exposure to interleukin-1 in vitro. J Orthop Res. 2008;26:504–12.
52. McNulty AL, Moutos FT, Weinberg JB, Guilak F. Enhanced integrative repair of the porcine meniscus in vitro by inhibition of interleukin-1 or tumor necrosis factor α. Arthritis Rheum. 2007;56:3033–43.
53. Riera KM, et al. Interleukin-1, tumor necrosis factor-alpha, and transforming growth factor-beta 1 and integrative meniscal repair: influences on meniscal cell proliferation and migration. Arthritis Res Ther. 2011;13:R187.

Regenerative Medicine in Tendon Pathologies

19

Stephen P. Abelow, Isabel Guillen-Garcia-Vicente,
Marta Guillen-Vicente,
Juan Manuel Lopez-Alcorocho, and Pedro Guillen

19.1 Introduction

Regenerative medicine and "orthobiology" have a tremendous and exciting promise in improving tendinopathy healing as well as other disorders of tendons. From the Greek suffix "pathy," indicating a disease or disorder, tendinopathy literally means a disease or disorder of tendon. Because there exists a multifactorial and complex pathology of the tendon causing tendinopathy, multiple new orthobiologic and regenerative therapeutic technologies are being investigated and developed. There are some promising preclinical and clinical results.

Tendinitis is an inflammatory process of the tendon. Tendinopathy is more of a degenerative type of process.

Factors causing tendinopathy can be repetitive microtrauma (overuse) causing microscopic damage to the tendon, lack of adequate tissue perfusion causing mitochondrial dysfunction and apoptosis, imbalance of sympathetic and parasympathetic signaling (increased glutamate and substance P), inflammation (release of proinflammatory cytokines such as IL-6 and IL-1), and metabolic/endocrine factors (diabetes, hypothyroidism, hypercholesterolemia, low vitamin D, excess weight) [1].

These diverse agents cause alteration in the structure and composition of the tendon cellular matrix causing altered cellular behavior. There is an increasing focus now on the role of immune cell population and inflammatory mediators (e.g., cytokines, nitric oxide, prostaglandins, and lipoxins) contributing to the development of tendinopathy) [2].

Regenerative techniques for tendinopathy should address matrix synthesis remodeling, cell proliferation, cell migration, angiogenesis, inflammatory mediators, production of nociceptive medium, and stimulation of intrinsic connective tissue progenitor cells (stem cell niche) [2].

Current regenerative and orthobiologic therapeutic strategies for tendinopathy include autologous blood products and cellular therapy.

Autologous blood products include platelet-rich plasma, autologous condition serum (ACS-Orthokine, Regenokine), autologous condition plasma (ACP), alpha-2-macroglobulin (alpha-2-M), platelet-rich fibrin matrix (PRFM), and autologous protein solution (APS) [2].

Current cell therapies include bone marrow (bone marrow aspirate (BMA) and/or bone marrow aspirate concentrate (BMAC)), adipose tissue, and perinatal tissue including umbilical cord blood, placenta, or amniotic membrane.

S. P. Abelow (✉) · I. Guillen-Garcia-Vicente
M. Guillen-Vicente · J. M. Lopez-Alcorocho
P. Guillen
Clinica CEMTRO, Madrid, Spain
e-mail: isabel.guillen@clinicacemtro.com; marta.guillen@clinicacemtro.com; jm.lopez@amplicel.com

A. Gobbi et al. (eds.), *Regenerative Medicine in Sports and Orthopaedics*,
https://doi.org/10.1007/978-3-031-84693-9_19

19.2 Platelet-Rich Plasma (PRP)

PRP is "a sample of autologous blood with concentrations of platelets above baseline values" (AAOS). It is created through a centrifugation process called plasmapheresis. Platelets release growth factors and cytokines as well as other polypeptides and proteins from alpha-granules of the platelets. Platelets have a half life of 7 days. Secretion of growth factors begins within 10 min, and 95% of the growth factors are released within 1 h. Many of the growth factors have a half life of 10 min.

PRP formulations may be plasma based (excludes red cells and white cells) or buffy coat based (both plasma and cellular layers). Buffy coat PRP usually has a higher platelet concentration and higher leukocyte concentration. There appear to be important signaling molecules in the plasma portion apart from the platelets.

PRP has been classified as "leukocyte-rich" versus "leukocyte-poor." Knowledge of which is better is still controversial. PRP may inhibit inflammation in a proinflammatory environment and be proinflammatory in the absence of inflammation.

Zang and Wang at the University at the University of Pittsburgh (Pennsylvania, USA) reported in 2010 that platelet-rich plasma releasate in rabbits promoted differentiation of "tendon stem cells" into active tenocytes exhibiting high proliferation rates and collagen production capability. During tendon healing platelet-derived growth factor (PDGF), epidermal growth factor, transforming growth factor beta-1 (TGF-B1), insulin-like growth factor (IGF-1), and certain receptors (example TGF-B1 receptor) are upregulated. Platelets appear to function as a natural reservoir for releasing of these growth factors that are essential to the repair of injured tissues [3].

A controlled laboratory study examining the difference in the mitotic, chemotactic, anabolic, and inflammatory effects between leukocyte-rich PRP (LR-PRP), platelet-rich plasma (PRP), and platelet-poor plasma (PPP) demonstrated there were stronger chemotactic factors and proliferation of proliferative properties with leukocyte-rich PRP and PRP. Platelet-poor plasma and platelet-rich plasma stimulated matrix metabolism. Leukocyte-rich PRP was more proinflammatory. It was felt that plasma formulations need to be tailored appropriately [4].

A randomized controlled trial (level of evidence 1) placebo-controlled study from Turkey reported that at 1 year follow-up a PRP injection was found to be no more effective in improving quality of life, pain, disability, and shoulder range of motion than placebo in patients with chronic rotator cuff tears than were treated with an exercise program [5].

Carr et al. (Oxford, UK) reported a randomized controlled trial (LOE 1) that PRP injection at the time of arthroscopic acromioplasty did not affect clinical outcomes. PRP significantly altered the tissue characteristics in tendons after surgery with residual cellularity and vascularity as well as increased levels of apoptosis. Tendon biopsy at 12 weeks showed increased apoptosis and decreased cellularity with PRP. The coapplication of PRP with acromioplasty did not improve clinical outcomes and may have potentially deleterious effects on healing tendons [6].

A double-blinded randomized controlled trial (LOE 1) by Schwitzguebel et al. (Switzerland) in 2019 concluded that PRP injections within interstitial supraspinatus tears did not improve tendon healing or clinical scores compared with saline injections. The incidence of adverse effects (pain greater than 48 h, frozen shoulder, extension of lesion) was significantly higher in the PRP group (54%) than the control group (26%) [7].

Meta-analysis and systematic review by Fitzpatrick et al. (Switzerland) reported good evidence to support a single injection of LR-PRP under ultrasound guidance in tendinopathy [8].

Systematic review of nonoperative treatment of rotator cuff disease with platelet-rich plasma by Hurley et al. (Ireland) in 2017 (LOE 2; systematic review of level 1 and 2 evidence) concluded that in the short-term PRP injections may not be beneficial when directly compared with exercise therapy. PRP did not result in superior functional outcome scores, pain scores, or range of motion. The literature was confounded by lack of reporting of the cytology and characteristics of the PRP [9].

Dragoo et al. (Stanford University, USA) reported a randomized controlled trial (LOE 1) for patellar tendinopathy comparing PRP with dry needling of the patellar tendon, a therapeutic regimen of standardized eccentric exercises and ultrasound-guided leukocyte-rich PRP injection with dry needling accelerated the recovery from patellar tendinopathy relative to exercise and ultrasound-guided dry needling alone. The benefits of PRP dissipated over time [10].

In 2014, Fitzpatrick et al. (Australia) reported on the effectiveness of platelet-rich plasma injections in gluteal tendinopathy (randomized controlled trial; LOE 1). Patients with chronic gluteal tendinopathy greater than 4 months achieved greater clinical improvement at 12 months when treated with a single PRP injection than those treated with a single corticosteroid injection (GPS III Kit; Zimmer Biomet) [11]. In 2019, Fitzpatrick et al. reported a double-blind randomized controlled trial (LOE 1) that leukocyte-rich plasma among patients with chronic gluteal tendinopathy and length of symptoms greater than 4 months, a single intratendinous LR-PRP under ultrasound guidance injection resulted in greater improvement in pain and function than a single corticosteroid injection. The improvement in pain and function after the LR-PRP injection was sustained at 2 years. The improvement from corticosteroid injection was maximal at 6 weeks and not maintained beyond 24 weeks [12].

de Vos et al. reported in 2010 a randomized controlled trial of platelet-rich plasma injection for chronic Achilles tendinitis. At 24 weeks PRP injection did not result in greater improvement in pain and activity compared with saline [13]. de Jong et al. (Netherlands) reported a randomized placebo-controlled trial of LR-PRP versus saline injection in addition to an eccentric training program (LOE 1). The randomized controlled trial showed "no clinical and ultrasonographic superiority to PRP injection over a placebo injection in chronic Achilles tendinitis at 1 year combined with an eccentric training program" [14].

Krogh et al. (Denmark) reported a randomized controlled trial (LOE 1) of Achilles tendinopathy with platelet-rich plasma or saline. PRP injection did not result in an improved VISA-A score over a 3-month period in patients with chronic Achilles tendinopathy compared to placebo. There was an increase in tendon thickness compared with saline injection [15].

A randomized double-blinded prospective study by Boesen et al. (Denmark) (LOE 1) concluded that treatment with high-volume injection of PRP in combination with eccentric training in chronic Achilles tendinopathy seems more effective in reducing pain, improving activity level, and reducing tendon thickness and intratendinous vascularity than eccentric training alone. It was felt that high-volume injection may be more effective in improving outcomes of chronic Achilles tendinopathy than PRP in the short term [16].

Peerbooms et al. performed a double-blinded randomized controlled trial (LOE 1) to try to determine the effectiveness of PRP as compared with corticosteroid injection of chronic plantar fasciitis. They concluded that the treatment of patients with chronic plantar fasciitis with PRP seems to reduce pain and increase function more as compared with no effect of corticosteroid injection [17].

The use of PRP in lateral epicondylitis of the elbow "plays to mixed reviews." Mishra and Pavelko reported in 2006 a 60% reduction in pain using leukocyte-rich PRP compared with 16% reduction in the control group at 8 weeks. At final follow-up, the LR-PRP group reported a 93% reduction in pain [18]. A prospective double-blinded randomized controlled trial found no difference comparing LR-PRP to a control group at 12 weeks. At 24 weeks, the success with LR-PRP was 82.1% compared with 60.3% for the control group [19].

Gosens and Peerbooms (Netherlands) (randomized controlled trial LOE 1) found that LR-PRP versus corticosteroid injection that the LR-PRP "reduced pain and increased function significantly exceeding the effect of corticosteroid injections even after a follow-up of 2 years."

A systematic review and meta-analysis by Vangsness et al. (LOE 1, prognostic) reported on 16 level 1 studies. They reported, "Considering the small number of comparable studies, lack of quantification of specific PRP contents, considerable heterogenicity between randomized controlled trials, and most effect sizes being equivocal within the framework of two estimated MCID values, the authors can neither scientifically support nor discourage the use of PRP for lateral epicondylitis despite finding significant improvement in pain and function outcomes" [20].

To further complicate the relevance of various regenerative and orthobiologic treatments for lateral epicondylitis, Drake and Ring reported in *the Journal of the American Academy of Orthopedic Surgeons*, "The natural history of lateral epicondylitis is spontaneous resolution but it can last for greater than 1 year," and "Despite the best efforts of medical science, no treatment, invasive or noninvasive, has proven to alter the natural history of the condition" [21].

Because of the heterogenicity of PRP formulations and design studies, there is tremendous variability in the reported outcomes for the use of regenerative medicine techniques in the treatment of lateral epicondylitis of the elbow. The question still remains whether PRP actually alters the natural history of lateral epicondylitis.

Rodeo et al. evaluated the effect of platelet-rich fibrin matrix (PRFM) on rotator cuff healing in a randomized clinical study (randomized controlled trial LOE 2). Platelet-rich fibrin matrix supplied to the tendon-bone interface at the time of rotator cuff repair had no demonstrable effect on tendon-to-bone healing, tendon vascularity, manual muscle strength, or clinical rating scales. Regression analysis suggested that PRFM may have a negative effect on healing [22].

19.3 Other Plasma-Based Modalities

Autologous condition serum (ACS) (Orthokine-Europe; Regenokine-US). Blood is harvested and concentrated overnight. IL-1Ra (IL-1R antagonist) is concentrated. IL-1Ra is a potent anti-inflammatory cytokine. This is not yet approved in the USA [2].

Autologous protein solution (APS) (nSTRIDE APS Kit Zimmer Biomet) concentrates growth factors and anti-inflammatory cytokines from whole blood. This concentrates the IL-1b receptor antagonist.

Alpha-2-macroglobulin (A2M) is an MMP inhibitor. It appears that matrix metalloproteinase (MMPs) and tissue inhibitors of matrix metalloproteinases (TIMP) play an important role in the pathogenesis of rotator cuff tears. A2M treatment in rats showed significantly greater fibrocartilage and collagen organization at the healing enthesis of the supraspinatus region and significant reduction in collagen degradation. This may offer a novel biologic pathway to augment tendon to bone healing of the rotator cuff [2].

19.4 Cell-Based Therapies for Tendinopathy

Currently available cell-based therapies include bone marrow aspirate (BMA), bone marrow aspirate concentrate (BMAC), peripheral blood-derived cells, adipose tissue (micro-fragmented "minimally manipulated" or stromal vascular fraction "enzymatic digestion"), and perinatal products (can only use under FDA-approved IND since June, 2021; cannot be used in the USA).

The concept of mesenchymal stem cells was introduced by Arnold Caplan, PhD, at Case Western Reserve University in Ohio (USA). The initial concept was that the stem cell progeny became committed to a specific phenotypic pathway. In 2011, Dr. Caplan postulated that the MSCs besides having a mesogenic line had a paracrine effect with immunomodulatory and trophic properties establishing a regenerative microenvironment and were essentially "medicinal signaling cells" [23]. These cells may be pericytes.

Criteria for stem cells dictate they are resting cells, unspecialized cells (not actively proliferating). They are capable of asymmetric cell division and self-renewal and have multiple differentiation properties.

In 2005, the International Society for Cellular and Genetic Therapy issued a position statement for minimal criteria for defining multipotent mesenchymal stromal cells. These multipotent stem cells adhere to plastic in standard culture and must express genes CD105, CD73, and CD90 and lack expression of CD45, CD34, CD14 or CD11b, CD79 or CD19, and HLA-DC. These cells are capable of differentiation into osteocytes, chondrocytes, and adipose tissue in vitro [24].

The classic definition of a mesenchymal stromal cell (MSC) is based on culture-expanded cells, which is different from cell populations in recently harvested tissue which are not non-culture expanded [2].

Currently used cell sources contain few, if any, "stem cells" as formal criteria. In bone marrow aspirate from 0.01 to 0.001% of the cells are represented by mesenchymal stromal cells.

Reporting cell therapy augmentation in rotator cuff repair, Hernigou et al. in 2014 (case controlled study) on biologic augmentation of rotator cuff repair with MSCs during arthroscopy improved healing and prevented further tears. At 10 years follow-up 87% of the MSC treated group had intact rotator cuffs while 44% of the control group had intact rotator cuffs. Long-term cuff integrity was better in those who received higher number of cells [25].

A cohort study (LOE 3) by Kim et al. using adipose-derived mesenchymal stem cells reported a 28.5% retear in MRI with conventional repair and 14.3% in the cell group [26].

Randelli et al. (Italy) in 2022 reported a prospective randomized controlled trial (LOE 2) demonstrating that intraoperative injection of autologous microfragmented adipose tissue (Lipogems) was safe and effective in improving short-term clinical functional results after a single row arthroscopic rotator cuff repair [27].

Cole et al. (USA) performed a randomized controlled trial (LOE 1) comparing arthroscopic rotator cuff repair with and without bone marrow aspirate concentrate augmentation. They reported that bone marrow aspirate concentration augmented arthroscopic rotator cuff repair may result in structurally superior repair by MRI. Arthroscopic rotator cuff repair without bone marrow aspirate concentrate had a 57% retear rate at 1 year compared to 18% retear rate in the BMAC augmented arthroscopic rotator cuff repair group. The bone marrow aspirate concentrate augmented rotator cuff repair group had no difference in patient-reported clinical outcomes [28].

19.5 Other Treatments for Tendinopathy

Several metabolic factors are associated with tendinopathy and impairing tendon healing, including hypercholesterolemia (hyperlipidemia), smoking, diabetes, osteoporosis, vitamin D deficiency, fluoroquinolone antibiotics, and thyroid disease [2].

Biological treatment with statins might provide protection against rotator cuff disease in patients with hyperlipidemia compared with no statin use [29].

Parathyroid hormone (teriparatide), a recombinant human parathyroid hormone, can be a systemic treatment option that significantly enhances the tendon-to-bone healing after arthroscopic rotator cuff repair with cuff tears greater than 2 cm [30].

Vitamin D affects osteoblastic activity and muscle function. Preclinical studies suggest vitamin D deficiency is related to impaired tendon healing [2].

19.6 Exosomes

Exosomes offer an exciting future potential for tendinopathy therapies. Exosomes are bioactive molecules that are contained in extracellular vesicles which can be purified from condition medium of human MSCs. Exosomes are quite small, with the size of exosomes being 35–150 nm (smaller than a bacteria and larger than a virus). There is an expression of specific exosome-associated markers (CD81, ALIX, CD9, and TSG101). The cargo of exosomes contains proteins and nucleic acids (m-RNA and mi-RNA) [31].

In a controlled laboratory study (n = 15), Wang et al. (Shanghai, China) reported adipose stem-cell-derived exosomes maintained metabolic homeostasis of human rotator cuff tendons to improve their histologic properties. The expression of MMP-9 and MMP-13 genes decreased, and there was an increased expression of type I and III collagen and increased type I/III rotator cuff collagen.

Exosomes are less immunogenic than viruses or stem-cell-based therapies. Exosomes from normal healing cells are involved in processes of wound healing and regeneration. Exosomes from injured cells are involved in the pathogenesis of many conditions including osteoarthritis.

Some key advantages of exosomes derived from stem cells are they are not perceived as foreign (and are therefore safe), can travel via systemic therapy, and deliver miRNA and mRNAm metabolites and proteins. Importantly, exosomes cannot transdifferentiate into other cells or into malignant cells. The dosage of exosomes can be controlled. It appears that exosomes can promote regeneration and immunomodulation, anti-inflammatory, and antifibrotic, and can home to inflammatory sites. Exosomes can also function as biomarkers for various diseases [31].

19.7 Conclusion

Regenerative medicine and "orthobiology" have a tremendous and exciting promise in improving tendinopathy healing as well as other disorders of tendons. We need to develop methods to select the desired cells and eliminate competing catabolic factors, giving us a so-called "designer orthobiology" for treatment of tendinopathies.

Finding the appropriate orthobiologic treatment for tendinopathy requires knowledge of the cellular and biomechanical modalities on its effect on targeted tendon disorder.

Because of the disease heterogenicity of tendinopathy and the heterogenicity of regenerative and orthobiological approaches, the interpretation of present clinical studies is difficult.

More randomized controlled trials with documentation of the quality and quantity of "orthobiologic" used is imperative to further define the importance of these various orthobiologic and regenerative therapies.

References

1. Mishra A, et al. Orthobiologics. Rosemont: American Academy of Orthopedic Surgeons; 2024.
2. Rodeo S. Instruction course lecture #304, American Orthopedic Society for Sports Medicine Annual Meeting, Denver, Colorado, July 2024.
3. Zhang J, et al. Platelet rich plasma release promotes differentiation of tendon stem cells into active tenocytes. Am J Sports Med. 2010;38(12):2477–88.
4. Rubio-Azpeitia E, et al. The properties of three different plasma formulas and their effect on tendinopathy cells. Am J Sports Med. 2016;44(8):1952–61.
5. Kesikburan S, et al. Platelet-rich plasma in the treatment of chronic rotator cuff tendinopathy: a randomized controlled trial with one year follow-up. Am J Sports Med. 2013;41(11):2609–16.
6. Carr AJ, et al. PRP injection with arthroscopic acromioplasty for chronic rotator cuff tendinopathy: a randomized controlled trial. Am J Sports Med. 2015;43(12):2891–7.
7. Schwitzguebel AJ, et al. Efficiency of platelet-rich plasma for the treatment of interstitial supraspinatus tears. Am J Sports Med. 2019;47(8):1885–92.
8. Fitzpatrick J, et al. The effectiveness of platelet-rich plasma in the treatment of tendinopathy: a meta-analysis of randomized controlled trials. Am J Sports Med. 2017;45(1):226–33.
9. Hurley ET, et al. Nonoperative treatment of rotator cuff disease with platelet-rich plasma: a systematic review of randomized controlled trials. Arthroscopy. 2019;35(5):1584–91.
10. Dragoo JL, et al. Platelet-rich plasma as treatment for patellar tendinopathy: a double-blind randomized controlled trial. Am J Sports Med. 2014;42(3):610–8.
11. Fitzpatrick J, et al. The effectiveness of platelet-rich plasma injections in gluteal tendinopathy: a randomized double-blind controlled trial comparing a single PRP injection with a single corticosteroid injection. Am J Sports Med. 2018;46(4):933–9.
12. Fitzpatrick J, et al. Leukocyte-rich platelet-rich plasma treatment of gluteus medius and minimus tendinopathy: a double-blind randomized controlled trial with two year follow-up. Am J Sports Med. 2019;47(5):1130–7.
13. de Vos RJ, et al. Platelet-rich plasma injection for chronic Achilles tendinopathy: a randomized controlled trial. JAMA J Am Med Assoc. 2010;303:144–9.
14. de Jange S, de Vos RJ, et al. One year follow-up of platelet-rich plasma treatment in chronic tendinopathy: a double-blind randomized placebo-controlled trial. Am J Sports Med. 2011;39(8):1623–9.
15. Krogh TP, et al. Ultrasound-guided injection therapy of Achilles tendinopathy with platelet-rich plasma or saline: a randomized blinded placebo-controlled trial. Am J Sports Med. 2016;44:1990–7.
16. Boesen AP, et al. Effect of high-volume injection, platelet-rich plasma, and sham treatment in chronic

midportion Achilles tendinopathy: a randomized double-blind prospective study. Am J Sports Med. 2017;45(9):1034–2043.
17. Peerbooms JC, et al. Positive effect of platelet-rich plasma on pain in plantar fasciitis: a double-blinded multicenter randomized controlled trial. Am J Sports Med. 2019;47:3238–46.
18. Mishra A, et al. Treatment of chronic elbow tendinitis with buffered platelet-rich plasma. Am J Sports Med. 2006;34:1774–8.
19. Mishra AK, et al. Efficacy of platelet-rich plasma for chronic tennis elbow: a double-blind prospective multicenter randomized controlled trial of 230 patients. Am J Sports Med. 2014;42:463–71.
20. Vangsness XT, et al. The efficacy of platelet-rich plasma for improving pain and function in lateral epicondylitis: a systematic review of meta-analysis with risk of bias assessment. Arthroscopy. 2021;37(9):2937–52.
21. Drake ML, Ring DC. Enthesopathy of the extensor carpi radialis brevis origin: effective common strategies. J Am Acad Orthop Surg. 2016;24(6):365–9.
22. Rodeo SA, et al. The effect of platelet-rich fibrin matrix on rotator cuff tendon healing: a prospective randomized clinical study. Am J Sports Med. 2012;40(6):1234–41.
23. Caplan C. The MSC and injury drugstore. Stem Cells. 2011;9(1):11–5.
24. International Society for Cellular and Genetic Therapy. Minimal criteria for defining multipotent mesenchymal stromal cells: the International Society for Cellular Therapy Position Statement. 2005.
25. Hernigou P, et al. Biologic augmentation of rotator cuff repair with mesenchymal stem cells during arthroscopy improves healing and prevents further tears: a case controlled study. Int Orthop. 2014;38(9):1811–8.
26. Kim YS, et al. Does an injection of adipose-derived mesenchymal stem cells loaded in fibrin glue influence rotator cuff repair outcomes? A clinical and magnetic resonance imaging study. Am J Sports Med. 2017;45(12):2010–8.
27. Randelli PS, et al. Arthroscopic rotator cuff repair augmented with autologous microfragmented lipoaspirate tissue is safe and effectively improves short-term clinical and functional results; a prospective randomized controlled trial with 24 month follow-up. Am J Sports Med. 2022;20(10):1–14.
28. Cole BJ, et al. Prospective randomized trial of biologic augmentation with bone marrow aspirate in patients undergoing arthroscopic rotator cuff repair. Am J Sports Med. 2023;51(5):1234–42.
29. Lin TT-L, et al. The effect of diabetes, hyperlipidemia, and statins on the development of rotator cuff disease. Am J Sports Med. 2015;43(9):2126–32.
30. Oh JH, et al. Effects of recombinant human parathyroid hormone on rotator cuff healing after arthroscopic repair. Arthroscopy. 2019;35(4):1064–71.
31. Ju H, et al. Preliminary investigation of the diagnosis of neonatal congenital small bowel atresia by ultrasound. Biomed Int. 2019;2019:7097159.

New Treatments for Musculoskeletal Diseases

20

Gabriel Ohana Marques Azzini,
Lucas Furtado da Fonseca, Fabio Ramos Costa,
Dra Silvia Visoni, Dra. Stephany Huber,
and José Fábio Lana

20.1 Principles of Regenerative Medicine

Regenerative medicine is an emerging interdisciplinary field that aims to develop new approaches to repair, regenerate and replace damaged cells, tissues and organs. It has the potential to offer permanent and often curative solutions to the most debilitating ailments by restoring the function of biological tissues and organs that have been impaired. Regenerative medicine in musculoskeletal diseases is of particular significance to patients because the structural changes that underlie degenerative processes in muscle, cartilage, bone and ligament increasingly preclude biologic reversal with conventional therapies that typically aim only to alleviate symptoms by reducing pain and inflammation. Musculoskeletal diseases encompass a broad spectrum of conditions, from spontaneous traumas and autoimmune or molecular aberrations to subtle ageing processes that may follow lifestyle-related behaviours, such as dietary habits or excess body weight, that have a long-term impact on musculoskeletal tissue structure and function. Regenerative medicine offers novel scientific approaches and clinical applications to treat patients with a wide range of debilitating conditions, from traumatic injuries and industrial diseases to spontaneously developed degenerative diseases and age-related tissue wear, including osteoarthritis, autoimmune diseases, osteoporosis and many other diverse musculoskeletal conditions [1–6].

The techniques involved in the field of regenerative medicine are manifold and ever evolving. Cellular therapy exploits the use of cells (often stem cells) to promote the growth and repair of tissue. Embryonic stem cells are pluripotent, meaning these cells possess the ability to differentiate into any type of body tissue. On the other hand, the group of cells known as adult stem cells includes mesenchymal stem cells (MSCs) that are progenitor cells, which can differentiate into multiple cell types, such as bone, cartilage and muscle, depending on the specific micro-environmental cues at the target site of implantation. Tissue engineering consists of the growth of biological tissues on scaffolds (biocompatible supports), seeded with either cells or a combination of cells and bioactive molecules. Although various types of biological tissues can be engineered in vitro, this methodology has been most widely employed to cultivate cartilage, bone and muscle. Cells generated under in vitro conditions

G. O. M. Azzini (✉) · L. F. da Fonseca · D. S. Visoni
D. S. Huber · J. F. Lana
Department of Orthopedics, Brazilian Institute of Regenerative Medicine (BIRM), Indaiatuba, Brazil
e-mail: contato@drlucasfonseca.med.br

F. R. Costa
Department of Orthopedics, FC Sports Traumatology Clinic, Salvador, Brazil

A. Gobbi et al. (eds.), *Regenerative Medicine in Sports and Orthopaedics*,
https://doi.org/10.1007/978-3-031-84693-9_20

can be easily implanted into patients, where they have the ability to enhance regeneration and repair.

Three-dimensional bioprinting has the potential to become a game-changer in the field of regenerative medicine, given that it enables assembling of tissue constructs layer by layer using 3D printers working with bioinks containing living cells. This technology seems to have great potential for the creation of complex tissues with spatial accuracy that, hopefully, could be translated into organ- or part-organ-scale reconstructions of musculoskeletal tissues with complicated anatomical configurations and mechanical and compositional properties, such as tendon-bone interfaces or articular cartilage. All these tools augment the therapeutic options for diseases initially regarded as incurable, as well as set the stage for the development of personalised and more effective therapies in a truly promising new era in musculoskeletal medicine.

20.2 Advances in Cellular Therapies

This feature would be important in regenerative medicine, and indeed, the interest in stem cells in general is due to the fact that they can differentiate into a variety of cell types and stimulate regeneration of tissue during the healing process. These cells can be divided into two main types: mesenchymal stem cells (MSCs) and induced pluripotent stem cells (iPSCs), both of which are of great interest in the treatment of osteoarticular diseases. Among these, MSCs have been extensively studied and are capable of differentiating into osteoblastic, chondrocytes and other mesenchymal cell lineages, which are responsible for the formation and regeneration of all types of musculoskeletal tissue. Although, beyond adipose tissue, MSCs are usually isolated from bone marrow, and more recently from the umbilical cord, their value lies in their immune modulatory activity and the release of bioactive factors that enhance tissue repair. Indeed, their capacity to reduce inflammation and promote tissue repair led to their use in osteoarthritis and bone lesion treatments [7, 8]. Moreover, MSCs are currently included in clinical trials for the repair of cartilage and sports injuries leading to promising outcomes in clinical trials, with the improvement of pain, while leading to an improvement in joint function [9].

iPSCs are embryonic-like cells obtained from the reprogramming of adult somatic cells by forcing their nucleus to overexpress a set of transcription factors. iPSC reprogramming represents a powerful technology for human cell modelling and regenerative medicine. These pluripotent cells can be differentiated into any somatic cell type, creating an important resource of specific cells that can be used for tissue regeneration purposes, as well as for research to understand disease pathogenesis and the development of new therapeutic approaches. For example, recent research has showed excellent cartilage repair efficacy of human iPSC-derived mesenchymal cells in animal models of human osteoarthritis, with the potential to be used in humans [10]. By generating cellular models derived from the most common osteoarticular diseases, it has been possible to explore new drugs and pharmacological agents, and gain deeper knowledge of the pathological pathways involved [11]. Applications of stem cells in the treatment of osteoarticular diseases include regeneration of cartilage in osteoarthritis patients, repair of bone defects, and immunomodulation of pain and inflammation. In degenerative joint diseases, the inflammatory process is very significant, and the paracrine activity of MSCs that are able to secrete potent growth and trophic factors effective in modulating inflammation, and also in the activation of endogenous progenitor cells in the damaged tissue, makes them an attractive new option in the management of degenerative joint diseases. iPSCs can be used to generate personalised tissues for autologous transplantation, thus reducing the risk of immunological rejection and improving the clinical outcomes [12]. So far, these approaches have been exploited in a preclinical setting, and it is urgent to bring them into clinical use to take advantage of the potential.

For instance, usage of such cells with immunomodulatory activity has become an important aspect of treating autoimmune and inflammatory diseases such as rheumatoid arthritis through their modification to alter the pro-inflammatory changes in the joint microenvironment. MSCs can be configured to release higher levels of anti-inflammatory cytokines and inhibitors of pro-inflammatory cytokines—possibly suppressing the dysregulated immune response in rheumatoid arthritis. MSC infusion induces such immunomodulation in animal models of rheumatoid arthritis, leading to a significant reduction of inflammation and resulting joint destruction [13]. Even though the potential of cellular therapies are vast, it's too early to celebrate this revolution. A plethora of issues need to be addressed to maximise clinical efficacy and safety. Uncertainties, such as growth of tumours after stem-cell implantation, infectious diseases and immunologic rejection—triggered by the 'foreign-ness' of the graft in the body—are still areas of concern. Well-conducted and long-term clinical trials must be in place to understand the actual safety of these therapies. Besides quality assurance (testing other parameters of stem cells such as growth, fitness for purpose, etc.), the heterogeneity in potency and character of stem cells in culture (expanded MSCs are not identical to native bone marrow cells) may affect their clinical efficacy to a large extent [14]. Standardising cultivation protocols and characterisation of the cells are the key [15]. The transnational nature of research and commercialisation complicates the regulation of cellular therapies. It varies considerably among different jurisdictions. In recent years, regulatory authorities, such as the FDA and the EMA, are developing specific guidelines for cellular therapies and outlining a framework for the development of regenerative medicine products [16].

Also, recent studies have also contributed to refine the optimised cultivation and application techniques of stem cells and strengthen the precision of the genetic editing technologies (e.g. CRISPR), enhancing cell specificity and safety of the cell therapies. In fact, by editing the DNA of iPSCs fibroblasts derived from the patient—before being differentiated and transplanted—it will become possible to correct genic mutations that could lead to ineffective or dangerous regenerative therapies, strengthening the precision of therapeutic outcomes. Mechanical stimulation, followed by a bath of osteogenic and angiogenic factors, induces the conversion of mesenchymal stem cells into an osteogenic phenotype [17]. Finally, new three-dimensional cultivation techniques and cell bioreactors will contribute to increase the efficiency and quality of the produced stem cells for transplantation purposes [18]. Based on these basic and translational findings, we will likely witness the outburst of completely new methods for the therapy of osteoarticular diseases, which will definitely enhance the paradigm of current treatment of these debilitating conditions.

20.3 Tissue Engineering and Bioprinting

Tissue engineering, a core discipline of regenerative medicine, involves the use of scaffolds, cells and bioactive molecules to repair or regenerate injured tissues. This approach is particularly useful for musculoskeletal disease, given its ability to intervene beyond the treatment of symptoms. For instance, a possible strategy for cartilage regeneration uses implantation of biomaterials, which are biodegradable scaffolds combined with cells that fit the niche of cartilage lesions (Fig. 20.2). Through this intervention, the cells are induced to create a new tissue and the biomaterials act as a temporary matrix that provides a structural scaffold, a temporary space for cell growth and migration, and a machinery that guides cell organisation and differentiation to mimic the structure of a native tissue. The balance between biocompatibility, porosity and degradation of implantable materials with native tissue formation is a key factor for the success of these implants [19].

In bone repair, scaffolds can be made using hydroxyapatite, biocompatible polymers or composite materials with specific osteoinductive properties and mechanical characteristics that are required to withstand loads during bone regeneration. Tissue engineering is developing these materials by integrating their capacities to provide physical support for anatomy reconstruction and, at the same time, inducing osteogenesis through growth factors that are released in a controlled fashion [20]. The emerging toolbox of regenerative medicine aims to move from the implantation of a scaffold as a physical cue that induced osteogenesis—as in bone- or cartilage-injecting injections, discussed earlier—to the 3D bioprinting of complex structures (Fig. 20.1). In this feasible approach, bioinks composed of living cells and feeding molecules are layered to build complex tissues in their natural architecture. Such 3D bioprinting is revolutionising regenerative medicine because it bypasses the limitations derived from the use of materials that are still synthetic and thus cause chronic, harmful inflammation derived from the formation of an immune barrier. An infinite level of anatomical complexity can be reconstructed by 3D bioprinting. Bioinks close to the mechanical and biological properties of cartilaginous tissue are used to build precise-size implants of articular cartilage using a laser to cure the structural matrix and then to increase the viability of the implanted cells. This technology is revolutionising the treatment of degenerative pathologies, such as osteoarthritis, due to the possibility of providing customised implants according to the patient's anatomy [21]. Targeting joint areas with specific needs based on the geometry of the damage is of huge clinical relevance. Instead of inserting artificial implants, joints could be repaired by creating the osteochondral interfaces, where bone meets with cartilage, for the first time in natural material gradients that allow [22].

However, there is still much to achieve before the numerous challenges associated with tissue engineering, and bioprinting can be overcome, such as sustained vascularisation of the printed tissues (to provide sufficient nutrients for long-term survival and efficient integration with the host tissue), as well as the complex mechanical properties recapitulating musculoskeletal tissues. Despite these hurdles, recent advancements in bioinks, printing techniques and post-processing strategies are markedly overcoming these obsta-

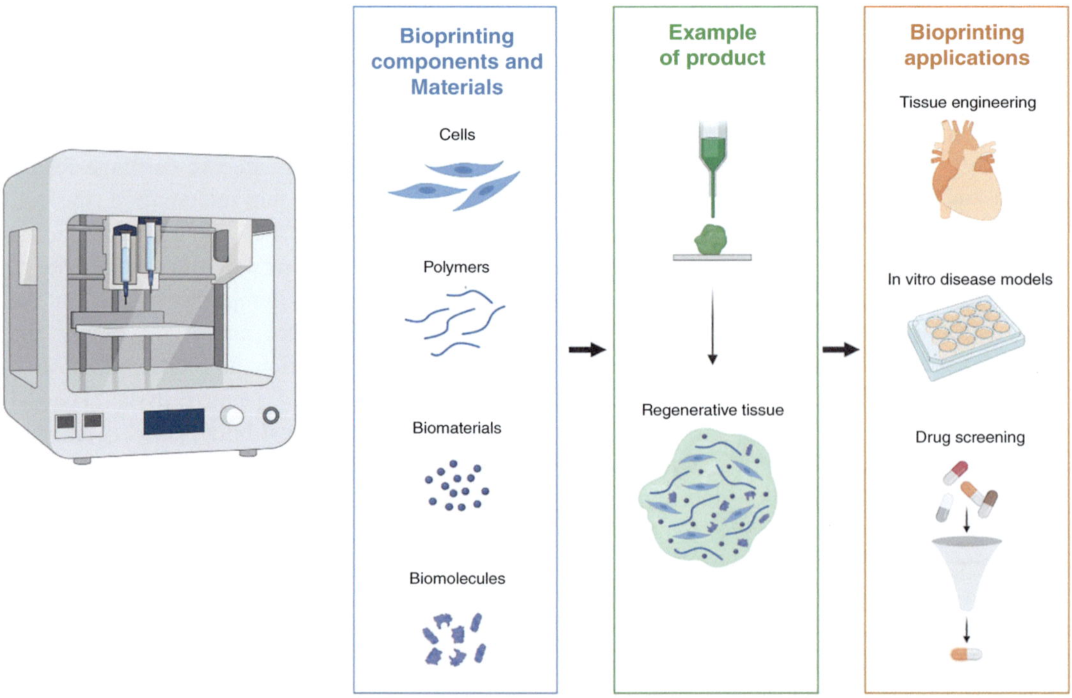

Fig. 20.1 3D bioprinting using bioinks to create cartilaginous tissues, a promising innovation in the treatment of osteoarticular diseases

cles, leading from the laboratory to the hospital bedside. These exponential developments reinforce the transformative potential of tissue engineering and bioprinting in musculoskeletal treatments and highlight a future where personalised and regenerative therapies will not be exceptions but the rule [23].

20.4 Molecular and Gene-Based Therapies

In addition to future strategies with a direct musculoskeletal target, gene therapy can directly treat musculoskeletal diseases by modulating the genetic information to repair or replace mutated or dysfunctional genes or to introduce new genes that help in the regeneration of tissues. This technique can either use viral vectors to deliver genetic material or non-viral vectors. Viral vectors are more widely used for carrying genetic materials into target cells. Some of the most studied viral vectors are adenoviruses, lentiviruses and adeno-associated viruses (AAVs). These are the most promising gene therapy vehicles because the viral structure allows them to enter the cells efficiently. This is particularly useful for musculoskeletal therapies, since they consider the different types of musculoskeletal cells and whether they are dividing or not [24]. These vectors are engineered to reduce the immunogenicity and toxicity that would prevent the use of viruses for gene transfer, making them safe enough for therapeutic use in humans.

Non-viral vectors, such as liposomes and nanoparticles, represent an interesting alternative. They are, in principle, safer than viral vectors, since they have a lower risk of immunogenicity and toxicity and can be safely reapplied in chronic musculoskeletal disease. Although they typically show less transduction efficiency than viral vectors, advances in their development to improve their transfection efficiency and specificity are still on the horizon [25].

Besides gene therapy, peptides and recombinant proteins can stimulate tissue regeneration directly, by promoting cellular processes including cellular proliferation, migration and differentiation. Excellent areas for biomaterial research and development are peptide and recombinant bioactive molecules that can behave as growth factors and stimulate tissue formation in a biomimetic, sustained and controlled way [26]. Many bioactive molecules, including BMP-2 (bone morphogenetic protein 2), have been studied and used in the clinic to stimulate bone formation and fracture healing [27]. These proteins can be encapsulated in scaffolds and biomaterials to enhance regenerative treatments by providing critical biological signals.

mRNA and RNAi (interfering RNA) may also enable tissue regeneration by controlling gene expression: it is possible to cause regeneration by administering mRNA that encodes growth factors or other therapeutic proteins instead of the genes for these proteins themselves. In doing so, we can enjoy the ease that mRNA offers in using nucleic acid biomolecules to control cell functions, without having to bear the concerns regarding direct gene therapy [28]. Furthermore, since mRNA-administered therapeutic proteins are translated quickly and used purely to treat damaged tissues, it is potentially much safer for temporary and controlled administration. mRNA molecules can be translated into a large amount of therapeutic proteins quickly in cells, allowing for efficient regeneration.

RNAi can be employed to silence gene pathways underlying pathological processes in musculoskeletal diseases, leading to a very selective modulation of gene expression in therapy for osteoarthritis, inflammatory rheumatic diseases and other musculoskeletal conditions [29]. Blockage of expression of genes responsible for the pathological inflammatory and degenerative processes (e.g. cytokines like TNFα) can be achieved by appropriate drugs that can reduce progression of disease and improve clinical outcomes. Such RNA strategies have been tested in preclinical assays and are beginning to be explored in clinical trials.

This type of molecule- and gene-based therapy represents a real step forward for regenerative medicine, offering new treatment approaches beyond classical options. Ongoing research and increasing technological development are continually enhancing their efficacy and safety, bringing them ever closer to clinical application.

20.5 Integration of Digital Technologies

Through the use of sophisticated software, digital technologies have enhanced how regenerative medicine studies diseases and develops new therapies. Digital theories of modelling and simulation create virtual access to musculoskeletal diseases and enable researchers and clinicians to gain a better understanding of pathological processes and tissue interactions. Experimental observations in individual patients can be extrapolated to large populations using simulation software, creating a digital archive of experimentation from which thousands of cases can be generated. In collaboration with researchers and clinicians, computer scientists and software engineers use simulation technologies to design and optimise new treatments to treat disease, such as scaffolds for tissue engineering, strategies for administration of stem cells and gene therapy protocols. Models predict how treatments will interact with human tissue and impact disease progression prior to human clinical trials [30]. This type of simulation ensures that potential failures are pinpointed and that treatments can be further refined to maximise efficacy and safety (Fig. 20.2).

Furthermore, treatment can be tailored to fit the characteristic biology of individual patients (e.g., details of joint anatomy) or disease (e.g., pathological progression kinetics) in order to improve therapy effectiveness and control side effects through personalization [31]. Personalisation aims to precision-engineer therapy to patient-specific conditions.

As the hallmark of musculoskeletal diseases is chronic inflammation, wearable devices and other monitoring technologies are indispensable tools towards continuous monitoring of the disease and response to therapy. They allow for a real-time assessment of the disease course and response to ongoing therapy and enable the timely adjustment of therapy should a patient not be responding or spontaneously improving. Sensors can be portable and take measurements of movement, joint pressure or biomarkers in the sweat, indicating inflammation and response to medications [32].

Combined with telemedicine platforms, such devices enable patients and doctors to remotely monitor the treatment effectiveness [33]. Such a collaboration between telemedicine and wearable gadgets will be particularly helpful to those who can't move around for some reason (for example, due to having a paraplegic condition or because they live far from a proper medical centre). A combination of telemedicine and wearable devices allows for a more proactive and preventive approach, supporting early treatment when warning signs of the condition-worsening are detected, all with a view to avoiding serious complications.

Besides improving the quality of treatment and life for patients, these technologies help make better use of medical resources by reducing the burden on healthcare personnel, who no longer need to physically see patients on a regular

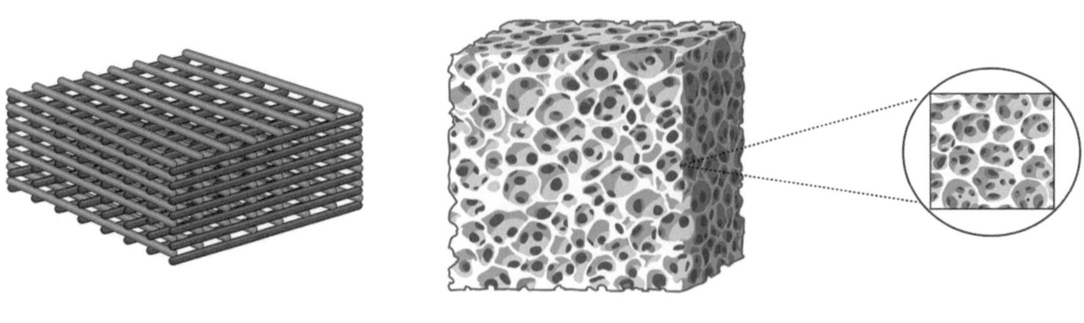

Fig. 20.2 Biodegradable scaffold used in tissue engineering to support cell growth and regenerate damaged tissues

basis, and by enabling more timely and proactive health management by medics. When huge amounts of data can be collected and processed in real time, doctors can discern patterns, make more informed and accurate decisions, and thus improve the efficacy of treatment and reduce intervention costs. Digital technologies therefore hold great promise for regenerative medicine to help improve patient outcomes and make the health system as a whole more efficient.

20.6 Regulatory and Ethical Challenges

With the advent of regenerative medicine, many people are raising valid ethical concerns about the manipulation of stem cells and the genome to cure or treat diseases. While stem-cell medicine is a very broad field, one of the most controversial issues is the use of embryonic stem cells. These cells are optimal for cell therapy, as they are pluripotent (they can differentiate into any cell type in the body). However, embryonic stem cells can be isolated only from embryos, and one has to consider where they come from to proceed with the research endeavour. Questions of origin inevitably lead to a discussion of the physical and intellectual capacities of the embryo, as well as its rights. This unsolved ethical problem is at the centre of intense debate in the scientific and general public communities [34].

These technologies appear to offer new hopes of genetic cure, in cases where a disease is caused by changes (mutations) in single genes: because the success of CRISPR/Cas9 and related technologies tends not to depend on knowing the precise details of the DNA change, small segments of DNA sequence can be targeted by the genome editing system, permitting precise enough correction of many disease-causing mutations [29]. However, these technologies also carry considerable risk of genetic damage, in cases where the edited segment of DNA is not quite right for the job at hand—a phenomenon that is still only partly understood. The molecules of CRISPR/Cas9 (bacterial in origin) frequently and spontaneously misfire and cut DNA where they shouldn't, leading to unwanted 'off-target' changes that can be just as harmful as the original disease. Also, there remains a clear possibility that use of these gene-editing systems, especially in embryos, could result in changes that are passed on to future generations, potentially altering not just the individual human genetic heritage but the entire human species—raising difficult questions about the ethics and prudence of such genetic interventions [35].

Informed consent is another key issue in the ethics of regenerative medicine. Patients need to be properly informed about the risks, benefits and experimental status of many regenerative therapies before they choose to accept treatment. The consent process must be sufficiently detailed to provide important information about the treatment, including uncertainties and potential side-effects. Transparency in novel medical technologies is vital to build and maintain public trust, and protect patients' rights [36].

Their use in new regenerative therapies involves multiple regulatory jurisdictions and sets of guidance, many of which need to be updated as new scientific knowledge emerges. For example, stem cell- and gene-based therapies are regulated differently between different countries and economic blocks, such as the European Union and the United States. This makes harmonisation difficult, but it is essential for clinical trials that cross multiple jurisdictions, which are essential for the safe and effective commercialisation of new therapies. As noted above, harmonised regulation promotes international collaboration and exchange of information and ensures that a drug or therapy is assessed in the same way by different national health authorities [37].

Quality standards and procedures on manufacturing need to be put in place in order to ensure the therapeutic safety and efficacy of regenerative therapies. Regulations are frequently updated to integrate new technologies and procedures that emerge at a breakthrough pace. Standardising manufacturing procedures minimises variability and ensures that products do not vary from one batch to another. For this purpose, the certification of laboratories, validation of procedures and quality assurance are crucial [38].

These regulatory and ethical issues signal that we need engagement and dialogue, not only among scientists and regulators, but also among doctors, patients and the general public, in order to bring new regenerative medicines to the clinic in the optimal and most ethical way. Progress will involve the engagement of all key stakeholders if the complexities and uncertainties are to be resolved. This kind of open and continued dialogue and engagement is fundamental for establishing a regulatory environment that will enable important new therapies to be made available safely and effectively, while preserving patients' rights.

20.7 Future of Regenerative Medicine in Treating Musculoskeletal Diseases

Regenerative medicine is changing rapidly, and there is a constant progression of new innovations that promise a better future for the treatment of musculoskeletal disease. Many of these innovations will streamline regenerative therapies, making them more effective, safer and more accessible to patients. The future of musculoskeletal disease holds unprecedented new hopes for human health.

Thanks to gene-editing techniques that eliminate or repair defective genes, such as CRISPR/Cas9, efforts to improve growth plate homeostasis will probably be attempted in the near future. In parallel with conventional approaches, attention is now focused on how to safely and effectively apply gene-editing techniques to stem cells that will eventually be administered as therapeutic products to correct the genetic defects [39].

Moreover, smart biomaterials able to respond to external signals, that is, on demand, according to pH, temperature or biological stimuli, are created to deliver growth factors or other therapeutic agents on demand. To be clear, this takes place in a controlled manner that provides new avenues for scaffold integration into the tissue, directing implants' performance. Imagine a smart biomaterial that, upon sensing a signal following tissue damage, is able to release growth factors and, in turn, aid the regeneration process in a more specific and effective way [40].

Organoids fabricated in a petri-dish from stem cells in the laboratory are now enabling detailed emulation of musculoskeletal disease. This provides a valuable platform for screening of novel drugs and therapies with the potential to be translated into new therapeutic interventions with optimal efficacy to mimic human disease. Organoids can replicate many of the 3D aspects of human tissue and allow for examination of disease and the creation of novel therapies [41].

Among potential applications, cells release exosomes—nanometric extracellular vesicles—to promotes therapeutic effects by transferring RNA, proteins and lipids. Preclinical studies suggest exosomes derived from stem cells could potentially be used to regulate inflammatory responses and promote tissue regeneration. As a natural messengers that transfer biological information among cells, exosomes can be exploited to enhance cell signalling and accelerate regeneration [42].

Indeed, comparatively few regenerative medicine studies are straightforward, and yet the development of therapies still requires a team effort involving scientists and engineers. Complex, unanswered problems often extend beyond the knowledge of any one researcher, so a multidisciplinary approach is essential. Various cell types are added to the tissue, some are integrated into the regenerated structure, and the rest become themselves isolated cells that can be extracted and cultured for possible use Relationships among scientists, clinicians and other healthcare professionals are important in helping to translate laboratory discoveries into safe and viable clinical therapies. For example, partnerships between bioengineers and molecular biologists are essential for advancing the development of new bioprinting devices, more functional scaffolds and genetic editing tools. This accelerates innovation so that better and more personalised solutions can be provided [43].

Academic institutions, as well as research institutes and hospitals, form the cornerstones of clinical trials testing the safety and efficacy of new regenerative therapies: scientific advances can be brought to the bedside only if research and practice are conducted in interaction. To do this, we must structure and perform clinical trials that identify valid new therapies and evaluate their safety and efficacy before bringing them to the bedside [44].

Third, partnerships with biotechnology and pharmaceutical companies are critical for developing and commercialising regenerative therapies. Biotechnology companies can provide essential financial resources and technical expertise to scale up production of regenerative treatments and help translate innovations from the laboratory to the clinic. Pharmaceutical companies have the resources to navigate regulatory and commercialisation pathways [45].

These innovations and collaborations from both basic and clinical disciplines make regenerative medicine a paradigm shifter to treat musculoskeletal diseases and improve patient quality of life and also transform our field.

20.8 Conclusion

This chapter provides updates on the progress and novel research applications of regenerative medicine in musculoskeletal diseases, with a focus on revealing the main research areas and technologies driving the future of regenerative medicine. The advances in using both mesenchymal stem cells and induced pluripotent stem cell science may facilitate tissue repair for patients who need new joints or biological joint replacements, achieve adipose-derived mesenchymal stem cell augmentation for cartilage repair and maintain a long-lasting supply of joint-relevant cell-based products that may particularly benefit patients with osteoarthritis or bone lesions, and prevent or delay the need for total joint replacement. Moreover, other immunomodulatory cells, such as adipose-derived mesenchymal stem cells, may act as conduits to modulate inflammation, offering new hope for treating these diseases. Tissue engineering and 3D bioprinting have reached the ability to create not only required composite tissues (cartilage, bone) but also recreate an osteochondral interface for repair or augmentation of articular joints; this capability offers us the possibility of more precise and personalised treatments.

Novel therapeutic opportunities are being created through gene therapy, peptides and recombinant proteins, messenger RNA (mRNA), and interfering RNA (RNAi), offering highly specific and physiologically regulated interventions to spur tissue regeneration. The integration of digital technologies that model and simulate, and that aid planning and decision-making, together with monitoring technologies including wearable devices, helps to personalise therapeutic approaches and provides real-time insight into disease progression and treatment outcomes.

These ethical and regulatory considerations as well as their effective application need to be discussed constantly between all the stakeholders to assure that the novel therapies can be applied in a safe and effective manner. The opportunities afforded by the latest innovations and the need for multidisciplinary collaborations have been emphasised to allow continued progress in the treatment of musculoskeletal diseases, offering glimpses of the road ahead with the promise of more effective and personalised therapies.

Regenerative medicine promises to truly revolutionise musculoskeletal disease treatment over the coming years. A few challenges remain, but there is a lot of hope that continued research will increase our knowledge, enhance existing technologies and see further advancement of regenerative therapies. Academic institutions, research centres, biotechnology industries and regulatory agencies must work together to remove barriers and get new therapies to patients and clinics safely and effectively.

Investments in research and development are crucial for discovering new options for these interventions and ensuring that scientific discoveries lead to clinical treatments in a timely manner. Additionally, clinicians who provide access to cell therapy must be trained or re-trained to use new technologies and therapies safely and effectively.

In conclusion, musculoskeletal regenerative medicine is in the pipeline to enhance the care for a wide range of patients suffering from various diseases. The potential of these advances that have been discussed in this chapter will transform patients' lives, provided that the research in this area continues and the global efforts for collaboration. A time will come when not just pain relief but true regeneration and cure will become reality for all patients.

References

1. Dallo I, Lane JG, Marin SV, Gobbi A. Evaluation of malalignment of the knee. In: The art of the musculoskeletal physical exam. Cham: Springer; 2023. p. 477–87.
2. Gobbi A, Whyte GP. Emerging orthobiologic approaches to ligament injury. In: Bio-orthopaedics: a new approach. Berlin: Springer; 2017. p. 313–24.
3. Gobbi A, Whyte GP. Biological augmentation in acute ACL repair. In: Bio-orthopaedics: a new approach. Berlin: Springer; 2017. p. 325–35.
4. Whyte GP, Gobbi A. Biologic knee arthroplasty for cartilage injury and early osteoarthritis. In: Bio-orthopaedics: a new approach. Berlin: Springer; 2017. p. 517–25.
5. Dallo I, Frank RM, Bradsell H, Piuzzi NS, Gobbi A. Overview of orthobiologics and joint function. In: Joint function preservation: a focus on the osteochondral unit. Cham: Springer; 2022. p. 21–31.
6. Jacob G, Nakamura N, Parker D, Lane J, Gobbi A. The use of ortho-biologics in sports medicine: an ISAKOS survey report. ISAKOS newsletter 2024 - current perspectives on arthroscopy, knee surgery & orthopaedic sports medicine, vol. 1, 2024, pp. 12–15.
7. Smith JR, et al. Role of mesenchymal stem cells in osteoarthritis treatment. J Orthop Res. 2017;35(3):487–95.
8. Caplan AI. Mesenchymal stem cells: time to change the name! Stem Cells Transl Med. 2017;6(6):1445–51.
9. Caplan AI. Mesenchymal stem cells: the past, the present, the future. Cartilage. 2010;1(1):6–9.
10. Yamashita A, et al. Cartilage tissue engineering using human iPS cells. BioRes Open Access. 2014;3(6):265–73.
11. Otsuka T, et al. Disease modeling of human familial hypercholesterolemia using induced pluripotent stem cells. Biochem Biophys Res Commun. 2010;394(3):603–7.
12. Yoshida Y, Yamanaka S. iPS cells: a source of cardiac regeneration. J Mol Cell Cardiol. 2011;50(2):327–32.
13. Wang L, et al. Human umbilical cord mesenchymal stem cell treatment alleviates collagen-induced arthritis by inhibiting Tfh cell expansion and reducing IL-21 production. Arthritis Res Ther. 2018;20(1):188.
14. Trounson A, McDonald C. Stem cell therapies in clinical trials: progress and challenges. Cell Stem Cell. 2015;17(1):11–22.
15. Wagner W, et al. Aging and replicative senescence have related effects on human stem and progenitor cells. PLoS One. 2009;4(6):e5846.
16. Bianco P, et al. Regulatory considerations for the clinical and research use of transplanted or transfused human cells, tissues, or cell- or tissue-based products. Health Policy. 2011;102(3):282–95.
17. Cox DBT, et al. Therapeutic genome editing: prospects and challenges. Nat Med. 2015;21(2):121–31.
18. Martin I, et al. Bioresorbable scaffolds in tissue engineering. Mater Sci Forum. 2016;250:124–30.
19. Hollister SJ. Porous scaffold design for tissue engineering. Nat Mater. 2005;4(7):518–24.
20. O'Brien FJ. Biomaterials & scaffolds for tissue engineering. Mater Today. 2011;14(3):88–95.
21. Murphy SV, Atala A. 3D bioprinting of tissues and organs. Nat Biotechnol. 2014;32(8):773–85.
22. Duan B, Hockaday LA, Kang KH, Butcher JT. 3D bioprinting of heterogeneous aortic valve conduits with alginate/gelatin hydrogels. J Biomed Mater Res A. 2013;101(5):1255–64.
23. Mironov V, Trusk T, Kasyanov V, Little S, Swaja R, Markwald R. Biofabrication: a 21st century manufacturing paradigm. Biofabrication. 2009;1(2):022001.
24. Evans CH, et al. Gene delivery to bone. Adv Drug Deliv Rev. 2012;64(12):1331–40.
25. Ramaswamy S, et al. Non-viral delivery systems in gene therapy. Gene Ther. 2014;21(2):247–59.
26. Martino MM, et al. Engineering the growth factor microenvironment with fibronectin domains to promote wound and bone tissue healing. Sci Transl Med. 2011;3(100):100ra89.
27. Bessa PC, et al. Bone morphogenetic proteins in tissue engineering: the road from the laboratory to the clinic, part I (basic concepts). J Tissue Eng Regen Med. 2009;3(1):1–13.
28. Sahin U, et al. mRNA-based therapeutics—developing a new class of drugs. Nat Rev Drug Discov. 2014;13(10):759–80.
29. Aagaard S, Rossi JJ. RNA interference therapeutics for cancer: challenges and opportunities. Mol Ther. 2012;20(7):1385–94.
30. Boccaccini AR, et al. 3D printing of composite tissue engineering scaffolds. Mater Today. 2018;21(8):861–74.
31. Hollister SJ. Porosity and pore size control in scaffold engineering. J Biomed Mater Res Part B. 2003;64(2):65–74.
32. Park S, et al. Wearable sensors for monitoring the physiological and biochemical profile of the athlete. NPJ Dig Med. 2019;2(1):72.
33. Mehta P, et al. Integration of wearable technologies into patients' electronic medical records for remote monitoring. J Med Internet Res. 2015;17(4):e97.
34. Lomax GP, et al. The ethical implications of human embryonic stem cell research. Sci Policy Forum. 2006;312(5774):1445–7.
35. Baltimore D, et al. Biotechnology: a prudent path forward for genomic engineering and germline gene modification. Science. 2015;348(6230):36–8.

36. King NMP. Ethical issues in stem cell research and therapy. Stem Cell Res Ther. 2014;5(4):85.
37. Sipp D, et al. Regulatory and ethical considerations for linking clinical and non-clinical data. Lancet. 2015;386(9992):2329–32.
38. Taylor PL. Regulation of clinical stem cell and exosome-based therapies. Nat Biotechnol. 2016;34(8):774–81.
39. Doudna JA, Charpentier E. The new frontier of genome engineering with CRISPR-Cas9. Science. 2014;346(6213):1258096.
40. Chen FM, Liu X. Advances in biomaterials for tissue engineering. Biomed Mater. 2016;11(1):014113.
41. Lancaster MA, et al. Cerebral organoids model human brain development and microcephaly. Nature. 2013;501(7467):373–9.
42. György B, et al. Therapeutic applications of extracellular vesicles: clinical promise and open questions. Annu Rev Pharmacol Toxicol. 2015;55:439–64.
43. Griffith LG, Swartz MA. Capturing complex 3D tissue physiology in vitro. Nat Rev Mol Cell Biol. 2006;7(3):211–24.
44. Robey PG, et al. Resource centers for human stem cell research. Nat Biotechnol. 2004;22(5):629–33.
45. Hynes RO, Lane TF. The road to commercialization. Nat Med. 2005;11(4):S24–7.

New Approach with Personalized Platelet-Rich Plasma

21

Melanio Acosta IV, Mikel Sánchez, Diego Delgado, Laura Rehak, Leandra Bizzoco, and Alberto Gobbi

21.1 Introduction

In recent years, there has been great interest in minimally invasive regenerative treatments, aimed to improve pain, enhance healing, and reduce recovery time of injuries. Platelet-rich plasma (PRP) has gained much traction and popularity in the world of orthopaedics as more studies and information have become available regarding the optimization and efficacy of use. Among the commonly used orthobiologics in orthopedic practice, PRP is the most popular and commonly used in joint disease, cartilage, tendon, muscle, ligament, and meniscus lesions [1]. PRP refers to the plasma fraction of centrifuged whole blood whose platelet concentration is above the baseline prior to centrifugation [2]. Within the α-granules of the platelets, there are cytokines and growth factors, such as epidermal growth factor, vascular endothelial growth factor (VEGF), endothelial cell growth factor, transforming growth factor–β (TGFβ), platelet-derived growth factor (PDGF), insulin-like growth factor (IGF-I, IGF-II), and fibroblast growth factor (FGF), that enhance tissue healing through angiogenesis, neovascularization, increased blood supply and nourishment to neighboring cells, cell proliferation, chemotaxis, and cell differentiation. By isolating and concentrating these beneficial cytokines and growth factors, healing can be enhanced when introducing them to sites of injury [3]. This chapter aims to discuss PRP and its specific uses in orthopaedic treatment.

21.2 Biology of PRP

When discussing PRP, platelets and their constituents have a vital role in driving the regenerative cascade. Platelets are discoid-shaped, non-nucleated cell elements (2–3 μm) that have hemostatic, inflammatory, and immunological properties [4] with a typical half-life of 7 days *in vivo*. Platelets are formed and released into the blood stream as hematopoietic progenitor cells called megakaryocytes that reside within the red marrow. A single megakaryocyte can produce thousands of platelets through a series of remodeling events.

Platelets have outer membranes covered with glycoprotein receptors and adhesion molecules

M. Acosta IV (✉) · L. Bizzoco · A. Gobbi
O.A.S.I. Bioresearch Foundation Gobbi N.P.O., Milan, Italy
e-mail: melanio.acosta@oasiortopedia.it

M. Sánchez
Arthroscopic Surgery Unit, Hospital Vithas Vitoria, Vitoria-Gasteiz, Spain

D. Delgado
Advanced Biological Therapy Unit, Hospital Vithas Vitoria, Vitoria-Gasteiz, Spain

L. Rehak
Athena Biomedical Innovations, Florence, Italy

A. Gobbi et al. (eds.), *Regenerative Medicine in Sports and Orthopaedics*,
https://doi.org/10.1007/978-3-031-84693-9_21

which are responsible for platelet adhesion and activation. Platelet activation can be induced rapidly through the binding of various molecules such as thrombin, calcium chloride, and calcium gluconate or more slowly and gradually when bound to molecules like type I collagen (typically seen in non-activated PRP).

Following platelet activation, platelets release many various molecules contained in their different types of granules that are important for hemostasis, immune function, inflammation, and regeneration. There are three important granules that characterize platelet function:

- α-granules: The most abundant and discussed platelet structure. These granules contain growth factors such as epidermal growth factor (EGF), fibroblastic growth factor (FGF), hepatocyte growth factor (HGF), platelet-derived growth factor (PDGF), transforming growth factor (TGF), and vascular endothelial growth factor (VEGF) and, in a lesser extent, angiogenesis regulators and coagulation factors, particularly fibrinogen, factor V, factor IX factor XIII, and von Willebrand factor (vWf). Other factors and molecules that make up the platelet secretome such as chemokines and cytokines are b-TG, CCL2 (MCP-1), CCL3 (MIP1a), CCL5 (RANTES), CTAP-III, CXCL1 (GRO-a), CXCL12 (SDF-1a), CXCL4 (PF4), CXCL5 (ENAÀ78), CXCL7 (PBP), CXCL8 (IL-8), NAP-2, and P-selectins. These granules are mainly involved in the recruitment and activation of immune cells [5–9].
- Dense granules: Contain high levels of calcium, magnesium, and potassium that have roles in the coagulation role of platelets. Through the release of various molecules such as histamine, ADP polyphosphates, serotonin, and epinephrine, these granules have immunomodulatory functions and help provide a link between innate and adaptive immune responses [5, 10, 11]. Components of these dense granules such as glutamate and serotonin can stimulate monocytes to differentiate into dendritic cells, as well as drive the migration of T lymphocytes.
- Lysosomes: There is a paucity of literature regarding the release of lysosomes and their contents from platelets. These lysosomes have been shown to play extracellular roles such as supporting vessel remodeling, degradation of extracellular matrix, and fibrinolysis. It is through these mechanisms lysosomes and their proteases act on the remodeling and maintenance of soft tissues such as tendons and ligaments as well as work jointly with macrophages in their bactericidal activity through phagocytosis [12, 13].

It is vital that the release of growth factors is consistent, slow, and long-lasting to maximize the recruitment and immunomodulatory effects of platelets on immune cells. These growth factors act differently depending on the site they are administered and the preparation method of PRP. For example, non-activated PRP is known to induce bone and cartilage development in vivo and in vitro, while pre-activated PRP reduces osteogenesis and chondrogenesis [14].

21.3 Leukocyte-Rich and Leukocyte-Poor PRP

One important, rarely discussed aspect is the PRP types used, mainly the use of leukocyte-rich (LR-PRP) and leukocyte-poor PRP (LP-PRP). The difference in leukocyte contents between the two types may elicit different physiologic effects on the body. For example, when LR-PRP is cultured with tendon stem cells from viable patellar tendons of rabbits, this induced catabolic and inflammatory changes in the differentiated tenocytes with increases in matrix metalloproteinase (MMP) 1, MMP-13, and pro-inflammatory cytokines IL-1b, IL-6, and TNF-a. While in the same models, LP-PRP induced anabolic effects, increasing the gene expression of alpha-smooth muscle actin and types I and III collagen. These outcomes suggest that LR-PRP may be detrimental in tendon injuries as they promote pro-inflammatory and catabolic activity, while LP-PRP may be more beneficial with their more anabolic- driven function [15, 16].

Mononuclear cells, comprised of natural killer cells, monocytes, and lymphocytes (B and T), have important roles in tissue healing. Monocytes in particular are vital as they have the ability to switch phenotypic expression and differentiate into particular tissue macrophages depending on their environment. These tissue macrophages act through phagocytosis of apoptotic cells and provide protection through innate immunity. While adaptive immunity is initiated through interactions between macrophages and activated T and B lymphocytes which promotes and enhances antimicrobial resistance [17–19].

Macrophages typically have two phenotypes—M1 and M2. M1 phenotypes typically function like proinflammatory cells, while M2 phenotypes have anti-inflammatory functions through type 2 immune responses. Type 2 immune responses act through activation and increasing numbers of IL-4, IL-5, IL-9, and IL-13 and depending on the setting may have a role in host protection, pathogenicity, as well as tissue healing and regeneration. The M1 phenotypes produce nitric oxide that may inhibit cell proliferation and have microbicidal activity through the release of TNF-α and IL-6 compared to M2 phenotypes where cell proliferation is promoted and enhanced driving tissue regeneration and repair through polyamine and collagen synthesis and the release of IL-4 and IL-10. M2 phenotypes have also been shown to express increased levels of cytokines and growth factors that stimulate angiogenesis [20–22]. It is also important to note that macrophages with a type 1 phenotype (M1) have the ability to polarize into the M2 phenotype through local environment signaling, further promoting tissue regeneration and healing. The use of PRP induces this environmental drive for M1 to M2 polarization [23].

21.4 No Place for Red Blood Cells in PRP

Red blood cells (RBCs) are anuclear cells with a bound heme protein that binds to iron. Through this heme and iron binding, they transport oxygen and carbon dioxide in the blood stream [24]. Damage to red blood cells and their membrane may release the contained iron, hemin, and hemoglobin, which may activate an inflammatory cascade and cause cytotoxicity due to oxidative stress, and loss of nitric oxide [25]. RBCs in PRP may also lead to eryptosis which can stimulate secretion of macrophage migration inhibitory factor (MMIF). MMIF inhibits the migration of monocytes and macrophages, as well as creating a pro-inflammatory environment impeding fibroblast proliferation, migration of stem cells, and eventually leading to cell dysfunction [24, 26]. The need to eliminate or minimize the presence of RBCs in PRP is vital in the centrifugation and processing steps when preparing PRP [27].

21.5 Reporting on PRP

There is a vast amount of literature regarding PRP; however, there is no standardization on the description, characterization, and preparation of PRP leading to heterogeneity between studies. In 2024 a collaboration of many worldwide experts in PRP and orthobiologics established a consensus to address some of these issues. Here, a strong consensus [28] was reached for the following recommendations:

1. PRP classification should be based on platelet count, leukocyte count, red blood count, activation method, and pure plasma versus fibrin matrix.
2. Important characteristics that should be reported in research studies are platelet count, leukocyte count, neutrophil count, red blood

cell count, total volume, the volume of injection, delivery method, and the number of injections.
3. The prognostic factors that should be reported are age, BMI, severity/grade of pathology, chronicity of pathology, prior injections and response, primary diagnosis (primary vs. postsurgery vs. post-trauma vs. psoriatic), comorbidities, and smoking history.
4. There is no minimum or maximum age or BMI, but clinical judgment should be used at the extremes.

PRP is typically described by absolute platelet concentration, shifting away from the original description of a platelet concentration simply above baseline values. It was first described by Marx et al. that a minimum platelet count of $1 \times 10^6/\mu L$ was required to exhibit improvement in soft tissue and bone healing [29]. Giusti et al. then expounded that a threshold of 1.5×10^9 platelets/mL is necessary to induce angiogenic responses in tissue repair mechanisms; however, excessively high concentrations reduced the angiogenic potential of platelets in perifollicular and follicular angiogenesis [30]. Thus, an optimal dose should be attained to stimulate cell proliferation, synthesize mesenchymal and neurotrophic factors, contribute to chemotactic cell migration, and stimulate immunomodulatory activities [31, 32].

21.6 PRP in Cartilage Disease

There are many different options when treating cartilage disease all with the same goal: to restore normal hyaline cartilage and restore normal mechanical properties [33, 34]. PRP is a biologic nonsurgical option toward this goal. However, PRP may have different biologic activity depending on the preferred site of action. With regard to cartilage regeneration, PRP contains various pro-chondrogenic growth factors, including BMP-2, bFGF, IGF-1, and TGF-β1, but may also contain antichondrogenic growth factors, such as IGF binding proteins, PDGFs, VEGF, and EGF [35]. Despite containing chondrogenic growth factors such as TGF-β1 that may stimulate type 1 collagen production in skin, synovium, and tendons, certain TGF-β1 concentrations in PRP may inhibit collagen production. The ideal platelet concentration may differ based on the intended lesion. With regard to cartilage regeneration, some authors noted that a "therapeutic dose" may fall within a range of 2–6 times higher than the baseline platelet count [36], whereas a lower platelet concentrations may produce suboptimal effects, while higher concentrations may have inhibitory effects. This emphasizes the intricate balance between anti-chondrogenic effects and pro-chondrogenic effects of platelets in PRP. Many authors and recent consensus also emphasize the importance of not just administering a single injection of PRP, but completing a full regimen cycle of repeated intra-articular injections over a course of time [37, 38].

Laver et al. [35] documented seven randomized controlled trials [39–45] showing significantly superior results in pain and functional scales for those patients given PRP compared to controls who were given hyaluronic acid, saline, or underwent an exercise program. Several studies [42, 45–48] also demonstrated better results when PRP is given to patients with early osteoarthritis and young age. However, there were other studies that showed improvement of clinical results in patients with osteoarthritis as high as Kellgren-Lawrence grade 3 [46].

In 2024, the European Society of Sports Traumatology, Knee Surgery and Arthroscopy (ESSKA) established the ORthoBIologics InitiaTive (ORBIT) which published a consensus on the use of blood-derived orthobiologics for the use in management of knee osteoarthritis [38]. Among the statements and questions discussed, they noted three grade A recommendations:

> (1) There is enough preclinical and clinical evidence to support the use of PRP in knee osteoarthritis. (2) Clinical evidence has shown the effectiveness of PRP in patients for mild to moderate degrees of knee osteoarthritis (Kellgren-Lawrence ≤3). (3) PRP injections have been shown to provide a longer effect in comparison to the short-term effect of CS injections.

The consensus also had many grade B recommendations: (1) There is support of the use of PRP over hyaluronic acid for knee osteoarthritis due to the evidence showing overall clinical improvement and expected long-lasting effects. (2) Leukocyte-poor and leukocyte-rich PRPs are both valid options for management of knee osteoarthritis, and there is no consensus support of use of one type over the other. (3) A range of 2–4 injections of PRP are recommended for treatment of knee osteoarthritis. (4) Interval ranges of 1–3 weeks are recommended for the interval between each injection of PRP as the main period of released growth factors occurs within the first 3 weeks from injection. (5) There is not enough evidence for the suggestion of combination of PRP and cell-based therapy over PRP or cell-based therapy used alone.

21.7 PRP in Ligament Injury

Ligament healing encompasses a series of events that results in a ligament which is more scar-like in character compared to native uninjured tissue. Ligament healing involves an inflammatory, proliferative, and remodeling phase. Immediately after injury, the inflammatory phase initiates and is characterized by an influx and accumulation of inflammatory cells, typically neutrophils, macrophages, monocytes, and lymphocytes. Here these immune cells clear injured tissue debris and secrete cytokines to start a healing cascade. Following the inflammatory phase, the proliferative phase consists of the recruitment of fibroblasts, endothelial cells, and additional macrophages which eventually form granulation tissue around the injured area which serves as a provisional scaffold for healing. The remodeling phase is the last phase of healing, where inflammatory cells, fibroblasts, and endothelial cells that were recruited to the site of injury now diminish and return to basal levels. This phase may last several months and the cellular profile of this healing ligamentous tissue may resemble more like normal tissue compared to those in the previous inflammatory and proliferative phases. Despite having a similar cellular profile to normal tissue, this tissue will more resemble a scar-like neo-ligamentous tissue, indicating healing may never healing may never result in regaining normal native tissue [49–51].

PRP has been used as a biologic agent to speed up the natural healing process and regeneration of tissue. Evidence shows that the actions of cellular components of PRP such as leukocytes, platelets, fibrinogen, growth factors, proteases, and hormones may provide pain relief, enhance and accelerate repair of ligaments, and allow patients to return to function faster [52–60]. This is through the actions which enhance cell proliferation, chemotaxis, migration, cellular differentiation, and extracellular matrix synthesis [61, 62].

Studies show mixed outcomes in regard to the use of PRP in ligament injuries. Some articles have found no significant benefit in the use of PRP and ligament healing [63, 64], seen in cases of ACL reconstruction [8, 65–67] and in *in vivo* skeletally immature animal models undergoing ACL repair [68, 69]. While others have demonstrated that growth factors in PRP, when combined with a synthetic scaffold or enhanced with bone marrow stimulation, can enhance ACL repair [70–72].

There are many different systems and protocols for the preparation and administration methods of PRP yielding different compositions. The lack of reporting standards including PRP preparation, composition, and administration may be one reason for the variability of results seen among studies [73–77]. Because of this, recent consensus has suggested the importance of recording baseline cellular and platelet compositions in whole blood and of the PRP after preparation to improve quality control measures in clinical research [38].

21.8 PRP in Tendinopathy

Tendon healing and regeneration undergoes a similar overlapping three-stage process as ligament healing—(1) inflammation, (2) proliferation, and (3) remodeling. After tendon injury, lasting approximately 48 hours, the inflammation stage comprises erythrocyte, leukocyte, platelet, and endothelial chemoattractant infiltration at the site of injury. Macrophages are at work consuming necrotic and non-viable tissue. The proliferation stage, which lasts a period of 7–21 days, consisting of macrophage and tenocyte-directed synthesis of new tendon tissue, results in less durable collagen III. During this stage, tenocytes also proliferate primarily in the epitenon. Following the proliferation stage, the remodeling phase typically lasts longer than 12 months. Collagen I replaces collagen III, resulting in more organized alignment of the extracellular matrix as collagen fibers undergo maturation and reorient themselves parallel to the direction of mechanical stress [78]. The stages of tendon healing occur around the mechanism of extrinsic and intrinsic factors. The extrinsic mechanisms of healing involve infiltration of inflammatory cells and fibroblasts which predominate early in the healing process, while intrinsic mechanisms involve cells originating from the tendon and epitenon which commence later [79]. Similar to ligament healing, the newly regenerated tissue is more scar-like in character with its formation of fibrovascular scar tissue having lower mechanical properties, making this tissue biomechanically inferior to original healthy tendon [80–82].

The use of PRP with the aim to provide a supraphysiological concentration of growth factors to the injury site, promoting tendon healing and potentially accelerating the recovery process, has been applied to various types of tendon injuries and pathologies such as rotator cuff tears, lateral epicondylitis, Achilles, and patellar tendinopathies [83–85]. There are no guidelines on use of PRP in tendinopathies as studies in the last decade show conflicting results on the benefit of PRP [86–92]. However, there have been recent studies and consensus showing promising results showing leukocyte-rich PRP having increased induction of growth factor release and tenocyte proliferation, as well as improved histological findings, molecular profiles, and mechanical function [4, 93–96].

21.9 PRP in Meniscal Lesions

The use of mechanical biologic enhancements as well as orthobiologics in meniscal injuries has grown in the past few years [62, 97]. Looking back half a decade ago, there was very little research on the efficacy of PRP in meniscal lesions, with only a single study being well-known and cited [98]. Due to the lack of investigations and studies, no specific guidelines or recommendation have been established [97]; however, this realization that there is a gap in our understanding of PRP and meniscal lesions sparked many studies thereafter. Since then, there have been many studies showing the effectiveness of PRP in improving clinical, functional, and radiographic outcomes when used in a non-surgical manner and when used in conjunction with surgical repair [76, 99–104, 105]. The use of ultrasound guidance to aid in percutaneous administration of intra- and perimeniscal PRP has also been investigated to further optimize the efficacy of PRP in meniscal lesions. The use of a more precise approach that targets the trigger tissue rather than the whole joint has proven to be an effective means of improving pain and functional scores while minimizing progression of subsequent osteoarthritis [102]. However, with regard to the use of leukocyte-poor versus leukocyte-rich preparations of PRP for meniscal lesions, there is no literature supporting one preparation over the other.

21.10 PRP in Muscle Healing

During skeletal muscle injury, skeletal muscles have a very high capability for repair through the presence of satellite cells or adult muscle stem cells. Healing takes place through three phases—(1) degeneration/inflammation phase, (2) regeneration phase, and (3) remodeling phase [106]. The degeneration/inflammation stage occurs immediately after injury caused by rupture and necrosis of myofibers. Here, there is a formation of a hematoma and infiltration of inflammatory cells. After muscle deterioration, the first inflammatory cells to enter the lesion are neutrophils. A large number of pro-inflammatory molecules such as chemokine (CCL2, CCL17), cytokines (IL-6, TNF-α), and growth factors (FGF, IGF-I, HGF, TGF-β1, VEGF) are secreted, attracting additional inflammatory cells such as monocytes and macrophages [107, 108]. During the regeneration phase, damaged tissue is taken up and phagocytized, followed by regeneration of myofibers and activation of satellite cells. Lastly, during the remodeling phase, there is maturation of the myofibers where there is recovery of the muscle's functional capacity. During this phase, there is also scar tissue and fibrosis formation in the area.

Muscle regeneration and myogenesis are dependent upon paracrine healing and growth factors, such as fibroblast growth factor 2 (FGF-2), hepatocyte growth factor (HGF), insulin-like growth factor-1 (IGF-1), platelet-derived growth factor (PDGF), prostaglandins (PG), transforming growth factor β 1 (TGFβ-1), and tumor necrosis factor-α (TNF-α). PRP supplies a concentrated release of these growth factors which can advance muscle healing. It has been shown that preparations of PRP releasate has helped improve skeletal muscle cell proliferation through the upregulation of protein such as cdk1, cdk2, PCNA, cyclin A1, and cyclin B1 [109]. Other studies have shown that PRP has improvement on pain control, inflammation, quicker full functional recovery, and significantly decreased time for return to sport [110–114]. There is currently no guideline or consensus with regard to superiority between leukocyte-poor and leukocyte-rich PRP; however, a study by Chen et al. may start the investigation. They investigate the similarly positive effects of leukocyte-poor and leukocyte-rich PRPs on a molecular level showing that both increase the differential gene expressions of myoblasts as well as promote their proliferation through the upregulated expression of cyclin A, cdk1, cdk2, and MyoD1 [115].

21.11 Personalizing PRP Uses

Platelet-rich plasma may be used in many different orthopedic pathologies; however, it is far from a one-size-fits-all remedy. As there are many various systems and preparations of PRP, one must take a personalized approach to its utilization by correlating pathology with preparation. Not every pathologic lesion or entity responds the same and not every preparation works equally on each pathology.

As in osteoarthritis, the ESSKA-ORBIT collaborative investigation has concluded that there is no superiority between leukocyte-rich and leukocyte-poor preparations of PRP. Conversely, in regard to tendinopathies, there are studies that demonstrate the superiority of leukocyte-rich over leukocyte-poor PRP in Achilles models [93], while the converse conclusions are seen with respect to rotator cuff [115–117]. There is still a paucity of data regarding the ideal preparation for each specific pathology, which calls for the need of additional investigation.

There are a limited number of studies that investigate the importance of platelet concentration, absolute platelet count, and platelet volume. This is exemplified by the discussion of PRP and degenerative cartilage disease by Laver et al. [35], as they discussed the paradoxical action of growth factors in relation to platelet concentration, where the platelet concentration of PRP preparations have crucial roles on their effect on different conditions and injuries. There is emphasis on a "therapeutic dose" range, as moderately elevated platelet concentrations have been suggested to induce optimal biological benefit, while lower platelet concentrations produce suboptimal effects and higher concentrations produce inhibitory effects. A recent study by Patel et al. [118] shows that with a fixed concentration of leukocyte-poor PRP, there was significant improvement in WOMAC, VAS, KOOS scores, and patient satisfaction in patients who received the higher volume PRP, thus higher absolute value of platelets.

Our own experience and philosophy with PRP is that it should not be a "one-size-fits-all" therapy. As mentioned in this chapter and further elaborated in the following chapter, there are many different types of preparations for PRP. Our personal goal is to unlock the potential that hides within each syringe and tube of extracted peripheral blood and to "personalize" each preparation to meet the needs of each patient for their individual and specific complaints at the time of visit. Through our in-house use of a personal hematology analyzer (Horiba, Kyoto, Japan), we can analyze the valuable products yielded from various preparation techniques and systems (Fig. 21.1). We do not utilize just one system or just one way to process and prepare our PRP solution, but take advantage of many different techniques and systems to yield the ideal type of PRP which we need to best manage our patients on a very personal and individual level (Fig. 21.2).

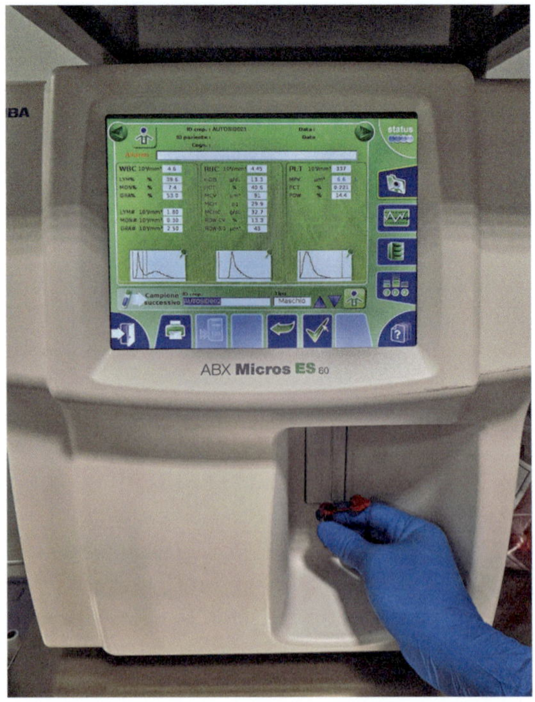

Fig. 21.1 Through the use of an in-house hematology analyzer, white blood cells, red blood cells, and platelet counts are determined from the peripheral blood extracted as well as the different products produced through various PRP preparation systems and methods

Through years of experience, we have developed an operational system of analyzing peripheral blood, as well as all the different types of products corresponding to the preparation technique and system used during each visit such as PPP layers, buffy coat layers, single-spin products, and products yielded from multiple-spin techniques. By analyzing the baseline characteristics of the peripheral blood and comparing it to the endless possibilities of products that can be produced, we can fine-tune each and every solution prior to administering it to our patients (Fig. 21.3). Through the preparation of a leukocyte-rich PRP, we can take advantage of the macrophages, specifically type 2 macrophages (M2), and control inflammation, prevent scar formation, and create an environment ideal for the regeneration and healing of connective tissue such as ligaments and tendons. While in patients with different pathologies such as cartilage and arthritic lesions, we may prefer a solution devoid of leukocytes with higher concentrations of platelets. Just as each and every patient is not the same, each and every preparation should not be the same as well. We hope to spread our philosophy of personalizing PRP and emphasize the importance of customizing each PRP preparation with regard to their use in different pathologies. Through future investigations on personalizing the different components of a PRP solution, we can better understand which preparation can best treat the different pathologies of our patients.

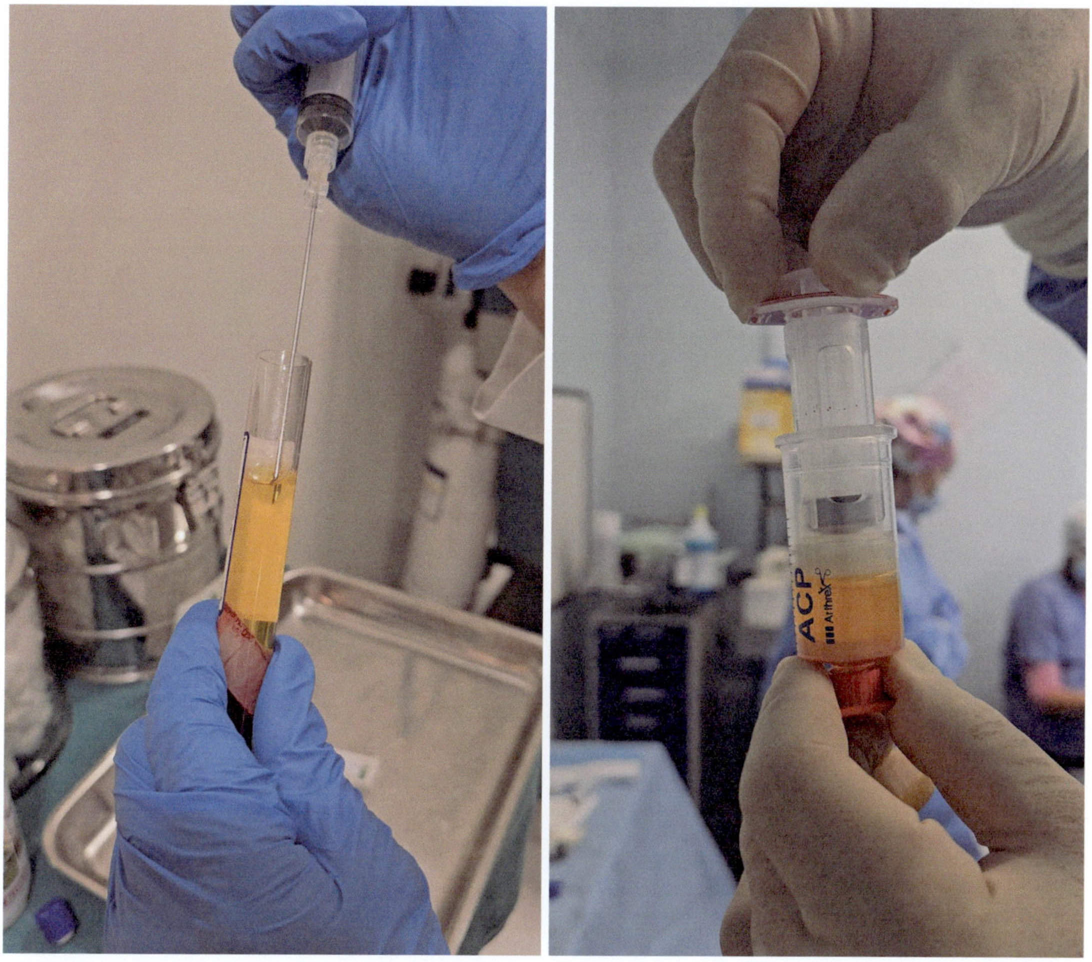

Fig. 21.2 Through the use of various preparation techniques and systems, personalized platelet- rich plasma and similar products can be achieved to suit each patient's individual needs

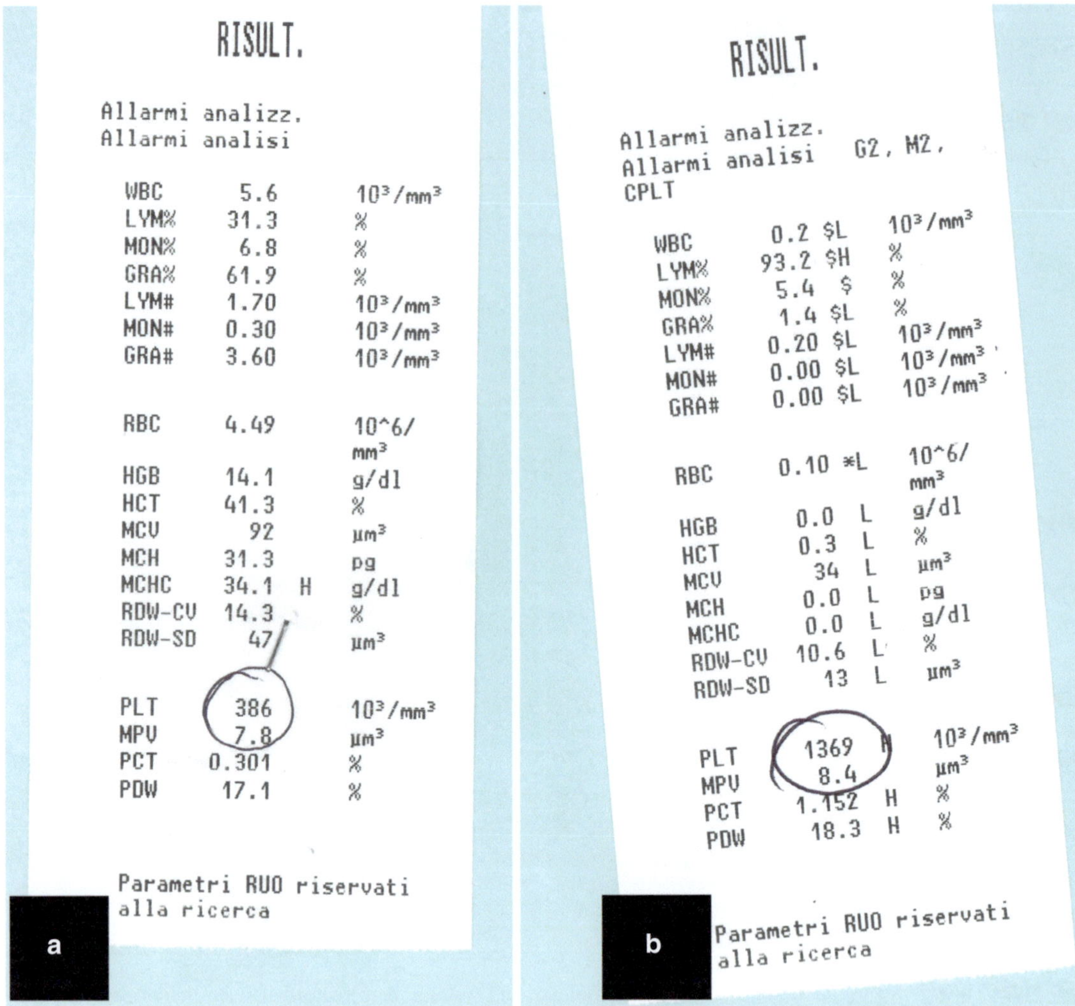

a

RISULT.

Allarmi analizz.
Allarmi analisi

WBC	5.6		10³/mm³
LYM%	31.3		%
MON%	6.8		%
GRA%	61.9		%
LYM#	1.70		10³/mm³
MON#	0.30		10³/mm³
GRA#	3.60		10³/mm³
RBC	4.49		10^6/mm³
HGB	14.1		g/dl
HCT	41.3		%
MCV	92		µm³
MCH	31.3		pg
MCHC	34.1	H	g/dl
RDW-CV	14.3		%
RDW-SD	47		µm³
PLT	386		10³/mm³
MPV	7.8		µm³
PCT	0.301		%
PDW	17.1		%

Parametri RUO riservati alla ricerca

b

RISULT.

Allarmi analizz.
Allarmi analisi G2, M2,
CPLT

WBC	0.2	$L	10³/mm³
LYM%	93.2	$H	%
MON%	5.4	$	%
GRA%	1.4	$L	%
LYM#	0.20	$L	10³/mm³
MON#	0.00	$L	10³/mm³
GRA#	0.00	$L	10³/mm³
RBC	0.10	*L	10^6/mm³
HGB	0.0	L	g/dl
HCT	0.3	L	%
MCV	34	L	µm³
MCH	0.0	L	pg
MCHC	0.0	L	g/dl
RDW-CV	10.6	L	%
RDW-SD	13	L	µm³
PLT	1369	H	10³/mm³
MPV	8.4		µm³
PCT	1.152	H	%
PDW	18.3	H	%

Parametri RUO riservati alla ricerca

Fig. 21.3 (**a**) Analysis of peripheral blood to determine baseline platelet levels. (**b**) Analysis of PRP to determine concentrated platelet counts, leukocyte-poor (or rich) preparations, and to confirm red blood cells are kept to a minimum

21.12 Future Trends of PRP

Since its beginnings, PRP has experienced a great proliferation both in the field of research and in its application in different medical specialties. Despite its rapid growth, this treatment has not seen major technical or technological advances. Their safety, ease of obtaining and application, and promising results may have masked the need for further scientific advancement in this field. However, as knowledge and experience in the use of the PRP increases, new challenges are being identified necessitating improvement in the techniques and systems of preparations. These new improvements or developments may affect different aspects of the PRP, some of which are described below.

21.13 Pharmaceutical Form

In addition to the aforementioned advantages of PRP, one of its best properties is versatility. This allows PRP to be applied in many different formulations, namely liquid and solid forms. The first is most widely used as it is the formulation of PRP injections most utlized in different pathologies and tissues. Furthermore, taking into account the coagulation process, PRP can be applied in a solid form by means of fibrin scaffolds, using clots or membranes. As a result, solid forms of PRP can be applied to pathologies where the action of liquid injections may be limited, such as in the treatment of wounds [119] or in surgical interventions [120].

In addition to these two main types of pharmaceutical forms, the possibility of formulating PRP in a nebulized form has recently been explored. Findings from *in vitro* laboratory studies showed that nebulization of platelet lysate, obtained from PRP, did not lead to decreased levels of growth factors. In addition, its bioactivity was also maintained, stimulating cell proliferation in lung fibroblasts [121]. This method of formulating PRP would allow it to act more effectively on pulmonary pathologies, as PRP is administered directly to the respiratory system. Thus, in a clinical study of patients with 25–50% burns and inhalation lung injury, the application of PRP shortened intubation time and hospital stay, as well as reducing the mortality rate [122]. The application of PRP in nebulized form could improve the responses similar to conventional PRP for the treatment of chronic respiratory diseases, through anti-inflammatory effects and stimulating tissue regeneration [123]. The application of PRP molecules through nebulization could stimulate anti-inflammatory actions through HGF and IGF-1 working directly on the intracellular pathways, stimulating polarization of M2 macrophages, and its action on oxidative stress. Through immunomodulation and anti-inflammation, nebulized PRP can have effects on viral pathologies such as COVID-19 as well as promote tissue regeneration. By modulating the inflammatory response and minimizing tissue destruction, PRP can have anti-fibrotic factors which may decrease or prevent pulmonary fibrosis [124–130].

21.14 Modulation of Composition

One of the main factors influencing the efficacy of PRP is the composition of the product. In fact, the term PRP encompasses a large number of products with different variations in composition, leading to results that may be contradictory and not comparable. The first solution to address the problem of ambiguity amongst PRP preparations and compositions effecting results and outcomes is the use of classification and coding systems. This allows the identification of the type and composition of PRP when it is used in the medical and scientific field [131]. The second solution is to further deepen our understanding of PRP and its different preparations and compositions in order to modify it according to the pathology or the patient for better results.

Platelest have been the main element attributed to the efficacy of PRP. Therefore, the goal of many PRP systems is to achieve high platelet concentrations in order to obtain a more effective PRP [132, 133]. However, although it may seem that the higher the concentration of platelets, results in greater the effect of PRP [134], high platelet concentrations do not guarantee optimal cell behavior [135]. In addition, it should be noted that platelets are not the only element present in PRP, but that there are also extra-platelet bioactive molecules circulating in the plasma [136]. The synergistic action of all the components of PRP is necessary for an optimal result of this treatment [137]. The modulation of PRP components requires the development of techniques and systems that not only act on cellular elements (leucocytes, erythrocytes, platelets) but should also optimize the levels of extra-platelet molecules [138].

Thus, in addition to the conventional centrifugation systems in which the kits for obtaining PRP are based, new systems have emerged in recent years that are capable of altering the levels of plasma content. Using different techniques such as absorption or filtration, these systems concentrate erythrocytes, leukocytes, and platelets as well as anabolic growth factors and anti-inflammatory cytokines. However, concentrating low molecular weight molecules still remains a challenge [139–141]. Other techniques are based on obtaining serum instead of PRP. They incubate blood under specific conditions to extract a plethora of cytokines and biomolecules from cellular elements such as monocytes [142, 143]. However, these enriched sera lack key coagulation molecules such as fibrinogen, so their versatility is limited by their inability to generate fibrin scaffolds.

Recently, new systems were described that focus on obtaining PRP by removing water to concentrate all components equally. For this purpose, procedures have been developed in evaporation [144], filtration [145], and absorption by means of hydrogels [146]. These systems are able to obtain plasma enriched in platelets and plasma proteins, although they may differ in the concentrations of certain growth factors. The degree of cell proliferation achieved with PRP obtained through these methods reached higher values than those achieved with conventional PRP.

Finally, modifying the levels of PRP molecules related to coagulation processes such as fibrinogen results in fibrin matrices that differ in their biomechanical properties [147]. As a result, customized fibrin matrices could be obtained depending on the indication or the needs of the physician and the patient.

21.15 Patient Influence: Toward Allogenic PRP

In addition to the method of application and the composition of PRP, the biology of the patient will also determine the success of treatment [148]. This factor could depend on both the molecular cocktail that defines PRP and the patient's tissue response to treatment due to interaction between biomolecules and cell receptors [149]. Several studies showed that aspects such as age and gender have an influence on the molecular composition of PRP [150–152]. For instance, young donors present higher levels of IGF-1 in PRP which leads to a greater stimulation of cell proliferation [153, 154]. Another study by Chowdhary et al. [155], where experimental investigations conducted *in vitro*, demonstrated platelet-rich plasma derived from younger donors promoted a more youthful phenotype in chondrocytes, as indicated by elevated levels of type II collagen and SOX-9 expression. However, this advantageous effect was markedly diminished when the cells were exposed to platelet-rich plasma obtained from older donors. Correspondingly, *in vivo* research indicated that animals receiving platelet-rich plasma from younger donors exhibited significantly enhanced cartilage integrity in comparison to those treated with plasma from older donors. So it is important to emphasize that the effect of PRP on tissue repair and clinical outcomes may depend on both the characteristics of the donor and the patient receiving the treatment [156, 157].

The findings suggest that employing allogenic PRP sourced from younger family members for the therapeutic treatment of older relatives could be a viable avenue for exploration. This raises the prospect for future research, particularly given the potential enhanced efficacy of allogenic PRP from younger individuals compared to "aged" autologous PRP in improving health outcomes for the elderly. However, this approach also invites an ethical discourse, focusing on critical issues such as the appropriate minimum age for the collection of allogenic PRP, the implications of obtaining such material from minors, the nuances of consent, the beneficiaries of this treatment, and the risks of possible exploitation of young donors.

Moreover, the patient's genetics may also play a key role in the response to PRP. Genetic polymorphisms present in an individual could determine the synthesis of both cellular receptors and PRP effector molecules, leading to a PRP with a defect or excess of certain molecules and conditioning its effect and response to treatment [158].

Therefore, a patient who shows a very positive response to PRP could present an ideal molecular cocktail and be a potential PRP donor. This could provide an alternative for patients who do not respond appropriately to PRP or who have contraindications to this treatment, as demonstrated in a study of patients with osteoarthritis and thrombocytopenia [159]. In other work, the application of allogenic PRP in patients with knee osteoarthritis achieved clinically superior results to the group of patients who received autologous PRP, without any adverse reactions to donor plasma [160]. A study on patients with rotator cuff pathology who received allogenic PRP also showed no adverse reactions and superior long-term results to corticosteroid therapy [161]. Thus, the use of allogenic PRP is increasingly being considered a safe option when autologous PRP is not feasible.

References

1. Jacob G, Nakamura N, Parker D, Lane J, Gobbi A. ISAKOS newsletter 2024; 2024.
2. Mazzocca AD, McCarthy MB, Chowaniec DM, Cote MP, Romeo AA, Bradley JP, Arciero RA, Beitzel K. Platelet-rich plasma differs according to preparation method and human variability. J Bone Joint Surg. 2012;94(4):308–16. https://doi.org/10.2106/JBJS.K.00430.
3. Foster TE, Puskas BL, Mandelbaum BR, Gerhardt MB, Rodeo SA. Platelet-rich plasma: from basic science to clinical applications. Am J Sports Med. 2009;37(11):2259–72. https://doi.org/10.1177/0363546509349921.
4. Morrell CN, Aggrey AA, Chapman LM, Modjeski KL. Emerging roles for platelets as immune and inflammatory cells. Blood. 2014;123(28):2758–67.
5. Sprugel KH, Mcpherson JM, Clowes AW, Ross R. Effects of growth factors in vivo I. Cell ingrowth into porous subcutaneous chambers. Am J Pathol. 1987;129:601.
6. Blair P, Flaumenhaft R. Platelet α–granules: basic biology and clinical correlates. Blood Rev. 2009;23:177.
7. Italiano JE, Battinelli EM. Selective sorting of alpha-granule proteins. J Thromb Haemost. 2009;7:173–6.
8. King SM, Reed GL. Development of platelet secretory granules. Semin Cell Dev Biol. 2002;13:293–302.
9. Sehgal S, Storrie B. Evidence that differential packaging of the major platelet granule proteins von Willebrand factor and fibrinogen can support their differential release. J Thromb Haemost. 2007;5:2009–16.
10. Ganor Y, Besser M, Ben-Zakay N, Unger T, Levite M. Human T cells express a functional ionotropic glutamate receptor GluR3, and glutamate by itself triggers integrin-mediated adhesion to laminin and fibronectin and chemotactic migration. J Immunol. 2003;170:4362–72.
11. Iberg CA, Hawiger D. Natural and induced tolerogenic dendritic cells. J Immunol. 2020;204:733–44.
12. Gray MA, et al. Phagocytosis enhances lysosomal and bactericidal properties by activating the transcription factor TFEB. Curr Biol. 2016;26:1955–64.
13. Wunderli SL, et al. Tendon response to matrix unloading is determined by the patho-physiological niche. Matrix Biol. 2020;89:11–26.
14. Han B, Woodell-May J, Ponticiello M, Yang Z, Nimni M. The effect of thrombin activation of platelet-rich plasma on demineralized bone matrix osteoinductivity. J Bone Joint Surg Am. 2009;91:1459–70.
15. Zhou Y, Zhang J, Wu H, Hogan MV, Wang JH. The differential effects of leukocyte containing and pure platelet-rich plasma (PRP) on tendon stem/progenitor cells – implications of PRP application for the clinical treatment of tendon injuries. Stem Cell Res Ther. 2015;6:173–86.
16. Lana JF, Huber SC, Purita J, Tambeli CH, Santos GS, Paulus C, Annichino-Bizzacchi JM. Leukocyte-rich PRP versus leukocyte-poor PRP - the role of monocyte/macrophage function in the healing cascade. J Clin Orthop Trauma. 2019;10(Suppl 1):S7–S12. https://doi.org/10.1016/j.jcot.2019.05.008.
17. Sundman EA, Cole BJ, Fortier LA. Growth factor and catabolic cytokine concentrations are influenced by the cellular composition of platelet-rich plasma. Am J Sports Med. 2011;39(10):2135–40.
18. Gratchev A, Khyshkowska J, Kothe K, et al. Mphi1 and Mphi2 can be repolarized by Th2 or Th1 cytokines, respectively, and respond to exogenous danger signals. Immunobiology. 2006;211:473–86.
19. Martinez FO, Gordon S. The M1 and M2 paradigm of macrophage activation: time for reassessment. F1000Prime Rep. 2014;3:6–13.
20. Nakayama T, Hirahara K, Onodera A, Endo Y, Hosokawa H, et al. Th2 cells in health and disease. Annu Rev Immunol. 2017;35:53–84.
21. Mills CD, Kincaid K, Alt JM, Heilman MJ, Hill AM. M-1/M-2 macrophages and the Th1/Th2 paradigm. J Immunol. 2000;164:6166–73.
22. Jetten N, Verbruggen S, Gikbels MJ, Post MJ, De Winther MPJ, Donner MMPC. Anti-inflammatory M1 macrophages promote angiogenesis in vivo. Angiogenesis. 2014;17(1):109–18.
23. Ferrante CJ, Leibovich SJ. Regulation of macrophage polarization and wound healing. Adv Wound Care. 2012;1:10–6.
24. Everts P, Onishi K, Jayaram P, Lana JF, Mautner K. Platelet-rich plasma: new performance understandings and therapeutic considerations in 2020. Int J Mol Sci. 2020;21:7794.
25. Schaer DJ, Buehler PW, Alayash AI, Belcher JD, Vercellotti GM. Hemolysis and free hemoglobin revisited: exploring hemoglobin and hem in scaven-

gers as a novel class of therapeutic proteins. Blood. 2013;21:1276–84.
26. Repsold L, Joubert AM. Eryptosis: an erythrocyte's suicidal type of cell death. Biomed Res Int. 2018;2018:9405617.
27. Gupta A, Maffulli N, Jain VK. Red blood cells in platelet-rich plasma: avoid if at all possible. Biomedicine. 2023;11:2425.
28. Hurley ET, Sherman SL, Stokes DJ, Rodeo SA, Shapiro SA, Mautner K, Buford DA, Dragoo JL, Mandelbaum BR, Zaslav KR, Cole BJ, Frank RM. Experts achieve consensus on a majority of statements regarding platelet-rich plasma treatments for treatment of musculoskeletal pathology. Arthroscopy. 2024;40(2):470–7. https://doi.org/10.1016/j.arthro.2023.08.020.
29. Marx RE. Platelet-rich plasma (PRP): what is PRP and what is not PRP? Implant Dent. 2001;10(4):225–8. https://doi.org/10.1097/00008505-200110000-00002.
30. Giusti I, Rughetti A, D'Ascenzo S, Millimaggi D, Pavan A, Dell'Orso L, Dolo V. Identification of an optimal concentration of platelet gel for promoting angiogenesis in human endothelial cells. Transfusion. 2009;49(4):771–8. https://doi.org/10.1111/j.1537-2995.2008.02033.x.
31. Zheng C, Zhu Q, Liu X, Huang X, He C, Jiang L, Quan D, Zhou X, Zhu Z. Effect of platelet-rich plasma (PRP) concentration on proliferation, neurotrophic function and migration of Schwann cells in vitro. J Tissue Eng Regen Med. 2016;10(5):428–36. https://doi.org/10.1002/term.1756.
32. Xu J, Gou L, Zhang P, Li H, Qiu S. Platelet-rich plasma and regenerative dentistry. Aust Dent J. 2020;65(2):131–42. https://doi.org/10.1111/adj.12754.
33. Gobbi A, Lane JG, Dallo I. Editorial Commentary: Cartilage restoration-what is currently available? Arthroscopy. 2020;36(6):1625–8. https://doi.org/10.1016/j.arthro.2020.04.001.
34. Scotti C, Gobbi A, Karnatzikos G, Martin I, Shimomura K, Lane JG, Peretti GM, Nakamura N. Cartilage repair in the inflamed joint: considerations for biological augmentation toward tissue regeneration. Tissue Eng Part B Rev. 2016;22(2):149–59. https://doi.org/10.1089/ten.TEB.2015.0297.
35. Laver L, Marom N, Dnyanesh L, Mei-Dan O, Espregueira-Mendes J, Gobbi A. PRP for degenerative cartilage disease: a systematic review of clinical studies. Cartilage. 2017;8(4):341–64. https://doi.org/10.1177/1947603516670709.
36. Sánchez M, Anitua E, Azofra J, Aguirre JJ, Andia I. Intra-articular injection of an autologous preparation rich in growth factors for the treatment of knee OA: a retrospective cohort study. Clin Exp Rheumatol. 2008;26(5):910–3.
37. Gobbi A, Lad D, Karnatzikos G. The effects of repeated intra-articular PRP injections on clinical outcomes of early osteoarthritis of the knee. Knee Surg Sports Traumatol Arthrosc. 2015;23(8):2170–7. https://doi.org/10.1007/s00167-014-2987-4.
38. Laver L, Filardo G, Sanchez M, Magalon J, Tischer T, Abat F, Bastos R, Cugat R, Iosifidis M, Kocaoglu B, Kon E, Marinescu R, Ostojic M, Beaufils P, de Girolamo L. The use of injectable orthobiologics for knee osteoarthritis: a European ESSKA-ORBIT consensus. Part 1-Blood-derived products (platelet-rich plasma). Knee Surg Sports Traumatol Arthrosc. 2024;32(4):783–97. https://doi.org/10.1002/ksa.12077.
39. Sánchez M, Fiz N, Azofra J, Usabiaga J, Aduriz Recalde E, Garcia Gutierrez A, et al. Randomized clinical trial evaluating plasma rich in growth factors (PRGF-Endoret) versus hyaluronic acid in the short-term treatment of symptomatic knee osteoarthritis. Arthroscopy. 2012;28(8):1070–8.
40. Patel S, Dhillon MS, Aggarwal S, Marwaha N, Jain A. Treatment with platelet-rich plasma is more effective that placebo for knee osteoarthritis: a prospective, double-blind, randomized trial. Am J Sports Med. 2013;41(2):356–64.
41. Rayegani SM, Raeissadat SA, Taheri MS, Babaee M, Bahrami MH, Eliaspour D, et al. Does intra articular platelet rich plasma injection improve function, pain and quality of life in patients with osteoarthritis of the knee? A randomized clinical trial. Orthop Rev. 2014;6(3):5405.
42. Kon E, Mandelbaum B, Buda R, Filardo G, Delcogliano M, Timoncini A, et al. Platelet-rich plasma intra-articular injection versus hyaluronic acid viscosupplementation as treatments for cartilage pathology: from early degeneration to osteoarthritis. Arthroscopy. 2011;27(11):1490–501.
43. Cerza F, Carnì S, Carcangiu A, Di Vavo I, Schiavilla V, Pecora A, et al. Comparison between hyaluronic acid and platelet-rich plasma, intra-articular infiltration in the treatment of gonarthrosis. Am J Sports Med. 2012;40(12):2822–7.
44. Vaquerizo V, Plasencia MÁ, Arribas I, Seijas R, Padilla S, Orive G, et al. Comparison of intra-articular injections of plasma rich in growth factors (PRGF-Endoret) versus Durolane hyaluronic acid in the treatment of patients with symptomatic osteoarthritis: a randomized controlled trial. Arthroscopy. 2013;29(10):1635–43.
45. Raeissadat SA, Rayegani SM, Hassanabadi H, Fathi M, Ghorbani E, Babaee M, et al. Knee osteoarthritis injection choices: platelet-rich plasma (PRP) versus hyaluronic acid (a one-year randomized clinical trial). Clin Med Insights Arthritis Musculoskelet Disord. 2015;8:1–8.
46. Filardo G, Kon E, Matteo B, Marino A, Sessa A, Merli ML, et al. Leukocyte-poor PRP application for the treatment of knee osteoarthritis. Joints. 2014;1(3):112–20.
47. Jang SJ, Kim JD, Cha SS. Platelet rich plasma (PRP) injections as an effective treatment for early osteoarthritis. Eur Orthop Surg Traumatol. 2013;23(5):573–80.

48. Torrero JI, Aroles F, Ferrer D. Treatment of knee chondropathy with platelet rich plasma. Preliminary results at 6 months of follow up with only one injection. J Biol Regul Homeost Agents. 2012;26(2 Suppl. 1):71s–8s.
49. Levenson SM, Geever EF, Crowley LV, Oates JF, Berard CW, Rosen H. The healing of rat skin wounds. Ann Surg. 1965;161(2):293–308. https://doi.org/10.1097/00000658-196502000-00019.
50. Frank C, Schachar N, Dittrich D. Natural history of healing in the repaired medial collateral ligament. J Orthop Res. 1983;1(2):179–88. https://doi.org/10.1002/jor.1100010209.
51. Chamberlain CS, Crowley E, Vanderby R. The spatio-temporal dynamics of ligament healing. Wound Repair Regen. 2009;17(2):206–15. https://doi.org/10.1111/j.1524-475X.2009.00465.x.
52. Whitman DH, Berry RL, Green DM. Platelet gel: an autologous alternative to fibrin glue with applications in oral and maxillofacial surgery. J Oral Maxillofac Surg. 1997;55(11):1294–9. https://doi.org/10.1016/s0278-2391(97)90187-7.
53. Dohan Ehrenfest DM, Rasmusson L, Albrektsson T. Classification of platelet concentrates: from pure platelet-rich plasma (P-PRP) to leucocyte- and platelet-rich fibrin (L-PRF). Trends Biotechnol. 2009;27(3):158–67. https://doi.org/10.1016/j.tibtech.2008.11.009.
54. Marx RE. Platelet-rich plasma: evidence to support its use. J Oral Maxillofac Surg. 2004;62(4):489–96. https://doi.org/10.1016/j.joms.2003.12.003.
55. Boswell SG, Cole BJ, Sundman EA, Karas V, Fortier LA. Platelet-rich plasma: a milieu of bioactive factors. Arthroscopy. 2012;28(3):429–39. https://doi.org/10.1016/j.arthro.2011.10.018.
56. Chen X, Jones IA, Park C, Vangsness CT Jr. The efficacy of platelet-rich plasma on tendon and ligament healing: a systematic review and meta-analysis with bias assessment. Am J Sports Med. 2018;46(8):2020–32. https://doi.org/10.1177/0363546517743746.
57. Lopez-Vidriero E, Goulding KA, Simon DA, Sanchez M, Johnson DH. The use of platelet-rich plasma in arthroscopy and sports medicine: optimizing the healing environment. Arthroscopy. 2010;26(2):269–78. https://doi.org/10.1016/j.arthro.2009.11.015.
58. Rodeo SA, Delos D, Weber A, Ju X, Cunningham ME, Fortier L, Maher S. What's new in orthopaedic research. J Bone Joint Surg. 2010;92(14):2491–501. https://doi.org/10.2106/JBJS.J.01174.
59. Chen X, Jones IA, Togashi R, Park C, Vangsness CT. Use of platelet-rich plasma for the improvement of pain and function in rotator cuff tears: a systematic review and meta-analysis with bias assessment. Am J Sports Med. 2020;48(8):2028–41. https://doi.org/10.1177/0363546519881423.
60. Gobbi A, Whyte GP. Emerging orthobiologic approaches to ligament injury. In: Gobbi A, Espregueira-Mendes J, Lane J, Karahan M, editors. Bio-orthopaedics. Berlin: Springer; 2017. https://doi.org/10.1007/978-3-662-54181-4_24.
61. Samy TM, Khater AH, Ahmed AK. Systematic review of literature on the efficiency of using PRP in arthroscopic rotator cuff repair. QJM. 2020;113(Supplement_1):13. https://doi.org/10.1093/QJMED/HCAA059.013.
62. Gobbi A, Herman K, Szwedowski D. Bio-orthopedics: a new approach to osteoarthritis and joint disorders. In: Cartilage disorders - recent findings and treatment. London: IntechOpen; 2023. https://doi.org/10.5772/intechopen.110845.
63. Kon E, Di Matteo B, Altomare D, et al. Biologic agents to optimize outcomes following ACL repair and reconstruction: a systematic review of clinical evidence. J Orthop Res. 2022;40(1):10–28. https://doi.org/10.1002/JOR.25011.
64. LaPrade RF, Goodrich LR, Phillips J, et al. Use of platelet-rich plasma immediately after an injury did not improve ligament healing, and increasing platelet concentrations was detrimental in an in vivo animal model. Am J Sports Med. 2018;46(3):702–12. https://doi.org/10.1177/0363546517741135.
65. Mirzatolooei F, Alamdari MT, Khalkhali HR. The impact of platelet-rich plasma on the prevention of tunnel widening in anterior cruciate ligament reconstruction using quadrupled autologous hamstring tendon: a randomised clinical trial. Bone Joint J. 2013;95-B(1):65–9. https://doi.org/10.1302/0301-620X.95B1.30487.
66. Magnussen RA, Flanigan DC, Pedroza AD, Heinlein KA, Kaeding CC. Platelet rich plasma use in allograft ACL reconstructions: two-year clinical results of a MOON cohort study. Knee. 2013;20(4):277–80. https://doi.org/10.1016/j.knee.2012.12.001.
67. Valentí Azcárate A, Lamo-Espinosa J, Aquerreta Beola JD, Hernandez Gonzalez M, Mora Gasque G, Valentí Nin JR. Comparison between two different platelet-rich plasma preparations and control applied during anterior cruciate ligament reconstruction. Is there any evidence to support their use? Injury. 2014;45(Suppl 4):S36–41. https://doi.org/10.1016/S0020-1383(14)70008-7.
68. Murray MM, Palmer M, Abreu E, Spindler KP, Zurakowski D, Fleming BC. Platelet-rich plasma alone is not sufficient to enhance suture repair of the ACL in skeletally immature animals: an in vivo study. J Orthop Res. 2009;27(5):639–45. https://doi.org/10.1002/JOR.20796.
69. Gobbi A, Whyte GP. Biological augmentation in acute ACL repair. In: Gobbi A, Espregueira-Mendes J, Lane J, Karahan M, editors. Bio-orthopaedics. Berlin: Springer; 2017. https://doi.org/10.1007/978-3-662-54181-4_25.
70. Xu H, Luo H, Chen J, Chen G, Yu X, Ye Z. BMP-2 releasing mineral-coated microparticle-integrated hydrogel system for enhanced bone regeneration. Front Bioeng Biotechnol. 2023;11:1217335. https://doi.org/10.3389/fbioe.2023.1217335.

71. Gobbi A, Bathan L, Boldrini L. Primary repair combined with bone marrow stimulation in acute anterior cruciate ligament lesions. Am J Sports Med. 2009;37(3):571–8. https://doi.org/10.1177/0363546508327141.
72. Gobbi A, Whyte GP. Long-term outcomes of primary repair of the anterior cruciate ligament combined with biologic healing augmentation to treat incomplete tears. Am J Sports Med. 2018;46(14):3368–77. https://doi.org/10.1177/0363546518805740.
73. Mishra A, Harmon K, Woodall J, Vieira A. Sports medicine applications of platelet rich plasma. Curr Pharm Biotechnol. 2012;13(7):1185–95. https://doi.org/10.2174/138920112800624283.
74. DeLong JM, Russell RP, Mazzocca AD. Platelet-rich plasma: the PAW classification system. Arthroscopy. 2012;28(7):998–1009. https://doi.org/10.1016/j.arthro.2012.04.148.
75. Magalon J, Chateau AL, Bertrand B, Louis ML, Silvestre A, Giraudo L, Veran J, Sabatier F. DEPA classification: a proposal for standardising PRP use and a retrospective application of available devices. BMJ Open Sport Exerc Med. 2016;2(1):e000060. https://doi.org/10.1136/bmjsem-2015-000060.
76. Dohan Ehrenfest DM, Andia I, Zumstein MA, Zhang CQ, Pinto NR, Bielecki T. Classification of platelet concentrates (platelet-rich plasma-PRP, platelet-rich fibrin-PRF) for topical and infiltrative use in orthopedic and sports medicine: current consensus, clinical implications and perspectives. Muscles Ligaments Tendons J. 2014;4(1):3–9.
77. Kunze KN, Pakanati JJ, Vadhera AS, Polce EM, Williams BT, Parvaresh KC, Chahla J. The efficacy of platelet-rich plasma for ligament injuries: a systematic review of basic science literature with protocol quality assessment. Orthop J Sports Med. 2022;10(2):23259671211066504. https://doi.org/10.1177/23259671211066504.
78. Voleti PB, Buckley MR, Soslowsky LJ. Tendon healing: repair and regeneration. Annu Rev Biomed Eng. 2012;14:47–71. https://doi.org/10.1146/annurev-bioeng-071811-150122.
79. Beredjiklian PK. Biologic aspects of flexor tendon laceration and repair. J Bone Joint Surg. 2003;85(3):539–50. https://doi.org/10.2106/00004623-200303000-00025.
80. Leadbetter WB. Cell-matrix response in tendon injury. Clin Sports Med. 1992;11(3):533–78.
81. Gelberman RH, Steinberg D, Amiel D, Akeson W. Fibroblast chemotaxis after tendon repair. J Hand Surg Am. 1991;16(4):686–93. https://doi.org/10.1016/0363-5023(91)90195-h.
82. Massimino ML, Rapizzi E, Cantini M, Libera LD, Mazzoleni F, Arslan P, Carraro U. ED2+ macrophages increase selectively myoblast proliferation in muscle cultures. Biochem Biophys Res Commun. 1997;235(3):754–9. https://doi.org/10.1006/bbrc.1997.6823.
83. De Vos RJ, Weir A, Van Schie HTM, et al. Platelet-rich plasma injection for chronic Achilles tendinopathy: a randomized controlled trial. JAMA. 2010;303(2):144–9. https://doi.org/10.1001/JAMA.2009.1986.
84. Randelli P, Arrigoni P, Ragone V, Aliprandi A, Cabitza P. Platelet rich plasma in arthroscopic rotator cuff repair: a prospective RCT study, 2-year follow-up. J Shoulder Elb Surg. 2011;20(4):518–28. https://doi.org/10.1016/J.JSE.2011.02.008.
85. Gosens T, Peerbooms JC, Van Laar W, Den Oudsten BL. Ongoing positive effect of platelet-rich plasma versus corticosteroid injection in lateral epicondylitis: a double-blind randomized controlled trial with 2-year follow-up. Am J Sports Med. 2011;39(6):1200–8. https://doi.org/10.1177/0363546510397173.
86. Yan R, Gu Y, Ran J, Hu Y, Zheng Z, Zeng M, Heng BC, Chen X, Yin Z, Chen W, Shen W, Ouyang H. Intratendon delivery of leukocyte-poor platelet-rich plasma improves healing compared with leukocyte-rich platelet-rich plasma in a rabbit achilles tendinopathy model. Am J Sports Med. 2017;45(8):1909–20. https://doi.org/10.1177/0363546517694357.
87. Fitzpatrick J, Bulsara MK, O'Donnell J, Zheng MH. Leucocyte-rich platelet-rich plasma treatment of gluteus medius and minimus tendinopathy: a double-blind randomized controlled trial with 2-year follow-up. Am J Sports Med. 2019;47(5):1130–7. https://doi.org/10.1177/0363546519826969.
88. Weber SC, Kauffman JI, Parise C, Weber SJ, Katz SD. Platelet-rich fibrin matrix in the management of arthroscopic repair of the rotator cuff: a prospective, randomized, double-blinded study. Am J Sports Med. 2013;41(2):263–70. https://doi.org/10.1177/0363546512467621.
89. Rodeo SA, Delos D, Williams RJ, Adler RS, Pearle A, Warren RF. The effect of platelet-rich fibrin matrix on rotator cuff tendon healing: a prospective, randomized clinical study. Am J Sports Med. 2012;40(6):1234–41. https://doi.org/10.1177/0363546512442924.
90. Mazzocca AD, McCarthy MB, Chowaniec DM, Dugdale EM, Hansen D, Cote MP, Bradley JP, Romeo AA, Arciero RA, Beitzel K. The positive effects of different platelet-rich plasma methods on human muscle, bone, and tendon cells. Am J Sports Med. 2012;40(8):1742–9. https://doi.org/10.1177/0363546512452713.
91. Jo CH, Kim JE, Yoon KS, Shin S. Platelet-rich plasma stimulates cell proliferation and enhances matrix gene expression and synthesis in tenocytes from human rotator cuff tendons with degenerative tears. Am J Sports Med. 2012;40(5):1035–45. https://doi.org/10.1177/0363546512437525.
92. Boswell SG, Schnabel LV, Mohammed HO, Sundman EA, Minas T, Fortier LA. Increasing platelet concentrations in leukocyte-reduced platelet-rich plasma decrease collagen gene synthesis in tendons.

Am J Sports Med. 2014;42(1):42–9. https://doi.org/10.1177/0363546513507566.
93. Jiang G, Wu Y, Meng J, Wu F, Li S, Lin M, Gao X, Hong J, Chen W, Yan S, Yan R, Feng G, Cheng Z. Comparison of leukocyte-rich platelet-rich plasma and leukocyte-poor platelet-rich plasma on achilles tendinopathy at an early stage in a rabbit model. Am J Sports Med. 2020;48(5):1189–99. https://doi.org/10.1177/0363546520906142.
94. Liu X, Zhang R, Zhu B, Li Y, Liu X, Guo S, Wang C, Wang D, Li S. Effects of leukocyte- and platelet-rich plasma on tendon disorders based on in vitro and in vivo studies. Exp Ther Med. 2021;21(6):639. https://doi.org/10.3892/etm.2021.10071.
95. Lin KY, Chen P, Chen AC, Chan YS, Lei KF, Chiu CH. Leukocyte-rich platelet-rich plasma has better stimulating effects on tenocyte proliferation compared with leukocyte-poor platelet-rich plasma. Orthop J Sports Med. 2022;10(3):23259671221084706. https://doi.org/10.1177/23259671221084706.
96. Rha DW, Park GY, Kim YK, Kim MT, Lee SC. Comparison of the therapeutic effects of ultrasound-guided platelet-rich plasma injection and dry needling in rotator cuff disease: a randomized controlled trial. Clin Rehabil. 2013;27(2):113–22. https://doi.org/10.1177/0269215512448388.
97. Kopf S, Beaufils P, Hirschmann MT, et al. Management of traumatic meniscus tears: the 2019 ESSKA meniscus consensus. Knee Surg Sports Traumatol Arthrosc. 2020;28(4):1177–94. https://doi.org/10.1007/S00167-020-05847-3.
98. Grifn JW, Hadeed MM, Werner BC, Diduch DR, Carson EW, Miller MD. Platelet-rich plasma in meniscal repair: does augmentation improve surgical outcomes? Clin Orthop Relat Res. 2015;473:1665–72.
99. Kaminski R, Kulinski K, Kozar-Kaminska K, Wielgus M, Langner M, Wasko MK, Kowalczewski J, Pomianowski S. A prospective, randomized, double-blind, parallel-group, placebo-controlled study evaluating meniscal healing, clinical outcomes, and safety in patients undergoing meniscal repair of unstable, complete vertical meniscal tears (bucket handle) augmented with platelet-rich plasma. Biomed Res Int. 2018;2018:9315815. https://doi.org/10.1155/2018/9315815.
100. Everhart JS, Cavendish PA, Eikenberry A, Magnussen RA, Kaeding CC, Flanigan DC. Platelet-rich plasma reduces failure risk for isolated meniscal repairs but provides no benefit for meniscal repairs with anterior cruciate ligament reconstruction. Am J Sports Med. 2019;47(8):1789–96. https://doi.org/10.1177/0363546519852616.
101. Li Z, Weng X. Platelet-rich plasma use in meniscus repair treatment: a systematic review and meta-analysis of clinical studies. J Orthop Surg Res. 2022;17(1):446. https://doi.org/10.1186/s13018-022-03293-0.
102. Di Matteo B, Altomare D, Garibaldi R, La Porta A, Manca A, Kon E. Ultrasound-guided meniscal injection of autologous growth factors: a brief report. Cartilage. 2021;13(1_suppl):387S–91S. https://doi.org/10.1177/1947603521l037390.
103. Guenoun D, Magalon J, de Torquemada I, Vandeville C, Sabatier F, Champsaur P, Jacquet C, Ollivier M. Treatment of degenerative meniscal tear with intrameniscal injection of platelets rich plasma. Diagn Interv Imaging. 2020;101(3):169–76. https://doi.org/10.1016/j.diii.2019.10.003.
104. Özyalvaç ON, Tüzüner T, Gürpinar T, Obut A, Acar B, Akman YE. Radiological and functional outcomes of ultrasound-guided PRP injections in intrasubstance meniscal degenerations. J Orthop Surg. 2019;27(2):2309499019852779. https://doi.org/10.1177/2309499019852779.
105. Sánchez M, Jorquera C, Bilbao AM, García S, Beitia M, Espregueira-Mendes J, González S, Oraa J, Guadilla J, Delgado D. High survival rate after the combination of intrameniscal and intraarticular infiltrations of platelet-rich plasma as conservative treatment for meniscal lesions. Knee Surg Sports Traumatol Arthrosc. 2023;31(10):4246–56. https://doi.org/10.1007/s00167-023-07470-4.
106. Laumonier T, Menetrey J. Muscle injuries and strategies for improving their repair. J Exp Orthop. 2016;3(1):15. https://doi.org/10.1186/s40634-016-0051-7.
107. Tidball JG. Inflammatory cell response to acute muscle injury. Med Sci Sports Exerc. 1995;27(7):1022–32. https://doi.org/10.1249/00005768-199507000-00011.
108. Toumi H, Best TM. The inflammatory response: friend or enemy for muscle injury? Br J Sports Med. 2003;37(4):284–6. https://doi.org/10.1136/bjsm.37.4.284.
109. Tsai WC, Yu TY, Lin LP, Lin MS, Wu YC, Liao CH, Pang JS. Platelet rich plasma releasate promotes proliferation of skeletal muscle cells in association with upregulation of PCNA, cyclins and cyclin dependent kinases. Platelets. 2017;28(5):491–7. https://doi.org/10.1080/09537104.2016.1227061.
110. Rossi LA, Molina Rómoli AR, Bertona Altieri BA, Burgos Flor JA, Scordo WE, Elizondo CM. Does platelet-rich plasma decrease time to return to sports in acute muscle tear? A randomized controlled trial. Knee Surg Sports Traumatol Arthrosc. 2017;25(10):3319–25. https://doi.org/10.1007/s00167-016-4129-7.
111. Delos D, Maak TG, Rodeo SA. Muscle injuries in athletes: enhancing recovery through scientific understanding and novel therapies. Sports Health. 2013;5(4):346–52. https://doi.org/10.1177/1941738113480934.
112. Hamid A, Ali MSM, Yusof A, George J, Lee LP. Platelet-rich plasma injections for the treatment of hamstring injuries: a randomized controlled trial. Am J Sports Med. 2014;42(10):2410–8. https://doi.org/10.1177/0363546514541540.

113. Mejia HA, Bradley JP. The effects of platelet-rich plasma on muscle: basic science and clinical application. Oper Tech Sports Med. 2011;19:149–53. https://doi.org/10.1053/j.otsm.2011.04.002.
114. Setayesh K, Villarreal A, Gottschalk A, Tokish JM, Choate WS. Treatment of muscle injuries with platelet-rich plasma: a review of the literature. Curr Rev Musculoskelet Med. 2018;11(4):635–42. https://doi.org/10.1007/s12178-018-9526-8.
115. Chen LS, Chen CK, Pang JS, Lin LP, Yu TY, Tsai WC. Leukocyte-poor platelet-rich plasma and leukocyte-rich platelet-rich plasma promote myoblast proliferation through the upregulation of cyclin A, cdk1, and cdk2. J Orthop Res. 2024;42(1):32–42. https://doi.org/10.1002/jor.25666.
116. Peng Y, Guanglan W, Jia S, Zheng C. Leukocyte-rich and leukocyte-poor platelet-rich plasma in rotator cuff repair: a meta-analysis. Int J Sports Med. 2022;43(11):921–30. https://doi.org/10.1055/a-1790-7982.
117. Hurley ET, Colasanti CA, Anil U, Luthringer TA, Alaia MJ, Campbell KA, Jazrawi LM, Strauss EJ. The effect of platelet-rich plasma leukocyte concentration on arthroscopic rotator cuff repair: a network meta-analysis of randomized controlled trials. Am J Sports Med. 2021;49(9):2528–35. https://doi.org/10.1177/0363546520975435.
118. Patel S, Gahlaut S, Thami T, Chouhan DK, Jain A, Dhillon MS. Comparison of conventional dose versus superdose platelet-rich plasma for knee osteoarthritis: a prospective, triple-blind, randomized clinical trial. Orthop J Sports Med. 2024;12(2):23259671241227863. https://doi.org/10.1177/23259671241227863.
119. Anitua E, Pino A. The management of postsurgical wound complications with plasma rich in growth factors: a preliminary series. Adv Skin Wound Care. 2020;33(4):202–8.
120. Sánchez M, Delgado D, Sánchez P, Fiz N, Azofra J, Orive G, et al. Platelet rich plasma and knee surgery. Biomed Res Int. 2014;2014:890630.
121. Beitia M, Delgado D, Sánchez P, Vallejo de la Cueva A, Cugat JR, Sánchez M. Platelet lysate nebulization protocol for the treatment of COVID-19 and its sequels: proof of concept and scientific rationale. Int J Mol Sci. 2021;22(4):1856.
122. Salama SM, Kamel IH, Ghanim M, Elsherif AF. The efficacy of autologous nebulized platelet rich plasma (PRP) as an early adjuvant therapeutic and prognostic treatment modality in the management of inhalation lung injury. Egypt J Plast Reconstr Surg. 2019;43(2):203–8.
123. Knight AD, Kacker S. Platelet-rich plasma treatment for chronic respiratory disease. Cureus. 2023;15(1):e33265.
124. Lo SK, Burhop KE, Kaplan JE, Malik AB. Role of platelets in maintenance of pulmonary vascular permeability to protein. Am J Phys. 1988;254:H763–71. https://doi.org/10.1152/ajpheart.1988.254.4.H763.
125. Bendinelli P, Matteucci E, Dogliotti G, Corsi MM, Banfi G, Maroni P, Desiderio MA. Molecular basis of anti-inflammatory action of platelet-rich plasma on human chondrocytes: mechanisms of NF-KB inhibition via HGF. J Cell Physiol. 2010;225:757–66.
126. Vasina EM, Cauwenberghs S, Feijge MAH, Heemskerk JWM, Weber C, Koenen RR. Microparticles from apoptotic platelets promote resident macrophage differentiation. Cell Death Dis. 2011;2:e211.
127. Haselton FR, Alexander JS. Platelets and a platelet-released factor enhance endothelial barrier. Am J Phys. 1992;263:L670–8.
128. Squecco R, Chellini F, Idrizaj E, Tani A, Garella R, Pancani S, Pavan P, Bambi F, Zecchi-Orlandini S, Sassoli C. Platelet-rich plasma modulates gap junction functionality and connexin 43 and 26 expression during TGF-B1-induced fibroblast to myofibroblast transition: clues for counteracting fibrosis. Cells. 2020;9:1199. https://doi.org/10.3390/cells9051199.
129. Sánchez M, Anitua E, Azofra J, Andía I, Padilla S, Mujika I. Comparison of surgically repaired achilles tendon tears using platelet-rich fibrin matrices. Am J Sports Med. 2007;35:245–51. https://doi.org/10.1177/0363546506294078.
130. Everts PAM, Devilee RJJ, Oosterbos CJM, Mahoney CB, Schattenkerk ME, Knape JTA, van Zundert A. Autologous platelet gel and fibrin sealant enhance the efficacy of total knee arthroplasty: improved range of motion, decreased length of stay and a reduced incidence of arthrofibrosis. Knee Surg Sports Traumatol Arthrosc. 2007;15:888–94. https://doi.org/10.1007/s00167-007-0296-x.
131. Kon E, Di Matteo B, Delgado D, Cole BJ, Dorotei A, Dragoo JL, et al. Platelet-rich plasma for the treatment of knee osteoarthritis: an expert opinion and proposal for a novel classification and coding system. Expert Opin Biol Ther. 2020;20(12):1447–60.
132. Muthu S, Krishnan A, Ramanathan KR. Standardization and validation of a conventional high yield platelet-rich plasma preparation protocol. Ann Med Surg. 2022;82:104593.
133. Rossi R, Garg AK, Kurtzman GM. Novel protocols for the production of autologous blood concentrates with high platelet volume. Compend Contin Educ Dent. 2022;43(3):140–5.
134. Berrigan WA, Bailowitz Z, Park A, Reddy A, Liu R, Lansdown D. A greater platelet dose may yield better clinical outcomes for platelet-rich plasma in the treatment of knee osteoarthritis: a systematic review. Arthrosc J Arthrosc Relat Surg. 2024. https://doi.org/10.1016/j.arthro.2024.03.018.
135. Giusti I, D'Ascenzo S, Mancò A, Di Stefano G, Di Francesco M, Rughetti A, et al. Platelet concentration in platelet-rich plasma affects tenocyte behavior in vitro. Biomed Res Int. 2014;2014:630870.
136. Beitia M, Delgado D, Mercader J, Sánchez P, López de Dicastillo L, Sánchez M. Action of platelet-rich plasma on in vitro cellular bioactivity: more than platelets. Int J Mol Sci. 2023;24(6):5367.
137. van der Bijl I, Vlig M, Middelkoop E, de Korte D. Allogeneic platelet-rich plasma (PRP) is superior

to platelets or plasma alone in stimulating fibroblast proliferation and migration, angiogenesis, and chemotaxis as relevant processes for wound healing. Transfusion. 2019;59(11):3492–500.

138. Ruiz JM, Beitia M, Delgado D, Sánchez P, Bilbao AM, Fiz N, et al. New formulations of platelet-rich plasma: the relevance of extraplatelet molecules. London: IntechOpen; 2024.
139. Higgins JC, Woodell-May J, Hoeppner JC. Use of a device for obtaining interleukin-1 receptor antagonist rich solutions [Internet]. EP2567692A1, 2013 [citado 9 de febrero de 2023]. Disponible en: https://patents.google.com/patent/EP2567692A1/en.
140. Muir SM, Reisbig N, Baria M, Kaeding C, Bertone AL. The concentration of plasma provides additional bioactive proteins in platelet and autologous protein solutions. Am J Sports Med. 2019;47(8):1955–63.
141. Kyle MR, Coull T. Fluid concentrator, autologous concentrated body fluids, and uses thereof [Internet]. WO2009002849A2, 2008 [citado 9 de febrero de 2023]. Disponible en: https://patents.google.com/patent/WO2009002849A2/en.
142. Schutte RJ, Parisi-Amon A, Reichert WM. Cytokine profiling using monocytes/macrophages cultured on common biomaterials with a range of surface chemistries. J Biomed Mater Res A. 2009;88(1):128–39.
143. Meijer H, Reinecke J, Becker C, Tholen G, Wehling P. The production of anti-inflammatory cytokines in whole blood by physico-chemical induction. Inflamm Res. 2003;52(10):404–7.
144. Mercader Ruiz J, Beitia M, Delgado D, Sánchez P, Sánchez MB, Oraa J, et al. Method to obtain a plasma rich in platelet- and plasma-growth factors based on water evaporation. Manuscr Submitt Publ. 2024.
145. Mercader Ruiz J, Beitia M, Delgado D, Sánchez P, Guadilla J, Pérez de Arrilucea C, et al. Method based on ultrafiltration to obtain a plasma rich in platelet and plasma growth factors. J Clin Med. 2023;12(18):5941.
146. Mercader Ruiz J, Beitia M, Delgado D, Sánchez P, Arnaiz MJ, López de Dicastillo L, et al. New formulation of platelet-rich plasma enriched in platelet and extraplatelet biomolecules using hydrogels. Int J Mol Sci. 2023;24(18):13811.
147. Delgado D, Beitia M, Mercader Ruiz J, Sánchez P, Montoya-Alzola M, Fiz N, et al. A novel fibrin matrix derived from platelet-rich plasma: protocol and characterization. Int J Mol Sci. 2024;25(7):4069.
148. Sánchez M, Jorquera C, de Dicastillo LL, Fiz N, Knörr J, Beitia M, et al. Real-world evidence to assess the effectiveness of platelet-rich plasma in the treatment of knee degenerative pathology: a prospective observational study. Ther Adv Musculoskelet Dis. 2022;14:1759720X221100304.
149. Padilla S, Sanchez M, Padilla I, Orive G, Anitua E. Healing or not healing. Curr Pharm Biotechnol. 2016;17(5):419–30.
150. Larsson A, Carlsson L, Gordh T, Lind AL, Thulin M, Kamali-Moghaddam M. The effects of age and gender on plasma levels of 63 cytokines. J Immunol Methods. 2015;425:58–61.
151. Bai MY, Chuang MH, Lin MF, Tang SL, Wong CC, Chan WP. Relationships of age and sex with cytokine content and distribution in human platelet fibrin gels. Sci Rep. 2018;8(1):10642.
152. O'Donnell C, Migliore E, Grandi FC, Koltsov J, Lingampalli N, Cisar C, et al. Platelet-rich plasma (PRP) from older males with knee osteoarthritis depresses chondrocyte metabolism and upregulates inflammation. J Orthop Res. 2019;37(8):1760–70.
153. Delgado D, Bilbao AM, Beitia M, Garate A, Sánchez P, González-Burguera I, et al. Effects of platelet-rich plasma on cellular populations of the central nervous system: the influence of donor age. Int J Mol Sci. 2021;22(4):1725.
154. Jiang Z, Huang C, Guo E, Zhu X, Li N, Huang Y, et al. Platelet-rich plasma in young and elderly humans exhibits a different proteomic profile. J Proteome Res. 2024;23(5):1788–800.
155. Chowdhary K, Sahu A, Iijima H, Shinde S, Borg-Stein J, Ambrosio F. Aging affects the efficacy of platelet-rich plasma treatment for osteoarthritis. Am J Phys Med Rehabil. 2023;102(7):597–604. https://doi.org/10.1097/PHM.0000000000002161.
156. Delgado D, Garate A, Sánchez P, Bilbao AM, García Del Caño G, Salles J, et al. Biological and structural effects after intraosseous infiltrations of age-dependent platelet-rich plasma: an in vivo study. J Orthop Res. 2020;38(9):1931–41.
157. Sánchez M, Jorquera C, López de Dicastillo L, Martínez N, Espregueira-Mendes J, Vergés J, et al. Women show a positive response to platelet-rich plasma despite presenting more painful knee osteoarthritis than men. Knee Surg Sports Traumatol Arthrosc. 2024;32(10):2516–25.
158. Niemiec P, Szyluk K, Balcerzyk A, Kalita M, Jarosz A, Iwanicka J, et al. Why PRP works only on certain patients with tennis elbow? Is PDGFB gene a key for PRP therapy effectiveness? A prospective cohort study. BMC Musculoskelet Disord. 2021;22(1):710.
159. Zhu X, Zhao L, Riva N, Yu Z, Jiang M, Zhou F, et al. Allogeneic platelet-rich plasma for knee osteoarthritis in patients with primary immune thrombocytopenia: a randomized clinical trial. iScience. 2024;27(5):109664.
160. Bottegoni C, Dei Giudici L, Salvemini S, Chiurazzi E, Bencivenga R, Gigante A. Homologous platelet-rich plasma for the treatment of knee osteoarthritis in selected elderly patients: an open-label, uncontrolled, pilot study. Ther Adv Musculoskelet Dis. 2016;8(2):35–41.
161. Jo CH, Lee SY, Yoon KS, Oh S, Shin S. Allogeneic platelet-rich plasma versus corticosteroid injection for the treatment of rotator cuff disease: a randomized controlled trial. J Bone Joint Surg Am. 2020;102(24):2129–37.

Essential Considerations in Platelet-Rich Plasma Preparations with Emphasis on Platelet Dosing and Bioformulations: There Is No One-Size-Fits-All Method

22

Peter A. Everts, José Fábio Lana, Melanio Acosta IV, Luyddy Pires, Anita van Domselaar, Alberto Gobbi, and Luga Podesta

22.1 Introduction

There is a pressing need for innovative and less invasive treatment options for patients suffering from a variety of musculoskeletal and spinal disorders [4]. In recent years, autologous biological cellular preparations, freshly prepared at the point of care, have emerged as a promising area of innovative medical healthcare settings. These preparations, including PRP, BMC, and ATC, offer a range of biological materials with potential therapeutic benefits [2]. By harnessing the power of platelets, MSCs, PGFs, leukocytes, signaling cells, cytokines, proteinases, chemokines, and interleukins, orthobiologics can contribute to the advancement of regenerative and orthobiological treatment plans, providing better outcomes for patients suffering from these debilitating conditions [5, 6]. Continued research and collaboration are necessary to further advance the field of orthobiologics and optimize its utilization in musculoskeletal and spinal treatments.

PRP orthobiological applications have shown promising results and are often used to meet the objectives of regenerative medicine therapy initiatives [7]. PRP preparations and treatment protocols have advanced significantly over the last two decades. However, there is currently a lack of consensus regarding standardized protocols for preparing different orthobiological bioformulations [8]. This lack of standardization has led to mixed results in some clinical studies, with some

P. A. Everts (✉)
The University of Queensland, School of Medicine, Brisbane, Australia

Max-Planck University, OrthoRegen Group, Indaiatuba, SP, Brazil

Medical School, Max Planck University Center (UniMAX), Indaiatuba, SP, Brazil
e-mail: peter@evertsscientific.com; peter@gulfcoastbiologics.com

J. F. Lana · L. Pires
Max-Planck University, OrthoRegen Group, Indaiatuba, SP, Brazil

Medical School, Max Planck University Center (UniMAX), Indaiatuba, SP, Brazil

M. Acosta IV · A. Gobbi
O.A.S.I. Bioresearch Foundation Gobbi N.P.O., Milan, Italy

A. van Domselaar
Relieve Clinics, Leuven, Belgium

L. Podesta
Bluetail Medical Group & Podesta Orthopedic Sports Medicine, Naples, FL, USA

Orlando College of Osteopathic Medicine, Orland, FL, USA

A. Gobbi et al. (eds.), *Regenerative Medicine in Sports and Orthopaedics*,
https://doi.org/10.1007/978-3-031-84693-9_22

studies reporting positive outcomes and others finding no benefit or even negative outcomes [9–11]. As a result, authors from these studies often discourage the use of orthobiological preparations for specific musculoskeletal (MSK) pathologies. It is essential to establish definitive and accepted standards for the preparation of orthobiological preparations to ensure reliable and reproducible results, ultimately leading to better outcomes for patients.

The orthobiological field requires significant changes to enhance the quality of orthobiological products and achieve more consistent and positive outcomes for patients. By adopting algorithms to determine optimal platelet dosing strategies combined with physiologically different PRP bioformulations, and fostering an unyielding awareness of product variability, we can overcome the limitations of the "one-size-fits-all" approach [1].

Ultimately, a better understanding of orthobiological cellular behavior in fascia, ligaments, tendons, cartilage, bone, muscle fibers, and nervous filaments will contribute to improved mechano-metabolic conditions to restore MSK function after injury or tissue degenerative processes. By tailoring treatments to specific tissue structures and pathologies, orthobiological research aims to revolutionize the management of musculoskeletal disorders. As further research is conducted, the hope of developing more effective and personalized treatments for these conditions grows.

22.2 Platelet-Rich Plasma Preparations Background Details

Autologous platelet-rich plasma (PRP) is a versatile, centrifuge-processed liquid fraction derived from harvested fresh peripheral blood. It is characterized by its heterogeneous and complex composition, containing a multitude of multicellular components within a small volume of plasma [7]. PRP is generally considered safe. No systemic adverse effects have been observed, making PRP an attractive treatment option. One of the key aspects of PRP is its significant increase in the number of platelets compared to baseline values, with some devices utilizing ingenious preparation techniques to ensure a high-quality PRP product [12, 13].

PRP therapies have been widely used for various indications for over 30 years. Numerous studies have reported clinical benefits and promising patient outcomes in treating various conditions such as osteoarthritis, tendinopathies, and other musculoskeletal disorders [14–16]. Despite this increasing popularity, there remains a significant challenge in standardizing the preparation, concentration, and formulations of PRP, leading to variations in its cellular properties, bioformulations, and clinical effectiveness. These profound dissimilarities pose sometimes challenges in establishing consistent clinical results, hindering its widespread adoption, and growing clinical treatment needs [17–20].

22.2.1 PRP Cellular Properties

The literature has mentioned that the formulation of PRP products differs significantly between devices with regard to platelet concentration (PLTc) and bioformulation, defined as the final cellular composition of PRP (e.g., leukocyte populations with different concentrations) [21, 22]. Consequently, the cellular properties vary significantly among PRP devices. PRP can contain a diverse range of cells, including platelets, neutrophils, monocytes, lymphocytes, and erythrocytes, which all potentially can contribute to the overall PRP regenerative potential [5, 7]. However, the ratios and distribution of these cells vary across different PRP preparations, impacting their biological activities and therapeutic outcomes [23]. Furthermore, the PRP growth factor content, total protein content, cytokine, and other biomolecules vary significantly [24, 25].

22.3 The Importance of PRP Platelet Dosing and Bioformulation in Angiogenesis, Immunomodulation, and Painkilling

PRP preparations have the potential to initiate and regulate critical pathways, like angiogenesis, inflammatory control, and immunomodulation, and accomplish painkilling, to achieve tissue repair, regeneration, and wound healing. During these pivotal processes, PRP platelet dosing and bioformulation strategies hold important roles to realize functional repair leading and positive patient outcomes.

22.3.1 Angiogenesis

In orthobiology, angiogenesis is an intricate and multifaceted process essential for the sprouting and organization of micro-vessels from pre-existing blood vessels in diseased and degenerative tissue, especially in areas with poor vascularity [1, 26]. One of the critical steps in angiogenesis is the migration and proliferation of endothelial cells. Platelet growth factors, in particular vascular endothelial growth factor-A (VEGF-A), adhesion molecules, and chemokines, serve as crucial mediators between cell membrane receptors and endothelial cells, facilitating and regulating cell-cell communication, and the promotion of cell migration [27, 28].

It is important to mention that Giusti and co-workers concluded in a dose-defining study that 1.5×10^6 platelets/μL was the optimal platelet dose for promoting angiogenesis [29]. Furthermore, cell-matrix interactions are instigated by the interaction of platelet growth factors with matrix proteins to promote cell adhesion, migration, and proliferation to stabilize newly formed blood vessels, crucially balancing the abundance of pro- and anti-angiogenic platelet factors [30, 31].

Apart from a high density of platelets, LR-PRP preparations are a rich source of leukocytes (neutrophils, monocytes, lymphocytes) and other biomolecules which are either directly, or indirectly, involved in angiogenesis [32–35]. Leukocytes in LR-PRP stimulate cell proliferation, differentiation, with monocyte secreting pro-angiogenic molecules, including the expression of VEGF [36]. Additionally, pro-inflammatory cytokines, TNF-α and interleukins, in combination with platelet growth factors VEGF and FGF incite angiogenesis [37].

22.3.2 Immunomodulation

The initiation of inflammatory pathways and the activation of the wound-healing cascade are closely linked to the involvement of leukocytes and platelets [38]. Leukocytes, particularly monocytes, macrophages, neutrophils, dendritic cells, and natural killer cells, possess specific functions in the innate immune system to detect and eliminate intruding pathogens or tissue fragments [39–41]. Lymphocytes and their subsets also play a role in the adaptive immune system [42, 43]. Platelets are among the first cells to recognize endothelial tissue injury and release platelet agonists, which contribute to the inflammatory response, regulating immunomodulatory processes to promote tissue repair and wound healing [44].

The presence of inflammatory stimuli in orthobiological PRP therapies causes a change in the surface expression of platelet receptors. These changes stimulate platelet–leukocyte interactions, leading to the formation of platelet–leukocyte aggregates [45]. Neutrophils and monocytes are both active participants in aggregate formations, contributing to the regulation of the innate immune response, and ultimately tissue repair [46, 47].

22.3.3 Painkilling

The exact mechanisms behind the painkilling effects of platelet-rich plasma (PRP) have not yet been fully elucidated. Everts et al. proposed that platelet-dense granules release high concentrations of serotonin (5-HT) in activated platelets [48]. They hypothesized that 5-HT may have a mechanistic role in painkilling. Evidence supporting this idea comes from studies that have demonstrated higher levels of 5-HT in peripheral blood after platelet activation. In addition to the release of 5-HT from platelets, mast cells and endothelial cells also release endogenous 5-HT in response to tissue injury or surgical trauma [49]. These endogenous 5-HT releasers can interfere with peripheral and multiple neuronal 5-HT receptors, potentially decreasing the extent of pain following noxious stimulation [50]. Another proposed analgesic mechanism involves a decrease in tissue inflammation and subsequent tissue repair. Araya et al. suggested that PRP may have anti-inflammatory properties that contribute to its analgesic effects [51].

Recently, a role for polarizing monocytes was described in pain relief [52]. Furthermore, it was demonstrated that intra-articular PRP injections decreased pain in neuropathic OA pain by down-regulating microglial activation and alleviating nerve pain [53].

22.4 PRP Properties in Tissue Repair and Regeneration

The literature indicates that the use of autologous PRP products is safe, with no known systemic adverse effects, compared to other non-autologous biologics. This safety profile makes PRP an attractive option for various clinical applications, especially in orthopedics, sports medicine, and spine pathologies, as well as other medical fields. The absence of systemic side effects reduces the risks associated with allogeneic products, such as the potential for the transmission of infectious diseases or the development of immune reactions. PRP employment in patients with tissue degenerative diseases, or after trauma, signifies complex multifactorial and synergistic biological processes. The therapeutic actions of PRP preparations are primarily due to the release of a large quantity of platelet factors, proteins, cytokines, adhesion molecules, and other biomolecules that are involved in tissue reparative processes leading to tissue repair and regeneration.

In the natural healing cascade, platelet activation is the crucially important first step as it triggers the release of various bioactive factors that are essential for tissue repair and regeneration [54]. Importantly, neutrophils, monocytes, and lymphocytes are active participants in the natural healing cascade, a series of well-orchestrated sequential cellular activities and extracellular signaling processes that are essential for tissue repair, integrity, remodeling, and ultimately functional improvement. Interestingly, many PRP devices and preparation protocols exclude the presence of these leukocytic cells, even though they are pivotal contributors in tissue repair as they are natural occurring cellular partners in the healing cascade [9, 17].

For PRP to mimic nature's healing cascades effectively, the platelet granular constituents, platelet proteins, and biomolecules must exhibit the ability to stimulate angiogenesis, express immunomodulatory activities, and contribute to painkilling [35]. To achieve this, PRP should be capable of inciting cell proliferation and cell differentiation, contributing to chemotactic cell migration, and facilitating the activities of mesenchymal and neurotrophic factors. Furthermore, the presence of high concentrations of leukocytes in an autologous multi-cellular PRP preparation can initiate the full complexity of biological cell signaling, including microenvironmental signaling [1].

22.5 PRP Preparation Considerations

Autologous PRP refers to the concentrated liquid fraction of freshly derived, anticoagulated peripheral blood that exceeds baseline values due to its high concentration of platelets, currently defined as PLTcs exceeding $1 \times 10^6/\mu L$ [55]. PRP specimens can be prepared using simple test-tube-like kits or more sophisticated PRP systems that are able to process larger whole blood volumes. PRP treatment specimen can have a meaningful impact on patient outcomes, as important PRP parameters differ greatly between the many available PRP devices, such as PLTc, and the presence, or absence, of leukocytes.

PRP formulations should be tailored to meet specific therapeutic requirements, such as tissue repair and regeneration, anti-inflammatory activities, and immunomodulatory therapies, based on tissue specificity [7]. For these reasons, PRP devices and their specific preparation protocols should be capable of producing different PRP formulations, including options for platelet dosing and bioformulation strategies.

Prior to PRP preparation, practitioners should be aware of the various, essential, preparation factors that have an effect on platelet dosing and bioformulation strategies, as shown in Table 22.2.

22.5.1 Whole Blood Centrifugation

When WB is subjected to centrifugation, the cells are suspended in a liquid medium, which can stress and damage them. Swing-out, horizontal whole blood (WB) centrifuges are most frequently used to process a unit of fresh anticoagulated WB as they allow for better, and more equal, separation of WB cellular and plasma components based on the densities of those components [56]. These devices utilize the principle of density gradient centrifugation, where the densities of different blood components determine the speed at which they are centrifuged. By adjusting the speed and duration of the centrifugation, the centrifuge can separate the components based on their individual cellular densities, red blood cells (RBCs), neutrophils, monocytes, lymphocytes, and platelets. Each component exhibits a different density, allowing separation to be achieved.

The horizontal orientation of the centrifuge allows for better distribution of the blood components, ensuring that each cell is evenly exposed to the centrifugal force [57]. This minimizes the risk of damage or clumping to the cells, leading to less trauma with a more homogenous PRP product.

22.5.2 Gravitational Cellular Density Separation

Dual-spin PRP devices leverage the principles of gravitational cellular density separation, as visualized in Fig. 22.1, to create a multicellular, layered, buffy-coat stratum [1]. After the first-spin, when dual-spin PRP devices are used, whole blood is separated in a red blood cell (RBC) fraction, and a plasma fraction, and in between these fractions an earliest buffy coat stratum. After the second-spin, cells are separated based on their individual size and density differences and are concentrated. The concentrated cells are arranged in an orderly manner in a volume of plasma at the bottom of the PRP device [25]. This arrangement should allow for the selective withdrawal of platelets and other cellular constituents, based on the practitioner's preferences.

As mentioned, dual-spin and large-volume PRP devices can optimize PRP preparations by producing a biological product with high PLTcs. In contrast, single-spin, low-volume devices create a product from the acellular plasma layer, which generally excludes all leukocytes and erythrocytes. This approach results in a PRP product with a low PLTc, lacking dense platelets and leukocytes.

Noteworthy, home-made, low-volume PRP preparations are utilized using the same principles as commercially available dual-spin PRP devices [58]. However, this requires foremost a dedicated and meticulously acting practitioner, with a "biological mindset," and patients will have to predonate more anticoagulated blood to prepare a therapeutically effective PRP product. Furthermore, depending on the PRP preparation disposable, such as syringes or Falcon tubes, the centrifugation parameters must be validated.

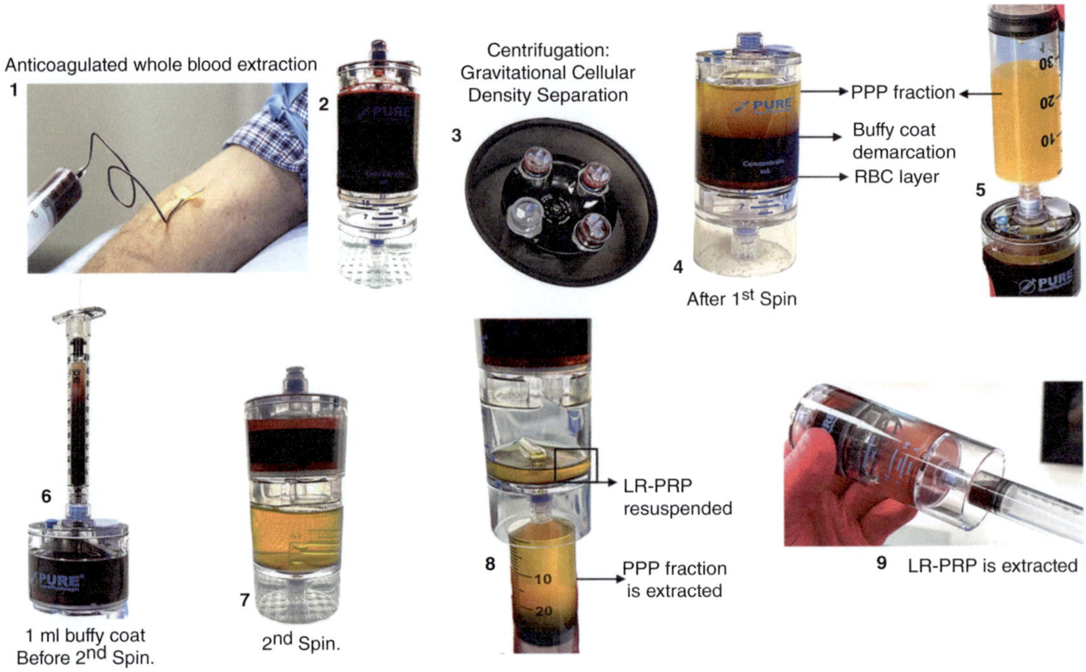

Fig. 22.1 Schematic representation from whole blood collection to dual-spin PRP processing
Gravitational density separation after a dual-spin PRP procedure (PurePRP®-SP, EmCyte Corporation®, Fort Myers, FL, USA)
Processing steps: (1) Whole blood is collected in a syringe containing sodium citrate anticoagulant; (2) Transferred to a PRP device; (3) Placed in a centrifuge syringe for the first spin; (4) After the first spin, the whole blood is separated in a PPP fraction and RBC layer, with in between the demarcation of the buffy coat layer; (5) The PPP fraction is extracted; (6) After collecting the PPP, an additional 1 ml of the buffy coat, to prepare LR-PRP, is extracted and injected in the bottom part of the PRP device, followed by the collected PPP; (7) After the second spin, the concentrated cells are at the bottom of the device; (8) The PPP fraction is removed till the remaining plasma volume reaches the desired total PRP volume, and the cells are resuspended; (9) After resuspension, the LR-PRP is extracted from the device

22.5.3 Single-Spin Versus Double-Spin PRP Preparations

PRP devices must meet rigorous validation standards to ensure the quality and consistency of the resulting formulations. These validation processes involve testing the efficacy and safety of different PRP formulations under controlled conditions. By adhering to validated protocols, practitioners can be confident that the PRP products they produce are safe and effective for their intended applications.

It has long been recognized that there are substantial differences in PRP cellular properties, PLTc, and bioformulation between single- and double-spin PRP devices [59, 60]. It should not be surprising that the variations in PRP quality and treatment volumes have a significant impact on patient outcomes [9, 11, 17]. These PRP quality differences can primarily be attributed to the differences in device design concepts, centrifugation parameters, WB predonation volumes, PRP preparation protocols, and finally PRP volumes [59, 61]. Additionally, the cellular characteristics of single- and double-spin PRP preparations differ significantly, with single-spin PRP preparations demonstrating a profuse variability in PLTc and bioformulation as they generate primarily a PRP product from the acellular plasma layer, excluding cellular components, and they demonstrate low platelet capture rates [21, 22]. The double-spin PRP device, on the other hand, consistently produces PRP profiles with more consistent PLTcs and optional bioformulation variances to include leukocytes, which play a crucial role in the classical healing cascade and therefore contribute to PRP's therapeutic potential. Generally, double-spin protocols provide superior PRP preparations in terms of quality, consistency, and PLTc [60].

Oh et al. recognized great variations in PLTcs, cellular composition, and biomolecular characteristics between single-spin and dual-spin preparation methods [3]. In addition, VEGF and PDGF concentrations were significantly higher following dual-spin PRP preparation than following single-spin methods [7, 62].

Marques et al. described in a comprehensive study the correlation between inferior PRP treatment outcomes and poor quality and inconsistent PRP preparations [63]. They concluded that inferior patient outcomes were directly related to the poor quality and inconsistent PRP preparations produced by single-spin devices. Furthermore, Fadadu et al. confirmed that single-spin test tube-type "PRP" systems generally produced platelet counts of 0.52 times baseline values, effectively generating platelet-poor plasma [21]. However, they also stated that double-spin PRP devices generate significantly higher PLTcs.

Practitioners should be aware of the technical capabilities and limitations of their own PRP device and preparation protocol in order to produce variable PRP formulations.

22.5.4 Failure to Classify PRP and Associated Biological Considerations

Lacking consensus on the standardization of PRP devices and their specific preparation protocols to create PRP formulations has contributed to the development and introduction of a wide variety of different PRP devices, with many devices demonstrating inconsistent patient outcomes [11, 17–20, 64, 65].

Analyzing published studies with negative or no positive outcomes revealed that the employed PRP devices produced dissimilar cell capture rates, with low PLTc and varying leukocyte concentrations.

Moreover, the absence of generally accepted standards presents a challenge in evaluating and comparing different preparation techniques, even though several groups published different classification systems [66–70], referring to platelet and leukocyte concentrations, platelet increase or decrease over baseline values, etc. Rossi et al. proposed three principal groups of PRP preparations for orthobiological applications: leukocyte-rich PRP (LR-PRP), leukocyte-poor PRP (LP-PRP), and pure platelet-rich plasma (P-PRP) and pure platelet-rich fibrin (P-PRF), which are more specific than the generic PRP definition [71]. However, the characteristics and cellular compositions of these groups may vary largely.

The complexity and diversity of PRP applications, combined with ongoing advancements in regenerative medicine, make it challenging to establish a standardized system that pleases the needs of all stakeholders.

22.5.5 PRP Device Operational Flexibility in Creating Multiple Formulations

Ideally, practitioners want to create different PRP formulations with the same PRP device, based on the treatment goals and pathoanatomic conditions. It is important to note that the different PRP formulations, P-PRP, LP-PRP, and LR-PRP (Fig. 22.2), even when prepared with the same device will significantly demonstrate varying platelet and other cell concentrations.

The same dual-spin PRP device was used to process blood from the same donor to prepare the three different PRP formulations. In Fig. 22.2, the differences in visual aspect between P-PRP, LP-PRP, and LR-PRP are noticeable. The PRP preparation parameters and hematology data for the three PRP formulations are shown in Table 22.1. PRP preparations, encompassing fewer leukocytes, revealed profoundly lower PCRs, and comprised of less TAP. LR-PRP contained the highest number of TAP, as well as a more than tenfold increase in reparative mono-

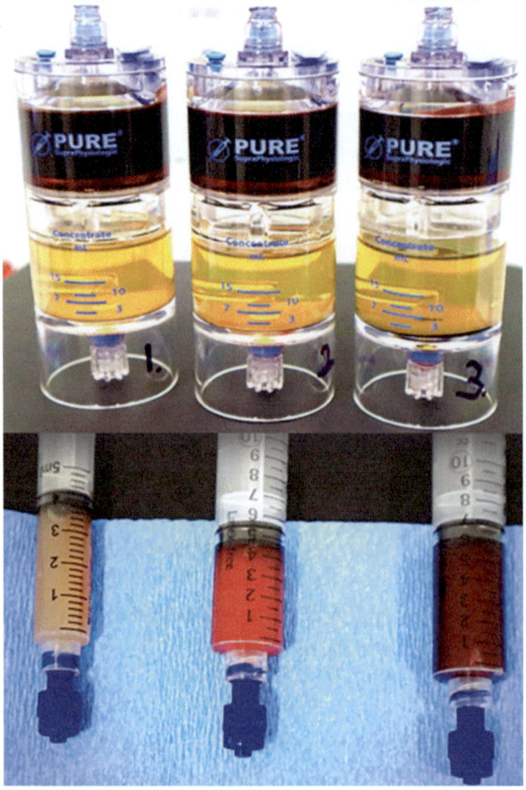

P-PRP **LP-PRP** **LR-PRP**

Fig. 22.2 Visual aspect of P-PRP, LP-PRP, and LR-PRP preparations with the same device. Abbreviations: *P-PRP* pure platelet-rich plasma, *LP-PRP* leukocyte-poor PRP, *LR-PRP* leukocyte-rich rich PRP

cytes, while the RBC concentration was the highest from all three preparations.

In Table 22.1, the PRP preparation parameters and hematology data (Beckman Coulter DxH 500–5-part differential, Brea, CA, USA) are presented. The PCR of the different PRP formulations is portrayed by the varying TAP numbers. The substantial increase in TAP from P-PRP to LP-PRP to LR-PRP (31.1%, and 12%, respectively) can be explained by the fact that denser platelets are captured as their density is closer to the densities of neutrophils and RBCs, with a high concentration of platelets present in the buffy coat stratum, as the TAP in the NR-PRP suspension were 65% higher than in the P-PRP.

In the event that practitioners do not have access to commercially available and dedicated PRP systems, home-made protocols can be created and utilized to produce safe and effective PRP products using alternative disposables, such as variable-size syringes or an equivalent alternative sterile disposable. Often, these disposables have to be manually re-shaped to be processed using tilting centrifuges. In Fig. 22.3, we present a dual-spin protocol to prepare LR-PRP (internal communication, Drs. Lana and Pires, Brazilian Institute Regenerative Medicine (BIRM), Indaiatuba SP, Brazil, 2024). This labor-intensive process requires meticulous processing tech-

Table 22.1 PRP preparation parameters and hematology data

Parameters and CBC data	Average WB	PurePRP	LP-PRP	LR-PRP
Volume (mL)	60	3.8	4.6	6.4
Platelet concentration ($\times 10^3/\mu L$)	224	1885	2174	1849
TAP ($\times 10^9$)	13.44	7.16	10.04	11.83
PCR (%)		53.3	77.4	88
Neutrophil ($\times 10^3/\mu L$)	3.4	0.61	3.97	11.76
Monocyte ($\times 10^3/\mu L$)	0.4	2	2.91	4.49
Lymphocyte ($\times 10^3/\mu L$)	1.15	9.27	8.95	12.47
RBC ($\times 10^9/\mu L$)	6.22	0.05	0.22	1.87

Abbreviations: *CBC* complete blood count, *WB* whole blood count, *PRO* platelet-rich plasma, *LP-PRP* leukocyte-poor, *PRP NR-PRP* leukocyte-rich rich PRP, *TAP* total available platelets, *PCR* platelet capture rate, *RBC* red blood cells

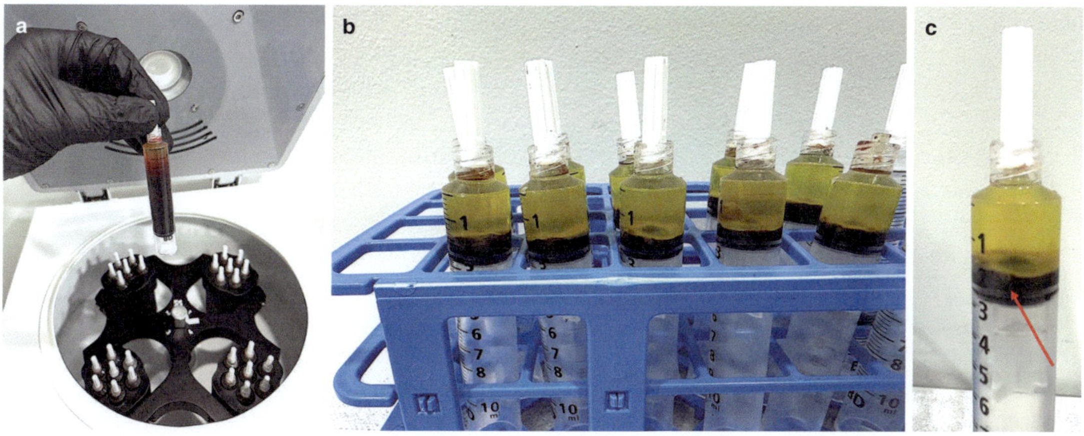

Fig. 22.3 Homemade dual-spin preparation method to create LR-PRP. (**a**) In a centrifuge, a predetermined number of re-shaped 10 mL syringes are loaded with anticoagulated whole blood for PRP preparation; (**b**) After a dual-spin centrifugation protocol, the individual PRP syringes can be consolidated for application; (**c**) The concentrated cell layer is clearly visible on top of the plunger (indicated by the red arrow). (Courtesy of Drs. Lana, Pires, BIRM, Indaiatuba SP, Brazil)

niques. Depending on the treatment plan, i.e., how many pathological sites will be treated and how much PRPv is needed, the number of syringes will be determined. This manual dual-spin preparation method can achieve a six- to sevenfold increase in PRP-PLTc over WB-PLTc (personal communications, Dr. Pires).

22.6 Defining Increase in Platelets Over Baseline, Total Available Platelets, and Platelet Dose

Many PRP devices are available on the market, with significantly varying differences in quality and efficacy. A number of performance parameters are described in the literature for quantifying and measuring the quality of PRPs.

22.6.1 PLTc

Strikingly, the literature cites frequently the increase of PLTc in the PRP above WB-PLTc as a quality parameter for PRP applications. Though, the fold increase over baseline values is only a limited quality marker for the efficacy of the employed PRP device and the correlated preparation procedure.

Importantly, the PRP-PLTc fold increase over baseline is an invalid parameter to correlate to patient outcomes following a PRP procedure. Therefore, the PLTc alone does not provide an accurate representation of the total number of reparative platelets (platelet dose) that have been administered. Different PRP devices and preparation methods produce different PLTcs [1, 72]. Astoundingly, Miron et al. described the preparation of a tenfold increase in PLTc using an injectable PRF preparation protocol; however, the volume of this platelet concentrate varied between 0.3 and 0.5 mL.

Formula Factor increase in platelets (or leukocytes):
"Platelet concentration in PRP/Platelet concentration in whole blood"

PRP-A prepared according to Everts et al. [1].
WB-PLTc: 195,000/µL; WBv: 61 mL.
PR-PLTc: 2,243,000/µL; PRPv: 4.5 mL.
PRP-PLT fold increase: 2,243,000: 195,000 = 11.50 times increase over WB-PLTc.

PRP-B prepared according to Everts et al. [1].
WB-PLTc: 211,000/µL; WBv: 62 mL.
PRP-PLTc: 1,753,000/µL; PRPv: 6.5 mL.
PRP-PLT fold increase: 1,753,000: 211,000 = 8.30 times increase over WB-PLTc.

PRP prepared according to Miron [72].
WB-PLTc: 220,000/µL—300,000/µL; WBv: 10 mL.
PRP-PLTc: 2,000,000—3,000,000/µL; PRPv: 0.3 ml–0.5 mL.
PRP-PLT fold increase: 2,000,000: 220,000 = 10 times increase over WB-PLTc.
PRP-PLT fold increase: 3,000,000: 300,000 = 10 times increase over WB-PLTc.

In the above example, using the same dual-spin PRP device, two different preparation protocols were utilized to prepare LP-PRP and LR-PRP (A and B, respectively) treatment specimens. The PRP-PLTc over WB baseline increase was 11.5 and 8.3 times, respectively. Furthermore, the method according to Miron produced a PRP with a PLTc 10 times over WB-PLTc.

A great variability is noted between PRP devices and accompanying preparation methods regarding WBv, PLTc, and PRPv. These published clinical data demonstrate that an increase in PLTc over WB baseline PLTc does not inform about the number of TAP to treat patients, let be that this parameter provides any information about the number of PLTs administered to multiple treatment sites using the same PRPv.

To provide more clarity about PRP formulations to treat different pathoanatomic conditions, we need to implement a calculation that informs about the total available platelets (TAP) to perform PRP applications.

22.6.2 Total Available Platelets (TAP)

The absolute number of platelets that are present in a particular volume is termed TAP. To calculate the TAP, the PLTc is multiplied by the volume in which the platelets reside, like in the patient's WB, to prepare PRP and the final PRP suspension. Note that to calculate the TAP, the unit of measures for PLTc and volumes (WB and PRP) need to be the same. Therefore, changing from µL to ml requires multiplying the PLTc with 1000.

Formula to calculate TAP (in whole blood, or PRP):
"Platelet concentration x volume of this platelet concentration"

PRP prepared according to Everts et al. [1].
WB-PLTc: 195,000/µL; WBv: 61 mL.
PRP-PLTc: 2,243,000/µL; PRPv: 4.5 mL.
TAP in prepared PRP: 2,243,000 × 4.8 × 1000 = 10.1 × 10^9 platelets (10.1 billion).

PRP prepared according to Everts et al. [1].
WB PLTc: 211,000/µL; WBv: 62 mL.
PRP PLTc: 1,753,000/µL; PRPv: 6.5 mL.
TAP in prepared PRP: 1,753,000 × 6.58 × 1000 = 11.4 × 10^9 platelets (11.4 billion).

PRP prepared according to Miron [72].
WB PLTc: 220,000/μL; WBv: 10 mL.
PRP PLTc: 2,000,000–3,000,000/μL PRPv: 0.3 mL–0.5 mL.
TAP in prepared PRP: $2,000,000 \times 0.5 \times 1000 = 1 \times 10^9$ *platelets (1 billion).*
TAP in prepared PRP: $2,000,000 \times 0.3 \times 1000 = 600 \times 10^6$ *platelets (600 million).*

TAP in prepared PRP: $3,000,000 \times 0.5 \times 1000 = 1.5 \times 10^9$ *platelets (1.5 billion).*
TAP in prepared PRP: $3,000,000 \times 0.3 \times 1000 = 900 \times 10^6$ *platelets (900 million).*

From this calculation, it is obvious that an increase over baseline of PRP-PLTc does not provide any useful information about the TAP to treat patients. The PRP specimen was only created by producing an extreme low PRPv, ranging from 0.3 to 0.5 mL. Consequently, the TAP to treat patients with the Miron method was significantly lower when compared to the A and B PRP preparations (ranging from 900 million to 1.5 billion, to 10.1 and 11.4 billion, respectively).

Noteworthy, when PRP is used to treat multiple pathological conditions, the number of TAP will decrease accordingly and need to be accounted for in the treatment plan and PRP production protocol.

22.6.3 Platelet Dose: Total Deliverable Platelets (TDP): Treatment Plan

The TAP can be considered as the "PRP platelet storage reservoir." The platelet dose is synonymous with the actual number of total deliverable platelets (TDP) to one treatment site. The platelet dose is calculated per treatment site by multiplying the applied PRPv with the PRP-PLTc. When multiple pathological conditions are treated, each platelet dose is separately calculated, based on guidelines addressing tissue specificity and pathology, in order to accomplish positive patient outcomes.

To have a sufficient platelet dose for each treatment site, it is imperative to develop a patient-specific treatment plan that includes the number of treatment sites with the expected PRP volumes per site, the platelet dose, and bioformulation per pathoanatomic condition.

Based on this PRP treatment plan, the total needed PRPv and TAP can be calculated. Accordingly, these parameters will determine the total WBv needed to prepare PRP, based on the WB-PLTc.

Formula to calculate platelet dose or total deliverable platelets:
"PRP platelet concentration x PRP volume of a single application"

PRP PLTc: 1,753,000/μL; PRPv: 6.5 mL.
TAP in prepared PRP: $1,753,000 \times 6.58 \times 1000 = 11.4 \times 10^9$ *platelets (11.4 billion).*

Platelet dose in a 4 mL treatment syringe: $4 \times 1,753,000 \times 1000 = 7.01$ *billion TDP.*
Platelet dose in a 2.5 mL treatment syringe: $2.5 \times 1,753,000 \times 1000 = 4.38$ *billion TDP.*

Cave: the limiting factor is the total prepared PRP volume when distributing the PRP volume in smaller aliquots to meet the individualized number of treatment sites and the anticipated injected PRPv per site.

22.7 Critical Factors Shaping PRP Platelet Dosing and Bioformulation

In order to calculate the platelet dose and bioformulation of any PRP preparation, it is essential to analyze both WB and PRP specimens using a hemoanalyzer, which measures, among other parameters, platelet and leukocyte counts. Importantly, calculating platelet dosages needs to be meticulously executed with emphasis on accurate volume readings for WBv and PRPv.

The creation of customized, pathoanatomic, and tissue-type specific PRP treatment vials requires a thorough understanding of some critical factors (Table 22.2) that influence the preparation of adequate and effective PRP formulations.

Table 22.2 Critical Factors in PRP Dosing and Bioformulation Strategies

Patient	PLTc	Variable: 100–400,000/μL
	WBv	Variable
PRP device	Number of spin cycles	Single vs. dual
	Size	Low vs. high volume
	Flexibility	Accurately extracting variable volumes
	Sterility	Use 1 device vs. multiple for the same PRPv
	PRPv	Variable preparation volume
	PRP-PLTc	Variable: High–low
	Platelet capture rate	High vs. low
	Buffy coat separation	Include no, low, high leukocytes

Abbreviations: *PLTc* platelet concentration, *WBv* whole blood volume, *PRPv* platelet-rich plasma volume, *PRP-PLTc* platelet-rich plasma-platelet concentration

22.7.1 Patient PLTc, WBv, and TAP

In order to meet accepted platelet dosing requirements for specific conditions, it is necessary to consider the WB-PLTc as a patient variable which must be addressed when preparing PRP, especially if the WB-PLTc is less than 100,000/μL.

The PRP-TAP can be increased in these circumstances by processing a higher WBv, since this will potentially lead to an increase in the number of WB-PLTs to prepare PRP. Additionally, the efficacy of the PRP device is important to capture a high concentration of the WB platelets (platelet capture rate, %) to create a PRP product with a high TAP number while using a minimal amount of patient's own blood.

WB-PLTc: 100,000/μL, WBv: 60 mL TAP: $100{,}000 \times 60 \times 1000 = 6 \times 10^9$.

WB-PLTc: 100,000/μL, WBv: 120 mL TAP: $100{,}000 \times 120 \times 1000 = 12 \times 10^9$.

A patient with a WB-PLTc of 100,000/μL has a low TAP number per unit of volume. When 60 mL of anticoagulated WB is extracted to produce PRP, the WB-TAP number is 6 billion platelets, whereas a blood draw of 120 mL would facilitate a PRP preparation where the WB-TAP is 12 billion platelets. Finally, the platelet capture rate (PCR) of the employed PRP will ultimately determine the PRP-TAP number.

22.7.2 PRP Device Size and Preparation Volume

The size of the PRP device allows for the WB volume that can be processed in a single preparation cycle. Different PRP devices have different maximum capacities to process a unit of WB, ranging from 7 mL to 180 mL. Generally, PRP devices can be categorized into low-volume (< 30 mL) and high-volume PRP devices. The maximum preparation volume determines how many PRP devices, or test tube-like kits need to be used for a treatment procedure.

WB-PLTc: 100,000/μL, WBv: 120 mL TAP: $100{,}000 \times 120 \times 1000 = 12 \times 10^9$.

PRP Device A: a high-volume device: maximum processing volume of 60 mL: 2 devices are needed.

PRP Device B: a low-volume device that has a maximum processing volume of 10 mL: 12 devices are needed.

It is fair to assume that the use of fewer devices to process the same WBv reduces the risk of bacterial contamination and decreases the overall processing time. Furthermore, the final PRP volume handling is simpler in high-volume devices.

Noteworthy, the PCR has still to be factored in regarding platelet dose and bioformulation specifics.

22.7.3 PRP Platelet Capture Rate

The effectiveness of PRP devices in concentrating platelets from WB can be measured objectively by the PCR and is expressed as a percentage of the WB-TAP.

Formula to determine Platelet Capture Rate Percentage:

"PRP-TAP / WB-TAP" × *100%*

PRP prepared according to Everts et al. [1].

WB PLTc: 211,000/μL; WBv: 62 mL.

TAP in WB: $211{,}000 \times 62 \times 1000 = 13.1 \times 10^9$ *platelets (11.4 billion).*

TAP in PRP: $1{,}753{,}000 \times 6.58 \times 1000 = 11.4 \times 10^9$ *platelets (11.4 billion).*

PCR: $11.4 \times 10^9 / 13.1 \times 10^9 \times 100\% = 87.0\%$.

PRP prepared according to Miron [72].

WB PLTc: 220,000/μL; WBv: 10 mL.

TAP in WBP: $220{,}000 \times 10 \times 1000 = 2.2 \times 10^9$ *platelets (2.2 billion).*

TAP in PRP: $2{,}000{,}000 \times 0.3 \times 1000 = 0.6 \times 10^9$ *platelets (0.6 billion).*

PCR: $0.6 \times 10^9 / 2.2 \times 10^9 \times 100\% = 27.0\%$.

As illustrated above, PRP devices and their contemporaneous operational preparation protocols differ greatly in extracting concentrated platelets, including the number of spin cycles, device-specific centrifugation parameters, PRP device design, and instructions for use. Here, the dual-spin PRP device demonstrated almost a 70% higher PCR when compared to the test-tube PRP device, despite its tenfold increase in PLTc above WB-PLTc. Therefore, PRP practitioners should consider using the PCR as a quality and performance indicator, as it has been reported that the PCR values can vary from more than 90% to less than 25% [59].

Accordingly, the PCR has a direct impact on the total volume of whole blood required to prepare adequate PRP volumes, used to meet PRP dosing and bioformulation strategies, and will also determine the number of PRP devices that need to be used to meet these requirements. Most importantly, the reparative and regenerative effects are negatively affected by low administered platelet dosages, as described clearly by Everts et al. [35].

22.7.4 Buffy Coat Separation and Extraction Variability

Dual-spin PRP devices leverage the principles of gravitational cellular density separation, as visualized in Fig. 22.4, to create a multicellular, layered, buffy-coat stratum. Only few PRP devices can pellet and resuspend platelets allowing for the creation of either P-PRP, LP-PRP, or LR-PRP suspensions. The concentration of monocytes, lymphocytes, and neutrophils will vary significantly, depending on the gravitational density separation and the extraction of P-PRP, LP-PRP, and LR-PRP suspensions, as graphically illustrated in Fig. 22.4.

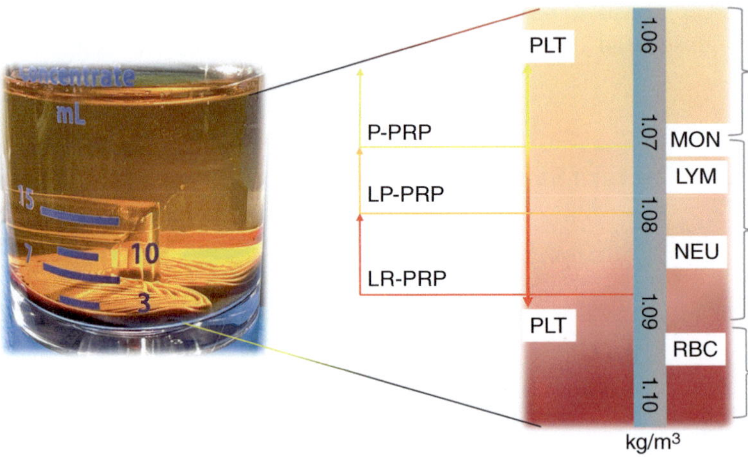

Fig. 22.4 Graphical illustration linking the buffy coat stratum to PRP formulary definitions
Gravitational cellular density separation of whole blood following a two-spin centrifugation procedure. (PurePRP®-SP device, EmCyte Corporation®, Fort Myers, FL, USA). After the first spin, WB is separated into basically two main layers—the platelet-poor plasma suspension and the RBC layer—with in between the inception of the buffy coat stratum. After the second spin, clearly visible at the bottom of the PRP device, a multicomponent, buffy coat layer has been created based, on the principles of gravitational cell density separation (in kg/m^3). The magnification indicates that the buffy coat layer is located between the pure plasma suspension and the RBC layer. The ruler shows the individual cell densities [73] in the buffy coat stratum and identifies which cells are present in P-PRP, LP-PRP, and LR-PRP, based on their cellular densities. Note the overlap in the range of cellular densities of the various cells, specifically the density of PLTs, can range from 1.040 to 1.091 kg/m^3. In this figure, a particular volume of PPP is aspirated after LR-PRP preparation, leaving approximately 6 ml of PPP behind to resuspend the multicellular buffy coat layer, containing high concentrations of platelets and monocytes, with elevated lymphocytes and neutrophil concentration, as illustrated in Fig. 22.5. Abbreviations: P-PRP pure platelet-rich plasma, LP-PRP leukocyte-poor PRP, LR-PRP leukocyte-rich PRP, PLT platelet, MON monocyte, LYM lymphocyte, NEU neutrophil, RBC red blood cell, PPP platelet-poor plasma, BCS buffy coat stratum

WB Baseline

Test	Result	Flags	Units	Low	High
WBC	4.85		x10³/µL	3.71	10.67
LY	26.96		%	18.94	46.71
MO	6.49		%	4.88	12.81
NE	63.06		%	40.62	71.65
EO	3.15		%	0.74	6.73
BA	0.34		%	0.05	0.48
LY#	1.31		x10³/µL	1.15	3.52
MO#	0.31		x10³/µL	0.25	0.99
NE#	3.06		x10³/µL	1.85	6.72
EO#	0.15		x10³/µL	0.04	0.48
BA#	0.02		x10³/µL	0.00	0.03
RBC	3.72	l	x10⁶/µL	3.87	5.68
HGB	11.61	l	g/dL	12.00	16.75
HCT	34.2	l	%	35.1	48.7
MCV	92.0		fL	78.4	97.6
MCH	31.2		pg	26.5	33.5
MCHC	33.9		g/dL	32.9	35.4
RDW	12.3	l	%	12.7	15.6
RDW-SD	37.8	l	fL	38.9	49.0
PLT	191.3		x10³/µL	150.5	366.8
MPV	7.96		fL	7.42	10.77

LR-PRP

Test	Result	Flags	Units	Low	High
WBC	28.20	h	x10³/µL	3.71	10.67
LY	47.79	h	%	18.94	46.71
MO	15.10	h	%	4.88	12.81
NE	36.18	l	%	40.62	71.65
EO	0.44	l	%	0.74	6.73
BA	0.50	h	%	0.05	0.48
LY#	13.48	H	x10³/µL	1.15	3.52
MO#	4.26	H	x10³/µL	0.25	0.99
NE#	10.20	h	x10³/µL	1.85	6.72
EO#	0.12		x10³/µL	0.04	0.48
BA#	0.14	h	x10³/µL	0.00	0.03
RBC	1.16	l	x10⁶/µL	3.87	5.68
HGB	3.87	L	g/dL	12.00	16.75
HCT	10.9	L	%	35.1	48.7
MCV	94.2		fL	78.4	97.6
MCH	33.4		pg	26.5	33.5
MCHC	35.5	h	g/dL	32.9	35.4
RDW	12.6	l	%	12.7	15.6
RDW-SD	37.3	l	fL	38.9	49.0
PLT	1951.3	H	x10³/µL	150.5	366.8
MPV	8.18		fL	7.42	10.77

WBv: 62 ml

TAP: 11,9 x 10^9

PRPv: 5.7 ml
PLTc 10.2 x over WB-PLTc
TAP: 11.1 x 10^9
PCR: 93 %
PLT dose: 4.0 ml syringe: = 7,8 x 10^9
1.7 ml syringe: = 3,3 x 10^9

Fig. 22.5 Hematology and LR-PRP data
LR-PRP was prepared from 54 mL of WB mixed and agitated with 8 mL of sodium citrate 3.8%. A total volume of exactly 62 mL was used to produce LR-PRP
After the second spin, 23 mL of PPP was removed and the buffy coat stratum was resuspended with the remaining PPP, extracting precisely 5.7 mL of LR-PRP
The PCR was 93%, resulting in 11.1 × 10^9 TAP. Subsequently, the 5.7 mL PRPv was divided in two syringes (4 mL and 1.7 mL), comprising each of 7.8 × 10^9, and 3.3 × 10^9 platelets, respectively. Abbreviations: WB whole blood, LR-PRP leukocyte-rich platelet-rich plasma, WBv whole blood volume, PRPv platelet-rich plasma volume, PLTc platelet concentration, WB-PLTc whole blood platelet concentration, TAP total available platelets, PCR platelet capture rate, PLT platelet

22.8 Patient Case Description and the Implementation of an Individualized PRP Treatment Plan

Pre-procedure the patient's WB-PLTc is 210,000/μL.

Multiple pathoanatomic lesions and injection sites are determined, requiring both LP-PRP and LR-PRP formulations, as detailed in Table 22.3.

The PRP device used is a two-spin 60 mL device, with a known PCR of 85%.

In this treatment plan, three different knee structures, each with multiple injection sites per structure, have been identified for PRP therapy.

Three separate syringes are prepared, with tailored platelet dosages, PRP volumes, and bioformulations.

Syringe 1: LP-PRP objective is 9×10^9 TDP, in 5 mL.

The required PRP-PLTc is calculated as follows:

(WBv × WB-PLTc) = (PRPv × PRP-PLTc),
PRP-PLTc = (60 × 210,000)/5 = 2520×10^3/μL, *if* all WB platelets can be captured.
However, capturing 100% of all WB platelets is impossible; the PCR of the used PRP device determines the factual PRP-PLTc = 2520×10^3/μL × 85% = 2142×10^3/μL.
Now, the TDP = 2142×10^3/μL × 5 mL = 10.71×10^9 platelets.

Conclusion: One PRP device can generate the predetermined LP-PRP platelet dose, in a 5 mL specimen.

Syringes 2 and 3: LR-PRP objective is each syringe has a TDP count of 4×10^9 platelets.

Needed is a TAP count of 8×10^9: ($4 \times 10^9 + 4 \times 10^9$); and a total volume of 4.5 mL (2.5 mL + and 2.0 mL).

The required PRP-PLTc is calculated as follows:

(WBv × WB-PLTc) = (PRPv × PRP-PLTc),
PRP-PLTc = (60 × 210,000)/4.5 = 2800×10^3/μL, *if* all WB platelets can be captured.
However, capturing 100% of all WB platelets is impossible; the PCR of the used PRP device determines the factual PRP-PLTc = 2800×10^3/μL × 85% = 2380×10^3/μL.
Now, the TAP = 2380×10^3/μL × 4.5 mL = 10.71×10^9 platelets.

Conclusion: One PRP device can generate the predetermined TAP count to be used to create two LR-PRP syringes with the predetermined TDP count for both, the 2 mL and 2.5 mL syringes.

This treatment plan can be executed using *two* 60 mL PRP devices with a known PCR of 85%, to create LP and LR-PRP formulations with the predetermined platelet dosages and treatment volumes, according to the detailed treatment plan.

Table 22.3 Detailed multi-injection orthobiological treatment plan using different platelet dosages and bioformulations to treat different tissue structures

Structures	PRP injection sites	Syringe	Platelet dose	Bioformulation	PRPv in mL
Knee OA	Intra-articular injection Peri-articular synovial capsule	1	9×10^9	LP-PRP	5
Medial meniscus tear	Intra-meniscus Meniscus fem lig Meniscus tib lig	2	4×10^9	LR-PRP	2.5
MCL	Intra-ligament MCL fem fibers MCL tib fibers	3	4×10^9	LR-PRP	2

Abbreviations: *OA* osteoarthritis, *LP-PRP* leukocyte-poor platelet-rich plasma, *PRPv* platelet-rich plasma volume, *LR-PRP* leukocyte-rich platelet-rich plasma, *MCL* medial collateral ligament

Alternatively, same patient simulation using a single-spin 30 ml device with a known PCR of 55%.

Syringe 1: LP-PRP syringe objective is 9×10^9 TDP, in 5 mL.

The required PRP-PLTc is calculated as follows:

(WBv × WB-PLTc) = (PRPv × PRP-PLTc),
PRP-PLTc = (30 × 210,000)/5 = 1260 × 10^3/μL, *if* all WB platelets can be captured.
However, capturing 100% of all WB platelets is impossible; the PCR of the used PRP device determines the factual PRP-PLTc = 1260 × 10^3/μL × 55% = 693 × 10^3/μL.
Now, the TDP = 693 × 10^3/μL × 5 mL = 3.47 × 10^9 platelets.

Conclusion: One PRP device cannot *generate the predetermined LP-PRP platelet dose of* 9×10^9.

To meet this dose, three PRP devices are needed!

Syringes 2 and 3: LR-PRP syringes objective is each syringe has a TDP number of 4×10^9.

Needed is a TAP of 8×10^9: ($4 \times 10^9 + 4 \times 10^9$); a total volume of 4.5 mL (2.5 mL + and 2.0 mL).

The required PRP-PLTc is calculated as follows:

(WBv × WB-PLTc) = (PRPv × PRP-PLTc),
PRP-PLTc = (30 × 210,000)/4.5 = 1400 × 10^3/μL, *if* all WB platelets can be captured.
However, capturing 100% of all WB platelets is impossible; the PCR of the used PRP device determines the factual PRP-PLTc = 1400 × 10^3/μL × 55% = 770 × 10^3/μL.
Now, the TAP = 770 × 10^3/μL × 4.5 mL = 3.47 × 10^9 platelets.

Conclusion: One PRP device cannot *generate the predetermined total LR-PRP platelet dose of* 8×10^9.

To meet this dose, three PRP devices are needed to produce two syringes of LR-PRP!

Overall, this treatment plan can *only* be executed using *six* 30 mL PRP devices with a known PCR of 55%, to create LP and LR-PRP formulations with predetermined platelet dosages and treatment volumes, according to the detailed treatment plan.

22.9 Discussion

PRP treatment protocols have evolved immensely over the past 10 years. Through experimental and clinical research, we now have a better understanding of the regenerative potential of PRP's cellular physiology and synergy between platelets, leukocytes, and other biomolecules. Furthermore, several high-quality systematic reviews, meta-analyses, and randomized controlled trials signify the effectiveness of PRP biotechnology in many medical fields, including orthopedic surgery, sports medicine, dermatology, cardiac surgery, plastic surgery, pain management, spinal disorders [14, 74–79], and recently as well in ENT [80] and infertility pathologies [81, 82].

The quality of PRP is often characterized by the fold increase in PRP-PLTc over WB-PLTc [83, 84]; it is measured by a PLTc greater than 1×10^6/μL [55], which is an approximate equivalent of a fivefold increase in platelets from baseline [79].

Fadadu et al. reviewed 33 PRP systems and protocols [21] and concluded that some systems produced PRP preparations with platelet concentrations less than that of the whole blood platelet concentration. They reported that some reviewed devices had a PRP platelet increase over baseline values as low as 0.52 with a single spin kit [85], whereas some dual-spin PRP systems produced the highest platelet concentration (1.6×10^6/μL). These data were in accordance with the work of Magalon et al. [59].

The heterogeneity of PRP formulations that are used can be attributed to the wide variety of PRP devices that are currently available on the market, each with its own preparation methods, and the lack of consensus regarding minimal criteria to define effective clinical PRP therapies.

Factors frequently cited include the volume of pre-donated whole blood, the type of anticoagulant, the geometrical mathematical and physical properties of the PRP device, and the platelet and leukocyte device capture rates [61]. In addition, the PRP centrifugal operational and performance variables, such as the number of spin cycles, centrifugation speed, spin cycle duration, as well as acceleration and deacceleration speeds, have rarely been mentioned as contributing factors. Therefore, with so many variables at play, it is difficult to determine the true biological potential of PRP therapy for tissue repair, regeneration, and functional recovery.

The regulation of PRP preparation methods, consensus on effective platelet dosing schemes, and tissue-specific bioformulations remain a challenge despite PRP's growing popularity. In the absence of a consensus on standardizing PRP protocols for manufacturing PRP formulations, new devices will continue to be introduced to the market, resulting in even greater variations in the number and characteristics of cells delivered to patients, including new devices that produce low platelet concentrations and varying levels of leukocytes. Due to the growing and non-regulated PRP device market, these profound and vital differences in cellular composition can make it increasingly difficult for practitioners to achieve positive, consistent clinical results.

A study by Everts et al. examined some of the MSK literature in which there was no evidence of positive outcomes for patients undergoing PRP therapy, and they concluded that in all failing studies, patients were suboptimally treated since they were given PRP with low platelet dosages and very low, to no, leukocyte presence. In the authors' opinion, the reasons for the delivered low platelet dosage were related to the use of low-volume, single-spin devices, which resulted in a low platelet count, and a low TAP number, consequently a low PRP platelet dose was administered. Furthermore, in several studies, when multiple pathoanatomic conditions were treated, the TDP per injection site was extracted from the limited TAP count. Accordingly, under these circumstances, the delivered PRP platelet dose was significantly lower per additional injection site. The most logical solution to process more blood to increase the PRP volume was never mentioned.

In spite of the fact that several medical fields have an unmet need for expanding clinical, nonsurgical, and biological treatments, the unawareness of how to design patient-specific treatment plans, including PRP formulation details and procedural calculations, will hinder PRP therapies from becoming widely accepted as a viable alternative therapeutic procedure that can be conveniently utilized both inside a clinical and office setting.

A treatment plan involving multiple injection sites requires a PRP volume with a high enough TAP count to prepare multiple treatment vials containing sufficient platelet dosages. The choice of a PRP device and/or preparation method therefore significantly impacts the TAP count and hence the individual TDP for each treatment site. The use of more advanced, dual-spin PRP systems will produce a larger PRP volume with a greater number of TAPs than simpler, single-spin devices. Furthermore, when injecting multiple pathological sites, the number of dual-spin, large-volume PRP devices necessary to produce sufficient TDP per injection is significantly less when compared to single-spin, low-volume test tube devices.

Harrison and co-workers reported that in single-spin test tube PRP kits, the migration of platelets to the buffy coat layer from both plasma and RBC layers was incomplete after centrifugation [86]. Even after 20 min of centrifugation, platelets remained in the RBC layer, approximately 30% of the total platelets [1]. They concluded that any single-spin PRP method will not have a PCR greater than approximately 70%. Furthermore, they reported that the denser and most regenerative PLTs are stuck in the RBC layer after the first spin.

In order to compensate for the limited PCRs when using single-spin test tube devices, some of the authors (JFL and LP) developed a labor-intensive PRP preparation method that included the use of larger predonated whole blood volumes, manually re-shaped syringes, a dual-spin validated centrifugation protocol, and refined preparation methods, with calibrated PRP extraction techniques. In this way, they realized a PCR ranging from 60% to 65%, producing approximately 25 mL of LR-PRP, prepared from 280 ml anticoagulated WB.

The importance of understanding the difference between PRP devices cannot be overstated. Practitioners who implement parameters such as PCR, TAP, TDP, and bioformulations while ignoring traditional quality parameters will not only improve the quality of their prepared PRP specimen, but they will most likely produce consistently better patient outcomes as well.

22.10 Conclusions

Globally, there is an unmet demand to treat a wide range of orthopedic and sports medicine disorders with non-surgical, innovative orthobiologic procedures that use autologous biological preparations such as PRP to treat a variety of MSK disorders. As more PRP products enter the market, differences in PRP preparation volumes, centrifugation methods, flexibility, and control of platelet and leukocyte extraction will continue to occur.

PRP practitioners can significantly improve the quality of their PRP treatments, and thus patient outcomes, by understanding the differences between devices and the clinical incorporation of innovative PRP parameters, such as PCR, TAP, TDP, and bioformulations into their treatment plan.

The need for more research is imperative in order to develop and consent to PRP formulations that guarantee consistent positive patient outcomes in the treatment of tissue-type specific pathologies in a minimum number of sessions.

References

1. Everts PA, Mazzola T, Mautner K, Randelli PS, Podesta L. Modifying orthobiological PRP therapies are imperative for the advancement of treatment outcomes in musculoskeletal pathologies. Biomedicines. 2022;10(11):2933.
2. Everts PA, Flanagan G, Podesta L. Autologous orthobiologics. In: Mostoufi SA, George TK, Tria Jr AJ, editors. Clinical guide to musculoskeletal medicine: a multidisciplinary approach. Cham: Springer International Publishing; 2022. p. 651–79. Available from: https://doi.org/10.1007/978-3-030-92042-5_62.
3. Oh JH, Kim W, Park KU, Roh YH. Comparison of the cellular composition and cytokine-release kinetics of various platelet-rich plasma preparations. Am J Sports Med. 2015;43(12):3062–70. Available from: http://journals.sagepub.com/doi/10.1177/0363546515608481
4. Manchikanti L, Navani A, Sanapati M. Evolution of regenerative medicine in managing musculoskeletal and spinal disorders. In: Essentials of regenerative medicine in interventional pain management. Springer; 2024. p. 3–34.
5. Mariani E, Pulsatelli L. Platelet concentrates in musculoskeletal medicine. IJMS. 2020;21(4):1328. Available from: https://www.mdpi.com/1422-0067/21/4/1328
6. Andia I, Maffulli N. New biotechnologies for musculoskeletal injuries. Surgeon. 2019;17(4):244–55. Available from: https://linkinghub.elsevier.com/retrieve/pii/S1479666X18300969
7. Everts P, Onishi K, Jayaram P, Lana JF, Mautner and K. Platelet-rich plasma: new performance understandings and therapeutic considerations in 2020. Med Pharmacol. 2020;21:7794. Available from: https://www.preprints.org/manuscript/202010.0069/v1
8. Chahla J, Cinque ME, Piuzzi NS, Mannava S, Geeslin AG, Murray IR, et al. A call for standardization in platelet-rich plasma preparation protocols and composition reporting: a systematic review of the clinical orthopaedic literature. J Bone Joint Surg. 2017;99(20):1769–79. Available from: http://Insights.ovid.com/crossref?an=00004623-201710180-00009
9. Korpershoek JV, Vonk LA, De Windt TS, Admiraal J, Kester EC, Van Egmond N, et al. Intra-articular injection with autologous conditioned plasma does not lead to a clinically relevant improvement of knee osteoarthritis: a prospective case series of 140 patients with 1-year follow-up. Acta Orthop. 2020;91(6):743–9. Available from: https://actaorthop.org/actao/article/view/813
10. Hurley ET, Lim Fat D, Moran CJ, Mullett H. The efficacy of platelet-rich plasma and platelet-rich fibrin in arthroscopic rotator cuff repair: a meta-analysis of randomized controlled trials. Am J Sports Med.

2019;47(3):753–61. Available from: http://journals.sagepub.com/doi/10.1177/0363546517751397
11. Bennell KL, Paterson KL, Metcalf BR, Duong V, Eyles J, Kasza J, et al. Effect of intra-articular platelet-rich plasma vs placebo injection on pain and medial tibial cartilage volume in patients with knee osteoarthritis: the RESTORE randomized clinical trial. JAMA. 2021;326(20):2021. Available from: https://jamanetwork.com/journals/jama/fullarticle/2786501
12. Gentile P, Garcovich S. Systematic review – the potential implications of different platelet-rich plasma (PRP) concentrations in regenerative medicine for tissue repair. IJMS. 2020;21(16):5702. Available from: https://www.mdpi.com/1422-0067/21/16/5702
13. Amable P, Carias RB, Teixeira MV, da Cruz PÍ, Corrêa do Amaral RJ, Granjeiro J, et al. Platelet-rich plasma preparation for regenerative medicine: optimization and quantification of cytokines and growth factors. Stem Cell Res Ther. 2013;4(3):67. Available from: http://stemcellres.com/content/4/3/67
14. Lutz C, Cheng J, Prysak M, Zukofsky T, Rothman R, Lutz G. Clinical outcomes following intradiscal injections of higher-concentration platelet-rich plasma in patients with chronic lumbar discogenic pain. Int Orthop. 2022;46(6):1381–5. Available from: https://link.springer.com/10.1007/s00264-022-05389-y
15. Berger DR, Centeno CJ, Steinmetz NJ. Platelet lysates from aged donors promote human tenocyte proliferation and migration in a concentration-dependent manner. Bone Joint Res. 2019;8(1):32–40. Available from: https://online.boneandjoint.org.uk/doi/10.1302/2046-3758.81.BJR-2018-0164.R1
16. Lin MT, Wei KC, Wu CH. Effectiveness of platelet-rich plasma injection in rotator cuff tendinopathy: a systematic review and meta-analysis of randomized controlled trials. Diagnostics. 2020;10(4):189. Available from: https://www.mdpi.com/2075-4418/10/4/189
17. Rodeo SA, Delos D, Williams RJ, Adler RS, Pearle A, Warren RF. The effect of platelet-rich fibrin matrix on rotator cuff tendon healing: a prospective, randomized clinical study. Am J Sports Med. 2012;40(6):1234–41. Available from: http://journals.sagepub.com/doi/10.1177/0363546512442924
18. Linnanmäki L, Kanto K, Karjalainen T, Leppänen OV, Lehtinen J. Platelet-rich plasma or autologous blood do not reduce pain or improve function in patients with lateral epicondylitis: a randomized controlled trial. Clin Orthop Relat Res. 2020;478(8):1892–900. Available from: https://journals.lww.com/10.1097/CORR.0000000000001185
19. Walsh MR, Nelson BJ, Braman JP, Yonke B, Obermeier M, Raja A, et al. Platelet-rich plasma in fibrin matrix to augment rotator cuff repair: a prospective, single-blinded, randomized study with 2-year follow-up. J Shoulder Elb Surg. 2018;27(9):1553–63. Available from: https://linkinghub.elsevier.com/retrieve/pii/S1058274618303318
20. Longo UG, Castricini R, De Benedetto M, Panfoli N, Pirani P, Zini R, et al. Paper # 117: platelet-rich fibrin matrix augmentation for arthroscopic rotator cuff repair: a randomized controlled trial. Arthroscopy. 2011;27(10):e145–6. Available from: https://linkinghub.elsevier.com/retrieve/pii/S0749806311008279
21. Fadadu PP, Mazzola AJ, Hunter CW, Davis TT. Review of concentration yields in commercially available platelet-rich plasma (PRP) systems: a call for PRP standardization. Reg Anesth Pain Med. 2019;44(6):652–9. Available from: http://rapm.bmj.com/lookup/doi/10.1136/rapm-2018-100356
22. Degen RM, Bernard JA, Oliver KS, Dines JS. Commercial separation systems designed for preparation of platelet-rich plasma yield differences in cellular composition. HSS Jrnl. 2017;13(1):75–80. Available from: http://link.springer.com/10.1007/s11420-016-9519-3
23. dos Santos RG, Santos GS, Alkass N, Chiesa TL, Azzini GO, da Fonseca LF, et al. The regenerative mechanisms of platelet-rich plasma: a review. Cytokine. 2021;144:155560. Available from: https://linkinghub.elsevier.com/retrieve/pii/S104346662100140X
24. Mazzucco L, Balbo V, Cattana E, Guaschino R, Borzini P. Not every PRP-gel is born equal evaluation of growth factor availability for tissues through four PRP-gel preparations: Fibrinet®, RegenPRP-kit®, Plateltex® and one manual procedure. Vox Sang. 2009;97(2):110–8. Available from: http://doi.wiley.com/10.1111/j.1423-0410.2009.01188.x
25. Everts PAM, Hoffmann J, Weibrich G, Mahoney CB, Schönberger JPAM, van Zundert A, et al. Differences in platelet growth factor release and leucocyte kinetics during autologous platelet gel formation. Transfus Med. 2006;16(5):363–8. Available from: http://doi.wiley.com/10.1111/j.1365-3148.2006.00708.x
26. Frey C, Shereff M, Greenidge N. Vascularity of the posterior tibial tendon. JBJS. 1990;72(6):884–8.
27. Lee S, Jilani SM, Nikolova GV, Carpizo D, Iruela-Arispe ML. Processing of VEGF-A by matrix metalloproteinases regulates bioavailability and vascular patterning in tumors. J Cell Biol. 2005;169(4):681–91.
28. Chen TT, Luque A, Lee S, Anderson SM, Segura T, Iruela-Arispe ML. Anchorage of VEGF to the extracellular matrix conveys differential signaling responses to endothelial cells. J Cell Biol. 2010;188(4):595–609.
29. Giusti I, D'Ascenzo S, Mancò A, Di Stefano G, Di Francesco M, Rughetti A, et al. Platelet concentration in platelet-rich plasma affects tenocyte behavior *in vitro*. Biomed Res Int. 2014;2014:1–12. Available from: http://www.hindawi.com/journals/bmri/2014/630870/
30. Blair P, Flaumenhaft R. Platelet alpha-granules: basic biology and clinical correlates. Blood Rev. 2009;23:177–89. https://doi.org/10.1016/j.blre.2009.04.001.
31. Kisucka J, Butterfield CE, Duda DG, Eichenberger SC, Saffaripour S, Ware J, et al. Platelets and platelet adhesion support angiogenesis while preventing

excessive hemorrhage. Proc Natl Acad Sci USA. 2006;103(4):855–60. Available from: https://pnas.org/doi/full/10.1073/pnas.0510412103
32. Romagnani P, Annunziato F, Liotta F, Lazzeri E, Mazzinghi B, Frosali F, et al. CD14+CD34 low cells with stem cell phenotypic and functional features are the major source of circulating endothelial progenitors. Circ Res. 2005;97(4):314–22. Available from: https://www.ahajournals.org/doi/10.1161/01.RES.0000177670.72216.9b
33. Spiller KL, Anfang RR, Spiller KJ, Ng J, Nakazawa KR, Daulton JW, et al. The role of macrophage phenotype in vascularization of tissue engineering scaffolds. Biomaterials. 2014;35(15):4477–88. Available from: https://linkinghub.elsevier.com/retrieve/pii/S0142961214001525
34. Sainson RCA, Johnston DA, Chu HC, Holderfield MT, Nakatsu MN, Crampton SP, et al. TNF primes endothelial cells for angiogenic sprouting by inducing a tip cell phenotype. Blood. 2008;111(10):4997–5007. Available from: https://ashpublications.org/blood/article/111/10/4997/24264/TNF-primes-endothelial-cells-for-angiogenic
35. Everts PA, Lana JF, Onishi K, Buford D, Peng J, Mahmood A, et al. Angiogenesis and tissue repair depend on platelet dosing and bioformulation strategies following orthobiological platelet-rich plasma procedures: a narrative review. Biomedicines. 2023;11(7):1922.
36. Lana JF. Leukocyte-rich PRP versus leukocyte-poor PRP – the role of monocyte/macrophage function in the healing cascade. J Clin Orthop Trauma. 2019;6:S7.
37. Korntner S, Lehner C, Gehwolf R, Wagner A, Grütz M, Kunkel N, et al. Limiting angiogenesis to modulate scar formation. Adv Drug Deliv Rev. 2019;146:170–89.
38. Vasina EM, Cauwenberghs S, Feijge MAH, Heemskerk JWM, Weber C, Koenen RR. Microparticles from apoptotic platelets promote resident macrophage differentiation. Cell Death Dis. 2011;2(9):e211. Available from: http://www.nature.com/articles/cddis201194
39. Acosta-Rodriguez EV, Napolitani G, Lanzavecchia A, Sallusto F. Interleukins 1β and 6 but not transforming growth factor-β are essential for the differentiation of interleukin 17–producing human T helper cells. Nat Immunol. 2007;8(9):942–9. Available from: http://www.nature.com/articles/ni1496
40. Kapur R, Zufferey A, Boilard E, Semple JW. Nouvelle cuisine: platelets served with inflammation. J Immunol. 2015;194(12):5579–87. Available from: http://www.jimmunol.org/lookup/doi/10.4049/jimmunol.1500259
41. Clark SR, Ma AC, Tavener SA, McDonald B, Goodarzi Z, Kelly MM, et al. Platelet TLR4 activates neutrophil extracellular traps to ensnare bacteria in septic blood. Nat Med. 2007;13(4):463–9. Available from: https://doi.org/10.1038/nm1565
42. Ubezio G, Ghio M. Bio-modulators in platelet-rich plasma: a comparison of the amounts in products from healthy donors and patients produced with three different techniques. Blood Transfus. 2014;12(Suppl 1):s214.
43. Martin CW, Muir IFK. The role of lymphocytes in wound healing. Br J Plast Surg. 1990;8:655.
44. Clemetson K, Clemetson J, Proudfoot A, Power C, Baggiolini M, Wells T. Functional expression of CCR1, CCR3, CCR4, and CXCR4 chemokine receptors on human platelets. Blood. 2000;96(13):4046–54. Available from: http://www.bloodjournal.org/cgi/content/full/96/13/4046
45. Zarbock A, Ley K, McEver RP, Hidalgo A. Leukocyte ligands for endothelial selectins: specialized glycoconjugates that mediate rolling and signaling under flow. Blood. 2011;118(26):6743–51. Available from: https://doi.org/10.1182/blood-2011-07-343566
46. Seta N, Kuwana M. Human circulating monocytes as multipotential progenitors. Keio J Med. 2007;56(2):41–7. Available from: http://joi.jlc.jst.go.jp/JST.JSTAGE/kjm/56.41?from=CrossRef
47. Ferrante CJ, Leibovich SJ. Regulation of macrophage polarization and wound healing. Adv Wound Care. 2012;1(1):10–6. Available from: http://www.liebertpub.com/doi/10.1089/wound.2011.0307
48. Everts PA, Devilee RJJ, Brown Mahoney C, van Erp A, Oosterbos CJM, Stellenboom M, et al. Exogenous application of platelet-leukocyte gel during open subacromial decompression contributes to improved patient outcome. Eur Surg Res. 2008;40(2):203–10. Available from: https://www.karger.com/Article/FullText/110862
49. Sommer C. Serotonin in pain and analgesia. Mol Neurobiol. 2004;30(2):117–25. Available from: https://doi.org/10.1385/MN:30:2:117
50. Wu WP, Hao JX, Xu XJ, Wiesenfeld-Hallin Z, Koek W, Colpaert FC. The very-high-efficacy 5-HT1A receptor agonist, F 13640, preempts the development of allodynia-like behaviors in rats with spinal cord injury. Eur J Pharmacol. 2003;478(2–3):131–7. Available from: https://doi.org/10.1016/j.ejphar.2003.08.047
51. Araya N, Miyatake K, Tsuji K, Katagiri H, Nakagawa Y, Hoshino T, et al. Intra-articular injection of pure platelet-rich plasma is the Most effective treatment for joint pain by modulating synovial inflammation and calcitonin gene-related peptide expression in a rat arthritis model. Am J Sports Med. 2020;48(8):2004–12. Available from: http://journals.sagepub.com/doi/10.1177/0363546520924011
52. Zhao W, Ma L, Deng D, Zhang T, Han L, Xu F, et al. M2 macrophage polarization: a potential target in pain relief. Front Immunol. 2023;14:1243149.
53. Yan X, Ye Y, Wang L, Xue J, Shen N, Li T. Platelet-rich plasma alleviates neuropathic pain in osteoarthritis by downregulating microglial activation. BMC Musculoskelet Disord. 2024;25(1):331.
54. Dobke M, Peterson DR, Mattern RH, Arm DM, Li WW. Microvascular tissue as a platform technology to modify the local microenvironment and influence the healing cascade. Regen Med. 2020;15(2):1313–28. Available from: https://www.futuremedicine.com/doi/10.2217/rme-2019-0139

55. Marx RE. Platelet-rich plasma (PRP): what is PRP and what is not PRP? Implant Dent. 2001;10(4):225–8. Available from: https://insights.ovid.com/crossref?an=00008505-200110000-00002
56. Jo CH, Roh YH, Kim JE, Shin S, Yoon KS. Optimizing platelet-rich plasma gel formation by varying time and gravitational forces during centrifugation. J Oral Implantol. 2013;39(5):525–32.
57. Arora G, Arora S. Platelet-rich plasma – where do we stand today? A critical narrative review and analysis. Dermatol Ther. 2020;34:e14343. Available from: https://onlinelibrary.wiley.com/doi/10.1111/dth.14343
58. Bausset O, Giraudo L, Veran J, Magalon J, Coudreuse JM, Magalon G, et al. Formulation and storage of platelet-rich plasma homemade product. BioRes Open Access. 2012;1(3):115–23. Available from: http://www.liebertpub.com/doi/10.1089/biores.2012.0225
59. Magalon J, Brandin T, Francois P, Degioanni C, De Maria L, Grimaud F, et al. Technical and biological review of authorized medical devices for platelets-rich plasma preparation in the field of regenerative medicine. Platelets. 2021;32(2):200–8. Available from: https://www.tandfonline.com/doi/full/10.1080/09537104.2020.1832653
60. Fitzpatrick J, Bulsara MK, McCrory PR, Richardson MD, Zheng MH. Analysis of platelet-rich plasma extraction: variations in platelet and blood components between 4 common commercial kits. Orthop J Sports Med. 2017;5(1):232596711667527. Available from: http://journals.sagepub.com/doi/10.1177/2325967116675272
61. Piao L, Park H, Jo CH. Theoretical prediction and validation of cell recovery rates in preparing platelet-rich plasma through a centrifugation. PLoS One. 2017;12(11):e0187509. Available from: https://dx.plos.org/10.1371/journal.pone.0187509
62. Oudelaar BW, Peerbooms JC, et al. Concentrations of blood components in commercial platelet-rich plasma separation systems: a review of the literature. Am J Sports Med. 2019;47(2):479–87. Available from: http://journals.sagepub.com/doi/10.1177/0363546517746112
63. Marques FP, Ingham SJM, Forgas A, da Franciozi CE, Sasaki PH, Abdalla RJ. A manual method to obtain platelet rich plasma. Acta Ortop Bras. 2014;22(2):75–7. Available from: http://www.scielo.br/scielo.php?script=sci_arttext&pid=S1413-78522014000200075&lng=en&tlng=en
64. Levy GM, Lucas P, Hope N. Efficacy of a platelet-rich plasma injection for the treatment of proximal hamstring tendinopathy: a pilot study. J Sci Med Sport. 2019;22(3):247–52. Available from: https://linkinghub.elsevier.com/retrieve/pii/S1440244018304791
65. Boesen AP, Hansen R, Boesen MI, Malliaras P, Langberg H. Effect of high-volume injection, platelet-rich plasma, and sham treatment in chronic Midportion Achilles tendinopathy: a randomized double-blinded prospective study. Am J Sports Med. 2017;45(9):2034–43. Available from: http://journals.sagepub.com/doi/10.1177/0363546517702862
66. Lana JFSD, Purita J, Paulus C, Huber SC, Rodrigues BL, Rodrigues AA, et al. Contributions for classification of platelet rich plasma – proposal of a new classification: MARSPILL. Regen Med. 2017;12(5):565–74. Available from: https://www.futuremedicine.com/doi/10.2217/rme-2017-0042
67. Magalon J, Chateau AL, Bertrand B, Louis ML, Silvestre A, Giraudo L, et al. DEPA classification: a proposal for standardising PRP use and a retrospective application of available devices. BMJ Open Sport Exerc Med. 2016;2(1):e000060. Available from: http://bmjopensem.bmj.com/lookup/doi/10.1136/bmjsem-2015-000060
68. Hurley ET, Sherman SL, Stokes DJ, Rodeo SA, Shapiro SA, Mautner K, et al. Experts achieve consensus on a majority of statements regarding platelet-rich plasma treatments for treatment of musculoskeletal pathology. Arthroscopy. 2024;40(2):470–7.
69. Ehrenfest DMD, Andia I, Zumstein MA, Zhang CQ, Pinto NR, Bielecki T. Classification of platelet concentrates (platelet-rich plasma-PRP, platelet-rich fibrin-PRF) for topical and infiltrative use in orthopedic and sports medicine: current consensus, clinical implications and perspectives. Muscles Ligaments Tendons J. 2014;4(1):3.
70. Mautner K, Malanga GA, Smith J, Shiple B, Ibrahim V, Sampson S, et al. A call for a standard classification system for future biologic research: the rationale for new PRP nomenclature. PM&R. 2015;7:S53–9. Available from: http://doi.wiley.com/10.1016/j.pmrj.2015.02.005
71. Rossi LA, Murray IR, Chu CR, Muschler GF, Rodeo SA, Piuzzi NS. Classification systems for platelet-rich plasma. Bone Joint J. 2019;101-B(8):891–6. Available from: https://online.boneandjoint.org.uk/doi/10.1302/0301-620X.101B8.BJJ-2019-0037.R1
72. Miron RJ, Chai J, Zhang P, Li Y, Wang Y, de Mourão CF, et al. A novel method for harvesting concentrated platelet-rich fibrin (C-PRF) with a 10-fold increase in platelet and leukocyte yields. Clin Oral Invest. 2020;24(8):2819–28. Available from: http://link.springer.com/10.1007/s00784-019-03147-w
73. Miron RJ, Chai J, Fujioka-Kobayashi M, Sculean A, Zhang Y. Evaluation of 24 protocols for the production of platelet-rich fibrin. BMC Oral Health. 2020;20(1):310. Available from: https://bmcoralhealth.biomedcentral.com/articles/10.1186/s12903-020-01299-w
74. Gupta AK, Versteeg SG, Rapaport J, Hausauer AK, Shear NH, Piguet V. The efficacy of platelet-rich plasma in the field of hair restoration and facial aesthetics – a systematic review and meta-analysis. J Cutan Med Surg. 2019;23(2):185–203. Available from: http://journals.sagepub.com/doi/10.1177/1203475418818073
75. Kirmani BH, Jones SG, Datta S, McLaughlin EK, Hoschtitzky AJ. A meta-analysis of platelet gel for

prevention of sternal wound infections following cardiac surgery. Blood Transfus. 2016;15:57.

76. Willemsen JCN, Van Dongen J, Spiekman M, Vermeulen KM, Harmsen MC, van der Lei B, et al. The addition of PRP to facial lipofilling: a double-blind placebo-controlled randomized trial. Plast Reconstr Surg. 2017;141:331. Available from: http://journals.lww.com/00006534-900000000-98336
77. Muchedzi TA, Roberts SB. A systematic review of the effects of platelet rich plasma on outcomes for patients with knee osteoarthritis and following total knee arthroplasty. Surgeon. 2018;16(4):250–8. Available from: https://linkinghub.elsevier.com/retrieve/pii/S1479666X1730118X
78. Tuakli-Wosornu YA, Terry A, Boachie-Adjei K, Harrison JR, Gribbin CK, LaSalle EE, et al. Lumbar Intradiskal platelet-rich plasma (PRP) injections: a prospective, double-blind, randomized controlled study. PM&R. 2016;8(1):1–10. Available from: http://doi.wiley.com/10.1016/j.pmrj.2015.08.010
79. Haunschild ED, Huddleston HP, Chahla J, Gilat R, Cole BJ, Yanke AB. Platelet-rich plasma augmentation in meniscal repair surgery: a systematic review of comparative studies. Arthroscopy. 2020;36(6):1765–74. Available from: https://linkinghub.elsevier.com/retrieve/pii/S0749806320301201
80. Yan CH, Jang SS, Lin HC, Ma Y, Khanwalkar AR, Thai A, et al. Use of platelet-rich plasma for COVID-19–related olfactory loss: a randomized controlled trial. Int Forum Allergy Rhinol. 2023;13(6):989–97. Available from: https://onlinelibrary.wiley.com/doi/10.1002/alr.23116
81. Sadiah YS, Dewanto A, Chandra LA. Efficacy of intraovarian autologous platelet-rich plasma administration in women with low ovarian reserve: a systematic review and meta-analysis. Asian Pac J Reprod. 2024;13(2):60–7. Available from: https://journals.lww.com/10.4103/apjr.apjr_141_23
82. Özyurt R, Bulutlar E, Yılmaz MB. Autologous platelet-rich plasma intrauterine perfusion improves the fertility outcome by correcting the thin endometrium due to clomiphene citrate. Med J Bakirkoy. 2024;20(1):72–8. Available from: https://bakirkoymedj.org/articles/doi/BMJ.galenos.2024.2023.10-3
83. Park MS, Moon SH, Kim TH, Oh JK, Yoon WY, Chang HG. Platelet-rich plasma for the spinal fusion. J Orthop Surg (Hong Kong). 2018;26(1):230949901875577. Available from: http://journals.sagepub.com/doi/10.1177/2309499018755772
84. de Mos M, van der Windt AE, Jahr H, van Schie HTM, Weinans H, Verhaar JAN, et al. Can platelet-rich plasma enhance tendon repair?: a cell culture study. Am J Sports Med. 2008;36(6):1171–8. Available from: http://journals.sagepub.com/doi/10.1177/0363546508314430
85. Kushida S, Kakudo N, Morimoto N, Hara T, Ogawa T, Mitsui T, et al. Platelet and growth factor concentrations in activated platelet-rich plasma: a comparison of seven commercial separation systems. J Artif Organs. 2014;17(2):186–92. Available from: http://link.springer.com/10.1007/s10047-014-0761-5
86. Harrison T, Bowler J, Cheng C, Reeves K. Optimizing platelet-rich plasma: spin time and sample source. Bioengineering. 2023;10:1270. Available from: https://doi.org/10.3390/bioengineering10111270

Use of New Formulation in Hyaluronic Acid in Regenerative Medicine

23

Melanio Acosta IV, Leandra Bizzoco, Camilla Boiocchi, and Alberto Gobbi

23.1 Introduction

Hyaluronic acid (HA) has been used to treat osteoarthritis through viscosupplementation as well as in the treatment of tendinopathies as early as the 1960s [1]. The concept of viscosupplementation is derived from the viscoelastic properties of synovial fluid and how they are altered in osteoarthritis [2, 3]. These changes in synovial fluid can be attributed to the qualitative and quantitative decline of hyaluronic acid. Hyaluronic acid is a major character in the properties of synovial fluid through its viscoelastic behavior, shock absorption, and lubrication. This is due to the hyaluronic acid molecule and protein-hyaluronic acid electrostatic interactions.

Molecular weight (MW) and the concentration of hyaluronic acid are major factors contributing to the viscoelastic properties of synovial fluid. These factors are mainly related to their ability to reduce mechanical stress in the joint [4–6]. Hyaluronic acid and its protective properties are related to the speed and direction at which the deforming forces are applied. When low shear forces are applied as seen in slow movement or walking, the linear chains of HA have time to slowly align themselves in the direction of deformation, allowing the viscous properties to predominate. However, when fast impact forces, seen in running and jumping, are applied, HA molecules do not have time to realign themselves and act more as a shock absorber, displaying more elastic properties.

Besides hyaluronic acid's advantages in decreasing mechanical stress, they also exhibit biologic properties involved in pain modulation, anti-inflammation, and enhancing synthesis of high MW hyaluronic acid [7–14]. After more than 60 years of clinical use and research on the safety and efficacy of hyaluronic acid treatment in orthopedics, there are new interests on how these mechanical, regenerative, anti-inflammatory, and pain reduction properties can be optimized and enhanced through the use of new preparations of HA as well as new formulations through combination with different biomaterials.

Despite tendon healing being a slow multifaceted process, it is common for the production of inferior quality fibrotic scar tissue or fibrous adhesions, which may modify the function and structure of tendons. Hyaluronic acid has not only been shown to be effective in arthritic pathologies but also may enhance tissue healing as well as pain reduction.

M. Acosta IV (✉) · L. Bizzoco · C. Boiocchi
A. Gobbi
O.A.S.I. Bioresearch Foundation Gobbi N.P.O., Milan, Italy
e-mail: melanio.acosta@oasiortopedia.it

A. Gobbi et al. (eds.), *Regenerative Medicine in Sports and Orthopaedics*,
https://doi.org/10.1007/978-3-031-84693-9_23

23.2 Biologic and Mechanical Properties of Hyaluronic Acid

Hyaluronic acid is a component of synovial fluid and the cartilage matrix as a glycosaminoglycan. Exogenous hyaluronic acid used in intra-articular injections is typically classified based on molecular weight preparations, categorized as high molecular weight (HMW), medium molecular weight (MMW), and low molecular weight (LMW). The molecular weight of hyaluronic acid is a major factor in the biologic properties it may produce within the body and the joints. Low molecular weight formulations are usually 500,000 to 730,000 Daltons (Da), medium molecular weight falls between 800,000 and 2,000,000 Da, and high molecular weight averaging 6,000,000 Da [15–17]. High molecular weight formulations are typically prepared through crosslinking, allowing the production of a highly viscous solution. In normal joint synovial fluid, the mass of the HA components is typically 5–7 million Daltons; however, this number drops to 1 million Daltons in osteoarthritic joints. This can be due to molecular fragmentation, abnormal synoviocyte production, and synovial fluid dilution secondary to effusion. With the use of exogenous HA, the goal is to replace the osteoarthritic synovial fluid with a higher concentration of HA, therefore, improving joint viscoelastic properties through enhancing the joint's ability for shock absorption and lubrication.

Other than mechanical properties, HA can bring about pharmacologic changes as well. CD44 receptors are the major players involved in hyaluronic acid's protective effects on the body. Through binding of CD44 receptors by HA molecules, there is inhibition of IL-1β expression, subsequently decreasing the production of matrix metalloproteases (MMPs) 1, 2, 3, 9, and 13, which impedes catabolic enzyme activity [18–21]. Through binding to these surface receptors, exogenous HA can also trigger intracellular cascades involving proliferation, differentiation, migration, and degradation of endogenous HA [22]. It is this affinity of HA to CD44 receptors that drives the potential biological effects within the body.

Synovial fibroblast's endogenous production of HA is directly influenced by the molecular weight and concentration of HA [23]. In low MW preparations, biosynthesis of HA may not be adequately induced due to weak binding to surface receptors. In medium MW preparations, strong bindings occur, allowing a high number of receptors to be stimulated, thus inducing and enhancing the production of endogenous HA. However, in high MW preparations, despite having maximal receptor binding, the number of sites actually occupied on a cell surface may be limited due to the sheer size of each molecule. Due to this crowding and lack of site binding, there may not be maximum stimulation of the production of endogenous HA [23].

Other than stimulation of endogenous HA, different MW formulations can also have other biologic effects. Higher MW formulations usually exhibit anti-inflammatory effects, while lower MW formulations promote angiogenesis, tissue remodeling, and wound healing, as well as exhibit pro-inflammatory properties in certain cell types such as chondrocytes [24, 25]. The administered HA forms a layer around cells and pain receptors, reducing pain induction, inciting anti-inflammatory effects by modulating pro-inflammatory mediators through the binding of cellular receptors, regulating cell proliferation and migration, and modulating gene expression [26–28]. Hyaluronic acid can also stimulate collagen production, most importantly type I collagen. This stimulates tissue recovery and regeneration as well as maintains cellular integrity [29]. Induction of the biosynthesis of endogenous HA and other extracellular matrix components are also important properties of administered exogenous HA. Through these actions, there is reduction of proteoglycan loss in joints and cartilage as well as reducing chondrocyte apoptosis [23, 26, 30].

23.3 Efficacy of Hyaluronic Acid Treatments

The various formulations used in the market and their popularity may be a reflection of different time periods, as low MW preparations were more prevalent before 2010, high MW preparations were more prevalent from 2010 to 2012, and medium MW preparations were more prevalent after 2012 [31]. Of course, these trends and their time periods may differ based on the availability of different products in different countries around the world. Despite having multiple formulations of various molecular weights on the market, there have been patterns seen in scientific literature of higher MW formulations having superior efficacy over those of lower MW [32–36].

Hummer et al. performed a meta-analysis that showed the superiority of high MW intra-articular injections over placebo and low molecular weight formulations. Here they demonstrated the statistical significance of high MW hyaluronic acid in improving WOMAC pain scale scores, exceeding the American Academy of Orthopaedic Surgeons (AAOS) defined minimal clinically important improvement (MCII) threshold. Low MW preparations showed lesser nonsignificant improvements, as well as having scores that did not exceed the MCII threshold [37].

Hyaluronic acid in tendinopathy has beneficial effects on the repair site as well as the synovial sheath. This is through its immunomodulatory effects in decreasing peripheral inflammatory response and enhancing contact healing by recruiting epitenon and endotenon cells [38]. Hyaluronic acid use can also lead to restoration of Type I collagen/Type III Collagen ratio, angiogenetic changes, reduced apoptosis of cells, and increased tenocyte regeneration [39–41]. Through these actions, it was shown that the use of HA has superior results in preventing adhesions with less interference on tendon healing when compared to corticosteroids [42].

Gobbi et al. [43] highlighted studies that HA had cytoprotective effects in tenosynovitis, where repeated peripatellar tendon injections in detrained rats showed maintenance of anabolic activity of tenocytes [44]. Safety and efficacy in providing significant clinical, functional, and pain outcomes of peritendinous HA injections in Achilles tendon and rotator cuff tendinopathies when compared to placebo has also been shown [45–49].

Merolla et al. compared physical therapy with ultrasound-guided subacromial injection of HA in rotator cuff tendinopathy. They showed superiority of HA to physical therapy in decreasing overall pain scores, as well as pain at night and during activity beyond 2 weeks up to 12 weeks [50]. Frizziero et al. investigated comparisons of low MW HA injections to extracorporeal shockwave therapy (ESWT) in rotator cuff tendinopathy. They not only confirmed Merolla's findings but also showed that low MW HA injections can be equally effective and safe when compared to ESWT. In fact, there was a faster trend to improve Constant-Murley scale scores and Dash scale scores in patients who underwent HA injections compared to ESWT. However, there was no significant difference in the final outcome scores by the last follow-up [51].

Hyaluronic acid has also been shown to be very effective in patellar and achilles tendinopathy as well. Many studies have shown improvement in functional, clinical, and pain score outcomes with the use of HA [45, 52–56].

23.4 New Formulations in Hyaluronic Acid

As the efficacy and safety in the clinical use of hyaluronic acid became established through years of research and study, new trends pushed towards how we can improve the effectiveness, durability, and longevity of HA, as well as finding new combinations and formulas that may afford an improved or even synergistic effect when applied to HA.

The longevity and effectiveness of exogenous HA administered into the joint can be related to how rapidly it is degraded or cleared by natural processes. This degradation and clearance typically involve HA synthase enzymes (HAS1, HAS2, HAS3), hydrolysis of hyaluronidases (HYALs), and reactive oxygen species (ROS)

[57]. These mechanisms contribute to the rapid clearance of intra-articular HA, limiting the time of effect to roughly 1 week for high MW cross-linked preparations to only a few days for linear HA preparations [58, 59]. For this reason, many stabilizing agents, such as polyols, sugars, amino acids, and salts, have been investigated to improve clearance delay and improved anti-inflammatory properties.

23.5 Sorbitol and Hyaluronic Acid

Sorbitol is a polyol molecule with potent scavenger characteristics for oxygen free radicals. Sorbitol acts by preventing IL-1β induced production of inflammatory, catabolic, and oxidative stress mediations, NADPH oxidase phosphorylation, release of nitric oxide, prostaglandins, and metalloproteinases, as well as caspase-3 activation, NF-κB activation, and DNA fragmentation [60–64]. These processes of catabolism, inflammation, oxidative stress, and apoptosis are not only characteristics of cartilage damage and joint inflammation in osteoarthritis but also contribute to the degradation and clearance of hyaluronic acid. In products that combine HA and sorbitol such as Synolis™ V-A (Aptissen, Plan-Les-Ouates, Switzerland), it has been shown to prevent tissue damage and cellular apoptosis caused by inflammation, catabolic, and oxidative processes as well as protect against endogenous and exogenous HA degradation, which are all key factors in the treatment of osteoarthritis.

23.6 Trehalose and Hyaluronic Acid

Trehalose is a nonreducing disaccharide found in plants, yeast, fungi, bacteria, and insects. However, as humans cannot produce trehalose naturally, it can be administered exogenously. Hyaluronic acid is easily degraded by strong acids, inorganic salts, organic solvents, and alkaline enzymes as their active end groups in the protein structure are vulnerable to binding which can lead to denaturation and loss of function. Trehalose is used as a stabilizing agent that binds to these active end groups preserving the structure and function of HA's protein molecule.

Trehalose not only has stabilization properties but has anti-inflammatory effects as well [65]. Through the mTOR (Mammalian Target of Rapamycin) independent pathway, trehalose acts as an autophagy activator [66, 67]. Trehalose also has cell-protective effects against oxidative damage, temperature denaturation, and dehydration through anti-apoptosis effects through BNP-3 upregulation which rescues mitochondria dysfunction. This, in turn, reverses age-related arterial stiffening which may be related to osteoarthritis senescence and metabolic-inflammatory syndrome [64, 68]. Studies investigating products with a combination of trehalose and hyaluronic acid showed significant functional, clinical, and pain scores.

Gobbi et al. conducted a prospective randomized, double-blinded, controlled trial comparing 1% trehalose hyaluronic acid (T-HA) versus non-trehalose hyaluronic acid. They found that at 3 months, IKDC, KOOS, and VAS scores improved for both groups but with no significant differences at that time between the two treatments. However, at 6 months, scores continued to improve in the group given the combination of trehalose and hyaluronic acid, while in the group given non-trehalose hyaluronic acid, scores decreased to the baseline, showing a significant decrease [65, 69].

23.7 Niacinamide and Hyaluronic Acid

Niacin is a generic name for nicotinic acid and nicotinamide (also known as niacinamide) which are water-soluble vitamin B complex vitamins that are components of enzyme cofactors nicotinamide adenine dinucleotide (NAD) and nicotinamide adenine dinucleotide phosphate (NADP). NAD and NADP are essential for maintaining cellular redox homeostasis and for controlling various biologic effects including cellular metabolism [70, 71]. Niacin is a semi-essential vitamin due to its endogenous formation from tryptophan. Niacinamide is an amide of nicotinic acid and consists of a pyridine ring to which a primary amide group is attached in the meta position [72].

Niacinamide provides resistance to oxidative degradation as well as protection against hyaluronidase activity, preventing hydrolysis of hyaluronic acid. When combined with hyaluronic acid, not only does it provide protection against oxidative processes and enzyme degradation but also preserves its viscosity through thermal stabilization.

In studies comparing high MW hyaluronic acid with addition of niacinamide versus standard medium MW HA, high MW hyaluronic acid with niacinamide preparations demonstrated superior results in efficacy and longer-lasting effects when compared to the HA group alone [73, 74]. N-HA group showed a statistically significant improvement at 3 and 6 months when compared to basal score with KOOS final score (P <0.05). HA group revealed an improvement at 3 months (P <0.01), but at 6 months there was a deterioration of the results, and the KOOS score was similar to pretreatment (P >0.05). At 3 months a statistically significant improvement was seen in SPORT, ADL, PAIN, and SYMPTOMS scores for N-HA group (P <0.05); furthermore SPORT scores maintained the improvement from month 3 to 6 (P <0.05). Finally, at 6 months N-HA had significantly better results [73].

23.8 Corticosteroids and Hyaluronic Acid

Corticosteroids in the treatment of osteoarthritis have been extensively studied. Since the 1950s, corticosteroids have been used in rheumatoid arthritis, osteoarthritis, and other rheumatic and musculoskeletal diseases [75]. Corticosteroids are mainly used in these inflammatory arthritic pathologies due to their anti-inflammatory effects within the joint. Many studies show its effectiveness in decreasing inflamed synovial tissue along with decreasing fluid volume in joints with effusion [75–78]. These effects usually translated to a decrease in pain and improved functional outcomes only in the short term. However, despite short-term improvements in pain and function, there is concern with repeated intra-articular corticosteroid injections and their dose-dependent deleterious effects on cartilage morphology, histology, and viability, causing cartilage damage and chondrocyte toxicity [79].

Preparations with corticosteroids and hyaluronic acid are not necessarily formulated to increase the stability of one another's structure; however, they are combined together to act in partnership with each other. Some comparative studies have shown that corticosteroids are more effective in reducing pain than HA due to their direct anti-inflammatory effect; however, pain relief is shorter in the long run. Through a combination of corticosteroids and hyaluronic acid, the advantages of short-term pain relief from the anti-inflammatory effects of corticosteroids are combined with the longer-lasting analgesic and biologic effects of hyaluronic acid [80–82].

23.9 Platelet-Rich Plasma and Hyaluronic Acid

As thoroughly discussed in the previous chapters, PRP and its use in osteoarthritis harnesses the power of growth factors and cytokines found in platelet granules to enhance tissue and cartilage healing through angiogenesis, neovascularization, increased blood supply and nourishment to neighboring cells, cell proliferation, chemotaxis, and cell differentiation as well as analgesic effects through the release of serotonin.

Combinations of PRP and hyaluronic acid aim to combine the therapeutic effects of each component in order to maximize and extend their efficacy for pain relief and improved function [83–87]. PRP does not negatively affect the mechanical, viscous, or elastic properties of hyaluronic acid, but hyaluronic acid, in fact, creates a bioactive scaffold in which platelets can attach and progressively release growth factors [88, 89]. Ciapini et al. demonstrated that at 3 and 6 months follow-up WOMAC and VAS scores were significantly lower in the group given the combination of PRP and HA compared to groups which had PRP or HA alone [90].

23.10 Conclusion

Exogenous hyaluronic acid not only enhances the viscoelastic behavior of the body's synovial fluid, while providing its additional shock absorption and lubrication within joints. Through binding of hyaluronic acid to various surface receptors, depending on the concentration and molecular weight, pharmacologic and biologic cascades can influence pain modulation, anti-inflammation, tissue regeneration, and stimulation of endogenous production of hyaluronic acid. As we understand the clinical benefits of hyaluronic acid on its own, we strive to find new combinations and formulations to enhance its longevity and efficacy. This is done by delaying degradation and turnover through action against enzymes and reactive oxygen species, as well as stabilization of the molecule itself. Hyaluronic acid can also be combined with anti-inflammatory agents such as corticosteroids or with other orthobiologic products for additive or synergistic effects. As further studies broaden our understanding of the body and orthobiologics, new formulations and combinations may be found to further improve our treatment of arthritic and other musculoskeletal disorders.

References

1. Rydell N, Balazs EA. Effect of intra-articular injection of hyaluronic acid on the clinical symptoms of osteoarthritis and on granulation tissue formation. Clin Orthop Relat Res. 1971;80:25–32. https://doi.org/10.1097/00003086-197110000-00006.
2. Balazs EA. Viscosupplementation for treatment of osteoarthritis: from initial discovery to current status and results. Surg Technol Int. 2004;12:278–89.
3. Balazs EA, Denlinger JL. Viscosupplementation: a new concept in the treatment of osteoarthritis. J Rheumatol Suppl. 1993;39:3–9.
4. Conrozier T, Mathieu P, Vignon E, Piperno M, Rinaudo M. Differences in the osteoarthritic synovial fluid composition and rheology between patients with or without flare: a pilot study. Clin Exp Rheumatol. 2012;30:729–34.
5. Milas M, Rinaudo M. Characterization and properties of hyaluronic acid (hyaluronan). In: Dimitriu S, editor. Polysaccharides: structural diversity and functional versatility. New York, NY: Marcel Dekker; 2004. p. 535–49.
6. Fam H, Bryant JT, Kontopoulou M. Rheological properties of synovial fluids. Biorheology. 2007;44:59–74.
7. Boettger MK, Kümmel D, Harrison A, Schaible HG. Evaluation of long-term antinociceptive properties of stabilized hyaluronic acid preparation (NASHA) in an animal model of repetitive joint pain. Arthritis Res Ther. 2011;13:R110.
8. Yasuda T. Hyaluronan inhibits prostaglandin E2 production via CD44 in U937 human macrophages. Tohoku J Exp Med. 2010;220:229–35.
9. Waddell DD, Kolomytkin OV, Dunn S, Marino AA. Hyaluronan suppresses IL1beta-induced metalloproteinase activity from synovial tissue. Clin Orthop Relat Res. 2007;465:241–8.
10. Li J, Gorski DJ, Anemaet W, et al. Hyaluronan injection in murine osteoarthritis prevents TGFbeta 1-induced synovial neovascularization and fibrosis and maintains articular cartilage integrity by a CD44-dependent mechanism. Arthritis Res Ther. 2012;14:R151.
11. Li P, Raitcheva D, Hawes M, et al. Hylan G-F 20 maintains cartilage integrity and decreases osteophyte formation in osteoarthritis through both anabolic and anticatabolic mechanisms. Osteoarthr Cartil. 2012;20:1336–46.
12. Wang CT, Lin YT, Chiang BL, Lin YH, Hou SM. High molecular weight hyaluronic acid down-regulates the gene expression of osteoarthritis-associated cytokines and enzymes in fibroblast-like synoviocytes from patients with early osteoarthritis. Osteoarthr Cartil. 2006;14:1237–47.
13. Greenberg DD, Stoker A, Kane S, Cockrell M, Cook JL. Biochemical effects of two different hyaluronic acid products in a co-culture model of osteoarthritis. Osteoarthr Cartil. 1993;1:97–103.
14. Bagga H, Burkhardt D, Sambrook P, March L. Long term effects of intra-articular hyaluronan on synovial fluid in osteoarthritis of the knee. J Rheumatol. 2006;33:946–50.
15. Kruel AVS, Ribeiro LL, Gusmão PD, Huber SC, Lana JFSD. Orthobiologics in the treatment of hip disorders. World J Stem Cells. 2021;13(4):304–16. https://doi.org/10.4252/wjsc.v13.i4.304.
16. Maheu E, Rannou F, Reginster JY. Efficacy and safety of hyaluronic acid in the management of osteoarthritis: evidence from real-life setting trials and surveys. Semin Arthritis Rheum. 2016;45(4 Suppl):S28–33. https://doi.org/10.1016/j.semarthrit.2015.11.008.
17. Costa FR, Costa Marques MR, Costa VC, Santos GS, Martins RA, Santos MDS, Santana MHA, Nallakumarasamy A, Jeyaraman M, Lana JVB, Lana JFSD. Intra-articular hyaluronic acid in osteoarthritis and tendinopathies: molecular and clinical approaches. Biomedicines. 2023;11(4):1061. https://doi.org/10.3390/biomedicines11041061.

18. Karna E, Miltyk W, Surazyński A, Pałka JA. Protective effect of hyaluronic acid on interleukin-1-induced deregulation of beta1-integrin and insulin-like growth factor-I receptor signaling and collagen biosynthesis in cultured human chondrocytes. Mol Cell Biochem. 2008;308(1–2):57–64. https://doi.org/10.1007/s11010-007-9612-5.
19. Julovi SM, Yasuda T, Shimizu M, Hiramitsu T, Nakamura T. Inhibition of interleukin-1beta-stimulated production of matrix metalloproteinases by hyaluronan via CD44 in human articular cartilage. Arthritis Rheum. 2004;50(2):516–25. https://doi.org/10.1002/art.20004.
20. Kalaci A, Yilmaz HR, Aslan B, Söğüt S, Yanat AN, Uz E. Effects of hyaluronan on nitric oxide levels and superoxide dismutase activities in synovial fluid in knee osteoarthritis. Clin Rheumatol. 2007;26(8):1306–11. https://doi.org/10.1007/s10067-006-0504-y.
21. Abate M, Pelotti P, De Amicis D, Di Iorio A, Galletti S, Salini V. Viscosupplementation with hyaluronic acid in hip osteoarthritis (a review). Ups J Med Sci. 2008;113(3):261–77. https://doi.org/10.3109/2000-1967-233.
22. Dicker KT, Gurski LA, Pradhan-Bhatt S, Witt RL, Farach-Carson MC, Jia X. Hyaluronan: a simple polysaccharide with diverse biological functions. Acta Biomater. 2014;10(4):1558–70. https://doi.org/10.1016/j.actbio.2013.12.019.
23. Smith MM, Ghosh P. The synthesis of hyaluronic acid by human synovial fibroblasts is influenced by the nature of the hyaluronate in the extracellular environment. Rheumatol Int. 1987;7(3):113–22. https://doi.org/10.1007/BF00270463.
24. Campo GM, Avenoso A, Campo S, D'Ascola A, Traina P, Rugolo CA, Calatroni A. Differential effect of molecular mass hyaluronan on lipopolysaccharide-induced damage in chondrocytes. Innate Immun. 2010;16(1):48–63. https://doi.org/10.1177/1753425909340419. (Retraction published Innate Immun. 2024 Feb;30(2–4):79. doi: 10.1177/17534259231224146).
25. Day AJ, de la Motte CA. Hyaluronan cross-linking: a protective mechanism in inflammation? Trends Immunol. 2005;26(12):637–43. https://doi.org/10.1016/j.it.2005.09.009.
26. Ghosh P, Guidolin D. Potential mechanism of action of intra-articular hyaluronan therapy in osteoarthritis: are the effects molecular weight dependent? Semin Arthritis Rheum. 2002;32(1):10–37. https://doi.org/10.1053/sarh.2002.33720.
27. Pozo MA, Balazs EA, Belmonte C. Reduction of sensory responses to passive movements of inflamed knee joints by hylan, a hyaluronan derivative. Exp Brain Res. 1997;116(1):3–9. https://doi.org/10.1007/pl00005742.
28. Bagga H, Burkhardt D, Sambrook P, March L. Longterm effects of intraarticular hyaluronan on synovial fluid in osteoarthritis of the knee. J Rheumatol. 2006;33(5):946–50.
29. Guidolin DD, Ronchetti IP, Lini E, Guerra D, Frizziero L. Morphological analysis of articular cartilage biopsies from a randomized, clinical study comparing the effects of 500-730 kDa sodium hyaluronate (Hyalgan) and methylprednisolone acetate on primary osteoarthritis of the knee. Osteoarthr Cartil. 2001;9(4):371–81. https://doi.org/10.1053/joca.2000.0398.
30. Díaz-Gallego L, Prieto JG, Coronel P, Gamazo LE, Gimeno M, Alvarez AI. Apoptosis and nitric oxide in an experimental model of osteoarthritis in rabbit after hyaluronic acid treatment. J Orthop Res. 2005;23(6):1370–6. https://doi.org/10.1016/j.orthres.2005.05.003.1100230619.
31. Shewale AR, Barnes CL, Fischbach LA, Ounpraseuth ST, Painter JT, Martin BC. Comparison of low-, moderate-, and high-molecular-weight hyaluronic acid injections in delaying time to knee surgery. J Arthroplast. 2017;32(10):2952–2957.e21. https://doi.org/10.1016/j.arth.2017.04.041.
32. Altman RD, Bedi A, Karlsson J, Sancheti P, Schemitsch E. Product differences in intra-articular hyaluronic acids for osteoarthritis of the knee. Am J Sports Med. 2016;44(8):2158–65. https://doi.org/10.1177/0363546515609599.
33. Rutjes AW, Jüni P, da Costa BR, Trelle S, Nüesch E, Reichenbach S. Viscosupplementation for osteoarthritis of the knee: a systematic review and meta-analysis. Ann Intern Med. 2012;157(3):180–91. https://doi.org/10.7326/0003-4819-157-3-201208070-00473.
34. Maheu E, Zaim M, Appelboom T, Jeka S, Trc T, Berenbaum F, Maasalu K, Berenbaum F. Comparative efficacy and safety of two different molecular weight (MW) hyaluronans F60027 and Hylan G-F20 in symptomatic osteoarthritis of the knee (KOA). Results of a non inferiority, prospective, randomized, controlled trial. Clin Exp Rheumatol. 2011;29(3):527–35.
35. Webb D, Naidoo P. Viscosupplementation for knee osteoarthritis: a focus on Hylan G-F 20. Orthop Res Rev. 2018;10:73–81. https://doi.org/10.2147/ORR.S174649.
36. Berenbaum F, Grifka J, Cazzaniga S, D'Amato M, Giacovelli G, Chevalier X, Rannou F, Rovati LC, Maheu E. A randomised, double-blind, controlled trial comparing two intra-articular hyaluronic acid preparations differing by their molecular weight in symptomatic knee osteoarthritis. Ann Rheum Dis. 2012;71(9):1454–60. https://doi.org/10.1136/annrheumdis-2011-200972.
37. Hummer CD, Angst F, Ngai W, Whittington C, Yoon SS, Duarte L, Manitt C, Schemitsch E. High molecular weight intraarticular hyaluronic acid for the treatment of knee osteoarthritis: a network meta-analysis. BMC Musculoskelet Disord. 2020;21(1):702. https://doi.org/10.1186/s12891-020-03729-w.
38. Amiel D, Ishizue K, Billings E Jr, Wiig M, Vande Berg J, Akeson WH, Gelberman R. Hyaluronan in flexor tendon repair. J Hand Surg Am. 1989;14(5):837–43. https://doi.org/10.1016/s0363-5023(89)80085-1.
39. Ciardulli MC, Marino L, Lovecchio J, Giordano E, Forsyth NR, Selleri C, Maffulli N, Porta GD. Tendon

and cytokine marker expression by human bone marrow mesenchymal stem cells in a hyaluronate/poly-lactic-co-glycolic acid (PLGA)/fibrin three-dimensional (3D) scaffold. Cells. 2020;9(5):1268. https://doi.org/10.3390/cells9051268.
40. Takahashi K, Goomer RS, Harwood F, Kubo T, Hirasawa Y, Amiel D. The effects of hyaluronan on matrix metalloproteinase-3 (MMP-3), interleukin-1beta(IL-1beta), and tissue inhibitor of metalloproteinase-1 (TIMP-1) gene expression during the development of osteoarthritis. Osteoarthr Cartil. 1999;7(2):182–90. https://doi.org/10.1053/joca.1998.0207.
41. Esenyel CZ, Tekin C, Cakar M, Bayraktar K, Saygili S, Esenyel M, Tekin ZN. Surgical treatment of the neglected Achilles tendon rupture with Hyalonect. J Am Podiatr Med Assoc. 2014;104(5):434–43. https://doi.org/10.7547/0003-0538-104.5.434.
42. Yuzawa K. [Experimental studies on the healing and restoration of gliding function of the injured digital flexor tendon. Part 9: the use of drugs to prevent adhesion formation of the injured tendon]. Nihon Seikeigeka Gakkai Zasshi. 1985;59(12):1107–18
43. Gobbi A, Morales M. Use of hyaluronic acid for rotator cuff tendinopathy. Am J Biomed Sci Res. 2021;14(4):334–5. https://doi.org/10.34297/AJBSR.2021.14.002007.
44. Salamanna F, Frizziero A, Pagani S, Giavaresi G, Curzi D, Falcieri E, Marini M, Abruzzo PM, Martini L, Fini M. Metabolic and cytoprotective effects of in vivo peri-patellar hyaluronic acid injections in cultured tenocytes. Connect Tissue Res. 2015;56(1):35–43. https://doi.org/10.3109/03008207.2014.979166.
45. Lynen N, De Vroey T, Spiegel I, Van Ongeval F, Hendrickx NJ, Stassijns G. Comparison of peritendinous hyaluronan injections versus extracorporeal shock wave therapy in the treatment of painful Achilles' tendinopathy: a randomized clinical efficacy and safety study. Arch Phys Med Rehabil. 2017;98(1):64–71. https://doi.org/10.1016/j.apmr.2016.08.470.
46. Blaine T, Moskowitz R, Udell J, Skyhar M, Levin R, Friedlander J, Daley M, Altman R. Treatment of persistent shoulder pain with sodium hyaluronate: a randomized, controlled trial. A multicenter study. J Bone Joint Surg Am. 2008;90(5):970–9. https://doi.org/10.2106/JBJS.F.01116.
47. Chou WY, Ko JY, Wang FS, Huang CC, Wong T, Wang CJ, Chang HE. Effect of sodium hyaluronate treatment on rotator cuff lesions without complete tears: a randomized, double-blind, placebo-controlled study. J Shoulder Elb Surg. 2010;19(4):557–63. https://doi.org/10.1016/j.jse.2009.08.006.
48. Moghtaderi A, Sajadiyeh S, Khosrawi S, Dehghan F, Bateni V. Effect of subacromial sodium hyaluronate injection on rotator cuff disease: a double-blind placebo-controlled clinical trial. Adv Biomed Res. 2013;2:89. https://doi.org/10.4103/2277-9175.122517.
49. Meloni F, Milia F, Cavazzuti M, Doria C, Lisai P, Profili S, Meloni GB. Clinical evaluation of sodium hyaluronate in the treatment of patients with supraspinatus tendinosis under echographic guide: experimental study of periarticular injections. Eur J Radiol. 2008;68(1):170–3. https://doi.org/10.1016/j.ejrad.2007.11.001.
50. Merolla G, Bianchi P, Porcellini G. Ultrasound-guided subacromial injections of sodium hyaluronate for the management of rotator cuff tendinopathy: a prospective comparative study with rehabilitation therapy. Musculoskelet Surg. 2013;97(Suppl 1):49–56. https://doi.org/10.1007/s12306-013-0259-y.
51. Frizziero A, Vittadini F, Barazzuol M, Gasparre G, Finotti P, Meneghini A, Maffulli N, Masiero S. Extracorporeal shockwaves therapy versus hyaluronic acid injection for the treatment of painful non-calcific rotator cuff tendinopathies: preliminary results. J Sports Med Phys Fitness. 2017;57(9):1162–8. https://doi.org/10.23736/S0022-4707.16.06408-2.
52. Kumai T, Muneta T, Tsuchiya A, Shiraishi M, Ishizaki Y, Sugimoto K, Samoto N, Isomoto S, Tanaka Y, Takakura Y. The short-term effect after a single injection of high-molecular-weight hyaluronic acid in patients with enthesopathies (lateral epicondylitis, patellar tendinopathy, insertional Achilles tendinopathy, and plantar fasciitis): a preliminary study. J Orthop Sci. 2014;19(4):603–11. https://doi.org/10.1007/s00776-014-0579-2.
53. Muneta T, Koga H, Ju YJ, Mochizuki T, Sekiya I. Hyaluronan injection therapy for athletic patients with patellar tendinopathy. J Orthop Sci. 2012;17(4):425–31. https://doi.org/10.1007/s00776-012-0225-9.
54. Fogli M, Giordan N, Mazzoni G. Efficacy and safety of hyaluronic acid (500-730 kDa) ultrasound-guided injections on painful tendinopathies: a prospective, open label, clinical study. Muscles Ligaments Tendons J. 2017;7(2):388–95. https://doi.org/10.11138/mltj/2017.7.2.388.
55. Frizziero A, Oliva F, Vittadini F, Vetrano M, Bernetti A, Giordan N, Vulpiani MC, Santilli V, Masiero S, Maffulli N. Efficacy of ultrasound-guided hyaluronic acid injections in Achilles and patellar tendinopathies. A prospective multicentric clinical trial. Muscles Ligaments Tendons J. 2019;9(3):305–13. https://doi.org/10.32098/mltj.03.2019.01. ISSN 2240-4554.
56. Gervasi M, Barbieri E, Capparucci I, Annibalini G, Sisti D, Amatori S, Carrabs V, Valli G, Donati Zeppa S, Rocchi MBL, Stocchi V, Sestili P. Treatment of Achilles tendinopathy in recreational runners with peritendinous hyaluronic acid injections: a viscoelastometric, functional, and biochemical pilot study. J Clin Med. 2021;10(7):1397. https://doi.org/10.3390/jcm10071397.
57. Berdiaki A, Neagu M, Spyridaki I, Kuskov A, Perez S, Nikitovic D. Hyaluronan and reactive oxygen species signaling-novel cues from the matrix? Antioxidants

(Basel). 2023;12(4):824. https://doi.org/10.3390/antiox12040824.

58. Larsen NE, Dursema HD, Pollak CT, Skrabut EM. Clearance kinetics of a hylan-based viscosupplement after intra-articular and intravenous administration in animal models. J Biomed Mater Res B Appl Biomater. 2012;100(2):457–62. https://doi.org/10.1002/jbm.b.31971.
59. Lindqvist U, Tolmachev V, Kairemo K, Aström G, Jonsson E, Lundqvist H. Elimination of stabilised hyaluronan from the knee joint in healthy men. Clin Pharmacokinet. 2002;41(8):603–13. https://doi.org/10.2165/00003088-200241080-00004.
60. Mongkhon JM, Thach M, Shi Q, Fernandes JC, Fahmi H, Benderdour M. Sorbitol-modified hyaluronic acid reduces oxidative stress, apoptosis and mediators of inflammation and catabolism in human osteoarthritic chondrocytes. Inflamm Res. 2014;63(8):691–701. https://doi.org/10.1007/s00011-014-0742-4.
61. Joshi S, Husain MM, Chandra R, Hasan SK, Srivastava RC. Hydroxyl radical formation resulting from the interaction of nickel complexes of L-histidine, glutathione or L-cysteine and hydrogen peroxide. Hum Exp Toxicol. 2005;24(1):13–7. https://doi.org/10.1191/0960327105ht493oa.
62. Chou LW, Wang J, Chang PL, Hsieh YL. Hyaluronan modulates accumulation of hypoxia-inducible factor-1 alpha, inducible nitric oxide synthase, and matrix metalloproteinase-3 in the synovium of rat adjuvant-induced arthritis model. Arthritis Res Ther. 2011;13(3):R90. https://doi.org/10.1186/ar3365.
63. Chang CC, Hsieh MS, Liao ST, Chen YH, Cheng CW, Huang PT, Lin YF, Chen CH. Hyaluronan regulates PPARγ and inflammatory responses in IL-1β-stimulated human chondrosarcoma cells, a model for osteoarthritis. Carbohydr Polym. 2012;90(2):1168–75. https://doi.org/10.1016/j.carbpol.2012.06.071.
64. Hashizume M, Mihara M. High molecular weight hyaluronic acid inhibits IL-6-induced MMP production from human chondrocytes by up-regulating the ERK inhibitor, MKP-1. Biochem Biophys Res Commun. 2010;403(2):184–9. https://doi.org/10.1016/j.bbrc.2010.10.135.
65. Morales M, Gobbi A. Trehalose: a promising stabilizer agent for hyaluronic acid (Chapter 1). In: Osteoarthritis, vol. 2. Reno: MedDocs Publishers; 2021. p. 1–7.
66. Zhang X, Chen S, Song L, Tang Y, Shen Y, Jia L, Le W. MTOR-independent, autophagic enhancer trehalose prolongs motor neuron survival and ameliorates the autophagic flux defect in a mouse model of amyotrophic lateral sclerosis. Autophagy. 2014;10(4):588–602. https://doi.org/10.4161/auto.27710.
67. Chen X, Li M, Li L, Xu S, Huang D, Ju M, Huang J, Chen K, Gu H. Trehalose, sucrose and raffinose are novel activators of autophagy in human keratinocytes through an mTOR-independent pathway. Sci Rep. 2016;6:28423. https://doi.org/10.1038/srep28423.
68. LaRocca TJ, Hearon CM Jr, Henson GD, Seals DR. Mitochondrial quality control and age-associated arterial stiffening. Exp Gerontol. 2014;58:78–82. https://doi.org/10.1016/j.exger.2014.07.008.
69. Gobbi A, Morales M, Avio G, D'Ambrosi R. Double-blinded prospective randomized clinical trial in knee joint osteoarthritis treatment: safety assessment and performance of trehalose hyaluronic acid versus standard infiltrative therapy based on medium-weight sodium hyaluronate. J Cartilage Joint Preserv. 2022;2(3):100043. https://doi.org/10.1016/j.jcjp.2022.100043.
70. Bogan KL, Brenner C. Nicotinic acid, nicotinamide, and nicotinamide riboside: a molecular evaluation of NAD+ precursor vitamins in human nutrition. Annu Rev Nutr. 2008;28:115–30. https://doi.org/10.1146/annurev.nutr.28.061807.155443.
71. Sauve AA. NAD+ and vitamin B3: from metabolism to therapies. J Pharmacol Exp Ther. 2008;324(3):883–93. https://doi.org/10.1124/jpet.107.120758.
72. Knip M, Douek IF, Moore WP, Gillmor HA, McLean AE, Bingley PJ, Gale EA, European Nicotinamide Diabetes Intervention Trial Group. Safety of high-dose nicotinamide: a review. Diabetologia. 2000;43(11):1337–45. https://doi.org/10.1007/s001250051536.
73. Gobbi A, Herman K, Bizzoco L, Avio G. Evaluation of safety and performance of hyaluronic acid combined with niacin-amide versus standard infiltrative therapy in the treatment of joint degenerative and post-traumatic diseases. J Orthop Muscular Syst. 2023;6(2):1024.
74. Gobbi A, Herman K, Szwedowski D. Bio-orthopedics: a new approach to osteoarthritis and joint disorders. London: IntechOpen; 2023. https://doi.org/10.5772/intechopen.110845.
75. Hollander JL, Brown EM Jr, Jessar RA, Brown CY. Hydrocortisone and cortisone injected into arthritic joints; comparative effects of and use of hydrocortisone as a local antiarthritic agent. J Am Med Assoc. 1951;147(17):1629–35. https://doi.org/10.1001/jama.1951.03670340019005.
76. Kumar A, Bendele AM, Blanks RC, Bodick N. Sustained efficacy of a single intra-articular dose of FX006 in a rat model of repeated localized knee arthritis. Osteoarthr Cartil. 2015;23(1):151–60. https://doi.org/10.1016/j.joca.2014.09.019.
77. Keen HI, Hensor EM, Wakefield RJ, Mease PJ, Bingham CO 3rd, Conaghan PG. Ultrasound assessment of response to intra-articular therapy in osteoarthritis of the knee. Rheumatology (Oxford). 2015;54(8):1385–91. https://doi.org/10.1093/rheumatology/keu529.
78. Wenham CY, Balamoody S, Grainger AJ, Hensor EM, Draycott S, Hodgson R, Conaghan PG. The responsiveness of novel, dynamic, contrast-enhanced mag-

netic resonance measures of total knee synovitis after intra-articular corticosteroid for painful osteoarthritis. Osteoarthr Cartil. 2014;22(10):1614–8. https://doi.org/10.1016/j.joca.2014.05.024.
79. Wernecke C, Braun HJ, Dragoo JL. The effect of intra-articular corticosteroids on articular cartilage: a systematic review. Orthop J Sports Med. 2015;3(5):2325967115581163. https://doi.org/10.1177/2325967115581163.
80. Wang SZ, Wu DY, Chang Q, Guo YD, Wang C, Fan WM. Intra-articular, single-shot co-injection of hyaluronic acid and corticosteroids in knee osteoarthritis: a randomized controlled trial. Exp Ther Med. 2018;16(3):1928–34. https://doi.org/10.3892/etm.2018.6371.
81. Smith C, Patel R, Vannabouathong C, Sales B, Rabinovich A, McCormack R, Belzile EL, Bhandari M. Combined intra-articular injection of corticosteroid and hyaluronic acid reduces pain compared to hyaluronic acid alone in the treatment of knee osteoarthritis. Knee Surg Sports Traumatol Arthrosc. 2019;27(6):1974–83. https://doi.org/10.1007/s00167-018-5071-7.
82. Wang CP, Lee WC, Hsieh RL. Effects of repeated co-injections of corticosteroids and hyaluronic acid on knee osteoarthritis: a prospective, double-blind randomized controlled trial. Am J Med. 2022;135(5):641–9. https://doi.org/10.1016/j.amjmed.2021.11.016.
83. Saturveithan C, Premganesh G, Fakhrizzaki S, Mahathir M, Karuna K, Rauf K, William H, Akmal H, Sivapathasundaram N, Jaspreet K. Intra-articular hyaluronic acid (HA) and platelet rich plasma (PRP) injection versus hyaluronic acid (HA) injection alone in patients with grade III and IV knee osteoarthritis (OA): a retrospective study on functional outcome. Malays Orthop J. 2016;10(2):35–40. https://doi.org/10.5704/MOJ.1607.007.
84. Karasavvidis T, Totlis T, Gilat R, Cole BJ. Platelet-rich plasma combined with hyaluronic acid improves pain and function compared with hyaluronic acid alone in knee osteoarthritis: a systematic review and meta-analysis. Arthroscopy. 2021;37(4):1277–1287.e1. https://doi.org/10.1016/j.arthro.2020.11.052.
85. Lana JF, Weglein A, Sampson SE, Vicente EF, Huber SC, Souza CV, Ambach MA, Vincent H, Urban-Paffaro A, Onodera CM, Annichino-Bizzacchi JM, Santana MH, Belangero WD. Randomized controlled trial comparing hyaluronic acid, platelet-rich plasma and the combination of both in the treatment of mild and moderate osteoarthritis of the knee. J Stem Cells Regen Med. 2016;12(2):69–78. https://doi.org/10.46582/jsrm.1202011.
86. Papalia R, Zampogna B, Russo F, Torre G, De Salvatore S, Nobile C, Tirindelli MC, Grasso A, Vadalà G, Denaro V. The combined use of platelet rich plasma and hyaluronic acid: prospective results for the treatment of knee osteoarthritis. J Biol Regul Homeost Agents. 2019;33(2 Suppl 1):21–8.
87. Yu W, Xu P, Huang G, Liu L. Clinical therapy of hyaluronic acid combined with platelet-rich plasma for the treatment of knee osteoarthritis. Exp Ther Med. 2018;16(3):2119–25. https://doi.org/10.3892/etm.2018.6412.
88. Madry H, Kon E, Condello V, Peretti GM, Steinwachs M, Seil R, Berruto M, Engebretsen L, Filardo G, Angele P. Early osteoarthritis of the knee. Knee Surg Sports Traumatol Arthrosc. 2016;24:1753–62.
89. Gobbi A, Dallo I, D'Ambrosi R. Autologous microfragmented adipose tissue and leukocyte-poor platelet-rich plasma combined with hyaluronic acid show comparable clinical outcomes for symptomatic early knee osteoarthritis over a two-year follow-up period: a prospective randomized clinical trial. Eur J Orthop Surg Traumatol. 2023;33(5):1895–904. https://doi.org/10.1007/s00590-022-03356-2.
90. Ciapini G, Simonettii M, Giuntoli M, Varchetta G, De Franco S, Ipponi E, Scaglione M, Parchi PD. Is the combination of platelet-rich plasma and hyaluronic acid the best injective treatment for grade II-III knee osteoarthritis? A prospective study. Adv Orthop. 2023;2023:1868943. https://doi.org/10.1155/2023/1868943.

Regenerative Medicine to Improve Skeletal Muscle Healing After Injury, Disease, and Aging

24

Ashley Groesbeck, Meredith Luhmann, Jacob Singer, Noah Knezic, Dane Lind, Luis Lopez, Sara Robinson, and Johnny Huard

24.1 Introduction

Skeletal muscle represents one of the most abundant tissues in the human body, encompassing approximately 40% of total body mass [1]. Its primary functions are generating forces for movement, sustaining body posture, maintaining body temperature, storing nutrients, and stabilizing joints [2]. Unlike smooth and cardiac muscle contraction, skeletal muscle contraction is predominantly voluntary, mediated by the somatic nervous system that sends signals to initiate muscle function.

Muscle fibers and connective tissue are the primary components of skeletal muscle. New skeletal muscle fibers form from the fusion of myoblasts, creating multinucleated cells through a process called myogenesis. Within these muscle fibers are numerous myofibrils, consisting of repeating protein structures called sarcomeres, which facilitate contraction. Sarcomeres, the fundamental contractile unit of the skeletal muscle, are organized into a unique striped pattern called striation [3]. Sarcomere interactions allow muscle fibers to contract and slide past each other, shortening the muscle. Connective tissue sheaths wrap thousands of muscle fibers together: the endomysium surrounds individual fibers, and the perimysium encases each bundle [3] (Fig. 24.1).

Skeletal muscle is highly vascularized and innervated, containing capillaries, nerve tissue, and mitochondria to supply oxygen, nutrients, and energy to the muscle fibers. Two of the most important myofilaments within muscle fibers are actin and myosin. Skeletal muscle contraction begins at the neuromuscular junction (NMJ), the synapse between a muscle fiber and a motor neuron [2]. The arrival of an action potential at the NMJ triggers a muscle contraction. Acetylcholine is released from the synaptic vesicles into the synaptic cleft and binds to postsynaptic receptors on the muscle fiber, opening the sodium ion channels on the motor end plate. Action potentials propagate along T-tubules through the sarcomeres and stimulate the sarcoplasmic reticulum to release calcium ions via voltage-gated channels, increasing intracellular calcium concentration. The influx of calcium ions causes a conformational change in the sarcomere, shortening the unit where myosin filaments pull actin filaments closer. The shortening of the sarcomeres is muscle contraction, described by the sliding filament model of muscle contraction. The resulting longitudinal force is transmitted through the extracellular matrix (ECM) to the bone via tendons.

Skeletal muscle fibers are categorized into slow-twitch (ST; type I) and fast-twitch (FT; type

A. Groesbeck · M. Luhmann · J. Singer · N. Knezic
D. Lind · L. Lopez · S. Robinson · J. Huard (✉)
Linda and Mitch Hart Regenerative and Personalized Medicine, Steadman Philippon Research Institute, Vail, CO, USA
e-mail: jhuard@sprivail.org

A. Gobbi et al. (eds.), *Regenerative Medicine in Sports and Orthopaedics*,
https://doi.org/10.1007/978-3-031-84693-9_24

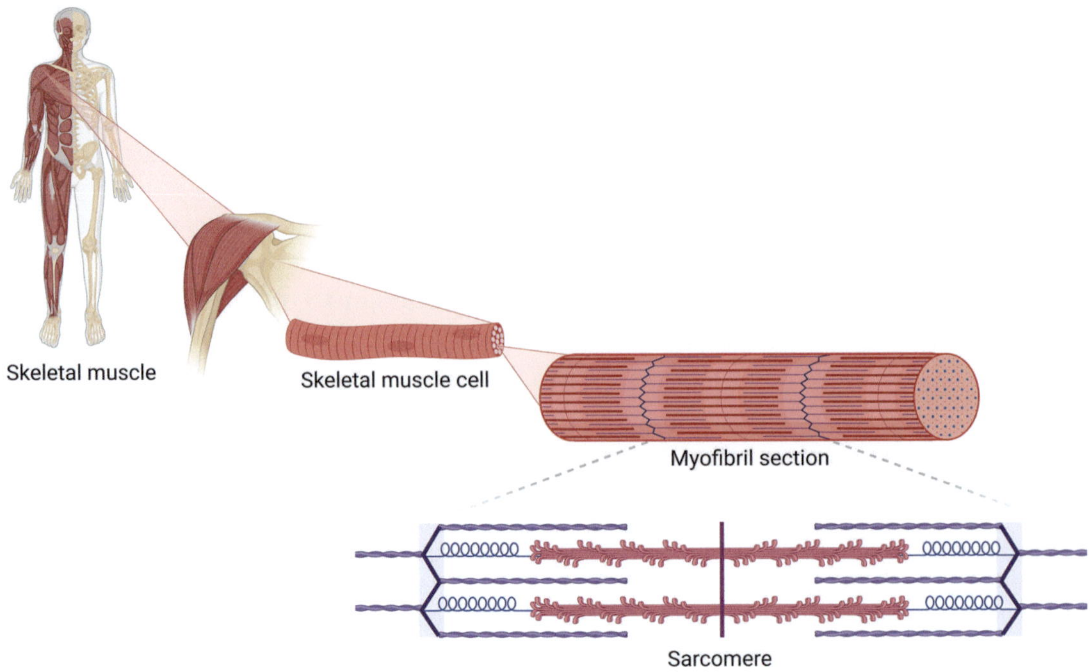

Fig. 24.1 Skeletal muscle structure. (*Created with Biorender*)

II) fibers [4]. ST fibers have minimal ATPase activity and are activated for high aerobic endurance activities, such as distance running or walking. In contrast, FT fibers have high ATPase activity and activate for short, quick bursts of muscle response. Due to their high ATPase activity and low glycolytic capacity, FT fibers are easily fatigued. FT fibers are further classified into type IIa and type IIx. Type IIa fibers are the intermediate between ST and type IIx fibers, specialized for moderately fast contractions. Type IIx fibers are optimal for rapid contractions in low aerobic conditions, making them ideal for quick bursts of energy in exercises such as sprinting or intense weightlifting [5].

Muscle damage frequently occurs in the body due to mechanical trauma, metabolic stress, or aging. Mechanical trauma involves the tearing of muscle fibers through hyperextension or abnormal muscle position. Metabolic stress is caused by anaerobic respiration, leading to the accumulation of lactate, phosphates, and hydrogen ions [6]. Metabolic stress and the resulting inflammation create microtears in muscle fibers, leading to hypertrophy. Aging leads to muscle mass loss due to decreased physical activity, hormonal decline, or nutritional decline. Substantial muscle loss or dysfunction can lead to musculoskeletal diseases such as sarcopenia, osteoporosis, and osteoarthritis [7]. The progressive loss of bone and muscle associated with age poses a significant threat to an individual's independence. Sarcopenia and osteoporosis are two conditions that substantially elevate the risk of fracture. Sarcopenia is a skeletal muscle disorder that is characterized by the progressive loss of muscle mass and function due to aging or immobility [8]. Symptoms of sarcopenia include weakness, loss of energy, issues with balance, and difficulty walking or standing. Osteoporosis is a bone disease that develops when bone mineral density and bone mass decrease, causing bones to become weak and brittle. This can cause a decrease in bone strength and an increased risk of bone fracture. Typically, bone undergoes continuous resorption and regeneration; however, in osteoporosis, new bone formation does not keep up with bone resorption, leading to a net loss of bone mass.

24.1.1 Sarcopenia

Sarcopenia is a musculoskeletal disease characterized by decreased muscle mass, strength, and function associated with aging [9]. Around 10–16% of the older adult population worldwide is diagnosed with sarcopenia, amounting to over 50 million people [10]. Sarcopenia is associated with increased functional decline, fractures, disability, poor mobility, and frailty [10, 11]. In sarcopenia, type II fibers are reduced and replaced by type I fibers and fat tissue [12]. Neuromuscular aging leads to NMJ deterioration, which contributes to a decrease in muscle strength [13]. Satellite cells (SCs), located between the plasma membrane of myofibers and the basal lamina, are skeletal muscle stem cells important for muscle growth, regeneration, and tissue repair [9]. SC function is reduced in sarcopenia muscles due to changes in regulatory factor levels [2]. Myostatin is a secreted growth factor upregulated in sarcopenia and a member of the TGF-β family, and it inhibits muscle growth by suppressing protein synthesis and promoting atrophy [14]. Forkhead box class O (FOXO) family member proteins, which induce muscle atrophy, are also upregulated in sarcopenia [13]. Other upregulated molecules include AMP-activated protein kinase (AMPK), an inhibitor of muscle anabolic processes, and inflammatory cytokines like tumor necrosis factor-alpha (TNF-α) and interleukin-6 (IL-6), both of which contribute to muscle wasting [15]. Downregulated factors include insulin-like growth factor-1 (IGF-1), which promotes muscle growth, and the mammalian target of rapamycin (mTOR), a key regulator of protein synthesis. Currently, there are no pharmacological agents approved for sarcopenia treatment. Management of sarcopenia includes maintaining a regular exercise regimen, which positively influences the neuromuscular system and increases the rate of protein synthesis [16]. There is interest in exploring pharmacological interventions, specifically with selective androgen receptor modulators (SARMs) that promote muscle growth and increase bone density [17].

24.1.2 Osteoporosis

Osteoporosis is a skeletal disease characterized by low bone mass and deterioration of bone tissue, causing fragility and higher susceptibility to fracture [18]. Osteoporosis affects over 200 million people worldwide, and it is estimated that 50% of postmenopausal women will have at least one osteoporosis-related fracture [19, 20]. Many factors contribute to the onset of osteoporosis, including estrogen deficiency, increased inflammation, natural aging, decreased osteogenesis, and reduced mitochondrial efficiency [21, 22]. Estrogen inhibits osteoclasts, but the hormone's deficiency during menopause removes this inhibition, resulting in bone loss [23]. Estrogen deficiency also increases the production of inflammatory cytokines, including TNF-α and IL-1ß, by activating T cells (Table 24.1). These cytokines stimulate osteoclast activity, inhibit osteoclast apoptosis, and inhibit osteoblast function. The imbalance of osteoblasts and osteoclasts leads to further bone loss [24]. Additionally, protection from reactive oxygen species (ROS) declines with age, which causes osteoblast and osteocyte apoptosis and decreased bone mass [23]. Non-pharmacological interventions include vitamin D and calcium supplementation [25]. Vitamin D improves calcium absorption and increases bone mass density [19]. Taking both together reduces the number of falls, increases quadricep strength, and decreases body sway [26]. Exercise increases homeostatic pressure and fluid flow within the bone marrow, leading to an increase in osteoblast activity and bone formation while also decreasing osteoclast activity [27]. Pharmacological interventions include antiresorptive agents, anabolic agents, and agents with both actions [19]. Antiresorptive agents such as hormones, selective estrogen receptor modulators, and bisphosphonates have roles in reducing bone resorption. Anabolic agents like teriparatide and abaloparatide increase bone formation. Dual-action agents like romosozumab have both antiresorptive and anabolic properties [28].

Table 24.1 Molecules involved in muscle disease and repair

Molecule	Type	Secreted by	Function
TNF-α	Cytokine	Immune cells (macrophages, T cells)	• Involved in sarcopenia – Muscle wasting • Involved in osteoporosis – Inhibit osteoclast apoptosis – Increase osteoclast activity – Inhibit osteoblast activity • Involved in muscle repair – Attract SCs to injury site – Promote SC proliferation • Involved in muscle atrophy – Activate catabolic pathways • Induces muscle inflammation leading to muscle breakdown
IL-6	Cytokine	Immune cells (T cells, B cells, macrophages, neutrophils, monocytes) Tissue formation cells (osteoblasts, chondrocytes, adipocytes)	• Involved in sarcopenia – Muscle wasting • Involved in muscle repair – Stimulate migration, proliferation, and differentiation of myoblasts • Involved in muscle atrophy – Activate catabolic pathways • Can be pro-inflammatory and anti-inflammatory, depending on the context
IL-1β	Cytokine	Immune cells (neutrophils, mast cells)	• Involved in osteoporosis – Inhibit osteoclast apoptosis – Increase osteoclast activity – Inhibit osteoblast activity • Involved in muscle repair – Recruit macrophages and T cells to injury site – Stimulate IL-6 production • Involved in muscle atrophy – Stimulate catabolic pathways • Contributes to muscle inflammation and damage during injury and disease
IFN-γ	Cytokine	Immune cells (neutrophils, T cells)	• Involved in muscle repair – Activate macrophages • Modulates immune response in muscle tissue – Pathogen clearance
IGF-1	Myokine	Muscle cells	• Involved in sarcopenia • Promotes muscle growth • Enhances muscle hypertrophy and regeneration – Activating SCs – Increase protein synthesis
VEGF	Growth factor	Hypoxic tissue	• Promotes new blood vessel formation • Involved in muscle regeneration – Improve blood supply to damaged tissue • Improves oxygen and nutrient delivery to muscle cells
TGF-β	Growth factor	Immune cells (lymphocytes and macrophages)	• Induces fibrosis in skeletal muscle • Regulates muscle repair – Modulate inflammation and fibroblast activity

24.1.3 Osteosarcopenia

The degeneration of muscle, as in sarcopenia, and bone, as in osteoporosis, are interconnected; skeletal muscle plays a crucial role in maintaining bone health through muscle-driven loading and biological interactions [29]. Mechanical loading is essential for maintaining bone mass and preventing osteoporosis. Conversely, bone health affects muscle function through paracrine signaling, creating a bidirectional relationship [30]. Muscles secrete myokines, such as IL-6, IGF-1, and myostatin, which act on nearby osteoblasts and osteoclasts to promote bone growth or remodeling (Table 24.1). Osteoblasts and osteocytes secrete osteokines, such as osteocalcin and sclerostin, which influence nearby muscle function and metabolism [30]. Because of the relationship between osteoporosis and sarcopenia, a new diagnosis of "osteosarcopenia" is used for patients who fit the diagnostic criteria for both diseases [31]. Therefore, patients with osteosarcopenia have the combined pathophysiology of both OP and SA discussed previously.

24.1.4 Muscular Dystrophies

Genetically inherited disorders of muscle generation are muscle dystrophies [32]. Over 30 different muscular dystrophy diseases vary in onset, severity, and affected muscles [33]. The congenital basis of muscular dystrophies has led to an increase in genetic diagnoses as more genes are associated with muscular dystrophies [34]. The most common muscular dystrophy is Duchenne Muscular Dystrophy (DMD), affecting 1 in every 5000 male births; females are not as impacted due to X-linked genetic variations [32]. Muscular dystrophy disorders involve damage to the protective membranes of muscle fibers or defects in the connective tissues [32]. Muscle damage and weakness in muscular dystrophies are primarily attributed to a defective protein, dystrophin, which anchors sarcomeres to cell membranes and the extracellular membrane [35]. Because dystrophin is also important for heart muscle function, muscular dystrophies can lead to organ failure and shortened lifespans. With modern treatments, patients with DMD can live up to 35–40 years [35]. Treatment options for muscular dystrophies have significantly improved over recent decades, leading to longer lifespans and improved quality of life for patients [35]. Medications for heart failure or hypertension can help manage cardiac complications. There is an increase in therapeutic research focused on understanding the disease mechanisms, specifically in heart muscles [36].

24.2 Biological Approaches to Improve Muscle Healing After Injuries

Skeletal muscle can naturally regenerate tissue in response to acute injuries. However, in instances of volumetric muscle loss (VML) exceeding 20%, the remaining muscle cannot fully restore its functional capacity, so interventions may be required to reestablish muscle integrity. Various strategies have been developed to promote muscle repair and regeneration, including physical therapy, surgical techniques, biomaterials, and muscular tissue engineering [37]. The functional loss of muscle can significantly impact the quality of life by limiting the body's capacity for movement. Contusions or strains are the most common forms of muscle injuries, accounting for greater than 90% of all sports-related injuries [38].

The formation of new skeletal muscle fibers involves the fusion of myoblasts into multinucleated muscle cells called myotubes. Satellite cells (SCs) play an important role in the regeneration process following muscle injury [39]. After injury, SCs are activated, proliferate, and differentiate into myogenic precursor cells known as myoblasts. Myoblasts form new myotubes that fuse with the damaged myofibers and mature into functional myofibers. Skeletal muscles originate from the paraxial mesoderm. Mesodermal cells differentiate into myogenic cells, which undergo mitosis to produce myoblasts. Myoblasts synthesize the proteins actin and myosin, which are important for muscle contraction. Once myoblasts align and fuse, they form multinucleated myotubes that are essential for functional muscle fiber formation. Myotubes continue to synthesize muscle-specific proteins like actin, myosin, troponin, tropomyosin, and other muscle proteins to allow muscle contraction and maintain structural integrity. These proteins organize into repeating units of sarcomeres to form myofibrils [2].

Muscle injuries are structural damage to muscle fibers caused by trauma, and they are the most common category of injuries in athletes [39]. Up to 50% of all sporting injuries are isolated to skeletal muscle [40]. Skeletal muscle injuries can stem from a variety of events, including direct trauma like muscle lacerations and contusions and indirect insults such as strains, ischemia, and denervation [39]. Muscle injury is followed by a healing response that consists of three phases: inflammation, regeneration, and fibrosis.

24.2.1 Muscle Regeneration Process

As soon as there is damage to skeletal muscle due to exercise, growth, or trauma, the muscle regeneration process is initiated. The process can be divided into phases: necrosis of injured muscle cells, activation of satellite cells, proliferation of activated satellite cells, differentiation of satellite cells, and maturation and remodeling of newly formed muscle fibers [41]. Muscle injury causes damage to muscle fiber membranes, leading to the release of cellular contents into the extracellular space. This release, in combination with cellular debris, triggers the immune system's response in three waves.

The first wave of the immune system response to muscle injury involves activation of the complement system, mast cells, and neutrophils. During this phase, muscle fiber necrosis, enlargement of the lesion, and debris clearance occur. The complement system, the first line of innate immune defense, is immediately activated after injury leading to neutrophil and macrophage infiltration to the site of injury [42]. Similarly, mast cells are immediately activated post-injury, leading to degranulation and release of proinflammatory cytokines, including TNF-α, IL-1ß, and histamine. These cytokines recruit more immune cells, such as mast cells and neutrophils, to the damaged tissue and promote inflammation [43]. Neutrophils secrete pro-inflammatory cytokines like TNF-α, interferon-γ (IFN-γ), and IL-1ß, as well as enzymes and oxidative factors. Specifically, IL-1 and IL-8 recruit macrophages to the injury site to remove necrotic muscle [41].

The second wave of the immune system response involves macrophages and T cells to activate satellite cell proliferation and migration. Neutrophils recruit macrophages to the injury site, where they are in their M1 proinflammatory phenotype. M1 macrophages are dominant in early muscle regeneration and release cytokines like TNF-α, IL-6, and IL-1ß [44]. TNF-α attracts satellite cells to the injury site and promotes their proliferation through nuclear factor-kappa B (NF-κB) signaling. TNF-α also activates the p38 signaling pathway to stimulate muscle cell differentiation. IL-6 stimulates the migration, proliferation, and differentiation of myoblasts. IL-1ß recruits macrophages and T cells to the damaged area and stimulates IL-6 production in skeletal muscle cells [45]. T cells secrete growth factors and cytokines, including TNF-α, IFN-γ, IL-1ß, IL-4, IL-12, and IL-13, to promote muscle repair, satellite cell proliferation, and migration [42].

The third wave of the immune system response is marked by a switch from pro-inflammatory (M1) macrophages to anti-inflammatory (M2) macrophages. M2 macrophages produce IL-4, IL-10, and IL-13 cytokines to suppress local inflammatory responses and promote satellite cell differentiation into myotubes [41]. Regulatory T cells (Tregs) suppress immune responses and secrete cytokines, including IL-10, which helps in the conversion of M1 macrophages to M2 macrophages and promotes myoblast differentiation [46]. Muscle Tregs express amphiregulin, an epidermal growth factor receptor (EGFR) which promotes the proliferation and differentiation of myoblasts [46].

24.2.2 Muscle Repair Transcription Factors

During the muscle repair process, many transcription factors have roles in regulating gene expression. Transcription factors such as Stat3, Stat5a, JunB, and Atf3 are involved in inflammation and are upregulated during muscle atrophy [47]. Atf3, Klf5, Stat3, Myf6, Myod1, and Myog are transcription factors contributing to the differentiation of muscle cells. Among these, the muscle regulatory factors (MRFs) Myf5, MyoD, Myog, and Mrf4 are important for myoblast determination, muscle differentiation, and determination of myogenic lineage [47]. Stat3 and Klf5 both contribute to muscle regeneration, highlighting their role in the muscle repair process post-injury [48].

24.2.3 PRICE and POLICE Protocols

The initial management of acute muscle injury commonly uses the protection, rest, ice, compression, and elevation (PRICE) protocol. The POLICE protocol is similar to the PRICE protocol but substitutes rest with optimal loading, suggesting that optimal loading can stimulate the healing process. A randomized controlled study conducted by Ankara University's Department of Emergency Medicine demonstrated that the POLICE protocol resulted in more effective and faster recovery in patients with ankle sprains compared to the PRICE protocol, indicating the therapeutic importance of optimal loading in expediting recovery [49].

Despite nonsteroidal anti-inflammatory drugs (NSAIDs) being commonly used to mitigate inflammation and pain associated with acute muscle injury, there is ongoing debate regarding the potential of NSAIDs to delay muscle healing by interfering with the body's natural anti-inflammatory response involved in the muscle repair process [50]. Acute inflammation, the body's immediate response to tissue damage, is essential to the healing process, increasing blood flow and activating neutrophil and macrophage infiltration. However, chronic inflammation is characterized as a prolonged inflammatory response, which can be detrimental to muscle healing and potentially lead to reinjury. Muscle reinjuries are common in athletes, ranging between 12% and 43% across different sports [51]. NS-398, a type of NSAID and a cyclooxygenase-2-specific inhibitor, has negative effects on skeletal muscle healing by obstructing the proliferation and maturation of differentiated myogenic precursor cells, showing that blocking acute inflammation further delays muscle healing [52].

Steroid injections are effective in reducing pain and inflammation in muscle injuries and can be utilized to determine if the injury is local or coming from another source [53]. Over time, however, consistent steroid injections can be detrimental to the affected joint. It is important to be cautious with the number of steroid injections given.

24.2.4 Surgical Intervention

For severe muscle trauma, such as VML or a complete muscle belly lesion, surgical intervention may be necessary. Surgical approaches for muscle injury include muscle repair surgery, tendon transfer, muscle grafting, and muscle resection. Additional surgical procedures include myotomy, myectomy, and myorrhaphy. Myotomy is a surgical incision or cut, a myectomy is the surgical removal of a part or all the muscle, and a myorrhaphy is the suturing of a muscle.

Muscle repair surgery aims to reattach torn muscles to restore their integrity and function. Surgery has remained a frequent method of muscle repair despite potential alterations to the local anatomy and underlying biomechanics. Suture repair techniques, including Kessler stitches, horizontal mattress stitches, and figure eight stitches, are commonly used, though the optimal suturing method remains undetermined [40, 54].

Tendon transfer surgery is necessary when muscle function is lost due to nerve damage. Muscles rely on signals from nerves, so when the nerve is injured, those muscles are paralyzed, and their function is lost. During the tendon transfer procedure, the origin of the muscle is left in place, and the tendon insertion into the bone is detached and reattached to a different area. Muscle grafting involves removing a portion of muscle tissue from one area of the body and transferring it to a defective or damaged location. Muscle grafts can be used to restore movement of a limb after trauma, denervation, or pathology.

Current treatments for severe muscle injury and myopathies, diseases that affect skeletal muscle structure, metabolism, or channel function, including autologous muscle transplantation and injection of ex vivo cultured muscle cells, have shown limited success [55]. Muscle laceration repair is technically demanding due to the low resistance of muscle tissue, which can cause sutures to pull out and lead to a higher rate of clinical failure. In addition to direct suture repair, work involving biologics and regenerative medicine has led to the emergence of biological scaffolds. These biomaterials, derived from various species, including bovine and porcine tissues, represent innovative approaches to improve muscle regeneration and repair [40].

Although these procedures have relatively high success rates, with any surgical procedure there are several risks, including infection, bleeding, pain, injury to nearby nerves, tendons, or blood vessels, swelling, rupture, poor incision healing, and blood clots. Surgery is typically considered a last resort approach for muscle injuries since it can introduce fibrotic scar tissue into the joint.

24.3 Unaddressed Components of Muscle Healing

Muscle healing is a complex process influenced by numerous biological factors. While there has been success with existing treatments, certain aspects of muscle healing are left unaddressed. These include angiogenesis, fibrosis, muscle atrophy, and cellular senescence, each of which contributes unique challenges to skeletal muscle repair. Addressing these factors comprehensively could lead to the development of a more efficacious treatment.

24.3.1 Angiogenesis

Angiogenesis, the formation of new blood vessels from existing ones, is critical for tissue formation and repair [56]. Many existing treatments for muscle injury fail to adequately promote angiogenesis, leading to insufficient blood flow to the injury site [39, 57]. The insufficiency of blood flow impedes the healing potential of the tissue due to inadequate vascularization, leading to hypoxia and nutrient deprivation [58]. These factors can result in incomplete or delayed muscle healing [59].

Vascular endothelial growth factor (VEGF) is important in promoting new blood vessel formation. Secreted by hypoxic tissues, VEGF directs vasculature formation to ensure proper nutrient supply [60, 61]. A challenge in establishing proper vasculature in injured muscle is regulating the amount of angiogenic factors present, like VEGF, as both excessive and insufficient levels can lead to abnormal vessel formation. Additionally, the timing of angiogenic factor secretion must coincide with the appropriate stages of repair to be effective [62].

24.3.2 Fibrosis

Fibrosis is a necessary part of the body's healing mechanism [63]. Post-injury, the repair process attempts to reconstruct muscle fibers and provide temporary structure support through scar tissue formation or fibrosis [64, 65]. However, fibrosis can become dysregulated, resulting in excessive ECM deposition, leading to stiff and mechanically inferior tissue [66, 67]. This tissue can cause pain or discomfort for the patient and result in muscle tissue that is less flexible and less capable of proper contraction and extension [68, 69]. Additionally, fibrotic tissue can impede the repair of surrounding muscle fibers and can become a frequent site for mechanical failure and reinjury in muscles [68, 70]. However, the development of scar tissue within the injured muscle can also predispose that muscle for reinjury which is defined as the recurrence of the same type of injury at the same site within 2 months up to 1 year after an athlete returns to full sports participation [51, 71].

Transforming growth factor beta (TGF-β) is known to induce a fibrosis cascade in skeletal muscle [72, 73]. TGF-β, secreted by immune cells such as lymphocytes and macrophages recruited to the injury site during the inflammatory response, creates a feedback loop of inflammation [73]. The elevated state of inflammation increases fibrotic tissue production and diminishes repair quality [74]. Zhu et al. [75] reported that myostatin, a protein that inhibits muscle growth, stimulates fibroblast proliferation post muscle injury [75]. The results suggest that myostatin and TGF-β1 work together to increase fibrosis by upregulating the expression level of each other [75]. It has also been observed that the release of TGF-β1 within the injured muscle induces the expression of TGF-β1 by local muscle cells, which can consequently differentiate into myofibroblasts that participate in muscle fibrosis [76, 77]. Using the CRISPR/Cas9 technology, Dr. Farshid Guilak has reported a novel approach that reprograms stem cells (called Stem Cells Modified for Autonomous Regenerative Therapy or SMART) to make it possible to deliver anti-inflammatory factor in an auto-regulated, feedback-controlled manner, and demonstrated its utility for musculoskeletal regenerative medicine [78]. They have reprogrammed muscle stem cells to be able to suppress TGF-β1 action locally around the cells by inducing TGF-β inhibitor from them whenever TGF-β1 is present in the injured skeletal muscle environment (i.e., autonomous suppression of fibrotic environment). These cells are currently being tested for musculoskeletal repair [79].

Maintaining or restoring muscle mass is a critical goal in muscle injury treatment, but it presents a significant challenge. Muscle injury often leads to offloading of the affected area to reduce strain, leading to muscle atrophy which is exacerbated by insufficient protein synthesis and chronic inflammation [45, 80]. Inflammatory cytokines such as TNF-α, IL-6, and IL-1β activate catabolic pathways that contribute to muscle atrophy [81, 82]. As such, the injured muscle becomes a conducive environment for atrophy, characterized by increased inflammation, lack of loading, and dysregulated nutrition. Muscle loss reduces tissue strength and function and increases the risk of prolonged recovery or reinjury [83].

24.4 Cellular Senescence in Muscle Healing

24.4.1 Senescence

Cellular senescence was first defined by Hayflick in the 1960s as a state of inhibited cell proliferation reached by human cells [84]. While cellular senescence is associated with aging [85], correlation does not imply causation. Various environmental factors can trigger cellular senescence, including oncogene activation, mitochondrial dysfunction, telomere attrition, irradiation, and epigenetic modification [86]. Most often, cellular senescence acts as an antitumor mechanism [87]. Senescent cells promote an inflammatory environment characterized by the senescence-associated secretory phenotype (SASP), releasing pro-inflammatory cytokines that lead to chronic inflammation [88].

The deterioration of skeletal muscle is closely related to aging, as sarcopenia decreases mobility and increases injury risk [89]. Given that the global population over 65 is expected to double in the next 30 years [90], understanding the relationship between senescence and skeletal muscle is important.

24.4.2 The Negative Impact of Senescence on Muscle and Muscle Injury

Senescence induces a deleterious, pro-inflammatory environment in skeletal muscle. Xu et al. [91] found that transplanting senescent cells into skeletal muscle increased the expression of senescent markers in muscle tissues and reduced grip strength [91]. Studies inducing senescence in myoblast cells reported that senescent cells promote apoptosis, atrophy, and the release of inflammatory factors contributing to sarcopenia [92–94].

Cellular senescence is particularly relevant to muscle injury and healing. Chiche et al. [95] demonstrated that muscle injury stimulates the release of IL-6, a cytokine associated with senescence and SASP [95, 96]. Furthermore, Xu et al. [91] found that clearing senescent cells using dasatinib and quercetin improves lifespan and physical function in mice, evidenced by improved walking speed, hanging endurance, and grip strength [91, 97].

Interestingly, some studies suggest that senescence may benefit muscle repair. Young et al. [98] identified "transient" senescent cells within mice that regenerate muscle and within the muscles of a muscular dystrophy mouse model [98]. They found that treatment with senolytics impaired muscle fiber growth, indicating that these "transient" senescent cells positively contribute to muscle growth and repair [98]. It is believed that senescent cells infiltrating the muscle early after injury are potentially beneficial for muscle repair, while senescent cells creating chronic inflammation at longer time points after injury might be detrimental.

24.5 Emerging Research and Innovations

Alternatives to these conventional therapies are under investigation, including platelet-rich plasma (PRP) and bone marrow aspirate concentrate (BMAC). Both PRP and BMAC are orthobiologic therapies that deliver biologically derived molecules to the damaged tissues to increase regeneration and repair. They contain a combination of cells, growth factors, and bioactive molecules. PRP has inconclusive results regarding its efficacy in improving muscle injury repair [99]. BMAC has yet to be tested on its benefit in potentially expediting muscle regeneration and repair processes [99, 100].

Aging is associated with a significant decline in musculoskeletal tissue function and a reduced capacity for regeneration. This decline can lead to muscle loss, weakness, and an increased risk of degenerative musculoskeletal diseases such as sarcopenia and osteoporosis. These conditions are interconnected; skeletal muscle significantly contributes to maintaining bone health through muscle-driven loading and biological interactions. It is well-established that loading is crucial for maintaining bone mass and preventing osteoporosis. Although exercise can mitigate age-related muscle weakness and slow the progression of conditions like sarcopenia and osteoporosis, adherence to exercise routines is often challenging for older individuals. The complex effects of whole-body exercise complicate the investigation of specific relationships between skeletal muscle and bone health throughout aging.

Current therapeutic approaches for skeletal muscle healing are not always effective, and there are ongoing challenges in clinical work to repair muscle. There remains a demand for the development of novel methods and materials to advance skeletal muscle repair and functional regeneration. Promising new methods include gene therapy, hydrogel injections, and exposure to high-altitude geographic environments.

24.5.1 Estrogen-Related Receptors

Estrogen is known to help improve muscle mass and strength and may help in muscle recovery and repair. Estrogen-related receptors (ERRs) are a subfamily of orphan nuclear receptors, a protein that has a similar structure to other identified receptors but has unidentified endogenous ligands involved in transcriptional processes and gene expression regulation [101, 102]. ERRs have sequence homology and share target genes with estrogen receptors (ERs) [103]. Compared to ERs, ERRs bind and regulate transcription through estrogen response elements (EREs) and extend ERE half-sites, known as ERR response elements (ERREs). However, they lack endogenous estrogen-binding capability [104]. Multiple papers have documented that ERs and ERRs can crosstalk between a shared binding site [105–107]. There are three known subtypes of estrogen-related receptors: estrogen-related receptor alpha (ERRα), beta (ERRβ), and gamma (ERRγ). ERRα and ERRγ are the two highly expressed ERR subtypes that regulate gene expression in skeletal muscle by their response to various extracellular signals, including exercise, hypoxia, fasting, and nutritional factors [108, 109] (Fig. 24.2).

ERRα is an important transcription factor involved in regulating mitochondrial biogenesis and oxidation of glucose and fatty acids during differentiation in skeletal myocytes. ERRα is present in tissues with high energy metabolism and mitochondrial biogenesis, such as the heart, kidney, intestinal tract, skeletal muscle, and brown adipose tissue [104]. The process of muscle regeneration involves several stages: necrosis of the injured muscle, inflammation, activation and proliferation of satellite cells, myogenesis and fusion, and remodeling of myofibers [110]. LaBarge et al. [111] showed that ERRα-deficient muscles showed delayed regeneration in response to injury, resulting in smaller myofibers with centralized nuclei [111]. The results of their experiment showed reduced muscle ATP content and enhanced AMP-activated protein kinase (AMPK) activation, a phylogenetically conserved fuel-sensing enzyme, in the ERRα-deficient muscles [111]. Additionally, it was demonstrated that the ERRα pathway, which is stimulated downstream of AMPK, regulated the mitochondrial biogenesis program required to support muscle regeneration [111].

Estrogen-related receptor gamma (ERRγ) is moderately expressed in the stomach, pancreas, kidney, spleen, prostate, and intestine and is highly expressed in skeletal muscle, regulating the perinatal switch to oxidative metabolism in the myocardium [104, 112] (Table 24.2). Exercise can induce ERRγ, which regulates slow twitch myofibers and promotes mitochondrial biogenesis and oxidative metabolism. Despite having smaller muscles, muscle-specific ERRγ transgenic mice have displayed an increase in exercise capacity, mitochondrial enzyme activity, and enlarged mitochondria [112]. The aging process adversely affects muscle function, impacting bone health and contributing to age-related musculoskeletal diseases. Transgenic (TG) mice overexpressing ERRγ (ERRGO mice) demonstrate that this overexpression activates AMPK, promoting mitochondrial biogenesis, oxidative capacity, and vascularization even without exercise (Fig. 24.3). Consequently, ERRGO mice display faster muscle healing and reduced cellular senescence.

Peripheral artery disease (PAD) is a circulatory condition characterized by narrowed blood vessels that reduce blood flow to the limbs, leading to cardiovascular complications [113]. It often results from atherosclerosis, which is the buildup of fatty deposits and calcium on arterial walls. A 2023 study found that ERRγ gene therapy can enhance muscle recovery in a preclinical model of PAD [114]. Intramuscular delivery of AAV9-Esrrg increases capillary density, improves blood flow, and restores oxidative metabolic capacity, demonstrating the potential of ERRγ gene therapy as a therapeutic approach for PAD

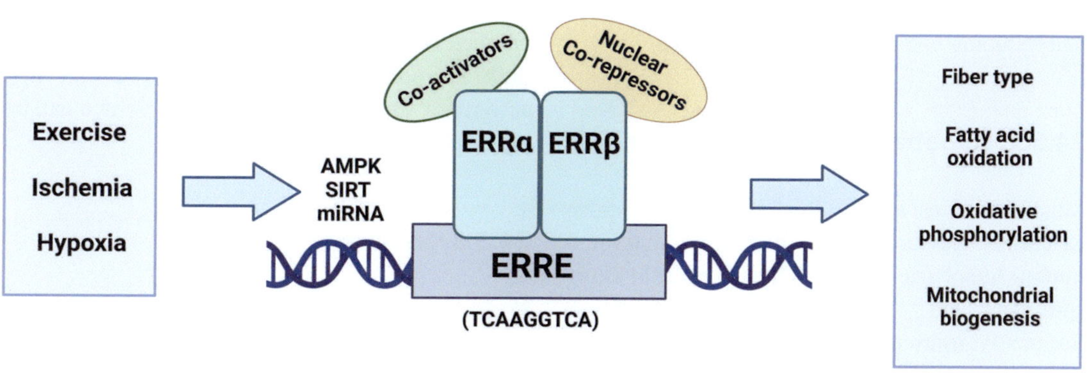

Fig. 24.2 Functions of ERRs on skeletal muscle. (*Created with Biorender*)

Fig. 24.3 Benefits of transgenic mouse model with an overexpression

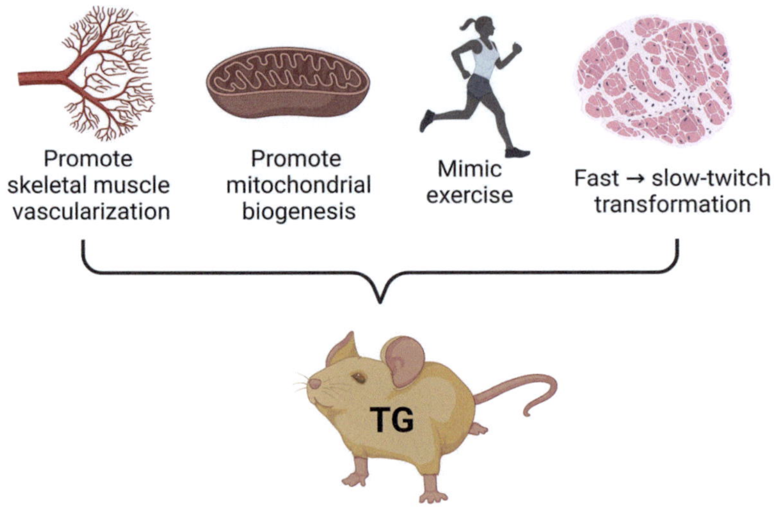

Table 24.2 Effect of estrogen on different tissues

Structure	Estrogen action	Estrogen deficiency/resistance
Brain	Energy balance	Obesity
Adipose	Adipose health	Obesity Adipose inflammation Altered secretory profile
Pancreas	β-cell function Survival	β-cell dysfunction Type 2 diabetes
Liver	Insulin sensitivity Lipid homeostasis	Insulin resistance Fatty liver
Muscle	Insulin sensitivity Energy homeostasis	Insulin resistance Impaired glucose homeostasis
Immune cells	Macrophage polarization	Inflammation Adiposity Atherosclerosis

and limb ischemia. Understanding the role of ERRγ in skeletal muscle can lead to novel pharmacological and rehabilitation approaches for improving musculoskeletal tissue regeneration and reducing age-related diseases [114].

24.5.2 Hypobaric Hypoxia Therapy

Muscle injuries are the leading cause of sports casualties. Research on rats has shown that intermittent hypobaric hypoxia (IHH) and cold exposure can aid skeletal muscle regeneration. Surgically injured muscles in rats regenerated histological structure and recovered function 21 days post-injury [115]. Elevated levels of phosphorylated AMPKα in the hypoxic group suggest the enzyme plays a key role in modulating the inflammatory response during early muscle regeneration [116]. Hypobaric hypoxia (HH), characterized by low oxygen concentration at high altitudes, influences the expression and activity of various cellular proteins, which can be beneficial for muscle repair.

24.5.3 Hydrogels

Hydrogels are cross-linked polymeric materials that are commonly used in drug delivery. Due to their ability to change structure in response to the environment, resemblance to the ECM of natural tissues, and capacity for minimally invasive administration, hydrogels present significant advantages for cell delivery in the treatment of muscle injuries [117, 118]. Hydrogels have the capacity to be customized to treat various injuries by facilitating cell attachment, proliferation, and differentiation. Hydrogels are engineered to gradually degrade within the body and completely integrate with host tissues. They can be used for muscle regeneration by inducing endogenous repair, delivering therapeutic agents, or scaffolding the formation of new tissue, which enhances the therapeutic process by providing a conducive environment for cell growth and tissue repair [117].

24.6 Conclusion

Skeletal muscle is the most common type of muscle in the human body, responsible for producing movement and stabilizing joints. Damage and injury to skeletal muscle impact daily functions and can inhibit an individual's independence. It is crucial to investigate the properties and interactions between genes and hormones expressed in skeletal muscle to fully understand and appreciate muscle regeneration and healing. Age-related decline in musculoskeletal tissue can result from reduced physical activity, hormonal decline, or nutritional deficiencies. This can lead to muscle loss, muscle weakness, and an increased risk of degenerative diseases, including sarcopenia and osteoporosis. Aging is associated with cellular senescence, a state of inhibited cell proliferation characterized by a pro-inflammatory environment. Aging and cellular senescence diminishes the capacity for muscle repair and regeneration. Current treatments for muscle injuries include various surgical techniques such as muscle repair, tendon transfer, and muscle grafting. Despite their popularity, these options are not optimal due to complications inherent in surgical procedures, including excessive bleeding and nerve damage and the potential introduction of fibrotic scar tissue. Given these limitations, there is a need for novel therapeutic methods to effectively promote skeletal muscle regeneration and repair. Promising new approaches, as described in this chapter, might be important in the future to improve muscle healing after injury, disease, and aging.

References

1. Frontera WR, Ochala J. Skeletal muscle: a brief review of structure and function. Calcif Tissue Int. 2015;96(3):183–95.
2. McCuller C, Jessu R, Callahan AL. Physiology, skeletal muscle. In: StatPearls. Treasure Island, FL: StatPearls Publishing; 2023.
3. Dave HD, Shook M, Varacallo M. Anatomy, skeletal muscle. In: StatPearls. Treasure Island, FL: StatPearls Publishing; 2023.
4. Schoenfeld BJ, et al. Loading recommendations for muscle strength, hypertrophy, and local endurance: a re-examination of the repetition continuum. Sports. 2021;9(2):32.
5. Plotkin DL, et al. Muscle fiber type transitions with exercise training: shifting perspectives. Sports. 2021;9(9):127.
6. de Freitas MC, et al. Role of metabolic stress for enhancing muscle adaptations: practical applications. World J Methodol. 2017;7(2):46–54.
7. Larsson L, et al. Sarcopenia: aging-related loss of muscle mass and function. Physiol Rev. 2019;99(1):427–511.
8. Zhao Y, et al. Advances in the role of leucine-sensing in the regulation of protein synthesis in aging skeletal muscle. Front Cell Dev Biol. 2021;9:646482.
9. Kim TN, Choi KM. Sarcopenia: definition, epidemiology, and pathophysiology. J Bone Metab. 2013;20(1):1.
10. Yuan S, Larsson SC. Epidemiology of sarcopenia: prevalence, risk factors, and consequences. Metabolism. 2023;144:155533.
11. Gielen E, et al. Sarcopenia, osteoporosis and frailty. Metabolism. 2023;145:155638.
12. Cho M-R, Lee S, Song S-K. A review of sarcopenia pathophysiology, diagnosis, treatment and future direction. J Korean Med Sci. 2022;37(18):e146.
13. Park SS, Kwon E-S, Kwon K-S. Molecular mechanisms and therapeutic interventions in sarcopenia. Osteoporos Sarcopenia. 2017;3(3):117–22.
14. Gumucio JP, Mendias CL. Atrogin-1, MuRF-1, and sarcopenia. Endocrine. 2013;43:12–21.
15. Zhou HH, et al. Association of muscle wasting with mortality risk among adults: a systematic review and meta-analysis of prospective studies. J Cachexia Sarcopenia Muscle. 2023;14(4):1596–612.
16. Koca TT. Bibliometric analysis of "sarcopenia" in web of science. Egypt Rheumatol Rehabil. 2023;50(1):26.
17. Dhillon RJ, Hasni S. Pathogenesis and management of sarcopenia. Clin Geriatr Med. 2017;33(1):17–26.

18. Edwards M, et al. Osteoporosis and sarcopenia in older age. Bone. 2015;80:126–30.
19. Munoz M, Robinson K, Shibli-Rahhal A. Bone health and osteoporosis prevention and treatment. Clin Obstet Gynecol. 2020;63(4):770–87.
20. Zhivodernikov IV, et al. Molecular and cellular mechanisms of osteoporosis. Int J Mol Sci. 2023;24(21):15772.
21. Papadopoulou SK, et al. Exercise and nutrition impact on osteoporosis and sarcopenia—the incidence of osteosarcopenia: a narrative review. Nutrients. 2021;13(12):4499.
22. Sözen T, Özışık L, Başaran NÇ. An overview and management of osteoporosis. Eur J Rheumatol. 2017;4(1):46.
23. Armas LA, Recker RR. Pathophysiology of osteoporosis: new mechanistic insights. Endocrinol Metab Clin. 2012;41(3):475–86.
24. Li H, et al. Osteoporosis: mechanism, molecular target and current status on drug development. Curr Med Chem. 2021;28(8):1489–507.
25. De Martinis M, et al. Vitamin D deficiency, osteoporosis and effect on autoimmune diseases and hematopoiesis: a review. Int J Mol Sci. 2021;22(16):8855.
26. Yu G, et al. A systematic review of cost-effectiveness analyses of sequential treatment for osteoporosis. Osteoporos Int. 2023;34(4):641–58.
27. Turner CH, Robling AG. Mechanisms by which exercise improves bone strength. J Bone Miner Metab. 2005;23:16–22.
28. Tabacco G, Bilezikian JP. Osteoanabolic and dual action drugs. Br J Clin Pharmacol. 2019;85(6):1084–94.
29. Ducher G, et al. Effects of repetitive loading on the growth-induced changes in bone mass and cortical bone geometry: a 12-month study in pre/peri-and postmenarcheal tennis players. J Bone Miner Res. 2011;26(6):1321–9.
30. Qin H, Jiao W. Correlation of muscle mass and bone mineral density in the NHANES US general population, 2017–2018. Medicine. 2022;101(39):e30735.
31. Hirschfeld H, Kinsella R, Duque G. Osteosarcopenia: where bone, muscle, and fat collide. Osteoporos Int. 2017;28:2781–90.
32. Mercuri E, Muntoni F. Muscular dystrophies. Lancet. 2013;381(9869):845–60.
33. Emery AE. The muscular dystrophies. Lancet. 2002;359(9307):687–95.
34. Butterfield RJ. Congenital muscular dystrophy and congenital myopathy. Continuum. 2019;25(6):1640–61.
35. Mogharehabed F, Czubryt MP. The role of fibrosis in the pathophysiology of muscular dystrophy. Am J Physiol Cell Physiol. 2023;325(5):C1326–35.
36. Goyenvalle A, et al. Therapeutic approaches to muscular dystrophy. Hum Mol Genet. 2011;20(R1):R69–78.
37. Tidball JG. Mechanisms of muscle injury, repair, and regeneration. Compr Physiol. 2011;1(4):2029–62.
38. Järvinen TA, et al. Muscle injuries: biology and treatment. Am J Sports Med. 2005;33(5):745–64.
39. Laumonier T, Menetrey J. Muscle injuries and strategies for improving their repair. J Exp Orthop. 2016;3(1):15.
40. Danna NR, et al. Therapeutic approaches to skeletal muscle repair and healing. Sports Health. 2014;6(4):348–55.
41. Yang W, Hu P. Skeletal muscle regeneration is modulated by inflammation. J Orthop Translat. 2018;13:25–32.
42. Watanabe S, et al. The role of macrophages in the resolution of inflammation. J Clin Invest. 2019;129(7):2619–28.
43. Forcina L, Cosentino M, Musarò A. Mechanisms regulating muscle regeneration: insights into the interrelated and time-dependent phases of tissue healing. Cells. 2020;9(5):1297.
44. Ziemkiewicz N, et al. The role of innate and adaptive immune cells in skeletal muscle regeneration. Int J Mol Sci. 2021;22(6):3265.
45. Howard EE, et al. Divergent roles of inflammation in skeletal muscle recovery from injury. Front Physiol. 2020;11:87.
46. Becker M, et al. Regulatory T cells require IL6 receptor alpha signaling to control skeletal muscle function and regeneration. Cell Metab. 2023;35(10):1736–1751.e7.
47. Yokoyama S, Asahara H. The myogenic transcriptional network. Cell Mol Life Sci. 2011;68(11):1843–9.
48. Yang X, et al. Changes of gene expression patterns of muscle pathophysiology-related transcription factors during denervated muscle atrophy. Front Physiol. 2022;13:923190.
49. Erdurmuş ÖY, et al. Comparison of the effects PRICE and POLICE treatment protocols on ankle function in patients with ankle sprain. Ulus Travma Acil Cerrahi Derg. 2023;29(8):920–8.
50. Shen W, et al. Inhibited skeletal muscle healing in cyclooxygenase-2 gene-deficient mice: the role of PGE2 and PGF2alpha. J Appl Physiol (1985). 2006;101(4):1215–21.
51. Palermi S, et al. Managing lower limb muscle reinjuries in athletes: from risk factors to return-to-play strategies. J Funct Morphol Kinesiol. 2023;8(4):155.
52. Shen W, et al. NS-398, a cyclooxygenase-2-specific inhibitor, delays skeletal muscle healing by decreasing regeneration and promoting fibrosis. Am J Pathol. 2005;167(4):1105–17.
53. Noonan TJ, Garrett WE Jr. Muscle strain injury: diagnosis and treatment. J Am Acad Orthop Surg. 1999;7(4):262–9.
54. Oliva F, et al. Surgical repair of muscle laceration: biomechanical properties at 6 years follow-up. Muscles Ligaments Tendons J. 2013;3(4):313–7.
55. Kwee BJ, Mooney DJ. Biomaterials for skeletal muscle tissue engineering. Curr Opin Biotechnol. 2017;47:16–22.

56. Veith AP, et al. Therapeutic strategies for enhancing angiogenesis in wound healing. Adv Drug Deliv Rev. 2019;146:97–125.
57. Liu J, et al. Current methods for skeletal muscle tissue repair and regeneration. Biomed Res Int. 2018;2018:1984879.
58. Nauta TD, van Hinsbergh VW, Koolwijk P. Hypoxic signaling during tissue repair and regenerative medicine. Int J Mol Sci. 2014;15(11):19791–815.
59. Yamamoto N, et al. VEGF and bFGF induction by nitric oxide is associated with hyperbaric oxygen-induced angiogenesis and muscle regeneration. Sci Rep. 2020;10(1):2744.
60. Frey SP, et al. VEGF improves skeletal muscle regeneration after acute trauma and reconstruction of the limb in a rabbit model. Clin Orthop Relat Res. 2012;470(12):3607–14.
61. Deasy BM, et al. Effect of VEGF on the regenerative capacity of muscle stem cells in dystrophic skeletal muscle. Mol Ther. 2009;17(10):1788–98.
62. DiPietro LA. Angiogenesis and wound repair: when enough is enough. J Leukoc Biol. 2016;100(5):979–84.
63. El Ayadi A, Jay JW, Prasai A. Current approaches targeting the wound healing phases to attenuate fibrosis and scarring. Int J Mol Sci. 2020;21(3):1105.
64. Gardner T, Kenter K, Li Y. Fibrosis following acute skeletal muscle injury: mitigation and reversal potential in the clinic. J Sports Med (Hindawi Publ Corp). 2020;2020:7059057.
65. Chapman MA, Meza R, Lieber RL. Skeletal muscle fibroblasts in health and disease. Differentiation. 2016;92(3):108–15.
66. Iwasaki Y, Kinoshita M. Thyroid function in patients with multiple sclerosis. Acta Neurol Scand. 1988;77(3):269.
67. Cholok D, et al. Traumatic muscle fibrosis: from pathway to prevention. J Trauma Acute Care Surg. 2017;82(1):174–84.
68. Mahdy MAA. Skeletal muscle fibrosis: an overview. Cell Tissue Res. 2019;375(3):575–88.
69. Brashear SE, et al. Passive stiffness of fibrotic skeletal muscle in mdx mice relates to collagen architecture. J Physiol. 2021;599(3):943–62.
70. Garg K, Corona BT, Walters TJ. Therapeutic strategies for preventing skeletal muscle fibrosis after injury. Front Pharmacol. 2015;6:87.
71. Ekstrand J, et al. Return to play after thigh muscle injury in elite football players: implementation and validation of the Munich muscle injury classification. Br J Sports Med. 2013;47(12):769–74.
72. Kim KK, Sheppard D, Chapman HA. TGF-β1 signaling and tissue fibrosis. Cold Spring Harb Perspect Biol. 2018;10(4):a022293.
73. Biernacka A, Dobaczewski M, Frangogiannis NG. TGF-β signaling in fibrosis. Growth Factors. 2011;29(5):196–202.
74. Kim J, Lee J. Role of transforming growth factor-β in muscle damage and regeneration: focused on eccentric muscle contraction. J Exerc Rehabil. 2017;13(6):621–6.
75. Zhu J, et al. Relationships between transforming growth factor-beta1, myostatin, and decorin: implications for skeletal muscle fibrosis. J Biol Chem. 2007;282(35):25852–63.
76. Huard J, Li Y, Fu FH. Muscle injuries and repair: current trends in research. J Bone Joint Surg Am. 2002;84(5):822–32.
77. Li Y, et al. Transforming growth factor-beta1 induces the differentiation of myogenic cells into fibrotic cells in injured skeletal muscle: a key event in muscle fibrogenesis. Am J Pathol. 2004;164(3):1007–19.
78. Brunger JM, et al. Genome engineering of stem cells for autonomously regulated, closed-loop delivery of biologic drugs. Stem Cell Rep. 2017;8(5):1202–13.
79. Pferdehirt L, et al. In vitro analysis of genome-engineered muscle-derived stem cells for autoregulated anti-inflammatory and antifibrotic activity. J Orthop Res. 2022;40(12):2937–46.
80. Gao Y, et al. Muscle atrophy induced by mechanical unloading: mechanisms and potential countermeasures. Front Physiol. 2018;9:235.
81. Tu H, Li YL. Inflammation balance in skeletal muscle damage and repair. Front Immunol. 2023;14:1133355.
82. Lieber RL, Ward SR. Skeletal muscle design to meet functional demands. Philos Trans R Soc Lond Ser B Biol Sci. 2011;366(1570):1466–76.
83. Appell HJ. Muscular atrophy following immobilisation. A review. Sports Med. 1990;10(1):42–58.
84. Hayflick L. The limited in vitro lifetime of human diploid cell strains. Exp Cell Res. 1965;37:614–36.
85. Rodier F, Campisi J. Four faces of cellular senescence. J Cell Biol. 2011;192(4):547–56.
86. Gorgoulis V, et al. Cellular senescence: defining a path forward. Cell. 2019;179(4):813–27.
87. Collado M, Serrano M. Senescence in tumours: evidence from mice and humans. Nat Rev Cancer. 2010;10(1):51–7.
88. Tchkonia T, Kirkland JL. Aging, cell senescence, and chronic disease: emerging therapeutic strategies. JAMA. 2018;320(13):1319–20.
89. Cruz-Jentoft AJ, Sayer AA. Sarcopenia. Lancet. 2019;393(10191):2636–46.
90. Cacchione PZ. World Health Organization leads the 2021 to 2030-decade of healthy ageing. Clin Nurs Res. 2022;31(1):3–4.
91. Xu M, et al. Senolytics improve physical function and increase lifespan in old age. Nat Med. 2018;24(8):1246–56.
92. Alcalde-Estévez E, et al. Endothelin-1 induces cellular senescence and fibrosis in cultured myoblasts. A potential mechanism of aging-related sarcopenia. Aging (Albany NY). 2020;12(12):11200–23.
93. Mijares A, Allen PD, Lopez JR. Senescence is associated with elevated intracellular resting [Ca(2+)] in

mice skeletal muscle fibers. An in vivo study. Front Physiol. 2020;11:601189.
94. Deschenes MR. Effects of aging on muscle fibre type and size. Sports Med. 2004;34(12):809–24.
95. Chiche A, et al. Injury-induced senescence enables in vivo reprogramming in skeletal muscle. Cell Stem Cell. 2017;20(3):407–414.e4.
96. Taguchi J, Yamada Y. Unveiling the role of senescence-induced cellular plasticity. Cell Stem Cell. 2017;20(3):293–4.
97. Hickson LJ, et al. Senolytics decrease senescent cells in humans: preliminary report from a clinical trial of Dasatinib plus quercetin in individuals with diabetic kidney disease. EBioMedicine. 2019;47:446–56.
98. Young LV, et al. Muscle injury induces a transient senescence-like state that is required for myofiber growth during muscle regeneration. FASEB J. 2022;36(11):e22587.
99. Yamaguchi FSM, et al. PRP and BMAC for musculoskeletal conditions via biomaterial carriers. Int J Mol Sci. 2019;20(21):5328.
100. Arnoczky SP, Shebani-Rad S. The basic science of platelet-rich plasma (PRP): what clinicians need to know. Sports Med Arthrosc Rev. 2013;21(4):180–5.
101. Deblois G, Giguère V. Functional and physiological genomics of estrogen-related receptors (ERRs) in health and disease. Biochim Biophys Acta. 2011;1812(8):1032–40.
102. Crevet L, Vanacker J-M. Regulation of the expression of the estrogen related receptors (ERRs). Cell Mol Life Sci. 2020;77:4573–9.
103. Madhu Krishna B, et al. Estrogen receptor α dependent regulation of estrogen related receptor β and its role in cell cycle in breast cancer. BMC Cancer. 2018;18(1):607.
104. Xu Z, et al. Research progress on the reproductive and non-reproductive endocrine tumors by estrogen-related receptors. J Steroid Biochem Mol Biol. 2016;158:22–30.
105. Cheung CP, et al. Expression and functional study of estrogen receptor-related receptors in human prostatic cells and tissues. J Clin Endocrinol Metab. 2005;90(3):1830–44.
106. Vanacker JM, et al. Transcriptional targets shared by estrogen receptor-related receptors (ERRs) and estrogen receptor (ER) alpha, but not by ERbeta. EMBO J. 1999;18(15):4270–9.
107. Cerutti C, Shi JR, Vanacker JM. Multifaceted transcriptional network of estrogen-related receptor alpha in health and disease. Int J Mol Sci. 2023;24(5):4265.
108. Sopariwala D, Nguyen H, Narkar V. Estrogen-related receptor signaling in skeletal muscle fitness. Int J Sports Med. 2023;44(9):609–17.
109. Eichner LJ, Giguère V. Estrogen related receptors (ERRs): a new dawn in transcriptional control of mitochondrial gene networks. Mitochondrion. 2011;11(4):544–52.
110. Ciciliot S, Schiaffino S. Regeneration of mammalian skeletal muscle. Basic mechanisms and clinical implications. Curr Pharm Des. 2010;16(8):906–14.
111. LaBarge S, et al. Estrogen-related receptor-α (ERRα) deficiency in skeletal muscle impairs regeneration in response to injury. FASEB J. 2014;28(3):1082–97.
112. Rangwala SM, et al. Estrogen-related receptor gamma is a key regulator of muscle mitochondrial activity and oxidative capacity. J Biol Chem. 2010;285(29):22619–29.
113. Conte SM, Vale PR. Peripheral arterial disease. Heart Lung Circ. 2018;27(4):427–32.
114. Sopariwala DH, et al. Estrogen-related receptor gamma gene therapy promotes therapeutic angiogenesis and muscle recovery in preclinical model of PAD. J Am Heart Assoc. 2023;12(16):e028880.
115. Santocildes G, et al. Simulated altitude is medicine: intermittent exposure to hypobaric hypoxia and cold accelerates injured skeletal muscle recovery. J Physiol. 2023;602:5855.
116. Chen X, et al. Activation of AMPK inhibits inflammatory response during hypoxia and reoxygenation through modulating JNK-mediated NF-κB pathway. Metabolism. 2018;83:256–70.
117. Lev R, Seliktar D. Hydrogel biomaterials and their therapeutic potential for muscle injuries and muscular dystrophies. J R Soc Interface. 2018;15(138):20170380.
118. Li X, et al. Functional hydrogels for the repair and regeneration of tissue defects. Front Bioeng Biotechnol. 2023;11:1190171.

Early OA Treatment Using Allogeneic Human Umbilical Cord MSCs

25

Bogusław Sadlik, Katarzyna Herman, and Dawid Szwedowski

25.1 Introduction

Treatment of early stages of knee osteoarthritis (OA) using mesenchymal stem cells (MSC) is a rapidly developing area in orthopedics. Various methods are used to induce reparative effects on the articular cartilage. The most popular methods are intra-articular injections of biological agents such as platelet-rich plasma (PRP) and concentrates of autologous stem cells derived from bone marrow or adipose tissue.

B. Sadlik
College of Physiotherapy in Wrocław, Wrocław, Poland

Biological Joint Reconstruction Department, ORTHOS Multidisciplinary Hospital, Komorowice, Poland

K. Herman
Department of Orthopaedics and Trauma Surgery, Bonifraters Medical Center, Katowice, Poland

Medical Rehabilitation Department, School of Health Sciences, Silesian Medical University in Katowice, Katowice, Poland

D. Szwedowski (✉)
Faculty of Pharmacy, Department of Organic Chemistry, Ludwik Rydygier Collegium Medicum in Bydgoszcz, Nicolaus Copernicus University in Toruń, Bydgoszcz, Poland

Department of Orthopaedics and Trauma Surgery, Provincial Polyclinical Hospital, Toruń, Poland
e-mail: dawid.szwedowski@cm.umk.pl

In the last decade, numerous clinical trials have been conducted using allogeneic stem cells derived from the umbilical cord. The results of these studies have shown mid-term pain reduction in patients over 40 years of age with early stages of knee osteoarthritis. Unfortunately, none of these studies confirmed cartilage regeneration through magnetic resonance imaging (MRI). The idea of biological restoration of the articular surface of the knee joint was announced in 1994 by Caplan and Goldberg [1].

The first bold clinical trial was conducted by Alberto Gobbi in 2011. Gobbi performed the reconstruction of degenerative lesions of the knee articular cartilage in patients with early OA, using a hyaluronic acid matrix soaked in autologous bone marrow concentrate, which was gelled before implantation [2].

These procedures were initially performed with open approach and later using the dry arthroscopy technique.

In the first decade of the twenty-first century, other authors attempted the implantation of stem cells derived from autologous adipose tissue, umbilical cord blood, and finally, from the umbilical cord [3].

Wharton's jelly is a gelatinous substance found within the umbilical cord. It provides structural support and protection to the blood vessels of the umbilical cord, preventing compression and ensuring the steady flow of blood between the placenta and the developing fetus.

A. Gobbi et al. (eds.), *Regenerative Medicine in Sports and Orthopaedics*,
https://doi.org/10.1007/978-3-031-84693-9_25

Named after the English anatomist Thomas Wharton, who first described it in 1656, Wharton's jelly is composed mainly of mucopolysaccharides, particularly hyaluronic acid and chondroitin sulfate, as well as fibroblasts, myofibroblasts, and various types of collagen fibers. Umbilical cord stem cells, specifically those derived from Wharton's jelly (WJ-MSCs), are a type of mesenchymal stem cell known for their high proliferation rate, multipotency, and immunomodulatory properties. They are capable of differentiating into various cell types, including osteoblasts, chondrocytes, and adipocytes, making them ideal for regenerative medicine applications. Additionally, WJ-MSCs exhibit low immunogenicity, which allows for their use in allogeneic transplantation without eliciting a strong immune response [4–8].

Two types of cells are found in the human umbilical cord. Human umbilical cord blood mesenchymal stem cells (UCB-MSCs) are harvested from the umbilical cord blood, requiring immediate collection postpartum and involving processes to separate MSCs from other blood components, resulting in lower concentrations of MSCs [9].

On the other hand, Wharton's jelly mesenchymal stem cells (WJ-MSCs) are sourced from the gelatinous Wharton's jelly material of the umbilical cord. This method generally yields higher concentrations and viability of MSCs due to less invasive isolation methods such as enzymatic digestion or explant culture techniques [10].

Both UCB-MSCs and WJ-MSCs express typical MSC markers like CD73, CD90, and CD105 and lack hematopoietic markers such as CD34 and CD45. Notably, WJ-MSCs may exhibit higher expression levels of these surface markers, indicating a more primitive and pluripotent stromal cell population [11].

WJ-MSCs display superior proliferation rates compared to UCB-MSCs, attributed to the more primitive nature of WJ-MSCs, which also grants them a broader differentiation potential into osteogenic, chondrogenic, adipogenic, and myogenic lineages [12].

UCB-MSCs demonstrate differentiation abilities along these pathways but generally show slower proliferation, requiring longer culture periods for significant yields.

Both cell types exhibit immunomodulatory effects crucial for therapeutic uses in allogeneic transplants. Studies suggest that WJ-MSCs may possess more potent immunosuppressive properties due to higher expressions of immunomodulatory factors, making them particularly useful for treating inflammatory and autoimmune disorders [13].

Our attempt to reconstruct degenerative lesions of the articular cartilage using allogeneic stem cells derived from Wharton's jelly of the umbilical cord began in 2015 at St. Luke's Hospital in Bielsko-Biała, Poland. The local Ethical Committee approved the trial for the first group of patients, who were eligible for primary total or unicompartmental knee arthroplasty but did not want to undergo the proposed therapy and consciously agreed to the experimental method. The safety of the new method was predicted based on previous implantations of WJUC-MSCs in hematological and neurodegenerative diseases in large groups of patients [14, 15]. The Bioethics Committee approved an age limit of up to 70 years at the time of implantation.

25.2 Clinical Application of Umbilical Cord MSCs

Mesenchymal stem cells (MSCs) have emerged as a potentially effective method for tissue regeneration in recent years. MSCs possess the ability to self-renew and differentiate into multiple cell types, while also playing a role in regulating immunity and enhancing tissue regeneration. These versatile cells can be harvested from various tissues including bone marrow, skeletal muscle, synovial membranes, periodontal ligaments, Wharton's jelly, umbilical cords, umbilical cord blood, amniotic fluid, placentae, and adipose tissue from different regions of the body. Originally, it was believed that MSCs migrate to damaged sites, engraft, and then differentiate into the desired cells for tissue regeneration to achieve therapeutic effects. Other research, however, has suggested that MSCs' therapeutic benefit comes from factors they secrete. Almost 30 years ago Haynesworth et al. identified MSCs' secretion of paracrine [16]. According to their findings, MSCs generate and secrete a wide range of chemokines, cytokines, and growth factors that affect the behavior of nearby cells. These secreted factors modulate immune responses, promote extracellular matrix remodeling, decrease apoptosis and fibrosis, and increase angiogenesis. Thus, MSCs either directly or through paracrine secretion promote regeneration to save damaged cells, lessen tissue damage, and ultimately speed up the healing of organs like skin, bone, liver, or kidney [17].

The process of wound healing in the skin involves several layers: the epidermis, the dermis, and the subcutaneous layer. Mesenchymal stem cells can migrate to the site of skin damage, suppress inflammation, and enhance the growth and differentiation of fibroblasts, endothelial, and epidermal cells [18]. Consequently, this can lead to reduced scarring. Preclinical studies have shown promising results in using umbilical cord blood mesenchymal stem cells (UCB-MSCs) for wound healing, paving the way for future clinical trials. For example, Zhang et al. showed reduction in scar formation and improvement in the regeneration of skin appendages, nerves, and vessels after UCB-MSCs treatment in rats [19]. The administration of UCB-MSCs to patients with skin wounds showed the potential in improved granulation tissue coverage and thickness [20].

The effects of MSCs in oncology have been a subject of debate, with conflicting reports of both inhibitory and proliferative actions observed in various types of cancers. In laboratory settings, MSCs have been shown to have tumoricidal effects on solid tumors like breast and lung cancer, while also promoting proliferation in melanoma [21, 22]. This dual role highlights the importance of fully understanding the impact of MSCs on tumor cell proliferation. The therapeutic potential of MSCs in hematologic malignancies is uncertain due to the contradictory inhibitory and promoting effects observed in both laboratory and animal studies [23].

While autologous nerve grafting is considered the standard treatment for repairing peripheral nerve damage, challenges such as limited donor nerve availability have prompted the exploration of alternative therapeutic approaches. The impact of various types of mesenchymal stem cells in peripheral nerve injury has been explored. Umbilical cord blood–derived MSCs have also been successfully seeded onto various biodegradable scaffolds, resulting in improved recovery rates relative to control groups [24].

25.3 Cartilage Regeneration Using MSCs

Cartilage is a unique type of connective tissue that originates from the mesoderm. It is characterized by its cellular component, which is surrounded by an extracellular matrix (ECM) consisting of ground substance, fibrous proteins, and interstitial fluid, primarily composed of water. Unlike other tissues, cartilage does not have a direct blood supply, lymphatics, or nerves. Instead, it relies on diffusion from the surrounding tissues for its nutrition. There are three distinct types of cartilaginous tissues based on their composition and function: hyaline, fibro-, and elastic cartilage. Articular cartilage possesses a restricted capacity for self-repair once it undergoes damage. In clinical settings, a reparative tissue is generated; however, it frequently exhibits mechanical deficiencies, being classified as fibrocartilage [25].

The management of early osteoarthritis has been a topic of interest in the medical field. For individuals with early osteoarthritic changes, in particular, regenerative therapy approaches appear to offer a significant potential to close the current treatment gap. In the field of surgical techniques for the treatment of chondral and osteochondral defects in early degenerative joints, therapies such as MSC transplantation and autologous or allogenous osteochondral transplant are encouraging [26].

Recent studies have shown that the utilization of articular chondrocytes as a cell source presents challenges due to age-related factors and variability among donors in their ability to form cartilage. To address these limitations, current research has shifted towards exploring alternative cell sources and experimental models for cartilage regeneration. The tissue source of MSCs seems to influence their therapeutic potential and chondrogenic capabilities [27]. Wakitani et al. involved the use of autologous MSCs from bone marrow to repair osteochondral defects in both animal and human models [28]. The results showed promising outcomes, with better arthroscopic and histological grading scores in the group that received MSC transplantation compared to the control group. This highlights the potential of autologous culture-expanded bone marrow mesenchymal cell transplantation for treating articular cartilage defects in patients with early osteoarthritis. Further studies have also supported the positive effects of MSC transplantation for cartilage repair, emphasizing the importance of utilizing autologous cells for optimal outcomes. Gobbi presented a novel approach for cartilage repair, utilizing mesenchymal stem cells derived from bone marrow during surgery. The bone marrow aspirate concentrate (BMAC) containing MSCs was concentrated to form a flexible clot. This clot was then combined with a hyaluronic acid-based material [29]. This one-stage procedure eliminates the need for cell cultivation, resulting in reduced costs. Long-term studies, spanning up to 10 years, have demonstrated positive clinical outcomes when evaluated using various clinical outcome measures, also in patients older than 45 years [30, 31]. It is important to note that the concentrated BMAC may contain not only MSCs but also hematopoietic progenitors.

Another cell type that has shown promising preclinical data and is currently undergoing a clinical trial is synovial tissue. Jorgenson et al. introduced a suspension bioreactor that integrates microcarrier technology to enhance the cultivation of synovial fluid-derived mesenchymal stem cells, offering a promising avenue for further investigations utilizing this cell source [32]. Shimomura et al. investigated the use of a scaffold-free tissue-engineered construct derived from autologous synovia-derived MSCs to repair knee chondral lesions [33]. Their findings indicated that this approach was both safe and successful in addressing articular cartilage defects within the knee joint [34].

Biomaterials play a crucial role in regenerative medicine, particularly in the context of articular cartilage repair. However, there is still ongoing debate regarding the necessity of a scaffold in this process [35]. The treatment of MSCs for cartilage repair is proposed as a less invasive alternative to metal implants, with the cells and their secretions expected to aid in tissue repair through their ability to trophic effects and modulate the immune response.

25.4 The Use of WJ-MSCs in Dry Arthroscopy

The efficacy of hUCB-MSCs in regenerative medicine, particularly for cartilage repair, has been extensively studied. These cells have demonstrated potential in improving function and alleviating pain in knee osteoarthritis, as evidenced by various scales such as the Visual Analogue Scale (VAS), Knee injury and Osteoarthritis Outcome Score (KOOS), and Lysholm score, which assess pain reduction and functional improvements following treatment.

One of the significant advantages of hUCB-MSCs is their young cellular age, which allows for a more robust proliferative capacity and pluripotency compared to adult stem cells, such as those derived from bone marrow. This feature makes hUCB-MSCs particularly suitable for older patients whose own stem cells may have diminished regenerative capabilities due to age [36, 37].

Moreover, hUCB-MSCs do not require the patient to undergo an invasive procedure for stem cell harvesting, unlike bone marrow-derived MSCs (BMAC-MSCs), which necessitate bone marrow extraction. This noninvasiveness is an added benefit, reducing the procedure's complexity and associated risks [37].

Arthroscopic approach allows for the precise placement of hUC-MSCs directly into the cartilage defects. This technique minimizes patient morbidity and accelerates recovery by reducing the surgical footprint compared to traditional open joint surgeries. Especially use of dry arthroscopy approach significantly reduces the risk of MSC washout by eliminating the use of fluid media commonly used in arthroscopic procedures, thereby ensuring a higher concentration of stem cells at the defect site [38].

Clinically, human umbilical cord mesenchymal stem cells (hUC-MSCs) are frequently embedded within biocompatible scaffolds that support cellular adhesion and help maintain their localization within the defect site. These scaffolds, often composed of materials such as hydrogels or silk fibroin, provide a structural framework that mimics the extracellular matrix, promoting cell growth and differentiation. The scaffolds are typically arthroscopically implanted into the lesion site, where they gradually degrade, leaving behind newly formed tissue that integrates with the existing cartilage [39].

Studies have consistently demonstrated that patients receiving human umbilical cord mesenchymal stem cell (hUC-MSC) treatments exhibit significant improvements in joint function and pain reduction. MRI analyses in these studies have shown increases in cartilage thickness and integrity post-intervention.

For instance, a study on knee osteoarthritis patients treated with a combination of hUC-MSCs and hyaluronic acid (HA) showed significant attenuation in disease progression, improved chondrocyte viability, and enhanced antioxidative capabilities. This combinatorial therapy not only reduced inflammation but also promoted the regeneration of cartilage, leading to improved joint function and reduced pain [40].

Despite promising clinical outcomes, the field continues to explore the optimization of scaffold materials and the mechanistic pathways through which hUC-MSCs exert their effects. Future research is also aimed at establishing standardized protocols for cell preparation and delivery, ensuring consistent and reproducible results across different clinical settings. Dry arthroscopy, being a less invasive approach, reduces the risk of perioperative trauma and postoperative complications, such as arthrofibrosis, making it a promising option for early OA treatment.

25.5 Current Clinical Results of the First Group of Patients

Preliminary results after 2 years of observation were very encouraging, as none of the 19 patients in the first group required knee arthroplasty, and only one patient needed to use a cane for mobility assistance. The remaining individuals functioned efficiently and independently. The clinical assessment of 19 after more than 6 years showed promising results.

The pilot group of the patients consists of 4 women and 15 men, aged between 40 and 74 years, with a mean age of 60.1 years. The average follow-up period was 6.2 years. The injury locations were distributed as follows: 13 medial femoral condyles (MFC), 3 lateral femoral condyles (LFC), 3 patellae, 3 trochleae, 2 tibial plateaus, and 1 lateral tibial condyle (LTC). Concomitant procedures performed during surgery included three high tibial osteotomies (HTO), three partial meniscectomies, and one anterior cruciate ligament reconstruction (ACLR).

Only four patients required additional operations postoperatively: one unicompartmental knee replacement (UKR), two high tibial osteotomies (HTO), and one synovectomy. Additionally, only one patient required the use of crutches postoperatively. Sixteen out of 19 patients reported an improvement in their quality of life following the surgery. Pain reduction was significant, with visual analog scale (VAS) scores decreasing from an average of 7.8 preoperatively to 2.3. Functional outcomes also showed marked improvement. Our results overcame previous prognosis in MRI observation as well (Figs. 25.1 and 25.2).

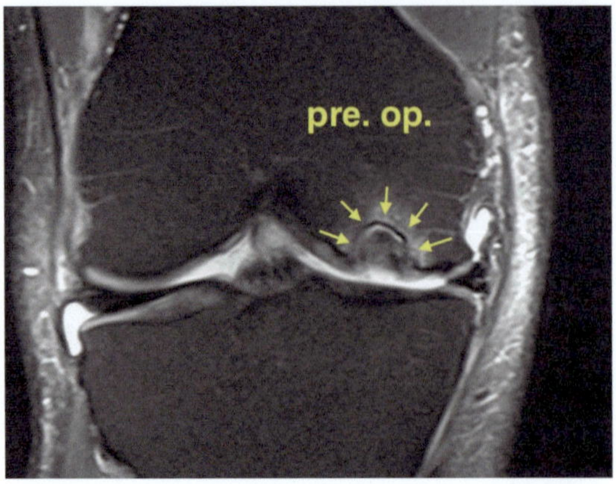

Fig. 25.1 Coronal MRI of the knee preoperatively

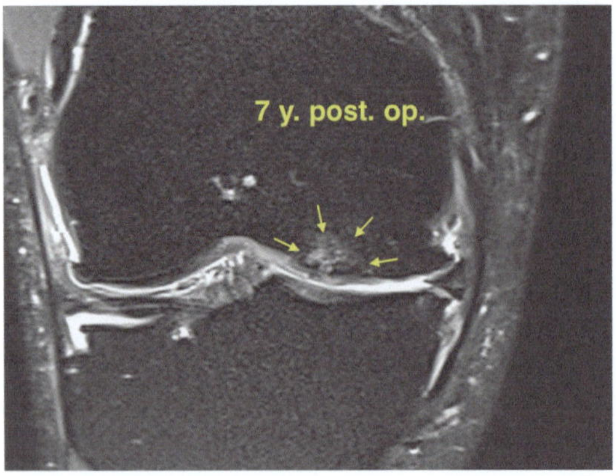

Fig. 25.2 Coronal MRI of the knee 7 years postoperatively

25.6 Prognosis and Future of Therapy

Attempts at biological joint reconstruction in the early stages of osteoarthritis remain an unresolved goal for researchers and clinicians, especially considering the aging population of physically active individuals. Treating gonarthrosis in this patient group with biological reconstruction methods is particularly desirable, as it is crucial to maintain good joint mobility into advanced age.

Enhancing the biological regenerative process using allogeneic stem cells derived from young human tissues, such as the umbilical cord, seems justified here. However, it is unknown what real impact these cells have on the regeneration of the articular surface when used with collagen matrices, which themselves induce such effects in younger patient groups. To date, no animal model studies have been conducted that could confirm the theoretical assumptions of the discussed therapeutic method.

There is no certainty that the cells enhance the effect of biological regeneration of the articular surface, considering the results of Palka's studies, where two groups of patients were compared. Patients with WJ-MSC transplants showed similar results to patients undergoing AMIC procedures without stem cells, although this was a slightly younger group of patients [37].

MRI observations clearly indicate that a regenerate forms on the articular surface undoubtedly. Our pilot study aimed to investigate two basic issues: whether any unforeseen adverse postoperative effects would occur after introducing WJ-MSCs into the knee joint, and whether patients treated with this method would extend the lifespan of their knee joint until total or unicompartmental knee arthroplasty became necessary. The answers to both questions were positive, but the method is not yet fully investigated [41, 42].

It is still unknown what the optimal concentration of stem cells should be implanted into the scaffold before implantation. It is also unclear what the actual concentration of cells is at the time of implantation into the joint after going through the entire preparation process. No studies have been conducted to determine the best scaffold for this type of implantation. Therefore, research on this therapy should be reverted to laboratory work for optimizing the proposed method.

The results of our experiment indicate that there is potential for biological regeneration of articular surfaces even in older individuals, but there is still much work to be done in this area. An undeniable advantage of our developed method is the possibility of treatment using dry arthroscopy, which significantly reduces perioperative trauma and decreases the risk of postoperative arthrofibrosis that occurs after performing arthrotomy. Looking to the future, minimally invasive techniques will undoubtedly be the preferred method for biomaterial implantation to reduce postoperative risks and avoid increasing knee joint stiffness. Performing arthrotomy for biological treatments of early osteoarthritis worsens the conditions for subsequent knee arthroplasty, which will likely be necessary for most patients.

In our opinion, the optimal method for treating early osteoarthritis of the locomotor system joints has not yet been fully developed, and the number of patients awaiting such therapies will increase. In vitro studies and animal models have shown that allogeneic stem cells derived from the umbilical cord, including WJ-MSCs, demonstrate significant potential for bioregeneration of bone and cartilage tissue. Therefore, their application in clinical studies could bring substantial changes, but it might also turn out that they do not significantly impact surgically induced tissue regeneration, which occurs with the involvement of the patient's cells despite their age. It is possible that WJ-MSCs exert a significant anti-inflammatory effect in the ongoing degenerative process, but this is difficult to study under clinical conditions. Thus, there is still much to be done regarding the implantation of WJ-MSCs in the treatment of early stages of osteoarthritis.

References

1. Goldberg VM, Caplan AI. Biological resurfacing: an alternative to total joint arthroplasty. Orthopedics. 1994;17(9):819–21.
2. Gobbi A, Karnatzikos G, Sankineani SR. One-step surgery with multipotent stem cells for the treatment of large full-thickness chondral defects of the knee. Am J Sports Med. 2014;42(3):648–57.
3. Lee DH, Kim SA, Song JS, Shetty AA, Kim BH, Kim SJ. Cartilage regeneration using human umbilical cord blood derived mesenchymal stem cells: a systematic review and meta-analysis. Medicina (Kaunas). 2022;58(12):1801. https://pubmed.ncbi.nlm.nih.gov/36557003/.
4. Wang H, Hung S, Peng S, Huang C, Wei H, Guo Y, et al. Mesenchymal stem cells in the Wharton's jelly of the human umbilical cord. Stem Cells. 2004;22(7):1330–7. https://pubmed.ncbi.nlm.nih.gov/15579650/.
5. Marmotti A, Peretti GM, Mattia S, Bonasia DE, Bruzzone M, Dettoni F, et al. A future in our past: the umbilical cord for orthopaedic tissue engineering. Joints. 2014;2(1):20. /pmc/articles/PMC4295663/.
6. Dhillon J, Kraeutler MJ, Belk JW, Scillia AJ. Umbilical cord–derived stem cells for the treatment of knee osteoarthritis: a systematic review. Orthop J Sports Med. 2022;10(7):23259671221104409. /pmc/articles/PMC9289921/.
7. Xia P, Wang X, Lin Q, Li X. Efficacy of mesenchymal stem cells injection for the management of knee osteoarthritis: a systematic review and meta-analysis. Int Orthop. 2015;39(12):2363–72. https://pubmed.ncbi.nlm.nih.gov/25944079/.
8. Song JS, Hong KT, Kim NM, Park HS, Choi NH. Human umbilical cord blood-derived mesenchymal stem cell implantation for osteoarthritis of the knee. Arch Orthop Trauma Surg. 2020;140(4):503–9. https://pubmed.ncbi.nlm.nih.gov/31980879/.
9. Bieback K, Kern S, Kocaömer A, Ferlik K, Bugert P. Comparing mesenchymal stromal cells from different human tissues: bone marrow, adipose tissue and umbilical cord blood. Biomed Mater Eng. 2008;18(1 Suppl):S71–6.
10. Davies JE, Walker JT, Keating A. Concise review: Wharton's jelly: the rich, but enigmatic, source of mesenchymal stromal cells. Stem Cells Transl Med. 2017;6(7):1620. /pmc/articles/PMC5689772/.
11. Sarugaser R, Lickorish D, Baksh D, Hosseini MM, Davies JE. Human umbilical cord perivascular (HUCPV) cells: a source of mesenchymal progenitors. Stem Cells. 2005;23(2):220–9. https://pubmed.ncbi.nlm.nih.gov/15671145/.
12. Kim DW, Staples M, Shinozuka K, Pantcheva P, Kang SD, Borlongan CV. Wharton's jelly-derived mesenchymal stem cells: phenotypic characterization and optimizing their therapeutic potential for clinical applications. Int J Mol Sci. 2013;14:11692–712. https://www.mdpi.com/1422-0067/14/6/11692/htm.
13. Weiss ML, Anderson C, Medicetty S, Seshareddy KB, Weiss RJ, VanderWerff I, et al. Immune properties of human umbilical cord Wharton's jelly-derived cells. Stem Cells. 2008;26(11):2865–74.
14. Krupa P, Vackova I, Ruzicka J, Zaviskova K, Dubisova J, Koci Z, et al. The effect of human mesenchymal stem cells derived from Wharton's jelly in spinal cord injury treatment is dose-dependent and can be facilitated by repeated application. Int J Mol Sci. 2018;19(5):1503. https://pubmed.ncbi.nlm.nih.gov/29772841/.
15. Barczewska M, Maksymowicz S, Zdolińska-Malinowska I, Siwek T, Grudniak M. Umbilical cord mesenchymal stem cells in amyotrophic lateral sclerosis: an original study. Stem Cell Rev Rep. 2020;16(5):922–32. https://pubmed.ncbi.nlm.nih.gov/32725316/.
16. Haynesworth SE, Baber MA, Caplan AI. Cytokine expression by human marrow-derived mesenchymal progenitor cells in vitro: effects of dexamethasone and IL-1α. J Cell Physiol. 1996;166:585.

17. Margiana R, Markov A, Zekiy AO, Hamza MU, Al-Dabbagh KA, Al-Zubaidi SH, et al. Clinical application of mesenchymal stem cell in regenerative medicine: a narrative review. Stem Cell Res Ther. 2022;13:366.
18. Hu MS, Borrelli MR, Lorenz HP, Longaker MT, Wan DC. Mesenchymal stromal cells and cutaneous wound healing: a comprehensive review of the background, role, and therapeutic potential. Stem Cells Int. 2018;2018:6901983.
19. Zhang Y, Pan Y, Liu Y, Li X, Tang L, Duan M, et al. Exosomes derived from human umbilical cord blood mesenchymal stem cells stimulate regenerative wound healing via transforming growth factor-β receptor inhibition. Stem Cell Res Ther. 2021;12:434.
20. Kim YJ, Yoo SM, Park HH, Lim HJ, Kim YL, Lee S, et al. Exosomes derived from human umbilical cord blood mesenchymal stem cells stimulates rejuvenation of human skin. Biochem Biophys Res Commun. 2017;493:1102.
21. Li L, Tian H, Yue W, Zhu F, Li S, Li W. Human mesenchymal stem cells play a dual role on tumor cell growth in vitro and in vivo. J Cell Physiol. 2011;226:1860.
22. Qiao L, Xu Z-L, Zhao T-J, Ye L-H, Zhang X-D. Dkk-1 secreted by mesenchymal stem cells inhibits growth of breast cancer cells via depression of Wnt signalling. Cancer Lett. 2008;269:67.
23. Studeny M, Marini FC, Dembinski JL, Zompetta C, Cabreira-Hansen M, Bekele BN, et al. Mesenchymal stem cells: potential precursors for tumor stroma and targeted-delivery vehicles for anticancer agents. J Natl Cancer Inst. 2004;96:1593.
24. Han Y, Li X, Zhang Y, Han Y, Chang F, Ding J. Mesenchymal stem cells for regenerative medicine. Cells. 2019;8:886.
25. Armiento AR, Alini M, Stoddart MJ. Articular fibrocartilage—why does hyaline cartilage fail to repair? Adv Drug Deliv Rev. 2019;146:289.
26. Angele P, Niemeyer P, Steinwachs M, Filardo G, Gomoll AH, Kon E, et al. Chondral and osteochondral operative treatment in early osteoarthritis. Knee Surg Sports Traumatol Arthrosc. 2016;24:1743.
27. Wakitani S, Mera H, Nakamura N, Gobbi A. Review of Caplan (1991) on cell-based therapeutic technology using mesenchymal stem cells. J ISAKOS. 2023;9:426.
28. Wakitani S, Imoto K, Yamamoto T, Saito M, Murata N, Yoneda M. Human autologous culture expanded bone marrow-mesenchymal cell transplantation for repair of cartilage defects in osteoarthritic knees. Osteoarthr Cartil. 2002;10:199.
29. Whyte GP, Gobbi A, Sadlik B. Dry arthroscopic single-stage cartilage repair of the knee using a hyaluronic acid-based scaffold with activated bone marrow-derived mesenchymal stem cells. Arthrosc Tech. 2016;5:e913.
30. Gobbi A, Scotti C, Karnatzikos G, Mudhigere A, Castro M, Peretti GM. One-step surgery with multipotent stem cells and Hyaluronan-based scaffold for the treatment of full-thickness chondral defects of the knee in patients older than 45 years. Knee Surg Sports Traumatol Arthrosc. 2017;25:2494.
31. Gobbi A, Whyte GP. Long-term clinical outcomes of one-stage cartilage repair in the knee with hyaluronic acid–based scaffold embedded with mesenchymal stem cells sourced from bone marrow aspirate concentrate. Am J Sports Med. 2019;47(7):1621–8.
32. Jorgenson KD, Hart DA, Krawetz R, Sen A. Production of adult human synovial fluid-derived mesenchymal stem cells in stirred-suspension culture. Stem Cells Int. 2018;2018:8431053.
33. Shimomura K, Ando W, Moriguchi Y, Sugita N, Yasui Y, Koizumi K, et al. Next generation mesenchymal stem cell (MSC)–based cartilage repair using scaffold-free tissue engineered constructs generated with synovial mesenchymal stem cells. Cartilage. 2015;6:13S.
34. Lee DH, Kim SA, Song JS, Shetty AA, Kim BH, Kim SJ. Cartilage regeneration using human umbilical cord blood derived mesenchymal stem cells: a systematic review and meta-analysis. Medicina (Lithuania). 2022;58:1801.
35. Wei ZJ, Wang QQ, Cui ZG, Inadera H, Jiang X, Wu CA. Which is the most effective one in knee osteoarthritis treatment from mesenchymal stem cells obtained from different sources?—A systematic review with conventional and network meta-analyses of randomized controlled trials. Ann Transl Med. 2021;9:452.
36. Huang J, Liu Q, Xia J, Chen X, Xiong J, Yang L, et al. Modification of mesenchymal stem cells for cartilage-targeted therapy. J Transl Med. 2022;20(1):1–15. https://translational-medicine.biomedcentral.com/articles/10.1186/s12967-022-03726-8.
37. Pałka K, Sadlik B, Kamiński P, Bułdak RJ, Dobrakowski M. Comparison of the efficacy of human umbilical cord-derived and bone marrow aspirate concentrate mesenchymal stem cells for cartilage repair defects of the knee via arthroscopic implementation on scaffolds in a retrospective study. J Pers Med. 2024;14(3):220. https://www.mdpi.com/2075-4426/14/3/220/htm.
38. Lee TJ, Jeong CD, Lee TH. Dry arthroscopic cartilage repair of the knee joint using umbilical cord mesenchymal stem cells: Kelly clamp technique. Arthrosc Tech. 2023;12(8):e1355. /pmc/articles/PMC10466224/.
39. Allardyce B, Nakazawa Y, Li G, Sun S. Silk fibroin-based biomaterials for tissue engineering applications. Molecules. 2022;27:2757. https://www.mdpi.com/1420-3049/27/9/2757/htm.
40. Wu JL, Wong PC, Ho CW, Chen CH, Liao KY, Lovel R, et al. Human umbilical cord mesenchymal stem cells in combination with hyaluronic acid ameliorate the progression of knee osteoarthritis. Appl Sci. 2021;11:6650. https://www.mdpi.com/2076-3417/11/14/6650/htm.
41. Sadlik B, Jaroslawski G, Puszkarz M, Blasiak A, Oldak T, Gladysz D, et al. Cartilage repair in the knee

using umbilical cord Wharton's jelly-derived mesenchymal stem cells embedded onto collagen scaffolding and implanted under dry arthroscopy. Arthrosc Tech. 2017;7(1):e57–63. https://pubmed.ncbi.nlm.nih.gov/29552470/.

42. Sadlik B, Jaroslawski G, Gladysz D, Puszkarz M, Markowska M, Pawelec K, et al. Knee cartilage regeneration with umbilical cord mesenchymal stem cells embedded in collagen scaffold using dry arthroscopy technique. Adv Exp Med Biol. 2017;1020:113–22. https://pubmed.ncbi.nlm.nih.gov/28243954/.

Meniscus Repair Using Synovial Stem Cells

26

Ichiro Sekiya

26.1 The Meniscus

The meniscus is a crescent-shaped fibrocartilage located between the femur and tibia in the knee joint. Each knee has two menisci—the medial and lateral menisci—and both play roles in shock absorption, load distribution, and lubrication. Trauma and repetitive loading can cause meniscal injuries and deteriorate their function. Also, as the body ages, the structure of the meniscus changes, leading to degeneration. An injured meniscus can cause a feeling of instability in the knee while also limiting extension and flexion and eliciting pain [1].

26.2 Investigating Osteoarthritis Progression with 3D MRI

Three-dimensional MRI is a powerful imaging tool for investigating the progression of knee osteoarthritis. Our research has utilized artificial intelligence to develop a fully automated 3D image analysis system for knee MRI [2]. This technology complements our ongoing Kanagawa Knee Study, a large-scale survey that has examined MRI scans from over 500 desk-bound workers in the Kanagawa Prefecture [3]. The study population includes government employees and retirees, with a balanced representation across genders and age groups (30–70 years old).

Analyzing the progression of osteoarthritis traditionally requires following subjects over a long period due to the gradual nature of the disease. To overcome this limitation, we employed a novel approach and focused on MRI data from women in their 70 s who were participating in the Kanagawa Knee Study. By selecting one knee with no cartilage defect and 14 knees with varying degrees of cartilage damage, we were able to create a sequence of 3D MRI images (Fig. 26.1), ordered from the mildest to the most severe cartilage defects. This approach offers valuable insights because it provides a visual demonstration of the tendency for cartilage defects to progress alongside medial meniscal extrusion. It also confirms that cartilage defects often initiate in the middle external subregion and expand outward without separating from the inner edge of the medial meniscus (MM) [4]. These findings further emphasize the critical role of the medial meniscus in maintaining joint health while also highlighting the association between meniscus injuries and medial meniscal extrusion—a known risk factor for osteoarthritis [5].

I. Sekiya (✉)
Center for Stem Cell and Regenerative Medicine, Department of Applied Regenerative Medicine, Institute of Science Tokyo, Tokyo, Japan
e-mail: sekiya.arm@tmd.ac.jp

A. Gobbi et al. (eds.), *Regenerative Medicine in Sports and Orthopaedics*,
https://doi.org/10.1007/978-3-031-84693-9_26

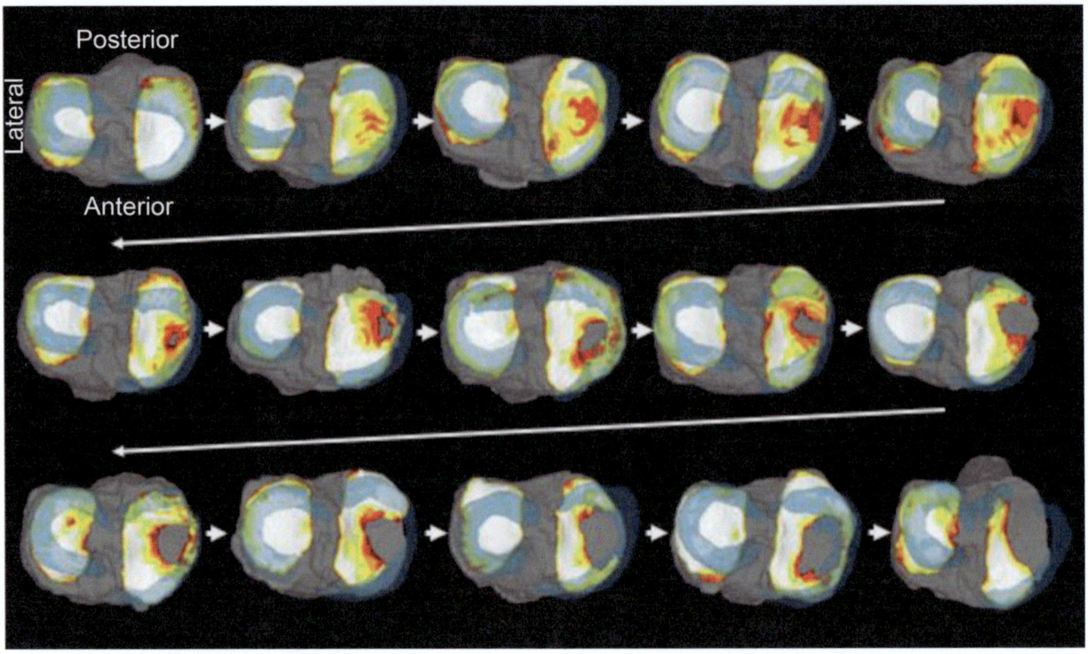

Fig. 26.1 Progression of osteoarthritis of the knee. 3D MRI images show bone in gray, meniscus in blue, and cartilage thickness in different colors (white indicates a thickness greater than 2 mm, yellow to red indicates thinning). From the Kanagawa Knee Study, one knee without a cartilage defect and 14 knees with cartilage defects were selected from women in their 70 s and arranged in order of the mildest to the most severe tibial cartilage defects

26.3 Limited Benefit and Potential Harm of Meniscectomy in Middle-Aged and Older Adults

Some studies have cast doubt on the effectiveness of meniscectomy in treating meniscal tears in middle-aged and older adults. For example, in a randomized controlled trial, Sihvonen et al. compared partial meniscectomy with a sham surgery in this population. Interestingly, the Lysholm scores showed no significant difference between the two groups after 1 year [6]. Similar findings have been reported by other research groups [7, 8]. An even more concerning finding is that meniscectomy in this age group may even be detrimental. Some evidence indicates an increased risk of developing osteoarthritis within a year after surgery [9], with a quadrupled risk at the 16-year mark [10]. Additionally, the need for total knee arthroplasty triples after 9 years in patients who undergo meniscectomy [11]. These findings suggest that meniscectomy for meniscal tears in middle-aged and older adults may not only be ineffective but could also accelerate the progression of osteoarthritis.

26.4 Current Landscape of Meniscus Surgery in Japan

Meniscus injuries are a common orthopedic issue, and surgical intervention typically involves two main approaches: meniscectomy and repair. The latest data obtainable on the NDB Open Data Japan national database, which contains receipt information released by the Ministry of Health, Labor, and Welfare [12], indicates that

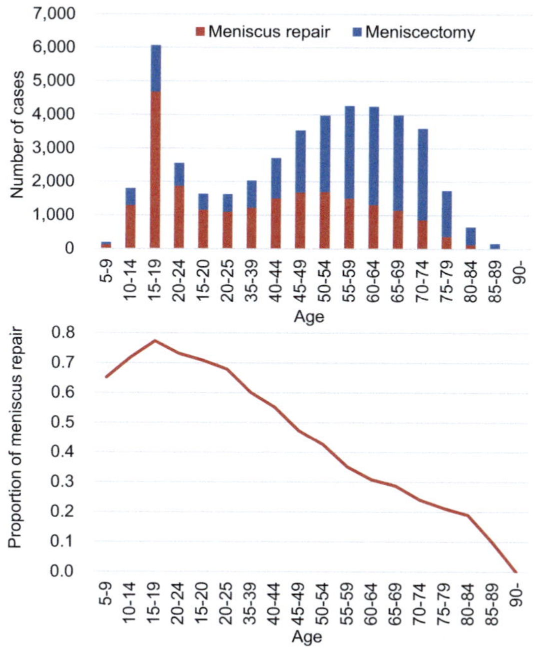

Fig. 26.2 Number of meniscus surgeries in Japan in fiscal year 2021. The number of surgeries by procedure (above) and the proportion of repair procedures by age group are shown

about 45,000 meniscus surgeries were performed in FY2021. The number of surgeries peaks sharply in patients in their late teens and shows a second, broader peak in patients in their 50–60 s (Fig. 26.2). The proportion of these surgeries that are meniscus repairs is 80% in patients in their late teens, but this proportion declines with age, dropping to about 30% in patients in their 50–60 s.

26.5 Meniscus Repair in Middle-Aged and Older Patients

The meniscus has poor blood circulation and low healing capacity; therefore, repair is limited to longitudinal tears along the circumference of the external third of the meniscus, where the blood supply is abundant. However, even after repair surgery, the risk of reinjury always remains. Meniscus injuries in middle-aged and older patients are particularly difficult to treat because they are often associated with complex injury patterns and age-related changes in the knee structure. Consequently, improvement is needed in the outcomes of meniscus repair, as is expansion of the indications for meniscus repair [13].

26.6 Spontaneous Repair of the Meniscus and the Role of the Synovium

When a cylindrical defect of 1 mm diameter is created in the meniscus of a rat, the synovium is induced to enter the meniscal defect, resulting in spontaneous repair (Fig. 26.3) because the synovium provides a blood supply and delivers cells to the defect. The induction of surrounding synovial tissue into the meniscal defect is crucial for meniscal repair, as reported by King more than 80 years ago [14]. Therefore, utilizing cells derived from the synovium to repair and regenerate the meniscus is a logical and valid method for promoting healing following meniscus injury.

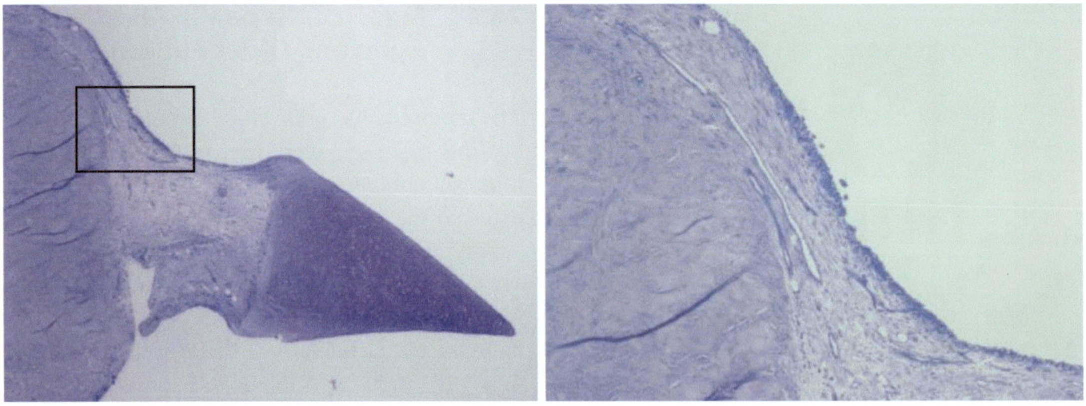

Fig. 26.3 Spontaneous repair process of the meniscus. Cylindrical defects 1 mm in diameter were created in the medial meniscus of rats and observed histologically as toluidine blue-stained sections 2 weeks later. The lower magnified image (left) shows that the 1 mm diameter cylindrical defect was filled with connective tissue, while the higher magnified image (right) shows that the connective tissue is continuous with the synovial membrane covering the surface of the meniscus

26.7 Synovial Stem Cells

The synovium, a membrane that lines the inside of the joint, consists of tissue that can regenerate quickly, even in vitro, following harvesting from the knee. Enzymatic treatment of harvested synovium releases synovial cells that can be cultured, and after 2 weeks of culture, approximately 10% of these cells form cell colonies. The colony-forming cells, when collected and cultured under specific conditions, demonstrate multipotency and can differentiate into cartilage, bone, and fat. These multipotent colony-forming cells express unique surface antigens that identify them as mesenchymal stem cells. We have previously reported that synovial mesenchymal stem cells (synovial stem cells) have high chondrogenic potential [15, 16].

26.8 Promotion of Meniscal Healing by Synovial Stem Cells in a Microminipig Model

A degenerative tear was created in the avascular region of the medial meniscus from the anterior horn to the posterior horn in both knees of microminipigs by repeatedly punching 1 mm diameter holes using a biopsy punch. After repairing the meniscus tear, a suspension of GFP-positive synovial stem cells was placed onto the repaired meniscal area for 10 min, resulting in the engraftment of GFP-positive cells in the meniscal defect and surrounding synovial tissue (Fig. 26.4). Macroscopic comparisons of the knees with and without synovial stem cell transplantation revealed that the tear became indistinct as early as 2 weeks in the transplanted knee, whereas the tear remained evident in the control knee. Analysis of synovial tissue induction into the meniscal tear after synovial stem cell transplantation revealed that a multilayered synovium-like tissue continuously covered the meniscal surface of the transplanted knees from the outer periphery to the defect site at 2 weeks, indicating synovial tissue induction into the meniscal defect. By contrast, in the control knees, the synovial tissue did not reach the meniscal defect even after 4 weeks [17]. These results in microminipigs demonstrate that placing a synovial stem cell suspension onto a repaired degenerative tear in the avascular region of the medial meniscus for 10 min led to the engraftment of synovial stem cells in the meniscal defect and surrounding synovium, promoted synovial tissue induction into the defect, and enhanced meniscal healing (Fig. 26.5).

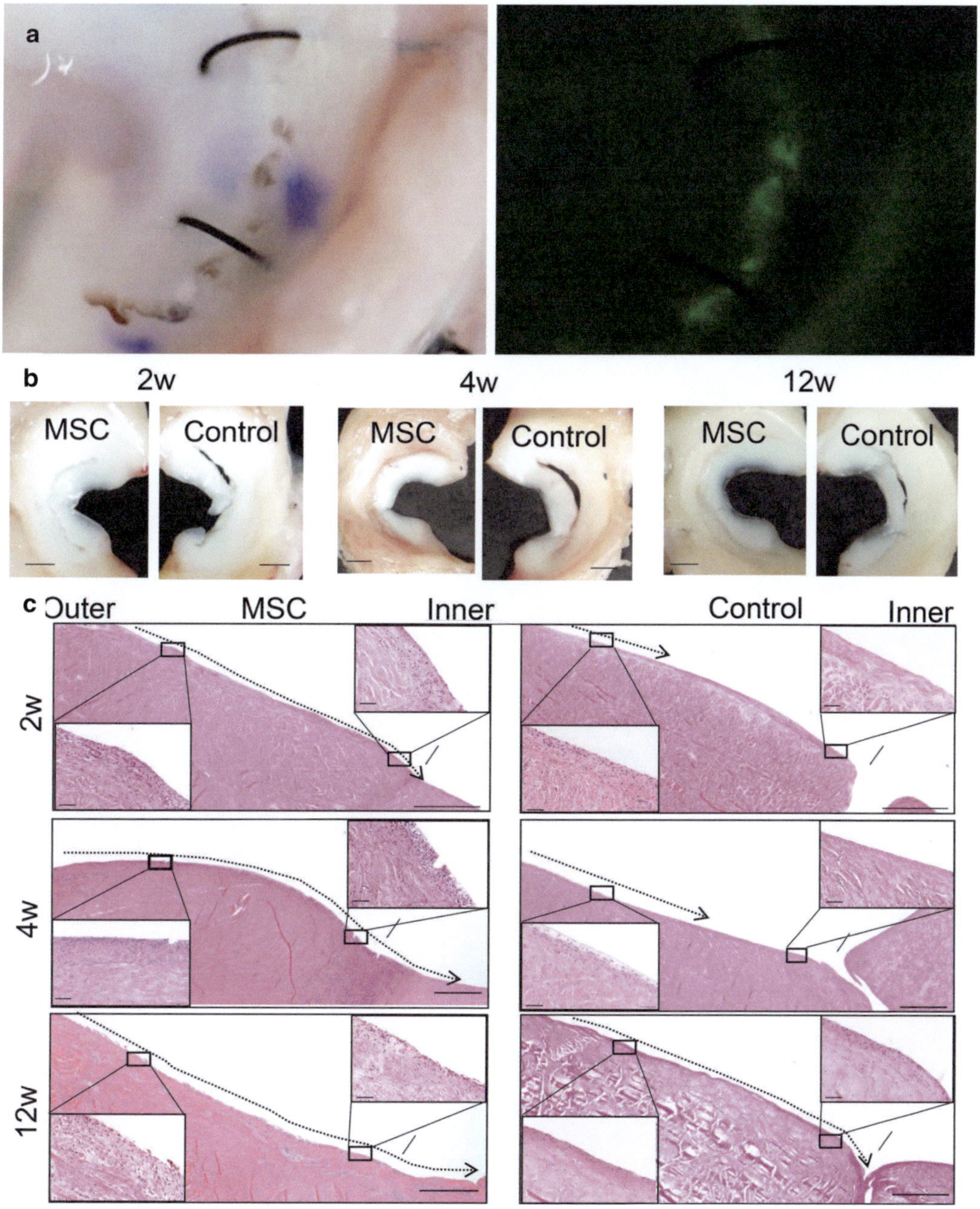

Fig. 26.4 Transplantation of synovial stem cells on the injured meniscus in microminipigs. (**a**) Distribution of synovial stem cells 10 min after transplantation. The medial meniscus was repeatedly punctured with a biopsy punch (1 mm diameter) in the avascular area to induce a degenerative tear. A suspension of synovial stem cells (derived from transgenic pigs expressing GFP) was placed over the injured meniscus, and the cells localized to the injured meniscus and surrounding synovium in the next 10 min. (**b**) The effect of synovial stem cells in promoting meniscus repair. Macroscopic images of the meniscus after suture removal are shown. On the control side, the tear still remains after 12 weeks, while on the cell-treated side, the tear is already obscured after 2 weeks. (**c**) Induction of synovial tissue in the injured meniscus after synovial stem cell transplantation. Histological sections of the meniscus stained with hematoxylin/eosin are shown. Arrows indicate the location of the meniscus tear, and dotted arrows indicate the location where synovial tissue was guided toward the meniscus tear. On the cell-treated side, the synovial tissue reached the meniscus tear earlier than on the control side. (Adapted from the reference by Nakagawa et al. [17])

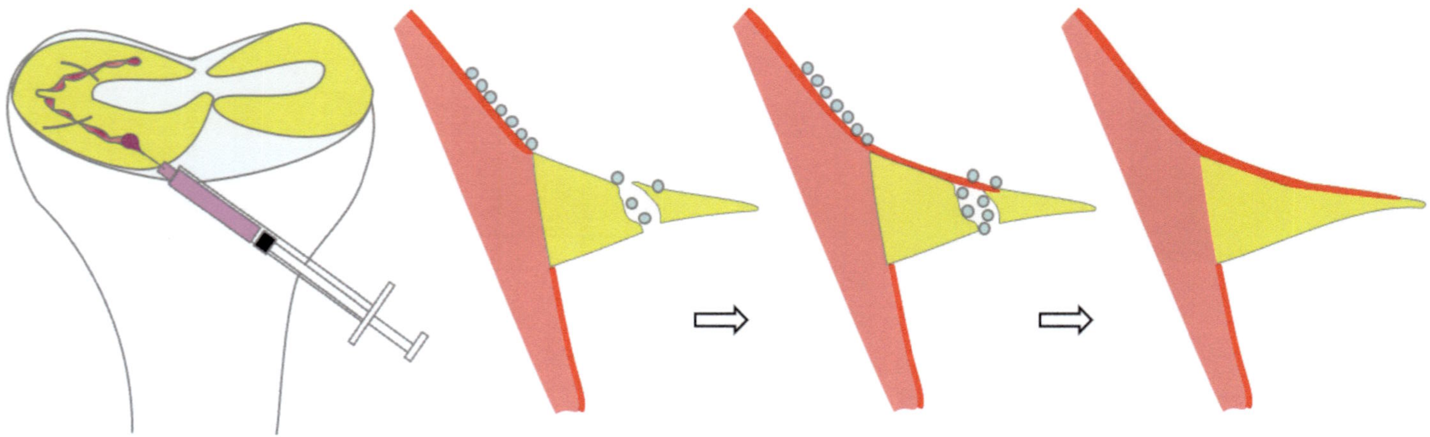

Fig. 26.5 The mechanism of synovial stem cells promoting meniscus repair. Ten minutes after placing a suspension of synovial stem cells on a sutured, degenerative torn meniscus, the cells adhere to the meniscus tear and surrounding synovial tissue. They then promote meniscal healing by guiding the surrounding synovial tissue into the meniscus tear

26.9 High-Resolution Imaging of Synovial Stem Cell Adhesion to the Meniscus

Scanning electron microscopy (SEM) provides a detailed view of the interaction between synovial stem cells and the meniscus. Observing a healthy pig meniscus using SEM reveals a smooth surface (Fig. 26.6). However, after abrasion to mimic degeneration, the surface becomes rough and frayed. When a suspension of synovial stem cells is introduced onto the abraded meniscus, they readily adhere to the surface. This adhesion is evident within 10 min, with the number of attached cells steadily increasing. By 6 h, numerous cells with altered morphology cover the meniscus surface. After 24 h, a complete layer of flattened cells is observed. At higher magnifica-

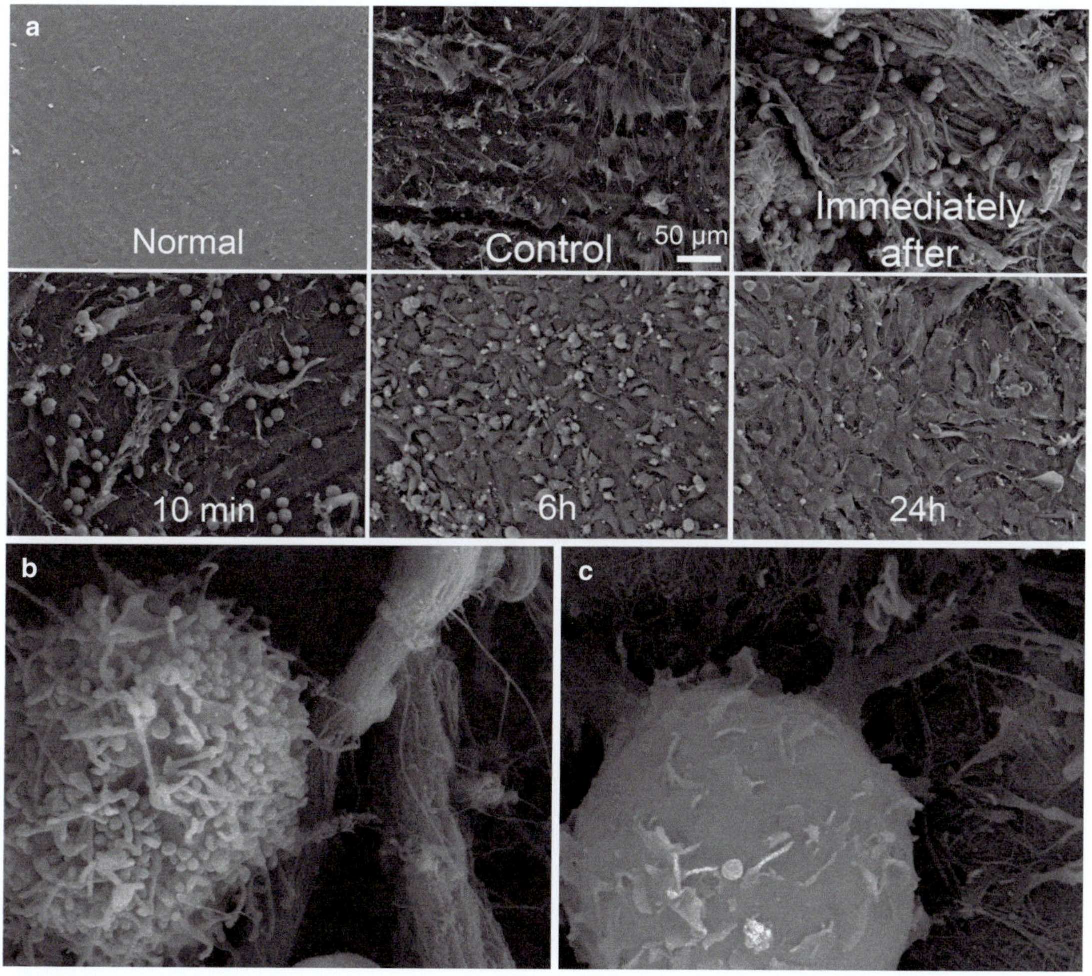

Fig. 26.6 Scanning electron microscopy observations after placement of a synovial stem cell suspension on the degenerated meniscus. (**a**) Morphological changes over time. The surface of a pig meniscus (normal) was abraded, and a suspension of synovial stem cells was placed and washed for observation immediately or after 10 min, 6 h, and 24 h. The cells adhered immediately after placement, and the number of adhering cells increased with increasing placement time. The cells became semi-flattened after 6 h, and flattened cells covered the entire meniscus after 24 h. (**b**) Adhesion by microspikes. Synovial stem cells that adhered immediately after placement had microspikes that became tangled in the collagen fibers of the meniscus. (**c**) Adhesion by pseudopodia. Cells that adhered 1 h after placement extended pseudopodia onto the meniscus. Adapted from the reference by Suzuki et al. [18]

tion, the cells adhering immediately after placement reveal microspikes that grasp the collagen fibers on the meniscus. The cells observed after 6 h display extended pseudopodia that further facilitate adhesion to the meniscus surface [18].

26.10 Essential Molecules for Meniscus Regeneration by Synovial Stem Cells

The essential molecules required for meniscus regeneration were identified by treating rat synovial stem cells with neutralizing antibodies against specific candidate molecules: integrin β1, PDGFRβ, or CD44. Other genes, namely Vcam1, Tnfr1, and Col2a1, were also deleted using the CRISPR/Cas9 system. Synovial MSCs treated with the integrin β1–neutralizing antibody showed inhibited adhesion to collagen-coated slides and failed to regenerate the meniscus when administered to a rat partial meniscectomy model (Fig. 26.7). Synovial MSCs treated with the PDGFRβ–neutralizing antibody exhibited suppressed proliferation in monolayer culture and impaired meniscus regeneration effects. Synovial MSCs lacking Col2a1 lost their in vitro chondrogenic differentiation ability and failed to promote meniscus regeneration in rats. Conversely, the suppression or deletion of CD44, Vcam1, or Tnfr1 expression had no effect on the meniscus regeneration-promoting effects of synovial MSCs. Expression of integrin β1, PDGFRβ, and Col2a1 genes is essential for meniscus regeneration by synovial MSCs in rats [19].

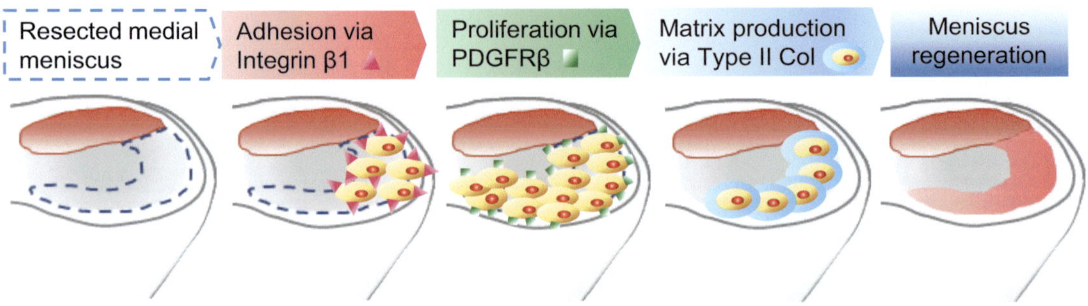

Fig. 26.7 Essential molecules for meniscus regeneration. The molecules responsible for meniscus regeneration were identified by treating rat synovial mesenchymal stem cells (MSCs) with neutralizing antibodies for integrin β1, PDGFRβ, or CD44, or with the CRISPR/Cas9 system to delete the Vcam1, Tnfr1, or Col2a1 genes. After partial meniscectomy, rat knees were injected with these MSCs, and the regenerated meniscus area was quantified 3 weeks later. The in vivo and in vitro functions of the treated and control MSCs were compared. The synovial MSCs injected into the knee joint promoted meniscus regeneration through adhesion to integrin β1 in the meniscectomized region, cell proliferation via PDGFRβ, and cartilage matrix production from type II collagen. (Adapted from the reference by Kitahashi et al. [19])

26.11 Clinical Application of Synovial Stem Cell Transplantation for Meniscus Repair

Based on the results of preclinical studies, we initiated a clinical study in 2014 to study the arthroscopic transplantation of synovial stem cells following meniscus repair surgery (Fig. 26.8). Specifically, we studied medial meniscal tears in middle-aged and older patients without malalignment or medial meniscus extrusion—a condition that would typically be an indication for meniscectomy. None of the five knees treated in this clinical study had any serious adverse outcomes during the 2-year observation period, and all showed improved MRI findings and clinical scores [20].

We also started a physician-initiated clinical trial in 2017. Clinical scores improved in all nine patients who received a transplant, and second-look arthroscopy was performed 1 year later for four patients with flap tears and two with radial tears. In a representative case of a flap tear, the posterior flap was severely displaced posteriorly and impinged into the femorotibial joint prior to treatment (Fig. 26.9). The meniscus was repaired using the tie-grip suture method to reduce the displaced flap. Two weeks later, an autologous synovial stem cell suspension was statically loaded onto the repaired area for 10 min, under arthroscopic observation and without joint irrigation. The second-look arthroscopy performed 1 year later confirmed that the repaired meniscal shape was maintained [21]. The inner margin of the repaired meniscus appeared irregular, and opinions may differ on whether it should have been trimmed or preserved. However, since this patient's clinical symptoms significantly improved, the author judged that preserving the inner margin was the appropriate choice.

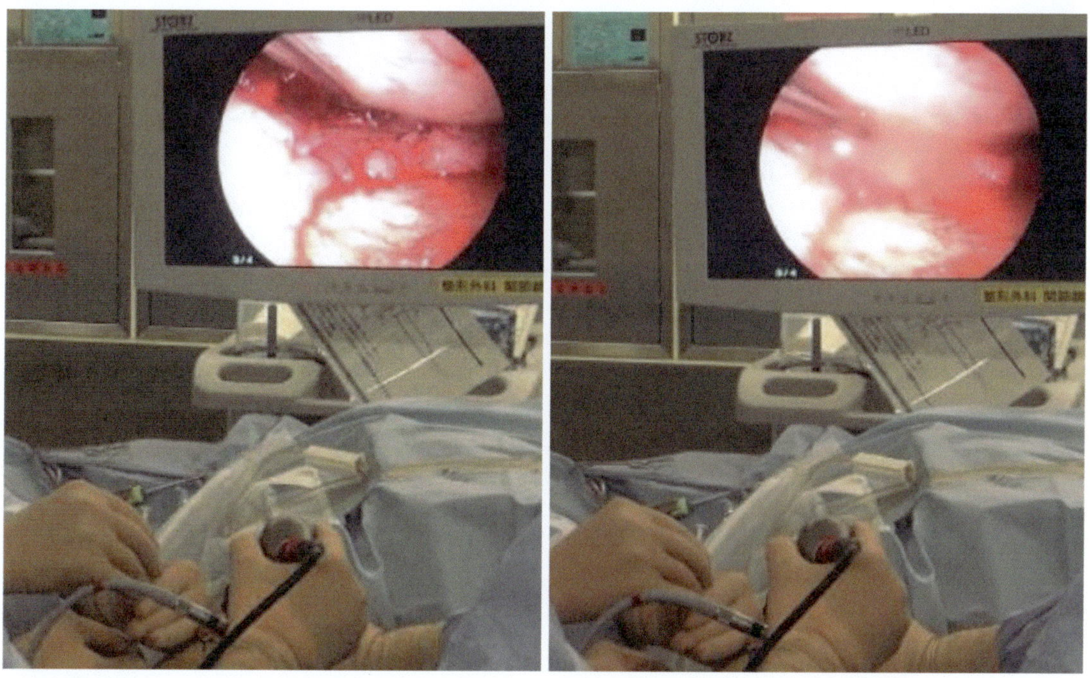

Fig. 26.8 Synovial stem cell transplantation into the repaired meniscus. Arthroscopy is performed without irrigation fluid (left), a synovial stem cell suspension is placed on the repaired meniscus using a syringe (right), and this position is maintained for 10 min

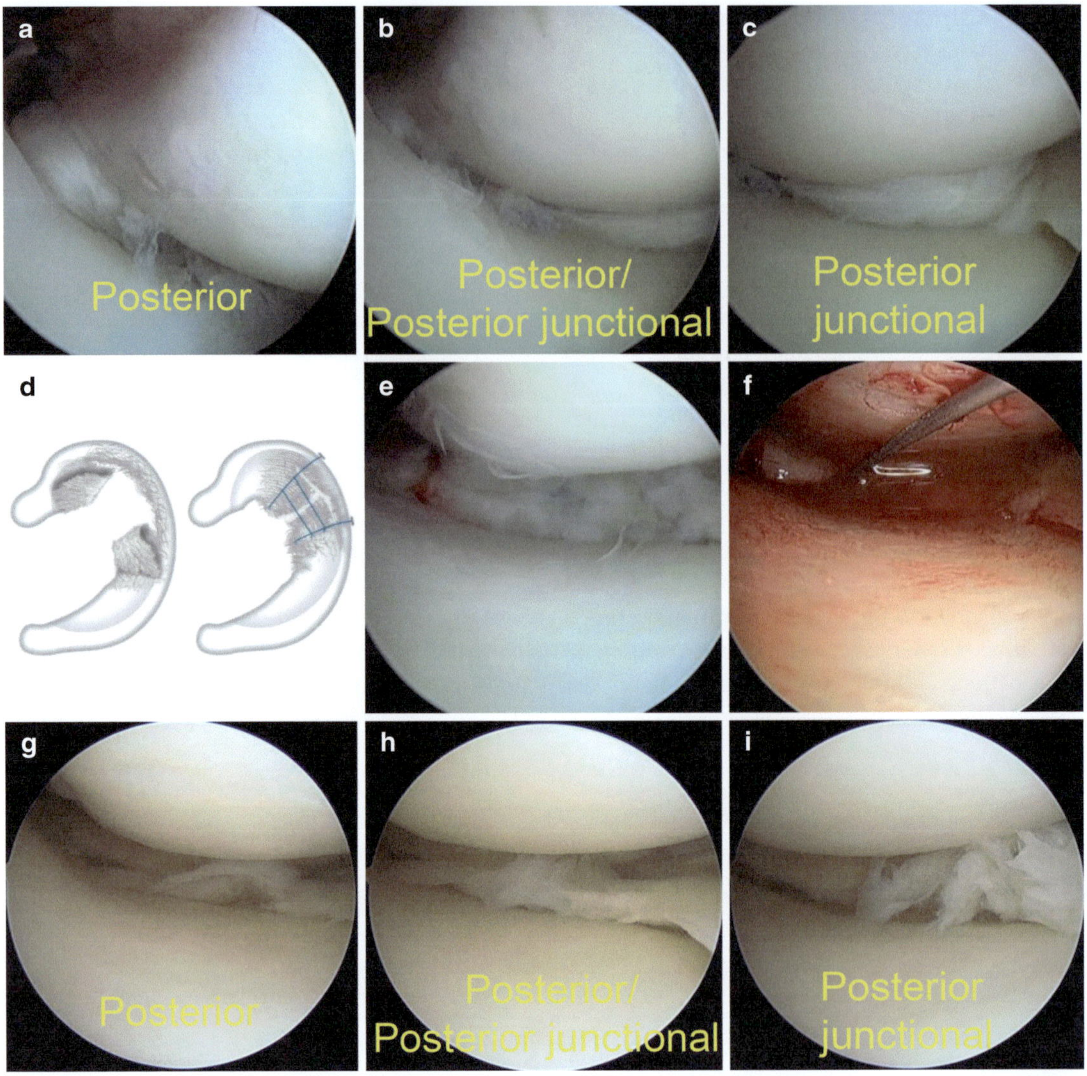

Fig. 26.9 Arthroscopic images of synovial stem cell transplantation for a flap tear of the medial meniscus in the right knee. (**a–c**) Before meniscus repair. (**d**) Illustrations of the meniscus before and immediately after meniscus repair. (**e**) Arthroscopic image immediately after repair. (**f**) Cell transplantation onto the repaired area. (**g–i**) Second-look arthroscopic images 1 year later. (Adapted from the reference by Sekiya et al. [21])

26.12 Multicenter Trial Evaluating the Arthroscopic Scoring System for Flap Tears

We are presently examining whether this treatment can be appropriately implemented at multiple medical institutions by conducting a multicenter trial, which was initiated in January 2023. The trial targets patients with flap tears, as the flap tear injury is considered particularly refractory among meniscal tears that would typically be indicated for meniscectomy. We currently use a novel arthroscopic scoring system that we have developed to enable objective quantitative analysis of flap tears for our assessment of whether clinical improvement is due to meniscal healing (Fig. 26.10) [21].

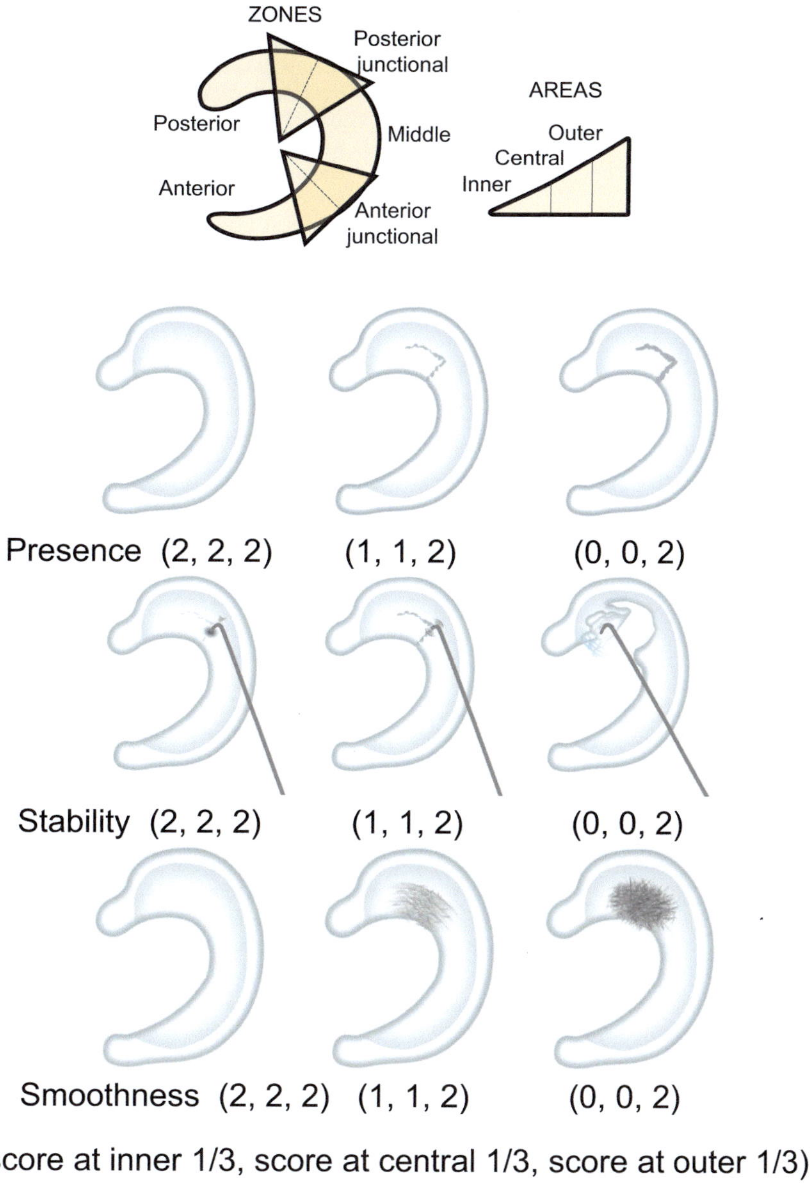

Fig. 26.10 Arthroscopic scoring system proposed by the author for flap tears. The presence, stability, and smoothness of the meniscus are scored as 0, 1, or 2 points, respectively. Scores are made at 15 locations, divided into five zones (anterior, anterior junctional, middle, posterior junctional, and posterior) and three areas (inner, central, and outer) within each zone. This figure shows the scores for the inner, middle, and outer areas of the posterior junctional zone. Second-look arthroscopic images 1 year later. (Adapted from the reference by Sekiya et al. [21])

26.13 Future Study

Our ongoing multicenter trial will continue to provide valuable data on the efficacy and safety of synovial stem cell transplantation for meniscus repair in a broad patient population. Future studies will focus on the optimization of cell culture conditions and transplantation timing, the use of synovial stem cells in combination with other surgical procedures (e.g., centralization for extruded menisci or osteotomy for malaligned knees), and investigations of the long-term durability of the repaired meniscus. Research into cost-effectiveness and the establishment of guidelines for patient selection will also be crucial for wider clinical adoption of this novel stem cell transplantation approach.

References

1. Duong V, Oo WM, Ding C, Culvenor AG, Hunter DJ. Evaluation and treatment of knee pain: a review. JAMA. 2023;330(16):1568–80.
2. Sekiya I, Katano H, Guermazi A, Miura Y, Okanouchi N, Tomita M, et al. Association of AI-determined Kellgren-Lawrence grade with medial meniscus extrusion and cartilage thickness by AI-based 3D MRI analysis in early knee osteoarthritis. Sci Rep. 2023;13(1):20093.
3. Sekiya I, Sasaki S, Miura Y, Aoki H, Katano H, Okanouchi N, et al. Medial Tibial osteophyte width strongly reflects medial meniscus extrusion distance and medial joint space width moderately reflects cartilage thickness in knee radiographs. J Magn Reson Imaging. 2022;56(3):824–34.
4. Katano H, Ozeki N, Koga H, Tomita M, Suzuki K, Masumoto J, Sekiya I. Three-dimensional MRI shows cartilage defect extension with no separation from the meniscus in women in their 70 s with knee osteoarthritis. Sci Rep. 2022;12(1):4198.
5. Costa CR, Morrison WB, Carrino JA. Medial meniscus extrusion on knee MRI: is extent associated with severity of degeneration or type of tear? AJR Am J Roentgenol. 2004;183(1):17–23.
6. Sihvonen R, Paavola M, Malmivaara A, Itälä A, Joukainen A, Nurmi H, et al. Arthroscopic partial meniscectomy versus sham surgery for a degenerative meniscal tear. N Engl J Med. 2013;369(26):2515–24.
7. Katz JN, Brophy RH, Chaisson CE, de Chaves L, Cole BJ, Dahm DL, et al. Surgery versus physical therapy for a meniscal tear and osteoarthritis. N Engl J Med. 2013;368(18):1675–84.
8. Yim JH, Seon JK, Song EK, Choi JI, Kim MC, Lee KB, Seo HY. A comparative study of meniscectomy and nonoperative treatment for degenerative horizontal tears of the medial meniscus. Am J Sports Med. 2013;41(7):1565–70.
9. Roemer FW, Kwoh CK, Hannon MJ, Hunter DJ, Eckstein F, Grago J, et al. Partial meniscectomy is associated with increased risk of incident radiographic osteoarthritis and worsening cartilage damage in the following year. Eur Radiol. 2017;27(1):404–13.
10. Englund M, Roos EM, Lohmander LS. Impact of type of meniscal tear on radiographic and symptomatic knee osteoarthritis: a sixteen-year followup of meniscectomy with matched controls. Arthritis Rheum. 2003;48(8):2178–87.
11. Rongen JJ, Rovers MM, van Tienen TG, Buma P, Hannink G. Increased risk for knee replacement surgery after arthroscopic surgery for degenerative meniscal tears: a multi-center longitudinal observational study using data from the osteoarthritis initiative. Osteoarthr Cartil. 2017;25(1):23–9.
12. Katano H, Koga H, Ozeki N, Otabe K, Mizuno M, Tomita M, et al. Trends in isolated meniscus repair and meniscectomy in Japan, 2011-2016. J Orthop Sci. 2018;23(4):676–81.
13. Ozeki N, Koga H, Sekiya I. Degenerative meniscus in knee osteoarthritis: from pathology to treatment. Life (Basel). 2022;12(4)
14. King D. The healing of semilunar cartilages. Clin Orthop Relat Res. 1936;1990(252):4–7.
15. Koga H, Muneta T, Nagase T, Nimura A, Ju YJ, Mochizuki T, Sekiya I. Comparison of mesenchymal tissues-derived stem cells for in vivo chondrogenesis: suitable conditions for cell therapy of cartilage defects in rabbit. Cell Tissue Res. 2008;333(2):207–15.
16. Sakaguchi Y, Sekiya I, Yagishita K, Muneta T. Comparison of human stem cells derived from various mesenchymal tissues: superiority of synovium as a cell source. Arthritis Rheum. 2005;52(8):2521–9.
17. Nakagawa Y, Muneta T, Kondo S, Mizuno M, Takakuda K, Ichinose S, et al. Synovial mesenchymal stem cells promote healing after meniscal repair in microminipigs. Osteoarthr Cartil. 2015;23(6):1007–17.
18. Suzuki S, Mizuno M, Sakamaki Y, Mimata A, Endo K, Kohno Y, et al. Morphological changes in synovial mesenchymal stem cells during their adhesion to the meniscus. Lab Investig. 2020;100(7):916–27.
19. Kitahashi T, Kogawa R, Nakamura K, Sekiya I. Integrin β1, PDGFRβ, and type II collagen are essential for meniscus regeneration by synovial mesenchymal stem cells in rats. Sci Rep. 2022;12(1):14148.
20. Sekiya I, Koga H, Otabe K, Nakagawa Y, Katano H, Ozeki N, et al. Additional use of synovial mesenchymal stem cell transplantation following surgical repair of a complex degenerative tear of the medial meniscus of the knee: a case report. Cell Transplant. 2019;28(11):1445–54.
21. Sekiya I, Koga H, Katano H, Mizuno M, Kohno Y, Otabe K, Ozeki N. Second-look arthroscopy after meniscus repair and synovial mesenchymal stem cell transplantation to treat degenerative flaps and radial tears of the medial meniscus: a case report. J Orthop Sci. 2022;27(4):821–34.

Regenerative Medicine for Osteoarthritis of the Knee Using Cell Sheets

27

Masato Sato, Eriko Toyoda, and Kosuke Hamahashi

Abbreviations

AC sheet	Adult knee joint chondrocyte sheet
ACAN	Aggrecan
ACKR4	Atypical chemokine receptor 4
ADAMTS5	ADAM metallopeptidase with thrombospondin type 1 motif 5
BMP6	Bone morphogenetic protein 6
CCN2	Cellular communication network factor 2
CNMD	Chondromoduline
COL1	Type I collagen
COL1A2	Collagen type I alpha 2 chain
COL2	Type II collagen
COL2A1	Collagen type II alpha 1 chain
COL27A1	Collagen type XXVII alpha 1 chain
COMP	Cartilage oligomeric matrix protein
CSGALNACT1	Chondroitin sulfate N-acetylgalactosaminyltransferase 1
CXCL6	C-X-C motif chemokine ligand 6
DKK1	Dickkopf-related protein 1
ESM1	Endothelial cell-specific molecule 1
FN	Fibronectin
GATA6	GATA binding protein 6
GDF5	Growth differentiation factor 5
GREM1	Gremlin 1
G0S2	G0/G1 switch 2
HAPLN1	Hyaluronan and proteoglycan link protein 1
HTO	High tibial osteotomy
KL	Kellgren–Lawrence
KOOS	Knee Injury and Osteoarthritis Outcome Score
LIPA	Laser-induced photoacoustic measurement
LKS	Lysholm Knee Score
MATN2	Matrilin 2
MHLW	Ministry of Health, Labour and Welfare
MIA	Melanoma inhibitory activity

M. Sato (✉) · E. Toyoda · K. Hamahashi
Department of Orthopaedic Surgery, Surgical Science, Tokai University School of Medicine, Isehara, Kanagawa, Japan

Center for Musculoskeletal innovative Research and Advancement (C-MiRA), Tokai University Graduate School, Isehara, Kanagawa, Japan
e-mail: sato-m@tokai.ac.jp; toyoda.eriko.r@tokai.ac.jp; hamako@tokai.ac.jp

A. Gobbi et al. (eds.), *Regenerative Medicine in Sports and Orthopaedics*,
https://doi.org/10.1007/978-3-031-84693-9_27

MSC	Mesenchymal stem cell
OAK	Osteoarthritis of the knee
OARSI	Osteoarthritis Research Society International
PD cell	Polydactyly-derived chondrocyte
PD sheet	Polydactyly-derived chondrocyte sheet
POU5F1	POU class 5 homeobox 1
PTGS2	Prostaglandin-endoperoxide synthase 2
RMSC	Treatment by resection of unhealthy tissue, marrow stimulating and covered by chondrocyte sheets
SOX9	SRY-box transcription factor 9
SP7	sp7 transcription factor
STC1	Stanniocalcin 1
TGF	Transforming growth factor
TIMP1	Tissue inhibitor of metallopeptidase 1
TKA	Total knee arthroplasty
VEGFA	Vascular endothelial growth factor A

27.1 Introduction

27.1.1 Applications of Cell Sheets in Regenerative Medicine

Regenerative medicine is a type of medical treatment to repair and regenerate tissues and organs that have lost their functions because of disease or trauma and requires the production and transplantation of the tissues and cells needed for treatment from the patient's own or other people's cells. The final form of regenerative medicine involves the construction of three-dimensional structures and the replacement of damaged organs by transplantation, although no organs created through regenerative medicine have reached this level.

The fabrication and application of three-dimensional structures, called multilayered tissues or organoids, produced in two-dimensional culture dishes have been attempted within regenerative medicine. Temperature-responsive culture dishes developed by Okano et al. can be used with cell sheet engineering to construct three-dimensional tissue without the use of scaffolds. This process has potential use in regenerative medicine because the cell sheets themselves have excellent adhesive properties and can be reliably transplanted to the target site without the need for sutures or patches [1].

Current research, clinical trials, and clinical applications of regenerative medicine are ongoing for use in the epidermis, cornea, myocardium, esophagus, middle ear, and periodontal tissue [2–4]. Approved and commercially available regenerative medicine products in Japan include epidermal-derived cell sheets for indications such as severe burns, corneal epithelial cell sheets derived from the corneal ring, and oral mucosa tissue for indications such as corneal epithelial stem cell exhaustion. In addition, a clinical trial has been initiated to transplant cardiomyocyte sheets produced from cardiomyocytes derived from induced pluripotent stem cells from healthy individuals into patients with dilated cardiomyopathy.

Regenerative medicine for osteoarthritis of the knee (OAK) is expected to have social, medical, and economic outcomes, but no regenerative medicine products are currently available for the treatment of OAK. We have applied cell sheet engineering to regenerative medicine in the treatment of cartilage defects caused by OAK and have conducted basic studies to examine its potential role in biological knee joint reconstruction [5–10].

27.1.2 Differences Between Autologous and Allogeneic Cell Sheets

Autologous cell sheets are adult knee joint chondrocyte cell sheets (AC sheets) made from the cartilage tissue of the healthy part of the patient's own knee joint. The proliferative capacity of chondrocytes in adults, especially the elderly, is low, and it has been difficult to culture these cells into a uniform cell sheet form. However, the co-culture method of seeding synovial membrane-derived cells collected at the same time as the cartilage tissue onto a companion plate and seeding the chondrocytes into a temperature-responsive insert promotes the proliferation of chondrocytes. Stacking the three resulting cell sheets and culturing them for 1 week enables the production of a flexible, highly adhesive sheet-like tissue in a short time [11] (Fig. 27.1).

Allogeneic cell sheets are made from polydactyly-derived chondrocyte cells (PD cells), which are isolated from cartilage waste tissue obtained from patients with polydactyly after excision of the extra digits. The cells are then treated with enzymes, expanded, and cryopreserved. The product produced from these cells is an allogeneic cell sheet, which we call the polydactyly-derived chondrocyte cell sheet (PD sheet) (Fig. 27.2). Because patients who undergo polydactyly surgery are mainly infants, the recovered PD cells show high proliferative potential and do not require co-culturing with synovial cells, as in the preparation of AC sheets. PD cells obtained from polydactyly tissue can be used to produce allogeneic cell sheets from 1/15th the number of cells needed when using AC sheets. Calculations show that with cells obtained from a single donor, 600 second-generation cell sheets and 3000 third-generation cell sheets can be produced for transplantation [12].

Because the cellular material for PD sheets is cryopreserved to form lots, some cells from the lot can be used to select donors for therapeutic use after confirming genetic stability by array comparative genomic hybridization analysis and karyotyping before and after long-term culture, and these cells can be adjusted and cultured when necessary. Another advantage in terms of quality control is that cell sheets can be prepared in advance, and their purity, production of factors, and efficacy can be evaluated and confirmed using animal models. Treatment with AC sheets requires surgery to harvest autologous tissue, and the number of cells recovered and their proliferative capacity can vary significantly between patients. Some patients with large areas of articular damage may not be able to produce sufficient AC sheets. However, treatment with PD sheets is superior because the required number of prequalified allogeneic cell sheets can be prepared in advance. On the other hand, because of the use of allogeneic cells, the patient's immune response

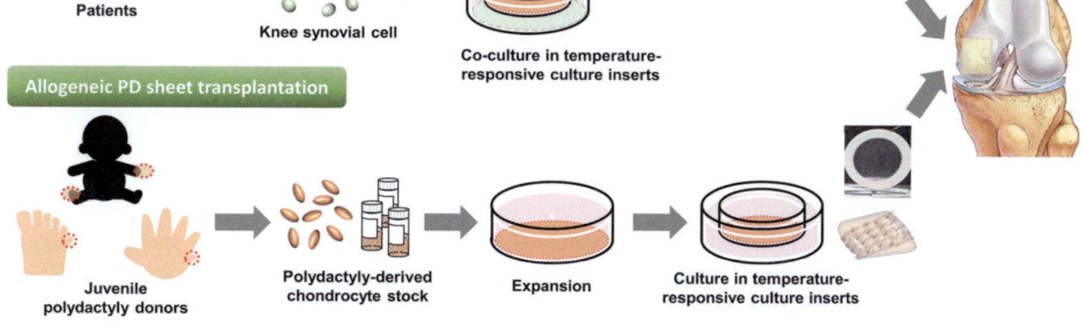

Fig. 27.1 Preparation of autologous or allogeneic cell sheets for treatment of osteoarthritis of the knee
AC sheet: Cell sheet prepared from chondrocyte obtained from joint tissue of OAK patient. Chondrocytes were co-cultured with synovial cells that were separated by the membrane of culture insert and layered. PD sheet: Cell sheet prepared from cryopreserved polydactyly-derived chondrocyte obtained from surgical waste of polydactyly patient

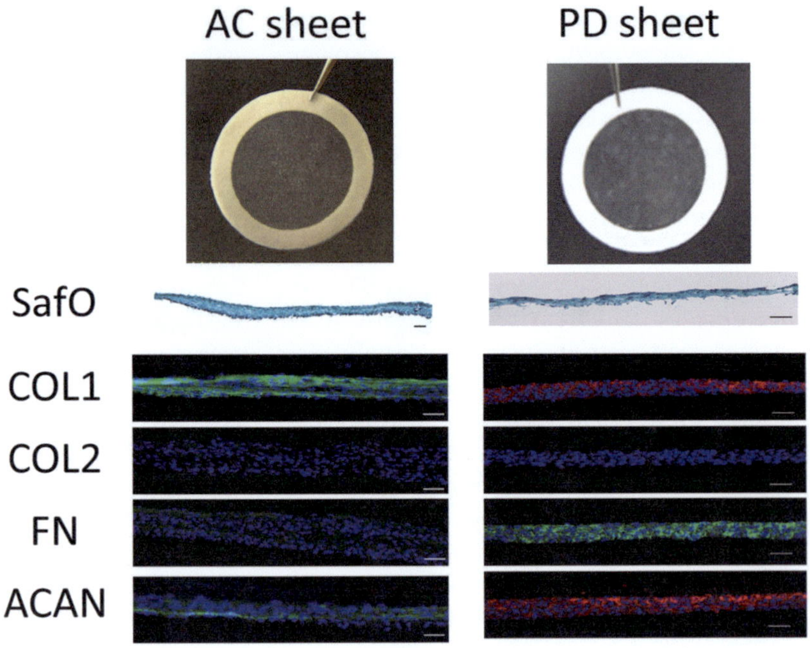

Fig. 27.2 Adult knee joint chondrocyte sheet (AC sheet) and polydactyly-derived chondrocyte sheet (PD sheet)

after transplantation must be assessed and followed up. We have confirmed the presence of antibodies to human leukocyte antigen in patients with allogeneic cell sheet transplants. However, we have also confirmed that this does not affect the clinical outcomes.

Macroscopic images show representative images of a PD sheet and AC sheet attached to a white PVDF membrane. Histological staining of cell sheet. Stained for Safranin O, immunostaining for type I collagen (COL1), type II collagen (COL2), fibronectin (FN), and aggrecan (ACAN) (scale bar = 50 μm).

Autologous cell sheets produced from adult knee cartilage tissue (AC sheets) and cell sheets produced from polydactyly patient cartilage tissue (PD sheets) differ in terms of the source of cartilage tissue but exhibit similar properties in many respects. Both cell sheets are CD31- and CD45-negative; comprise cells positive for the mesenchymal cell markers CD44, CD81, and CD90; and contain some CD146-, CD49a-, and GD2-positive cells. These cell sheets also produce cartilage anabolic and other related factors such as melanoma inhibitory activity (MIA), transforming growth factor β (TGFβ), tissue inhibitor of metallopeptidase 1 (TIMP1), stanniocalcin 1 (STC1), and hyaluronan and proteoglycan link protein 1 (HAPLN1). However, PD sheets tend to produce more TGFβ, MIA, and TIMP1, whereas AC sheets produce more STC1 and HAPLN1 [10, 12]. Cell sheets are thought to physically cover the injured area, to prevent the outflow of factors such as proteoglycans from the injured area matrix, and to protect the cartilage defect from catabolic factors in joint fluid. At the same time, the cell sheet continuously supplies many fluid factors and mobilizes cells involved in tissue repair to the injured area [10].

Cross-sections of both AC and PD sheets show a multilayered cell structure that expresses type I collagen, aggrecan, and fibronectin. These sheets show little staining in Col2A1 immunohistochemical staining and have characteristics similar to those of cells in the superficial layer of cartilage and dedifferentiated chondrocytes (Fig. 27.2). However, AC and PD sheets have been observed to repair and regenerate cartilage defects with hyaline cartilage-like tissue that exhibits type II collagen-positive and safranin O-positive properties when implanted in the knee joint in rabbit and rat models of total cartilage defects [13–15] and OAK patients [16, 17].

We have previously identified MIA, Dickkopf-related protein 1 (DKK1), endothelial cell-specific molecule 1 (ESM1), and gremlin 1 (GREM1) as humoral factors whose gene expression levels and protein production levels correlate with hyaline cartilage repair action in animal models of PD sheets [15]. MIA, DKK1, and GREM1 are known to be associated with chondrogenic differentiation and are involved in repair as chondrotrophic factors. A clinical study has reported the potential production of these factors by PD sheets implanted in patients with osteoarthritis (OAK) [16].

Two previous clinical studies have analyzed gene expression in cell sheets and attempted to identify a set of genes useful for predicting clinical outcome. In a clinical study of autologous cell sheets transplanted with AC sheets, the genes predictive of the Knee Injury and Osteoarthritis Outcome Score (KOOS) at 1 year postoperatively were aggrecan (*ACAN*), chondroitin sulfate N-acetyl galactosaminyl transferase 1 (*CSGALNACT1*), bone morphogenetic protein 6 (*BMP6*), prostaglandin-endoperoxide synthase 2 (*PTGS2*), POU class 5 homeobox 1(*POU5F1*), and C-X-C motif chemokine ligand 6 (*CXCL6*). GATA binding protein 6 (*GATA6*), chondromodculine (*CNMD*), growth differentiation factor 5 (*GDF5*), ADAM metallopeptidase with thrombospondin type 1 motif 5 (*ADAMTS5*), Sp7 transcription factor (*SP7*), and matrilin 2 (*MATN2*) were identified as postoperative predictors of the Lysholm Knee Score (LKS). *ACAN*, *BMP6*, *POU5F1*, *CXCL6*, vascular endothelial growth factor A (*VEGFA*), and *PTGS2* were identified as candidate genes of the Osteoarthritis Research Society International (OARSI) histological score [17]. By contrast, clinical studies of allogeneic cell sheets implanted with PD sheets identified *PTGS2*, *TGFB1*, *MIA*, and G0/G1 switch 2(*G0S2*) as postoperative KOOS predictor genes, and collagen type I alpha 2 chain (*COL1A2*), *MATN2*, SRY-box transcription factor 9 (*SOX9*), and *TGFB1* were identified as postoperative predictors of the LKS. Cellular communication network factor 2 (*CCN2*), cartilage oligomeric matrix protein (*COMP*), atypical chemokine receptor 4 (*ACKR4*), *ESM1*, and *GREM1* were identified as predictors of the OARSI histological score for histological repair. Collagen type II alpha 1 chain (*COL2A1*), collagen type XXVII alpha 1 chain (*COL27A1*), *ACKR4*, and *ESM1* were identified as predictors of the International Cartilage Research Society's overall assessment [16]. Although validation of the predictive performance of these gene groups is an issue for the future, the two types of sheets share specific genes. Confirmation that the clinical outcomes can be predicted according to the expression of these genes may also provide a method for assessing the efficacy of the cell sheets.

27.1.3 Current Status of OAK and Its Significance as Joint Preservation Surgery

Japan is now a super-aging society with an unprecedented proportion of older people. According to the World Health Organization's 2023 World Health Statistics, life expectancy in Japan is 81.5 years for men, the second highest in the world, and 86.9 years for women, the highest in the world [18]. The number of people with joint disease is increasing with the aging of society as a whole. It is estimated that 10 million people have subjective symptoms of OAK, a typical joint disease, and that 30 million patients have been radiologically diagnosed with OAK, which makes this a national disease in Japan [19].

Although joint replacement surgery can achieve stable results as a surgical treatment for OAK, the durability of current knee joint prostheses is estimated at about 20 years. Considering the future need for revision surgery, one indication for arthroplasty in Japanese patients with OAK should generally be age 65 years or older, and the indication for young people should be carefully considered. According to a market survey, the estimated annual numbers in Japan are about 100,000 for total knee arthroplasty (TKA), 14,000 for unilateral knee arthroplasty, and 10,000 for high tibial osteotomy (HTO). The indication for HTO surgery is appropriate up to Kellgren–

Lawrence (KL) grade 3 in people with OAK, when the articular cartilage is still intact. In advanced disease (KL grade 4), where the articular cartilage has disappeared and the subchondral bone is in contact, an artificial joint is indicated. The number of patients in Japan who refuse surgical treatment and who are undergoing long-term, repeated hyaluronic acid injections as conservative treatment covered by the universal health insurance system is expected to be considerable, which means that the number of patients with potential indications for surgery is probably far greater than the actual number of surgical cases.

Knee joint preservation surgery, mainly HTO, is being reviewed in Japan, and HTO is being performed in adolescents and in elderly patients who wish to continue heavy work and sports activities. When combined with appropriate postoperative rehabilitation, HTO is a beneficial treatment that is expected to improve activities of daily living and quality of life to a level equal to or better than before surgery [20, 21]. HTO corrects lower limb alignment and decompresses the tibiofemoral joint, which allows for repair of damaged cartilage, although most of the repaired cartilage is fibrocartilage, which is different in nature from hyaline cartilage [22]. A systematic review of treatments combining HTO and biological intra-articular augmentation (e.g., osteotomy, stem cell transplantation) found no superiority in all three criteria (histological findings, imaging findings, and clinical symptoms) compared with a control group and that no reports have shown superiority [23].

Osteochondral column transplantation [24] and cultured autologous chondrocyte cell transplantation [25] are the most common methods for treating cartilage injuries, but these are limited to relatively small cartilage defects, such as traumatic chondral defects and osteochondritis dissecans. By contrast, OAK is often surrounded by degenerated and thinned cartilage, which makes it practically impossible to treat OAK cartilage damage in the same way (Table 27.1).

In a report of treatment results using TKA surgery as the endpoint, HTO survivorship was about 80% at 5 years and 60% at 10 years [26]. It is essential to both correct the lower limb alignment and regenerate cartilage defects in the loading zone lost because of OAK using hyaline cartilage to achieve long-term results. We believe that correcting the lower limb alignment as well as regenerating the cartilage defects in the loading zone lost because of OAK with hyaline cartilage is crucial for improving long-term results. We have applied cell sheet engineering using temperature-responsive culture dishes in the regeneration of articular cartilage and have devised a combination procedure with HTO. To date, we have transplanted autologous chondrocyte cell sheets in eight patients [17] and allogeneic chondrocyte cell sheets in 10 patients as clinical studies and have reported regeneration in hyaline cartilage and good clinical results [16]. We have repaired and regenerated cartilage defects in the loading zone as hyaline cartilage by transplanting cell sheets at the same time as optimizing the intra-articular environment by correcting alignment with HTO. We believe that this combined technique may help to avoid the future need for arthroplasty.

Table 27.1 Treatment for cartilage defect of osteoarthritis of the knee

Operative procedure	OAK indication	Target disease	Method	Cartilage repair
Microfracture	(off-label) Recommendation for 35~40 years[a]	Traumatic cartilage defect or osteochondritis dissecans within 1 year after injury Size 2 cm^2 or less	Remove unhealthy cartilage and create multiple small holes to allow bleeding from bone marrow at 3–4 mm intervals	Repair with fibrous cartilage by mobilization of mesenchymal stem cells or chondroprogenitor cells
Autologous osteochondral mosaicplasty	(off-label) Under 40 years[a]	Osteochondritis dissecans in the knee joint Cartilage defect of the patella Cartilage defects caused by ligament or meniscal damage Early stage of arthritis, Osteonecrosis Size 1 ~ 4 cm^2	Osteochondral plugs are harvested from nonweight-bearing site of the same joint and transplanted to the cartilage defect	Repair with mixture of transplanted autologous osteochondral plugs and regenerated fibrous cartilage
Autologous chondrocyte implantation	(off-label) 14~45 years[a]	The last resort for traumatic cartilage defect or osteochondritis dissecans with defect size 4 cm^2 and more (excludes knee osteoarthritis	Autologous periosteum and cartilage need to be harvested Autologous chondrocytes are cultured in collagen gel and transplanted under sutured periosteum or collagen membrane	Repair by periosteum or collagen membrane and autologous chondrocytes
RMSC[b] with autologous AC sheets	(clinical research) 20 ~ 60 years	Cartilage defect caused by trauma or degeneration Defect size 4.2 cm^2 and less	Autologous chondrocyte sheet prepared by co-culture with autologous synovial cells are transplanted by RMSC method	Repair with hyaline cartilage regeneration induced by paracrine effect and recruitment of MSCs or chondroprogenitor cells by factors secreted from AC sheets
	(advanced medical care B program) 20 ~ 80 years	Knee osteoarthritis indicated for high tibial osteotomy Defect size 1.0 ~ 8.4 cm^2		
RMSC[b] with allogeneic PD sheets	(clinical research) 20 ~ 60 years	Knee osteoarthritis indicated for high tibial osteotomy No limitation for defect size	Allogeneic chondrocyte sheet prepared from tissue obtained from polydactyly patients are transplanted by RMSC method	Repair with hyaline cartilage regeneration induced by paracrine effect and recruitment of MSCs or chondroprogenitor cells by factors secreted from PD sheets
RMSC[b] with allogeneic PD sheets	(phase III) 20 ~ 80 years	Knee osteoarthritis indicated for high tibial osteotomy No limitation for defect size	Allogeneic chondrocyte sheet prepared from tissue obtained from polydactyly patients transplanted by RMSC method	Repair with hyaline cartilage regeneration induced by paracrine effect and recruitment of MSCs or chondroprogenitor cells by factors secreted from PD sheets

[a]Modified from Report of working group for establishing standards for use of autologous cultured cartilage 2013, The Japanese Orthopaedic Association

[b]RMSC: Treatment by resection of unhealthy tissue, marrow stimulating and covered by chondrocyte sheets

27.2 Autologous Cartilage Cell Sheets

27.2.1 Clinical Studies

Autologous chondrocyte sheets (AC sheets) were transplanted into eight patients in a clinical study between 2011 and 2013 (Table 27.2). The age at the time of surgery was 30–59 years, and the area of cartilage damage was relatively small (2.3–4.1 cm^2). This was because of restrictions by the Ministry of Health, Labour and Welfare (MHLW); that is, their viewpoint was that safety should be assessed first, and therefore, the size of the defect was required to be covered by a single cell sheet. For tissue collection, arthroscopic surgery was performed first, and normal cartilage was harvested from the synovial tissue and nonweight-bearing areas. Chondrocytes and synovial cells were isolated enzymatically and co-cultured over 3 weeks to produce chondrocyte sheets, which were stacked in triplicate and transplanted into the cartilage defect simultaneously with anterior cruciate ligament reconstruction or HTO (Fig. 27.1).

No serious adverse events were observed during the study, and a second look and biopsy of the cartilage-regenerated area were obtained 1 year after surgery. Histological examination showed characteristics of hyaline cartilage that stained with Safranin-O and expressed type 2 collagen in samples from all patients. Both the KOOS and LKS had improved significantly from their preoperative levels, and the clinical results were well maintained up to 3 years postoperatively [17]. All patients treated with autologous chondrocyte sheet transplantation have been followed for 10 years after transplantation, and the long-term results will be reported in the near future.

Table 27.2 Comparison between autologous chondrocyte sheets and allogeneic chondrocyte sheets

Summary of therapy	Joint treatment by cell sheet	
Target disease	Cartilage defects in osteoarthritis of the knee (OAK)	
Outcome assessment	Primary outcome measure: Safety; frequency and content of adverse events Secondary outcome measure: Efficacy; ① clinical evaluations ② X-ray photographs ③ magnetic resonance imaging (MRI) ④ arthroscopic evaluation ⑤ LIPA[a] for viscoelastic properties of cartilage ⑥ biopsy (histological evaluations)	
Combination therapy	High tibial osteotomy + RMSC method[b]	
Characteristics	Includes patients with OAK No use of periosteum or sutures during transplantation RMSC method	
Summary of clinical studies	**Autologous AC sheets** **NPJ Regen Med 2019**	**Allogeneic PD sheets** **NPJ Regen Med 2022**
Risk classification[c]	Class II Regenerative medicine	Class I Regenerative medicine
Cell source	Autologous chondrocytes	Allogeneic polydactyly-derived cells
Limitations and advantages of each treatment	Requires two surgeries Limitation to size of cartilage defects Donor site morbidity, not re-operable Effectiveness of chondrocyte sheets is limited by the quality of patients' own cells	Single surgery No limitation to size of cartilage defects Re-operation is possible Consistency in quality and quantity of chondrocyte sheets (lot management/quality assurance)
Phase	Advanced medical care B In progress	Phase III clinical trial In progress

[a]Laser-induced photoacoustic measurement

[b]Removal of unhealthy tissue, marrow stimulation, and covered by chondrocyte sheets (RMSC method) has the potential to regenerate hyaline cartilage

[c]Risk classification under the Act on the Safety of Regenerative Medicine (Japan)

Problems with the autologous chondrocyte sheets include the need for arthroscopic surgery for tissue collection before transplantation and individual differences in the quantity and quality of the chondrocytes that can be collected.

27.2.2 Advanced Medical Treatment

In April 2019, the Ministry of Health, Labour and Welfare (MHLW) accepted notification of autologous cartilage cell sheets as an Advanced Medical Care. However, it was approved only for patients with medial-type OAK when HTO is performed at the same time. The size of the cartilage defect was also expanded to 8.4 cm^2 (the area of two cell sheets), based on the experience of producing two or more autologous cell sheets in all patients in the preceding clinical study. To date, 18 patients with OAK have been transplanted with autologous cell sheets at Tokai University Hospital. The cost is high, about 4 million yen, but this copayment is waived if the patient has an advanced medical treatment clause in a private medical insurance policy, and the patient must pay only for the concomitant HTO. In Japan, the universal health insurance system applies to HTO, which significantly reduces the patient's cost. To ensure that all patients who wish to undergo joint-sparing surgery can receive treatment with autologous cell sheets, we will continue using this advanced medical treatment and aim to have it covered by insurance.

27.3 Allogeneic Cartilage Cell Sheets

27.3.1 Clinical Research

By using allogeneic cells, cell sheets of stable quality can be produced in the required number of cell sheets according to the HTO surgery schedule without the need to harvest the patient's own healthy tissue and can be adapted to a wide range of cartilage defects. We focused on waste tissue from polydactyly surgery as a source of allogeneic cells.

Chondrocytes were isolated, cultured, and cryopreserved from the waste tissue from polydactyly surgery, and the frozen cell stock was used to produce allogeneic chondrocyte cell sheets. The chondrocytes derived from polydactyly tissues had high cell proliferative capacity and did not require co-culture with synovial cells, enabling the production of more than 600 cell sheets from a single waste tissue [12].

After repeated basic studies on the safety and efficacy, the implementation of clinical research using allogeneic cartilage cell sheets (PD sheets) was approved under the Type 1 Regenerative Medicine Provision Plan based on the Law for Ensuring the Safety of Regenerative Medicine (Table 27.2). Only patients with medial-type OAK for whom HTO was indicated were eligible. The area of the cartilage defect was no longer restricted, and patients with extensive cartilage defects, such as the medial femoral condyle, tibial kissing lesion, and femoral glenoid, totaling more than 20 cm^2, were also included.

Ten patients underwent allogeneic PD sheet transplantation. Postoperative X-rays and intraoperative photographs of all patients at the time of transplantation are provided (Fig. 27.3).

No serious adverse events were observed during the procedure, and biopsies were taken during the second look of the regenerated cartilage area at 1 year postoperatively. All patients showed features of hyaline cartilage on histological examination, and both the KOOS and LKS were significantly improved from their preoperative levels (Fig. 27.4) The viscoelastic properties and cartilage thickness were measured by laser-induced photoacoustics [27]; setting the value for normal surrounding cartilage as 1.0, the relative preoperative mean was 0.46 and the postoperative mean was 0.91 [16]. The average cartilage thickness was 1.37 mm preoperatively and 3.53 mm postoperatively. To date, 3–6 years of postoperative follow-up has

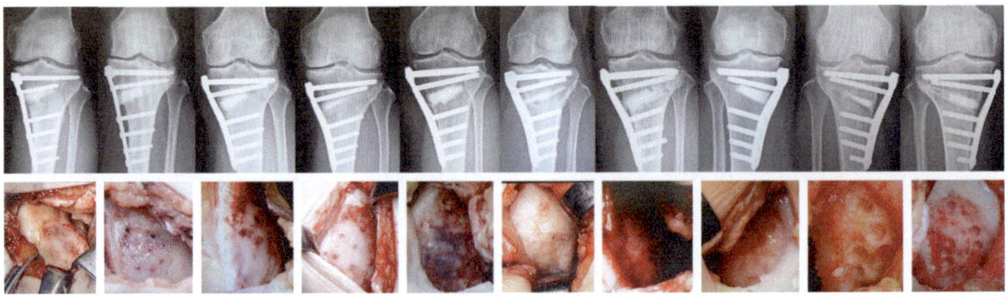

Fig. 27.3 Ten cases of allogeneic cell sheet transplantation after HTO
Upper row: postoperative radiographs. Lower row: Photographs after cell sheet transplantation using the RMSC* method
*After Removing the unhealthy cartilage tissue on the femoral side, bone Marrow Stimulation was performed on the cartilage defect and covered by Chondrocyte sheets. Because of its excellent tissue adhesion, no suture or patch is required, but external fixation is performed for 2 weeks postoperatively

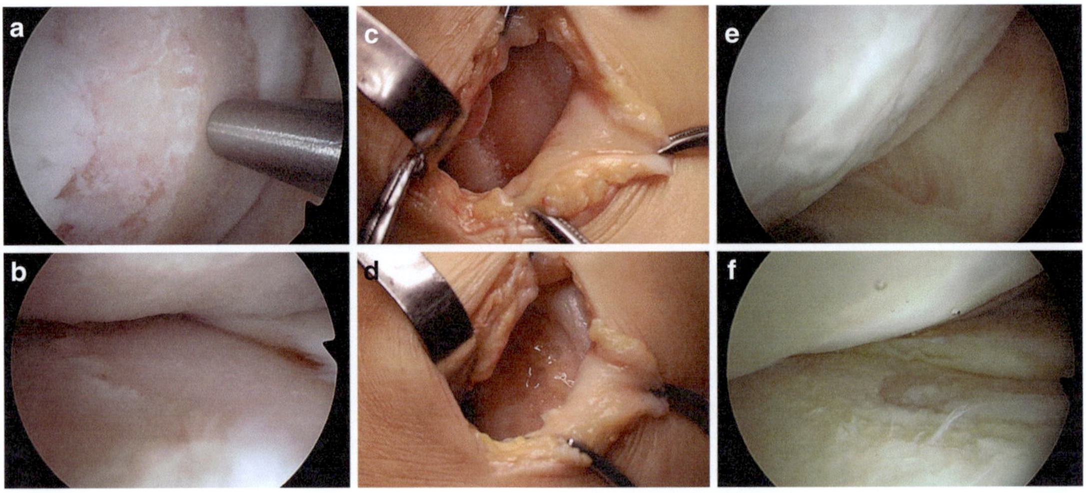

Fig. 27.4 Case 7
Preoperative arthroscopic findings showed extensive exposure of subchondral bone at the femoral medial condyle (**a**) and tibial medial condyle (**b**). The joint was incised, the unhealthy cartilage tissue on the femoral side was removed (**c**), and after abrasion, an allogeneic chondrocyte sheet was transplanted (**d**). A second look 1 year later showed cartilage regeneration on both the femoral (**e**) and tibial (**f**) sides

been completed, and all patients have a good postoperative course.

27.3.2 Company Clinical Trials and the Future

Efficacy and safety were confirmed in all ten patients in the aforementioned clinical study of allogeneic cell sheets [16], and the clinical course was good. As a result of face-to-face advice at the Pharmaceuticals and Medical Devices Agency (Japan), the clinical study was approved as a Phase II study, and a Phase III study was approved for implementation as a corporate clinical trial under the Pharmaceuticals and Medical Devices Act. The clinical trial notification was submitted in September 2023.

In future, while conducting clinical trials, we will develop a mass culture method using automated culture equipment to provide OAK for more patients, develop a cryopreservation method for cell sheets to make it easier to adjust the timing of transplantation, and improve the shape and product packaging to make it easier to use during surgery and to provide clinically. We are working with related companies to develop products that are easy to use in the clinical field.

References

1. Okano T, Yamada N, Sakai H, et al. A novel recovery system for cultured cells using plasma-treated polystyrene dishes grafted with poly (N-isopropylacrylamide). J Biomed Mater Res. 1993;27:1243–51.
2. Sawa Y, Yoshikawa Y, Toda K, et al. Safety and efficacy of autologous skeletal myoblast sheets (TCD-51073) for the treatment of severe chronic heart failure due to ischemic heart disease. Circ J. 2015;79:991–9.
3. Nishida K, Yamato M, Hayashida Y, et al. Corneal reconstruction with tissue-engineered cell sheets composed of autologous oral mucosal epithelium. New Engl J Med. 2004;351:1187–96.
4. Kobayashi S, Kanai N, Ohki T, et al. Prevention of esophageal strictures after endoscopic submucosal dissection. World J Gastroenterol. 2014;20:15098–109.
5. Kaneshiro N, Sato M, Ishihara M, et al. Bioengineered chondrocyte sheets may be potentially useful for the treatment of partial thickness defects of articular cartilage. Biochem Biophys Res Commun. 2006;349:723–31.
6. Kaneshiro N, Sato M, Ishihara M, et al. Cultured articular chondrocytes sheets for partial thickness cartilage defects utilizing temperature-responsive culture dishes. Eur Cell Mater. 2007;13:87–92.
7. Mitani G, Sato M, Lee JI, et al. The properties of bioengineered chondrocyte sheets for cartilage regeneration. BMC Biotechnol. 2009;9:17.
8. Ebihara G, Sato M, Yamato M, et al. Cartilage repair in transplanted scaffold-free chondrocyte sheets using a minipig model. Biomaterials. 2012;33:3846–51.
9. Sato M, Yamato M, Hamahashi K, et al. Articular cartilage regeneration using cell sheet technology. Anat Red (Hoboken). 2014;297:36–43.
10. Hamahashi K, Sato M, Yamato M, et al. Studies of the humoral factors produced by layered chondrocyte sheets. J Tissue Eng Regen Med. 2012;9:24–30.
11. Kokubo M, Sato M, Yamato M, et al. Characterization of chondrocyte sheets prepared using a co-culture method with temperature-responsive culture inserts. J Tissue Eng Regen Med. 2013;10:486–95.
12. Maehara M, Sato M, Toyoda E, et al. Characterization of polydactyly-derived chondrocyte sheets versus adult chondrocyte sheets for articular cartilage repair. Inflamm Regener. 2017;37:22.
13. Takahashi T, Sato M, Toyoda E, et al. Rabbit xenogeneic transplantation model for evaluating human chondrocyte sheets used in articular cartilage repair. J Tissue Eng Regen Med. 2018;12:2067–76.
14. Takizawa D, Sato M, Okada E, et al. Regenerative effects of human chondrocyte sheets in a xenogeneic transplantation model using immune-deficient rats. J Tissue Eng Regen Med. 2020;14:1296–306.
15. Toyoda E, Sato M, Takahashi T, et al. Transcriptomic and proteomic analyses reveal the potential mode of action of chondrocyte sheets in hyaline cartilage regeneration. Int J Mol Sci. 2019;21:E149.
16. Hamahashi K, Toyoda E, Ishihara M, et al. Polydactyly-derived allogeneic chondrocyte cell-sheet transplantation with high tibial osteotomy as regenerative therapy for knee osteoarthritis. NPJ Regen Med. 2022;7:71.
17. Sato M, Yamato M, Mitani G, et al. Combined surgery and chondrocyte cell-sheet transplantation improves clinical and structural outcomes in knee osteoarthritis. NPJ Regen Med. 2019;4:4.
18. World Health Organization. World health statistics 2023: Monitoring health for the SDGs, sustainable development goals. 2023. https://www.who.int/publications/i/item/9789240074323
19. Ministry of Health, Labour and Welfare. https://www.mhlw.go.jp/shingi/2008/07/dl/s0701-5a.pdf
20. Takeuchi R, Ishikawa H, Aratake M, et al. Medial opening wedge high tibial osteotomy with early full weight bearing. Arthroscopy. 2009;25:46–53.
21. Takeuchi R, Ishikawa H, Miyasaka Y, et al. A novel closed-wedge high tibial osteotomy procedure to treat osteoarthritis of the knee: hybrid technique and rehabilitation measures. Arthrosc Tech. 2014;3:e431–7.
22. Jung WH, Takeuchi R, Chun CW, et al. Second-look arthroscopic assessment of cartilage regeneration after medial opening-wedge high tibial osteotomy. Arthroscopy. 2014;30:72–9.
23. Reale D, Feltri P, Franceschini M, et al. Biological intra-articular augmentation for osteotomy in knee osteoarthritis: strategies and results: a systematic review of the literature from the ESSKA Orthobiologics initiative. Knee Surg Sports Traumatol Arthrosc. 2023;31:4327–46.
24. Hangody L, Kish G, Kárpáti Z, et al. Arthroscopic autogenous osteochondral mosaicplasty for the treatment of femoral condylar articular defects. A preliminary report. Knee Surg Sports Traumatol Arthrosc. 1997;5:262–7.
25. Brittberg M, Lindahl A, Nilsson A, et al. Treatment of deep cartilage defects in the knee with autologous chondrocyte transplantation. New Engl J Med. 1994;331:889–95.
26. Pannell WC, Heidari KS, Mayer EN, et al. High tibial osteotomy survivorship: a population-based study. Orthop J Sports Med. 2019;7:2325967119890693.
27. Sato M, Ishihara M, Kikuchi M, Mochida J. A diagnostic system for articular cartilage using non-destructive pulsed laser irradiation. Lasers Surg Med. 2011;43:421–32.

28 Adipose-Derived Cell-Based Products for Regenerative Medicine

Laura de Girolamo, Giulio Grieco, Vincenzo Raffo, Christopher J. Rogers, and Alberto Gobbi

28.1 Adipose-Derived Products: What Are They?

Adipose tissue is primarily composed of mature adipocytes (67.6%) but also contains adipose-derived mesenchymal stem cells (ASCs), pericytes, leukocytes, fibroblasts, macrophages, and pre-adipocytes within a vascularized stromal network known as stromal vascular fraction (SVF) [1–4]. Mesenchymal stem cells (MSCs) were first identified in adipose tissue in 2001 and were described as an adequate cell pool for regenerative medicine approaches [5]. Traditionally, MSCs, regardless of their origin, have been utilized as a "cell replacement tool" due to their capacity to differentiate into several cell lineages, making them ideal candidates for regenerative medicine purposes. More recently, this role has expanded to view MSCs as "sensing cells" that can influence their environment. It is now known that MSCs manage the healing process through a paracrine action, which stimulates antiapoptotic, immunomodulatory, trophic, and angiogenic properties by interacting with resident cells.

In fact, this cell type has garnered growing clinical interest due to their combined paracrine effects and multipotent differentiation capacity [6]. Nevertheless, identifying the distinct cell types and their true phenotypes in their native environment remains contentious [7]. Currently, two main procedures are used to deliver ASC-containing products: one involves the use of freshly isolated SVF or micro-fragmented adipose tissue (microfat), and the other employs culture-expanded ASCs [8, 9]. ASCs, themselves, can rapidly proliferate and can be collected in substantial quantities and cryopreserved for future clinical use. Both methods are used in regenerative medicine strategies and have demonstrated anti-inflammatory, immunomodulatory, and reparative effects in vivo, with safety validated in numerous clinical trials (NCT01953523), including those for musculoskeletal diseases.

The first approach, involving the use of SVF or microfat, facilitates one-step procedures that preserve the cells within their niche. This method has shown potent therapeutic effects but involved complex cell mixtures. The second approach, using culture-expanded ASCs, allows for the isolation and enrichment of a more homogeneous

L. de Girolamo (✉) · G. Grieco · V. Raffo
Orthopaedic Biotechnology Laboratory, Galeazzi Sant'Ambrogio Hospital, Milan, Italy
e-mail: laura.degirolamo@grupposandonato.it; giulio.grieco@grupposandonato.it; vincenzo.raffo@grupposandonato.it

C. J. Rogers
San Diego Orthobiologics Medical Group, Carlsbad, CA, USA
e-mail: rogers@sdomg.com

A. Gobbi
O.A.S.I. Bioresearch Foundation Gobbi N.P.O., Milan, Italy
e-mail: gobbi@cartilagedoctor.it

A. Gobbi et al. (eds.), *Regenerative Medicine in Sports and Orthopaedics*,
https://doi.org/10.1007/978-3-031-84693-9_28

cell population, potentially enhancing therapeutic outcomes through a paracrine mechanism [10–12]. However, this method requires a GMP facility increasing cost for patients and healthcare systems. The composition and effectiveness of SVF are influenced by donor characteristics, tissue quality, and isolation protocols, which affect the proportions and profiles of expanded cell populations [13, 14]. Given that not all SVF cells may have satisfactory therapeutic effects [15], precise characterization of ASCs is crucial. Identifying specific cell subsets and their association to particular functions could enable the development of tailored cell-based protocols to achieve desired outcomes.

In 2006, the Society for Cellular Therapy established guidelines for developing and standardizing stem cell products, identifying adherence to plastic, tri-lineage differentiation capacity (adipocytes, chondrocytes, osteocytes), and specific surface markers (CD73, CD90, CD105) as minimal criteria for MSCs, which also apply to ASCs. These cells should lack expression of CD45, CD34, CD14 or CD11b, CD79a or CD19, and HLA-DR 14. More recently, additional markers such as CD13, CD29, CD44 (positive), CD31, and CD235a (negative) have been introduced [16]. Assessing cell size, granularity, telomere length, senescence status, secretome profile, and immunomodulation is also essential to predict the regenerative potential of MSCs [17].

28.2 Adipose-Derived Products: Rationale for Their Use in Inflammatory Diseases

The rationale behind the therapeutic use of MSCs in treating inflammatory diseases, such as osteoarthritis (OA) and tendinopathy, relies on their immunomodulatory capacity. This facilitates a shift from an inflammatory microenvironment to a pro-regenerative one, thereby promoting the restoration of tissue homeostasis. MSCs are known to operate at the level of the adaptive immune system characterized by inflammation through inhibiting T cell proliferation, suppressing B cells, attracting regulatory T cells, and inducing the production of anti-inflammatory factors [18]. Additionally, MSCs support the transition from pro-inflammatory M1 macrophages to anti-inflammatory M2 macrophages through the secretion of prostaglandin E2. This prostaglandin, by interacting with the EP4 receptor on stimulated M1 macrophages, suppresses the secretion of pro-inflammatory cytokines, such as tumor necrosis factor-α (TNF-α), and interleukin-1β (IL-1β) and increases the expression of anti-inflammatory mediators like IL-10 by M2 macrophages [19] (Fig. 28.1b).

For these reasons, MSCs can prevent macrophage-mediated cartilage degeneration in early OA by inhibiting the activation of matrix-degrading enzymes, such as matrix metalloproteinases (MMPs) and aggrecanases, and by promoting the synthesis of matrix proteins by chondrocytes [20]. MSCs also exhibit immune-evasive characteristics due to their low expression of MHC class I molecules and lack of MHC class II expression, rendering them minimally immunogenic and advantageous for the use of both autologous and allogeneic MSC-based therapies [21, 22]. Notably, the restoration process is ensured through the initial interaction between MSCs and resident cells, thus their prolonged survival at the injection site is not essential for their efficacy [23]. These observations suggest that the benefits observed after MSC transplantation are primarily attributed to the secretion of soluble factors rather than direct cell transdifferentiation. Consequently, research is increasingly focused on characterizing the secreted factors, collectively known as "secretomes" which include a wide array of cytokines, chemokines, and lipids with trophic and immunomodulatory activities [24].

Caplan's description of MSCs as "drugstores," which release bioactive molecules in response to injury signals to modulate the local immune response and establish a regenerative microenvironment, highlights the therapeutic potential of MSC secretome [25]. Among the factors identi-

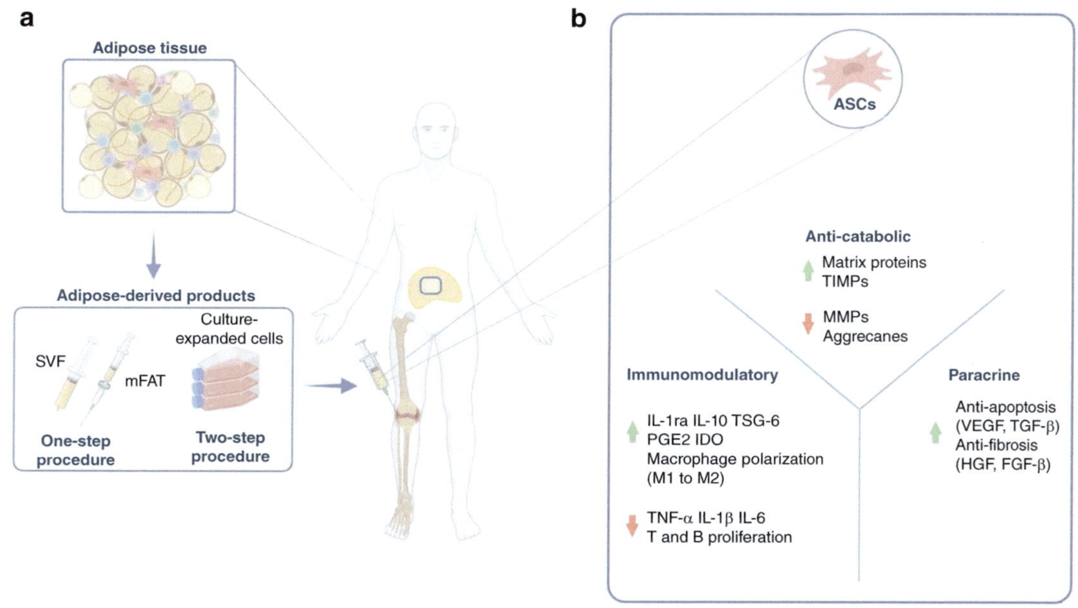

Fig. 28.1 Harvesting, application, and therapeutic mechanisms of adipose-derived products for regenerative medicine. (**a**) Source and processing of adipose-derived products. One-step procedure includes stromal vascular fraction (SVF) and micro-fragmented adipose tissue (mFAT); two-step procedure relies on culture-expanded cells. (**b**) Biological effects and molecular mechanisms of cell-based products

fied within secretome, extracellular vesicles (EVs) have been recognized as active entities [26, 27], encapsulating diverse active molecules such as DNA, mRNAs, miRNAs, pre-miRNAs, ncRNAs, and proteins [28]. ASCs exhibit strong secretory activity [29], producing various bioactive molecules that deliver protective and supportive factors, which can reduce apoptosis, fibrosis, and inflammation [25], thereby contributing to the treatment of early OA.

Several studies have demonstrated the efficacy of ASCs, and their paracrine actions have been proven to restore tissue homeostasis by exerting both anti-inflammatory and protective activities in chondrocytes and tenocytes exposed to pathological conditions, allowing their use as a therapeutic option in treating joint diseases like OA [21, 30, 31]. For example, local intra-articular injections of autologous ASCs have shown beneficial effects and improved cartilage quality, likely due to their immunomodulatory and immunosuppressive activities on the synovial microenvironment [32].

28.3 Types of Adipose-Derived Products

While the most traditional method of delivering cell therapy involves isolating in vitro cultivated cells, intraoperative approaches that do not require cell expansion have been recently developed as one-step approaches to balance clinical efficacy and practicality. One-step procedures usually involve minimally manipulated products that do not require extensive tissue and cell processing while preserving their original characteristics. Among minimally manipulated products derived from adipose tissue, SVF is isolated through mechanical digestion often followed by a centrifugation step [33], resulting in a liquid cell suspension containing heterogeneous cell populations. Alternatively, adipose tissue can be used at the point-of-care in the form of microfat, where the tissue is purified, washed out from residual blood and oil, and resized to allow passage through an injection needle (Fig. 28.1a). The main difference between SVF and microfat is that SVF is a suspension of various cell populations with minimal tissue matrix, whereas microfat retains the native adipose tissue micro-architecture, thus preserving its stem cell niches. So far, no direct comparative studies have been conducted to compare these two techniques, and therefore it is not possible to claim superiority of one approach over the other.

Overall, one-step procedures exploiting the regenerative potential of adipose tissue are convenient in terms of time and cost savings but lack the precision to selectively isolate ASC populations and yield a higher number of pre-regenerative cells [34]. However, encouraging results from initial clinical trials together with its ease of use have resulted in its increasing usage.

On the other hand, culture-expanded cells consist of a more homogeneous cell-based product, where cells are extensively characterized and produced in an adequate number for that given treatment, ensuring that the treatment is tailored to the individual patient's requirements.

Both minimally manipulated products and expanded ASCs can be combined with scaffolds to treat focal lesions such as bone abnormalities, tendon ruptures, or chondral lesions. These constructs are usually delivered surgically, and in this case, they can help the regeneration process both by direct differentiation into tissue-specific cells and by paracrine modulation of the microenvironment.

Characteristics including age, body mass index (BMI), gender, lifestyle, and harvest sites may operate as confounding variables, as indicated by the variety of outcomes. Research has shown that the total number of nucleated cells produced decreases with age [35, 36] and that the ability of ASCs to proliferate and differentiate declines significantly [37], however at a lower extent than MSCs from bone marrow. Current evidence does not demonstrate clear limitations or associations with regard to cell quantity/cell viability and product efficacy; therefore, a recommended patient age range cannot be determined. Therefore, for elderly patients, adipose-derived products are still a feasible choice despite these limitations, provided that physicians inform such patients that outcomes may not be as favorable or as durable as for younger grades of knee OA [38].

Higher BMI is linked to decreased cell migratory, angiogenic, and proliferative abilities, as well as fewer viable mature adipocytes per gram of tissue. It is interesting to note that BMI can have the opposite effect on cell function. Long-term calorie restriction achieved with diet and bariatric surgery can improve the ASC profile, lowering DNA damage, increasing viability, and prolonging the replicative lifespan [39]. This data is consistent with research showing a beneficial relationship between decreased inflammation and weight loss [40]. However, there is a lack of literature and clinical data regarding the direct impact of diet, fasting, and other lifestyle recommendations on the therapeutic effects of adipose-derived products. Since basic science literature reports negative effects of these behaviors on MSC performances, it may be important to adopt dietary and lifestyle restrictions (including alcohol consumption and smoking) a few weeks before the treatment [38].

Gender and donor site effects are still debatable. While some studies have indicated no difference in the proliferation and yield of ASCs across genders, others have suggested that men may have a larger yield of ASCs than women [41]. In a comparable way, the best location for fat harvesting remains to be determined.

28.4 Adipose-Derived Products: Preclinical and Clinical Findings in Joint Degeneration

Numerous preclinical models of OA in a variety of animals, including rabbits [42], dogs [43, 44], goats [45], mice [46], horses [47], and rats [48], have demonstrated the powerful regenerative and immunomodulatory capabilities of adipose-derived products. These models translate similar OA traits to the human condition. Anterior cruciate ligament (ACL) transection or meniscectomy are two surgical methods used to generate OA, as well as chemically induced OA with the use of collagenase and papain is typically used in animal models [49]. While early tissue changes and the effects of early therapeutic procedures can be properly studied using chemical induction, the surgical method aims to replicate a more natural development of OA. The evaluation of outcomes in animal studies is complicated because these animal studies frequently do not identify the disease stage, which is crucial as the timing of cell therapy administration is crucial for treatment. For example, delivering an ASC injection 7 days after chemical induction reduces synovial thickening in a murine model of OA, while an injection 14 days after induction offers no discernible protection against synovial activation or joint degeneration [46].

The method of cell delivery is also a crucial factor as ASCs or any other adipose-derived products can be suspended in a variety of media in preclinical research—animal serum [46, 50], phosphate-buffered saline solution [48], and cell culture medium [42]. ASCs and platelet-rich plasma (PRP), most importantly the growth factors contained in the latter, can be occasionally mixed together to promote cell growth and chondrogenic differentiation [45, 51]. A different approach consists of distributing cells on 3D absorbable scaffolds, which are intended to stimulate chondrogenic development and sustain cell replacement at the defect site [52, 53]. Nonetheless, injectable therapies are more popular than cell-scaffold constructions due to the easier administration and the reduced costs.

A significant number of investigations have demonstrated that preclinical applications of adipose-derived products are safe overall, with no reports of systemic problems such as altered weight, behavior, or organ function, nor of local adverse effects like edema or stiffness in the joints [32]. This safety profile is consistent even in the case of xenografts and allogeneic ASCs, suggesting their low immunogenicity [45, 48]. Animal studies have also shown positive results at the macroscopic, histological, and immunohistochemical levels as well as in the assessment of biomarkers, clinical, and imaging outcomes, supporting possible disease-modifying effects at the synovial and cartilage level [46, 54–56].

Interestingly, a study in dogs with spontaneous OA showed that intra-articular injections of microfat, probably due to its potent analgesic, anti-inflammatory, and trophic properties, considerably enhanced joint function and decreased pain and symptoms for at least 6 months [57]. Despite not having the mechanical characteristics or longevity of typical hyaline cartilage, the tissue that was created demonstrates that MSCs support a regenerative milieu as an alternative to directly differentiating into cartilage.

In recent years, the European Society of Sports Traumatology Knee Surgery and Arthroscopy (ESSKA) has published a consensus with regard to the injectable use of orthobiologics in knee OA, developed following a strict modified Delphi Methodology [38], which utilized a large amount of literature [58–62], including level I and II clinical trials and meta-analysis. Among the main statements, they conclude that the current scientific evidence has shown that the use of point-of-care cell-based therapy (POC-CBT) products (including adipose-derived ones) as well as culture-expanded MSCs (including ASCs) can provide clinical benefit and are a safe treatment option for knee OA. Of course, certain limitations of current evidence exist, mainly related to heterogeneity of products and lack of studies on larger populations. However, clinical improvement has been shown at both short (6 months) and long (12 months) follow-ups in most of the studies available in literature. Therefore, the consensus group concluded that there was sufficient clinical evidence to support the use of POC-CBT and expanded ASCs for knee OA; however, due to the limitations mentioned above, the use of these products should not be considered as a first-line injectable treatment for knee OA. While most preclinical research investigates expanded ASCs, most clinical trials concentrate on one-step procedures using SVF or microfat [63–68]. Among studies regarding the use of expanded cells, some focused on cell dose. Despite the fact that high doses of expanded ASCs for intra-articular injection have been shown to be safe and beneficial in three early OA studies on patients (KL I–II), doses of 50–100 × 10^6 cells have been demonstrated to yield considerable clinical effects that have been maintained for up to 2 years [69–71] with maintained improvements on MRI for up to 5 years [72]. Furthermore, significant improvements in pain, knee function, and cartilage volume were observed with recurrent ASC injections at 3, 6, and 48 weeks [70, 71]. In a recently published prospective clinical trial, significant improvements in pain and knee function were reported when lower doses of GMP-manufactured SVF were used for moderately severe knee OA [73]. In a retrospective study, intra-articular injections of SVF (38 patients, 69 knees) and ASCs (42 patients, 59 knees) were compared for knee OA (KL II–IV). The results showed that the ASC group had a higher proportion of responders based on OMERACT-OARSI criteria and that their pain and symptoms improved more quickly. As OA severity increased, the ASC group's 100% response rate in KL II patients decreased to zero at 6 months [34].

Concerning OA severity, the ESSKA consensus group reports that current evidence has shown the clinical benefit of CBT in knee OA Kellgren-Lawrence (KL) grades 1–4; however, with most studies involving populations with KL grades 2–3, the use of adipose-derived products is mainly indicated for knee OA mainly in grades 1–3. However, some clinical benefits have also been shown in KL grade 4, albeit lower in comparison to lower KL grades. For this reason, these treatments should be considered in patients with late-stage OA mainly if they have failed other nonoperative strategies and who are not yet indicated for, are not willing to undergo knee replacement surgery, or cannot undergo such surgery due to comorbidities. However, it is important to inform these patients about a possible less favorable treatment outcome [38].

28.5 Adipose-Derived Products: Preclinical and Clinical Findings in Tendon and Muscle Disorders with Special Focus in Sports Medicine

Tendon injury and tendinopathy are among the most frequently encountered musculoskeletal diseases and represent a challenging issue with a global socioeconomic burden. Rotator cuff tendon tears (RCT) impact 25% of individuals in their sixth decade and nearly half of individuals in their eighth decade with 40% of asymptomatic partial-thickness tendon tears eventually progressing to a full-thickness RCT [74, 75]. Similar to osteoarthritis, tendon injuries are often nonresponsive to conservative care, and in some cases, surgical repair may not be an option. Surgical treatment carries the risk of potential complications and lengthy recovery, and the clinical outcomes may not be superior to conservative management [76]. Even when successful, post-surgical retear rates for full-thickness RCTs are as high as 13.1% [77]. Yet, the currently available conservative [78–83] and surgical [84–86] treatments are unable to restore a tendon's original biomechanical attributes, structure, and function [87, 88]. Therefore, treatments aimed to improve healthy tendon regeneration are necessary.

Rotator cuff tendon tear progression and retear rates following surgical repair are mainly the result of failed tendon healing, scar tissue formation, adhesions, and matrix disorganization [89]. Chronic rotator cuff tendon tears often lead to a decreased muscle fiber count and muscle volume with increased fatty infiltration [90, 91]. Patients with stage 3 or 4 Goutallier grade RCTs have been shown to have lower success rates following surgical tendon repair [92, 93]. Insufficient tissue vascularization reduces the availability of oxygen, growth factors, and nutrients necessary for tendon regeneration. Inflammatory mediators have been shown to be intensified in tendinopathy with several studies demonstrating the ability of adipose-derived stromal cells to improve tendon healing [94–101]. In an equine study of collagenase-induced tendinitis, treatments with SVF revealed significant improvements in tendon fiber architecture, reduction in vascularity, inflammatory cell infiltrate, and collagen type III formation with improvements in tendon fiber density and alignment [102]. Similarly, studies using ASCs have shown improved tendon tensile strength, reduced edema, neutrophil count, muscle fibrosis, and atrophy with a shift of the inflammatory environment to a more regenerative state in murine models and canine patients [103–108]. Multipotent stromal cell-derived exosomes also promote tendon regeneration by facilitating the migration and proliferation of endogenous tendon stem and progenitor cells resulting in decreased fatty infiltration, inflammation, atrophy, and muscle vascularization [96–101, 109].

Significant improvements in pain and function as well as structural healing on ultrasound and MRI have been reported in human clinical case series that used ASCs for lateral epicondylitis and chronic patellar tendinopathy [110–113], with results seen as early as 6 weeks and maintained up to 30 months. Two randomized controlled trials of non-insertional Achilles tendinopathy demonstrated improved functional outcomes following SVF injection earlier than PRP injection with similar results at each study's endpoint [114, 115]. Microfat has been reported in clinical case series of partial and full thickness rotator cuff tendon tears with or without glenohumeral osteoarthritis. No clinically significant adverse events were reported and statistically significant improvements were seen in pain and function up to 12 months follow-up [116–119]. In a prospective, randomized, controlled study of partial thickness RTCs, statistically significant improvements in shoulder pain and function were seen at 52-week follow-up in patients who received microfat tendon injection versus methylprednisolone and bupivacaine subacromial space injection [120]. No significant differences in retear rate by MRI at 18 months were seen in a prospective, randomized controlled trial of arthroscopic rotator cuff repair augmented with microfat [121]. Yet, lower retear rates were seen on MRI at 1 year follow-up when rotator cuff repair was augmented with ASC and fibrin glue injection in a retrospective clinical trial [122]. In a prospective dose-ranging study of intratendinous ASC injection for partial-thickness RCTs, better shoulder pain reduction was seen at 1 year follow-up in the high dose group. At 6 month follow-up, MRI revealed 90% reduction in the tendon defect. Further pain improvements were seen in all groups at 2 year follow-up with near complete healing in the high dose group [70, 123].

28.6 Summary

With promising results reported in both animal and human clinical trials, the safe and effective use of adipose-derived orthobiologics seems assured in the near future. However, randomized clinical trials with long-term follow-up combined with real-world data collection will be necessary to inform physicians of the most cost-effective use of these therapies. Proper patient selection, as well as optimized cellular products and treatment protocols will need to be determined in order to assure significant advances are made in the care of joint and tendon disease.

References

1. Pettersson P, Cigolini M, Sjöström L, Smith U, Björntorp P. Cells in human adipose tissue developing into adipocytes. Acta Med Scand. 1984;215:447–51.
2. Tsuji W. Adipose-derived stem cells: implications in tissue regeneration. World J Stem Cells. 2014;6:312.
3. Peinado JR, Pardo M, De La Rosa O, Malagón MM. Proteomic characterization of adipose tissue constituents, a necessary step for understanding adipose tissue complexity. Proteomics. 2012;12:607–20.
4. Lin C-S, Xin Z-C, Deng C-H, Ning H, Lin G, Lue TF. Defining adipose tissue-derived stem cells in tissue and in culture. Histol Histopathol. 2010:807–15.
5. Zuk PA, et al. Multilineage cells from human adipose tissue: implications for cell-based therapies. Tissue Eng. 2001;7:211–28.
6. Mazini L, Rochette L, Amine M, Malka G. Regenerative capacity of adipose derived stem cells (ADSCs), comparison with mesenchymal stem cells (MSCs). Int J Mol Sci. 2019;20:2523.

7. Dubey N, et al. Revisiting the advances in isolation, characterization and secretome of adipose-derived stromal/stem cells. Int J Mol Sci. 2018;19:2200.
8. Aronowitz JA, Lockhart RA, Hakakian CS. Mechanical versus enzymatic isolation of stromal vascular fraction cells from adipose tissue. Springerplus. 2015;4:713.
9. Shah FS, Wu X, Dietrich M, Rood J, Gimble JM. A non-enzymatic method for isolating human adipose tissue-derived stromal stem cells. Cytotherapy. 2013;15:979–85.
10. Domergue S, et al. Comparison between stromal vascular fraction and adipose mesenchymal stem cells in remodeling hypertrophic scars. PLoS One. 2016;11:e0156161.
11. Nyberg E, Farris A, O'Sullivan A, Rodriguez R, Grayson W. Comparison of stromal vascular fraction and passaged adipose-derived stromal/stem cells as point-of-care agents for bone regeneration. Tissue Eng Part A. 2019;25:1459–69.
12. Pittenger MF, et al. Mesenchymal stem cell perspective: cell biology to clinical progress. Npj Regen Med. 2019;4:22.
13. Peng Q, Alipour H, Porsborg S, Fink T, Zachar V. Evolution of ASC immunophenotypical subsets during expansion in vitro. Int J Mol Sci. 2020;21:1408.
14. Dominici M, Paolucci P, Conte P, Horwitz EM. Heterogeneity of multipotent mesenchymal stromal cells: from stromal cells to stem cells and vice versa. Transplantation. 2009;87:S36–42.
15. Nielsen FM, et al. Discrete adipose-derived stem cell subpopulations may display differential functionality after in vitro expansion despite convergence to a common phenotype distribution. Stem Cell Res Ther. 2016;7:177.
16. Bourin P, et al. Stromal cells from the adipose tissue-derived stromal vascular fraction and culture expanded adipose tissue-derived stromal/stem cells: a joint statement of the International Federation for Adipose Therapeutics and Science (IFATS) and the International Society for Cellular Therapy (ISCT). Cytotherapy. 2013;15:641–8.
17. Samsonraj RM, et al. Concise review: multifaceted characterization of human mesenchymal stem cells for use in regenerative medicine. Stem Cells Transl Med. 2017;6:2173–85.
18. English K. Mechanisms of mesenchymal stromal cell immunomodulation. Immunol Cell Biol. 2013;91:19–26.
19. YlÖstalo JH, Bartosh TJ, Coble K, Prockop DJ. Human mesenchymal stem/stromal cells cultured as spheroids are self-activated to produce prostaglandin E2 that directs stimulated macrophages into an anti-inflammatory phenotype. Stem Cells. 2012;30:2283–96.
20. Benito MJ. Synovial tissue inflammation in early and late osteoarthritis. Ann Rheum Dis. 2005;64:1263–7.
21. Lopa S, Colombini A, Moretti M, De Girolamo L. Injective mesenchymal stem cell-based treatments for knee osteoarthritis: from mechanisms of action to current clinical evidences. Knee Surg Sports Traumatol Arthrosc. 2019;27:2003–20.
22. Gimble JM, Katz AJ, Bunnell BA. Adipose-derived stem cells for regenerative medicine. Circ Res. 2007;100:1249–60.
23. Yorukoglu AC, Kiter AE, Akkaya S, Satiroglu-Tufan NL, Tufan AC. A concise review on the use of mesenchymal stem cells in cell sheet-based tissue engineering with special emphasis on bone tissue regeneration. Stem Cells Int. 2017;2017:1–13.
24. Madrigal M, Rao KS, Riordan NH. A review of therapeutic effects of mesenchymal stem cell secretions and induction of secretory modification by different culture methods. J Transl Med. 2014;12:260.
25. Caplan AI, Correa D. The MSC: an injury drugstore. Cell Stem Cell. 2011;9:11–5.
26. Murphy MB, Moncivais K, Caplan AI. Mesenchymal stem cells: environmentally responsive therapeutics for regenerative medicine. Exp Mol Med. 2013;45:e54.
27. Théry C, et al. Minimal information for studies of extracellular vesicles 2018 (MISEV2018): a position statement of the International Society for Extracellular Vesicles and update of the MISEV2014 guidelines. J Extracell Vesicles. 2018;7:1535750.
28. Hong P, Yang H, Wu Y, Li K, Tang Z. The functions and clinical application potential of exosomes derived from adipose mesenchymal stem cells: a comprehensive review. Stem Cell Res Ther. 2019;10:242.
29. Niada S, et al. Adipose-derived stromal cell secretome reduces TNFα-induced hypertrophy and catabolic markers in primary human articular chondrocytes. Stem Cell Res. 2019;38:101463.
30. Denkovskij J, et al. Paracrine potential of the human adipose tissue-derived stem cells to modulate balance between matrix metalloproteinases and their inhibitors in the osteoarthritic cartilage in vitro. Stem Cells Int. 2017;2017:1–13.
31. Viganò M, et al. Autologous microfragmented adipose tissue reduces the catabolic and fibrosis response in an in vitro model of tendon cell inflammation. Stem Cells Int. 2019;2019:1–10.
32. Perdisa F, et al. Adipose-derived mesenchymal stem cells for the treatment of articular cartilage: a systematic review on preclinical and clinical evidence. Stem Cells Int. 2015;2015:1–13.
33. Trivisonno A, et al. Intraoperative strategies for minimal manipulation of autologous adipose tissue for cell- and tissue-based therapies: concise review. Stem Cells Transl Med. 2019;8:1265–71.
34. Yokota N, et al. Comparative clinical outcomes after intra-articular injection with adipose-derived cultured stem cells or noncultured stromal vascular fraction for the treatment of knee osteoarthritis. Am J Sports Med. 2019;47:2577–83.
35. Varghese J, Griffin M, Mosahebi A, Butler P. Systematic review of patient factors affecting adipose stem cell viability and function: implica-

tions for regenerative therapy. Stem Cell Res Ther. 2017;8:45.
36. Gonzalez-Garza MT, Cruz-Vega DE. Regenerative capacity of autologous stem cell transplantation in elderly: a report of biomedical outcomes. Regen Med. 2017;12:169–78.
37. Girolamo LD, et al. Human adipose-derived stem cells isolated from young and elderly women: their differentiation potential and scaffold interaction during in vitro osteoblastic differentiation. Cytotherapy. 2009;11:793–803.
38. The use of injectable orthobiologics for knee osteoarthritis. Part 2: cell-based therapy (CBT). A formal ESSKA-ORBIT Consensus. https://cdn.ymaws.com/www.esska.org/resource/resmgr/docs/consensus_projects/2024_orbit_complete_report.pdf.
39. Mitterberger MC, Mattesich M, Zwerschke W. Bariatric surgery and diet-induced long-term caloric restriction protect subcutaneous adipose-derived stromal/progenitor cells and prolong their life span in formerly obese humans. Exp Gerontol. 2014;56:106–13.
40. Moschen AR, et al. Anti-inflammatory effects of excessive weight loss: potent suppression of adipose interleukin 6 and tumour necrosis factor expression. Gut. 2010;59:1259–64.
41. Faustini M, et al. Nonexpanded mesenchymal stem cells for regenerative medicine: yield in stromal vascular fraction from adipose tissues. Tissue Eng Part C Methods. 2010;16:1515–21.
42. Toghraie F, et al. Scaffold-free adipose-derived stem cells (ASCs) improve experimentally induced osteoarthritis in rabbits. Arch Iran Med. 2012;15:495–9.
43. Kriston-Pál É, et al. Characterization and therapeutic application of canine adipose mesenchymal stem cells to treat elbow osteoarthritis. Can J Vet Res. 2017;81:73–8.
44. Guercio A, et al. Production of canine mesenchymal stem cells from adipose tissue and their application in dogs with chronic osteoarthritis of the humeroradial joints. Cell Biol Int. 2012;36:189–94.
45. Ko J-Y, Lee J, Lee J, Im G-I. Intra-articular xenotransplantation of adipose-derived stromal cells to treat osteoarthritis in a goat model. Tissue Eng Regen Med. 2017;14:65–71.
46. Ter Huurne M, et al. Antiinflammatory and chondroprotective effects of intraarticular injection of adipose-derived stem cells in experimental osteoarthritis. Arthritis Rheum. 2012;64:3604–13.
47. Frisbie DD, Kisiday JD, Kawcak CE, Werpy NM, McIlwraith CW. Evaluation of adipose-derived stromal vascular fraction or bone marrow-derived mesenchymal stem cells for treatment of osteoarthritis. J Orthop Res. 2009;27:1675–80.
48. Mei L, et al. Culture-expanded allogenic adipose tissue-derived stem cells attenuate cartilage degeneration in an experimental rat osteoarthritis model. PLoS One. 2017;12:e0176107.
49. Cucchiarini M, et al. Basic science of osteoarthritis. J Exp Orthop. 2016;3:22.
50. Desando G, et al. Intra-articular delivery of adipose derived stromal cells attenuates osteoarthritis progression in an experimental rabbit model. Arthritis Res Ther. 2013;15:R22.
51. Van Pham P, et al. Activated platelet-rich plasma improves adipose-derived stem cell transplantation efficiency in injured articular cartilage. Stem Cell Res Ther. 2013;4:91.
52. Jurgens WJFM, et al. One-step surgical procedure for the treatment of osteochondral defects with adipose-derived stem cells in a caprine knee defect: a pilot study. Biores Open Access. 2013;2:315–25.
53. Jurgens WJ, Kroeze RJ, Bank RA, Ritt MJPF, Helder MN. Rapid attachment of adipose stromal cells on resorbable polymeric scaffolds facilitates the one-step surgical procedure for cartilage and bone tissue engineering purposes. J Orthop Res. 2011;29:853–60.
54. Hsu T-H, et al. Accuracy of critical shoulder angle and acromial index for predicting supraspinatus tendinopathy. Diagnostics. 2022;12:283.
55. Cheng J-H, et al. Autologous adipose-derived mesenchymal stem cells combined with shockwave therapy synergistically ameliorates the osteoarthritic pathological factors in knee joint. Pharmaceuticals. 2021;14:318.
56. Perucca Orfei C, et al. Cell-based therapies have disease-modifying effects on osteoarthritis in animal models. A systematic review by the ESSKA Orthobiologic Initiative. Part 1: adipose tissue-derived cell-based injectable therapies. Knee Surg Sports Traumatol Arthrosc. 2023;31:641–55.
57. Zeira O, et al. Intra-articular administration of autologous micro-fragmented adipose tissue in dogs with spontaneous osteoarthritis: safety, feasibility, and clinical outcomes. Stem Cells Transl Med. 2018;7:819–28.
58. Xu S, et al. Effect of mesenchymal stromal cells for articular cartilage degeneration treatment: a meta-analysis. Cytotherapy. 2015;17:1342–52.
59. Kim SH, et al. Intra-articular injection of mesenchymal stem cells for clinical outcomes and cartilage repair in osteoarthritis of the knee: a meta-analysis of randomized controlled trials. Arch Orthop Trauma Surg. 2019;139:971–80.
60. Kim K-I, Kim M-S, Kim J-H. Intra-articular injection of autologous adipose-derived stem cells or stromal vascular fractions: are they effective for patients with knee osteoarthritis? A systematic review with meta-analysis of randomized controlled trials. Am J Sports Med. 2023;51:837–48.
61. Lu L, et al. Treatment of knee osteoarthritis with intra-articular injection of autologous adipose-derived mesenchymal progenitor cells: a prospective, randomized, double-blind, active-controlled, phase IIb clinical trial. Stem Cell Res Ther. 2019;10:143.
62. Gupta PK, et al. Efficacy and safety of stempeucel in osteoarthritis of the knee: a phase 3 randomized, double-blind, multicenter, placebo-controlled study. Am J Sports Med. 2023;51:2254–66.

63. Dallo I, Morales M, Gobbi A. Platelets and adipose stroma combined for the treatment of the arthritic knee. Arthrosc Tech. 2021;10(11):e2407–14.
64. Gobbi A, Dallo I, Rogers C, Striano RD, Mautner K, Bowers R, et al. Two-year clinical outcomes of autologous microfragmented adipose tissue in elderly patients with knee osteoarthritis: a multi-centric, international study. Int Orthop. 2021;45:1179–88.
65. Dallo I, Szwedowski D, Mobasheri A, Irlandini E, Gobbi A. A prospective study comparing leukocyte-poor platelet-rich plasma combined with hyaluronic acid and autologous microfragmented adipose tissue in patients with early knee osteoarthritis. Stem Cells Dev. 2021;30(13):651–9.
66. Gobbi A, Dallo I, D'Ambrosi R. Autologous microfragmented adipose tissue and leukocyte-poor platelet-rich plasma combined with hyaluronic acid show comparable clinical outcomes for symptomatic early knee osteoarthritis over a two-year follow-up period: a prospective randomized clinical trial. Eur J Orthop Surg Traumatol. 2023;33(5):1895–904.
67. Gobbi A, de Girolamo L, Whyte GP, Sciarretta FV. Clinical applications of adipose tissue-derived stem cells. In: Bio-orthopaedics: a new approach. Berlin, Heidelberg: Springer Berlin Heidelberg; 2017. p. 553–9.
68. Herman K, Gobbi A. Evidence-based approach to orthobiologics for osteoarthritis and other joint disorders. Phys Med Rehabil Clin N Am. 2023;34(1):71–81.
69. Jo CH, et al. Intra-articular injection of mesenchymal stem cells for the treatment of osteoarthritis of the knee: a proof-of-concept clinical trial. Stem Cells. 2014;32:1254–66.
70. Jo CH, et al. Intra-articular injection of mesenchymal stem cells for the treatment of osteoarthritis of the knee: a 2-year follow-up study. Am J Sports Med. 2017;45:2774–83.
71. Song Y, et al. Human adipose-derived mesenchymal stem cells for osteoarthritis: a pilot study with long-term follow-up and repeated injections. Regen Med. 2018;13:295–307.
72. Kim KI, et al. Safety and efficacy of the intra-articular injection of mesenchymal stem cells for the treatment of osteoarthritic knee: a 5-year follow-up study. Stem Cells Transl Med. 2022;11(6):586–96.
73. Rogers CJ, et al. Clinical evaluation of safety and efficacy of a central current good manufacturing practices laboratory produced autologous adipose-derived stromal vascular fraction cell therapy product for the treatment of knee osteoarthritis. Stem Cells Dev. 2024;33(7–8):168–76.
74. Yamamoto A, et al. Prevalence and risk factors of a rotator cuff tear in the general population. J Shoulder Elb Surg. 2010;19:116–20.
75. Mall NA, et al. Symptomatic progression of asymptomatic rotator cuff tears: a prospective study of clinical and sonographic variables. J Bone Joint Surg Am. 2010;92:2623–33.
76. Kukkonen J, et al. Treatment of non-traumatic rotator cuff tears. Bone Joint J. 2014;96-B:75–81.
77. Kim I-B, Kim M-W. Risk factors for retear after arthroscopic repair of full-thickness rotator cuff tears using the suture bridge technique: classification system. Arthroscopy. 2016;32:2191–200.
78. Childress MA, Beutler A. Management of chronic tendon injuries. Am Fam Physician. 2013;87:486–90.
79. Gerdesmeyer L, et al. Current evidence of extracorporeal shock wave therapy in chronic Achilles tendinopathy. Int J Surg. 2015;24:154–9.
80. Bannuru RR, Flavin NE, Vaysbrot E, Harvey W, McAlindon T. High-energy extracorporeal shock-wave therapy for treating chronic calcific tendinitis of the shoulder. Ann Intern Med. 2014;160:542.
81. Couppé C, Svensson RB, Silbernagel KG, Langberg H, Magnusson SP. Eccentric or concentric exercises for the treatment of tendinopathies? J Orthop Sports Phys Ther. 2015;45:853–63.
82. Barker-Davies RM, et al. Study protocol: a double blind randomised control trial of high volume image guided injections in Achilles and patellar tendinopathy in a young active population. BMC Musculoskelet Disord. 2017;18:204.
83. Chaudhry FA. Effectiveness of dry needling and high-volume image-guided injection in the management of chronic mid-portion Achilles tendinopathy in adult population: a literature review. Eur J Orthop Surg Traumatol. 2017;27:441–8.
84. Brockmeyer M, Diehl N, Schmitt C, Kohn DM, Lorbach O. Results of surgical treatment of chronic patellar tendinosis (jumper's knee): a systematic review of the literature. Arthroscopy. 2015;31:2424–2429.e3.
85. Ahmad Z, et al. Lateral epicondylitis. Bone Joint J. 2013;95-B:1158–64.
86. Baltes TPA, Zwiers R, Wiegerinck JI, van Dijk CN. Surgical treatment for midportion Achilles tendinopathy: a systematic review. Knee Surg Sports Traumatol Arthrosc. 2017;25:1817–38.
87. Shapiro E, Grande D, Drakos M. Biologics in Achilles tendon healing and repair: a review. Curr Rev Musculoskelet Med. 2015;8:9–17.
88. Palumbo Piccionello A, et al. Adipose micro-grafts enhance tendinopathy healing in ovine model: an in vivo experimental perspective study. Stem Cells Transl Med. 2021;10:1544–60.
89. Liu C-F, et al. What we should know before using tissue engineering techniques to repair injured tendons: a developmental biology perspective. Tissue Eng Part B Rev. 2011;17:165–76.
90. Kang JR, Gupta R. Mechanisms of fatty degeneration in massive rotator cuff tears. J Shoulder Elb Surg. 2012;21:175–80.
91. Gupta R, et al. Muscle stem cells and rotator cuff injury. JSES Rev Rep Tech. 2021;1:186–93.
92. Somerson JS, Hsu JE, Gorbaty JD, Gee AO. Classifications in brief: Goutallier classification of fatty infiltration of the rotator cuff musculature. Clin Orthop. 2016;474:1328–32.

93. Gerber C, Wirth SH, Farshad M. Treatment options for massive rotator cuff tears. J Shoulder Elb Surg. 2011;20:S20–9.
94. Shi L, et al. Alterations of tendons in diabetes mellitus: what are the current findings? Int Orthop. 2015;39:1465–73.
95. Soleymaninejadian E, Pramanik K, Samadian E. Immunomodulatory properties of mesenchymal stem cells: cytokines and factors. Am J Reprod Immunol. 2011;67:1–8.
96. Connor DE, et al. Therapeutic potential of exosomes in rotator cuff tendon healing. J Bone Miner Metab. 2019;37:759–67.
97. Zhang M, et al. Tendon stem cell-derived exosomes regulate inflammation and promote the high-quality healing of injured tendon. Stem Cell Res Ther. 2020;11:402.
98. Yu H, et al. Bone marrow mesenchymal stem cell-derived exosomes promote tendon regeneration by facilitating the proliferation and migration of endogenous tendon stem/progenitor cells. Acta Biomater. 2020;106:328–41.
99. Wang C, Hu Q, Song W, Yu W, He Y. Adipose stem cell–derived exosomes decrease fatty infiltration and enhance rotator cuff healing in a rabbit model of chronic tears. Am J Sports Med. 2020;48:1456–64.
100. Sevivas N, et al. Mesenchymal stem cell secretome improves tendon cell viability in vitro and tendon-bone healing in vivo when a tissue engineering strategy is used in a rat model of chronic massive rotator cuff tear. Am J Sports Med. 2017;46:449–59.
101. Zhang X, et al. Adipose stem cell–derived exosomes recover impaired matrix metabolism of torn human rotator cuff tendons by maintaining tissue homeostasis. Am J Sports Med. 2021;49:899–908.
102. Nixon AJ, Dahlgren LA, Haupt JL, Yeager AE, Ward DL. Effect of adipose-derived nucleated cell fractions on tendon repair in horses with collagenase-induced tendinitis. Am J Vet Res. 2008;69:928–37.
103. Costa-Almeida R, Reis RL, Gomes ME. Metabolic disease epidemics: emerging challenges in regenerative medicine. Trends Endocrinol Metab. 2019;30:147–9.
104. Nichols A, et al. Effects of type II diabetes mellitus on tendon homeostasis and healing. J Orthop Res. 2020;38:13–22.
105. Mora MV, Antuña SA, Arranz MG, Carrascal MT, Barco R. Application of adipose tissue-derived stem cells in a rat rotator cuff repair model. Injury. 2014;45:S22–7.
106. Chen H-S, et al. Human adipose-derived stem cells accelerate the restoration of tensile strength of tendon and alleviate the progression of rotator cuff injury in a rat model. Cell Transplant. 2015;24:509–20.
107. Gumucio JP, et al. Stromal vascular stem cell treatment decreases muscle fibrosis following chronic rotator cuff tear. Int Orthop. 2016;40:759–64.
108. Canapp SO Jr, et al. The use of adipose-derived progenitor cells and platelet-rich plasma combination for the treatment of supraspinatus tendinopathy in 55 dogs: a retrospective study. Front Vet Sci. 2016;3:61. Doornaert MAJ, et al. Intrinsic dynamics of the fat graft. Plast Reconstr Surg. 2012;130:1001–9.
109. Wang T, et al. Load-induced regulation of tendon homeostasis by SPARC, a genetic predisposition factor for tendon and ligament injuries. Sci Transl Med. 2021;13:eabe5738.
110. Lee YH, Kim S, Lim D, Song H-T, Suh J-S. MR quantification of the fatty fraction from T2*-corrected dixon fat/water separation volume-interpolated breathhold examination (VIBE) in the assessment of muscle atrophy in rotator cuff tears. Acad Radiol. 2015;22:909–17.
111. Freitag J, Shah K, Wickham J, Tenen A. Effect of autologous adipose-derived mesenchymal stem cell therapy in combination with autologous platelet-rich plasma in the treatment of elbow tendinopathy. BMJ Case Rep. 2020;13:e234592.
112. Khoury MA, et al. Expanded adipose derived mesenchymal stromal cells are effective in treating chronic insertional patellar tendinopathy: clinical and MRI evaluations of a pilot study. J Exp Orthop. 2021;8:49.
113. Khoury M, et al. Promising improvement of chronic lateral elbow tendinopathy by using adipose derived mesenchymal stromal cells: a pilot study. J Exp Orthop. 2021;8:6–6.
114. Albano D, et al. Magnetic resonance and ultrasound in Achilles tendinopathy: predictive role and response assessment to platelet-rich plasma and adipose-derived stromal vascular fraction injection. Eur J Radiol. 2017;95:130–5.
115. Usuelli FG, et al. Intratendinous adipose-derived stromal vascular fraction (SVF) injection provides a safe, efficacious treatment for Achilles tendinopathy: results of a randomized controlled clinical trial at a 6-month follow-up. Knee Surg Sports Traumatol Arthrosc. 2017;26:2000–10.
116. Striano RD, Battista V, Bilboo N. Non-responding knee pain with osteoarthritis, meniscus and ligament tears treated with ultrasound guided autologous, micro-fragmented and minimally manipulated adipose tissue. Open J Regen Med. 2017;06:17–26.
117. Ferrell JL, Dodson A, Martin J. Microfragmented adipose tissue in the treatment of a full-thickness supraspinatus tear: a case report. Regen Med. 2023;18:773–80.
118. Hogaboom N, Malanga G, Cherian C, Dyson-Hudson T. A pilot study to evaluate micro-fragmented adipose tissue injection under ultrasound guidance for the treatment of refractory rotator cuff disease in wheelchair users with spinal cord injury. J Spinal Cord Med. 2021;44:886–95.
119. Marathe A, Song B, Jayaram P. Microfragmented adipose tissue with adjuvant platelet-rich plasma combination therapy for partial-thickness supraspinatus tear. Cureus. 2021;13:e15583.
120. Hurd JL, et al. Safety and efficacy of treating symptomatic, partial-thickness rotator cuff tears with fresh, uncultured, unmodified, autologous adipose-

derived regenerative cells (UA-ADRCs) isolated at the point of care: a prospective, randomized, controlled first-in-human pilot study. J Orthop Surg. 2020;15:122.
121. Randelli PS, et al. Arthroscopic rotator cuff repair augmentation with autologous microfragmented lipoaspirate tissue is safe and effectively improves short-term clinical and functional results: a prospective randomized controlled trial with 24-month follow-up. Am J Sports Med. 2022;50:1344–57.
122. Kim YS, Sung CH, Chung SH, Kwak SJ, Koh YG. Does an injection of adipose-derived mesenchymal stem cells loaded in fibrin glue influence rotator cuff repair outcomes? A clinical and magnetic resonance imaging study. Am J Sports Med. 2017;45:2010–8.
123. Jo CH, Chai JW, Jeong EC, Oh S, Yoon KS. Intratendinous injection of mesenchymal stem cells for the treatment of rotator cuff disease: a 2-year follow-up study. Arthroscopy. 2020;36:971–80.

Orthobiologically Augmented Scaffolds in Cartilage Repair

29

Fabio Valerio Sciarretta and Melanio Acosta IV

29.1 Introduction

Focal articular cartilage defects are very common in the knee joint, with prevalence ranging approximately 60%. In 2007, Widuchowski et al. [1] studied 25,124 patients, analyzing the previously reported 60% prevalence of knee chondral lesions in patients undergoing knee arthroscopies. In their own investigation, they found that 67% of patients had either focal chondral or osteochondral lesions following the Outerbridge classification, where most were localized in the patella (36%) and in the medial femoral condyle (34%). Curl et al. [2] also looked at 31,516 patients and found an incidence of 63% for chondral lesions encountered in patients undergoing knee arthroscopies. With much evidence we can see that cartilage injuries are very frequent; however, being avascular, aneural, and alymphatic, articular cartilage tissue has limited capacity for intrinsic repair or regeneration. The goal of any cartilage restoration procedure is to restore the articular surface by matching the histological, biochemical, and biomechanical properties of normal hyaline cartilage, improve patient symptoms and function, and impede lesion degeneration and predisposition to osteoarthritis (OA), consequently slowing down or preventing progression of focal cartilage injuries to end-stage arthritis which, if untreated, may eventually necessitate joint replacement surgeries by unicompartmental or total joint arthroplasty.

Over the last decades several treatment options have been proposed for the treatment of focal cartilage defects. For smaller cartilage lesions, treatment modalities include debridement, marrow stimulation through microfracture, abrasion, drilling and nano-fracture, advanced scaffold augmented marrow stimulation techniques (AMIC, BST-Cargel, GelrinC), and osteochondral autograft transplantation (OATs)/mosaicplasty. For larger lesions, osteochondral allograft (OCA) transplantation and autologous chondrocyte implantation (ACI) have been introduced and appear to be highly effective. In recent years, orthobiologic treatments with cells, mesenchymal stem cells (MSCs), and growth factors have grown in interest as a treatment option in an attempt to repair or regenerate articular cartilage. These treatments have also been used to treat different pathologies regarding muscle, tendon, bone, and especially cartilage, in order to improve clinical outcome scores, prevent osteoarthritis progression, and reduce recovery times, thus, enabling patients a quicker return to their daily lives and sport activities.

F. V. Sciarretta (✉)
Nostra Signora Della Mercede Clinic, Rome, Italy

M. Acosta IV
O.A.S.I. Bioresearch Foundation Gobbi N.P.O., Milan, Italy
e-mail: melanio.acosta@oasiortopedia.it

A. Gobbi et al. (eds.), *Regenerative Medicine in Sports and Orthopaedics*,
https://doi.org/10.1007/978-3-031-84693-9_29

29.2 Mesenchymal Stem Cells, Orthobiologics, and Scaffolds

Arnold Caplan, considered the father of clinical therapeutics of adult MSCs, has demonstrated that MSCs derive from perivascular cells called "pericytes," located in the periphery of blood vessels [3, 4]. It is these MSCs that are the most studied and investigated type of cell for regenerative medicine and play a crucial role in many studies and novel application protocols. The latest discoveries, based on the initial isolation and subsequent studies conducted by Arnold Caplan, around pericytes belong to the so-called perivascular niche [5], where they live in a quiescent state. When a vessel is damaged, typically occurring during injury, there is a release of pericytes that initially exist in a quiescent phase which pass to an activated phase, where they start to exhibit the MSC phenotype. Activated MSCs start to release a cascade of bioactive molecules to counteract the overaggressive immune response [6] and trophic factors that are used to establish a regenerative microenvironment, promote angiogenesis, and stimule the proliferation of tissue-specific progenitor cells [6, 7]. The updated interpretation of MSC activity emphasizes the fact that these cells show a very effective *in vivo* function and attribute their therapeutic action mostly to paracrine and trophic actions. For this reason MSCs have been recently defined as a "drugstore," due to their ability to release a wide range of growth factors and cytokines in the surrounding microenvironment, acting as a very powerful drug administered locally [6, 8].

MSCs reside in many types of tissues, including bone marrow, adipose tissue, and synovium, which are all fairly easy to isolate. Due to the fact that MSCs of various origins have been demonstrated the ability to differentiate into various mesenchymal lineages, including chondrocytes, there has been an explosive trend and a growing interest focusing on different strategies to regenerate damaged articular cartilage through MSC activity. Clinical applications of MSCs should meet the minimal criteria established by the International Society for Cellular Therapy including [9] being plastic-adherent in culture conditions; expressing cluster of differentiation 105 (CD105), CD73, and CD90; lacking expression of CD45, CD34, CD14 or CD11b, CD79 or CD19, and human leukocyte antigen-DR isotype (HLA-DR) surface molecules; while possessing tri-lineage differentiation into osteoblasts, adipocytes, and chondroblasts.

In orthobiology, the innovative field of modern biomedical technology that provides biologic therapies for reconstructing damaged tissues, mesenchymal stem cells are, in most cases, obtained from bone marrow or adipose tissue. Orthobiologics include bone marrow aspirate concentrate (BMAC), adipose tissue-derived stroma cells (ADSCs), platelet-rich plasma (PRP), and micronized allogeneic cartilage. Parallel to the use of orthobiologics, tissue engineering advancements have drawn attention as another promising strategy that has been recently introduced and extensively studied. Literature on orthobiologics augmented cartilage repair procedures have shown that different emerging orthobiologics therapies of recent popular interest have the potential to improve the long-term results offered in the treatment of articular cartilage lesions and slow down the progression to OA joint degeneration. However, these complex surgical procedures rely on different components, each one contributing to their success. It is therefore essential to utilize MSCs of proven chondrogenic capacity, obtained through processing kits with proven efficacy that will be seeded on scaffolds whose biologic and chemical composition perfectly adapts to the needs of a specific tissue engineering procedure. The use of either scaffolds and matrices alone or scaffolds seeded with MSCs helps to enhance differentiation in chondrocytes and create hyaline cartilage tissue. One-step cartilage repair techniques using different MSC sources have become a new potential avenue for the use of three-dimensional scaffold augmentation surgical regeneration techniques.

29.3 Hyaluronic Acid–Based Scaffold Embedded with MSCs Sourced from Bone Marrow Aspirate Concentrate

The multilineage potential of MSCs derived from bone marrow, their chondrogenic potential, the availability of these cells, and the fact that they can be harvested from various sites have drawn the attention of many researchers [10–14]. Further interest stemmed from works demonstrating bone marrow–derived MSCs possessed superior proliferation rates than chondrocytes. There are also studies demonstrating MSC transplantation had similar results to autologous chondrocyte implantation [15].

The use of scaffold-associated multipotent cells from bone marrow aspirate concentrate (BMAC) in one-step cell-based repair techniques has shown excellent results over the medium term in treating sizable lesions, comparable to those managed with two-step- cell-based techniques. The option of single-stage cell-based procedures is appealing due to the potential cost savings and the need for the patient to undergo only one surgical procedure instead of two [16–19]. The use of multipotent cells from BMAC combined with biologic scaffolding for one-stage repair has shown similar short- to medium-term clinical results as autologous chondrocyte implantation (ACI) in treating chondral lesions sized 0.4 cm^2 [17, 18, 20].

Gobbi et al. [21] described the long-term results of patients who had undergone one-stage cartilage repair in the knee with hyaluronic acid–based scaffold embedded with MSCs sourced from bone marrow aspirate concentrate (HA-BMAC). Their technique started with a diagnostic arthroscopy (Fig. 29.1) to confirm the areas of chondral injury and once the need for HA-BMAC cartilage repair was established, 60 mL of bone marrow was aspirated from the ipsilateral iliac crest and processed with a commercial system (BMAC Harvest Smart PreP2 System; Harvest Technologies) to isolate bone marrow aspirate with a total nucleated marrow cell concentration approximately six times the baseline concentration [22]. Batroxobin enzyme (Plateltex Act; Plateltex SRO) was used as a catalyst to activate the BMAC, producing a malleable clot.

Through a mini-arthrotomy (Fig. 29.2) or arthroscopically, the areas of chondral injury were exposed. Unstable chondral fragments were removed and lesions were appropriately prepared using chondrotomes, creating vertical cartilage walls around the periphery of each defect. The base of each defect underwent debridement to remove the calcified cartilage layer, taking care not to damage the underlying subchondral plate [19, 20]. A hyaluronic acid–based scaffold (Hyalofast; Anika Therapeutics) prepared to the exact size of the defect was then combined with the activated BMAC clot and implanted onto the cartilage defects. The implanted patch was then secured using sutures or fibrin glue. They concluded after long-term follow-ups, as long as 10 years, that there was significantly improved Tegner, visual analog scale, IKDC, and KOOS scores in those patient who had underwent HA-BMAC. Good to excellent outcomes were achieved across a wide range of patient ages, articular cartilage lesion sizes, lesion location, and when cartilage repair was performed concurrently with treatment of associated conditions. These successful outcomes are comparable with those of cell-based therapy combined with ACI after long-term follow-up of similar duration [23].

Through the use of the same hyaluronic acid-based scaffold, it has been demonstrated that clinical outcomes are comparable to those with high-grade knee chondral lesions treated with matric-induced ACI (MACI) or HA-BMAC after 3 years follow-up [24].

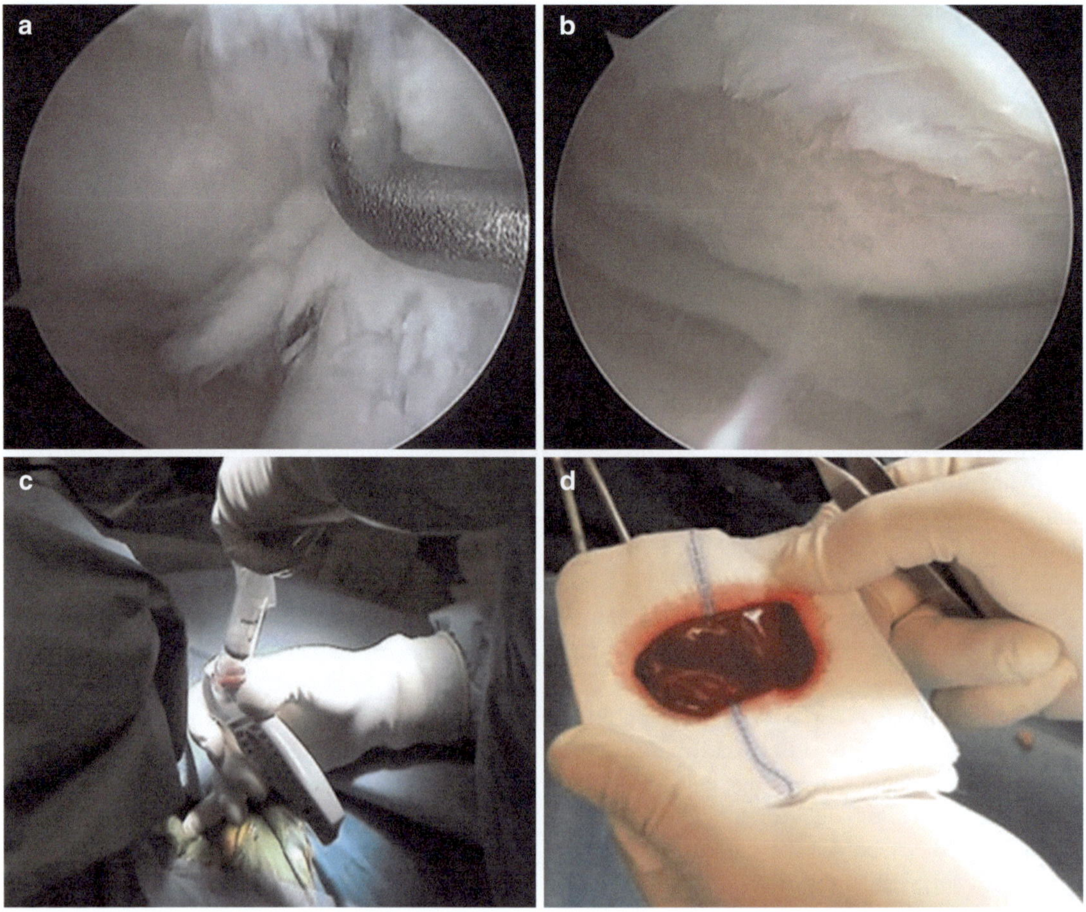

Fig. 29.1 Diagnostic arthroscopic surgery was initially performed to evaluate the condition of the joint and confirm magnetic resonance imaging findings: grade 4 cartilage lesion on the medial femoral condyle (**a**) before and (**b**) after debridement of the lesion (also shown in Fig. 29.2). (**c**) Bone marrow aspiration and (**d**) a bone marrow aspirate concentrate clot after activation [16]

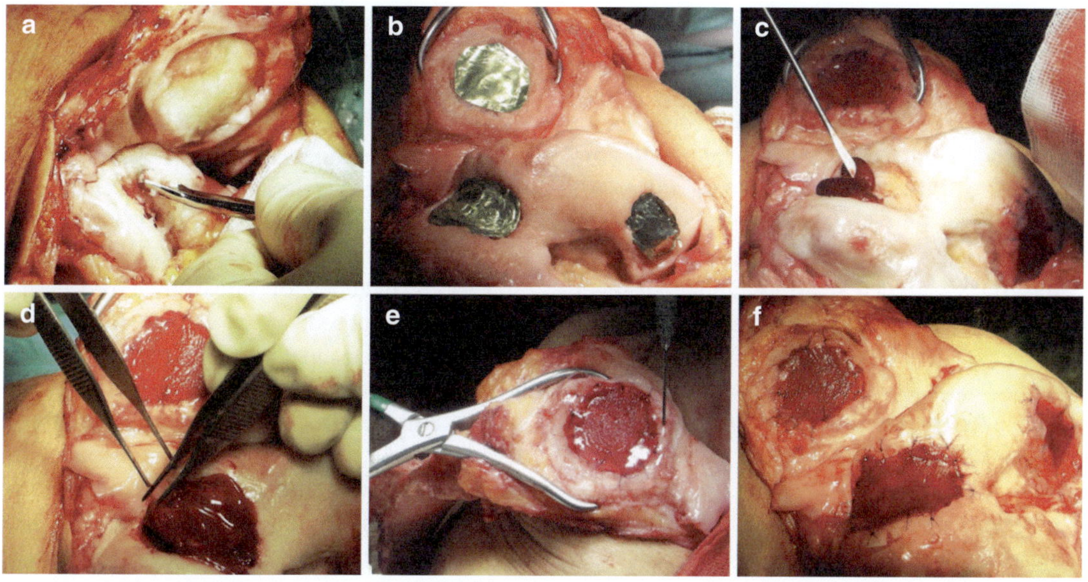

Fig. 29.2 Lesions were debrided, (**a**) templated, and the collagen membrane was fashioned according to the defect size (**b**). The bone marrow aspirate concentrate clot was pasted into the lesions (**c**) and covered with a collagen-based matrix (**d**), which was anchored to the surrounding cartilage and sealed with fibrin glue (**e**, **f**) [16]

29.4 Adipose Tissue and Adipose-Derived Stem Cells

Adipose tissue has become globally recognized as a source of increasing interest for adult stem cells not only for plastic surgery but also for regenerative medicine. Adipose tissue contains a high number of adipose-derived stem cells (ADSCs), 500-fold greater than MSCs from bone marrow. In fact, adipose tissue represents an ideal source of MSCs that can be easily collected in abundance through less invasive means and fewer age-related restrictions. Adipose tissue grafts and adipose-derived stem cells (ADSCs) are commonly isolated from the upper arm, medial thigh, buttocks, trochanteric region, superficial abdominal depots, and infrapatellar fat pad.

Nowadays, subcutaneous adipose tissue is the first choice for cell isolation as it can be easily obtained via lipoaspiration. It is relatively abundant in many patients and can be harvested by a minimally invasive procedure and safely and effectively transplanted in an autologous or allogeneic setting [25–27]. This type of tissue provides an abundant source of stromal vascular fraction (SVF) cells for immediate administration.

The SVF fraction contains multiple noncultured cell types, including preadipocytes, MSCs and endothelial progenitor cells (EPCs), fibroblasts, pericytes, and vascular smooth muscle cells. Upon processing and administration, the adipose-derived SVF cells can differentiate into different tissue types, supporting neovascularization, replacing cells, and repairing injured tissue. Each adipocyte is completely surrounded by a capillary system, explaining how the amount of MSCs in adipose tissue is five times higher than those found in bone marrow. The identification of the stroma and the possibility to immediately use this stromal vascular fraction with its high prevalence of stem/stromal cells for therapeutic use have made adipose tissue a suitable source for clinical applications.

ADSCs represent a population of adult MSCs that possess the ability to self-renew and multipotentially differentiate into different lineages such as adipocytes, chondrocytes, myocytes, and osteoblasts, therefore playing a key role in regenerative and tissue engineering medicine. This can be employed alone by injection or in combination with complex reconstructive surgeries through seeding on different types of scaffolds, mainly composed of collagen or hyaluronic acid. According to published studies, cartilage repair with scaffold augmentation has improved clinical outcomes, radiological fill, and histological repair compared to microfracture alone. In 2017, Pot et al. [28] conducted a systematic review and meta-analysis on cellular and cellular scaffolds. The research found that cartilage regeneration using ADSCs-seeded scaffolds improved regeneration compared to acellular scaffolds.

29.5 Adipose-Derived Stem Cells in Cartilage Repair

For regenerative purposes, the main interest in adipose tissue–derived mesenchymal cells stems from the fact that these cells are multipotent, therefore, capable of differentiating into different cell types including chondrocytes. Over the years, the chondrogenic potential of human ADSCs has been demonstrated by various studies. In 2002, after an *in vitro* differentiation process, Erickson et al. [29] showed that human ADSCs maintain the chondrocyte phenotype and form cartilage tissue when subcutaneously implanted *in vivo* in immunodeficient mice for up to 12 weeks. In several studies [30–32], it has been demonstrated that under certain conditions, ADSCs can express genes and proteins of several specific cartilage molecules, including type II collagen and aggrecan, without the expression of hypertrophic chondrocyte markers, such as type X collagen. However, under several defined conditions, ADSCs can be induced to synthesize type I and type II collagen, suggesting that a fibrocartilaginous phenotype is also possible. When kept in pellet culture or encapsulated in alginate beads and cultured with 10 ng/mL of TGF-β1 growth factor, ascorbate, and dexamethasone, ADSCs have been shown to express a chondrocyte-like phenotype and synthesize type II collagen, aggrecan, link proteins, and chondroitin sulfate in a time-dependent manner. Based on the analysis through immunohistochemistry and incorporation of mRNA radiotracers, there was an observed significant increase in the synthesis of proteoglycans and proteins under chondrogenic conditions. Winter et al. [31] also reported that the gene expression profile of ADSCs under chondrogenic conditions is similar to that of mesenchymal cells. Therefore, the mesenchymal cells from adipose tissue can definitely be considered capable of forming cartilage tissue.

Various methods of lipoaspirate processing have been proposed, which can be divided into two broad lines: enzymatic and nonenzymatic processing techniques. The former uses enzymes such as collagenase, trypsin, or dispase to digest adipose tissue, with a certain variability as to the number of washes, the concentration of enzymes, the centrifugation and filtration parameters, and the erythrocyte lysis methods used. The fraction capable of adhering to the plastic, which contains the ADSCs, is obtained once the stages of processing are concluded, and it can be further made homogeneous if grown by expansion.

However, enzymes can change the safety and effectiveness of the process, raise costs, and may not allow full compliance with the local law and policies. For this reason, attention has shifted to nonenzymatic methods of cell separation that exploit mechanical steps, such as centrifugation, pressure, and filtration to separate cells from adipose tissue. These innovative systems allow minimal manipulation of the tissue (only through mechanical action), as permitted by law, and are designed to maintain the niche of the stromal vascular fraction and promote the removal of impurities, allowing a regenerative product to be immediately available in the same setting in which it can be applied to a defect and scaffold. Also, as previously mentioned, the specific type and characteristics of these scaffolds are very important to achieve a successful clinical result. Kramer et al. [33], in 2006, demonstrated in an *in vitro* study that a collagen membrane is able to retain mesenchymal cells derived from bone marrow after microfractures and these cells have the ability to differentiate toward the chondrogenic, adipogenic, and osteogenic lines. Dickhut et al. [34] carried out *in vitro* research that demonstrated that a biphasic carrier of type I and III collagen is able to promote chondrogenesis of mesenchymal cells. Compared with a collagen-free membrane, it provides greater stability of the repair tissue. Gille et al. [35], in a 12-month study conducted on sheep, demonstrated that when a collagen membrane is applied to a microfractured area being repaired, a thicker repair tissue is obtained compared to that obtained with microfractures alone.

Recent interest and research have established positive results achieved through the use of ADSCs as intra-articular injections in the hip and knee, adopted as an OA treatment, and as an orthobiologic complex scaffold used in reconstruction surgeries. This has brought about the need to validate one-step procedures based on adipose tissue graft containing adipose tissue MSCs. The aim was to improve outcomes by providing a stronger chondrogenic environment and exploiting MSCs paracrine action.

29.6 Autologous Matrix-Induced Chondrogenesis Modified by Adipose Tissue Mesenchymal Cells Augmentation

We personally developed the LIPO-AMIC procedure [36], where in a single procedure, a cartilage defect is subjected to careful debridement, accurate microfractures, then covered by a collagen membrane, where the 3D-matrix bilayer scaffold is carefully cut into the exact size and shape of the defect. The scaffold then is treated and soaked for 10–15 min in the microfragmented adipose tissue transfer graft that contains ADSCs and the stromal vascular fraction. The adipose tissue graft, which is a small quantity of lipoaspirate (about 50–80 mL) obtained from the abdomen adipose tissue reservoir, is treated in a closed system through a dedicated single-use kit for harvesting and subsequent processing (filtration and microfragmentation). Adipose tissue grafting, which utilizes mechanical disruption of the sample, enables the discarding of oil and debris, maintains the cellular niche and secretomes, and allows the eventual transfer of the graft to a syringe for use.

According to the classification of the International Cartilage Regeneration and Joint Preservation Society (ICRS), the LIPO-AMIC technique is used for the treatment of knees with symptomatic degree III and IV with focal cartilage defects greater than 1 cm^2 in size (with mean lesion size of 3.1 cm^2 in studies [36]), affecting the medial or lateral femoral condyles, the retrosurface of the patella, or femoral trochlea in patients between 30 and 60 years of age.

29.7 Details of LIPO-AMIC Surgical Technique

Once the presence of a suitable defect for the treatment is confirmed by arthroscopic analysis, patients undergo a one-step biological resurfacing procedure using microfracture, adipose tissue grafts, ADSCs, and collagen membrane coverage for full-thickness defects (Fig. 29.3).

All surgical procedures are performed in a single surgical setting, under loco-regional anesthesia, and after routine preparation of the sterile

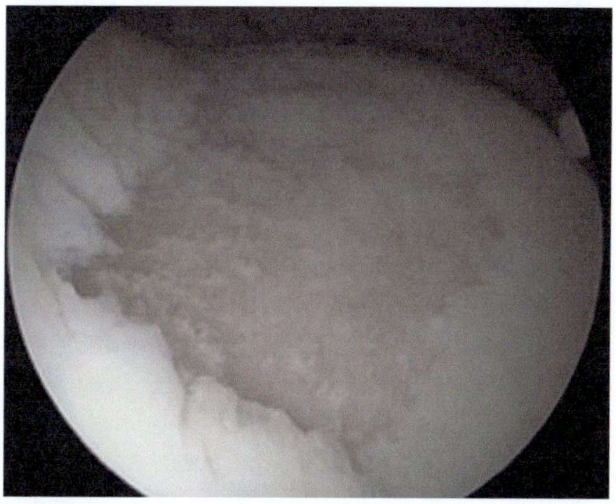

Fig. 29.3 Large full-thickness defect on the femoral trochlea seen arthroscopically suitable for one-step biological resurfacing procedure using microfracture, adipose tissue grafts, ADSCs, and collagen membrane coverage [37]

operating field. The first phase of the procedure involves performing a normal diagnostic arthroscopic examination through the anteromedial and anterolateral standard portals, in order to confirm magnetic resonance imaging findings with respect to location and size of the lesions. Once the diagnostic stage is completed, the extent and depth of the chondral or osteochondral defect and any associated meniscal or synovial pathologies are identified and treated during the standard arthroscopic surgical procedure.

Next, infiltration of Klein's solution is done under local anesthesia of the abdomen and 50–80 ml of lipoaspirate is harvested by lipoaspiration from the periumbilical adipose tissue. Once the adipose graft harvesting process has been completed and the lipoaspirate graft has been processed, filtrated, and microfragmented for preparation to be incorporated into the scaffold, the following step of the surgery is the repair of the focal chondral or osteochondral defect.

Once the cartilage defect is identified, an extensive debridement of the defect is completed. At this stage, it is extremely important to focus on removing all damaged and unstable cartilage until surrounding healthy viable cartilage tissue is reached, creating solid and stable margins around the defect. To accomplish this, straight or curve curettes, specifically designed chondrotomes, and an arthroscopic full radius cutter are normally used to carefully remove partially or fully degenerated cartilage that is still connected to the defect's edges. It is crucial to take care and always remove the calcified cartilage layer with a curette without violating the subchondral plate.

The defect is then measured and an exact figure is created through a sterile aluminum template that will be used to repair the full-thickness cartilage defect. This process helps to produce an exact imprint of the defect that will be utilized to cut the scaffold to the exact size and shape needed which will later be used for the biological resurfacing. While the cut scaffold is soaking in the adipose tissue graft, microfractures in the subchondral bed of the lesion are done using different angled drills, until subchondral bone is reached. Once the microfracture procedure is completed, the dry collagen bilayer membrane is cut to the exact size and shape of the defect and then laid on with care to match the size and shape perfectly.

The next step is the repair of the defect. This phase involves several sequential steps: the first step being the injection of ADSCs in the defect bed, followed by the insertion of the enriched membrane into the joint directly and accurately covering the entire defect. The remaining part of the cells obtained from the adipose tissue is then infiltrated on the site of the lesion. Then the membrane, adipose graft, and MSCs are sealed and secured to surrounding cartilage with fibrin glue (Fig. 29.4). After complete adhesion of the scaffold, the joint is cycled in flexion and extension several times and the stability of the applied membrane is checked.

The postoperative protocol utilizes progressive immediate partial weightbearing assisted with crutches. In the first postoperative day, the patient starts immediately to regain joint range of motion (ROM). The careful evaluation of the compliance characteristics of each patient helps determine in favor or not to use continuous passive motion (CPM), which promotes cartilage healing by favoring joint nutrition and reducing

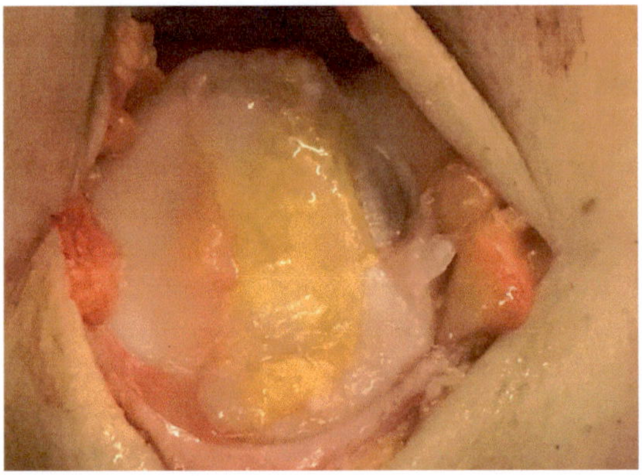

Fig. 29.4 Sealed and secured scaffold membrane, adipose graft, and MSCs over defect and surrounding cartilage using fibrin glue [37]

intra-articular adhesion formation. Range of motion and weightbearing protocols are differentiated depending on the defect location, if located on the femoral condyle or the patello-femoral joint. In case of patello-femoral joint defects, progressive weightbearing with crutches is immediately allowed, limiting ROM from 0° to 60° of flexion for the first 3 weeks. In femoral condyle defects, weightbearing is restricted for the first 4 weeks postoperatively, achieving unrestricted complete weightbearing generally by 6 weeks. Emphasis on the early achievement of normal gait patterns of patients is also of great importance. Once muscle strengthening is completed and joint movement is fully recovered, specific sport activity exercises are initiated under strict guidance of physiotherapists and trainers.

We recently published the 5 years follow-up results of LIPO-AMIC technique [37] that confirmed that all operated patients showed significant improvements in all scores at 6- and 12-month follow-ups. Furthermore, during the check-ups at 2 and 5 years, patients did not manifest any adverse reactions or intra- and postoperative complications (Table 29.1). Patients showed significant progressive improvement of all scores starting as early as 6 months. Additionally, further increased values were noted up until the last follow-up at 60 months. Mean subjective preoperative IKDC score of 36.1 significantly increased to 86.4 at 24 months and to 87.2 at 60 months. Mean preoperative Lysholm score of 44.4 reached 93.5 at 2 years and 93.5 at 5 years. MRI examination showed early subchondral lamina regrowth and progressive maturation of repair tissue and filling of defects. The mean total MOCART score showed a significant improvement at 2-year follow-ups (mean score of 69.1 points) to final follow-ups, having a mean score of 81.9 points (range, 30–100 points; SD 24). Complete filling of the defect at the level of the surrounding cartilage was found in 77.8% of the cases.

The main finding was that the LIPO-AMIC technique achieved statistically significant and stable outcomes for the treatment of full-thickness knee cartilage defects. Improved outcomes were seen compared to those previously reported in recent literature on the standard AMIC technique when utilizing mesenchymal cells from adipose tissue graft attached to a type I and III double-layer collagen membrane. The positive effects of LIPO-AMIC procedure were observed early during the 6-month postoperative visit, with clinical and statistical improvement in pain and functional outcomes.

Table 29.1 Statistical evaluation of clinical outcome scores (IKDC, LYSHOLM, KOOS, and VAS) pre-op and at 6-month, 12-month, 2-year, and 5-year follow-ups [37]

	Follow-up					*P* value	
	Preoperative	6 months	12 months	24 months	60 months	Preoperative/6–12–24–60 months	24/60 months
VAS						$p < 0.05$	$p = 0.52$
Mean value, SD	6.5 ± 0.7	1.1 ± 0.7	0.8 ± 0.7	0.7 ± 0.7	0.8 ± 08		
(Range)	(5–8)	(0–3)	(0–3)	(0–3)	(0–2)		
IKDC						$p < 0.05$	$p = 0.75$
Mean value, SD	36.1 ± 7.1	78.7 ± 6.4	82 ± 4.9	86.4 ± 6.7	87.2 ± 7		
(Range)	(24–44)	(60–90)	(66–92)	(68–94)	(67–94)		
LYSHOLM						$p < 0.05$	$p = 0.98$
Mean value, SD	44.4 ± 10.5	83.2 ± 3.5	87.7 ± 2.8	93.5 ± 0.7	93.5 ± 0.7		
(Range)	(26–66)	(65–93)	(66–96)	(69–98)	(68–98)		
KOOS pain						$p < 0.05$	$p = 0.84$
Mean value, SD	64.1 ± 9.2	88.9 ± 8.4	93.1 ± 4.9	94.2 ± 3.5	93.7 ± 3.5		
(Range)	(51–72)	(73–100)	(80–100)	(80–100)	(80–100)		
KOOS sympt						$p < 0.05$	$p = 0.78$
Mean value, SD	66 ± 4.7	86.8 ± 5.6	90.9 ± 3.6	92.5 ± 5	92.1 ± 1.4		
(Range)	(54–72)	(68–95)	(82–95)	(82–100)	(85–100)		
KOOS adl						$p < 0.05$	$p = 0.81$
Mean value, SD	68.1 ± 1.4	90.9 ± 2.8	94.7 ± 2.8	95.1 ± 2.8	94.8 ± 4.2		
(range)	(56–77)	(78–100)	(85–100)	(85–100)	(88–100)		
KOOS sport						$p < 0.05$	$p = 0.80$
Mean value, SD	43.6 ± 7	68.5 ± 4.9	72.9 ± 8.5	74.4 ± 8.8	75.2 ± 7.9		
(Range)	(28–55)	(45–85)	(60–93)	(60–90)	(65–93)		
KOOS qol						$p < 0.05$	$p = 0.94$
Mean value, SD	38.7 ± 3.5	76.3 ± 12	77–9 ± 14.1	82.4 ± 3.5	82.1 ± 3.5		
(Range)	(30–53)	(63–100)	(64–95)	(60–100)	(65–100)		

29.8 Conclusion

In conclusion, taking into consideration preclinical and clinical studies, the current literature shows encouraging results on the use of orthobiologics in cartilage repair. PRP, BMAC, and ADSCs are the most frequent orthobiologic products used as injection therapy and in new scaffold augmented surgical procedures. MSCs can be obtained from various tissues, such as bone marrow, adipose, and synovial tissues with each source-derived MSC having its own unique advantages and disadvantages. For example, with respect to the number of cells collected, adipose-derived MSCs (ASCs) yield the most, while bone marrow-derived MSCs yield the least. MSC therapies have already been used in clinical phase trials. The respective reviews have shown that the MSC injection of various sources is able to improve the quality of cartilage tissue and delay OA progression.

Cartilage repair is undoubtedly a field that is continuously progressing. In recent decades great advances have been made and the use of orthobiologics is showing great potential for further improvement. The use of cells, growth factors, and biomaterials has revolutionized cartilage regeneration with significant promising prospects. Cartilage repair and regeneration with its evolutive scheme and future next-generation treatment modalities will certainly better define orthobiologics' usefulness, qualities, and specifics and introduce novel strategies that will further improve patient care.

References

1. Widuchowski W, Widuchowski J, Trzaska T. Articular cartilage defects: study of 25,124 knee arthroscopies. Knee. 2007;14:177–82.
2. Curl WW, Krome J, Gordon ES, Rushing J, Smith BP, Poehling GG. Cartilage injuries: a review of 31,516 knee arthroscopies. Arthroscopy. 1997;13:456–60.
3. Caplan AI. MSCs: the sentinel and safe-guards of injury. J Cell Physiol. 2016;231(7):1413–6.
4. Chen WC, Péault B, Huard J. Regenerative translation of human blood-vessel-derived MSC precursors. Stem Cells Int. 2015;2015:375187.
5. Chen WC, Corselli M, Péault B, Huard J. Human blood-vessel-derived stem cells for tissue repair and regeneration. J Biomed Biotechnol. 2012;2012:597439.
6. Caplan AI, Correa D. The MSC: an injury drugstore. Cell Stem Cell. 2011;9(1):11–5.
7. Kean TJ, Lin P, Caplan AI, Dennis JE. MSCs: delivery routes and engraftment, cell-targeting strategies, and immune modulation. Stem Cells Int. 2013;2013:732742.
8. de Girolamo L, Lucarelli E, Alessandri G, Avanzini MA, Bernardo ME, Biagi E, Brini AT, D'Amico G, Fagioli F, Ferrero I, Locatelli F, Maccario R, Marazzi M, Parolini O, Pessina A, Torre ML, Italian Mesenchymal Stem Cell Group. Mesenchymal stem/stromal cells: a new "cells as drugs" paradigm. Efficacy and critical aspects in cell therapy. Curr Pharm Des. 2013;19(13):2459–73.
9. Dominici M, Le Blanc K, Mueller I, Slaper-Cortenbach I, Marini F, et al. Minimal criteria for defining multipotent mesenchymal stromal cells. The international society for cellular therapy position statement. Cytotherapy. 2006;8:315–7.
10. Caplan AI. Mesenchymal stem cells: cell-based reconstructive therapy in orthopedics. Tissue Eng. 2005;11(7–8):1198–211.
11. Robey PG, Bianco P. The use of adult stem cells in rebuilding the human face. J Am Dent Assoc. 2006;137(7):961–72.
12. Fortier LA, Potter H, Rickey E, et al. Concentrated bone marrow aspirate improves full-thickness cartilage repair compared with micro- fracture in the equine model. J Bone Joint Surg Am. 2010;92(10):1927–37.
13. Jones E, McGonagle D. Human bone marrow mesenchymal stem cells in vivo. Rheumatology (Oxford). 2008;47(2):126–31.
14. Caplan AI. Mesenchymal stem cells: the past, the present, the future. Cartilage. 2010;1(1):6–9.
15. Nejadnik H, Hui JH, Feng Choong EP, et al. Autologous bone marrow–derived mesenchymal stem cells versus autologous chondrocyte implantation: an observational cohort study. Am J Sports Med. 2010;38(6):1110–6.
16. Gobbi A, Karnatzikos G, Sankineani SR. One-step surgery with multipotent stem cells for the treatment of large full-thickness chondral defects of the knee. Am J Sports Med. 2014;42(3):648–57.
17. Gobbi A, Scotti C, Karnatzikos G, Mudhigere A, Castro M, Peretti GM. One-step surgery with multipotent stem cells and hyaluronanbased scaffold for the treatment of full-thickness chondral defects of the knee in patients older than 45 years. Knee Surg Sports Traumatol Arthrosc. 2017;25(8):2494–501.
18. Gobbi A, Whyte GP. One-stage cartilage repair using a hyaluronic acid–based scaffold with activated bone marrow–derived mesenchymal stem cells compared with microfracture: five-year follow-up. Am J Sports Med. 2016;44(11):2846–54.

19. Sadlik B, Matlak A, Blasiak A, Klon W, Puszkarz M, Whyte GP. Arthroscopic cartilage lesion preparation in the human cadaveric knee using a curette technique demonstrates clinically relevant histologic variation. Arthroscopy. 2018;34(7):2179–88.
20. Blasiak A, Whyte GP, Matlak A, Brzo'ska R, Sadlik B. Morphologic properties of cartilage lesions in the knee arthroscopically prepared by the standard curette technique are inferior to lesions prepared by specialized chondrectomy instruments. Am J Sports Med. 2018;46(4):908–14.
21. Gobbi A, Whyte GP. Long-term clinical outcomes of one-stage cartilage repair in the knee with hyaluronic acid-based scaffold embedded with mesenchymal stem cells sourced from bone marrow aspirate concentrate. Am J Sports Med. 2019;47(7):1621–8. https://doi.org/10.1177/0363546519845362.
22. Hermann PC, Huber SL, Herrler T, et al. Concentration of bone marrow total nucleated cells by a point-of-care device provides a high yield and preserves their functional activity. Cell Transplant. 2008;16:1059–69.
23. Aldrian S, Zak L, Wondrasch B, et al. Clinical and radiological longterm outcomes after matrix-induced autologous chondrocyte transplantation. Am J Sports Med. 2014;42(11):2680–8.
24. Gobbi A, Chaurasia S, Karnatzikos G, Nakamura N. Matrix-induced autologous chondrocyte implantation versus multipotent stem cells for the treatment of large patellofemoral chondral lesions: a nonrandomized prospective trial. Cartilage. 2015;6(2):82–97.
25. Zuk PA, Zhu M, Mizuno H, et al. Multilineage cells from human adipose tissue: implications for cell-based therapies. Tissue Eng. 2001;7(2):211–28.
26. Zuk PA, Zhu M, Ashjian P, De Ugarte DA, Huang JI, Mizuno H, Alfonso ZC, Fraser JK, Benhaim P, Hedrick MH. Human adipose tissue is a source of multipotent stem cells. Mol Biol Cell. 2002;13(12):4279–95. https://doi.org/10.1091/mbc.e02-02-0105. PMID: 12475952; PMCID: PMC138633.
27. Bunnell BA, Flaat M, Gagliardi C, Patel B, Ripoll C. Adipose-derived stem cells: isolation, expansion and differentiation. Methods. 2008;45(2):115–20. https://doi.org/10.1016/j.ymeth.2008.03.006.
28. Pot MW, van Kuppevelt TH, Gonzales VK, Buma P, IntHout J, de Vries RBM, Daamen WF. Augmented cartilage regeneration by implantation of cellular versus acellular implants after bone marrow stimulation: a systematic review and meta-analysis of animal studies. PeerJ. 2017;5:e3927. https://doi.org/10.7717/peerj.3927. Published online 2017 Oct 27.
29. Erickson GR, Gimble JM, Franklin DM, Rice HE, Awad H, Guilak F. Chondrogenic potential of adipose tissue-derived stromal cells in vitro and in vivo. Biochem Biophys Res Commun. 2002;290:763–9.
30. Gronthos S, Franklin DM, Leddy HA, Robey PG, Storms RW, Gimble JM. Surface protein characterization of human adipose tissue-derived stromal cells. J Cell Physiol. 2001;189:54–63.
31. Winter A, Breit S, Parsch D, Benz K, Steck E, Hauner H, Weber RM, Ewerbeck V, Richter W. Cartilage-like gene expression in differentiated human stem cell spheroids: a comparison of bone mar-row-derived and adipose tissue-derived stromal cells. Arthritis Rheum. 2003;48:418–29.
32. McIntosh K, Zvonic S, Garrett S, Mitchell JB, Floyd ZE, Hammill L, et al. The immunogenicity of human adipose derived cells: temporal changes in vitro. Stem Cells. 2006;24:1245–53.
33. Kramer J, Böhrnsen F, Lindner U, Behrens P, Schlenke P, Rohwedelet J. In vivo matrix-guided human mesenchymal stem cells. Cell Mol Life Sci. 2006;63(5):616–26.
34. Dickhut A, Gottwald E, Steck E, et al. Chondrogenesis of mesenchymal stem cells in gel-like bio-materials in vitro and in vivo. Front Biosci. 2008;13(12):4517–28.
35. Gille J, Kunow J, Boisch L, et al. Cell-laden and cell-free matrix-induced chondrogenesis versus microfracture for the treatment of articular cartilage defects: a histological and biomechanical study in sheep. Cartilage. OnlineFirst. 2010. https://doi.org/10.1177/1947603509358721.
36. Sciarretta FV, Ascani C, Fossati C, Campisi S. LIPO-AMIC: technical description and eighteen pilot patients report on AMIC® technique modified by adipose tissue mesenchymal cells augmentation. GIOT. 2017;43:156–62.
37. Sciarretta FV, Ascani C, Sodano L, et al. One-stage cartilage repair using the autologous matrix-induced chondrogenesis combined with simultaneous use of autologous adipose tissue graft and adipose tissue mesenchymal cells technique: clinical results and magnetic resonance imaging evaluation at five-year follow-up. Int Orthop. 2024;48:267–77. https://doi.org/10.1007/s00264-023-05921-8.

Part IV

Future Trends

Biophysical Stimulation in Regenerative Medicine

30

Antonio Marmotti, Francesca Veronesi, Laura Mangiavini, Nicolò Rossi, Riccardo Giorgino, Martina Coco, and Stefania Setti

30.1 Introduction

Electromagnetic fields (EMFs) represent a dynamic interaction between magnetic and electrical forces, exerting a mutual influence characterized by varying signal frequencies, waveforms, and intensities, each of which is capable of eliciting beneficial or adverse biological effects. In clinical practice, EMFs can be used in a dual role: as diagnostic tools and as potential therapeutic interventions, where they are promising therapeutic agents, with emerging medical evidence suggesting their efficacy in modulating cellular responses and promoting tissue repair [1]. On cell membranes they exert influence on cellular processes, potentially affecting homeostasis and cellular function. Their biological interactions and potential therapeutic implications underscore the importance of vigilant surveillance and ongoing research. By advancing our understanding of EMFs and their effects on biological systems, we can harness their benefits while mitigating their potential risks, thereby optimizing their utility in medical practice.

Among the different subtypes of EMFs, pulsed electromagnetic fields (PEMFs) stand out as promising therapeutic agents. These noninvasive and cost-effective physical stimulations have demonstrated therapeutic potential in treating a wide range of diseases [2, 3]. PEMFs have been shown to influence cellular processes such as cell cycle regulation, apoptosis, proliferation, differentiation, and interaction with extracellular matrix components. PEMFs can alter cell function and trigger a cascade of biological responses [4]. Moreover, the effects of PEMFs may propagate along signal transduction pathways, ultimately influencing cell behavior and restoring homeostatic functions. PEMFs induce biological changes that can affect Ca^{2+} efflux, nitric oxide signaling, and growth factor secretion [5].

A. Marmotti (✉) · M. Coco
Department of Orthopaedics and Traumatology, University of Turin, Torino, Italy

Department of Molecular Biotechnology and Health Sciences, Molecular Biotechnology Center "Guido Tarone", University of Torino, Torino, Italy
e-mail: antonio.marmotti@inwind.it

F. Veronesi
Surgical Sciences and Technologies, IRCCS Istituto Ortopedico Rizzoli, Bologna, Italy

L. Mangiavini
IRCCS Istituto Ortopedico Galeazzi, Milan, MI, Italy

Department of Biomedical Sciences for Health, University of Milan, Milan, Italy

N. Rossi
IRCCS Ospedale Galeazzi Sant'Ambrogio, Milan, MI, Italy

R. Giorgino
Residency Program in Orthopaedics and Traumatology, University of Milan, Milan, Italy

S. Setti
IGEA SpA, Clinical Biophysics, Carpi, Italy
e-mail: s.setti@igeamedical.com

A. Gobbi et al. (eds.), *Regenerative Medicine in Sports and Orthopaedics*,
https://doi.org/10.1007/978-3-031-84693-9_30

The biological effects of PEMFs are influenced by various factors, including the signal waveform, amplitude and frequency, and the dosage of the treatment [6]. Parameter optimization in preclinical studies is essential to deliver therapeutic PEMFs in clinical practice. Moreover, different cell types exhibit varying degrees of sensitivity to PEMFs. Elucidating the true biological implications of PEMF exposure requires careful consideration of dosimetry alongside an understanding of the complex interplay between PEMF and biological systems. As with pharmacological research, dose-response effects are of fundamental importance for the development of clinical applications.

Biophysics represents an interdisciplinary field of study that bridges physics and biology, employing the methods and theories of physics to investigate biological systems. Biophysics studies how nonionizing physical stimuli, such as electrical, magnetic, and mechanical forces, interact with biological systems, changing their behavior: "biophysical stimulation." Biophysical stimulation techniques offer versatile applications in clinical practice, serving as stand-alone therapies to strengthen reparative and anabolic processes within tissues. Moreover, they can synergistically complement traditional pharmaceutical approaches when used in conjunction with drugs or growth factors. This combination not only increases their effectiveness but also mitigates the potential side effects of drugs by decreasing their intake, making biophysical stimulation a powerful adjunct to conventional drug treatments.

In this chapter we will address the effectiveness of biophysical stimulation with PEMFs in the field of regenerative medicine for skeletal tissues in sports and orthopedics.

30.2 Biophysical Stimulation in Regenerative Medicine for Osteo/Cartilaginous Lesions

Biophysical stimulation in regenerative medicine for osteo/cartilaginous lesions represents a promising approach to enhance tissue repair and regeneration. In this ambit, PEMF stimulation offers a noninvasive and promising therapeutic strategy, because it works by delivering low-frequency electromagnetic waves to the affected area.

PEMFs influence cellular signaling pathways, increasing the proliferation and differentiation of osteoblasts and chondrocytes and enhancing extracellular matrix (ECM) component synthesis, such as collagens and aggrecan.

In addition, PEMFs can improve local blood circulation and reduce inflammation, creating a favorable environment for tissue regeneration. By optimizing the microenvironment and mechanical conditions, PEMFs not only accelerate the repair process but also improve the quality and functionality of the regenerated tissue.

This approach is particularly valuable in treating osteochondral lesions, where both bone and cartilage require coordinated regeneration to restore joint integrity and function, potentially leading to improved clinical outcomes and faster recovery times.

In this chapter the main potentials of PEMFs, used in the preclinical and clinical settings for the regeneration of bone and cartilage, will be shown, focusing both on the cellular and molecular aspects.

30.2.1 Preclinical Evidence

30.2.1.1 Bone Tissue

Regarding bone tissue, human mesenchymal stem cells (hMSCs), derived from different sources, have been extensively used for tissue engineering. PEMFs induce proliferation and an early and high osteogenic differentiation of human MSCs derived from bone marrow (hBMSCs) [7]. Similarly, PEMFs increase also Alkaline Phosphatase (ALP), mineralization Collagen I (COLL I), and Osteocalcin (OCN) gene expression and Akt and mTOR pathways in human MSCs derived from adipose tissue (hADSCs) [8–10]. ALP is highly expressed in the cells of mineralized tissue and plays a critical function in the formation of hard tissue; COLL I and OCN are the main components of bone ECM, and Akt/mTOR pathway plays a pivotal role in osteogenic differentiation of MSCs [11, 12]. In both hBMSCs and hADSCs, osteogenic differentiation is induced by PEMFs by increasing ALP, and OCN protein production and matrix mineralization. The combination with Bone Morphogenetic Protein-2 (BMP-2) further increases hBMSC osteogenic differentiation, while in hADSCs this effect is independent by BMP-2 [13]. In detail, the combination of BMP-2 and PEMF stimulation on hBMSCs increases BMP-6, BMP type I receptor gene expression, as well as SMAD 1/5/8 and P38 MAPK activation, showing that PEMFs induce osteogenic differentiation through BMP signaling components [14]. Finally, making a comparison between hBMSCs and hADSCs, PEMFs seem to promote more cell proliferation, and more bone ECM deposition in hBMSCs than in hADSCs [15].

As regards other molecular pathways implicated in PEMF effect, also calcium-related osteogenic pathways (permeation, and regulation of cytosolic concentration) are responsible for PEMF ability in promoting MSC osteogenic differentiation [16]. Finally, PEMFs up-modulate the expression of Micro-RNA (miR)-26a and miR-29b, the inductors of osteogenic differentiation, and promote the expression and release of miRNAs also involved in angiogenesis, showing their potential to regulate the osteogenic-angiogenic coupling, important in bone tissue regeneration [17].

PEMF stimulation increases osteogenic features of hBMSCs, seeded onto TiO2 nanostructured surface, interfering with selective calcium-related osteogenic pathways and enhancing the expression of osteogenic genes and their relative protein production, such as ALP, COLL I, OCN, Osteopontin (OPN), BMP2, and Decorin (DCN) [18]. Finally, also in murine MSC cell lines, C3H10T1/2, PEMFs show the same results observed in hMSCs: they increase ALP activity, mineralization, and intracellular Ca2+ concentration and upregulate Wnt1, which plays a key role in bone mass regulation, phosphor-Lrp6, involved in skeletal development, and β-catenin, important for maintenance of bone homeostasis and mechanotransduction [19].

Since PEMFs are efficacious in inducing the osteogenic effect of PEMFs on MSCs, from different sources, they could be proposed as new strategies for bone tissue engineering optimization. PEMFs could be considered a possible tool to improve autologous cell-based regeneration of bone defects in clinic.

Always in the context of bone regeneration, PEMF stimulation shows its effects in counteracting osteolysis. Wear-mediated osteolysis is a common complication occurring around implanted prosthesis, which ultimately leads to bone loss with mechanical instability. In this context, PEMFs, in combination with platelet-rich plasma (PRP), are able to increase the amount of the anabolic factor transforming growth factor β1 (TGFβ1) and to reduce the inflammatory cytokine tumor necrosis factor α (TNFα) and the number of osteoclasts. In addition, PEMFs increase bone ingrowth and osseointegration of titanium alloy pins implanted in distal femurs of rats [20, 21].

30.2.1.2 Cartilage Tissue

Articular cartilage injuries are a common source of joint pain and dysfunction. As articular cartilage is avascular, it exhibits a poor intrinsic healing capacity of self-repair. In summary, PEMFs stimulate chondrocyte proliferation, differentiation, and ECM release of anabolic factors and of anti-inflammatory cytokines [22].

In detail, PEMFs are able to differentiate bovine BMSCs into a chondrogenic phenotype, and when treated with interleukin 1β (IL1β) to simulate an in vitro osteoarthritic microenvironment, PEMFs are also able to counteract the IL1β-induced inhibition of chondrogenesis, to increase proteoglycans (PGs) content and Aggrecan (ACAN) and COLL II gene expression. These results suggest PEMFs as a therapeutic strategy for improving clinical outcome of cartilage engineering repair procedures, based on the use of MSCs [23].

Other important cells involved in cartilage regeneration are synovial fibroblasts (FLS). PEMF stimulation promotes bovine FLS migration and increases the recruitment of FLS from bovine synovium explants to the site of cartilage wound injury recreated in vitro. In addition, an increase in glycosaminoglycans (GAGs) and collagens are also observed [24]. Also in canine chondrocytes, embedded in an agarose hydrogel scaffold, PEMFs induce GAG production in vitro [20].

In dog focal cartilage defects, PEMFs improve the regeneration after the implantation of osteochondral construct, showing an early repair after 3 months from the treatment, with normal chondrocytes and PG content [25]. Similarly, in rabbit osteochondral defects, treated with collagen scaffold and bone marrow concentrate (BMC), PEMFs increase the quality of the regenerated tissue [26].

30.2.2 Clinical Evidence

On the clinical side, in patients affected by knee chondral lesions of ICRS grade III and IV, that undergo arthroscopic matrix-assisted autologous chondrocyte implantation (MACI), PEMF stimulation improves SF-36 and EuroQol and pain, with effects also observed after 60 months from the treatment [27]. Similarly, in patients affected by talus osteochondral defects and who undergo BMSC transplantation, PEMF stimulation significantly increases AOFAS score and decreases pain at 6 and 12 months of follow-up in comparison to not stimulated patients [28]. In a prospective, randomized, double-blind study, Zorzi et al. [29] recruited patients with Outerbridge grade II, III, and IV osteoarthritic lesions treated surgically by arthroscopic microfractures. The results showed that the percentage of patients taking NSAIDs was significantly lower in patients in the active group than in the placebo group and that functional scores were significantly better in the PEMF-treated group 90 days after surgery. Knee function recovery time was significantly reduced in the short term, and 87.5% of patients in the placebo group were unable to return to full sports activity 3 years after surgery, compared with 37.5% in the active group ($p < 0.05$).

Another important aspect of regenerative medicine in this ambit is the inflammatory front. It is well known that adenosine is a powerful endogenous inhibitor of inflammatory processes and that it acts through interaction with four membrane receptors. In particular, the A_{2A} receptors are able to modulate the inflammatory process. In chondrocytes and synoviocytes PEMFs are able to increase the number of bonds with A_{2A} receptors in stimulated bovine and human chondrocytes, synoviocytes, and human OBs [30, 31].

30.3 Conclusions and Future Perspectives

In conclusion, PEMFs are currently used as a safe and noninvasive treatment to enhance bone healing and to provide joint protection. PEMFs enhance cell proliferation and both osteogenic and chondrogenic differentiation of MSCs, isolated from different sources, suggesting a synergic effect with other regenerative actors, such as scaffolds or growth factors.

PEMFs ameliorate bone ingrowth and osseointegration of a bone implant and regeneration of cartilaginous tissues. Finally, PEMF stimulation is able to control the joint inflammatory process, enabling better tissue regeneration, especially following regenerative medicine procedures, allowing restoration of proper joint homeostasis. It also allows better functional recovery and standard rehabilitation with less pain, maintaining stable results over time (Table 30.1).

Table 30.1 PEMF relating to MSCs, Osteoclasts, Chondrocytes and synoviocytes along with experimental preclinical and clinical evidences

Cells/Tissues	PEMF effects			
MSCs	↑	Osteogenic and chondrogenic differentiation ALP protein production COLL I, OCN gene expression Akt/mTOR pathway Mineralization Calcium-related osteogenic pathway MiR-26a and MiR-29b Wnt1, p-Lrp6, β-catenin		
MSCs seeded onto TiO2 nanostructured surface	↑	A ALP, COLL I. OCN, OPN, BMP2, DCN gene expression and protein production		
Osteoclasts	↑	TGFβ1	↓	TNFa, N° of osteoclasts
Ti Alloy in distal femurs of rats	↑	Osseointegration		
MSCs+IL1β	↑	PG ACAN and COLL II gene expression		
FLS	↑	Migration GAGs		
Chondrocytes in agarose hydrogel scaffold	↑	GAGs		
Dog focal cartilage defect treated with osteochondral construct	↑	Regeneration after 3 months		
Rabbit osteochondral defect treated with collagen scaffold and BMC	↑	Quality of regenerated tissue		
Patients affected by knee chondral lesion treated with MACI	↑	SF-36, EuroQoL	↓	Pain
Patients affected by talus osteochondral defect treated with BMSC transplantation	↑	AOFAS score	↓	Pain
Chondrocytes, synoviocytes and osteoblasts	↑	N° of A2A receptors	↓	Inflammation

30.4 Biophysical Stimulation in Regenerative Medicine for Muscular Lesions

Skeletal muscle is one of the largest organs in the human body, essential for respiration and movement. Due to its complex organization, this tissue has a remarkable regenerative potential, primarily supported by muscle stem cells. However, in cases of severe injuries, this self-repair capability can be compromised. Recent preclinical research is exploring biophysical stimulation techniques to enhance muscle regeneration. Biophysical stimulation is emerging as a significant innovation in regenerative medicine for the treatment of muscular lesions.

This chapter will examine the preclinical evidence on the effectiveness of these innovative approaches in regenerative medicine for muscle injuries, discussing the mechanisms of action and the results obtained so far. Subsequently, the initial clinical evidence supporting the efficacy of these techniques will be reviewed.

30.4.1 Preclinical Evidence

Skeletal muscle is considered one of the largest organs of human body and its sophisticated organization is responsible for the efficient function of generating forces that guarantee breathing and movement [32]. The contractile capacity is related to the myofibers organization, which originates from the fusion of myoblasts, forming multinucleated myotubes. Furthermore, this tissue is known for the outstanding regenerative potential after minor traumas or damage [33]. The main supporter of this self-regenerative capacity is related to muscle stem cells, also called satellite cells, which lie in a quiescent state but are ready to activate when repair and regeneration are needed allowing for fast recovery and maintenance of functionality. However, this self-healing capacity can be compromised in case of more serious injuries or pathological conditions. In addition to satellite cells and myoblasts, progenitor cells with high myogenic potential have been found within adult skeletal muscle. These cell types include CD133+ progenitor cells, muscle-derived stem cells (MDSCs), multipotent perivascular progenitor cells, and muscle-derived side-population (SP) cells. Moreover, multipotent cells with myogenic potential have also been found in tissues other than skeletal muscle such as bone marrow (BM), adipose, and umbilical cord [33, 34].

Tissue engineering and regenerative medicine approaches to augment the skeletal response to injury have been widely investigated and can be resumed in three main families [34]:

1. A cell-therapy approach employs the use of exogenous myogenic progenitor cells in sites of skeletal muscle injury with intramuscular injection. One of the most used cell therapies is the stromal vascular fraction derived from adipose tissue. This approach has shown the ability to differentiate not only into adipogenic, chondrogenic, osteogenic, and myogenic cells but also to have a myogenic potential in vitro. This potential has been also examined in vivo [35]. After injection at the injured muscle site, ADSCs are able to downregulate inflammation, upregulate angiogenesis, and restore dystrophin expression in a mouse model of muscular dystrophy [36]. Kang et al. [37] studied the ability of ADSCs to repair tissue damage in a murine ischemic hindlimb model. At 4 weeks post injury and implantation, ADSC-transplanted limbs showed increased vascular density and reduced muscle atrophy.
2. The second approach aims to create an implantable tissue-engineered muscle construct (TEMC) composed of exogenous myogenic cells coupled with a biomaterial scaffold.
3. The acellular approach employs the implantation of biologic scaffolds composed of decellularized mammalian tissue, which upon their degradation recruit endogenous myogenic progenitor cells. These approaches share the same goal to restore site-appropriate and functional skeletal muscle tissue. On the other hand, these strategies have been limited by poor cell viability, engraftment into host tissue, and regulatory challenges.

Moreover, it has been demonstrated that biophysical stimulation can play a pivotal role in enhancing the regenerative potential of the muscular unit due to the contractile ability of this tissue. Mechanical stimulation, electrical stimulation, and their synergistic effect have been explored to investigate the ability of stimulating the muscular regenerative niche [38]. While electrical stimulation and the administration of chemical compounds have been widely used in tissue regeneration studies, both approaches present significant drawbacks. These limitations have driven scientists to seek alternative approaches. In this context, pulsed electromagnetic fields (PEMF) are emerging as a potential therapeutic option to stimulate regeneration of skeletal muscle cells. PEMF are recognized to stimulate osteoblastogenesis and enhance the paracrine role of mesenchymal stem cells for cartilage regeneration [39]. Maiullari et al. showed in their in vitro study that PEMF can induce skeletal muscle cell repair by sustaining the expression of proteins involved in the response to cellular damage and oxidative stress. SkMC stimulated with 1.5 mT PEMF for 4 h and repeated for 2 days led to cell proliferation without inducing cell apoptosis or significant impairment of the metabolic activity. Interestingly, PEMF treatment could speed up the regenerative process, inducing a more complete cell migration to close the scratch and wound healing. These data support the possibility of using PEMF to increase SkMC regeneration [40].

30.4.2 Clinical Evidence

Biophysical stimulation is emerging as a significant innovation within regenerative medicine for the treatment of muscular lesions. Biophysical stimulation refers to a series of interventions aimed at stimulating the body's inherent healing processes through various biological mechanisms. This approach has shown promising preliminary results, although their true clinical efficacy still requires further confirmatory studies.

PEMF is a noninvasive technique to promote tissue regeneration and muscle recovery. Recent studies suggest that PEMF can modulate mitochondrial activities, fostering muscle regeneration and tissue healing. This therapeutic approach could be particularly useful in the context of postoperative rehabilitation, for example, patients with anterior cruciate ligament reconstruction (ACLR) injuries.

In a randomized controlled clinical trial [41], the use of PEMF during the postoperative period showed potential benefits in increasing quadriceps muscle strength compared to traditional rehabilitation alone. Patients treated with PEMF demonstrated an improvement in muscle volume through strength measurements using isokinetic evaluations, along with imaging techniques such as ultrasound and magnetic resonance imaging (MRI).

Another recent study [42] investigated the effects of PEMF on reducing delayed onset muscle soreness (DOMS) in elbow flexors. This study involved 30 healthy male students and was designed as a randomized, double-blind, placebo-controlled trial. Primary outcomes measured included muscle pain, peak torque, median frequency (MDF), and electromechanical delay (EMD) during isometric contraction at 24, 48, and 72 h after DOMS induction. The results demonstrated that PEMF application was effective in reducing physiological deficits associated with DOMS, improving perceived muscle pain recovery, MDF, and EMD during isometric contraction. However, no greater efficacy of PEMF treatment over the placebo group was observed in generating peak isometric torque.

In yet another study [43], the effect of PEMF on muscle activity and metabolism during exercise in semi-professional cyclists was investigated. Twenty cyclists performed constant load exercise with active (ON) or inactive (OFF) PEMF stimulation in a randomized manner. During the load-free warm-up phase, PEMF stimulation caused a significant increase in muscle activity. Additionally, blood lactate concentration was higher during PEMF stimulation, suggesting that PEMF may influence glycolytic metabolism. This study concluded that PEMF stimulation can increase muscle fiber activity and metabolism during physical exercise, particularly at low intensity.

30.5 Conclusions and Future Perspectives

While PEMFs have shown potential benefits in muscle regeneration and injury treatment, it is clear that further high-quality studies are necessary to confirm their clinical efficacy. Current evidence is sometimes conflicting and limited by methodological quality and the number of studies available.

An additional innovative aspect that shows great promise is the integration of biophysical stimulatory therapy with regenerative cellular therapy as a synergistic approach to treating muscle injuries. Cellular therapy, which includes the use of mesenchymal stem cells (MSCs) and other progenitor cells, has demonstrated the ability to promote tissue regeneration through the release of local growth factors, modulation of immune responses, and differentiation into muscle cells. Cellular therapy could potentially amplify tissue regeneration efficacy when combined with biophysical stimulation techniques such as pulsed electromagnetic field (PEMF).

Biophysical stimulation can create a favorable microenvironment by improving vascularization, reducing inflammation, and stimulating mitochondrial activity, which in turn can enhance the effectiveness of implanted regenerative cells. Preclinical studies have begun exploring these combined therapeutic approaches, suggesting that the integrated use of the two methodologies may accelerate functional recovery, reduce healing times, and improve muscle regeneration quality. However, well-designed clinical studies are needed to confirm these benefits and optimize combined treatment protocols.

Therefore, the future of biophysical stimulation in regenerative medicine requires not only continuous engagement in clinical and preclinical research but also a comprehensive understanding of the underlying biological mechanisms and the standardization of treatment protocols, which will be crucial to fully exploit the potential of these innovative technologies.

30.6 Biophysical Stimulation in Regenerative Medicine for Tendon Lesions

Tendon may be considered one of the latest frontiers for PEMFs stimulation, and, at present, the goal of a widespread clinical use of PEMF in the treatment of tendinopathy and tendon lesions has not yet been achieved. Several modern in vitro pieces of evidence are promising, and surprisingly, a long history of little-known clinical and preclinical studies dates back to the 1980s showing a real possible role of PEMFs for conservative therapy of tendon disorders.

30.6.1 Clinical Evidence

It is since 1984 that PEMFs have shown a promising effect on tendinopathy, in a pilot study published in *Lancet* by Binder et al. [44] focused on persistent rotator cuff tendinitis. In a double-blind randomized controlled trial, they described the beneficial use of PEMFs in rotator cuff tendinitis

Following those observations, several different clinical studies were performed. In 1985, Devereaux et al. [45] treated patients with chronic lateral humeral epicondylitis but no clinical improvement was detected, while several years after, in 2007, a significant improvement in pain during rest, activity, and nighttime was described by Uzunca et al. [46] comparing PEMFs to corticosteroid injections and sham PEMF. This study allows for admitting a real clinical benefit of PEMFs for a common pathology such as the "tennis elbow." Subsequently, in 2015 Osti et al. [47] showed a positive effect of PEMFs in reducing postoperative pain, analgesic use, and stiffness in the short-term period after rotator cuff repair. This showed also the clinical safety of postoperative PEMF application and the early clinical benefit, albeit, at 2 years, no difference was seen in outcomes in patients who did or did not undergo treatment with pulsed electromagnetic fields. A concept, thus, appeared related to the treatment with PEMFs: the possible acceleration of the repair process driven by PEMF exposure. The positive potential of PEMFs toward rotator cuff pathology was also shown in another study in which patients were treated conservatively and exposed to a highly energetic pulsed electromagnetic field therapy in association with extracorporeal shock wave therapy (ESWT). Kluter et al. [48] demonstrated that the combination of high-energy PEMFs and ESWT significantly improves pain and function compared to ESWT alone after 24 weeks, showing a possible efficacy of PEMFs on rotator cuff disease even if not associated with surgical therapy. Thus, a possible clinical benefit of PEMFs on rotator cuff was unquestionably observed in literature, even if a recent study [49] failed to show a consistent benefit of PEMFs associated with diathermy (US) and electrotherapy (TENS) compared to US and TENS alone, probably due to the brief exposure to PEMFs lasting only 25 min in the conservative program chosen by the authors that implied one session a day, 5 days/week for 2 weeks.

A very recent study in 2023 proposed PEMF exposure to improve Achilles tendinopathy. The authors from the Chinese University of Hong Kong have suggested a possible role of PEMFs as an adjunct to eccentric exercise, suggesting the potential of PEMFs in obtaining pain relief and improved response to eccentric exercise by modulating the mechanical properties of the Achilles tendon. This prospective randomized double-blinded, placebo-controlled trial still ongoing may reveal another precious benefit that could be easily and safely obtained by PEMF exposure for the conservative treatment of Achilles tendinopathy.

30.6.2 Preclinical Evidence

Clinical research is necessarily biased by different patients' characteristics and nonhomogeneous clinical picture, thus multiple factors may be involved in the low-powered evidence present in the literature. Conversely, preclinical studies may offer a more reproducible setting due to the possibility of choosing animal features and creating "artificially" pathological pictures. And in this landscape, results are very promising. The first animal study was performed in the 1980s, when Watkins et al. [50] created defects in the superficial digital flexor tendons in a horse model and exposed the animals to PEMFs, albeit for only 2 h per day, showing a delayed maturation of the tissue formed within the defect. During the following decades, the small animal models were the standard for tendon disorders, likely for economic issues and more reproducible clinical pictures. Lee et al. [51], in 1997, showed the possible role of PEMFs in the healing process of experimental Achilles tendonitis in the rat, describing a reduction of inflammation, with a better return of the tendon to histological normality. In a similar rat model, in 2006 Strauch et al. [52] analyzed the effect of PEMF after Achilles transection and repair and observed an increase in tensile strength of up to 69% at the repair site at 3 weeks after compared with nonstimulated control animals. This started to suggest the potential "accelerating" effect of PEMFs on tendon repair, hypothesizing in humans an early rehabilitation protocol during PEMF exposure. The improvement in early tendon healing was also demonstrated by Tucker et al. [53] in 2017 in a rat rotator cuff acute detachment and repair model. The authors described very interesting aspects concerning PEMF potential. Briefly, PEMF improves early tendon-to-bone healing through an improvement of tendon mechanical and histological properties, thus validating again the "early" action of PEMFs in the repair process. Specifically, mechanical properties were improved at the early follow-up period (4–8 weeks) with an increase in the elastic modulus, and histologically, tendon cells showed a more rounded shape, suggesting that the cells were metabolically more active due to the PEMF exposure, and collagen fiber in the mid substance showed also a better spatial organization. This study was certainly a cardinal point in the field of preclinical research. Similar observations were described by Huegel et al. [54] in 2018 in a rat model of acute bilateral supraspinatus injury and repair, where the authors showed again improvements in tendon mechanical properties (both modulus and stiffness) at 4 weeks and in collagen organization after PEMFs exposure up to 6 h/day. A recent study, also, by Dolkart et al. [55] in 2021 showed similar results in the same rat model, again observing the positive effect on the early stages of healing. Furthermore, Huegel et al. [56] showed recently the potential mechanisms triggered by PEMFs during rotator cuff healing as the upregulation in the BMP2 signaling pathway, the increase in pro-osteogenic genes at the insertion, promoting the tendon-bone interface, the downregulation of genes related to a fibrotic healing response and some anti-inflammatory effects as the shift from M1 to M2 macrophages. The same research group of Huegel [57], surprisingly, did not find the same positive effect of PEMF stimulation in a rat Achilles tendon tear model (complete with repair, partial without repair) after PEMF exposure up to 3 h/day. The setting was different because an Achilles tendon tear is a soft tissue lesion, while a supraspinatus tear is mainly a tendon-bone lesion, and the histological structure of the enthesis is far more complex than the body of the tendon itself. This conflicting evidence emphasizes the need to better clarify preclinically the potential benefit of PEMF therapy that seemed to be ineffective toward complete tear and repair but effective for a partial tear at the early time point (1-week post-injury) suggesting a possible positive effect in early tendon healing following partial width, full thickness Achilles injury. These results, however, were impaired due to the rapid baseline healing of the experimental rats used in the research; thus, potentially, a greater effect could be observed in a clinical human setting. Achilles tendinopathy has been recently studied by the research group of De Girolamo, and, in a recent work, they showed a positive anabolic effect of PEMF exposure 8 h/day. In their study, Perucca et al. [58] induced tendinopathy by type 1 collagenase injection, and they found the best results when PEMFs were applied in the mid-acute phase (7–14 days after the collagenase injection). Indeed, PEMF exposure significantly improved the restoration of physiological fiber structure and cell morphology and decreased the inflammatory infiltrates. A similar result was obtained in a rabbit Achilles tendon tear and repair model by the Turkish research group of Uzun et al. [59]. In this study, the authors exposed the rabbits for 21 days after surgery, and they obtained positive results opposite to the previously mentioned study of Huegel, and they concluded that PEMF are efficient for Achilles tendon recovery, although their mechanisms of action are still not fully understood. Taken together, these observations are surely promising for a possible clinical human application in a complex context such as Achilles tendinopathy.

30.7 In Vitro Modern Evidence

The clinical and preclinical results suggest that the main concept of PEMF stimulation is an early anabolic effect on the overall repair process, ultimately influencing the mechanical properties of the tendon tissue.

As suggested by a previous review in the German language [60], therapy with electromagnetic fields is a very old classic in conservative medicine, but little is known about the real effects of PEMF on tendon metabolism. This is the field of in vitro studies that are aimed to find a correlation between PEMF exposure and cellular changes in receptor and gene activation. Some precious information is available in the present state of the art, and they may be summarized in several focal points as recently described by Rosso et al. [61]:

- The key role of adenosine receptor A_{2A} and A_3; this has been described by the research group of De Girolamo in several works; in a model in which tendon cells were treated with PEMFs (48 h exposure) and IL-1β, Colombini et al. [62] observed an increase in higher adenosine receptor A_{2A} surface exposition/ligand affinity along with an early anabolic and reparative response typical of the physiological first inflammatory phase of tendon healing, as higher secretion of IL-6 and PGE2 and higher expression of SCX and COL3A1; previously, De Girolamo et al. [63, 64] showed also the increase of IL-10, VEGF-A, and TGF-β.
- The expression of anti-inflammatory genes, associated with the MAPK (ERK1/2) pathway; the work of Vinhas et al. showed in several studies [65, 66] the potential anti-inflammatory effect of PEMFs on magnetically assisted cell sheets constructed with human tendon-derived cells and magnetic nanoparticles to study inflammation activity upon exposure to IL-1β; they showed [67] similar anti-inflammatory potential of a brief PEMFs exposure on a tenocyte monolayer culture treated with exogenous IL-1β, describing the decreased production of IL-6 and tumor necrosis factor-α, IL-8, COX-2, MMP-1, MMP-2, and MMP-3, and the increased expression of IL-4, IL-10, and TIMP-1; Gehwolf et al. [68] showed also an increase in the expression of Il-1β decoy receptor Il1r2 that works as a decoy receptor for IL-1 in both a membrane-bound and soluble form.
- Increased expression of collagen type I, collagen type VI, and scleraxis; in a recent study,

Marmotti et al. [69] observed the anabolic effect of PEMFs (21 days prolonged exposure) on tenocyte from hamstring graft harvested during anterior cruciate ligament reconstruction; they showed the anabolic and tenogenic effect of PEMFs on healthy tendon cells as well as the increased expression of collagen type VI chat has a role as mechanoceptors interacting with the extracellular matrix molecules; similar effects were described by Marmotti et al. [70] after prolonged exposure of umbilical cord-derived mesenchymal stem cells to PEMFs, suggesting a possible effect of PEMFs on mesenchymal precursor cells, according to the role of MSCs as therapeutic target of currently applied biophysical treatments described by Viganò et al. [71]; interestingly, Liu et al. [72] showed, in a human rotator cuff tenocytes monolayer culture, a similar anabolic effect of PEMFs but only in presence of IL-1β stimulation; indeed they demonstrated that, under standard culture condition, PEMF did not significantly change the expression of tenocyte growth and differentiation genes, but, under inflammatory conditions (10 ng/ml of IL-1), PEMF significantly increased the gene expression of collagen I TGFb-1, PDGFb, BMP12, and TIMP4, suggesting the potential of PEMFs' anabolic exposure in an inflammatory setting as the postoperative state after rotator cuff repair.

30.8 Future Developments

The present knowledge of PEMF effects on tendon cells is still incomplete. The need for preclinical and clinical studies focusing on the different sources of tendons, i.e., rotator cuff vs Achilles tendon vs patellar tendon, will be crucial to determine the field of possible future widespread clinical application of PEMFs for treating tendon disorders in light of the present results that, albeit sometimes conflicting, are surely promising. The well-known early role of PEMFs toward tendon anabolism is also a key point that needs further research; recent studies have related PEMF exposure to an increased c-fos expression, and this concept may represent the early anabolic stimulus triggered by PEMFs that may be directed toward multiple differentiation pathways depending on the cell-niche in which the target cell is located. Thus, the well-known early action of PEMFs toward the osteogenic, chondrogenic, and tenogenic cell lines may be explained by the activation of early differentiation genes as c-fos that can "translate" the PEMFs differentiation potential toward multiple cell lines depending on the "context" in which the cells reside.

Furthermore, following the recent discovery of the mechanosensitive ion channels PIEZO1 and PIEZO2 receptors by the 2021 Nobel Prize winner Ardem Patapoutian, and the very recent study of Matsushima et al. [73] describing the central role of PIEZO1 in sensing mechanical stimuli in tenocytes, what if PEMFs effect is mediated primarily by the interaction with these specific cell mechano-sensors that would initiate the conversion of mechanical forces into biological signals? This is certainly one of the future frontiers of in vitro and preclinical studies that will allow for understanding the mechanism and the potential of PEMFs for treating tendon disorders.

References

1. Cadossi R, Massari L, Racine-Avila J, Aaron RK. Pulsed electromagnetic field stimulation of bone healing and joint preservation: cellular mechanisms of skeletal response. J Am Acad Orthop Surg Glob Res Rev. 2020;4(5):e1900155. https://doi.org/10.5435/JAAOSGlobal-D-19-00155. PMID: 33970582; PMCID: PMC7434032.
2. Peng L, Fu C, Xiong F, Zhang Q, Liang Z, Chen L, He C, Wei Q. Effectiveness of pulsed electromagnetic fields on bone healing: a systematic review and meta-analysis of randomized controlled trials. Bioelectromagnetics. 2020;41(5):323–37. https://doi.org/10.1002/bem.22271. Epub 2020 Jun 3. PMID: 32495506.
3. Gobbi A, Lad D, Petrera M, Karnatzikos G. Symptomatic early osteoarthritis of the knee treated with pulsed electromagnetic fields: two-year follow-up. Cartilage. 2014;5(2):78–85.
4. Varani K, Vincenzi F, Pasquini S, Blo I, Salati S, Cadossi M, De Mattei M. Pulsed electromagnetic field stimulation in osteogenesis and chondrogenesis:

signaling pathways and therapeutic implications. Int J Mol Sci. 2021;22(2):809. https://doi.org/10.3390/ijms22020809. PMID: 33467447; PMCID: PMC7830993.
5. Brighton CT, Wang W, Seldes R, Zhang G, Pollack SR. Signal transduction in electrically stimulated bone cells. J Bone Joint Surg Am. 2001;83(10):1514–23. https://doi.org/10.2106/00004623-200110000-00009. PMID: 11679602.
6. Mansourian M, Shanei A. Evaluation of pulsed electromagnetic field effects: a systematic review and meta-analysis on highlights of two decades of research in vitro studies. Biomed Res Int. 2021;2021:6647497. https://doi.org/10.1155/2021/6647497. PMID: 34368353; PMCID: PMC8342182.
7. Esposito M, Lucariello A, Riccio I, Riccio V, Esposito V, Riccardi G. Differentiation of human osteoprogenitor cells increases after treatment with pulsed electromagnetic fields. In Vivo. 2012;26(2):299–304. PMID: 22351673.
8. Poh PSP, Seeliger C, Unger M, Falldorf K, Balmayor ER, van Griensven M. Osteogenic effect and cell signaling activation of extremely low-frequency pulsed electromagnetic fields in adipose-derived mesenchymal stromal cells. Stem Cells Int. 2018;2018:5402853. https://doi.org/10.1155/2018/5402853. PMID: 30123287; PMCID: PMC6079332.
9. Heydari Asl S, Hosseinpoor H, Parivar K, Hayati Roodbari N, Hanaee-Ahvaz H. Physical stimulation and scaffold composition efficiently support osteogenic differentiation of mesenchymal stem cells. Tissue Cell. 2018;50:1–7. https://doi.org/10.1016/j.tice.2017.11.001. Epub 2017 Nov 22. PMID: 29429509.
10. Ferroni L, Gardin C, Dolkart O, Salai M, Barak S, Piattelli A, Amir-Barak H, Zavan B. Pulsed electromagnetic fields increase osteogenetic commitment of MSCs via the mTOR pathway in TNF-α mediated inflammatory conditions: an in-vitro study. Sci Rep. 2018;8(1):5108. https://doi.org/10.1038/s41598-018-23499-9. PMID: 29572540; PMCID: PMC5865106.
11. Gao S, Chen B, Zhu Z, Du C, Zou J, Yang Y, Huang W, Liao J. PI3K-Akt signaling regulates BMP2-induced osteogenic differentiation of mesenchymal stem cells (MSCs): a transcriptomic landscape analysis. Stem Cell Res. 2023;66:103010. https://doi.org/10.1016/j.scr.2022.103010. Epub 2022 Dec 23. PMID: 36580886.
12. Vimalraj S. Alkaline phosphatase: structure, expression and its function in bone mineralization. Gene. 2020;754:144855. https://doi.org/10.1016/j.gene.2020.144855. Epub 2020 Jun 6. PMID: 32522695.
13. Ongaro A, Pellati A, Bagheri L, Fortini C, Setti S, De Mattei M. Pulsed electromagnetic fields stimulate osteogenic differentiation in human bone marrow and adipose tissue derived mesenchymal stem cells. Bioelectromagnetics. 2014;35(6):426–36. https://doi.org/10.1002/bem.21862. Epub 2014 Aug 6. PMID: 25099126.
14. Martini F, Pellati A, Mazzoni E, Salati S, Caruso G, Contartese D, De Mattei M. Bone morphogenetic protein-2 signaling in the osteogenic differentiation of human bone marrow mesenchymal stem cells induced by pulsed electromagnetic fields. Int J Mol Sci. 2020;21(6):2104. https://doi.org/10.3390/ijms21062104. PMID: 32204349; PMCID: PMC7139765.
15. Ceccarelli G, Bloise N, Mantelli M, Gastaldi G, Fassina L, De Angelis MG, Ferrari D, Imbriani M, Visai L. A comparative analysis of the in vitro effects of pulsed electromagnetic field treatment on osteogenic differentiation of two different mesenchymal cell lineages. Biores Open Access. 2013;2(4):283–94. https://doi.org/10.1089/biores.2013.0016. Erratum in: Biores Open Access. 2013 Oct;2(5):397. PMID: 23914335; PMCID: PMC3731679.
16. Petecchia L, Sbrana F, Utzeri R, Vercellino M, Usai C, Visai L, Vassalli M, Gavazzo P. Electro-magnetic field promotes osteogenic differentiation of BM-hMSCs through a selective action on Ca(2+)-related mechanisms. Sci Rep. 2015;5:13856. https://doi.org/10.1038/srep13856. PMID: 26364969; PMCID: PMC4568470.
17. De Mattei M, Grassilli S, Pellati A, Brugnoli F, De Marchi E, Contartese D, Bertagnolo V. Pulsed electromagnetic fields modulate miRNAs during osteogenic differentiation of bone mesenchymal stem cells: a possible role in the osteogenic-angiogenic coupling. Stem Cell Rev Rep. 2020;16(5):1005–12. https://doi.org/10.1007/s12015-020-10009-6. PMID: 32681233.
18. Bloise N, Petecchia L, Ceccarelli G, Fassina L, Usai C, Bertoglio F, Balli M, Vassalli M, Cusella De Angelis MG, Gavazzo P, Imbriani M, Visai L. The effect of pulsed electromagnetic field exposure on osteoinduction of human mesenchymal stem cells cultured on nano-TiO2 surfaces. PLoS One. 2018;13(6):e0199046. https://doi.org/10.1371/journal.pone.0199046. PMID: 29902240; PMCID: PMC6002089.
19. Wu S, Yu Q, Lai A, Tian J. Pulsed electromagnetic field induces Ca2+-dependent osteoblastogenesis in C3H10T1/2 mesenchymal cells through the Wnt-Ca2+/Wnt-β-catenin signaling pathway. Biochem Biophys Res Commun. 2018;503(2):715–21. https://doi.org/10.1016/j.bbrc.2018.06.066. Epub 2018 Jun 18. PMID: 29909008.
20. Tschon M, Veronesi F, Contartese D, Sartori M, Martini L, Vincenzi F, Ravani A, Varani K, Fini M. Effects of pulsed electromagnetic fields and platelet rich plasma in preventing osteoclastogenesis in an in vitro model of osteolysis. J Cell Physiol. 2018;233(3):2645–56. https://doi.org/10.1002/jcp.26143. Epub 2017 Sep 4. PMID: 28786478.
21. Veronesi F, Fini M, Sartori M, Parrilli A, Martini L, Tschon M. Pulsed electromagnetic fields and platelet rich plasma alone and combined for the treatment of wear-mediated periprosthetic osteolysis: an in vivo study. Acta Biomater. 2018;77:106–15. https://doi.

org/10.1016/j.actbio.2018.07.012. Epub 2018 Jul 6. PMID: 29981946.

22. Iwasa K, Reddi AH. Pulsed electromagnetic fields and tissue engineering of the joints. Tissue Eng Part B Rev. 2018;24(2):144–54. https://doi.org/10.1089/ten.TEB.2017.0294. Epub 2017 Nov 17. PMID: 29020880; PMCID: PMC5905856.
23. Ongaro A, Pellati A, Setti S, Masieri FF, Aquila G, Fini M, Caruso A, De Mattei M. Electromagnetic fields counteract IL-1β activity during chondrogenesis of bovine mesenchymal stem cells. J Tissue Eng Regen Med. 2015;9(12):E229–38. https://doi.org/10.1002/term.1671. Epub 2012 Dec 17. PMID: 23255506.
24. Sakhrani N, Stefani RM, Setti S, Cadossi R, Ateshian GA, Hung CT. Pulsed electromagnetic field therapy and direct current electric field modulation promote the migration of fibroblast-like synoviocytes to accelerate cartilage repair in vitro. Appl Sci (Basel). 2022;12(23):12406. https://doi.org/10.3390/app122312406. Epub 2022 Dec 4. PMID: 36970107; PMCID: PMC10035757.
25. Stefani RM, Barbosa S, Tan AR, Setti S, Stoker AM, Ateshian GA, Cadossi R, Vunjak-Novakovic G, Aaron RK, Cook JL, Bulinski JC, Hung CT. Pulsed electromagnetic fields promote repair of focal articular cartilage defects with engineered osteochondral constructs. Biotechnol Bioeng. 2020;117(5):1584–96. https://doi.org/10.1002/bit.27287. Epub 2020 Feb 5. PMID: 31985051; PMCID: PMC8845061.
26. Veronesi F, Cadossi M, Giavaresi G, Martini L, Setti S, Buda R, Giannini S, Fini M. Pulsed electromagnetic fields combined with a collagenous scaffold and bone marrow concentrate enhance osteochondral regeneration: an in vivo study. BMC Musculoskelet Disord. 2015;16:233. https://doi.org/10.1186/s12891-015-0683-2. PMID: 26328626; PMCID: PMC4557597.
27. Collarile M, Sambri A, Lullini G, Cadossi M, Zorzi C. Biophysical stimulation improves clinical results of matrix-assisted autologous chondrocyte implantation in the treatment of chondral lesions of the knee. Knee Surg Sports Traumatol Arthrosc. 2018;26(4):1223–9. https://doi.org/10.1007/s00167-017-4605-8. Epub 2017 Jun 17. PMID: 28624853.
28. Cadossi M, Buda RE, Ramponi L, Sambri A, Natali S, Giannini S. Bone marrow-derived cells and biophysical stimulation for talar osteochondral lesions: a randomized controlled study. Foot Ankle Int. 2014;35(10):981–7. https://doi.org/10.1177/1071100714539660. Epub 2014 Jun 10. PMID: 24917648.
29. Zorzi C, Dall'Oca C, Cadossi R, Setti S. Effects of pulsed electromagnetic fields on patients' recovery after arthroscopic surgery: prospective, randomized and double-blind study. Knee Surg Sports Traumatol Arthrosc. 2007;15(7):830–4. https://doi.org/10.1007/s00167-007-0298-8. Epub 2007 Feb 28. PMID: 17333120.
30. Varani K, Vincenzi F, Ravani A, Pasquini S, Merighi S, Gessi S, Setti S, Cadossi M, Borea PA, Cadossi R. Adenosine receptors as a biological pathway for the anti-inflammatory and beneficial effects of low frequency low energy pulsed electromagnetic fields. Mediat Inflamm. 2017;2017:2740963. https://doi.org/10.1155/2017/2740963. Epub 2017 Feb 1. PMID: 28255202; PMCID: PMC5309410.
31. Fini M, Pagani S, Giavaresi G, De Mattei M, Ongaro A, Varani K, Vincenzi F, Massari L, Cadossi M. Functional tissue engineering in articular cartilage repair: is there a role for electromagnetic biophysical stimulation? Tissue Eng Part B Rev. 2013;19(4):353–67. https://doi.org/10.1089/ten.TEB.2012.0501. Epub 2013 Mar 19. PMID: 23339690.
32. Frontera WR, Ochala J. Skeletal muscle: a brief review of structure and function. Calcif Tissue Int. 2015;96(3):183–95. https://doi.org/10.1007/s00223-014-9915-y. Epub 2014 Oct 8. PMID: 25294644.
33. Dumont NA, Bentzinger CF, Sincennes MC, Rudnicki MA. Satellite cells and skeletal muscle regeneration. Compr Physiol. 2015;5(3):1027–59. https://doi.org/10.1002/cphy.c140068. PMID: 26140708.
34. Sicari BM, Dearth CL, Badylak SF. Tissue engineering and regenerative medicine approaches to enhance the functional response to skeletal muscle injury. Anat Rec (Hoboken). 2014;297(1):51–64. https://doi.org/10.1002/ar.22794. Epub 2013 Dec 2. PMID: 24293290.
35. Di Rocco G, Iachininoto MG, Tritarelli A, Straino S, Zacheo A, Germani A, Crea F, Capogrossi MC. Myogenic potential of adipose-tissue-derived cells. J Cell Sci. 2006;119(Pt 14):2945–52. https://doi.org/10.1242/jcs.03029. PMID: 16825428.
36. Pinheiro CH, de Queiroz JC, Guimarães-Ferreira L, Vitzel KF, Nachbar RT, de Sousa LG, de Souza AL Jr, Nunes MT, Curi R. Local injections of adipose-derived mesenchymal stem cells modulate inflammation and increase angiogenesis ameliorating the dystrophic phenotype in dystrophin-deficient skeletal muscle. Stem Cell Rev Rep. 2012;8(2):363–74. https://doi.org/10.1007/s12015-011-9304-0. PMID: 21874281.
37. Kang Y, Park C, Kim D, Seong CM, Kwon K, Choi C. Unsorted human adipose tissue-derived stem cells promote angiogenesis and myogenesis in murine ischemic hindlimb model. Microvasc Res. 2010;80(3):310–6. https://doi.org/10.1016/j.mvr.2010.05.006. Epub 2010 May 25. PMID: 20510252.
38. Terrie L, Burattini M, Van Vlierberghe S, Fassina L, Thorrez L. Enhancing myoblast fusion and myotube diameter in human 3D skeletal muscle constructs by electromagnetic stimulation. Front Bioeng Biotechnol. 2022;10:892287. https://doi.org/10.3389/fbioe.2022.892287. PMID: 35814025; PMCID: PMC9256958.
39. Parate D, Kadir ND, Celik C, Lee EH, Hui JHP, Franco-Obregón A, Yang Z. Pulsed electromagnetic fields potentiate the paracrine function of mesenchymal stem cells for cartilage regeneration. Stem Cell

Res Ther. 2020;11(1):46. https://doi.org/10.1186/s13287-020-1566-5. PMID: 32014064; PMCID: PMC6998094.
40. Maiullari S, Cicirelli A, Picerno A, Giannuzzi F, Gesualdo L, Notarnicola A, Sallustio F, Moretti B. Pulsed electromagnetic fields induce skeletal muscle cell repair by sustaining the expression of proteins involved in the response to cellular damage and oxidative stress. Int J Mol Sci. 2023;24(23):16631. https://doi.org/10.3390/ijms242316631. PMID: 38068954; PMCID: PMC10706358.
41. Ong MT, Man GC, Lau LC, He X, Qiu J, Wang Q, Chow MC, Choi BC, Yu M, Yung PS. Effect of pulsed electromagnetic field as an intervention for patients with quadriceps weakness after anterior cruciate ligament reconstruction: a double-blinded, randomized-controlled trial. Trials. 2022;23(1):771. https://doi.org/10.1186/s13063-022-06674-2. PMID: 36096886; PMCID: PMC9465849.
42. Jeon HS, Kang SY, Park JH, Lee HS. Effects of pulsed electromagnetic field therapy on delayed-onset muscle soreness in biceps brachii. Phys Ther Sport. 2015;16(1):34–9. https://doi.org/10.1016/j.ptsp.2014.02.006. Epub 2014 Mar 7. PMID: 24906295.
43. Trofè A, Piras A, Muehsam D, Meoni A, Campa F, Toselli S, Raffi M. Effect of pulsed electromagnetic fields (PEMFs) on muscular activation during cycling: a single-blind controlled pilot study. Healthcare (Basel). 2023;11(6):922. https://doi.org/10.3390/healthcare11060922. PMID: 36981580; PMCID: PMC10048902.
44. Binder A, Parr G, Hazleman B, Fitton-Jackson S. Pulsed electromagnetic field therapy of persistent rotator cuff tendinitis. A double-blind controlled assessment. Lancet. 1984;1(8379):695–8. https://doi.org/10.1016/s0140-6736(84)92219-0. PMID: 6143039.
45. Devereaux MD, Hazleman BL, Thomas PP. Chronic lateral humeral epicondylitis—a double-blind controlled assessment of pulsed electromagnetic field therapy. Clin Exp Rheumatol. 1985;3(4):333–6. PMID: 4085165.
46. Uzunca K, Birtane M, Taştekin N. Effectiveness of pulsed electromagnetic field therapy in lateral epicondylitis. Clin Rheumatol. 2007;26(1):69–74. https://doi.org/10.1007/s10067-006-0247-9. Epub 2006 Apr 22. PMID: 16633709.
47. Osti L, Buono AD, Maffulli N. Pulsed electromagnetic fields after rotator cuff repair: a randomized, controlled study. Orthopedics. 2015;38(3):e223–8. https://doi.org/10.3928/01477447-20150305-61. PMID: 25760511.
48. Klüter T, Krath A, Stukenberg M, Gollwitzer H, Harrasser N, Knobloch K, Maffulli N, Hausdorf J, Gerdesmeyer L. Electromagnetic transduction therapy and shockwave therapy in 86 patients with rotator cuff tendinopathy: a prospective randomized controlled trial. Electromagn Biol Med. 2018;37(4):175–83. https://doi.org/10.1080/15368378.2018.1499030. Epub 2018 Sep 5. PMID: 30183430.
49. Özdemir M, Yaşar MF, Yakşi E. Effect of pulsed electromagnetic field therapy in patients with supraspinatus tendon tear. Rev Assoc Med Bras (1992). 2021;67(2):282–6. https://doi.org/10.1590/1806-9282.67.02.20200730. PMID: 34406254.
50. Watkins JP, Auer JA, Morgan SJ, Gay S. Healing of surgically created defects in the equine superficial digital flexor tendon: effects of pulsing electromagnetic field therapy on collagen-type transformation and tissue morphologic reorganization. Am J Vet Res. 1985;46(10):2097–103. PMID: 4062013.
51. Lee EW, Maffulli N, Li CK, Chan KM. Pulsed magnetic and electromagnetic fields in experimental Achilles tendonitis in the rat: a prospective randomized study. Arch Phys Med Rehabil. 1997;78(4):399–404. https://doi.org/10.1016/s0003-9993(97)90232-x. PMID: 9111460.
52. Strauch B, Patel MK, Rosen DJ, Mahadevia S, Brindzei N, Pilla AA. Pulsed magnetic field therapy increases tensile strength in a rat Achilles' tendon repair model. J Hand Surg Am. 2006;31(7):1131–5. https://doi.org/10.1016/j.jhsa.2006.03.024. PMID: 16945715.
53. Tucker JJ, Cirone JM, Morris TR, Nuss CA, Huegel J, Waldorff EI, Zhang N, Ryaby JT, Soslowsky LJ. Pulsed electromagnetic field therapy improves tendon-to-bone healing in a rat rotator cuff repair model. J Orthop Res. 2017;35(4):902–9. https://doi.org/10.1002/jor.23333. Epub 2016 Jun 22. PMID: 27282093; PMCID: PMC5554861.
54. Huegel J, Choi DS, Nuss CA, Minnig MCC, Tucker JJ, Kuntz AF, Waldorff EI, Zhang N, Ryaby JT, Soslowsky LJ. Effects of pulsed electromagnetic field therapy at different frequencies and durations on rotator cuff tendon-to-bone healing in a rat model. J Shoulder Elb Surg. 2018;27(3):553–60. https://doi.org/10.1016/j.jse.2017.09.024. Epub 2017 Nov 22. PMID: 29174271; PMCID: PMC5835831.
55. Dolkart O, Kazum E, Rosenthal Y, Sher O, Morag G, Yakobson E, Chechik O, Maman E. Effects of focused continuous pulsed electromagnetic field therapy on early tendon-to-bone healing. Bone Joint Res. 2021;10(5):298–306. https://doi.org/10.1302/2046-3758.105.BJR-2020-0253.R2. PMID: 33934605; PMCID: PMC8160030.
56. Huegel J, Boorman-Padgett JF, Nuss CA, Raja HA, Chan PY, Kuntz AF, Waldorff EI, Zhang N, Ryaby JT, Soslowsky LJ. Effects of pulsed electromagnetic field therapy on rat Achilles tendon healing. J Orthop Res. 2020;38(1):70–81. https://doi.org/10.1002/jor.24487. Epub 2019 Oct 15. PMID: 31595543; PMCID: PMC6917903.
57. Huegel J, Chan PYW, Weiss SN, Nuss CA, Raja H, Waldorff EI, Zhang N, Ryaby JT, Soslowsky LJ, Kuntz AF. Pulsed electromagnetic field therapy alters early healing in a rat model of rotator cuff injury and repair: potential mechanisms. J Orthop Res. 2022;40(7):1593–603. https://doi.org/10.1002/jor.25185. Epub 2021 Oct 17. PMID: 34657312; PMCID: PMC9013393.

58. Perucca Orfei C, Lovati AB, Lugano G, Viganò M, Bottagisio M, D'Arrigo D, Sansone V, Setti S, de Girolamo L. Pulsed electromagnetic fields improve the healing process of Achilles tendinopathy: a pilot study in a rat model. Bone Joint Res. 2020;9(9):613–22. https://doi.org/10.1302/2046-3758.99.BJR-2020-0113.R1. PMID: 33072305; PMCID: PMC7533373.
59. Uzun C, Erdal N, Gürgül S, Kalaycı D, Yılmaz ŞN, Özdemir AA, Yetkin D, Yılmaz C. Comparison of the effects of pulsed electromagnetic field and extracorporeal shockwave therapy in a rabbit model of experimentally induced Achilles tendon injury. Bioelectromagnetics. 2021;42(2):128–45. https://doi.org/10.1002/bem.22314. Epub 2020 Dec 25. PMID: 33368423.
60. Pieber K, Schuhfried O, Fialka-Moser V. Magnetfeldtherapie—Ergebnisse hinsichtlich evidence based medicine [Pulsed electromagnetic fields (PEMF)—results in evidence based medicine]. Wien Med Wochenschr. 2007;157(1–2):34–6. German. https://doi.org/10.1007/s10354-006-0369-3. PMID: 17471830.
61. Rosso F, Bonasia DE, Marmotti A, Cottino U, Rossi R. Mechanical stimulation (pulsed electromagnetic fields "PEMF" and extracorporeal shock wave therapy "ESWT") and tendon regeneration: a possible alternative. Front Aging Neurosci. 2015;7:211. https://doi.org/10.3389/fnagi.2015.00211. PMID: 26617513; PMCID: PMC4637423.
62. Colombini A, Perucca Orfei C, Vincenzi F, De Luca P, Ragni E, Viganò M, Setti S, Varani K, de Girolamo L. A_{2A} adenosine receptors are involved in the reparative response of tendon cells to pulsed electromagnetic fields. PLoS One. 2020;15(9):e0239807. https://doi.org/10.1371/journal.pone.0239807. PMID: 32998161; PMCID: PMC7527253.
63. de Girolamo L, Stanco D, Galliera E, Viganò M, Colombini A, Setti S, Vianello E, Corsi Romanelli MM, Sansone V. Low frequency pulsed electromagnetic field affects proliferation, tissue-specific gene expression, and cytokines release of human tendon cells. Cell Biochem Biophys. 2013;66(3):697–708. https://doi.org/10.1007/s12013-013-9514-y. PMID: 23345006.
64. de Girolamo L, Viganò M, Galliera E, Stanco D, Setti S, Marazzi MG, Thiebat G, Corsi Romanelli MM, Sansone V. In vitro functional response of human tendon cells to different dosages of low-frequency pulsed electromagnetic field. Knee Surg Sports Traumatol Arthrosc. 2015;23(11):3443–53. https://doi.org/10.1007/s00167-014-3143-x. Epub 2014 Jun 24. PMID: 24957914.
65. Vinhas A, Almeida AF, Rodrigues MT, Gomes ME. Prospects of magnetically based approaches addressing inflammation in tendon tissues. Adv Drug Deliv Rev. 2023;196:114815. https://doi.org/10.1016/j.addr.2023.114815. Epub 2023 Mar 30. PMID: 37001644.
66. Vinhas A, Gonçalves AI, Rodrigues MT, Gomes ME. Human tendon-derived cell sheets created by magnetic force-based tissue engineering hold tenogenic and immunomodulatory potential. Acta Biomater. 2021;131:236–47. https://doi.org/10.1016/j.actbio.2021.06.036. Epub 2021 Jun 27. PMID: 34192569.
67. Vinhas A, Rodrigues MT, Gonçalves AI, Reis RL, Gomes ME. Pulsed electromagnetic field modulates tendon cells response in IL-1β-conditioned environment. J Orthop Res. 2020;38(1):160–72. https://doi.org/10.1002/jor.24538. Epub 2019 Dec 10. PMID: 31769535.
68. Gehwolf R, Schwemberger B, Jessen M, Korntner S, Wagner A, Lehner C, Weissenbacher N, Tempfer H, Traweger A. Global responses of Il-1β-primed 3D tendon constructs to treatment with pulsed electromagnetic fields. Cells. 2019;8(5):399. https://doi.org/10.3390/cells8050399. PMID: 31052237; PMCID: PMC6562657.
69. Marmotti A, Mattia S, Mangiavini L, Bellato E, Ragni E, de Girolamo L, Peretti GM, Blonna D, Bonasia D, Setti S, Castoldi F. Hamstring grafts are tenogenic constructs for ACL reconstruction and Pulsed Electromagnetic Fields improve tendon specific markers expression. An in-vitro study. J Biol Regul Homeost Agents. 2020;34(4 Suppl. 3):363–76. Congress of the Italian Orthopaedic Research Society.
70. Marmotti A, Peretti GM, Mattia S, Mangiavini L, de Girolamo L, Viganò M, Setti S, Bonasia DE, Blonna D, Bellato E, Ferrero G, Castoldi F. Corrigendum to "pulsed electromagnetic fields improve tenogenic commitment of umbilical cord-derived mesenchymal stem cells: a potential strategy for tendon repair-an in vitro study". Stem Cells Int. 2019;2019:9761573. https://doi.org/10.1155/2019/9761573. Erratum for: Stem Cells Int. 2018 Jul 30;2018:9048237. https://doi.org/10.1155/2018/9048237. PMID: 31191691; PMCID: PMC6525792.
71. Viganò M, Sansone V, d'Agostino MC, Romeo P, Perucca Orfei C, de Girolamo L. Mesenchymal stem cells as therapeutic target of biophysical stimulation for the treatment of musculoskeletal disorders. J Orthop Surg Res. 2016;11(1):163. https://doi.org/10.1186/s13018-016-0496-5. PMID: 27986082; PMCID: PMC5162101.
72. Liu M, Lee C, Laron D, Zhang N, Waldorff EI, Ryaby JT, Feeley B, Liu X. Role of pulsed electromagnetic fields (PEMF) on tenocytes and myoblasts-potential application for treating rotator cuff tears. J Orthop Res. 2017;35(5):956–64. https://doi.org/10.1002/jor.23278. Epub 2017 Apr 7. PMID: 27138553.
73. Matsushima T, Hiroshi A. Molecular mechanisms of mechanosensing and plasticity of tendons and ligaments. J Biochem. 2024:mvae039. https://doi.org/10.1093/jb/mvae039. Epub ahead of print. PMID: 38729213.

Biological Products and Biomaterials for Tendon Regeneration

31

Giuseppe M. Peretti and Nicolò Rossi

31.1 Introduction

Over the last decades, tissue engineering has gained popularity as an emerging discipline that combines the use of biomaterials, bioengineering, and cell biology to repair or regenerate biological tissues [1]. Tissue engineering techniques involve the application of materials used to create scaffolds that can be combined with cells (generally stem cells) and biologically active molecules to heal, regenerate, or replace parts or whole-body tissues. This discipline is challenging due to the need to create a complex structure whose functionality, biomechanical, and structural behavior resembles more closely the biological structures they replace or heal [2]. The principal characteristics of suitable biomaterials are hydrophilicity, wettability, biocompatibility, biodegradability, less cytotoxicity, and good mechanical strength. Moreover, among parameters that should be measured for a scaffold include porosity, pore size, degradation rates, and elasticity [3].

The main biomaterials used for tendon/ligament tissue engineering include collagen, silk, alginate, chitosan, polycaprolactone, polyglycolic acid, and poly-lactic acid [4].

In the context of preclinical research, the use of scaffolds for tendon regeneration in animal models plays a crucial role in evaluating the efficacy and safety of these technologies before moving on to clinical studies with human patients. Animal models provide a controlled experimental system that allows scientists to closely study the mechanisms of tissue healing and regeneration. Studies conducted on animal models, such as mice, rats, and pigs, have shown that tendon scaffolds can promote the regeneration of damaged tendons and improve joint function. These studies allow the evaluation of different characteristics of scaffolds, such as biocompatibility, degradability, and mechanical properties, in order to optimize their design for clinical applications. During preclinical studies, scaffolds are implanted into the tendons of animals to evaluate their effectiveness in promoting tissue regenera-

G. M. Peretti (✉)
IRCCS Ospedale Galeazzi Sant'Ambrogio, Milan, Italy

Department of Biomedical Sciences for Health, University of Milan, Milan, Italy
e-mail: giuseppe.peretti@unimi.it

N. Rossi
IRCCS Ospedale Galeazzi Sant'Ambrogio, Milan, MI, Italy

A. Gobbi et al. (eds.), *Regenerative Medicine in Sports and Orthopaedics*,
https://doi.org/10.1007/978-3-031-84693-9_31

tion. Researchers analyze the inflammatory response, extracellular matrix deposition, and integration of the scaffold with the surrounding tissue to evaluate the success of tendon regeneration. Information from animal model studies is critical to guide the development and optimization of tendon scaffolds for clinical use. These studies provide valuable scientific evidence on the safety and efficacy of scaffolds, allowing researchers to progress toward clinical trials with greater confidence in expected results. In conclusion, the use of scaffolds for tendon regeneration in animal models is a critical step in the process of developing new therapies for tendon injuries. Preclinical animal research provides a solid scientific basis to demonstrate the effectiveness of these technologies, paving the way for new treatment opportunities for patients suffering from tendon injuries.

In the clinical setting, scaffolds are most of the time three-dimensional structures designed to provide mechanical support and promote cell growth and regeneration of damaged tissue. In the context of tissue engineering, scaffolds are fundamental as they can mimic the characteristics of native tissue, thus facilitating the healing process. Essentially, scaffolds must possess adequate mechanical properties to withstand physical stresses at the injury site. Materials commonly used for tendon scaffolds include biopolymers such as polylactic acid (PLA), polyglycolic acid (PGA), and polyurethane. In the clinical context, scaffolds are surgically implanted at the site of tendon injury to provide structural support and promote tissue regeneration. These scaffolds can be formed into different geometric patterns and structures to suit specific patient and lesion needs.

Clinical studies have shown that using scaffolds for tendon regeneration can improve long-term outcomes compared to traditional treatment methods. Furthermore, the use of scaffolds combined with other therapies, such as stem cells, has shown promising results in tendon regeneration and restoration of joint function [5].

31.2 Current Application of Tendon Tissue Engineering

The biocompatibility of a scaffold is crucial for supporting cellular activity by providing an ideal physical environment for cell growth, proliferation, and neovascularization. Factors like mechanical strength, elasticity, and biodegradation influence the scaffold's biostability. Matching a scaffold's mechanical properties to the native extracellular matrix (ECM) helps preventing mechanical failure and supports tissue growth. Controlling scaffold biodegradation and new ECM formation is essential to maintain mechanical strength and architecture. Ideal scaffolds act as templates, offering biochemical cues for cellular functions mimicking the native ECM [6].

Various biomaterials like collagen, silk, alginate, and polylactic acid have been used to create scaffolds for tissue engineering through techniques like electrospinning, knitting, and freeze-drying. Current tissue-engineered tendon/ligament substitutes show promising short-term outcomes but face challenges in maintaining long-term efficiency. Several studies explore promising scaffold biomaterials for tendon and ligament repair in animal models, with the ACL being a common focus due to sports-related injuries.

Research demonstrates successful regeneration of ligaments using scaffold-MSC composites in animal models like rabbits and pigs [7]. Studies using materials like polycaprolactone (PCL) show varying outcomes in terms of mechanical strength and biocompatibility. Incorporating growth factors like basic fibroblast growth factor (bFGF) can enhance scaffold performance [8].

Decellularized porcine tendons re-cellularized with human MSCs in a bioreactor setting showed promising results in enhancing ACL mechanical strength [7]. Various scaffold types such as PGA/PLA and chitosan-based scaffolds have been studied for Achilles tendon regeneration in animal models, emphasizing the importance of incorporating stem cells for successful tissue maturation and regeneration [9].

Overall, stem cells play a crucial role in enhancing the success of engineered tendon/ligament grafts post-transplantation. Studies highlight the significance of scaffold material selection, mechanical properties, and cellular components in optimizing scaffold performance for tissue regeneration.

31.3 Types of Scaffolds

Scaffolds in tissue engineering are classified as biologic, synthetic, or composite (Table 31.1) based on the materials used [1]. *Biologic scaffolds*, like decellularized tendon matrices and natural polymers, offer biomimetic properties conducive to faster cell response and mimicry of native tissue composition. *Synthetic scaffolds*, often polyester derivatives, allow precise customization of architecture, mechanical properties, and degradation characteristics, although they may evoke host responses. *Composite scaffolds*, combining natural and synthetic materials, are increasingly favored for enhanced structural complexity resembling natural tendons. Combinations like PCL with chitosan/hyaluronic acid, PCL with collagen I, and PLGA with fibrin are among the studied formulations. The versatility of composite scaffolds lies in leveraging the advantageous properties of both materials to create tailored structures for improved tissue regeneration [3].

Table 31.1 Types of scaffold biomaterials in tendon/ligament tissue engineering

Source	Biomaterial	Advantages
Biologic	Collagen	Slow degradation rate, main component of native ECM, good mechanical strength
	Silk	Slow degradation rate, good mechanical strength
	Aliginate	Biocompatible and with hydrophilic proprieties
Synthetic	PCL	Easy to process
	PLGA	Low cost and mass production
	PLLA	High mechanical strength
Synthetic + biologic	PCL/CHT	High mechanical strength
	PLC/Collagen	Combination of the proprieties of natural and synthetic scaffolds
	PCL/CHT/HA	Combination of the proprieties of natural and synthetic scaffolds
	PLLA/CHT-collagen/aliginate	Combination of the proprieties of natural and synthetic scaffolds

PCL polycaprolactone, *PLGA* poly (lactic-co-glycolic acid), *CHT* chitosan, *PLCL* poly (L-lactide-co-e-caprolactone), *PLLA* poly-L-lactic acid, *HA* hydroxyapatite

31.4 Promising Techniques for Scaffolds Development in Tendon Tissue Engineering

31.4.1 3D Printing

3D printing enables the creation of detailed three-dimensional structures layer by layer, with materials referred to as inks in this process. In tissue engineering, two approaches are commonly used: traditional 3D printing for scaffold fabrication followed by cell and growth factor integration, and the more advanced 3D bioprinting where bioinks containing cells and growth factors are utilized [10]. Bioprinting offers highly customizable scaffolds that closely mimic natural tissue characteristics, allowing precise control over structure geometry, pore size, and material composition. The benefits of 3D bioprinting include automation, material versatility, incorporation of proteins and growth factors, and reproducibility in fabricating complex tissue structures. Notably, this technology shows potential for tendon tissue engineering by enabling the use of diverse materials, promoting cell integration, and adjusting mechanical properties to reflect those of tendons. However, challenges exist, such as time-consuming processes and potential cell damage during printing. Vascular network formation remains a critical hurdle for tissue survival, but advancements in the field aim to address this limitation [11]. The process of 3D bioprinting involves pre-processing, processing, and post-processing stages, crucial for designing, printing, and maturing functional tissue constructs. Various bioprinting techniques, including extrusion-based, inkjet-based, and laser-assisted, offer distinctive approaches to scaffold fabrication [10]. While 3D bioprinting for tendon tissue engineering remains a budding area with complexities to overcome, researchers are making encouraging progress by experimenting with diverse materials, optimizing printing parameters, and refining printing methods for promising outcomes in this field.

31.4.2 Amniotic Membrane

Recent advancements in tissue engineering have focused on utilizing human amniotic membrane (AM) for its regenerative potential, serving as a scaffold due to its unique properties. AM, sourced from the inner layer of the placenta adjacent to the fetus, contains growth factors beneficial for tissue engineering applications without ethical concerns. Decellularization is a crucial step in preparing AM scaffolds, enhancing biocompatibility for recellularization with target cells like tenocytes for tendon engineering. Decellularized AM retains essential ECM proteins and growth factors that support cellular functions, adhesion, and proliferation, making it a valuable scaffold material [12].

AM exhibits antimicrobial, anti-inflammatory, and anti-fibrotic properties, making it suitable for tissue engineering applications. Its viscoelastic characteristics are attributed to collagen and elastin content, influencing mechanical strength and elasticity. AM is used in various fields, such as ophthalmology for ocular surface reconstruction and nerve conduits in peripheral nerve regeneration, showcasing its versatility and potential in different regenerative medicine applications [13]. The utilization of amniotic membrane (AM) in tendon/ligament tissue engineering has shown promising outcomes in enhancing mechanical strength and reducing inflammation. AM, when combined with other biomaterials, provides structural support capable of withstanding physiological loads at a damaged site [14]. Studies have demonstrated that AM wrapping around injured tendons can enhance collagen fiber alignment, prevent peritendinous adhesion, and improve tendon healing. However, mixed results have been observed in clinical studies regarding the efficacy of AM in tendon repair.

Research by Seo and colleagues showcased the effectiveness of combining AM with silk scaffolds in Achilles tendon regeneration, leading to improved tissue organization, increased neo-angiogenesis, and enhanced mechanical strength [14]. Another case study utilizing dehydrated

AM as a patch in ACL reconstruction reported positive outcomes with early vascularization and improved patient mobility. Additionally, a randomized controlled trial using micronized AM for plantar fasciitis treatment demonstrated pain relief and improved functionality, attributed to its anti-inflammatory and growth factor properties [15].

These findings highlight the potential of AM in tissue engineering applications, suggesting that combining AM with complementary scaffolding materials can optimize therapeutic outcomes in tendon/ligament injuries and other musculoskeletal conditions.

31.4.3 Electrospinning

Electrospinning is a widely studied technique for producing nanofibrous scaffolds in tendon/ligament tissue engineering. This method generates fine fibers ranging from 15 nm to 10 μm through electrostatic forces, creating a non-woven mesh with high surface area and interconnected porosity. The process involves subjecting a biomaterial solution in a syringe to a strong electric field, resulting in fibers of micrometers to nanometers in diameter deposited on a rotating collector [16]. Various parameters like viscosity, distance between needle tip and collector, voltage, flow rate, temperature, and humidity affect the characteristics of the electrospun fibers. Increased viscosity of the polymer solution leads to larger fiber diameter control. The distance between the needle and rotating collector can affect the electrospun fiber diameter, with a shorter collecting distance typically resulting in smaller fibers due to faster stretching in the electric field. However, some studies suggest conflicting results, with claims of increased distance leading to larger fiber diameters or no effect. Applied voltage influences fiber diameter; higher voltage strengthens the electrical field and stretching force, reducing fiber diameter, yet increased mass flow rate may create thicker fibers. Higher flow rates result in thicker fibers, impacting the scaffold's surface-to-volume ratio crucial for cell proliferation and adhesion. Various successful studies have fabricated aligned nanofibers through electrospinning, enhancing tenocyte proliferation and gene expression. Techniques like braiding and multilayered scaffold development have shown promising mechanical mimicry of native tendons/ligaments and enhanced cell adhesion and differentiation markers [17]. Recent studies using silk/collagen scaffolds and poly-(L-lactic acid)/collagen scaffolds have also shown potential in supporting stem cell attachment, proliferation, and favorable biocompatibility for tendon/ligament regeneration efforts.

31.5 Stem Cells for Tendon Tissue Engineering

Stem cells play a vital role in tendon and ligament tissue engineering, offering the potential for regenerative therapy through their self-renewal and multilineage differentiation capabilities. Various stem cell types, including induced pluripotent stem cells (iPSCs), embryonic stem cells (ESCs), and mesenchymal stem cells (MSCs), have shown promise in tendon repair, with MSCs being widely studied due to their accessibility and multilineage differentiation potential [18]. Studies have demonstrated that co-culture of different cell types can enhance tenogenic markers and tissue formation in tendon-like structures. Moreover, these cells have shown promising results if associated with pulsed electromagnetic fields [19].

ESCs and iPSCs, being pluripotent, offer unlimited proliferative capacity and differentiation potential but come with ethical concerns and potential teratoma risks. On the other hand, MSCs, derived from multiple sources like bone marrow and adipose tissue, show effectiveness in promoting tendon healing without teratoma risk. Manipulating the culture environment during MSC expansion is crucial to ensure safe and efficient therapy, with factors like medium supplements influencing cell differentiation potential. Studies in animal models and clinical settings have highlighted the therapeutic potential of MSCs in tendon repair, with positive outcomes observed in various tendon sites such as rotator cuff and superficial digital flexor tendon [20].

Mechanisms involving specific gene expressions like EGR1, BMP12, and Mkx have been identified to promote tenogenic differentiation of MSCs and improve tendon matrix synthesis. Augmentation of these factors prior to MSC administration has shown to enhance collagen synthesis and mechanical properties in tendon repair. The integration of stem cells in tissue engineering holds promise for advancing regenerative medicine strategies in treating tendon and ligament injuries effectively [21].

31.6 Unipotent Tenocytes and Ligamentocytes for Tendon Tissue Engineering

Previous discussions highlight tenocytes and ligamentocytes as key cells driving tendon and ligament repair through ECM synthesis and growth factor production. While using these cells for regeneration seems feasible, their limited proliferation and differentiation abilities pose challenges. Autologous transplantation raises concerns regarding donor site morbidity and inadequate healing post-biopsy. Studies demonstrated that it is possible to isolate TSPCs from very small fragments of tissue and that they show the typical features of MSCs and multi-differentiation potential, above all toward osteogenic and chondrogenic lineages [22]. Wang et al. on chronic lateral epicondylitis patients treated with autologous tenocytes showed significant long-term improvements in pain, function, grip strength, and MRI findings. Animal studies support the efficacy of autologous tenocyte therapy in enhancing tendon healing and collagen synthesis without adverse immune responses. Implementation of collagen scaffolds with tenocytes has also demonstrated improved tendon healing outcomes [23].

31.7 Cell-Secreted Products for Tendon Tissue Engineering

New research suggests that stem cells may benefit tendon/ligament repair not only through cellular regeneration but also via their paracrine activity, secreting growth factors, cytokines, and extracellular vesicles. Exosomes, microvesicles rich in proteins and genetic materials, have shown potential in tissue repair by modulating the environment and cell responses. Stem cell-derived exosomes exhibit promise in treating various diseases and have implications for tendon/ligament regeneration, although more studies are needed [24]. Additionally, secretomes/growth factors play a crucial role in cell signaling and differentiation relevant for tendon/ligament healing. Various growth factors like PDGF have been studied in scaffold treatments, showing mixed outcomes in mechanical strength and cellular responses, emphasizing the importance of optimal dosing [25]. The use of conditioned medium from different cell types has demonstrated improvements in tendon healing, enhancing tissue organization, cellular density, and collagen production, suggesting a promising avenue for future tissue engineering approaches.

31.8 Bioreactors in Tendon Tissue Engineering

The tendon and ligament structure comprises dense collagen fibrils arranged in an organized manner to withstand tension and compressive forces. Tenocytes and ligamentocytes synthesize ECM and collagen influenced by mechanical factors, with studies demonstrating increased collagen secretion in response to mechanical stretch [26]. Mimicking the native tissue environment, controlled bioreactors provide mechanical and biological cues for tissue maturation, essential for tendon/ligament engineering [27]. Bioreactors designed for ACL grafts and ligament tissue simulate physiological conditions accurately, promoting cell adherence, gene expression, and tendon healing [28]. Cyclic loading within bioreactors enhances graft maturation, cellular phenotype, and expression of tenogenic differentiation markers like SCX and ECM-related genes such as COL-1 and COL-III. Bioreactor-induced mechanical stimulation fosters the maturation of engineered ligaments by promoting ECM gene activation and tendon-specific elements critical for tendon healing and repair.

31.9 Future Perspectives

Advancements in technology and knowledge have led to the development of engineered tissues closely resembling native tissue, aided by bioreactors with mechanical stimulation for in vitro maturation of tendons and ligaments. These bioreactors offer insights into cellular behavior post-transplantation and facilitate the precise development of engineered tissues. Future possibilities include culturing cells within bioreactors to create tendon and ligament organoids with complex stimuli and multiple cell types for functional tissue formation without solely relying on scaffolds. Understanding the biological and mechanical factors influencing cell behavior in organoid production could overcome limitations in tendon and ligament tissue engineering.

References

1. Ruiz-Alonso S, Lafuente-Merchan M, Ciriza J, Saenz-Del-Burgo L, Pedraz JL. Tendon tissue engineering: cells, growth factors, scaffolds and production techniques. J Control Release. 2021;333:448–86.
2. Thomas D, Gaspar D, Sorushanova A, Milcovich G, Spanoudes K, Mullen AM, et al. Scaffold and scaffold-free self-assembled systems in regenerative medicine. Biotechnol Bioeng. 2016;113(6):1155–63.
3. Lim WL, Liau LL, Ng MH, Chowdhury SR, Law JX. Current progress in tendon and ligament tissue engineering. Tissue Eng Regen Med. 2019;16(6):549–71.
4. Leong NL, Kator JL, Clemens TL, James A, Enamoto-Iwamoto M, Jiang J. Tendon and ligament healing and current approaches to tendon and ligament regeneration. J Orthop Res. 2020;38(1):7–12.
5. Scotti C, Gobbi A, Nakamura N, Peretti GM. Stem cells for cartilage regeneration: a roadmap to the clinic. Stem Cells Int. 2018;2018:7348560.
6. Law JX, Liau LL, Saim A, Yang Y, Idrus R. Electrospun collagen nanofibers and their applications in skin tissue engineering. Tissue Eng Regen Med. 2017;14(6):699–718.
7. Fan H, Liu H, Toh SL, Goh JC. Anterior cruciate ligament regeneration using mesenchymal stem cells and silk scaffold in large animal model. Biomaterials. 2009;30(28):4967–77.
8. Leong NL, Kabir N, Arshi A, Nazemi A, Wu B, Petrigliano FA, et al. Evaluation of polycaprolactone scaffold with basic fibroblast growth factor and fibroblasts in an athymic rat model for anterior cruciate ligament reconstruction. Tissue Eng Part A. 2015;21(11–12):1859–68.
9. Lee KI, Lee JS, Kang KT, Shim YB, Kim YS, Jang JW, et al. In vitro and in vivo performance of tissue-engineered tendons for anterior cruciate ligament reconstruction. Am J Sports Med. 2018;46(7):1641–9.
10. Matai I, Kaur G, Seyedsalehi A, McClinton A, Laurencin CT. Progress in 3D bioprinting technology for tissue/organ regenerative engineering. Biomaterials. 2020;226:119536.
11. Souness A, Zamboni F, Walker GM, Collins MN. Influence of scaffold design on 3D printed cell constructs. J Biomed Mater Res B Appl Biomater. 2018;106(2):533–45.
12. Fairbairn NG, Randolph MA, Redmond RW. The clinical applications of human amnion in plastic surgery. J Plast Reconstr Aesthet Surg. 2014;67(5):662–75.
13. Henry FP, Goyal NA, David WS, Wes D, Bujold KE, Randolph MA, et al. Improving electrophysiologic and histologic outcomes by photochemically sealing amnion to the peripheral nerve repair site. Surgery. 2009;145(3):313–21.
14. Seo YK, Kim JH, Eo SR. Co-effect of silk and amniotic membrane for tendon repair. J Biomater Sci Polym Ed. 2016;27(12):1232–47.
15. Zelen CM, Poka A, Andrews J. Prospective, randomized, blinded, comparative study of injectable micronized dehydrated amniotic/chorionic membrane allograft for plantar fasciitis—a feasibility study. Foot Ankle Int. 2013;34(10):1332–9.
16. Jun I, Han HS, Edwards JR, Jeon H. Electrospun fibrous scaffolds for tissue engineering: viewpoints on architecture and fabrication. Int J Mol Sci. 2018;19(3):745.
17. Khorshidi S, Solouk A, Mirzadeh H, Mazinani S, Lagaron JM, Sharifi S, et al. A review of key challenges of electrospun scaffolds for tissue-engineering applications. J Tissue Eng Regen Med. 2016;10(9):715–38.
18. Stanco D, Viganò M, Perucca Orfei C, Di Giancamillo A, Peretti GM, Lanfranchi L, et al. Multidifferentiation potential of human mesenchymal stem cells from adipose tissue and hamstring tendons for musculoskeletal cell-based therapy. Regen Med. 2015;10(6):729–43.
19. Marmotti A, Peretti GM, Mattia S, Mangiavini L, de Girolamo L, Viganò M, et al. Pulsed electromagnetic fields improve tenogenic commitment of umbilical

cord-derived mesenchymal stem cells: a potential strategy for tendon repair-an in vitro study. Stem Cells Int. 2018;2018:9048237.
20. Xu W, Wang Y, Liu E, Sun Y, Luo Z, Xu Z, et al. Human iPSC-derived neural crest stem cells promote tendon repair in a rat patellar tendon window defect model. Tissue Eng Part A. 2013;19(21–22):2439–51.
21. Wu S, Wang Y, Streubel PN, Duan B. Living nanofiber yarn-based woven biotextiles for tendon tissue engineering using cell tri-culture and mechanical stimulation. Acta Biomater. 2017;62:102–15.
22. Stanco D, Viganò M, Orfei CP, Giancamillo ADI, Thiebat G, Peretti G, et al. In vitro characterization of stem/progenitor cells from semitendinosus and gracilis tendons as a possible new tool for cell-based therapy for tendon disorders. Joints. 2014;2(4):159–68.
23. Chen JM, Willers C, Xu J, Wang A, Zheng MH. Autologous tenocyte therapy using porcine-derived bioscaffolds for massive rotator cuff defect in rabbits. Tissue Eng. 2007;13(7):1479–91.
24. Khalyfa A, Gozal D. Exosomal miRNAs as potential biomarkers of cardiovascular risk in children. J Transl Med. 2014;12:162.
25. Uggen C, Dines J, McGarry M, Grande D, Lee T, Limpisvasti O. The effect of recombinant human platelet-derived growth factor BB-coated sutures on rotator cuff healing in a sheep model. Arthroscopy. 2010;26(11):1456–62.
26. Raimondi MT, Laganà M, Conci C, Crestani M, Di Giancamillo A, Gervaso F, et al. Development and biological validation of a cyclic stretch culture system for the ex vivo engineering of tendons. Int J Artif Organs. 2018;41(7):400–12.
27. Sun L, Qu L, Zhu R, Li H, Xue Y, Liu X, et al. Effects of mechanical stretch on cell proliferation and matrix formation of mesenchymal stem cell and anterior cruciate ligament fibroblast. Stem Cells Int. 2016;2016:9842075.
28. Hohlrieder M, Teuschl AH, Cicha K, van Griensven M, Redl H, Stampfl J. Bioreactor and scaffold design for the mechanical stimulation of anterior cruciate ligament grafts. Biomed Mater Eng. 2013;23(3):225–37.

iPS Cell-Based Technology for Joint Resurfacing

32

Takahiro Negishi and Noriyuki Tsumaki

32.1 Introduction

The discovery of induced pluripotent stem cells (iPSCs) presented a paradigm shift in the field of regenerative medicine and cell therapy. iPSCs are a type of cells that can be generated from virtually any type of somatic cells through reprogramming, and proliferated almost indefinitely keeping its pluripotent state, providing an ideal source for allogeneic cell-based therapeutics [1, 2]. Research addressing the various clinical challenges of iPSC-based technology is underway, and the regeneration of cartilage in this context is certainly no exception. A substantial number (15+) of clinical studies are currently being conducted across the world, covering a variety of tissues and diseases [3].

As a cell source, iPSCs have several benefits in comparison to other types of cells used in the regeneration of articular cartilage. Since these cells are generated from somatic cells such as peripheral blood cells, dermal fibroblasts, or cryopreserved cord blood cells simply by enhancing expression of reprogramming factors, ethical issues are negligible compared to embryonic stem cells that require destruction of human embryo. The self-renewal activity of iPSCs is also advantageous over other stem cells such as human mesenchymal stromal cells (MSCs), and the proliferative character warrants consistent manufacturing from the same cell bank. The use of the same cell source for manufacturing ensures reproducible product quality, and the allogeneic nature, although a risk of immunogenic reaction remains, lends itself to an advantage over autologous products in that the product quality is independent of the patient and one-step operation with an off-the-shelf product is possible. Taken together, these attributes make iPSCs a prominent cell source for allogeneic therapy, and

T. Negishi
Department of Cartilage Regeneration, Graduate School of Medicine/Faculty of Medicine, Osaka University, Osaka, Japan
e-mail: t.negishi.docr@osaka-u.ac.jp

N. Tsumaki (✉)
Department of Tissue Biochemistry, Graduate School of Medicine (Graduate School of Frontier Biosciences/Premium Research Institute for Human Metaverse Medicine), The University of Osaka, Osaka, Japan
e-mail: tsumaki.noriyuki.med@osaka-u.ac.jp

A. Gobbi et al. (eds.), *Regenerative Medicine in Sports and Orthopaedics*,
https://doi.org/10.1007/978-3-031-84693-9_32

several chondrogenesis differentiation protocols showed successful construction of cell suspension, pellets, and cartilage-like tissues containing chondrocyte-like cells with efficacy in vivo. Chondrogenesis is achieved generally using families of growth factors such as TGF (Transforming Growth Factor) and BMP (Bone Morphogenetic Protein), and canonical Wnt signal activation. Embryonic bodies (EB), monolayers, micromass, or scaffolds are used during the differentiation [4, 5]. These protocols of chondrogenesis provide researchers with a range of options for generating functional transplants to treat cartilage defects. iPSC-derived chondrogenic transplants are well-characterized in vitro, including extensive gene expression analysis [4, 6]. The safety and efficacy of the transplants are evaluated in various animal models, including mice, rats, rabbits, sheep, and minipigs [7].

Despite much effort in the advancement of iPSC-based technology, there remain risks in clinical applications mainly with respect to manufacturing and potential safety. Regarding the manufacturing issue, the differentiation protocols of iPSCs to articular chondrocytes with cartilage-like structures are often long and intricate, which present certain commercial manufacturing obstacles such as robustness, reproducibility, and cost, which could be hindrance to clinical application [8]. Regarding the safety issue, in addition to generally accepted risks in cell therapy (e.g., donor eligibility and communicable diseases), transient expression of reprogramming genes during iPSCs' generation and long cultivation time with repetitive passaging impose risks of genomic alteration that may cause tumorigenicity and undesirable cell structure [9]. Conventional usage of viral vectors such as retro/lentivirus and Sendai virus for reprogramming is not employed for clinical purposes anymore and methods using episomal plasmid, synthetic RNA, and recombinant proteins are introduced to minimize the risk [10].

This chapter is dedicated to iPSC-based technology for joint resurfacing with a specific emphasis on the differentiation and manufacturing processes, characteristics of iPSC-derived chondrogenic transplants, the safety and efficacy of the technology, and the future direction of this field of research. These are considered current topics of research in achieving clinical application of iPSC-based technology in joint repair.

32.2 iPSCs Differentiation to Chondrocytes and Chondrogenic Tissues

Although an exceeding number of differentiation protocols from iPSCs to articular chondrocyte-like cells and chondrogenic tissues have been reported, standardized protocols for clinical application have yet to be established [4, 6, 11]. A significant number of studies have aimed to elucidate the pathophysiology of diseases such as osteoarthritis using iPSC-derived chondrocytes as a disease model [5, 12]. There are a few concepts behind each of the differentiation protocols coined so far which are described below.

32.2.1 Coculturing

It was previously shown that coculturing of pluripotent stem cells including iPSCs with primary chondrocytes can induce chondrogenic differentiation [13, 14]. The primary advantage of this method is the utilization of paracrine effects of innate chondrocytes to stimulate chondrogenesis, and the trans-well system has been applied [13]. However, it was also shown that it is not suited to produce hyaline cartilage extensively [13].

32.2.2 EB Formation Preceding Cartilage Differentiation

Embryonic bodies (EB) are frequently used to obtain cells of three germ layers in large amounts from iPSCs. Some chondrogenic differentiation protocols employed EB formation between iPSCs to chondrogenesis, initially by using undirected EB [15–17]. Generation of EB and subsequent monolayer outgrowth and 3D pelleting have successfully granted 3D structure of the iPSC-derived chondrogenic transplants [18].

Given the 3D structure, the method is combined with decellularization method to show efficacy [19]. Differentiation through EB is further combined with genetic manipulation using nonviral minicircle vectors carrying *TGFβ3* and *BMP2*, presenting options not to rely thoroughly on the constituent of differentiation media [20].

32.2.3 Mesenchymal Stromal Cell Induction Preceding Cartilage Differentiation

Human mesenchymal stromal cells (MSCs) are cells found in bone marrow, adipose tissue, umbilical cord blood, and dental pulp, which are said to have osteogenic, chondrogenic, myogenic, and adipogenic potential in vitro. It was previously reported that MSCs are differentiated into chondrocytes for transplantation aiming for cartilage repair by the addition of BMP2 and TGFβ3 [21, 22]. Therefore, common differentiation protocols aim to obtain MSC-like cells from iPSCs before the chondrogenic differentiation [23, 24]. Early studies paved the path of iPSCs differentiation through MSCs to chondrocyte pellet structure mainly by utilizing TGFβ3 as the chondroprogenitor inducer, which presumably functions as the modulator of *SOX9* resulting in type II collagen synthesis [25, 26]. These approaches were further augmented to simplify the protocol by resigning the use of BMPs to avoid hypertrophic cartilage formation and to optimize the yield by employing monolayer expansion of cells for MSC-like state [27, 28]. However, MSCs do not mature into chondrocytes in vivo, and it remains elusive that transplantation of MSC-like cells forms articular cartilage-like tissue but repairs the defect through paracrine effects.

32.2.4 Directed Chondrogenic Differentiation

In terms of developmental lineage, human articular cartilage originates from mesoendoderm and mesoderm lineage. As such, a number of scientists employed mesoderm lineage differentiation through canonical Wnt signaling activation of the pluripotent stem cells. Use of cytokines and growth factors in chondrogenic differentiation was modified using reporter lines under the control of the promoter/enhancer sequences of *COL2A1*, or *COL11A2*, the genes encoding cartilage-specific collagen proteins [29–31] and serial addition of WNT3A, FGF2, BMP4, follistatin, neurotrophin 4 (NT4), and GDF5 showed prospective in articular cartilage-like tissue generation through mesodermal lineage [32]. Mesodermal lineage was also shown to be an effective path for chondrogenic differentiation as activation of canonical Wnt signaling by CHIR99021, a GSK3β inhibitor, promoted chondrogenesis in vitro and following retinoic acid treatment to mimic the limb bud formation is also effective [30, 33, 34]. Additionally, use of iPSC-aggregate prior to direct chondrogenesis was shown to be successful [31]. Indeed, iPSC-aggregate formation prior to chondrogenesis is a promising technology to manufacture chondrogenic tissue-like structure, which could directly be used as a potential source of allogeneic cartilage tissue transplant (Fig. 32.1).

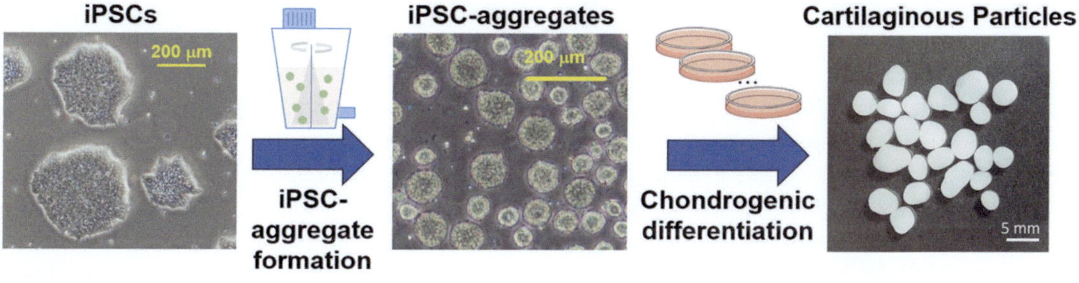

Fig. 32.1 iPSC-derived chondrogenic tissue via iPSC-aggregate formation. Images are provided by Tsumaki lab

32.2.5 Toward Clinical Manufacturing

Chondrogenic differentiation technologies are at the heart of the manufacturing process development of cartilage repair by iPSC-based technology. Yet, difficult requirement criteria for the manufacturing process hinder the transfer from bench to bedside [35]. One such impediment is the robustness and reproducibility of the process, suggesting that the differentiation protocol should rather be simple when considering clinical application. Critical materials such as cytokines and growth factors essential for differentiation of iPSCs are often not available for clinical use, not to mention equipment such as EB forming dishes/reactors necessary for large-scale production. Efforts have been made to reduce the risks of using growth factors by replacing them with small molecules and establishing a serum-free differentiation protocol [33]; however, the challenges for commercial manufacturing still remain. Another concern specific to iPSC-based technology is the choice of iPSC bank as the starting material for the chondrogenic product. There are reports of genetic background such that specific SNPs are associated with degenerative cartilage diseases [36, 37], and iPSCs of OA patients affect chondrogenesis [38]. Therefore, in addition to donor eligibility criteria of iPSCs, genetic background may have to be considered when selecting the starting material.

32.3 Quality of iPSC-Derived Chondrogenic Transplants

As a result of chondrogenesis, cells are expected to be functional chondrocytes that are safe and effective as joint cartilage therapeutics. Extensive characterization of cellular and extracellular matrix (ECM) constituents is conducted in most iPSC-based studies [4, 6, 12, 39], evaluating their characterization, safety, and efficacy. In vitro characterization along with safety and efficacy studies using animal models together define the quality of the iPSC-based articular cartilage repair. The potential specification criteria more specific to iPSC-derived chondrogenic transplants are shown in Table 32.1.

32.3.1 In Vitro Characterization

When iPSC-derived chondrogenic transplants are expected to be engrafted in the host tissue, it is important to show that they functionally resemble healthy chondrogenic tissue or articular cartilage in ECM and cellular constituents. Articular cartilage ECM is known for its hyaline property that ensures the mechanical properties of viscoelasticity, lubrication, and smooth surface. Hyaline cartilage property is shown by Safranin O or Alcian blue positive staining, which stains acidic glycosaminoglycans (GAGs) and proteoglycans, and the histological characterization is frequently employed to show hyaline property of iPSC-derived chondrogenic tissue [15, 17, 19–21, 24, 26–28, 30–32, 40–44]. In addition, histological immunostaining of type II collagen, Aggrecan, and SOX9 also characterize hyaline cartilage representing abundant collagen, GAGs, and chondrogenic transcription factor, respectively [15–17, 20, 24, 26, 28, 31, 40, 41, 43, 45, 46]. On the contrary, deposition of type I collagen and type X collagen is representative of fibrocartilage and hypertrophic cartilage, respectively, which are also characterized by immunostaining [15, 20, 26, 30, 31, 40]. On a more quantitative note, GAGs are easily quantified by the most widely used method of Dimethylmethylene Blue (DMMB) Assay to detect sulfated GAGs such as chondroitin sulfate and heparan sulfate within the ECM of the derived chondrogenic transplants [30, 41, 43]. Regarding its cellular characterization, gene expression plays an essential role. In almost all the studies, *COL2A1*, *SOX9*, and *ACAN* are often shown for their upregulation in comparison to the iPSC counterparts, and *COL1A2* and *COL10A1* are minimally expressed [15, 17, 20, 21, 24, 26, 29–33, 41, 42, 44, 45]. Size and the mechanical properties of the derived chondrogenic pellets or tissues, especially when using a scaffold, are also characterized [43, 45].

Table 32.1 Specification and characterization criteria of iPSC-derived chondrogenic transplants which should be considered in addition to the cases of general cellular therapies

Quality	Characterization criteria	Examples of measured targets or procedures
In vitro characterization	Hyaline cartilage	Histochemistry (Safranin O, Alcian blue, HE)
		Immunohistochemistry Positive staining: Type II collagen, Aggrecan, SOX9 Negative staining: Type I collagen, Type X collagen
		GAG contents (DMMB assay of ECM)
	Chondrogenesis	Chondrogenic gene expression (*COL2A1*, *SOX9*, *ACAN*)
	Fibrous cartilage	Fibrocartilage gene expression (*COL1A2*)
	Hypertrophic cartilage	Hypertrophic marker (*COL10A1*)
	Shape and size	Images, weight
	Mechanical property	Compressive and tensile strength measurement
	Pluripotency	Gene expression of pluripotent markers (*OCT3/4*, *NANOG*, *SOX2*, *SSEA4*)
	Unintended cell type	Osteogenic gene expression (*BMP4*, *RUNX2*), and other potential cell-type marker expression
	Mineralization	ALP activity
	Cellular heterogeneity	Single-cell RNAseq
Safety (preclinical)	Tumorigenicity	Subcutaneous transplants in immunodeficient mice or rats, orthotopic (i.e., knee joint) transplants in rats
	Immunogenicity	Use of allogeneic and large animal models
Efficacy (preclinical)	Graft survival	Anti-human specific immunohistochemistry
	Hyaline cartilage in vivo	Histochemistry (Safranin O, Alcian blue, HE)
		Immunohistochemistry Positive staining: Type II collagen, Aggrecan, SOX9 Negative staining: Type I collagen, Type X collagen

There are at least two main focuses that are more specific to iPSC-based technologies regarding in vitro characterization of the derived chondrogenic transplants. One is the assessment of pluripotent markers which is critical in assessing the contamination of undifferentiated cells, and *OCT3/4*, *NANOG*, *SOX2,* and *SSEA4* pluripotent markers are often selected for the evaluation [15, 20, 31–33, 41–43]. Another focus is the homogeneous terminal differentiation to minimize the contamination of unintended cell types in the differentiated cell population such as mesodermal cells, osteoblastic cells, or melanocytes, of which cell type is expected to depend largely on the method of differentiation. *BMP4*, osteocalcin, and *RUNX2* are the candidates [42, 43]. Regarding the evaluation of contamination with unintended cells, precise expression markers should be selected according to the differentiation protocol of choice and the histological analysis. It is worth noting that the iPSC-derived chondrocytes exhibit lower alkaline phosphatase (ALP) activity and are more resistant to mineralization in vitro in comparison to MSCs [27]. The risk of cellular heterogeneity is considered higher in iPSC-based technology because elaborate pluripotent to chondrogenic differentiation is inevitable [47]. To this end, single-cell RNA-seq analysis will be a powerful tool to characterize the heterogeneity in a systematic manner [46, 48, 49]. In the future, spatial transcriptomic approach should also be employed to overcome the practical difficulty of isolating chondrocytes from their ECM [50, 51].

Evaluation of the cellular characteristics during chondrogenic differentiation also presents a wealth of scientific knowledge regarding lineage of the chondrogenesis. Gene expression of differentiation markers in a time-course manner is evaluated in order to ensure the intended cellular lineage [31, 33, 41, 49]. Indeed, not only the specification of terminal chondrogenic tissue but also the in-process analysis will be the key for establishing a manufacturing protocol for commercialization. In order to realize a precise process control for commercial manufacturing, a concept such as Process Analytical Technology (PAT) will be useful [52, 53]. Such gene expression analysis is invaluable for the real-time understanding, monitoring, analysis, and control of the manufacturing process to establish a robust operation protocol for product development.

32.3.2 Safety

iPSC-based technology poses additional safety issues to the risks associated with other allogeneic cell therapies. These issues must be carefully evaluated in vivo using appropriate animal models.

One of the issues is the unintended differentiation of the transplanted cells in vivo. iPSC-derived chondrogenic transplants are subcutaneously placed in immunodeficient mice and rats to evaluate teratoma and other unintended tissue formation such as endochondral ossification, and most of the studies show stable hyaline cartilage tissue formation represented by positive Safranin O staining [31, 33, 41] except for one study that shows teratoma formation [32]. Malignant tumor formation was never observed. So far, iPSC-derived chondrogenic transplants have little tumorigenicity in vivo [54]; however, utmost care must be taken in evaluating the risks of tumorigenicity of iPSC-derived chondrogenic transplants when considering clinical application. It is often difficult to conduct a life-long experiment using these animal models, making it incomplete to be a preclinical safety test. In this regard, international collaboration is underway to standardize tumorigenicity testing to develop standardized analytical procedure for the assessment, which should be a valuable resource [55].

Another issue is the immunogenicity of the transplants since iPSC-based technology is intended to be an allogeneic cell therapy. Allogeneic cell therapies of articular cartilage are already in practice without serious reports on immunological rejection [56, 57], presumably due to the relatively immuno-privileged nature of cartilage supported by avascularity and the ECM. Chondrocytes are not only residing in immuno-privileged environment but they are also immuno-suppressive or immuno-modulating [58, 59]. However, it is still vastly controversial, and evidence presents immune response in articular cartilage repair [60]. The current understanding is that the immunological response to allogeneic cell therapy of articular cartilage depends on factors such as host/systemic conditions and local/surgical factors. The immuno-privileged nature of the host cartilage is accommodated for uncomplicated chondral defects; however, abundant blood flow in such cases of osteochondral defects abolishes the characteristics. It was clearly shown, using a primate model, that a surgically induced chondral defect with an intact subchondral bone is a suitable indication for iPSC-based allogeneic cell therapy without eliciting immunological response [46]. In addition, regarding immunogenicity of the graft, the exposure of chondrocytes to the host immune cells should elicit immunogenic response, suggesting cellular injection may be more immunogenic than tissue-like transplants in which chondrocytes are embedded in the ECM preventing access to the immune cells. A more fundamental approach in developing hypoimmunogenic iPSC line by genome editing of *HLA-A*, *HLA-B*, and *CIITA* has been reported which should largely reduce the immunogenicity of the iPSC-derived chondrogenic transplants [61].

32.3.3 Efficacy

Although not necessarily specific to iPSC-based technology, in vivo efficacy of iPSC-derived chondrogenic transplants is also a focus of research because interaction with the host environment such as sustained hyaline cartilage property after implantation can only be evaluated in animal studies.

Choice of the animal model is one of the most crucial aspects of efficacy testing, and mice, rats, rabbits, and mini-pigs are frequently used [6, 7]. In addition, the defect size/location and the study period are the decisive factors of a study. So far, many studies using immunodeficient rats for 8–12 weeks transplant period have been reported [15, 16, 18, 20, 26, 31, 42], presumably due to the availability of immunodeficient model, the affordability, the accessibility, and the fact that their joints are larger than those of mice. When evaluating the efficacy in vivo, first, the residual graft is detected by anti-human specific antibody such as human vimentin antibody immunohistochemistry, and then the hyaline property is shown using staining such as Safranin O, Alcian blue, and type II collagen immunohistochemistry [15, 16, 18, 20, 26, 31, 42]. When considering clinical representation, a larger animal model with a longer transplant period is desirable [7]. Only a few studies using rabbits up to 12 weeks [28, 45, 62] and mini-pigs up to 4 weeks [31] are conducted, despite their successful outcomes. Difficulty in conducting efficacy studies in larger animal models is due to the lack of immunodeficient models, which limits the transplant period without graft rejection [7].

In principle, the clinical target indication should be the decisive factor in which the animal model is selected for the efficacy study. Until now, most of the studies focus on chondral or osteochondral defects of articular cartilage of the knee joint, of which model is created using drills or tissue punches to essentially mimic traumatic defects of the articular cartilage. More interestingly, some of the studies use osteoarthritis (OA) models to suggest efficacy in OA treatment [16, 28]. Pathological histologies induced by treatments such as single intra-articular injection of monosodium iodoacetate (MIA) in rat model or anterior cruciate ligament transection (ACLT) in rabbit model were rescued by intra-articular injection of iPSC-derived chondrocytes [16, 28].

32.4 Clinical Application and Future Direction

Articular cartilage regeneration using iPSC-based technology is advancing rapidly and it holds immense possibilities in revolutionizing the treatment of cartilage defects and osteoarthritis. There are a number of chondrogenic differentiation protocols for iPSCs reported, which gives us clues in establishing manufacturing processes. Ongoing effort must be put into optimization and standardization of the differentiation protocols to generate mature and functional chondrocytes in concert with three-dimensional ECM scaffold by refinements of culture conditions, use of growth factors, and bioreactors. Technologies are already available in combining biomaterial scaffolds to improve the biomechanical properties of the transplants [45, 63, 64].

One of the critical characteristics attributing to the safety and efficacy of iPSC-derived chondrocytes is the cellular heterogeneity of the iPSC-derived chondrogenic transplants. Precise assessment using innovative analytical procedures should be introduced, such as single-cell omics technologies. Regarding safety and efficacy, improved animal models should increase the expected clinical success of the technology. In order to mimic the clinical indications and treatments, use of larger animals with longer observation periods is suited, especially that articular cartilage in vivo has little turnover. Longer observation should let us observe the integration of the transplants with the host articular cartilage tissue, and the formation of zonal structure of superficial, transitional, and deep zones.

Lastly, the translation of iPSC-based articular cartilage repair from preclinical studies to clinical trials requires collaborative efforts among academic researchers, clinicians, industries, and regulatory agencies [65, 66]. To the best of our knowledge, there is only one clinical trial currently underway using iPSCs in the field of cartilage repair, which targets knee OA with iPSC-derived MSCs [39] and one clinical study [67]. Regulatory framework is still under development; however, in addition to cGMP (current Good Manufacturing Practice), 21 CFR 1271 Subpart C—Donor Eligibility, REGULATION (EC) No 1394/2007, and Japanese Standards for Biological Ingredients/Raw Materials (MHLW Public Notice No. 210 2003) should be applied to iPSC-based technology in FDA, EMA, and MHLW governing regions of the world, respectively. There are guidelines published specific to iPSC-based technology [68] and the international harmonization is an ongoing effort.

In conclusion, ongoing advancements in stem cell biology are pushing iPSC-based cartilage repair toward translation to clinical applications. Researchers have uncovered ways for harnessing regenerative potential of iPSCs to address articular cartilage-related conditions and iPSC-based technology holds great promise in the field of joint repair in the future.

Funding This work was supported by AMED under Grant Number JP24ym0126127.

References

1. Takahashi K, Yamanaka S. Induction of pluripotent stem cells from mouse embryonic and adult fibroblast cultures by defined factors. Cell. 2006;126(4):663–76.
2. Takahashi K, Tanabe K, Ohnuki M, Narita M, Ichisaka T, Tomoda K, et al. Induction of pluripotent stem cells from adult human fibroblasts by defined factors. Cell. 2007;131(5):861–72.
3. Kim JY, Nam Y, Rim YA, Ju JH. Review of the current trends in clinical trials involving induced pluripotent stem cells. Stem Cell Rev Rep. 2022;18(1):142–54.
4. Kamaraj A, Kyriacou H, Seah KTM, Khan WS. Use of human induced pluripotent stem cells for cartilage regeneration in vitro and within chondral defect models of knee joint cartilage in vivo: a preferred reporting items for systematic reviews and meta-analyses systematic literature review. Cytotherapy. 2021;23(8):647–61.
5. Lach MS, Rosochowicz MA, Richter M, Jagiełło I, Suchorska WM, Trzeciak T. The induced pluripotent stem cells in articular cartilage regeneration and disease modelling: are we ready for their clinical use? Cells. 2022;11(3):529.
6. Castro-Viñuelas R, Sanjurjo-Rodríguez C, Piñeiro-Ramil M, Hermida-Gómez T, Fuentes-Boquete IM, de Toro-Santos FJ, et al. Induced pluripotent stem cells for cartilage repair: current status and future perspectives. Eur Cell Mater. 2018;36:96–109.
7. Yamashita A, Tsumaki N. Recent progress of animal transplantation studies for treating articular cartilage damage using pluripotent stem cells. Develop Growth Differ. 2021;63(1):72–81.
8. Kim JH, Kawase E, Bharti K, Karnieli O, Arakawa Y, Stacey G. Perspectives on the cost of goods for hPSC banks for manufacture of cell therapies. NPJ Regen Med. 2022;7(1):54.
9. Dashnau JL, Xue Q, Nelson M, Law E, Cao L, Hei D. A risk-based approach for cell line development, manufacturing and characterization of genetically engineered, induced pluripotent stem cell-derived allogeneic cell therapies. Cytotherapy. 2023;25(1):1–13.
10. Karami Z, Moradi S, Eidi A, Soleimani M, Jafarian A. Induced pluripotent stem cells: generation methods and a new perspective in COVID-19 research. Front Cell Dev Biol. 2022;10:1050856.
11. Nakayama N, Pothiawala A, Lee JY, Matthias N, Umeda K, Ang BK, et al. Human pluripotent stem cell-derived chondroprogenitors for cartilage tissue engineering. Cell Mol Life Sci. 2020;77(13):2543–63.
12. Eremeev A, Pikina A, Ruchko Y, Bogomazova A. Clinical potential of cellular material sources in the generation of iPSC-based products for the regeneration of articular cartilage. Int J Mol Sci. 2023;24(19):14408.
13. Qu C, Puttonen KA, Lindeberg H, Ruponen M, Hovatta O, Koistinaho J, et al. Chondrogenic differentiation of human pluripotent stem cells in chondrocyte co-culture. Int J Biochem Cell Biol. 2013;45(8):1802–12.
14. Wei Y, Zeng W, Wan R, Wang J, Zhou Q, Qiu S, et al. Chondrogenic differentiation of induced pluripotent stem cells from osteoarthritic chondrocytes in alginate matrix. Eur Cell Mater. 2012;23:1–12.
15. Ko JY, Kim KI, Park S, Im GI. In vitro chondrogenesis and in vivo repair of osteochondral defect with human induced pluripotent stem cells. Biomaterials. 2014;35(11):3571–81.
16. Zhu Y, Wu X, Liang Y, Gu H, Song K, Zou X, et al. Repair of cartilage defects in osteoarthritis rats with induced pluripotent stem cell derived chondrocytes. BMC Biotechnol. 2016;16(1):78.
17. Li Y, Liu T, Van Halm-Lutterodt N, Chen J, Su Q, Hai Y. Reprogramming of blood cells into induced pluripotent stem cells as a new cell source for cartilage repair. Stem Cell Res Ther. 2016;7:31.
18. Rim YA, Nam Y, Park N, Lee J, Park SH, Ju JH. Repair potential of nonsurgically delivered induced pluripotent stem cell-derived chondrocytes in a rat osteochondral defect model. J Tissue Eng Regen Med. 2018;12(8):1843–55.
19. Choi SH, Lee K, Han H, Mo H, Jung H, Ryu Y, et al. Prochondrogenic effect of decellularized extracellular matrix secreted from human induced pluripotent stem cell-derived chondrocytes. Acta Biomater. 2023;167:234–48.
20. Rim YA, Nam Y, Park N, Jung H, Lee K, Lee J, et al. Chondrogenic differentiation from induced pluripotent stem cells using non-viral minicircle vectors. Cells. 2020;9(3):582.
21. Guzzo RM, Gibson J, Xu RH, Lee FY, Drissi H. Efficient differentiation of human iPSC-derived mesenchymal stem cells to chondroprogenitor cells. J Cell Biochem. 2013;114(2):480–90.
22. Shen B, Wei A, Tao H, Diwan AD, Ma DD. BMP-2 enhances TGF-beta3-mediated chondrogenic differentiation of human bone marrow multipotent mesen-

chymal stromal cells in alginate bead culture. Tissue Eng Part A. 2009;15(6):1311–20.
23. Villa-Diaz LG, Brown SE, Liu Y, Ross AM, Lahann J, Parent JM, et al. Derivation of mesenchymal stem cells from human induced pluripotent stem cells cultured on synthetic substrates. Stem Cells. 2012;30(6):1174–81.
24. Liu Y, Goldberg AJ, Dennis JE, Gronowicz GA, Kuhn LT. One-step derivation of mesenchymal stem cell (MSC)-like cells from human pluripotent stem cells on a fibrillar collagen coating. PLoS One. 2012;7(3):e33225.
25. Umeda K, Zhao J, Simmons P, Stanley E, Elefanty A, Nakayama N. Human chondrogenic paraxial mesoderm, directed specification and prospective isolation from pluripotent stem cells. Sci Rep. 2012;2:455.
26. Nejadnik H, Diecke S, Lenkov OD, Chapelin F, Donig J, Tong X, et al. Improved approach for chondrogenic differentiation of human induced pluripotent stem cells. Stem Cell Rev Rep. 2015;11(2):242–53.
27. Diederichs S, Klampfleuthner FAM, Moradi B, Richter W. Chondral differentiation of induced pluripotent stem cells without progression into the endochondral pathway. Front Cell Dev Biol. 2019;7:270.
28. Chang YH, Wu KC, Ding DC. Induced pluripotent stem cell-differentiated chondrocytes repair cartilage defect in a rabbit osteoarthritis model. Stem Cells Int. 2020;2020:8867349.
29. Yang SL, Harnish E, Leeuw T, Dietz U, Batchelder E, Wright PS, et al. Compound screening platform using human induced pluripotent stem cells to identify small molecules that promote chondrogenesis. Protein Cell. 2012;3(12):934–42.
30. Adkar SS, Wu CL, Willard VP, Dicks A, Ettyreddy A, Steward N, et al. Step-wise chondrogenesis of human induced pluripotent stem cells and purification via a reporter allele generated by CRISPR-Cas9 genome editing. Stem Cells. 2019;37(1):65–76.
31. Yamashita A, Morioka M, Yahara Y, Okada M, Kobayashi T, Kuriyama S, et al. Generation of scaffoldless hyaline cartilaginous tissue from human iPSCs. Stem Cell Reports. 2015;4(3):404–18.
32. Saito T, Yano F, Mori D, Kawata M, Hoshi K, Takato T, et al. Hyaline cartilage formation and tumorigenesis of implanted tissues derived from human induced pluripotent stem cells. Biomed Res. 2015;36(3):179–86.
33. Kawata M, Mori D, Kanke K, Hojo H, Ohba S, Chung UI, et al. Simple and robust differentiation of human pluripotent stem cells toward chondrocytes by two small-molecule compounds. Stem Cell Reports. 2019;13(3):530–44.
34. Kreuser U, Buchert J, Haase A, Richter W, Diederichs S. Initial WNT/β-catenin activation enhanced mesoderm commitment, extracellular matrix expression, cell aggregation and cartilage tissue yield from induced pluripotent stem cells. Front Cell Dev Biol. 2020;8:581331.
35. McKenna DH, Perlingeiro RCR. Development of allogeneic iPS cell-based therapy: from bench to bedside. EMBO Mol Med. 2023;15(2):e15315.
36. Boer CG, Hatzikotoulas K, Southam L, Stefánsdóttir L, Zhang Y, Coutinho de Almeida R, et al. Deciphering osteoarthritis genetics across 826,690 individuals from 9 populations. Cell. 2021;184(24):6003–5.
37. Aubourg G, Rice SJ, Bruce-Wootton P, Loughlin J. Genetics of osteoarthritis. Osteoarthr Cartil. 2022;30(5):636–49.
38. Khan NM, Diaz-Hernandez ME, Chihab S, Priyadarshani P, Bhattaram P, Mortensen LJ, et al. Differential chondrogenic differentiation between iPSC derived from healthy and OA cartilage is associated with changes in epigenetic regulation and metabolic transcriptomic signatures. elife. 2023;12:e83138.
39. Evenbratt H, Andreasson L, Bicknell V, Brittberg M, Mobini R, Simonsson S. Insights into the present and future of cartilage regeneration and joint repair. Cell Regen. 2022;11(1):3.
40. Craft AM, Rockel JS, Nartiss Y, Kandel RA, Alman BA, Keller GM. Generation of articular chondrocytes from human pluripotent stem cells. Nat Biotechnol. 2015;33(6):638–45.
41. Lee J, Taylor SE, Smeriglio P, Lai J, Maloney WJ, Yang F, et al. Early induction of a prechondrogenic population allows efficient generation of stable chondrocytes from human induced pluripotent stem cells. FASEB J. 2015;29(8):3399–410.
42. Cheng A, Kapacee Z, Peng J, Lu S, Lucas RJ, Hardingham TE, et al. Cartilage repair using human embryonic stem cell-derived chondroprogenitors. Stem Cells Transl Med. 2014;3(11):1287–94.
43. Chijimatsu R, Ikeya M, Yasui Y, Ikeda Y, Ebina K, Moriguchi Y, et al. Characterization of mesenchymal stem cell-like cells derived from human iPSCs via neural crest development and their application for osteochondral repair. Stem Cells Int. 2017;2017:1960965.
44. Xu L, Hanamatsu H, Homan K, Onodera T, Miyazaki T, Furukawa JI, et al. Alterations of glycosphingolipid glycans and chondrogenic markers during differentiation of human induced pluripotent stem cells into chondrocytes. Biomolecules. 2020;10(12):1622.
45. Liu J, Nie H, Xu Z, Niu X, Guo S, Yin J, et al. The effect of 3D nanofibrous scaffolds on the chondrogenesis of induced pluripotent stem cells and their application in restoration of cartilage defects. PLoS One. 2014;9(11):e111566.
46. Abe K, Yamashita A, Morioka M, Horike N, Takei Y, Koyamatsu S, et al. Engraftment of allogeneic iPS cell-derived cartilage organoid in a primate model of articular cartilage defect. Nat Commun. 2023;14(1):804.
47. Diederichs S, Gabler J, Autenrieth J, Kynast KL, Merle C, Walles H, et al. Differential regulation of SOX9 protein during chondrogenesis of induced pluripotent stem cells versus mesenchymal stromal cells: a shortcoming for cartilage formation. Stem Cells Dev. 2016;25(8):598–609.
48. Kamatani T, Hagizawa H, Yarimitsu S, Morioka M, Koyamatsu S, Sugimoto M, et al. Human iPS cell-derived cartilaginous tissue spatially and func-

tionally replaces nucleus pulposus. Biomaterials. 2022;284:121491.
49. Wu CL, Dicks A, Steward N, Tang R, Katz DB, Choi YR, et al. Single cell transcriptomic analysis of human pluripotent stem cell chondrogenesis. Nat Commun. 2021;12(1):362.
50. Vandereyken K, Sifrim A, Thienpont B, Voet T. Methods and applications for single-cell and spatial multi-omics. Nat Rev Genet. 2023;24(8):494–515.
51. Fan X, Sun AR, Young RSE, Afara IO, Hamilton BR, Ong LJY, et al. Spatial analysis of the osteoarthritis microenvironment: techniques, insights, and applications. Bone Res. 2024;12(1):7.
52. Yu LX, Amidon G, Khan MA, Hoag SW, Polli J, Raju GK, et al. Understanding pharmaceutical quality by design. AAPS J. 2014;16(4):771–83.
53. Kovács B, Péterfi O, Kovács-Deák B, Székely-Szentmiklósi I, Fülöp I, Bába LI, et al. Quality-by-design in pharmaceutical development: from current perspectives to practical applications. Acta Pharma. 2021;71(4):497–526.
54. Takei Y, Morioka M, Yamashita A, Kobayashi T, Shima N, Tsumaki N. Quality assessment tests for tumorigenicity of human iPS cell-derived cartilage. Sci Rep. 2020;10(1):12794.
55. Watanabe T, Yasuda S, Kusakawa S, Kuroda T, Futamura M, Ogawa M, et al. Multisite studies for validation and improvement of a highly efficient culture assay for detection of undifferentiated human pluripotent stem cells intermingled in cell therapy products. Cytotherapy. 2021;23(2):176–83.
56. Farr J, Tabet SK, Margerrison E, Cole BJ. Clinical, radiographic, and histological outcomes after cartilage repair with particulated juvenile articular cartilage: a 2-year prospective study. Am J Sports Med. 2014;42(6):1417–25.
57. Dawkins BJ, Shubin Stein BE, Mintz DN, Fabricant PD, Gomoll AH, Strickland SM, et al. Patellofemoral joint cartilage restoration with particulated juvenile allograft in patients under 21 years old. Knee. 2022;36:120–9.
58. Adkisson HD, Milliman C, Zhang X, Mauch K, Maziarz RT, Streeter PR. Immune evasion by neocartilage-derived chondrocytes: implications for biologic repair of joint articular cartilage. Stem Cell Res. 2010;4(1):57–68.
59. Osiecka-Iwan A, Hyc A, Radomska-Lesniewska DM, Rymarczyk A, Skopinski P. Antigenic and immunogenic properties of chondrocytes. Implications for chondrocyte therapeutic transplantation and pathogenesis of inflammatory and degenerative joint diseases. Cent Eur J Immunol. 2018;43(2):209–19.
60. Arzi B, DuRaine GD, Lee CA, Huey DJ, Borjesson DL, Murphy BG, et al. Cartilage immunoprivilege depends on donor source and lesion location. Acta Biomater. 2015;23:72–81.
61. Kitano Y, Nishimura S, Kato TM, Ueda A, Takigawa K, Umekage M, et al. Generation of hypoimmunogenic induced pluripotent stem cells by CRISPR-Cas9 system and detailed evaluation for clinical application. Mol Ther Methods Clin Dev. 2022;26:15–25.
62. Xu X, Shi D, Liu Y, Yao Y, Dai J, Xu Z, et al. repair of full-thickness cartilage defect with human iPSC-derived mesenchymal progenitor cells in a rabbit model. Exp Ther Med. 2017;14(1):239–45.
63. Aisenbrey EA, Bilousova G, Payne K, Bryant SJ. Dynamic mechanical loading and growth factors influence chondrogenesis of induced pluripotent mesenchymal progenitor cells in a cartilage-mimetic hydrogel. Biomater Sci. 2019;7(12):5388–403.
64. Agarwal T, Chiesa I, Presutti D, Irawan V, Vajanthri KY, Costantini M, et al. Recent advances in bioprinting technologies for engineering different cartilage-based tissues. Mater Sci Eng C Mater Biol Appl. 2021;123:112005.
65. Moradi S, Mahdizadeh H, Šarić T, Kim J, Harati J, Shahsavarani H, et al. Research and therapy with induced pluripotent stem cells (iPSCs): social, legal, and ethical considerations. Stem Cell Res Ther. 2019;10(1):341.
66. Nordberg RC, Otarola GA, Wang D, Hu JC, Athanasiou KA. Navigating regulatory pathways for translation of biologic cartilage repair products. Sci Transl Med. 2022;14(659):eabp8163.
67. https://jrct.niph.go.jp/en-latest-detail/jRCTa050190104.
68. Hayakawa T, Aoi T, Umezawa A, Ozawa K, Sato Y, Sawa Y, et al. A study on ensuring the quality and safety of pharmaceuticals and medical devices derived from processing of allogeneic human induced pluripotent stem(-Like) cells. Regen Ther. 2015;2:95–108.

33 Nanotechnologies, Biomaterials, and Scaffolds

Travis Peng, Apratim Maity, and Daniel A. Grande

33.1 Introduction to Bioscaffolds

Tissue regeneration was initially thought to be achievable through stem cell transplantation, where stem cells are harvested, expanded, and transplanted into the wound site [1]. However, this approach was not very successful. Proper tissue regeneration requires stem cells to know the exact position and time to proliferate and differentiate into the correct cell types [2, 3]. These instructions come in the form of chemical signals, cell–cell interactions, and cell–matrix interactions. While chemical signals can be provided through the addition of signaling molecules, simple stem cell transplantation would still lack proper cell–matrix interactions [4]. To address this problem, the use of bioscaffolds can be employed [5].

Bioscaffolds are engineered structures that support cell–matrix interactions by mimicking the ECM of natural tissues and aiding in the growth and differentiation of cells during tissue regeneration [5]. They are a cornerstone in the triad of regenerative medicine. For bioscaffolds to be successful, they must provide a three-dimensional environment where seeded cells can attach, differentiate, and function as they would in native tissue. Depending on the biomaterials used, they can be classified as either natural, synthetic, or composite [5, 6]. Natural bioscaffolds are made from molecules found in the body or nature, such as collagen, fibrin, or chitosan. Synthetic bioscaffolds are made from man-made polymers like polylactic acid or polyethylene glycol. Composite bioscaffolds combine natural and synthetic materials [7]. Later in this chapter, we will discuss how the choice of biomaterials impacts the success of bioscaffolds in facilitating tissue regeneration.

Interestingly, the use of bioscaffolds in tissue regeneration dates all the way back to 1500 BC [8]. The Papyrus of Ebers, an ancient Egyptian medical document, describes treating skin wounds with honey, grease, and lint. The honey acted as an antibiotic, the grease as a barrier to pathogens, and the lint as a fibrous scaffold promoting tissue regeneration. Other historical evidence of biomaterial use in general includes a

T. Peng · A. Maity · D. A. Grande (✉)
Northwell Health, New Hyde Park, NY, USA

Donald and Barbara Zucker School of Medicine at Hofstra/Northwell, Hempstead, NY, USA

Feinstein Institute for Medical Research, Manhasset, NY, USA
e-mail: dgrande@northwell.edu

A. Gobbi et al. (eds.), *Regenerative Medicine in Sports and Orthopaedics*,
https://doi.org/10.1007/978-3-031-84693-9_33

mummified body with a wooden toe from ancient Egypt, a skull with a dental implant from the Gallo-Roman period, and the Mayan use of blue nacre shells as dental implants. Although some medical knowledge from ancient times was lost, a significant amount was preserved by various cultures and expanded upon. During the Age of Enlightenment, philosophers began investigating phenomena associated with reproduction and regeneration and it was out of this time period that the foundational principles of tissue engineering and regenerative medicine were born.

In 1817, German biologist Christian H. Pander proposed that tissue development depends on the dynamic interplay between cells and their microenvironment [8]. Unbeknownst at the time, this hypothesis would serve as the basis for the creation of the first bioscaffold. In 1910, on his way to designing a standardized protocol for cell culture, renowned French surgeon and biologist Alexis Carrel used thin layers of clotted plasma to embed tissue fragments onto, which he would maintain under fluid phase for several weeks at a time. Once confluent, Carrel would use pieces of the clot to seed new flasks, and his observations that these tissues could maintain a state of active growth for long periods of time supported Pander's hypothesis regarding the interplay between cells and their microenvironments. Carrel's use of plasma clots in tissue culture served as the first use of a modern scaffold.

The following years saw an increase in the understanding of the nature of cell growth. In the 1970s, Judah Folkman, a pioneer in angiogenesis research, made three key observations: (1) tissue development requires adequate transport of nutrients and waste, (2) cell growth and proliferation are heavily influenced by the physical properties of the substrate on which they are grown, and (3) cells form complex structures when stimulated by environmental cues found in their natural environments [8]. These observations necessitated the use of bioscaffolds and morphogens to supplement the tissue regeneration process. Around the same time, an orthopedic surgeon named William T. Green utilized these theories, using spicules of bone to seed chondrocytes into rabbit articular cartilage [9]. Though unsuccessful at the time, these techniques are still used in tissue regeneration today.

In 1985, transplant surgeon Paul S. Russell wrote a review highlighting selective cell transplantation after which a notable rise in classic tissue engineering experiments using bioscaffolds was seen [10]. It was around this time that leaders in the field, Joseph Vacanti and Robert Langer, began to design bioscaffolds for cell delivery rather than simple cell seeding [11]. A bioscaffold designed for cell delivery not only provides structural support but also acts as a dynamic and active structure that facilitates the delivery of the seeded cells to their natural positions within the native tissue. Vacanti and Langer's work led to today's sophisticated approach to scaffold design, which emphasizes properties such as porosity, stiffness, and functionalization. In this chapter, we will discuss the various types of bioscaffolds, their properties, and their fabrication techniques. We will also discuss current applications of bioscaffolds in both research and medicine, as well as important clinical challenges and considerations regarding their use. Lastly, we will end with a discussion of future directions in bioscaffold fabrication and design.

33.2 Types of Bioscaffolds

Bioscaffolds are critical in tissue engineering, providing structures that support cell attachment, proliferation, and differentiation to regenerate damaged tissues. These scaffolds can be broadly categorized into natural, synthetic, and composite types, each with distinct properties and applications.

33.2.1 Natural Bioscaffolds

For soft tissue repair, natural, polymer-based scaffolds have become a widely popular, and in some cases commercially available [12], choice among clinicians and scientists alike. There are, however, limitations such as host immune response and poor mechanical properties must be considered [13]. Manufacturing natural scaffolds from their native state requires careful processing, either chemically or physically, where some leftover biologic materials may trigger a host immune response [14]. However, in many cases where processing techniques are nonspecific, overprocessed materials may result in lower mechanical strength [15]. Thus, when considering the usefulness of the natural scaffolds discussed in this section, it is important to be mindful not only of their advantageous properties but the techniques necessary for their procurement.

Natural bioscaffolds are derived from biological sources and typically include extracellular matrix (ECM)-derived scaffolds and decellularized tissues or organs. ECM-derived scaffolds, such as collagen, fibrin, gelatin, chitin, chitosan, coralline, and silk, are widely used due to their excellent biocompatibility and ability to promote cell adhesion and growth [14]. Collagen, for example, is a vital structural protein in the ECM of mammals composed of a long-chained macromolecule that forms a triple-helix structure. This structure provides a high tensile strength and structural integrity, making it a popular choice for wound healing, cartilage repair, and bone regeneration applications [16, 17]. Similarly, fibrin, derived from fibrinogen in the blood through a polymerization process triggered by thrombin during the clotting process, supports cell proliferation and differentiation, suitable for vascular grafts, skin regeneration, and nerve repair [18–20].

Another naturally derived and widely popular material is gelatin, a hydrolyzed form of collagen. Gelatin's denatured protein structure retains many of collagen's beneficial properties while being easier to process, making it useful for bone, cartilage, and skin regeneration [21]. Chitin and chitosan, found in the exoskeletons of crustaceans and insects, are polysaccharides composed of repeating units of N-acetylglucosamine and glucosamine respectively [22]. These materials, being biocompatible, biodegradable, and naturally antimicrobial, make them suitable for wound dressings, drug delivery systems, and bone regeneration. Coralline scaffolds, derived from coral and consisting primarily of calcium carbonate arranged in a porous structure that mimics natural bone, support osteoconduction primarily used in bone grafting and regeneration [23–25]. Finally silk fibroin, obtained from silkworms, is characterized by its β-sheet structure, offers excellent mechanical properties and biocompatibility, supporting cell adhesion, proliferation, and differentiation in applications such as bone and cartilage regeneration, ligament repair, and wound healing [26, 27].

In addition, tissues or even whole organs can also be decellularized, leaving behind the biochemical and structural cues of the native extracellular matrix (ECM). The tissue-specific extracellular matrix (ECM) is a key component of the cellular microenvironment, mediating cellular processes such as survival, adhesion, migration, proliferation, and differentiation [28, 29]. Cells interact bidirectionally with their ECM, sensing and responding to biochemical and biomechanical signals while actively remodeling the ECM [30]. Given the critical role of ECM in directing cell behavior, there is growing interest in developing three-dimensional (3D) in vitro models and cell delivery systems that replicate the native ECM properties [28].

Decellularized tissues represent an abundant source of tissue-specific ECM for developing cell-instructive scaffolds that promote tissue regeneration. These organs can be sourced from allogenic or xenogenic donors but must be removed of all cellular antigens [31]. After thorough processing, they are used in applications like heart valve replacement, vascular grafts, and whole-organ engineering. The decellularization process involves applying physical, chemical, and enzymatic treatments to eliminate immunogenic cellular components from tissues while preserving the structural, biochemical, and biomechanical properties of the ECM [32, 33]. Decellularized matrices have been prepared from various tissues and investigated for use as cell culture and delivery platforms. Customized decellularization protocols for tissues such as human adipose tissue and porcine auricular cartilage generate scaffolds with well-preserved ECM structure and composition [34].

These decellularized ECM scaffolds can provide naturally inductive microenvironments that promote specific cellular differentiation. For example, scaffolds derived from decellularized adipose tissue (DAT) have been shown to enhance the adipogenic differentiation of human adipose-derived stem/stromal cells (ASCs) in vitro and promote fat formation in vivo [35, 36]. Similarly, decellularized cartilage tissue (DCT) scaffolds support chondrogenic differentiation [37]. The inherent bioactivity of decellularized tissues mediates cell function, but these biomaterials offer limited flexibility in tuning scaffold properties to control cell–cell and cell–ECM interactions. To address this limitation, a strategy has been developed to incorporate synthetic and natural components to form a composite bioscaffold, which will be further discussed later.

Another popular type of bioscaffold is hydrogels, the most common of which are alginate and hyaluronic acid [38]. As their name implies, hydrogels have excellent water retention properties and can provide a hydrated environment for cell growth [39]. Cell encapsulation, the entrapment of living cells within a semipermeable membrane or homogeneous solid mass using hydrogels, is known for its ease of handling and homogenous cell distribution [40–42]. However, it should be noted that hydrogels are seldom used for load-bearing tissues due to poor mechanical properties. Alginate, derived from brown seaweed, is a polysaccharide composed of β-D-mannuronic acid and α-L-guluronic acid units that form hydrogels upon contact with divalent cations like calcium. It is used in wound dressings, cell encapsulation, and drug delivery systems. Hyaluronic acid, a naturally occurring glycosaminoglycan found in connective tissues composed of glucuronic acid and N-acetylglucosamine, forms hydrogels with excellent biocompatibility and is used in tissue engineering, wound healing, and ophthalmic surgeries [43].

33.2.2 Synthetic Bioscaffolds

The most widely utilized synthetic bioscaffolds are various forms of polymers and hydrogels. Polymers such as polylactic acid (PLA), polyglycolic acid (PGA), and polycaprolactone (PCL) are popular due to their controlled physical and mechanical properties. PLA is a biodegradable thermoplastic derived from renewable resources like cornstarch or sugarcane. Its long-chained macromolecule forms a crystalline or semi-crystalline structure, providing it with the mechanical strength suitable and widely used for bone screws, plates, and tissue engineering scaffolds [44]. PGA is a biodegradable polymer with a similar crystalline structure made up of glycolic acid units. Its high degradation rate is suitable for temporary scaffold applications and is often used in conjunction with other polymers to enhance mechanical properties such as in sutures, drug delivery systems, and tissue engineering scaffolds [45, 46]. In contrast, PCL is a biodegradable polyester with a slower degradation rate. PCL consists of repeating caprolactone units in a semi-crystalline structure, providing long-term mechanical support useful for load-bearing applications, such as bone regeneration, cartilage repair, and drug delivery systems.

33.2.3 Composite Bioscaffolds

Composite bioscaffolds combine natural and synthetic materials to leverage the advantages of both. They aim to more accurately mimic the complex structure and functionality of natural tissues. Examples include combining collagen with hydroxyapatite with hydroxyapatite for bone regeneration or incorporating synthetic polymers with natural hydrogels for enhanced mechanical properties and bioactivity [47]. These scaffolds are used in a variety of applications, including bone and cartilage regeneration, wound healing, and soft tissue repair.

As previously mentioned, singular material sources are limited in their ability to control cell–cell interactions. To address this limitation, a strategy has been developed to incorporate processed decellularized ECM in the form of micronized particles within naturally derived or synthetic hydrogel carriers, generating composite bioscaffolds with tissue-specific instructive properties. Hydrogels provide a tunable platform that can mimic the mechanical properties of native soft tissues and allow the permeability of oxygen, nutrients, and other metabolites, making them effective scaffolds for cell encapsulation. Hydrogels can be crosslinked ex vivo within a mold to generate scaffolds with a defined shape and volume or designed to crosslink in situ for minimally invasive cell delivery at a target site for tissue repair.

This composite hydrogel approach offers versatility, as various hydrogels can be utilized to enable cell encapsulation with high viability. Examples include methacrylated glycol chitosan, thiol-modified hyaluronic acid, alginate, and polyethylene glycol-based materials [48–50]. These hydrogels provide flexibility depending on the properties required for the target application. For instance, methacrylated chondroitin sulfate (MCS) has been used as a hydrogel carrier, showing high cell viability and enhanced adipogenic differentiation when combined with DAT [51].

From the perspective of cartilage regeneration, chondroitin sulfate (CS), a major structural and bioactive component of native ECM, supports the viability of encapsulated chondrocytes and promotes the chondrogenic differentiation of stem cells [40]. The composite approach using DAT and DCT demonstrates how this strategy can be applied with other ECM sources to create tissue-specific instructive biomaterials. The ECM particle size and concentration, as well as cell types and seeding densities, can be refined to promote desired cellular responses. For example, smaller ECM particles can enhance cell–cell contact and improve differentiation outcomes.

Composite bioscaffolds represent a powerful tool in tissue engineering, combining the bioactivity of natural ECM with the tunability of synthetic hydrogels to create scaffolds that can be tailored for specific applications in the ever-advancing field of regenerative medicine (Table 33.1).

Table 33.1 Common materials used in bioscaffolds [52–58]

Material	Advantages	Drawbacks	Uses
Collagen	Biocompatible, promotes cell adhesion	Rapid degradation, limited mechanical strength	Bone, skin, and cartilage tissue engineering
Chitosan	Biocompatible, biodegradable, antimicrobial	Poor mechanical properties	Wound healing, drug delivery
Alginate	Biocompatible, easy gelation	Low mechanical strength, rapid degradation	Cartilage and wound healing
Silk fibroin	High mechanical strength, biocompatible	Variable degradation rates, potential immunogenicity	Bone and ligament tissue engineering
Hyaluronic acid	Biocompatible, promotes cell migration	Rapid degradation, low mechanical strength	Cartilage and wound healing
Poly(lactic-co-glycolic acid) (PLGA)	Biodegradable, tunable degradation rate	Potential inflammatory response, acidic degradation products	Bone and cartilage tissue engineering, drug delivery
Polycaprolactone (PCL)	Biocompatible, biodegradable, good mechanical properties	Slow degradation, potential cytotoxicity	Bone and cartilage tissue engineering
Polyethylene glycol (PEG)	Biocompatible, hydrophilic	Nondegradable, potential toxicity	Drug delivery, tissue engineering
Polyvinyl alcohol (PVA)	Biocompatible, hydrophilic, good mechanical properties	Nondegradable, potential toxicity	Drug delivery, wound dressing
Polyurethane (PU)	Elastic, biocompatible, durable	Potential cytotoxicity, complex synthesis	Soft tissue engineering, vascular grafts
Hydroxyapatite (HA)	Osteoconductive, bioactive	Brittle, low mechanical strength	Bone tissue engineering
Tricalcium phosphate (TCP)	Osteoconductive, biodegradable	Brittle, low mechanical strength	Bone tissue engineering
Calcium phosphate cements (CPC)	Injectable, osteoconductive	Low mechanical strength	Bone tissue engineering
Bioglass	Bioactive, promotes bone regeneration	Brittle, low mechanical strength	Bone tissue engineering
Carbon nanotubes (CNTs)	High mechanical strength, electrical conductivity	Potential cytotoxicity, complex synthesis	Bone and nerve tissue engineering
Collagen/HA composite	Improved mechanical properties, osteoconductive	Potential immunogenicity, complex fabrication	Bone tissue engineering
Chitosan/alginate composite	Enhanced biocompatibility, easy to fabricate	Poor mechanical properties	Wound healing, drug delivery
PLGA/PCL composite	Controlled degradation, enhanced mechanical properties	Complex fabrication, potential inflammatory response	Bone and cartilage tissue engineering
PCL/HA composite	Enhanced mechanical properties, osteoconductive	Complex fabrication	Bone tissue engineering
Silk fibroin/CNT composite	Enhanced mechanical properties, biocompatible	Complex synthesis, potential cytotoxicity	Bone and nerve tissue engineering

33.3 Properties of Bioscaffolds

Bioscaffolds are essential for tissue engineering, serving as temporary structures that support cell attachment, proliferation, and differentiation. The effectiveness of bioscaffolds in mimicking the native tissue environment and promoting tissue regeneration depends on several key properties. By understanding and optimizing these properties, researchers can develop scaffolds that better mimic the native tissue environment, enhancing tissue regeneration and integration with host tissues. Here we will explore these parameters that are critical to their effectiveness in tissue engineering applications.

33.3.1 Microarchitecture

The microarchitecture of a scaffold includes porosity, pore size, and pore interconnectivity. Proper pore size ensures efficient mass transfer of nutrients and waste, facilitating tissue penetration and vascularization. Optimal pore sizes range from 75 to 100 μm, which helps promote cell attachment for in vitro applications, up to 200–500 μm, which helps support tissue penetration for in vivo applications [59]. An interconnected porous system provides a larger surface area for cell attachment, enhancing cell-to-cell interactions and promoting integration with native tissues. Scaffolds have been suggested to ideally have a pore network comprising more than 60 percent of pores with diameters between 150 and 400 μm and at least 20% smaller than 20 μm. Additionally, porosity must be high to allow for sufficient nutrient transport and cellular infiltration. For various cell types, the optimal pore sizes differ: for human skin fibroblasts (HSF), less than 160 μm; for smooth muscle cells, 60–150 μm; and for osteogenic cells, 100–150 μm. High porosity and appropriate pore sizes are crucial for scaffolds intended for different tissue engineering applications [60, 61]. The structural design must balance porosity and mechanical stability to ensure functionality (Table 33.2).

33.3.2 Mechanical Properties

Scaffold materials must possess mechanical properties similar to those of native tissues at the implantation site to provide structural support and shape stability. This minimizes the risk of stress shielding, implant-related osteopenia, and subsequent re-fracture [47]. Mechanical properties include elastic modulus, tensile strength, fracture toughness, fatigue resistance, and elongation percentage [64]. Tensile and compressive testing are conventional methods to characterize these properties, with parameters such as compressive/tensile strength, toughness, and Young's modulus being critical.

For bone tissue engineering, the scaffold should mimic the native structure of cancellous bone [65]. Understanding the mechanical behavior of bone, which is a nanometer-scale composite of brittle apatite and ductile collagen, provides insights into designing scaffolds with appropriate mechanical properties [66]. Cancellous bone's compressive stress-strain curve exhibits linear elasticity, a long plateau of cell wall collapse, and a densification regime [67]. Scaffolds should exhibit similar behavior, with mechanical properties tailored to match those of the surrounding tissue to avoid bone resorption or mechanical failure.

33.3.3 Biodegradability

Biodegradability refers to the scaffold's ability to be broken down chemically or enzymatically over time within a living organism. The degradation rate should match the rate of new tissue formation, allowing the scaffold to be gradually replaced by native tissue through "creeping substitution" [68, 69], where the scaffold is reabsorbed and replaced over time. The degradation products should be nontoxic, either being recycled as metabolites in other biochemical reactions or exiting the body without interfering with other organs and tissues. This property ensures that the scaffold provides temporary support

Table 33.2 Ideal pore size based on cell/tissue type [59, 62, 63]

Cell/tissue type	Pore size (μm)
Endothelial cells	<80
Keratinocytes	20–125
Smooth muscle	60–150
Fibroblasts	<160
Osteogenic cells	100–150
Fibrocartilagenous tissue	150–300

while being replaced by natural tissue over time. For instance, scaffolds fabricated from chitosan and hydroxyapatite (HA) can exhibit controlled degradation properties, making them suitable for bone tissue engineering.

33.3.4 Biocompatibility

Biocompatibility is the ability of a scaffold to support cell adhesion and proliferation without eliciting significant chronic immune responses [70]. Minimal inflammatory reactions should occur, resolving within 2 weeks to prevent adverse effects on healing or tissue integration. Scaffold materials must be compatible with the cellular components of engineered tissues and endogenous cells in the host tissue. This compatibility ensures that the scaffold does not provoke an immune response that could compromise the integration and functionality of the newly formed tissue [71, 72]. Materials like polyhydroxybutyrate (PHB) and polyhydroxybutyrate-co-valerate (PHBV) have demonstrated good biocompatibility, making them suitable for various tissue engineering applications [73].

33.3.5 Bioactivity

Bioactivity involves the scaffold's interaction with surrounding cellular components to enhance cell migration, differentiation, tissue regeneration, and integration with the host tissue [74]. Bioactive scaffolds can be functionalized with cell-adhesive ligands or with physical indicators like topography to improve cell attachment and morphology. For example, scaffolds with surface roughness have been shown to improve osteoblast function in bone tissue engineering, highlighting the importance of bioactivity [75, 76]. Surface properties, including surface chemistry and roughness, significantly influence cell attachment, proliferation, and differentiation. Surface roughness improves osteoblast function in bone tissue engineering, with variations in cellular behavior dependent on texture size. Scaffolds with hydrophilic surfaces (contact angle less than 90°) facilitate better cell adhesion and proliferation than hydrophobic surfaces. Surface topography can be categorized into macroroughness (100 μm–mm), microroughness (100 nm–100 μm), and nanoroughness (less than 100 nm), each influencing different cellular responses [77, 78]. Proper surface design enhances the scaffold's interaction with cells, promoting better integration and functionality.

Scaffolds may also serve as carriers for growth factors, enhancing tissue regeneration. Hydrogels synthesized by covalent or ionic crosslinking can entrap proteins by a mechanism controlled by the swelling of the hydrogels for later release. This active interaction promotes the desired cellular responses and tissue regeneration processes.

33.4 Fabrication Techniques

The development of three-dimensional (3D) scaffolds has become a fundamental aspect of tissue engineering, providing the necessary support for cell attachment, growth, and differentiation. Various methods have been devised to produce scaffolds with specific characteristics tailored to different applications. In this section, we will explore the principles, advantages, and limitations of some of the most prominent scaffold fabrication techniques: salt-leaching, electrospinning, freeze-drying, 3D printing, and self-assembly.

33.4.1 Salt-Leaching

Salt-leaching is a traditional and widely used technique for creating porous scaffolds. The process involves mixing a polymer solution with salt particles, which act as a porogen. After the polymer solidifies, the salt is leached out using water or another solvent, leaving behind a porous structure [79]. This method allows for precise control over pore size and porosity by adjusting the size and concentration of the salt particles used.

In the salt-leaching process, a polymer-salt mixture is first prepared by combining a polymer solution with salt particles of a desired size. The mixture is then cast into a mold and allowed to solidify, typically by evaporating the solvent or cooling the polymer melt. Once the polymer has solidified, the salt is dissolved by immersing the composite in water or another suitable solvent, resulting in a porous scaffold.

Salt-leaching offers several advantages, including simplicity, cost-effectiveness, and the ability to control pore size and porosity [80]. However, it also has limitations, such as being restricted to water-soluble porogens and potentially resulting in nonuniform pore structures. Despite these drawbacks, salt-leaching remains a valuable method for fabricating scaffolds for various tissue engineering applications.

33.4.2 Electrospinning

Electrospinning is a versatile technique for producing fiber-based scaffolds with a nanoscale architecture similar to the extracellular matrix (ECM). The process involves applying a high-voltage electric field to draw a polymer solution or melt into fine fibers, which are collected on a grounded surface [81, 82]. Electrospun scaffolds exhibit a high surface area to volume ratio, enhancing cell attachment and providing a biomimetic environment for tissue regeneration [83].

The electrospinning process begins with the preparation of a polymer solution, which is loaded into a syringe with a needle connected to a high-voltage power supply [84]. When a high voltage is applied, the polymer solution is ejected from the needle tip, forming fibers that are collected on a grounded surface. The diameter of the fibers can be controlled by adjusting parameters such as solution viscosity, electrical field strength, and distance between the needle tip and the collector [85].

Electrospinning is widely used in soft tissue repair, vascular tissue engineering, and cardiac tissue engineering. For instance, electrospun scaffolds composed of natural polymers like collagen can support the formation of blood vessels, while scaffolds with aligned fibers can guide the orientation of cardiomyocytes, mimicking the anisotropic structure of cardiac tissue [86, 87]. While electrospinning allows for the production of fibers with diameters ranging from nanometers to micrometers, achieving uniform fiber diameter and scaling up for large tissue constructs remain challenging. Despite these challenges, electrospinning continues to be a powerful technique for fabricating ECM-mimic nanofibrous scaffolds with tunable properties.

33.4.3 Freeze-Drying

Freeze-drying, or lyophilization, is a process that involves freezing a polymer solution and then sublimating the solvent under vacuum, leaving behind a porous structure. Freeze-drying is particularly suitable for producing scaffolds from natural polymers, as it preserves the bioactivity of incorporated biomolecules [88].

The freeze-drying process consists of three major steps: freezing, primary drying, and secondary drying. First, a polymer solution is poured into a mold and frozen at low temperatures. In the primary drying phase, the frozen solution is placed under a vacuum, causing the ice to sublimate. Finally, in the secondary drying or adsorption phase, a higher temperature is used to remove any ionically bound residual moisture, resulting in a porous scaffold [62].

Freeze-dried scaffolds with high porosity and interconnected pore structures facilitate cell attachment, proliferation, and mass transport. This makes freeze-drying useful for fabricating scaffolds for wound healing, nerve regeneration, and other applications.

While freeze-drying offers several advantages, such as generating scaffolds with high porosity and avoiding high temperatures, it also has limitations. The process is time-consuming, and there is a risk of pore structure collapse during drying [79]. Despite these challenges, freeze-drying remains a valuable method for fabricating natural polymer-based scaffolds with tailored properties.

33.4.4 3D Printing

3D printing, also known as additive manufacturing, is a cutting-edge technique for fabricating scaffolds with precise control over their architecture. The process involves building structures layer by layer based on a digital design, enabling the production of scaffolds with complex shapes and tailored properties.

There are several 3D printing techniques used in scaffold fabrication, including stereolithography (SLA), selective laser sintering (SLS), and bioprinting. Each technique has its unique principles and applications.

Stereolithography (SLA) uses ultraviolet (UV) light to cure photopolymer resins layer by layer through photopolymerization. The UV light induces the polymerization of the resin, forming a solid material. SLA offers control over the internal and external geometry of the scaffold, including pore size, porosity, and patterns. However, the technique has limitations, such as the need for photoinitiators, potential cytotoxicity, and relatively long processing times.

Selective laser sintering (SLS) involves the use of a laser to sinter powdered materials, forming solid layers. This technique is primarily used for bone tissue engineering due to its ability to produce scaffolds with high mechanical strength. However, SLS has limitations, including poor surface finishing, presence of defects, and the need for post-processing treatments.

Bioprinting combines 3D printing with living cells or other biological materials to create tissue constructs. The process involves depositing cells loaded in bioink through nozzle-based or laser-assisted techniques. Nozzle-based bioprinting uses pneumatic or mechanical systems to dispense bioink, while laser-assisted bioprinting uses a pulsed laser to transfer cells to a substrate. Bioprinting enables the fabrication of cell-laden scaffolds that mimic native tissues, supporting tissue regeneration. 3D printing offers several advantages, including high precision, control over scaffold geometry, and the ability to fabricate complex structures. However, material limitations, potential cytotoxicity of some components, and scalability challenges need to be addressed for broader applications.

33.4.5 Self-Assembly

As its name suggests, self-assembly leverages the natural tendency of certain molecules to form nanofibers and other nanostructures, autonomously organizing them into structured arrangements to mimic ECM [89, 90]. The self-assembly process involves designing molecules, such as peptides, with amphiphilic properties. These molecules can self-assemble into nanofibers or other structures under specific conditions, such as changes in pH or ionic concentration. For example, peptide amphiphiles can form nanofibers via pH-induced self-assembly, creating a scaffold that supports cell growth and differentiation [89].

Self-assembled scaffolds have been used in cartilage regeneration and neural tissue engineering. For instance, peptide amphiphiles can form nanofibers that promote chondrocyte differentiation and cartilage formation. In neural tissue engineering, self-assembling peptides create scaffolds that support neurite outgrowth and synaptogenesis [91, 92].

While self-assembly offers the advantage of mimicking natural biological processes, it also has limitations. Self-assembled nanostructures often have poor mechanical properties and stability, and the high cost of synthesizing biomaterials can limit their applications. Despite these challenges, self-assembly remains a promising approach for creating scaffolds that closely resemble the ECM.

33.5 Applications in Research

The development of bioscaffolds has revolutionized biomedical research, providing innovative platforms for various applications. This section delves into the applications of bioscaffolds in cell culture studies, drug delivery systems, disease modeling, and understanding cell–material interactions, emphasizing their potential to enhance research outcomes and clinical translations.

33.5.1 Cell Culture Studies

Bioscaffolds have significantly advanced cell culture studies by providing more physiologically relevant environments for cells. Traditional 2D cell culture methods often fail to replicate the complex three-dimensional (3D) architecture and microenvironment of tissues, leading to differences in cell behavior, gene expression, and functionality [93, 94]. Bioscaffolds, particularly those derived from decellularized tissues, offer a solution by mimicking the native extracellular matrix (ECM) and its biochemical and structural cues. As an example, decellularized brain tissue scaffolds have been used to study neural cell behavior. Koh et al. developed a patient-specific 3D brain tumor model using decellularized human brain biopsies reseeded with glioblastoma cells [95]. This model facilitated a more accurate representation of tumor invasiveness and cell morphology compared to traditional collagen-based models. Similarly, Sood et al. created a model mimicking both gray and white matter of the brain, using decellularized porcine brain ECM combined with silk scaffolds [96]. This setup supported neural network formation and axon elongation, highlighting the importance of tissue-specific ECM in neural studies.

In cardiac research, decellularized heart scaffolds have been employed to study cell behavior and cardiac tissue engineering. Nguyen et al. created a "humanized organ model" using decellularized rat hearts recellularized with human cells [97]. This model allowed the study of human cell integration and cardiac tissue responses in a more physiologically relevant context. Such studies demonstrate the potential of bioscaffolds to improve the accuracy and relevance of cell culture studies, providing better insights into cellular processes and interactions. In addition, hydrogels derived from decellularized tissues have shown promise in enhancing tissue culture. These hydrogels retain the bioactive components of the ECM, which can enhance cell behavior [98]. For example, decellularized liver ECM hydrogels have been used to culture hepatocytes, demonstrating improved cell viability and function compared to traditional collagen hydrogels [99].

33.5.2 Drug Delivery Systems

Bioscaffolds also play a crucial role in developing advanced drug delivery systems. These scaffolds can be designed to provide controlled and sustained release of therapeutic agents, improve drug efficacy, and reduce side effects [100]. The use of decellularized ECM-based scaffolds in drug delivery is particularly promising due to their biocompatibility and ability to mimic the natural tissue environment [101–103]. To this end, electrospun nanofibers made from natural and synthetic polymers have been widely explored for drug delivery applications. These nanofibers offer high surface area and porosity, allowing for efficient drug loading and controlled release. Electrospun scaffolds can be designed to respond to specific stimuli, such as pH or temperature, to release drugs in a controlled manner. Similarly, researchers have developed injectable hydrogels containing chemotherapeutics specifically designed to release when in proximity of the tumor microenvironment. These "smart hydrogels" allow for localized and safe delivery of otherwise highly toxic chemicals.

Three-dimensional (3D) printing has emerged as a powerful tool for fabricating drug delivery scaffolds with precise control over geometry and drug distribution. 3D-printed scaffolds can be customized to deliver drugs locally to specific tissues or organs, enhancing therapeutic outcomes [104]. For example, 3D-printed patches have been used for local delivery of anticancer drugs in pancreatic cancer treatment, and transdermal drug delivery systems have been developed to treat conditions like acne and wounds [105].

33.5.3 Disease Modeling

Bioscaffolds provide a powerful platform for disease modeling, enabling researchers to more accurately replicate disease conditions in vitro [106, 107]. Decellularized tissue scaffolds retain the complex ECM architecture and biochemical cues of the original tissue, making them ideal for studying disease mechanisms and testing therapeutic interventions.

In cancer research, decellularized tissue scaffolds have been used to model various types of tumors. Chen et al. have developed a colorectal cancer model using decellularized human colon tissue reseeded with primary human colonic epithelial cells carrying cancer-associated mutations [108]. This model successfully replicated the invasive and proliferative behavior of colorectal cancer, providing a valuable tool for studying cancer progression and identifying therapeutic targets. Similarly, Zhao et al. created a tongue squamous cell carcinoma model using decellularized tongue ECM reseeded with cancer cells. This model enabled the study of cancer cell migration, proliferation, and response to chemotherapy, offering insights into tumor biology and potential treatments [109].

In addition to cancer, bioscaffolds have been used to model other diseases, such as kidney fibrosis and liver diseases. Fu et al. used decellularized fibrotic kidney ECM to study the role of tenascin-C in fibrosis [110]. Their model showed increased fibroblast proliferation and metabolic activity in the fibrotic environment, highlighting the impact of the ECM on disease progression. In liver disease modeling, decellularized liver scaffolds have been used to study non-alcoholic fatty liver disease (NAFLD) and its progression to non-alcoholic steatohepatitis (NASH). These models have provided valuable insights into the mechanisms of disease development and potential therapeutic targets [111–113].

33.5.4 Understanding Cell–Material Interactions

Understanding cell–material interactions is crucial for developing effective biomaterials and tissue engineering strategies. Bioscaffolds, particularly those derived from decellularized tissues, offer a unique opportunity to study these interactions in a physiologically relevant context.

Decellularized tissue scaffolds provide a native-like ECM environment that can influence cell behavior, including adhesion, proliferation, differentiation, and migration. Studies using decellularized skeletal muscle ECM, for example, have shown how the ECM composition and mechanical properties can direct muscle stem cell behavior. Zhang et al. demonstrated that arsenic-exposed skeletal muscle ECM induced fibroblast differentiation of muscle stem cells, highlighting the impact of environmental factors on cell–ECM interactions [114].

In another study, Stearns-Reider et al. investigated the effect of aging on muscle stem cell behavior using decellularized skeletal muscle ECM from young and old mice [115]. Their findings revealed that aged ECM promoted fibrogenesis over myogenesis, suggesting that age-related changes in the ECM can influence stem cell fate. These studies underscore the importance of the ECM in regulating cell behavior and the potential of bioscaffolds to provide insights into cell–material interactions.

Moreover, decellularized tissue scaffolds have been used to study drug–ECM interactions. Wassenaar et al. used decellularized skeletal muscle to model drug absorption after intramuscular injection [116]. Their study revealed that drug-binding affinities to the ECM influenced absorption rates, providing valuable information for optimizing drug delivery strategies. Such models can help in understanding how drugs interact with the ECM, aiding in the design of more effective therapeutic agents.

33.5.5 Future Directions

Bioscaffolds have emerged as powerful tools in biomedical research, offering innovative platforms for cell culture studies, drug delivery systems, disease modeling, and understanding cell–material interactions. By providing a more physiologically relevant environment, bioscaffolds derived from decellularized tissues enhance the accuracy and relevance of in vitro models, facilitating better insights into cellular processes, disease mechanisms, and therapeutic responses. The continued development and application of bioscaffolds hold great promise for advancing research and improving clinical outcomes.

In the future, advancements in 3D bioprinting and organ-on-a-chip technologies are expected to further enhance the capabilities of bioscaffolds, enabling the creation of even more sophisticated and personalized in vitro models. These developments will undoubtedly play a crucial role in bridging the gap between laboratory research and clinical applications, ultimately leading to more effective treatments and improved patient care.

33.6 Applications in Medicine

The ability of bioscaffolds to facilitate natural cell–matrix interactions makes them essential in various medical applications including tissue engineering, organ regeneration, wound healing, and surgical implants. This section will provide an overview of the significance of bioscaffolds within each of these areas.

33.6.1 Tissue Engineering

Tissue engineering is an interdisciplinary field that integrates biology, engineering, and materials science to create functional tissues to repair or replace damaged or diseased tissues in the body. The application of tissue engineering has been broadly used in a variety of tissue types such as skin, bone, cartilage, blood vessels, bladder, and heart valves. This discussion will primarily explore the use of bioscaffolds in skin, bone, and cartilage engineering.

Tissue-engineered skin (TES) has been widely used to treat both acute and chronic skin wounds [5]. Since the first isolation of human epidermal keratinocytes in 1975, TES has evolved significantly, moving from simple epidermal substitutes to full-thickness skin constructs containing various types of seed cells. These seed cells, either skin-derived or non-skin-derived, rely on various types of bioscaffolds that can alter their survival and growth. For instance, skin-derived cells such as keratinocytes thrive on bioscaffolds made from natural materials like collagen, gelatin, fibrin, and chitosan. A notable challenge in the development of TES has also been achieving angiogenesis. Non-skin-derived seed cells, such as adipose-derived stem cells (ADSC), have been shown to have enhanced angiogenesis and wound healing rates when seeded on decellularized extracellular matrix scaffolds such as small intestinal submucosa and acellular dermal matrix (ADM). By coculturing various seed cells with endothelial cells on collagen sponge bioscaffolds, scientists were able to successfully develop TES constructs with complex vascular networks.

The design of TES depends heavily on bioscaffolds to ensure successful engraftment of the construct. These scaffolds promote granulation tissue formation, fibroblast-driven remodeling, angiogenesis, and re-epithelialization, with each type of scaffold offering unique advantages and drawbacks. Collagen scaffolds are the most extensively studied and clinically used natural scaffolds in TES engineering due to their good biocompatibility, porosity, and low immunogenicity [117]. However, they have poor mechanical strength and degrade rapidly. Chemical and biophysical cross-linking techniques, like cross-linking collagen with chitosan or heparin, can enhance mechanical properties, although high purification costs and variability in physicochemical properties remain challenges in their use for TES [118].

Non-collagenous biomaterials such as hyaluronic acid, fibrin, laminin, and elastin have also been explored in TES engineering [5, 119]. While these biomaterials promote good cell interactions, signaling contributions, and an ability to take on multiple phases such as suspensions, gels, sponges, films, or sheets, they still lack strong mechanical properties in comparison to synthetic materials. Synthetic scaffolds offer controllable mechanical properties and diverse plasticity, but their role in TES has not been extensively studied, resulting in limited knowledge regarding their optimal use. Decellularized ECM scaffolds on the other hand, such as acellular dermal matrix (ADM), maintain natural mechanical stability because they retain structural and functional proteins that constitute the native ECM [120]. Studies have shown that ADM enhances cell adhesion, proliferation, angiogenesis, cell recruitment, and cell division, making it particularly effective for designing TES to treat burn injuries. In summary, the selection and optimization of bioscaffolds are critical to the success of TES, influencing the survival and growth of seed cells, tissue integration, and overall healing outcomes.

The gold standard for repairing and regenerating diseased bones has traditionally been the use of autologous bone grafts [121]. Since the graft originates from the patient's body, it eliminates the risk of immune rejection, and the live bone cells within the graft directly contribute to new bone formation. However, autologous bone grafting has several limitations, including donor site morbidity, graft site complications, surgical risks, and constraints on the amount of bone that can be harvested. Specific complications at the donor site include infection, prolonged wound drainage, pain, and sensory loss. Allografts, while an alternative, carry similar concerns with the added risk of immune reactions. Consequently, bone tissue engineering (BTE) has been proposed as a promising alternative for the repair of bone defects.

The primary goal of bone tissue engineering is to use three-dimensional scaffolds to create bone grafts for implantation at the defect site. Typically, immune responses to these scaffolds are triggered by the preferential adsorption of fibrin onto the implant's surface [122]. The fibrin deposition attracts neutrophils, which in turn draw monocytes to the fibrin-coated surface. The monocytes then differentiate into macrophages and secrete proinflammatory cytokines. Therefore, in addition to osteoinductivity and osteoconductivity, biocompatibility is a critical consideration in BTE scaffold design. Moreover, the architecture of the scaffold is crucial as it influences the extent of angiogenesis, which has been shown to directly impact the success of the final graft. Key physical attributes of successful BTE scaffolds include pore size, porosity, pore structure, and interconnectivity. Recently, "intelligent" vascularized scaffolds that can spontaneously and programmatically respond to specific stimuli throughout the regeneration process have gained attention to enhance angiogenesis in bone repair tissue [123].

In terms of scaffold types, natural, synthetic, ceramic, and composite scaffolds all play vital roles in BTE. Among natural polymer scaffolds, collagen bioscaffolds are the most extensively studied. They are frequently combined with more robust biomaterials, such as hydroxyapatite, to enhance mechanical properties and support load-bearing applications. More recently, decellularized bone matrix (DBM) scaffolds have attracted attention for their strong potential in bone regeneration [124]. They are often fabricated into hydrogels and provide protection against infection, inflammation, and excessive wound secretion, thereby addressing the donor site morbidity concerns associated with autologous bone grafts. Their excellent biocompatibility and osteoinductive properties significantly enhance their regenerative potential. For instance, a clinical trial conducted by Karalashvili et al. successfully used DBM bioscaffolds to reconstruct zygomatic bone defects in motor vehicle accident patients. Another clinical study involved seeding autologous mesenchymal stem cells onto a DBM scaffold to complete a distal tibia reconstruction, which demonstrated active and novel bone formation 6 months post-surgery. DBM has also been used to create bioinks for constructing 3D bioprinted scaffolds with complex geometries and interconnected porosity. Lastly, bioglass is a type of biocompatible and bioactive material composed primarily of silicon dioxide, calcium oxide, sodium oxide, and phosphorous pentoxide [125, 126]. When implanted into the body, bioglass forms a hydrocarbonate apatite layer on its surface which is similar to the mineral phase of bone.

Beyond their structural and regenerative functions, bioscaffolds are increasingly being explored as drug delivery systems for bone regeneration [127, 128]. In the 1990s, researchers tested bone morphogenetic proteins (BMPs) and other growth factors to enhance bone healing in fractures and spinal fusions [129]. However, they quickly recognized the challenges in delivering these growth factors effectively to the injury site. The short duration of action and systemic toxicity, coupled with the inability to deliver growth factors in a contained and sustained-release manner impeded the development of growth factors into effective regenerative treatments. Functionalizing bioscaffolds, however, enables control over the localization, timing, and kinetics of signaling molecule release. Examples of molecules used to functionalize scaffolds in BTE include BMPs, TGF-β, RGD peptides, osteoinductive peptides, and more recently, exosomes [127, 128].

33.6.2 Tissue-Engineered Cartilage

The regeneration of articular cartilage has been of particular interest within the field of orthopedics since the early 1900s [130]. Due to its avascular nature and lack of a perichondrium, articular cartilage demonstrates poor healing and regenerative properties. Its reliance on passive diffusion for obtaining nutrients from synovial fluid is slow and ineffective, which hampers the robust repair of natural cartilage injuries and lesions [131]. Damage to the collagenous framework from overactivity of matrix metalloproteinases (MMPs) is also thought to be a significant factor in the progression of cartilage fibrillation and osteoarthritis [132]. Although articular cartilage lacks innervation, subchondral bone is richly innervated, leading to substantial pain once cartilage defects have degraded to expose the subchondral bone [131]. Additional complications include synovitis, joint effusions, avascular necrosis, subchondral sclerosis, and the formation of osteophytes.

In recent years, numerous tissue engineering approaches have emerged to address cartilage degradation [133, 134]. By restoring the structure and function of articular cartilage, these advanced therapies hold promise for improving patient outcomes in conditions like osteoarthritis, traumatic cartilage injuries, and joint instability. One such approach is matrix-induced autologous chondrocyte repair (MACI). In MACI, chondrocytes are harvested from a non-weight-bearing zone of the knee and expanded over a bioresorbable membrane, usually a collagen scaffold [133]. After preparation of the cartilage defect, the chondrocyte-loaded membrane is implanted into the defect and fixed using a fibrin glue. Patients treated with MACI have consistently shown significant improvements in pain and knee function. The Knee injury and Osteoarthritis Outcome Scores (KOOS) for pain and function were notably higher in MACI-treated patients compared to those who underwent microfracture marrow stimulation procedures alone. While MACI is a two-step process, autologous matrix-induced chondrogenesis (AMIC) is a single-step procedure that combines microfracture abrasion techniques with scaffold-based regeneration [134]. Immediately after inducing microfractures within the cartilage defect, an acellular collagen scaffold is placed into the defect space to enable infiltration of blood and bone marrow stromal cells. Clinical studies of AMIC have also shown improved clinical scores and patient satisfaction, but imaging results were variable.

In addition to collagen scaffolds, hyaluronic acid-based scaffolds have also been used in surgical procedures for repairing articular cartilage lesions. Hyalograft C is a bioscaffold composed of a hyaluronic ester, which has been shown to have positive effects on cartilage healing and regeneration, including long-term stable glycosaminoglycan content in knee repair tissue [135]. Similarly, chitosan-based scaffolds, such as BST-CarGel, also demonstrated significantly improved imaging results and clinical outcomes when used to repair cartilage defects. As for osteochondral defects, natural, synthetic, bioceramic, and composite biomaterials have been explored [136]. These scaffolds are generally multiphasic, meaning they are designed with multiple distinct layers or phases, each mimicking a different part of the osteochondral unit. Acellular multiphasic scaffolds such as MaioRegen and Agili-C have all shown significantly improved pain reduction and functional scores in addition to good integration and defect filling when used to treat osteochondral defects [137, 138]. Agili-C in particular received FDA approval on March 29, 2022, making it one of the latest off-the-shelf cartilage repair implants.

33.6.3 Organ Regeneration

Organogenesis in vivo is a highly intricate process that encompasses tissue self-organization, cell–cell interactions, regulated cell signaling, and precise cell migration [139]. This process typically commences with the induction of specific embryonic regions to differentiate into "organ-forming fields" which serve as the microenvironments where organs and tissues develop. These organ-forming fields are characterized by unique cellular environments, signaling cues, and gene expression patterns. The formation of most of the body's organs is driven by reciprocal interactions between organ-inductive potential stem cells within the embryonic epithelium and mesenchyme in each organ-forming field. These collections of organ-inductive potential stem cells are referred to as "organ germs."

Efforts to regenerate organs in vitro have been ongoing for decades; however, these studies have often been hindered by the inadequate incorporation of stem cells into the regeneration process, primarily due to the failure to establish proper cell–cell interactions. In 2007, a pioneering strategy for 3D organ-inductive stem cell manipulation emerged, leading to the successful generation of bioengineered organ germs. This innovative approach utilized collagen scaffolds to facilitate the appropriate compartmentalization of epithelial and mesenchymal stem cells, thereby mimicking the self-organization and multicellular assembly observed in in vivo organogenesis. This technique has successfully been used to bioengineer ectodermal organ germs, including teeth, hair follicles, and secretory glands.

Another prominent technique in the field of organ regeneration is whole organ decellularization [140]. This process involves the removal of all cellular components from an organ, leaving behind the extracellular matrix and forming a whole organ-derived decellularized ECM scaffold. Because the decellularized scaffold retains the complex architecture and biochemical cues of the original organ, it provides an ideal structure for recellularization with new cells to form a complete large organ. Decellularization protocols are generally organ-specific due to the variation in physico-chemical properties between different organs. A combination of chemical and mechanical techniques must therefore be employed to achieve the desired result. Additionally, bioreactors are often utilized to provide a controlled environment for the recellularized organ to allow for accurate cell growth and differentiation. Whole organ decellularization has been successfully applied to various organs, including the heart, lungs, liver, kidneys, and pancreas, and in each case, has shown limited functionality within the bioartificial organs.

Although not equivalent to the regeneration of large organs, bioscaffolds play a crucial role in the development of mini-organs, known as organoids. Organoids serve as versatile and powerful tools that bridge the gap between traditional cell cultures and whole organisms, providing a more accurate and ethical model for studying human biology and developing novel therapies. However, as mini-organs, organoids inherently reflect the limitations of in vitro systems incapable of generating large, fully functional organs. A method known as clustering-dependent embryoid body transplantation, however, was developed to help in forming organ systems [140]. This technique involves the spatial arrangement of embryoid bodies onto collagen scaffolds to facilitate the generation of functional organ systems. This approach has been successfully applied to the integumentary organ system, resulting in the formation of skin appendages, hair follicles, sebaceous glands, and subcutaneous adipose tissue. But despite the significant progress achieved in organ replacement regenerative therapies, organoid technologies remain at a stage where only mini-organs with limited structural and functional recovery can be generated.

33.6.4 Wound Healing

Proper wound healing necessitates a comprehensive approach that encompasses adequate blood supply, a moist wound environment, effective infection control, nutritional support, inflammation management, and optimal cellular activity [141]. Recent advancements in bioscaffold technology have significantly elevated the standard of wound care. These innovative scaffolds provide essential structural support, efficiently manage exudate, protect against infection, and deliver therapeutic agents directly to the wound site.

Together, these features contribute to improved and accelerated wound healing outcomes, offering a sophisticated and highly effective solution for modern wound care.

Multifunctional hydrogel scaffolds are among the most widely used bioscaffold types in wound healing [142]. These highly porous 3D constructs are based on polysaccharides, allowing them to be functionalized with cells, bioactive molecules, secretomes, and antibacterial agents. This versatility is crucial as different wound types have unique requirements for optimal healing. Acute skin wounds, for example, require effective hemostasis and biocompatibility. Their healing depends on multiple overlapping phases involving interactions between platelets, keratinocytes, fibroblasts, endothelial cells, neutrophils, macrophages, and progenitor cells. Polysaccharide matrix hydrogels provide excellent hemostasis, biocompatibility, antioxidant, anti-inflammatory, and antimicrobial properties, making them ideal for treating acute skin wounds.

In contrast, the healing of diabetic skin wounds necessitates a more nuanced approach [143]. The wound healing process in diabetes is significantly more chaotic compared to normal healing. Microvascular disease leads to reduced blood supply to the wound site, while recurrent bacterial infections, oxidative stress, and chronic inflammation further delay healing. Loading polysaccharide hydrogels with drugs and nanoparticles has proven to be an effective strategy for treating diabetic wounds [141, 142]. Additionally, the treatment of burn wounds has also greatly benefited from the advent of polysaccharide hydrogels and their ability to be functionalized. Excessive exudate in burn wounds can lead to moisture and nutrient loss, promoting bacterial growth. Polysaccharide matrices themselves can disrupt bacterial cell walls, and when loaded with metal-organic frameworks, they release metal ions slowly, enhancing their antibacterial effectiveness.

In addition to hydrogels, collagen-based scaffolds have also emerged as a cornerstone in modern wound care. These scaffolds are available in various forms, each tailored to specific wound types and healing needs, offering unique benefits and applications. Sheet collagen scaffolds, for instance, are flat and dense, providing a uniform surface well suited for supporting superficial wounds, surgical incisions, and burns. Fractionated (powdered) collagen acts as a bioactivator, using collagen fragments to signal macrophage and fibroblast recruitment [143]. The lack of crosslinking in fractionated collagen enhances its incorporation into the wound site, and it can be compounded with sterile saline, mixed in a syringe, and injected into deep tunneling wounds, as commonly seen in diabetic ulcers. Composite collagen bioscaffolds typically come as bilayer wound dressings, consisting of a bovine collagen soft tissue matrix bonded to a silicone sheet. These scaffolds offer multifunctional properties, with the silicone component helping to retain moisture and facilitate healing. Lastly, the role of collagen bioscaffolds in fat pad augmentation is also worth mentioning. In fat pad augmentation, collagen bioscaffolds and other biocompatible substances are implanted into tissue adjacent to bony prominences to help reduce focal pressures that can lead to ulcers.

33.6.5 Surgical Implants

Bioscaffolds have a wide range of applications as surgical implants due to their ability to support tissue regeneration, provide structural integrity, and integrate with host tissues. Orthopedic surgical implants that incorporate bioscaffolds are designed to support bone regeneration, repair defects, and enhance healing outcomes. INFUSE® Bone Graft is an absorbable collagen sponge functionalized with recombinant human Bone Morphogenetic Protein-2 (rhBMP-2) [144]. Actifuse™ is a synthetic bone graft substitute containing silicate-substituted calcium phosphate [145]. Vitoss® is another synthetic bone graft substitute but is made from a combination of beta-tricalcium phosphate (β-TCP) and collagen [146]. These and many other scaffold-based implants are used in spinal fusion, trauma, and reconstructive surgeries to fill bone voids and promote bone regeneration.

As previously mentioned, bioscaffolds are also extensively used in cartilage repair procedures. MACI and AMIC both utilize collagen bioscaffolds to help regenerate articular cartilage at the site of cartilage defects. Meniscal repair procedures also make use of collagen-based bioscaffolds, with Collagen Meniscus Implant (CMI®) serving as one of the most widely used meniscal implants on the market today [147, 148].

Reconstructive surgical applications outside of orthopedics have also greatly benefited from the use of bioscaffold-based implants. AlloDerm®, Strattice™, and Matriderm® are commonly used scaffolds in breast reconstruction surgery that promote tissue regeneration and improve the structural integrity of reconstructed breasts [149, 150]. Gore-Tex® and DuraGraft® have been used to enhance the viability and function of vascular grafts in coronary artery bypass grafting (CABG) and peripheral artery disease treatments [151]. NeuraGen®, Axoguard®, and Reaxon® are commonly used nerve guides designed to repair peripheral nerves by providing a conduit that supports axonal growth and enhances functional recovery [152, 153].

These examples illustrate the diverse applications of bioscaffold-based surgical implants and highlight the variation in structural and biological support that these implants provide, ultimately showcasing their critical role in advancing medical treatments and improving patient outcomes.

33.7 Clinical Challenges and Considerations

Bioscaffolds have emerged as a promising solution in tissue engineering and regenerative medicine. However, the immunogenicity and host response to these scaffolds pose significant challenges, affecting their clinical translation and long-term success. Understanding these challenges is crucial for developing more effective and biocompatible bioscaffolds.

33.7.1 Host Response to Bioscaffolds

The host response to bioscaffolds includes both innate and adaptive immune mechanisms. Upon implantation, the innate immune system initiates the foreign body response (FBR), characterized by the recruitment of neutrophils and macrophages to the site of implantation [154]. This response is followed by the production of inflammatory cytokines, leading to the formation of foreign body giant cells and the encapsulation of the scaffold in a fibrotic capsule.

The innate immune response is primarily mediated by damage-associated molecular patterns (DAMPs), which are released upon cell and ECM damage during scaffold processing or implantation. DAMPs are recognized by pattern recognition receptors (PRRs) on various immune cells, leading to the transcription of genes involved in the inflammation process and the recruitment of immune cells [154, 155]. Pro-inflammatory cytokines such as interleukin-1 (IL-1), IL6, and tumor necrosis factor (TNF) are released, promoting the polarization of macrophages toward the M1 phenotype, which is associated with graft degradation [156–158].

The adaptive immune response is driven by T cells, which require the presence of antigens, co-stimulatory signals, and pro-inflammatory cytokines for activation [159]. Decellularization processes often leave residual antigens, such as major histocompatibility complex (MHC) molecules and minor histocompatibility antigens, within the scaffold. These antigens are recognized by T cells, leading to the production of antibodies and further activation of the immune response [160]. The activation of complement cascades by DAMPs also contributes to the polarization of T cells toward pro-inflammatory phenotypes, exacerbating the immune response and scaffold degradation. In the next subsection, we will explore this process further [161].

33.7.2 Immunogenicity of Bioscaffolds

Immunogenicity refers to the ability of a substance to provoke an immune response. For bioscaffolds, this immune response can lead to rejection and failure of the implanted material. The primary drivers of immunogenicity in bioscaffolds are the presence of major histocompatibility complex (MHC) molecules and other antigenic determinants.

The immunologic rejection of transplanted tissues is primarily attributed to MHC molecules. MHC I molecules, present on all nucleated cells, and MHC II molecules, found on specialized antigen-presenting cells (APCs), play pivotal roles in immune recognition [162]. MHC I molecules activate CD8+ T cells, leading to direct invasion of immunogenic cellular targets. MHC II molecules, on the other hand, stimulate CD4+ T cells, which recruit other inflammatory cells and produce cytokines that further propagate the immune response. The adaptive immune response against bioscaffolds is not limited to mismatched MHC molecules. Minor histocompatibility antigens, which interact with MHC molecules, can also activate T cells and lead to graft rejection, even in MHC-matched transplants [163]. The recognition of these antigens by T cells occurs through three major pathways: direct, indirect, and semi-direct.

In the direct pathway, donor MHC molecules interact directly with recipient T cells, provoking a robust immune response [164, 165]. This response diminishes as donor APCs are depleted over time. In contrast, the indirect pathway uses recipient APCs to process donor MHC molecules or minor antigens and present them to T cells [166, 167]. This pathway induces a less vigorous but prolonged immune response. The semi-direct pathway is a hybrid of both pathways, where recipient dendritic cells acquire and present donor MHC molecules without processing them through trogocytosis (cell-cell transfer of membrane) or exosome-mediated transfer, leading to transient immune responses [168, 169].

33.7.3 Strategies to Mitigate Immunogenicity

Several strategies have been developed to mitigate the immunogenicity of bioscaffolds and enhance their biocompatibility.

Decellularization is the process of removing cellular components from tissues while preserving the ECM structure. Effective decellularization reduces the immunogenic content of the scaffold. However, incomplete decellularization can leave behind immunogenic cellular remnants, provoking severe immune responses. Ensuring the complete removal of DNA and other cellular materials is critical for minimizing immunogenicity.

In addition to decellularization, antigen removal techniques such as the use of α-galactosidase can eliminate specific antigens like α-gal epitopes, reducing the scaffold's immunogenicity [170]. Solubilization methods using salts and detergents can also be employed to remove hydrophilic and hydrophobic antigens, further decreasing the scaffold's immunogenicity [171].

Crosslinking is a technique used to stabilize the ECM structure and mask antigenic sites. Chemical crosslinkers like glutaraldehyde (GA) and natural crosslinkers like genipin can enhance the mechanical stability of the scaffold and reduce its immunogenicity [172, 173]. However, heavy crosslinking can impair biodegradability and promote chronic inflammation, necessitating a balance between stability and biocompatibility.

Sterilization is essential for eliminating pathogens and endotoxins that can enhance the immunogenicity of bioscaffolds. Techniques such as peracetic acid (PAA) treatment, γ-irradiation, and ethylene oxide sterilization are commonly used [101, 174]. Each method has its advantages and drawbacks, and the choice of sterilization technique must consider the preservation of ECM integrity and the elimination of immunogenic contaminants.

33.8 Current Research and Future Directions

33.8.1 Incorporation of Bioactive Molecules

Incorporating bioactive molecules into bioscaffolds is a critical strategy for enhancing their functionality and promoting tissue regeneration. Bioactive molecules, such as growth factors, cytokines, and peptides, can be embedded within scaffolds to provide biological cues that guide cellular behaviors, including proliferation, differentiation, and migration.

Growth factors are potent molecules that play essential roles in tissue development and repair. Incorporating growth factors like bone morphogenetic proteins (BMPs), vascular endothelial growth factor (VEGF), and transforming growth factor-beta (TGF-β) into bioscaffolds can significantly enhance their regenerative potential [175, 176]. For example, BMP-2 has been widely used to promote osteogenesis in bone tissue engineering, while VEGF facilitates angiogenesis, essential for supplying nutrients and oxygen to regenerating tissues [177].

Peptides and small molecules can also be integrated into scaffolds to provide specific biological signals. For instance, the RGD peptide, a well-known cell adhesion motif, can be incorporated to enhance cell attachment and spreading on scaffolds [178, 179]. Similarly, small molecules like dexamethasone can be used to induce osteogenic differentiation of stem cells seeded on the scaffolds [180].

The controlled release of bioactive molecules from scaffolds is crucial for maintaining their bioactivity over the required duration of tissue regeneration. Techniques such as encapsulation in nanoparticles, coating on scaffold surfaces, or embedding within the scaffold matrix can be used to achieve sustained and controlled release of bioactive agents. This approach ensures that the molecules remain effective over time, reducing the need for repeated administration and improving therapeutic outcomes.

33.8.2 Personalized Medicine Approaches

Personalized medicine aims to tailor medical treatments to individual patient characteristics. In the context of bioscaffold design, this approach involves creating scaffolds that are specifically designed to match the patient's unique anatomical and physiological conditions.

Using imaging techniques like MRI or CT scans, patient-specific scaffolds can be designed and fabricated to fit the exact dimensions of the defect site. This precision ensures better integration and function of the scaffold once implanted. 3D printing is particularly well-suited for this purpose, allowing the creation of custom scaffolds with complex geometries and internal structures [181, 182].

Personalized scaffolds can be seeded with autologous cells, such as the patient's own stem cells, to minimize the risk of immune rejection and enhance tissue regeneration. These cells can be harvested, expanded in vitro, and then incorporated into the scaffold before implantation. This approach leverages the patient's biological material to promote healing and regeneration more effectively.

Recent advances in genetic engineering allow for the customization of cell-laden scaffolds at the molecular level. By incorporating genes that encode for specific growth factors or signaling molecules, scaffolds can be designed to produce these bioactive agents in situ. This strategy provides a continuous supply of necessary factors directly at the site of regeneration, enhancing the scaffold's effectiveness [183, 184].

33.8.3 Nanotechnology in Bioscaffold Design

Nanotechnology offers innovative solutions for enhancing the properties and performance of bioscaffolds. The integration of nanomaterials can improve scaffold strength, promote cell interactions, and facilitate the controlled release of therapeutic agents [185]. Nanofibers, produced through electrospinning, create a highly porous scaffold structure that closely mimics the natural ECM. These nanofibrous scaffolds support cell attachment and proliferation, providing a conducive environment for tissue regeneration [186]. Additionally, incorporating nanoparticles into scaffolds can enhance their mechanical properties and functionality. For example, hydroxyapatite nanoparticles can be added to polymer matrices to create nanocomposites with improved osteoconductivity for bone tissue engineering [187, 188].

Nanoparticles can be used to deliver bioactive molecules directly to the scaffold. These particles can be designed to release their cargo in response to specific environmental triggers, such as pH changes or enzymatic activity. This targeted delivery system ensures that bioactive agents are released at the optimal time and location, maximizing their therapeutic impact [189, 190].

Surface functionalization of scaffolds with nanoscale features can enhance cell adhesion, proliferation, and differentiation. Techniques such as nanopatterning or coating with bioactive molecules create surfaces that mimic the natural ECM, providing physical and chemical cues that guide cellular behavior [104, 191]. This approach improves the biocompatibility and efficacy of scaffolds in tissue engineering applications.

References

1. McKee C, Chaudhry GR. Advances and challenges in stem cell culture. Colloids Surf B Biointerfaces. 2017;159:62–77. https://doi.org/10.1016/j.colsurfb.2017.07.051.
2. Baptista CA, Hatten ME, Blazeski R, Mason CA. Cell-cell interactions influence survival and differentiation of purified Purkinje cells in vitro. Neuron. 1994;12(2):243–60. https://doi.org/10.1016/0896-6273(94)90268-2.
3. Frisch SM, Francis H. Disruption of epithelial cell-matrix interactions induces apoptosis. J Cell Biol. 1994;124(4):619–26. https://doi.org/10.1083/jcb.124.4.619.
4. De Pieri A, Rochev Y, Zeugolis DI. Scaffold-free cell-based tissue engineering therapies: advances, shortfalls and forecast. NPJ Regen Med. 2021;6(1):1–15. https://doi.org/10.1038/s41536-021-00133-3.
5. Bi H, Jin Y. Current progress of skin tissue engineering: seed cells, bioscaffolds, and construction strategies. Burns Trauma. 2013;1(2):63–72. https://doi.org/10.4103/2321-3868.118928.
6. Costa A, Naranjo JD, Londono R, Badylak SF. Biologic scaffolds. Cold Spring Harb Perspect Med. 2017;7(9):a025676. https://doi.org/10.1101/cshperspect.a025676.
7. "Polymers | Free Full-Text | A comparative review of natural and synthetic biopolymer composite scaffolds." Accessed 18 June 2024. [Online]. Available: https://www.mdpi.com/2073-4360/13/7/1105.
8. Kaul H, Ventikos Y. On the genealogy of tissue engineering and regenerative medicine. Tissue Eng Part B Rev. 2015;21(2):203–17. https://doi.org/10.1089/ten.teb.2014.0285.
9. Green WT Jr, Ferguson RJ. Histochemical and electron microscopic comparison of tissue produced by rabbit articular chondrocytes in VIVO and in VITRO. Arthritis Rheum. 1975;18(3):273–80. https://doi.org/10.1002/art.1780180313.
10. Russell PS. Selective transplantation. An emerging concept. Ann Surg. 1985;201(3):255–62. https://doi.org/10.1097/00000658-198503000-00001.
11. Langer R, Vacanti JP. Tissue engineering. Science. 1993;260(5110):920–6. https://doi.org/10.1126/science.8493529.
12. Koons GL, Diba M, Mikos AG. Materials design for bone-tissue engineering. Nat Rev Mater. 2020;5(8):584–603. https://doi.org/10.1038/s41578-020-0204-2.
13. O'Brien FJ. Biomaterials & scaffolds for tissue engineering. Mater Today. 2011;14(3):88–95. https://doi.org/10.1016/S1369-7021(11)70058-X.
14. Asadi N, Del Bakhshayesh AR, Davaran S, Akbarzadeh A. Common biocompatible polymeric materials for tissue engineering and regenerative medicine. Mater Chem Phys. 2020;242:122528. https://doi.org/10.1016/j.matchemphys.2019.122528.
15. "Scaffolds from biomaterials: advantages and limitations in bone and tissue engineering | Biologia." Accessed 18 June 2024. [Online]. Available: https://link.springer.com/article/10.1515/biolog-2016-0056.
16. Suesca E, Dias AMA, Braga MEM, de Sousa HC, Fontanilla MR. Multifactor analysis on the effect of collagen concentration, cross-linking and fiber/pore orientation on chemical, microstructural, mechanical and biological properties of collagen type I scaffolds. Mater Sci Eng C Mater Biol Appl. 2017;77:333–41. https://doi.org/10.1016/j.msec.2017.03.243.
17. Raftery RM, et al. Multifunctional biomaterials from the sea: assessing the effects of chitosan incorporation into collagen scaffolds on mechanical and biological functionality. Acta Biomater. 2016;43:160–9. https://doi.org/10.1016/j.actbio.2016.07.009.
18. Gribova V, et al. Construction and myogenic differentiation of 3D myoblast tissues fabricated by

fibronectin-gelatin nanofilm coating. Biochem Biophys Res Commun. 2016;474(3):515–21. https://doi.org/10.1016/j.bbrc.2016.04.130.
19. Moncion A, et al. Design and characterization of fibrin-based acoustically responsive scaffolds for tissue engineering applications. Ultrasound Med Biol. 2016;42(1):257–71. https://doi.org/10.1016/j.ultrasmedbio.2015.08.018.
20. Barsotti MC, Felice F, Balbarini A, Di Stefano R. Fibrin as a scaffold for cardiac tissue engineering. Biotechnol Appl Biochem. 2011;58(5):301–10. https://doi.org/10.1002/bab.49.
21. Sharma C, Dinda AK, Potdar PD, Chou C-F, Mishra NC. Fabrication and characterization of novel nano-biocomposite scaffold of chitosan-gelatin-alginate-hydroxyapatite for bone tissue engineering. Mater Sci Eng C Mater Biol Appl. 2016;64:416–27. https://doi.org/10.1016/j.msec.2016.03.060.
22. Aranaz I, et al. Chitosan: an overview of its properties and applications. Polymers. 2021;13(19):3256. https://doi.org/10.3390/polym13193256.
23. Sheehy EJ, Lemoine M, Clarke D, Gonzalez Vazquez A, O'Brien FJ. The incorporation of marine coral microparticles into collagen-based scaffolds promotes osteogenesis of human mesenchymal stromal cells via calcium ion signalling. Mar Drugs. 2020;18(2):74. https://doi.org/10.3390/md18020074.
24. "Seaweeds and corals from the Brazilian coast: review on biotechnological potential and environmental aspects – PubMed." Accessed 18 June 2024. [Online]. Available: https://pubmed.ncbi.nlm.nih.gov/37298760/.
25. "Advanced polymeric membranes as biomaterials based on marine sources envisaging the regeneration of human tissues – PubMed." Accessed 18 June 2024. [Online]. Available: https://pubmed.ncbi.nlm.nih.gov/36975696/.
26. Sun W, Gregory DA, Tomeh MA, Zhao X. Silk fibroin as a functional biomaterial for tissue engineering. Int J Mol Sci. 2021;22(3):1499. https://doi.org/10.3390/ijms22031499.
27. "Evolution of arthropod silks – PubMed." Accessed 18 June 2024. [Online]. Available: https://pubmed.ncbi.nlm.nih.gov/15012314/.
28. Kazemi T, Mohammadpour AA, Matin MM, Mahdavi-Shahri N, Dehghani H, Kazemi Riabi SH. Decellularized bovine aorta as a promising 3D elastin scaffold for vascular tissue engineering applications. Regen Med. 2021;16(12):1037–50. https://doi.org/10.2217/rme-2021-0062.
29. Zouhair S, et al. A comprehensive comparison of bovine and porcine decellularized pericardia: new insights for surgical applications. Biomol Ther. 2020;10(3):371. https://doi.org/10.3390/biom10030371.
30. Fidalgo C, et al. A sterilization method for decellularized xenogeneic cardiovascular scaffolds. Acta Biomater. 2018;67:282–94. https://doi.org/10.1016/j.actbio.2017.11.035.
31. Dehghani S, Aghaee Z, Soleymani S, Tafazoli M, Ghabool Y, Tavassoli A. An overview of the production of tissue extracellular matrix and decellularization process. Cell Tissue Bank. 2024;25(1):369–87. https://doi.org/10.1007/s10561-023-10112-1.
32. Khakpour E, Tavassoli A, Mahdavi-Shahri N, Matin MM. Assessing the biocompatibility of bovine tendon scaffold, a step forward in tendon tissue engineering. Cell Tissue Bank. 2023;24(1):11–24. https://doi.org/10.1007/s10561-022-10012-w.
33. Xiang J-X, et al. Liver regeneration using decellularized splenic scaffold: a novel approach in tissue engineering. Hepatobiliary Pancreat Dis Int. 2015;14(5):502–8. https://doi.org/10.1016/s1499-3872(15)60423-4.
34. Niveditha K, Ca V, Joseph J, Arun U, John A, Abraham A. Mesenchymal stem cells seeded decellularized tendon scaffold for tissue engineering. Curr Stem Cell Res Ther. 2021;16(2):155–64. https://doi.org/10.2174/1574888X15666200723123901.
35. Mohiuddin OA, et al. Decellularized adipose tissue: biochemical composition, in vivo analysis and potential clinical applications. Adv Exp Med Biol. 2020;1212:57–70. https://doi.org/10.1007/5584_2019_371.
36. Morissette Martin P, Shridhar A, Yu C, Brown C, Flynn LE. Decellularized adipose tissue scaffolds for soft tissue regeneration and adipose-derived stem/stromal cell delivery. Methods Mol Biol. 2018;1773:53–71. https://doi.org/10.1007/978-1-4939-7799-4_6.
37. Kim YS, Majid M, Melchiorri AJ, Mikos AG. Applications of decellularized extracellular matrix in bone and cartilage tissue engineering. Bioeng Transl Med. 2018;4(1):83–95. https://doi.org/10.1002/btm2.10110.
38. Dutta SD, Hexiu J, Patel DK, Ganguly K, Lim K-T. 3D-printed bioactive and biodegradable hydrogel scaffolds of alginate/gelatin/cellulose nanocrystals for tissue engineering. Int J Biol Macromol. 2021;167:644–58. https://doi.org/10.1016/j.ijbiomac.2020.12.011.
39. Krishani M, Shin WY, Suhaimi H, Sambudi NS. Development of scaffolds from bio-based natural materials for tissue regeneration applications: a review. Gels. 2023;9(2):100. https://doi.org/10.3390/gels9020100.
40. "Injectable decellularized cartilage matrix hydrogel encapsulating urine-derived stem cells for immunomodulatory and cartilage defect regeneration | NPJ Regen Med." Accessed 18 June 2024. [Online]. Available: https://www.nature.com/articles/s41536-022-00269-w.
41. Maharjan B, et al. Regenerated cellulose nanofiber reinforced chitosan hydrogel scaffolds for bone tissue engineering. Carbohydr Polym. 2021;251:117023. https://doi.org/10.1016/j.carbpol.2020.117023.
42. Goodarzi H, Jadidi K, Pourmotabed S, Sharifi E, Aghamollaei H. Preparation and in vitro characterization of cross-linked collagen-gelatin hydro-

gel using EDC/NHS for corneal tissue engineering applications. Int J Biol Macromol. 2019;126:620–32. https://doi.org/10.1016/j.ijbiomac.2018.12.125.
43. "Polymeric materials, advances and applications in tissue engineering: a review – PMC." Accessed 18 June 2024. [Online]. Available: https://www.ncbi.nlm.nih.gov/pmc/articles/PMC9952269/.
44. "PLA/Hydroxyapatite scaffolds exhibit in vitro immunological inertness and promote robust osteogenic differentiation of human mesenchymal stem cells without osteogenic stimuli | Sci Rep." Accessed 18 June 2024. [Online]. Available: https://www.nature.com/articles/s41598-022-05207-w.
45. Toosi S, et al. PGA-incorporated collagen: toward a biodegradable composite scaffold for bone-tissue engineering. J Biomed Mater Res A. 2016;104(8):2020–8. https://doi.org/10.1002/jbm.a.35736.
46. "Fabrication of PGA/PLA scaffold with the shape of human nose | IntechOpen." Accessed 18 June 2024. [Online]. Available: https://www.intechopen.com/chapters/43775.
47. Perić Kačarević Ž, et al. An introduction to bone tissue engineering. Int J Artif Organs. 2020;43(2):69–86. https://doi.org/10.1177/0391398819876286.
48. Amsden BG, Sukarto A, Knight DK, Shapka SN. Methacrylated glycol chitosan as a photopolymerizable biomaterial. Biomacromolecules. 2007;8(12):3758–66. https://doi.org/10.1021/bm700691e.
49. Wirostko B, Mann BK, Williams DL, Prestwich GD. Ophthalmic uses of a thiol-modified hyaluronan-based hydrogel. Adv Wound Care. 2014;3(11):708–16. https://doi.org/10.1089/wound.2014.0572.
50. "Drug delivery systems based on polyethylene glycol hydrogels for enhanced bone regeneration – PMC." Accessed 18 June 2024. [Online]. Available: https://www.ncbi.nlm.nih.gov/pmc/articles/PMC9923112/.
51. Brown CFC, Yan J, Han TTY, Marecak DM, Amsden BG, Flynn LE. Effect of decellularized adipose tissue particle size and cell density on adipose-derived stem cell proliferation and adipogenic differentiation in composite methacrylated chondroitin sulphate hydrogels. Biomed Mater. Bristol Engl. 2015;10(4):045010. https://doi.org/10.1088/1748-6041/10/4/045010.
52. Woodruff M, Hutmacher D. The return of a forgotten polymer-polycaprolactone in the 21st century. Prog Polym Sci. 2010;35. https://doi.org/10.1016/j.progpolymsci.2010.04.002.
53. Ramakrishna S, Mayer J, Wintermantel E, Leong KW. Biomedical applications of polymer-composite materials: a review. Compos Sci Technol. 2001;61(9):1189–224. https://doi.org/10.1016/S0266-3538(00)00241-4.
54. Rezwan K, Chen QZ, Blaker JJ, Boccaccini AR. Biodegradable and bioactive porous polymer/inorganic composite scaffolds for bone tissue engineering. Biomaterials. 2006;27(18):3413–31. https://doi.org/10.1016/j.biomaterials.2006.01.039.
55. Hench LL. Bioceramics. J Am Ceram Soc. 1998;81(7):1705–28. https://doi.org/10.1111/j.1151-2916.1998.tb02540.x.
56. Middleton JC, Tipton AJ. Synthetic biodegradable polymers as orthopedic devices. Biomaterials. 2000;21(23):2335–46. https://doi.org/10.1016/s0142-9612(00)00101-0.
57. Bhat S, Kumar A. Biomaterials and bioengineering tomorrow's healthcare. Biomatter. 2013;3(3):e24717. https://doi.org/10.4161/biom.24717.
58. Hutmacher DW. Scaffolds in tissue engineering bone and cartilage. Biomaterials. 2000;21(24):2529–43. https://doi.org/10.1016/s0142-9612(00)00121-6.
59. Murphy CM, O'Brien FJ. Understanding the effect of mean pore size on cell activity in collagen-glycosaminoglycan scaffolds. Cell Adhes Migr. 2010;4(3):377–81. https://doi.org/10.4161/cam.4.3.11747.
60. O'Brien FJ, Harley BA, Yannas IV, Gibson LJ. The effect of pore size on cell adhesion in collagen-GAG scaffolds. Biomaterials. 2005;26(4):433–41. https://doi.org/10.1016/j.biomaterials.2004.02.052.
61. O'Brien FJ, Harley BA, Waller MA, Yannas IV, Gibson LJ, Prendergast PJ. The effect of pore size on permeability and cell attachment in collagen scaffolds for tissue engineering. Technol Health Care. 2007;15(1):3–17.
62. Yang J, et al. Fabrication and surface modification of macroporous poly(L-lactic acid) and poly(L-lactic-co-glycolic acid) (70/30) cell scaffolds for human skin fibroblast cell culture. J Biomed Mater Res. 2002;62(3):438–46. https://doi.org/10.1002/jbm.10318.
63. "Mechanical and biological properties of scaffold materials – ScienceDirect." Accessed 18 June 2024. [Online]. Available: https://www.sciencedirect.com/science/article/abs/pii/B978008100979600001X.
64. Chan BP, Leong KW. Scaffolding in tissue engineering: general approaches and tissue-specific considerations. Eur Spine J. 2008;17(Suppl 4):467–79. https://doi.org/10.1007/s00586-008-0745-3.
65. Roseti L, et al. Scaffolds for bone tissue engineering: state of the art and new perspectives. Mater Sci Eng C Mater Biol Appl. 2017;78:1246–62. https://doi.org/10.1016/j.msec.2017.05.017.
66. Wang X, et al. Topological design and additive manufacturing of porous metals for bone scaffolds and orthopaedic implants: a review. Biomaterials. 2016;83:127–41. https://doi.org/10.1016/j.biomaterials.2016.01.012.
67. Rege A. Constitutive modeling of the densification behavior in open-porous cellular solids. Materials. 2021;14(11):2731. https://doi.org/10.3390/ma14112731.
68. Roberts TT, Rosenbaum AJ. Bone grafts, bone substitutes and orthobiologics. Organogenesis. 2012;8(4):114–24. https://doi.org/10.4161/org.23306.

69. "The biology of bone graft repair – PubMed." Accessed 18 June 2024. [Online]. Available: https://pubmed.ncbi.nlm.nih.gov/6339139/.
70. Nikolova MP, Chavali MS. Recent advances in biomaterials for 3D scaffolds: a review. Bioact Mater. 2019;4:271. https://doi.org/10.1016/j.bioactmat.2019.10.005.
71. Mostafa S, et al. The wonders of BMP9: from mesenchymal stem cell differentiation, angiogenesis, neurogenesis, tumorigenesis, and metabolism to regenerative medicine. Genes Dis. 2019;6(3):201–23. https://doi.org/10.1016/j.gendis.2019.07.003.
72. Zhao X, et al. Applications of biocompatible scaffold materials in stem cell-based cartilage tissue engineering. Front Bioeng Biotechnol. 2021;9:603444. https://doi.org/10.3389/fbioe.2021.603444.
73. Wei Y-H, Chen W-C, Wu H-S, Janarthanan O-M. Biodegradable and biocompatible biomaterial, polyhydroxybutyrate, produced by an indigenous Vibrio sp. BM-1 isolated from marine environment. Mar Drugs. 2011;9(4):615–24. https://doi.org/10.3390/md9040615.
74. "Bioactivity – an overview | ScienceDirect Topics." Accessed 18 June 2024. [Online]. Available: https://www.sciencedirect.com/topics/engineering/bioactivity.
75. Tabatabaei Mirakabad FS, et al. Optimization of topography and surface properties of polyacrylonitrile-based electrospun scaffolds via nonoclay concentrations and its effect on osteogenic differentiation of human mesenchymal stem cells. Iran J Pharm Res. 2021;20(4):385–504. https://doi.org/10.22037/ijpr.2021.115119.15208.
76. Chi J, et al. Topographic orientation of scaffolds for tissue regeneration: recent advances in biomaterial design and applications. Biomimetics. 2022;7(3):131. https://doi.org/10.3390/biomimetics7030131.
77. Zamani F, Amani-Tehran M, Latifi M, Shokrgozar MA. The influence of surface nanoroughness of electrospun PLGA nanofibrous scaffold on nerve cell adhesion and proliferation. J Mater Sci Mater Med. 2013;24(6):1551–60. https://doi.org/10.1007/s10856-013-4905-6.
78. Parikh SD, Wang W, Nelson MT, Sulentic CEW, Mukhopadhyay SM. Bioinspired hierarchical carbon structures as potential scaffolds for wound healing and tissue regeneration applications. Nanomaterials. 2023;13(11):1791. https://doi.org/10.3390/nano13111791.
79. Heijkants RGJC, et al. Preparation of a polyurethane scaffold for tissue engineering made by a combination of salt leaching and freeze-drying of dioxane. J Mater Sci. 2006;41(8):2423–8. https://doi.org/10.1007/s10853-006-7065-y.
80. "Polymers | Free Full-Text | Comparison of scaffolds fabricated via 3D printing and salt leaching: in vivo imaging, biodegradation, and inflammation." Accessed 18 June 2024. [Online]. Available: https://www.mdpi.com/2073-4360/12/10/2210.
81. Blaker JJ, Knowles JC, Day RM. Novel fabrication techniques to produce microspheres by thermally induced phase separation for tissue engineering and drug delivery. Acta Biomater. 2008;4(2):264–72. https://doi.org/10.1016/j.actbio.2007.09.011.
82. Soliman S, et al. Multiscale three-dimensional scaffolds for soft tissue engineering via multimodal electrospinning. Acta Biomater. 2010;6(4):1227–37. https://doi.org/10.1016/j.actbio.2009.10.051.
83. Zulkifli MZA, Nordin D, Shaari N, Kamarudin SK. Overview of electrospinning for tissue engineering applications. Polymers. 2023;15(11):2418. https://doi.org/10.3390/polym15112418.
84. Research Centre for Sustainable Process Technology, Faculty of Engineering & Built Environment, Universiti Kebangsaan Malaysia, Research Centre for Sustainable Process Technology, Nordin D, Zulkifli MZA, Kamarudin NHN. Polycaprolactone/chlorophyllin sodium copper salt nanofibrous mats prepared by electrospinning for soft tissue engineering. J Kejuruter. 2019;si2(1):69–76. https://doi.org/10.17576/jkukm-2019-si2(1)-09.
85. Bhardwaj N, Kundu SC. Electrospinning: a fascinating fiber fabrication technique. Biotechnol Adv. 2010;28(3):325–47. https://doi.org/10.1016/j.biotechadv.2010.01.004.
86. Barcena AJR, et al. Bioresorbable mesenchymal stem cell-loaded electrospun polymeric scaffold inhibits neointimal hyperplasia following arteriovenous fistula formation in a rat model of chronic kidney disease. Adv Healthc Mater. 2023;12(26):e2300960. https://doi.org/10.1002/adhm.202300960.
87. Zaszczyńska A, et al. Enhanced electroactive phases of poly (vinylidene fluoride) fibers for tissue engineering applications. Int J Mol Sci. 2024;25(9):4980. https://doi.org/10.3390/ijms25094980.
88. Sultana N, Wang M. PHBV/PLLA-based composite scaffolds fabricated using an emulsion freezing/freeze-drying technique for bone tissue engineering: surface modification and in vitro biological evaluation. Biofabrication. 2012;4(1):015003. https://doi.org/10.1088/1758-5082/4/1/015003.
89. Nagahama K, Ouchi T, Ohya Y. Temperature-induced hydrogels through self-assembly of cholesterol-substituted star PEG-b-PLLA copolymers: an injectable scaffold for tissue engineering. Adv Funct Mater. 2008;18(8):1220–31. https://doi.org/10.1002/adfm.200700587.
90. Peck M, Dusserre N, McAllister TN, L'Heureux N. Tissue engineering by self-assembly. Mater Today. 2011;14(5):218–24. https://doi.org/10.1016/S1369-7021(11)70117-1.
91. Castelletto V, Hamley IW, Adamcik J, Mezzenga R, Gummel J. Modulating self-assembly of a nanotape-forming peptide amphiphile with an oppositely charged surfactant. Soft Matter. 2011;8(1):217–26. https://doi.org/10.1039/C1SM06677C.
92. Andrique L, et al. A model of guided cell self-organization for rapid and spontaneous formation

of functional vessels. Sci Adv. 2019;5(6):eaau6562. https://doi.org/10.1126/sciadv.aau6562.
93. Pampaloni F, Reynaud EG, Stelzer EHK. The third dimension bridges the gap between cell culture and live tissue. Nat Rev Mol Cell Biol. 2007;8(10):839–45. https://doi.org/10.1038/nrm2236.
94. Rijal G, Li W. 3D scaffolds in breast cancer research. Biomaterials. 2016;81:135–56. https://doi.org/10.1016/j.biomaterials.2015.12.016.
95. "The mode and dynamics of glioblastoma cell invasion into a decellularized tissue-derived extracellular matrix-based three-dimensional tumor model | Sci Rep." Accessed 18 June 2024. [Online]. Available: https://www.nature.com/articles/s41598-018-22681-3.
96. Sood D, et al. Fetal brain extracellular matrix boosts neuronal network formation in 3D bioengineered model of cortical brain tissue. ACS Biomater Sci Eng. 2016;2(1):131–40. https://doi.org/10.1021/acsbiomaterials.5b00446.
97. Nguyen DT, Althage M, Magnone MC, Heydarkhan-Hagvall S. Translational strategy: humanized mini-organs. Drug Discov Today. 2018;23(11):1812–7. https://doi.org/10.1016/j.drudis.2018.05.039.
98. Qiu Y, Park K. Environment-sensitive hydrogels for drug delivery. Adv Drug Deliv Rev. 2001;53(3):321–39. https://doi.org/10.1016/s0169-409x(01)00203-4.
99. Ma J, et al. Hydrogels for localized chemotherapy of liver cancer: a possible strategy for improved and safe liver cancer treatment. Drug Deliv. 29(1):1457–76. https://doi.org/10.1080/10717544.2022.2070299.
100. Liu B, Chen K. Advances in hydrogel-based drug delivery systems. Gels. 2024;10(4):262. https://doi.org/10.3390/gels10040262.
101. Yugatama A, et al. Oral delivery of photopolymerizable nanogels loaded with gemcitabine for pancreatic cancer therapy: formulation design, and in vitro and in vivo evaluations. Int J Nanomedicine. 2024;19:3753–72. https://doi.org/10.2147/IJN.S443610.
102. Sun Z, Song C, Wang C, Hu Y, Wu J. Hydrogel-based controlled drug delivery for cancer treatment: a review. Mol Pharm. 2020;17(2):373–91. https://doi.org/10.1021/acs.molpharmaceut.9b01020.
103. Khan F, et al. Synthesis, classification and properties of hydrogels: their applications in drug delivery and agriculture. J Mater Chem B. 2022;10(2):170–203. https://doi.org/10.1039/d1tb01345a.
104. Vayachuta L, et al. Printable-microencapsulated ascorbic acid for personalized topical delivery. ACS Appl Bio Mater. 2023;6(12):5385–98. https://doi.org/10.1021/acsabm.3c00648.
105. Yi H-G, et al. A 3D-printed local drug delivery patch for pancreatic cancer growth suppression. J Control Release. 2016;238:231–41. https://doi.org/10.1016/j.jconrel.2016.06.015.
106. Abuwatfa WH, Pitt WG, Husseini GA. Scaffold-based 3D cell culture models in cancer research. J Biomed Sci. 2024;31(1):7. https://doi.org/10.1186/s12929-024-00994-y.
107. Dalir Abdolahinia E, Han X. The three-dimensional in vitro cell culture models in the study of oral cancer immune microenvironment. Cancers. 2023;15(17):4266. https://doi.org/10.3390/cancers15174266.
108. Chen HJ, et al. A recellularized human colon model identifies cancer driver genes. Nat Biotechnol. 2016;34(8):845–51. https://doi.org/10.1038/nbt.3586.
109. Zhao L, Huang L, Yu S, Zheng J, Wang H, Zhang Y. Decellularized tongue tissue as an in vitro model for studying tongue cancer and tongue regeneration. Acta Biomater. 2017;58:122–35. https://doi.org/10.1016/j.actbio.2017.05.062.
110. Fu H, et al. Tenascin-C is a major component of the fibrogenic niche in kidney fibrosis. J Am Soc Nephrol. 2017;28(3):785–801. https://doi.org/10.1681/ASN.2016020165.
111. Aasadollahei N, Rezaei N, Golroo R, Agarwal T, Vosough M, Piryaei A. Bioengineering liver microtissues for modeling non-alcoholic fatty liver disease. EXCLI J. 2023;22:367–91. https://doi.org/10.17179/excli2022-5892.
112. Dai Q, Jiang W, Huang F, Song F, Zhang J, Zhao H. Recent advances in liver engineering with decellularized scaffold. Front Bioeng Biotechnol. 2022;10:831477. https://doi.org/10.3389/fbioe.2022.831477.
113. Mazza G, et al. Decellularized human liver as a natural 3D-scaffold for liver bioengineering and transplantation. Sci Rep. 2015;5:13079. https://doi.org/10.1038/srep13079.
114. Zhang C, et al. Arsenic promotes NF-Kb-mediated fibroblast dysfunction and matrix remodeling to impair muscle stem cell function. Stem Cells. 2016;34(3):732–42. https://doi.org/10.1002/stem.2232.
115. Stearns-Reider KM, et al. Aging of the skeletal muscle extracellular matrix drives a stem cell fibrogenic conversion. Aging Cell. 2017;16(3):518–28. https://doi.org/10.1111/acel.12578.
116. Wassenaar JW, Boss GR, Christman KL. Decellularized skeletal muscle as an in vitro model for studying drug-extracellular matrix interactions. Biomaterials. 2015;64:108–14. https://doi.org/10.1016/j.biomaterials.2015.06.033.
117. Del Bakhshayesh AR, et al. An overview of advanced biocompatible and biomimetic materials for creation of replacement structures in the musculoskeletal systems: focusing on cartilage tissue engineering. J Biol Eng. 2019;13(1):85. https://doi.org/10.1186/s13036-019-0209-9.
118. Dong C, Lv Y. Application of collagen scaffold in tissue engineering: recent advances and new perspectives. Polymers. 2016;8(2):42. https://doi.org/10.3390/polym8020042.
119. Morwood AJ, El-Karim IA, Clarke SA, Lundy FT. The role of extracellular matrix (ECM) adhesion motifs in functionalised hydrogels. Molecules. 2023;28(12):4616. https://doi.org/10.3390/molecules28124616.

120. Rosadas M, Silva IV, Costa JB, Ribeiro VP, Oliveira AL. Decellularized dermal matrices: unleashing the potential in tissue engineering and regenerative medicine. Front Mater. 2024;10. https://doi.org/10.3389/fmats.2023.1285948.
121. Schmidt AH. Autologous bone graft: is it still the gold standard? Injury. 2021;52(Suppl 2):S18–22. https://doi.org/10.1016/j.injury.2021.01.043.
122. Koushik TM, Miller CM, Antunes E. Bone tissue engineering scaffolds: function of multi-material hierarchically structured scaffolds. Adv Healthc Mater. 2023;12(9):e2202766. https://doi.org/10.1002/adhm.202202766.
123. Han X, et al. Intelligent vascularized 3D/4D/5D/6D-printed tissue scaffolds. Nanomicro Lett. 2023;15(1):239. https://doi.org/10.1007/s40820-023-01187-2.
124. Amirazad H, Dadashpour M, Zarghami N. Application of decellularized bone matrix as a bioscaffold in bone tissue engineering. J Biol Eng. 2022;16(1):1. https://doi.org/10.1186/s13036-021-00282-5.
125. Yu H, Peng J, Xu Y, Chang J, Li H. Bioglass activated skin tissue engineering constructs for wound healing. ACS Appl Mater Interfaces. 2016;8(1):703–15. https://doi.org/10.1021/acsami.5b09853.
126. Cannio M, Bellucci D, Roether JA, Boccaccini DN, Cannillo V. Bioactive glass applications: a literature review of human clinical trials. Materials. 2021;14(18):5440. https://doi.org/10.3390/ma14185440.
127. Rahmati S, Khazaei M, Nadi A, Alizadeh M, Rezakhani L. Exosome-loaded scaffolds for regenerative medicine in hard tissues. Tissue Cell. 2023;82:102102. https://doi.org/10.1016/j.tice.2023.102102.
128. Oliveira ÉR, et al. Advances in growth factor delivery for bone tissue engineering. Int J Mol Sci. 2021;22(2):903. https://doi.org/10.3390/ijms22020903.
129. Poon B, Kha T, Tran S, Dass CR. Bone morphogenetic protein-2 and bone therapy: successes and pitfalls. J Pharm Pharmacol. 2016;68(2):139–47. https://doi.org/10.1111/jphp.12506.
130. "History of regenerative medicine in the field of orthopedics – J Arthrosc Surg Sports Med." Accessed 18 June 2024. [Online]. Available: https://jassm.org/history-of-regenerative-medicine-in-the-field-of-orthopedics/.
131. Liu Y, Shah KM, Luo J. Strategies for articular cartilage repair and regeneration. Front Bioeng Biotechnol. 2021;9. https://doi.org/10.3389/fbioe.2021.770655.
132. Evenbratt H, Andreasson L, Bicknell V, Brittberg M, Mobini R, Simonsson S. Insights into the present and future of cartilage regeneration and joint repair. Cell Regen. 2022;11(1). https://doi.org/10.1186/s13619-021-00104-5.
133. Smith BD, Grande DA. The current state of scaffolds for musculoskeletal regenerative applications. Nat Rev Rheumatol. 2015;11(4):213–22. https://doi.org/10.1038/nrrheum.2015.27.
134. LaPorta TF, Richter A, Sgaglione NA, Grande DA. Clinical relevance of scaffolds for cartilage engineering. Orthop Clin North Am. 2012;43(2):245–54, vi. https://doi.org/10.1016/j.ocl.2012.02.002.
135. Brix MO, Stelzeneder D, Trattnig S, Windhager R, Domayer SE. Cartilage repair of the knee with Hyalograft C:® magnetic resonance imaging assessment of the glycosaminoglycan content at mid-term. Int Orthop. 2013;37(1):39–43. https://doi.org/10.1007/s00264-012-1700-9.
136. Li X, Ding J, Wang J, Zhuang X, Chen X. Biomimetic biphasic scaffolds for osteochondral defect repair. Regen Biomater. 2015;2(3):221–8. https://doi.org/10.1093/rb/rbv015.
137. D'Ambrosi R, Valli F, De Luca P, Ursino N, Usuelli FG. MaioRegen osteochondral substitute for the treatment of knee defects: a systematic review of the literature. J Clin Med. 2019;8(6):783. https://doi.org/10.3390/jcm8060783.
138. Boffa A, Solaro L, Poggi A, Andriolo L, Reale D, Di Martino A. Multi-layer cell-free scaffolds for osteochondral defects of the knee: a systematic review and meta-analysis of clinical evidence. J Exp Orthop. 2021;8(1):56. https://doi.org/10.1186/s40634-021-00377-4.
139. Takeo M, Tsuji T. Organ regeneration based on developmental biology: past and future. Curr Opin Genet Dev. 2018;52:42–7. https://doi.org/10.1016/j.gde.2018.05.008.
140. Yesmin S, Paget MB, Murray HE, Downing R. Bio-scaffolds in organ-regeneration: clinical potential and current challenges. Curr Res Transl Med. 2017;65(3):103–13. https://doi.org/10.1016/j.retram.2017.08.002.
141. Fadilah NIM, et al. Functionalised-biomatrix for wound healing and cutaneous regeneration: future impactful medical products in clinical translation and precision medicine. Front Bioeng Biotechnol. 2023;11:1160577. https://doi.org/10.3389/fbioe.2023.1160577.
142. Ding C, et al. Multifunctional hydrogel bioscaffolds based on polysaccharide to promote wound healing: a review. Int J Biol Macromol. 2024;259(Pt 2):129356. https://doi.org/10.1016/j.ijbiomac.2024.129356.
143. Landsman A, Taft D, Riemer K. The role of collagen bioscaffolds, foamed collagen, and living skin equivalents in wound healing. Clin Podiatr Med Surg. 2009;26(4):525–33. https://doi.org/10.1016/j.cpm.2009.08.012.
144. McKay WF, Peckham SM, Badura JM. A comprehensive clinical review of recombinant human bone morphogenetic protein-2 (INFUSE Bone Graft). Int Orthop. 2007;31(6):729–34. https://doi.org/10.1007/s00264-007-0418-6.
145. Lerner T, Liljenqvist U. Silicate-substituted calcium phosphate as a bone graft substitute in surgery for adolescent idiopathic scoliosis. Eur Spine J. 2013;22

Suppl 2(Suppl 2):S185–94. https://doi.org/10.1007/s00586-012-2485-7.
146. Sinha R, Menon P, Chakranarayan A. Vitoss synthetic cancellous bone (void filler). Med J Armed Forces India. 2009;65(2):173. https://doi.org/10.1016/S0377-1237(09)80136-6.
147. Monibi FA, Cook JL. Tissue-derived extracellular matrix bioscaffolds: emerging applications in cartilage and meniscus repair. Tissue Eng Part B Rev. 2017;23(4):386–98. https://doi.org/10.1089/ten.TEB.2016.0431.
148. Grassi A, Lucidi GA, Agostinone P, di Paolo S, dal Fabbro G, Zaffagnini S. Lateral collagen meniscus implant (CMI): techniques and outcomes—a narrative review. Ann Jt. 2022;7:18. https://doi.org/10.21037/aoj-21-2.
149. Jansen LA, Macadam SA. The use of AlloDerm in postmastectomy alloplastic breast reconstruction: part I. A systematic review. Plast Reconstr Surg. 2011;127(6):2232–44. https://doi.org/10.1097/PRS.0b013e3182131c56.
150. Dikmans REG, et al. Single-stage breast reconstruction using Strattice™: a retrospective study. J Plast Reconstr Aesthetic Surg. 2016;69(2):227–33. https://doi.org/10.1016/j.bjps.2015.11.008.
151. Sandner S, et al. Clinical outcomes and quality of life after contemporary isolated coronary bypass grafting: a prospective cohort study. Int J Surg. 2023;109(4):707–15. https://doi.org/10.1097/JS9.0000000000000259.
152. di Summa PG, Kingham PJ, Campisi CC, Raffoul W, Kalbermatten DF. Collagen (NeuraGen®) nerve conduits and stem cells for peripheral nerve gap repair. Neurosci Lett. 2014;572:26–31. https://doi.org/10.1016/j.neulet.2014.04.029.
153. Thomson SE, et al. Bioengineered nerve conduits and wraps for peripheral nerve repair of the upper limb. Cochrane Database Syst Rev. 2017;2017(3):CD012574. https://doi.org/10.1002/14651858.CD012574.
154. Janeway CA Jr, Medzhitov R. Innate immune recognition. Annu Rev Immunol. 2002;20. https://doi.org/10.1146/annurev.immunol.20.083001.084359.
155. "The inflammasomes: guardians of the body – PubMed." Accessed 18 June 2024. [Online]. Available: https://pubmed.ncbi.nlm.nih.gov/19302040/.
156. Tidball JG, Villalta SA. Regulatory interactions between muscle and the immune system during muscle regeneration. Am J Physiol Regul Integr Comp Physiol. 2010;298(5):R1173–87. https://doi.org/10.1152/ajpregu.00735.2009.
157. Janeway CA Jr, et al. Immunobiology. 5th ed. New York: Garland Science; 2001.
158. Spiller KL, et al. Sequential delivery of immunomodulatory cytokines to facilitate the M1-to-M2 transition of macrophages and enhance vascularization of bone scaffolds. Biomaterials. 2015;37. https://doi.org/10.1016/j.biomaterials.2014.10.017.
159. Zakrzewski JL, van den Brink MRM, Hubbell JA. Overcoming immunological barriers in regenerative medicine. Nat Biotechnol. 2014;32(8). https://doi.org/10.1038/nbt.2960.
160. Alberts B, Johnson A, Lewis J, Raff M, Roberts K, Walter P. T cells and MHC proteins. In: Molecular biology of the cell. 4th ed. Garland Science; 2002. Accessed 18 June 2024. [Online]. Available: https://www.ncbi.nlm.nih.gov/books/NBK26926/.
161. Matzinger P. Tolerance, danger, and the extended family. Annu Rev Immunol. 1994;12. https://doi.org/10.1146/annurev.iy.12.040194.005015.
162. Harper SJF, et al. CD8 T-cell recognition of acquired alloantigen promotes acute allograft rejection. Proc Natl Acad Sci USA. 2015;112(41):12788. https://doi.org/10.1073/pnas.1513533112.
163. Tostanoski LH, Gosselin EA, Jewell CM. Engineering tolerance using biomaterials to target and control antigen presenting cells. Discov Med. 2016;21(117). Accessed 18 June 2024. [Online]. Available: https://pubmed.ncbi.nlm.nih.gov/27355336/.
164. Siu JHY, Surendrakumar V, Richards JA, Pettigrew GJ. T cell allorecognition pathways in solid organ transplantation. Front Immunol. 2018;9:2548. https://doi.org/10.3389/fimmu.2018.02548.
165. Elkins WL, Guttmann RD. Pathogenesis of a local graft versus host reaction: immunogenicity of circulating host leukocytes. Science. 1968;159(3820):1250–1. https://doi.org/10.1126/science.159.3820.1250.
166. Gallon L, Watschinger B, Murphy B, Akalin E, Sayegh MH, Carpenter CB. The indirect pathway of allorecognition. The occurrence of self-restricted T cell recognition of allo-MHC peptides early in acute renal allograft rejection and its inhibition by conventional immunosuppression. Transplantation. 1995;59(4). Accessed 18 June 2024. [Online]. Available: https://pubmed.ncbi.nlm.nih.gov/7878766/.
167. "How T cells recognize alloantigen: evidence for two pathways of allorecognition – PubMed." Accessed 18 June 2024. [Online]. Available: https://pubmed.ncbi.nlm.nih.gov/8559468/.
168. Afzali B, Lombardi G, Lechler RI. Pathways of major histocompatibility complex allorecognition. Curr Opin Organ Transplant. 2008;13(4):438–44. https://doi.org/10.1097/MOT.0b013e328309ee31.
169. Nakayama M, Hori A, Toyoura S, Yamaguchi S-I. Shaping of T cell functions by trogocytosis. Cells. 2021;10(5). https://doi.org/10.3390/cells10051155.
170. Kasravi M, et al. Immunogenicity of decellularized extracellular matrix scaffolds: a bottleneck in tissue engineering and regenerative medicine. Biomater Res. 2023;27. https://doi.org/10.1186/s40824-023-00348-z.
171. White LJ, et al. The impact of detergents on the tissue decellularization process: a ToF-SIMS study. Acta

Biomater. 2017;50:207. https://doi.org/10.1016/j.actbio.2016.12.033.

172. "Graft-specific immune tolerance is determined by residual antigenicity of xenogeneic extracellular matrix scaffolds – PMC." Accessed 18 June 2024. [Online]. Available: https://www.ncbi.nlm.nih.gov/pmc/articles/PMC6349227/.
173. Gao M, et al. Comparative evaluation of decellularized porcine liver matrices crosslinked with different chemical and natural crosslinking agents. Xenotransplantation. 2019;26(1). https://doi.org/10.1111/xen.12470.
174. Yoganarasimha S, et al. Peracetic acid: a practical agent for sterilizing heat-labile polymeric tissue-engineering scaffolds. Tissue Eng Part C Methods. 2014;20(9):714. https://doi.org/10.1089/ten.tec.2013.0624.
175. Kim J, et al. Bone regeneration using hyaluronic acid-based hydrogel with bone morphogenic protein-2 and human mesenchymal stem cells. Biomaterials. 2007;28(10). https://doi.org/10.1016/j.biomaterials.2006.11.050.
176. Gaharwar AK, et al. Nanoclay-enriched poly (ε-caprolactone) electrospun scaffolds for osteogenic differentiation of human mesenchymal stem cells. Tissue Eng Part A. 2014;20(15–16):2088. https://doi.org/10.1089/ten.tea.2013.0281.
177. "BMP-2 and VEGF-A modRNAs in collagen scaffold synergistically drive bone repair through osteogenic and angiogenic pathways – PMC." Accessed 19 June 2024. [Online]. Available: https://www.ncbi.nlm.nih.gov/pmc/articles/PMC7815925/.
178. Bellis SL. Advantages of RGD peptides for directing cell association with biomaterials. Biomaterials. 2011;32(18):4205. https://doi.org/10.1016/j.biomaterials.2011.02.029.
179. Kumar VB, Tiwari OS, Finkelstein-Zuta G, Rencus-Lazar S, Gazit E. Design of functional RGD peptide-based biomaterials for tissue engineering. Pharmaceutics. 2023;15(2). https://doi.org/10.3390/pharmaceutics15020345.
180. Sun Z, et al. A dexamethasone-eluting porous scaffold for bone regeneration fabricated by selective laser sintering. ACS Appl Bio Mater. 2020;3(12):8739–47. https://doi.org/10.1021/acsabm.0c01126.
181. "3D-printed scaffold engineered to grow complex tissues," National Institutes of Health (NIH). Accessed 19 June 2024. [Online]. Available: https://www.nih.gov/news-events/nih-research-matters/3d-printed-scaffold-engineered-grow-complex-tissues.
182. Li H, Yin Y, Xiang Y, Liu H, Guo R. A novel 3D printing PCL/GelMA scaffold containing USPIO for MRI-guided bile duct repair. Biomed Mater. 2020;15(4). https://doi.org/10.1088/1748-605X/ab797a.
183. Place ES, Evans ND, Stevens MM. Complexity in biomaterials for tissue engineering. Nat Mater. 2009;8(6):457–70. https://doi.org/10.1038/nmat2441.
184. Wang M, Deng Z, Guo Y, Xu P. Designing functional hyaluronic acid-based hydrogels for cartilage tissue engineering. Mater Today Bio. 2022;17. https://doi.org/10.1016/j.mtbio.2022.100495.
185. Vasita R, Katti DS. Nanofibers and their applications in tissue engineering. Int J Nanomedicine. 2006;1(1):15. https://doi.org/10.2147/nano.2006.1.1.15.
186. Jiang Z, et al. Nanofiber scaffolds as drug delivery systems promoting wound healing. Pharmaceutics. 2023;15(7). https://doi.org/10.3390/pharmaceutics15071829.
187. "Recent developments in polymer nanocomposites for bone regeneration – PMC." Accessed 19 June 2024. [Online]. Available: https://www.ncbi.nlm.nih.gov/pmc/articles/PMC9959928/.
188. "The application of hyaluronic acid in bone regeneration – PubMed." Accessed 19 June 2024. [Online]. Available: https://pubmed.ncbi.nlm.nih.gov/31751713/.
189. Czekanska EM, et al. Combinatorial delivery of bioactive molecules by a nanoparticle-decorated and functionalized biodegradable scaffold. J Mater Chem B. 2018;6(27). https://doi.org/10.1039/c8tb00474a.
190. Li Y, Xu C, Lei C. The delivery and activation of growth factors using nanomaterials for bone repair. Pharmaceutics. 2023;15(3). https://doi.org/10.3390/pharmaceutics15031017.
191. Sun Y, Jallerat Q, Szymanski JM, Feinberg AW. Conformal nanopatterning of extracellular matrix proteins onto topographically complex surfaces. Nat Methods. 2015;12(2):134. https://doi.org/10.1038/nmeth.3210.

Scaffold-Free Tissue Engineering in Osteochondral Restoration

34

Kazunori Shimomura, Wataru Ando,
David A. Hart, and Norimasa Nakamura

34.1 Introduction

Osteoarthritis (OA) is a common disease that causes joint pain, joint deformity, and functional disability, impacting the quality of life of both young and elderly patients worldwide [1]. Current treatment strategies can be divided into nonsurgical (conservative) and surgical therapies, depending on the severity of OA [2, 3]. In the early stages of OA, pharmacologic and/or physical therapies are typically selected as conservative treatments to reduce pain and, in some cases, to attempt to delay the progressive structural deterioration in affected joints. Surgical therapies, such as joint replacement and osteotomy, are available for patients who fail to respond to more conservative measures. These treatments are well established and effective for reducing pain and improving quality of life. However, despite these therapeutic options, there is no method available that facilitates the complete healing of the articular cartilage [4–9]. Recently, several biological approaches, such as the use of tissue-engineered materials, have been tested to overcome these challenges. This chapter will focus on the feasibility of tissue-engineered materials in osteochondral repair and highlight recent advances in the biological repair of osteochondral lesions.

K. Shimomura (✉)
Department of Rehabilitation, Kansai University of Welfare Sciences, Osaka, Japan

Department of Orthopaedic Surgery, Osaka University Graduate School of Medicine, Osaka, Japan
e-mail: kazunori-shimomura@umin.net

W. Ando
Department of Orthopaedic Surgery, Osaka University Graduate School of Medicine, Osaka, Japan

Department of Orthopaedic Surgery, Kansai Rosai Hospital, Hyogo, Japan
e-mail: w-ando@umin.ac.jp

D. A. Hart
McCaig Institute for Bone & Joint Health, University of Calgary, Calgary, AB, Canada
e-mail: hartd@ucalgary.ca

N. Nakamura
Department of Orthopaedic Surgery, Osaka University Graduate School of Medicine, Osaka, Japan

Institute for Medical Science in Sports, Osaka Health Science University, Osaka, Japan

Global Center for Medical Engineering and Informatics and Informatics, Osaka University, Osaka, Japan
e-mail: norimasa.nakamura@ohsu.ac.jp

34.2 Osteochondral Unit

The osteochondral complex consists of articular cartilage and underlying subchondral bone. Both articular cartilage and subchondral bone are tightly connected and interact with each other biologically as well as biomechanically [10, 11].

A. Gobbi et al. (eds.), *Regenerative Medicine in Sports and Orthopaedics*,
https://doi.org/10.1007/978-3-031-84693-9_34

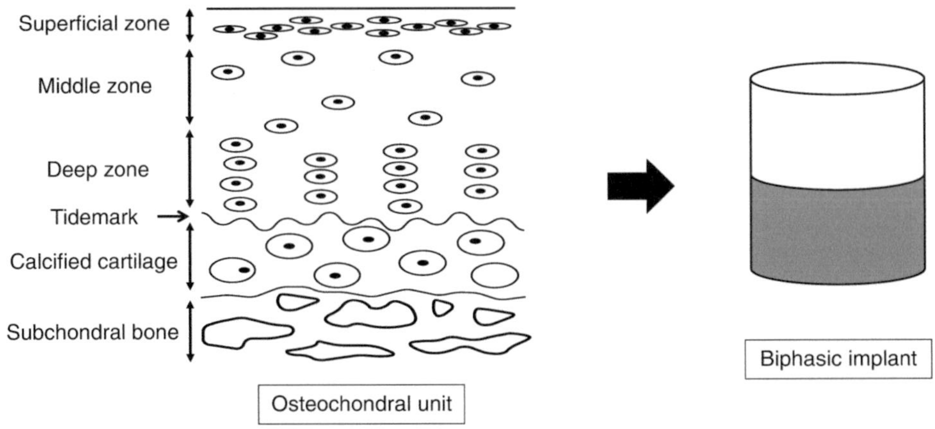

Fig. 34.1 Schematic drawing of osteochondral unit composed of the four zones of articular cartilage and subchondral bone. The biphasic implant facilitates zonal restoration of cartilage and subchondral bone, layer-by-layer, mimicking the natural articular structure

Therefore, these structures could be considered as one osteochondral unit.

Articular cartilage is primarily composed of water, chondrocytes, type II collagen, and proteoglycan [12, 13]. Cartilage can be differentiated into four distinct zones: the superficial, middle, deep, and calcified cartilage zones [14] (Fig. 34.1). Each zone is characterized by a specific composition and organization of cells and extracellular matrix (ECM) molecules. The varying proportions of ECM components influence the mechanical properties of each cartilage zone. Specifically, the superficial zone is strong in tension along the alignment of its collagen fibrils, aiding in the resistance of shear forces at the surface. In contrast, the deep zone exhibits greater compressive strength.

Subchondral bone is a complex tissue composed of water, type I collagen, and hydroxyapatite, with the two latter components providing the tissue's stiffness and compressive strength [12, 15]. The compressive modulus of subchondral bone is higher than that of cartilage. The different morphological compositions and mechanical properties of subchondral bone and cartilage highlight the complexity of the tissue interface [16].

34.3 Pathology of Osteoarthritis

OA is characterized by degenerative changes, such as articular cartilage loss, subchondral bone thickening, and osteophyte formation [17, 18]. The primary morphologic changes include thinning, fissuring, and fragmentation of articular cartilage. As the disease progresses, there is a continuous loss of articular cartilage, accompanied by a decrease in type II collagen and aggrecan, leading to the exposure of subchondral bone [19, 20]. Secondary changes are frequently observed in the underlying bone including fibrosis, cystic change, and new bone formation. These changes are considered to be triggered by a multitude of factors, including aging, trauma, obesity, mechanical overload, congenital disorders, genetics, and infection, which fail to heal spontaneously once damage has started.

On the molecular level, OA development is associated with loss of homeostatic balance between degradation and repair mechanisms in the articular cartilage [19, 21]. This imbalance induces senescence, differentiation, proliferation, and death in joint cells through gene and/or protein expression networks that shift from anabolic to catabolic outcomes [22]. Cartilage-degrading enzymes, such as a disintegrin and metalloproteinases with thrombospondin motifs (ADAMTS)-4 and ADAMTS-5, and matrix metalloproteinase (MMP)-13, play critical roles in OA pathogenesis [19]. Periarticular bone formation like osteophyte is generated by a process of endochondral ossification, which recapitulates the cellular mechanisms of bone growth that occur during skeletal growth and development, where new bone is formed by replacement of a cartilaginous matrix [23].

34.4 Development of Biphasic Implant

Considering the characteristics of osteochondral unit, it is reasonable to facilitate zonal restoration of cartilage and subchondral bone, layer-by-layer, mimicking the natural articular structure [10, 24–28] (Fig. 34.1). As a strategy to regenerate these structures in a layer-by-layer fashion, biphasic (or triphasic) implants have been developed for both mechanical and biological reasons. These reasons include the acquisition of initial mechanical strength, mimicking the natural articular structure, creating a uniform tidemark at the osteochondral junction, and integrating the biphasic implant with host tissue to sustain biological function. Thus, an ideal osteochondral implant should possess a rigid osseous layer to support the overlying cartilage and integrate with the native bone, and a chondral layer to allow the seeding and proliferation of chondrocytes or mesenchymal stem cells (MSCs) and the subsequent deposition of cartilaginous ECM [29]. Taken together, the selection of cells and materials is crucial for a successful osteochondral repair. The details are discussed below.

34.4.1 Cells

As a cell source, mature osteoblasts and chondrocytes could be a good candidate for osteochondral repair. However, the number of such cells obtained is usually limited, which is typically not enough to allow seeding onto scaffolds. Additionally, the expansion of primary cells like chondrocytes can result in a loss of differentiation capacity due to their de-differentiation into fibroblast [30, 31]. To overcome such potential problems related to de-differentiation, three-dimensional (3D) culture can be used to retain the cellular phenotype and avoid de-differentiation [32]. The most common method involves the use of various scaffolds to create a 3D culture environment [33, 34]. This may be combined with the supplementation of growth factors [35], the use of bioreactors [36], mechanical stimulation of the cells [37, 38], and the use of low oxygen tension [39, 40] during cultivation.

As an alternative, MSCs could be promising cell sources [41, 42]. MSCs have the capability to differentiate into a variety of connective tissue cell types, including bone, cartilage, tendon, muscle, and adipose tissue [5, 43]. These cells can be isolated from various tissues, such as bone marrow, skeletal muscle, synovial membrane, adipose tissue, and umbilical cord blood [7, 8, 44–46]. More recently, induced pluripotent stem cells (iPSCs) have been tested for their feasibility in cartilage repair [47, 48].

34.4.2 Materials

To date, several methods have been tested in development of biphasic scaffolds, which combine two distinct biomaterials, for cartilage and bone regeneration [49]. Natural or synthetic polymer-based scaffolds are commonly utilized for the cartilage layer. In contrast, for the scaffold of the subchondral bone layer, it is crucial to select materials with initial mechanical strength, favorable bone ingrowth, and integration with the surrounding native bone. Ceramics, glasses, and metallic materials are frequently employed for this purpose. Additionally, natural or synthetic polymers, similar to those used in the cartilage layer, could be utilized alone or in combination with ceramics [50–55].

Natural polymers offer a naturally conducive environment for cells and tissues, potentially promoting cell proliferation and differentiation [56, 57]. Additionally, they typically feature specific molecular domains that support and guide cells throughout their development, enhancing the biological interaction of the scaffold with host tissue [12, 49]. However, they generally exhibit lower biomechanical strength and stiffness compared to other materials [12]. Collagen, gelatin, glycosaminoglycan, chitosan, starch, hyaluronic acid, and alginate are commonly employed as materials.

Biodegradable synthetic polymers offer several advantages over other materials in scaffold development for tissue engineering. They allow for precise control over mechanical properties such as strength and stiffness, as well as degradation rate [58]. Synthetic polymers are also appealing because they can be shaped into various forms with desired pore structures to accommodate cell migration and tissue ingrowth rates [59]. Furthermore, advancements in techniques like electrospinning and 3D printing have simplified scaffold design and fabrication, enabling the replication of original tissue structures [60–62]. However, synthetic polymers may lack bioactivity due to their hydrophobic surfaces, which hinder cell attachment and proliferation [63–65]. Surface modifications using compounds like chondroitin sulfate [66], silicate [67], and alkaline [68] can enhance hydrophilicity, making them more suitable for tissue engineering applications. Commonly used biodegradable synthetic polymers include poly(glycolic acid) (PGA), poly(D,L-lactic-co-glycolic acid) (PLGA), poly(L-lactic acid) (PLLA), poly(caprolactone) (PCL), and poly(ethylene glycol) (PEG).

Ceramics, such as hydroxyapatite (HA) or other calcium phosphates like tricalcium phosphate (TCP), find extensive use in bone tissue engineering [69–72]. These materials facilitate the formation of bone-like tissue and enhance scaffold integration with host tissue due to their excellent osteoconductivity and osteoinductivity. However, they possess low structural integrity, being brittle and unsuitable for applications under mechanical stress, despite their appropriate stiffness [12]. Controlling the degradation behavior of these scaffolds is possible by adjusting the porous structures, which can be tailored to achieve degradation kinetics suitable for bone tissue engineering. It is known that increasing porosity can further compromise the mechanical properties of bioceramic scaffolds. This issue can be addressed by modifying porous scaffolds through infiltration or coating with biodegradable polymers [73–75].

34.5 Clinically Available Techniques

As an osteochondral solution must satisfy many functions, recent research has focused on combining different materials to meet all the necessary criteria. The goal of these biphasic implants is zonal restoration of the chondral and subchondral layers in a layer-by-layer fashion, and several such implants currently exist [10, 29]. Current concerns when employing polymers and synthetics for osteochondral restoration include the biocompatibility of the materials and their long-term safety implications. Additionally, effective cell delivery and integration of the implants have been challenging, although some selected implants have shown promise [53, 76].

To date, only a few biphasic implants have been utilized in clinical trials, using materials such as collagen, HA, various synthetics, and a novel material, aragonite [77]. The most studied implant has been Maioregen™ (Fin-Ceramica, Faenza, Italy), which is composed of collagen and HA organized in three layers.

Clinical trials have shown good integration and filling of the chondral defect, demonstrating the implant to be an efficacious option with a low complication rate [78–80]. Another reported implant is the synthetic biphasic implant consisting of polylactic acid, polyglycolic acid, and calcium phosphate, marketed as TruFit™ (Smith and Nephew, Andover, MA). A few studies reported good results; however, the literature has also reported delayed integration, with longer follow-ups noting subchondral cysts and bone edema [81–84]. A more recent bilayer scaffold under study is a crystalline coral aragonite-hyaluronic acid implant (Agili-C™, Cartiheal (2009) Ltd, Israel), which showed good integration and clinical improvement in a study with up to 12 months of follow-up [85] and further improvement at 24 months [86]. Additionally, patients with talar osteochondral lesions who were treated with microfracture and cell-free scaffold (Chondrotissue®: hyaluronic acid impregnated woven polyglycolic acid) implantation in a single-step arthroscopic surgery showed improved clinical and MOCART scores [87]. At present, there is not enough evidence to deem any of these implants superior or more efficient than the others, except for TruFit™, which is considered less favorable due to adverse reactions.

34.6 Biphasic Implant Composed of a Scaffold-Free MSC-Derived TEC Combined with Artificial Bone

As a scaffold-free technique, we originally developed a 3D scaffold-free tissue-engineered construct (TEC) through simple-cell culture methods and demonstrated its feasibility for cartilage repair and regeneration in the first-in-human clinical trial [88, 89]. To extend its application to osteochondral injuries, we have developed a new biphasic implant composed of a scaffold-free TEC derived from synovial MSCs combined with artificial bone (Fig. 34.2a) and tested its feasibility for osteochondral repair in a rabbit osteochondral defect model [90] (Fig. 34.2b, c). Our experimental details are described below.

Currently, artificial bones like HA or beta-tricalcium phosphate (bTCP) are widely utilized in clinical practice for treating bone defects resulting from fractures or the resection of bone tumors [69, 91, 92]. We have developed a novel fully interconnected HA artificial bone with a sufficient initial strength, as well as an excellent bone-formation capacity [69, 70]. We previously reported its feasibility in subchondral bone repair [69]. In addition, we have developed a scaffold-free 3D TEC composed of MSCs derived from

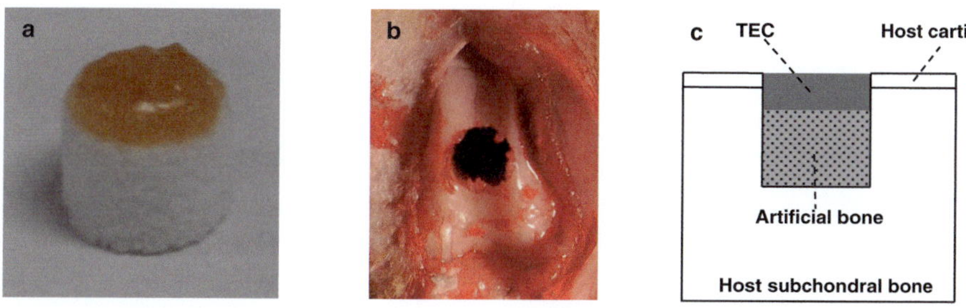

Fig. 34.2 (**a**) The biphasic implant generated with a tissue-engineered construct (TEC) and an artificial bone. (**b**) Osteochondral defects in the femoral groove of the rabbit knee. (**c**) Schematic representation of the implantation of TEC and artificial bone. Quoted and modified in accordance with the publisher reuse guideline from Ref. [90]. (Copyright © 2014, Mary Ann Liebert, Inc.)

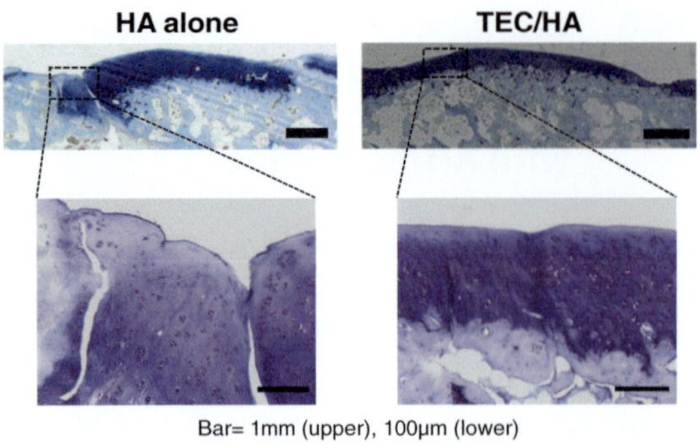

Fig. 34.3 Toluidine blue staining of repair tissues in the HA-alone control and TEC/HA groups. Note that the repair tissue in defects treated with the TEC/HA implant demonstrated good integration with the adjacent host tissue, while the control group showed poor integration. Bar = 1 mm (upper images). Bar = 100 um (lower images). Quoted and modified in accordance with the publisher reuse guideline from Ref. [90]. (Copyright © 2014, Mary Ann Liebert, Inc.)

the synovium and ECMs synthesized by the cells, as mentioned above [8]. We demonstrated its feasibility for cartilage repair in a large animal model as well as clinical trials [5, 7, 88]. These TECs, without artificial scaffolds, minimize potential side effects from extrinsic materials. They adhere well to the cartilage matrix and integrate securely with adjacent tissue. We hypothesized that combining TEC with fully interconnected HA-based artificial bone would effectively repair osteochondral lesions. We tested this in a rabbit osteochondral defect model. At 6 months post-implantation, defects treated with biphasic implants showed faster subchondral bone repair and well-integrated cartilaginous tissue compared to HA-alone controls (Fig. 34.3). Control defects exhibited delayed subchondral bone repair and poorly integrated cartilaginous tissue, suggesting early degenerative changes (Fig. 34.3). Biomechanically, repaired tissue with the biphasic implants at 6 months had stiffness similar to normal tissue, indicating significant improvement in osteochondral repair [90].

To enhance repair quality and expedite maturation, we assessed alternative options for the artificial bone component compared to HA. bTCP, being highly biocompatible and resorbable, presents a viable alternative [70]. Its faster in vivo resorption may lead to more efficient subchondral bone remodeling post-implantation. Thus, a biphasic implant combining bTCP with synovial MSC-derived TEC might offer superior osteochondral repair compared to an HA-based one, as tested in the rabbit model mentioned earlier. Osteochondral defects treated with TEC/bTCP implants showed faster subchondral bone repair at 1 month (Fig. 34.4), but deteriorated cartilaginous tissue over 6 months (Fig. 34.5). Conversely, defects treated with TEC/HA exhibited delayed subchon-

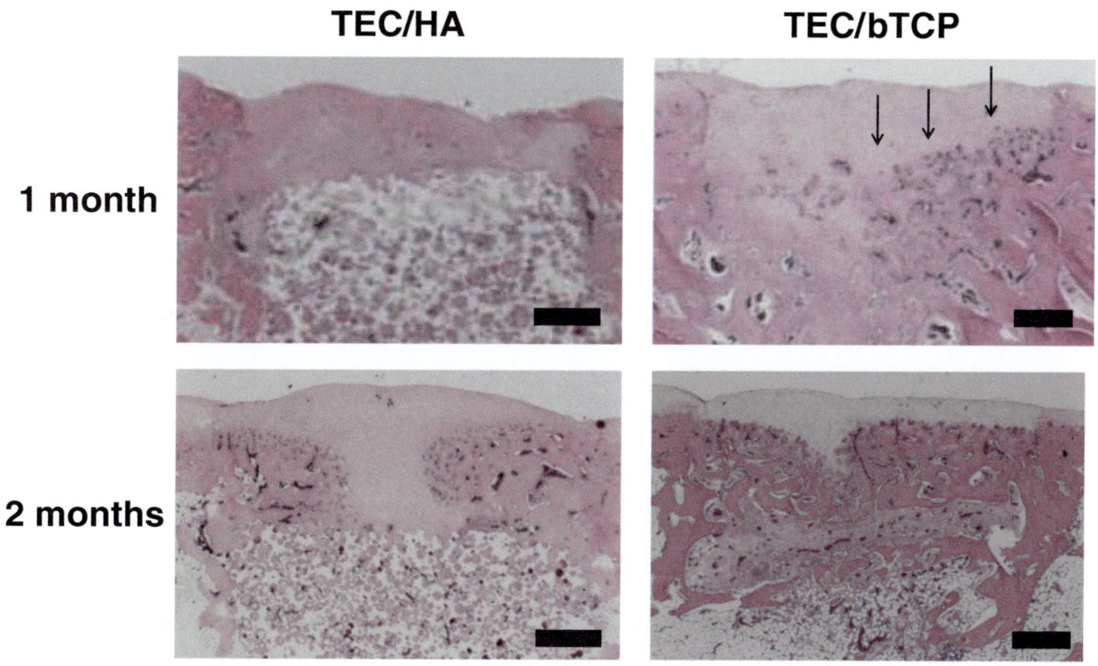

Fig. 34.4 Hematoxylin and eosin (H&E) staining of repair tissue resulting from the implantation of TEC/HA or TEC/bTCP biphasic implants. The osteochondral defect treated with TEC/bTCP showed rapid subchondral bone repair at 1 month. The osteochondral defect treated with TEC/HA exhibited delayed subchondral bone repair at 1 month but achieved a similar quality of subchondral bone repair to TEC/bTCP at 2 months. Bar = 1 mm. Quoted and modified in accordance with the publisher reuse guideline from Ref. [11]. (Copyright © 2017, © SAGE Publications)

dral bone repair at 1 month but similar quality by 2 months (Fig. 34.4). Notably, the TEC/HA implants maintained good histological quality (Fig. 34.5) and demonstrated better biomechanical properties at 6 months compared to TEC/bTCP (Fig. 34.6). Contrary to expectations, TEC/HA biphasic implants facilitated superior osteochondral repair. This suggests the importance of stable subchondral bone restoration for effective long-term repair over rapid remodeling [11].

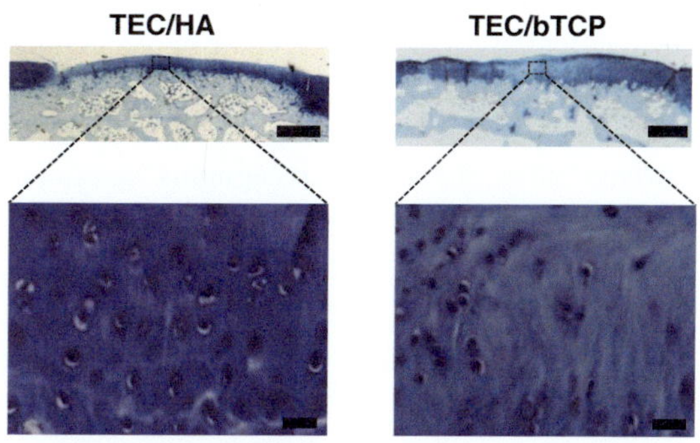

Fig. 34.5 Toluidine blue staining of repair tissue resulting from the implantation of TEC/HA or TEC/bTCP biphasic implants at 6 months after surgery. The repair tissue exhibited fibrocartilaginous-like tissue with weak toluidine blue staining in the TEC/bTCP group, while the TEC/HA group showed hyaline cartilage-like features, with chondrocytes arranged in longitudinal columns. Bar = 1 mm (upper images). Bar = 20 μm (lower images). Quoted and modified in accordance with the publisher reuse guideline from Ref. [11]. (Copyright © 2017, © SAGE Publications)

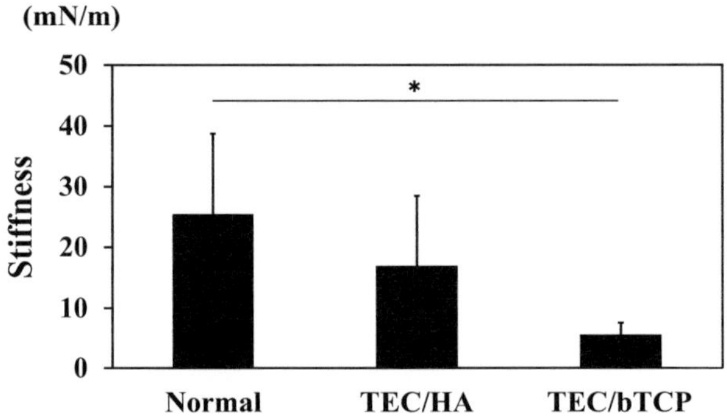

Fig. 34.6 The osteochondral tissue stiffness of a healthy rabbit and after implantation of a TEC/HA or TEC/bTCP hybrid implant at 6 months after surgery. The TEC/HA biphasic implant exhibited enhanced mechanical properties compared to the TEC/bTCP hybrid implant. *; $p < 0.05$. Quoted and modified in accordance with the publisher reuse guideline from Ref. [11]. (Copyright © 2017, © SAGE Publications)

34.7 Hybrid Implant Containing Both Scaffold-Free Human MSC-Derived TEC and Human iPSC-Cartilaginous Particles

Both MSCs and iPSCs offer the potential for repairing damaged connective tissues. The use of hybrid implants containing both human MSCs and iPSCs was investigated to assess their combined potential for enhanced repair of osteochondral defects [93]. Prior to the experiment, human iPSCs were chondrogenically differentiated to produce iPSC-derived cartilaginous particles (iPSC-CPs) using a previously described method [94]. Human iPSC-CPs, wrapped with TEC containing human MSCs, attained secure defect filling with good integration into adjacent tissue in a rat osteochondral injury model (Fig. 34.7a–c). The presence of MSCs in the hybrid implants was required for effective biphasic osteochondral repair (Fig. 34.8). Thus, the TEC component of such hybrid implants serves several critical functions, including adhesion to the defect site via the matrix, facilitation of the repair via MSCs, and enhanced angiogenesis and neovascularization. Based on these encouraging results, such hybrid implants may offer an effective future intervention for the repair of complex osteochondral defects.

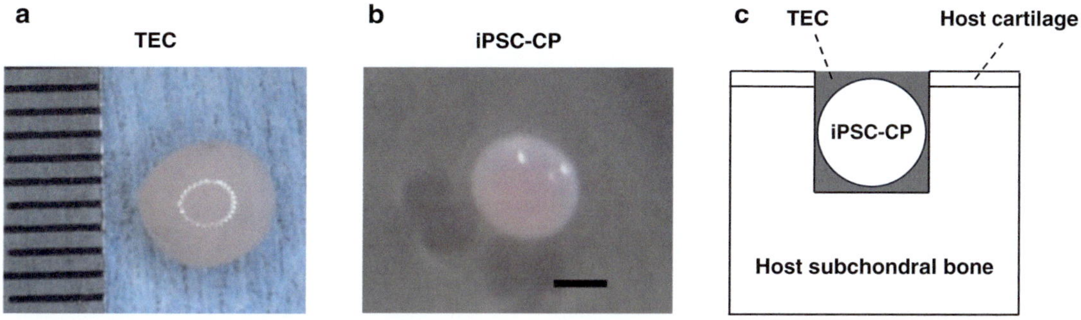

Fig. 34.7 Macroscopic view of a TEC (**a**) and an iPSC-CP (**b**). Bars = 1 mm. (**c**) Schematic representation of an iPSC-CP/TEC implantation for an osteochondral defect. Quoted and modified in accordance with the publisher reuse guideline from Ref. [93]. (Copyright © 2023, The Author(s))

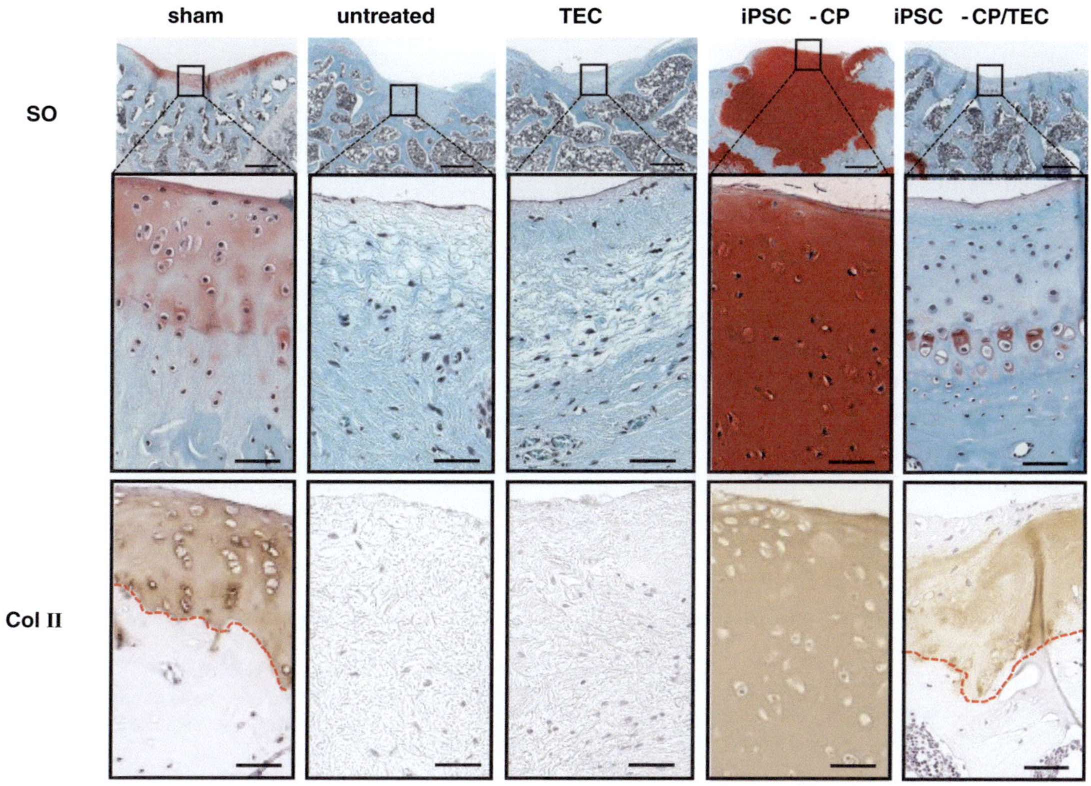

Fig. 34.8 Histological images and immunohistochemical analysis for col II in implantation sites at 24 weeks postoperatively. (Upper) Histology of the tissue stained with Safranin O (SO) (Bar = 500 μm), and (Middle) high-magnification images of articular surface stained with SO (Bar = 100 μm). (Lower) Immunohistochemical analysis of articular surface with col II (Bar = 100 μm). The red dotted lines demonstrated tidemarks. At 24 weeks post-implantation, the defects in both the untreated group and the TEC group were filled with fibrous tissue. In contrast, the osteochondral defects in the iPSC-CP group were completely filled with hyaline-like cartilaginous tissue, without subchondral bone repair. Notably, the repair tissue in the iPSC-CP/TEC group exhibited biphasic osteochondral repair. Quoted and modified in accordance with the publisher reuse guideline from Ref. [93]. (Copyright © 2023, The Author(s))

34.8 Future Perspective and Conclusion

The management of OA remains challenging and contentious. With the steady advancement of tissue engineering and cell-based technologies in the past decade, new therapeutic options for osteochondral repair in clinical practice may emerge. This chapter focuses on current techniques and recent advancements in tissue-engineered biomaterial scaffolds, along with our novel MSC-based biphasic implant for treating osteochondral lesions. Numerous promising scaffolds have been developed, some demonstrating effective osteochondral repair in vivo. Recent research has explored not only the efficacy of materials or cells but also novel concepts and techniques such as mechanical [95–101], microstructural [59], and local microenvironment modification [102] for the design and fabrication of scaffolds. Thus, the introduction of new implants for osteochondral lesions is foreseeable in the near future. However, the most suitable biomaterials for cartilage or subchondral bone layers have not been thoroughly investigated, despite the availability of many biomaterials for osteochondral repair. Comparative studies of these materials are necessary to identify the optimal choice. Additionally, current methods, when applied in clinical practice, can be highly invasive. Hence, minimally invasive approaches (e.g., arthroscopic implantation) should be considered. Further studies should be conducted rigorously to advance our understanding of this field.

Acknowledgments This work was supported by a Health and Labour Sciences Research grant from the Ministry of Health, Labour, and Welfare of Japan; a grant from the New Energy and Industrial Technology Development Organization, Japan; and a Grant-in-Aid for Scientific Research, Japan Society for the Promotion of Science; Centers for Clinical Application Research on Specific Disease/Organ (type B) Grant NO. 17bm0304004h005 from the Japan Agency for Medical Research and Development.

References

1. GBD 2021 Osteoarthritis Collaborators. Global, regional, and national burden of osteoarthritis, 1990–2020 and projections to 2050: a systematic analysis for the Global Burden of Disease Study 2021. Lancet Rheumatol. 2023;5(9):e508–e22.
2. Zhang W, Moskowitz RW, Nuki G, Abramson S, Altman RD, Arden N, et al. OARSI recommendations for the management of hip and knee osteoarthritis, part I: critical appraisal of existing treatment guidelines and systematic review of current research evidence. Osteoarthr Cartil. 2007;15(9):981–1000.
3. McAlindon TE, Bannuru RR, Sullivan MC, Arden NK, Berenbaum F, Bierma-Zeinstra SM, et al. OARSI guidelines for the non-surgical management of knee osteoarthritis. Osteoarthr Cartil. 2014;22(3):363–88.
4. Kandel RA, Grynpas M, Pilliar R, Lee J, Wang J, Waldman S, et al. Repair of osteochondral defects with biphasic cartilage-calcium polyphosphate constructs in a sheep model. Biomaterials. 2006;27(22):4120–31.
5. Shimomura K, Ando W, Tateishi K, Nansai R, Fujie H, Hart DA, et al. The influence of skeletal maturity on allogenic synovial mesenchymal stem cell-based repair of cartilage in a large animal model. Biomaterials. 2010;31(31):8004–11.
6. Brittberg M, Lindahl A, Nilsson A, Ohlsson C, Isaksson O, Peterson L. Treatment of deep cartilage defects in the knee with autologous chondrocyte transplantation. N Engl J Med. 1994;331(14):889–95.
7. Ando W, Tateishi K, Hart DA, Katakai D, Tanaka Y, Nakata K, et al. Cartilage repair using an in vitro generated scaffold-free tissue-engineered construct derived from porcine synovial mesenchymal stem cells. Biomaterials. 2007;28(36):5462–70.
8. Ando W, Tateishi K, Katakai D, Hart DA, Higuchi C, Nakata K, et al. In vitro generation of a scaffold-free tissue-engineered construct (TEC) derived from human synovial mesenchymal stem cells: biological and mechanical properties and further chondrogenic potential. Tissue Eng Part A. 2008;14(12):2041–9.
9. Makris EA, Gomoll AH, Malizos KN, Hu JC, Athanasiou KA. Repair and tissue engineering techniques for articular cartilage. Nat Rev Rheumatol. 2015;11(1):21–34.
10. Gomoll AH, Madry H, Knutsen G, van Dijk N, Seil R, Brittberg M, et al. The subchondral bone in articular cartilage repair: current problems in the surgical management. Knee Surg Sports Traumatol Arthrosc. 2010;18(4):434–47.

11. Shimomura K, Moriguchi Y, Nansai R, Fujie H, Ando W, Horibe S, et al. Comparison of 2 different formulations of artificial bone for a hybrid implant with a tissue-engineered construct derived from synovial mesenchymal stem cells. Am J Sports Med. 2017;45(3):666–75. https://doi.org/10.1177/0363546516668835.
12. Nooeaid P, Salih V, Beier JP, Boccaccini AR. Osteochondral tissue engineering: scaffolds, stem cells and applications. J Cell Mol Med. 2012;16(10):2247–70.
13. Keeney M, Pandit A. The osteochondral junction and its repair via bi-phasic tissue engineering scaffolds. Tissue Eng Part B Rev. 2009;15(1):55–73.
14. Madry H, van Dijk CN, Mueller-Gerbl M. The basic science of the subchondral bone. Knee Surg Sports Traumatol Arthrosc. 2010;18(4):419–33.
15. Arvidson K, Abdallah BM, Applegate LA, Baldini N, Cenni E, Gomez-Barrena E, et al. Bone regeneration and stem cells. J Cell Mol Med. 2011;15(4):718–46.
16. Yang PJ, Temenoff JS. Engineering orthopedic tissue interfaces. Tissue Eng Part B Rev. 2009;15(2):127–41.
17. Primorac D, Molnar V, Rod E, Jeleč Ž, Čukelj F, Matišić V, et al. Knee osteoarthritis: a review of pathogenesis and state-of-the-art non-operative therapeutic considerations. Genes (Basel). 2020;11(8):854.
18. Tong L, Yu H, Huang X, Shen J, Xiao G, Chen L, et al. Current understanding of osteoarthritis pathogenesis and relevant new approaches. Bone Res. 2022;10(1):60.
19. Jiang Y. Osteoarthritis year in review 2021: biology. Osteoarthr Cartil. 2022;30(2):207–15.
20. Giorgino R, Albano D, Fusco S, Peretti GM, Mangiavini L, Messina C. Knee osteoarthritis: epidemiology, pathogenesis, and mesenchymal stem cells: what else is new? An update. Int J Mol Sci. 2023;24(7):6405.
21. Fujii Y, Liu L, Yagasaki L, Inotsume M, Chiba T, Asahara H. Cartilage homeostasis and osteoarthritis. Int J Mol Sci. 2022;23(11):6316.
22. Muthu S. Osteoarthritis, an old wine in a new bottle! World J Orthop. 2023;14(1):1–5.
23. Goldring SR. Alterations in periarticular bone and cross talk between subchondral bone and articular cartilage in osteoarthritis. Ther Adv Musculoskelet Dis. 2012;4(4):249–58.
24. Kon E, Delcogliano M, Filardo G, Busacca M, Di Martino A, Marcacci M. Novel nano-composite multilayered biomaterial for osteochondral regeneration: a pilot clinical trial. Am J Sports Med. 2011;39(6):1180–90.
25. Jiang CC, Chiang H, Liao CJ, Lin YJ, Kuo TF, Shieh CS, et al. Repair of porcine articular cartilage defect with a biphasic osteochondral composite. J Orthop Res. 2007;25(10):1277–90.
26. Schek RM, Taboas JM, Segvich SJ, Hollister SJ, Krebsbach PH. Engineered osteochondral grafts using biphasic composite solid free-form fabricated scaffolds. Tissue Eng. 2004;10(9–10):1376–85.
27. Orth P, Meyer HL, Goebel L, Eldracher M, Ong MF, Cucchiarini M, et al. Improved repair of chondral and osteochondral defects in the ovine trochlea compared with the medial condyle. J Orthop Res. 2013;31(11):1772–9.
28. Orth P, Cucchiarini M, Kohn D, Madry H. Alterations of the subchondral bone in osteochondral repair—translational data and clinical evidence. Eur Cell Mater. 2013;25:299–316; discussion 4–6.
29. Shimomura K, Moriguchi Y, Murawski CD, Yoshikawa H, Nakamura N. Osteochondral tissue engineering with biphasic scaffold: current strategies and techniques. Tissue Eng Part B Rev. 2014;20(5):468–76.
30. Schnabel M, Marlovits S, Eckhoff G, Fichtel I, Gotzen L, Vecsei V, et al. Dedifferentiation-associated changes in morphology and gene expression in primary human articular chondrocytes in cell culture. Osteoarthr Cartil. 2002;10(1):62–70.
31. Benya PD, Shaffer JD. Dedifferentiated chondrocytes reexpress the differentiated collagen phenotype when cultured in agarose gels. Cell. 1982;30(1):215–24.
32. Takahashi T, Ogasawara T, Asawa Y, Mori Y, Uchinuma E, Takato T, et al. Three-dimensional microenvironments retain chondrocyte phenotypes during proliferation culture. Tissue Eng. 2007;13(7):1583–92.
33. Zheng MH, Willers C, Kirilak L, Yates P, Xu J, Wood D, et al. Matrix-induced autologous chondrocyte implantation (MACI): biological and histological assessment. Tissue Eng. 2007;13(4):737–46.
34. Marcacci M, Berruto M, Brocchetta D, Delcogliano A, Ghinelli D, Gobbi A, et al. Articular cartilage engineering with Hyalograft C: 3-year clinical results. Clin Orthop Relat Res. 2005;435:96–105.
35. Chubinskaya S, Segalite D, Pikovsky D, Hakimiyan AA, Rueger DC. Effects induced by BMPS in cultures of human articular chondrocytes: comparative studies. Growth Factors. 2008;26(5):275–83.
36. Forsey RW, Tare R, Oreffo RO, Chaudhuri JB. Perfusion bioreactor studies of chondrocyte growth in alginate-chitosan capsules. Biotechnol Appl Biochem. 2012;59(2):142–52.
37. Kawanishi M, Oura A, Furukawa K, Fukubayashi T, Nakamura K, Tateishi T, et al. Redifferentiation of dedifferentiated bovine articular chondrocytes enhanced by cyclic hydrostatic pressure under a gas-controlled system. Tissue Eng. 2007;13(5):957–64.
38. El-Ayoubi R, DeGrandpre C, DiRaddo R, Yousefi AM, Lavigne P. Design and dynamic culture of 3D-scaffolds for cartilage tissue engineering. J Biomater Appl. 2011;25(5):429–44.
39. Kurz B, Domm C, Jin M, Sellckau R, Schunke M. Tissue engineering of articular cartilage under the influence of collagen I/III membranes and low oxygen tension. Tissue Eng. 2004;10(7–8):1277–86.

40. Yasui Y, Chijimatsu R, Hart DA, Koizumi K, Sugita N, Shimomura K, et al. Preparation of scaffold-free tissue-engineered constructs derived from human synovial mesenchymal stem cells under low oxygen tension enhances their chondrogenic differentiation capacity. Tissue Eng Part A. 2016;22(5–6):490–500.
41. Sundelacruz S, Kaplan DL. Stem cell- and scaffold-based tissue engineering approaches to osteochondral regenerative medicine. Semin Cell Dev Biol. 2009;20(6):646–55.
42. Shimomura K, Ando W, Moriguchi Y, Sugita N, Yasui Y, Koizumi K, et al. Next generation mesenchymal stem cell (MSC)–based cartilage repair using scaffold-free tissue engineered constructs generated with synovial mesenchymal stem cells. Cartilage. 2015;6(Suppl 2):13S–29S.
43. De Bari C, Dell'Accio F, Tylzanowski P, Luyten FP. Multipotent mesenchymal stem cells from adult human synovial membrane. Arthritis Rheum. 2001;44(8):1928–42.
44. Martin MJ, Muotri A, Gage F, Varki A. Human embryonic stem cells express an immunogenic nonhuman sialic acid. Nat Med. 2005;11(2):228–32.
45. Sakaguchi Y, Sekiya I, Yagishita K, Muneta T. Comparison of human stem cells derived from various mesenchymal tissues: superiority of synovium as a cell source. Arthritis Rheum. 2005;52(8):2521–9.
46. Koga H, Shimaya M, Muneta T, Nimura A, Morito T, Hayashi M, et al. Local adherent technique for transplanting mesenchymal stem cells as a potential treatment of cartilage defect. Arthritis Res Ther. 2008;10(4):R84.
47. Tsumaki N, Okada M, Yamashita A. iPS cell technologies and cartilage regeneration. Bone. 2015;70:48–54.
48. Abe K, Yamashita A, Morioka M, Horike N, Takei Y, Koyamatsu S, et al. Engraftment of allogeneic iPS cell-derived cartilage organoid in a primate model of articular cartilage defect. Nat Commun. 2023;14(1):804.
49. Mano JF, Reis RL. Osteochondral defects: present situation and tissue engineering approaches. J Tissue Eng Regen Med. 2007;1(4):261–73.
50. Deng T, Lv J, Pang J, Liu B, Ke J. Construction of tissue-engineered osteochondral composites and repair of large joint defects in rabbit. J Tissue Eng Regen Med. 2012;8:546.
51. Zhang S, Chen L, Jiang Y, Cai Y, Xu G, Tong T, et al. Bi-layer collagen/microporous electrospun nanofiber scaffold improves the osteochondral regeneration. Acta Biomater. 2013;9(7):7236–47.
52. Zhang W, Chen J, Tao J, Hu C, Chen L, Zhao H, et al. The promotion of osteochondral repair by combined intra-articular injection of parathyroid hormone-related protein and implantation of a bi-layer collagen-silk scaffold. Biomaterials. 2013;34(25):6046–57.
53. Alhadlaq A, Mao JJ. Tissue-engineered osteochondral constructs in the shape of an articular condyle. J Bone Joint Surg Am. 2005;87(5):936–44.
54. Kon E, Delcogliano M, Filardo G, Fini M, Giavaresi G, Francioli S, et al. Orderly osteochondral regeneration in a sheep model using a novel nano-composite multilayered biomaterial. J Orthop Res. 2010;28(1):116–24.
55. Chen J, Chen H, Li P, Diao H, Zhu S, Dong L, et al. Simultaneous regeneration of articular cartilage and subchondral bone in vivo using MSCs induced by a spatially controlled gene delivery system in bilayered integrated scaffolds. Biomaterials. 2011;32(21):4793–805.
56. Tan W, Twomey J, Guo D, Madhavan K, Li M. Evaluation of nanostructural, mechanical, and biological properties of collagen-nanotube composites. IEEE Trans Nanobioscience. 2010;9(2):111–20.
57. Hsu FY, Hung YS, Liou HM, Shen CH. Electrospun hyaluronate-collagen nanofibrous matrix and the effects of varying the concentration of hyaluronate on the characteristics of foreskin fibroblast cells. Acta Biomater. 2010;6(6):2140–7.
58. Gunatillake PA, Adhikari R. Biodegradable synthetic polymers for tissue engineering. Eur Cell Mater. 2003;5:1–16; discussion.
59. Duan P, Pan Z, Cao L, He Y, Wang H, Qu Z, et al. The effects of pore size in bilayered poly(lactide-co-glycolide) scaffolds on restoring osteochondral defects in rabbits. J Biomed Mater Res A. 2013;102:180.
60. Li WJ, Cooper JA Jr, Mauck RL, Tuan RS. Fabrication and characterization of six electrospun poly(alpha-hydroxy ester)-based fibrous scaffolds for tissue engineering applications. Acta Biomater. 2006;2(4):377–85.
61. Li WJ, Mauck RL, Cooper JA, Yuan X, Tuan RS. Engineering controllable anisotropy in electrospun biodegradable nanofibrous scaffolds for musculoskeletal tissue engineering. J Biomech. 2007;40(8):1686–93.
62. Lin H, Zhang D, Alexander PG, Yang G, Tan J, Cheng AW, et al. Application of visible light-based projection stereolithography for live cell-scaffold fabrication with designed architecture. Biomaterials. 2013;34(2):331–9.
63. Shafiee A, Soleimani M, Chamheidari GA, Seyedjafari E, Dodel M, Atashi A, et al. Electrospun nanofiber-based regeneration of cartilage enhanced by mesenchymal stem cells. J Biomed Mater Res A. 2011;99(3):467–78.
64. Sarasam AR, Krishnaswamy RK, Madihally SV. Blending chitosan with polycaprolactone: effects on physicochemical and antibacterial properties. Biomacromolecules. 2006;7(4):1131–8.
65. Bhattarai SR, Bhattarai N, Viswanathamurthi P, Yi HK, Hwang PH, Kim HY. Hydrophilic nanofibrous structure of polylactide; fabrication and cell affinity. J Biomed Mater Res A. 2006;78(2):247–57.

66. Chang KY, Cheng LW, Ho GH, Huang YP, Lee YD. Fabrication and characterization of poly(gamma-glutamic acid)-graft-chondroitin sulfate/polycaprolactone porous scaffolds for cartilage tissue engineering. Acta Biomater. 2009;5(6):1937–47.
67. Chouzouri G, Xanthos M. In vitro bioactivity and degradation of polycaprolactone composites containing silicate fillers. Acta Biomater. 2007;3(5):745–56.
68. Pena J, Corrales T, Izquierdo-Barba I, Serrano MC, Portoles MT, Pagani R, et al. Alkaline-treated poly(epsilon-caprolactone) films: degradation in the presence or absence of fibroblasts. J Biomed Mater Res A. 2006;76(4):788–97.
69. Tamai N, Myoui A, Hirao M, Kaito T, Ochi T, Tanaka J, et al. A new biotechnology for articular cartilage repair: subchondral implantation of a composite of interconnected porous hydroxyapatite, synthetic polymer (PLA-PEG), and bone morphogenetic protein-2 (rhBMP-2). Osteoarthr Cartil. 2005;13(5):405–17.
70. Yamasaki N, Hirao M, Nanno K, Sugiyasu K, Tamai N, Hashimoto N, et al. A comparative assessment of synthetic ceramic bone substitutes with different composition and microstructure in rabbit femoral condyle model. J Biomed Mater Res B Appl Biomater. 2009;91(2):788–98.
71. Martin I, Miot S, Barbero A, Jakob M, Wendt D. Osteochondral tissue engineering. J Biomech. 2007;40(4):750–65.
72. Hutmacher DW. Scaffolds in tissue engineering bone and cartilage. Biomaterials. 2000;21(24):2529–43.
73. Chen QZ, Boccaccini AR. Poly(D,L-lactic acid) coated 45S5 Bioglass-based scaffolds: processing and characterization. J Biomed Mater Res A. 2006;77(3):445–57.
74. Miao X, Tan DM, Li J, Xiao Y, Crawford R. Mechanical and biological properties of hydroxyapatite/tricalcium phosphate scaffolds coated with poly(lactic-co-glycolic acid). Acta Biomater. 2008;4(3):638–45.
75. Ren J, Zhao P, Ren T, Gu S, Pan K. Poly (D,L-lactide)/nano-hydroxyapatite composite scaffolds for bone tissue engineering and biocompatibility evaluation. J Mater Sci Mater Med. 2008;19(3):1075–82.
76. Ahn JH, Lee TH, Oh JS, Kim SY, Kim HJ, Park IK, et al. Novel hyaluronate-atelocollagen/beta-TCP-hydroxyapatite biphasic scaffold for the repair of osteochondral defects in rabbits. Tissue Eng Part A. 2009;15(9):2595–604.
77. Kon E, Filardo G, Perdisa F, Venieri G, Marcacci M. Clinical results of multilayered biomaterials for osteochondral regeneration. J Exp Orthop. 2014;1(1):10.
78. Filardo G, Kon E, Di Martino A, Busacca M, Altadonna G, Marcacci M. Treatment of knee osteochondritis dissecans with a cell-free biomimetic osteochondral scaffold: clinical and imaging evaluation at 2-year follow-up. Am J Sports Med. 2013;41(8):1786–93.
79. Kon E, Filardo G, Di Martino A, Busacca M, Moio A, Perdisa F, et al. Clinical results and MRI evolution of a nano-composite multilayered biomaterial for osteochondral regeneration at 5 years. Am J Sports Med. 2014;42(1):158–65.
80. Brix M, Kaipel M, Kellner R, Schreiner M, Apprich S, Boszotta H, et al. Successful osteoconduction but limited cartilage tissue quality following osteochondral repair by a cell-free multilayered nano-composite scaffold at the knee. Int Orthop. 2016;40(3):625–32.
81. Carmont MR, Carey-Smith R, Saithna A, Dhillon M, Thompson P, Spalding T. Delayed incorporation of a TruFit plug: perseverance is recommended. Arthroscopy. 2009;25(7):810–4.
82. Bedi A, Foo LF, Williams RJ 3rd, Potter HG. The maturation of synthetic scaffolds for osteochondral donor sites of the knee: an MRI and T2-mapping analysis. Cartilage. 2010;1(1):20–8.
83. Bugelli G, Ascione F, Dell'Osso G, Zampa V, Giannotti S. Biphasic bioresorbable scaffold (TruFit(®)) in knee osteochondral defects: 3-T MRI evaluation of osteointegration in patients with a 5-year minimum follow-up. Musculoskelet Surg. 2018;102(2):191–9.
84. Verhaegen J, Clockaerts S, Van Osch GJ, Somville J, Verdonk P, Mertens P. TruFit plug for repair of osteochondral defects-where is the evidence? Systematic review of literature. Cartilage. 2015;6(1):12–9.
85. Kon E, Robinson D, Verdonk P, Drobnic M, Patrascu JM, Dulic O, et al. A novel aragonite-based scaffold for osteochondral regeneration: early experience on human implants and technical developments. Injury. 2016;47(Suppl 6):S27–s32.
86. Kon E, Di Matteo B, Verdonk P, Drobnic M, Dulic O, Gavrilovic G, et al. Aragonite-based scaffold for the treatment of joint surface lesions in mild to moderate osteoarthritic knees: results of a 2-year multicenter prospective study. Am J Sports Med. 2021;49(3):588–98.
87. Kanatlı U, Eren A, Eren TK, Vural A, Geylan DE, Öner AY. Single-step arthroscopic repair with cell-free polymer-based scaffold in osteochondral lesions of the talus: clinical and radiological results. Arthroscopy. 2017;33(9):1718–26.
88. Shimomura K, Yasui Y, Koizumi K, Chijimatsu R, Hart DA, Yonetani Y, et al. First-in-human pilot study of implantation of a scaffold-free tissue-engineered construct generated from autologous synovial mesenchymal stem cells for repair of knee chondral lesions. Am J Sports Med. 2018;46(10):2384–93.
89. Shimomura K, Ando W, Hart DA, Yonetani Y, Horibe S, Nakamura N. Five-year outcomes after implantation of a scaffold-free tissue-engineered construct generated from autologous synovial mesenchymal stromal cells for repair of knee chondral lesions. Orthop J Sports Med. 2023;11(8):23259671231189474.

90. Shimomura K, Moriguchi Y, Ando W, Nansai R, Fujie H, Hart DA, et al. Osteochondral repair using a scaffold-free tissue-engineered construct derived from synovial mesenchymal stem cells and a hydroxyapatite-based artificial bone. Tissue Eng Part A. 2014;20(17–18):2291–304. https://doi.org/10.1089/ten.tea.2013.0414.
91. Tamai N, Myoui A, Kudawara I, Ueda T, Yoshikawa H. Novel fully interconnected porous hydroxyapatite ceramic in surgical treatment of benign bone tumor. J Orthop Sci. 2010;15(4):560–8.
92. Shen C, Ma J, Chen XD, Dai LY. The use of beta-TCP in the surgical treatment of tibial plateau fractures. Knee Surg Sports Traumatol Arthrosc. 2009;17(12):1406–11.
93. Nakagawa S, Ando W, Shimomura K, Hart DA, Hanai H, Jacob G, et al. Repair of osteochondral defects: efficacy of a tissue-engineered hybrid implant containing both human MSC and human iPSC-cartilaginous particles. NPJ Regen Med. 2023;8(1):59. https://doi.org/10.1038/s41536-023-00335-x.
94. Yamashita A, Morioka M, Yahara Y, Okada M, Kobayashi T, Kuriyama S, et al. Generation of scaffoldless hyaline cartilaginous tissue from human iPSCs. Stem Cell Reports. 2015;4(3):404–18.
95. Dallo I, Gobbi A. Knee osteochondral lesions treatments. In: Joint function preservation: a focus on the osteochondral unit. Cham: Springer; 2022. p. 337–44.
96. Gobbi A, Alvarez R, Irlandini E, Dallo I. Current concepts in subchondral bone pathology. In: Joint function preservation: a focus on the osteochondral unit. Cham: Springer; 2022. p. 173–80.
97. Szwedowski D, Dallo I, Irlandini E, Gobbi A. Osteocore plasty: a minimally invasive approach for subchondral bone marrow lesions of the knee. Arthrosc Tech. 2020;9(11):e1773–7.
98. Herman K, Whyte GP, Montagna A, Bizzoco L, Shabshin N, Szwedowski D, Gobbi A. Osteocore plasty: minimally invasive approach for subchondral pathologies. J Cartilage Joint Preserv. 2023;3(1):100101.
99. Gobbi A, Arbas AJ, Dallo I. Proximal tibial subchondral cystic lesion treatment with osteo-core-plasty. In: Joint function preservation: a focus on the osteochondral unit. Cham: Springer; 2022. p. 237–46.
100. Dallo I, D'Ambrosi R, Szwedowski D, Mobasheri A, Gobbi A. Minimally invasive cell-based therapy for symptomatic bone marrow lesions of the knee: a prospective clinical study at 1 year. Stem Cells Dev. 2022;31(15–16):488–97.
101. Gobbi A, Dallo I. Osteo-Core-Plasty technique for the treatment of a proximal tibial subchondral cystic lesion. Bone. 2021;1(7):27–8.
102. Reyes R, Delgado A, Solis R, Sanchez E, Hernandez A, San Roman J, et al. Cartilage repair by local delivery of TGF-beta1 or BMP-2 from a novel, segmented polyurethane/polylactic-co-glycolic bilayered scaffold. J Biomed Mater Res A. 2014;102(4):1110–20.

Biomarkers in Regenerative Medicine and Bio-orthopedics

35

Graeme P. Whyte and Eric J. Strauss

35.1 Introduction to Biomarkers

Cytokines, proteases, growth factors, and a vast array of other metabolically active substances are molecular biomarkers that are becoming increasingly studied in the field of regenerative medicine and orthobiologics. Cytokines are small proteins involved in cell signaling. This term was first proposed by Stanley Cohen in 1974 [1] to describe protein molecules that modulate cellular differentiation, migration, growth, and function. The word chemokine is used to describe chemotactic cytokines that are especially active in mediating cellular chemotaxis. The term growth factor is generally applied to secreted proteins involved in cellular proliferation and differentiation, although this term is sometimes used interchangeably to describe certain cytokines that are associated with anabolic activity.

G. P. Whyte (✉)
New York University Grossman School of Medicine, NYU Langone Orthopedic Hospital, New York, NY, USA

Cornell University, Weill Medical College, New York, NY, USA
e-mail: info@whytemd.com

E. J. Strauss
New York University Grossman School of Medicine, NYU Langone Orthopedic Hospital, New York, NY, USA
e-mail: eric.strauss@nyulangone.org

Pathophysiology of articular cartilage degradation is a complex process involving pro-inflammatory and anti-inflammatory cytokines, proteases, matrix metalloproteinases, proteoglycans, and peptides [2]. Some medical treatments designed to alter the expression of pro-inflammatory cytokines associated with tissue degradation and pain have been well studied, including nonsteroidal anti-inflammatory drugs (NSAIDs). These medications have been shown to affect pro-inflammatory cytokines involved in chondrocyte turnover and degradation of supportive connective tissue matrix [3].

The ability of cytokines and other bioactive substances to mediate anabolic and catabolic processes critically impacts tissue structure after joint injury and also affects the durability and health of articular cartilage long after the acute incident. Impairment of normal joint kinematics, such that occurs in knee ligament or meniscal injury, alters stress and strain patterns on articular cartilage and can contribute to the progression of articular cartilage injury. In the event of acute injury, however, there are also substantial changes to the microenvironment profile of numerous cytokines and proteases that impact short- and long-term function of chondrocytes and the supportive connective tissue matrix.

Detailing and understanding the complex actions of biomarkers in anabolic and catabolic cascades is an important frontier in medicine that is expected to drive therapeutic innovation, and

A. Gobbi et al. (eds.), *Regenerative Medicine in Sports and Orthopaedics*,
https://doi.org/10.1007/978-3-031-84693-9_35

enable precise manipulation of pathophysiologic pathways to encourage tissue preservation and regeneration.

35.2 Synovial Fluid Collection and Biomarker Analysis

Synovial fluid samples are collected in the office setting, or in the operating room prior to surgical incision. An 18-gauge needle is inserted into the joint space under sterile conditions. In the case of the knee joint, a superolateral approach is preferred. Fluid is aspirated and transferred to a sterile tube containing a protease inhibitor solution (Halt Protease Inhibitor Cocktail, EDTA free; Pierce Biotechnology). A ratio of protease inhibitor volume to synovial fluid of 1:100 is used. A larger volume of synovial fluid is typically obtained than the microliters required for biomarker analysis. Samples are centrifuged at 3820 rpm for 10 min, and then stored at −80 C after aliquoting the supernatant into sterile tubes for later analysis.

Biomarkers of interest are identified and the appropriate multiplex biomarker assay is chosen (select biomarkers of interest are summarized in Table 35.1). Samples are thawed and assayed using multiplex enzyme-linked immunosorbent assay plates, according to the relevant manufacturer's protocol. Computer software is then used to calculate the concentrations of each analyte.

Table 35.1 Summary of select biomarkers of interest

Biomarker	Function
Pro-inflammatory	
IL-6	Interleukin 6. Pro-inflammatory cytokine secreted by macrophages and T lymphocytes. Stimulates release of other inflammatory cytokines, and modulates inflammatory responses by stimulating release of anti-inflammatory cytokines IL-Ra and IL-10
IL-8	Interleukin 8. Pro-inflammatory chemokine secreted by macrophages. Promotes migration of neutrophils to sites of inflammation and injury
IL-1β	Interleukin 1 beta. Pro-inflammatory cytokine secreted by macrophages. Modulates inflammatory response. Promotes chondrocyte protease production. Cyclooxygenase-2 (COX-2) activator
TNF-α	Tumor necrosis factor alpha. Pro-inflammatory cytokine released by monocytes, lymphocytes, and granulocytes. Involved in acute inflammatory response and can promote cellular apoptosis. Capable of impairing proteoglycan synthesis and degrading cartilage matrix
MIPs	Macrophage inflammatory proteins. Family of chemokines released from macrophages that recruit monocytes, lymphocytes, and granulocytes. Promote release of numerous pro-inflammatory cytokines. Involved in bone resorption and remodeling
MCPs	Monocyte chemotactic proteins. Pro-inflammatory chemokines that promote monocyte, macrophage, and other leukocyte migration to sites of tissue injury and inflammation
RANTES	"Regulated on activation, normal T cell expressed and secreted"; also known as CCL5. Chemokine released by T lymphocytes and monocytes. Chemotactic factor for leukocytes. Activity increases several days after injury and functions to maintain inflammatory response
MMPs	Matrix metalloproteinases. Protease enzymes that degrade extracellular matrix, including collagen, elastin, and fibronectin. Involved in cellular recruitment and apoptosis at sites of tissue injury and inflammation
Anti-inflammatory	
IL-1Ra	Interleukin 1 receptor antagonist. Anti-inflammatory cytokine that modulates inflammatory response. Competitively inhibits the action of pro-inflammatory IL-1β
TIMPs	Tissue inhibitors of metalloproteinases. Anti-inflammatory protein molecules that modulate inflammation by inhibition of MMPs. Have a chondroprotective role and can promote chondrogenesis

(continued)

Table 35.1 (continued)

Biomarker	Function
Growth factor	
VEGF	Vascular endothelial growth factor. Growth factor involved in cellular signaling to promote angiogenesis. Stimulates migration and proliferation of endothelial cells, and is a chemotactic factor for leukocytes
bFGF	Basic fibroblast growth factor; also known as FGF-2. Cell-signaling growth factor that is expressed in vascular smooth muscle cells and endothelial cells. Involved in tissue differentiation and development, including angiogenesis and chondrogenesis. Promotes wound healing and formation of granulation tissue
PDGF	Platelet-derived growth factor. Family of glycoprotein dimers involved in cellular proliferation, migration, and differentiation. Act as mitogens and chemokines, and capable of promoting tissue growth and inhibiting catabolic processes associated with inflammatory cytokines

35.3 Specific Biomarkers of Interest

35.3.1 Interleukins

IL-6 is a prominent pro-inflammatory cytokine of the interleukin family that is secreted by macrophages and T lymphocytes. It is active after traumatic injury and is associated with other inflammatory processes such as infection and neoplasia. This cytokine is capable of modulating inflammatory cascades through negative feedback, by stimulating release of the anti-inflammatory cytokines IL-Ra and IL-10. Clair et al. [4] compared IL-6 concentrations in meniscus-injured knees to asymptomatic controls. Increased concentrations of IL-6 were associated with meniscus injury, as well as increased age. Cuellar et al. [5] evaluated synovial fluid samples at the time of knee arthroscopy and in contralateral knees. There was a strong positive correlation between IL-6 concentration and intra-operative ICRS score, and of IL-6 with postoperative pain intensity at final follow-up. Additionally, IL-6 concentrations were associated with more severe articular cartilage injury. IL-6 concentrations have been studied in the pathogenesis of knee osteoarthritis (OA). Monibi et al. [6] identified significantly higher IL-6 concentrations in the synovial fluid of patients suffering from knee OA compared with normal patients. Furthermore, radiographic evaluation demonstrated that increased degenerative change had a strong positive correlation with IL-6 levels. In cases of anterior cruciate ligament (ACL) injury, Kingery et al. [7] determined that IL-6 concentration decreased in synovial samples taken from knees at the time of ACL reconstruction, compared to synovial fluid concentrations in synovial fluid samples taken from the acutely ACL-injured knees a month prior. Kaplan et al. [8] evaluated synovial fluid of ACL-injured knees at the time of ligament reconstruction, and compared knees with and without articular cartilage injury to control samples. It was found that ACL-injured knees with or without articular cartilage injury had significantly higher concentrations of IL-6 compared with contralateral controls. Anil et al. [9] examined knee synovial fluid at the first postoperative visit, at a mean of 10 days after isolated ACL reconstruction. Synovial fluid concentrations of IL-6 were significantly increased at the first postoperative visit compared to the time of surgery, demonstrating the impact of surgical intervention on the release of the pro-inflammatory IL-6 cytokine.

The role of numerous other pro-inflammatory cytokines of the interleukin family has been studied for their role in modulating inflammatory processes. IL-8 is a pro-inflammatory chemokine that is secreted by macrophages and promotes migration of neutrophils by acting as a chemotactic agent. IL-8 concentrations have been shown to be significantly higher in the synovial fluid of OA patients compared with normal patients, and radiographic findings of increased degenerative change have shown a strong positive correlation with IL-8 [6]. Increased concentrations of IL-8 are also seen in response to mechanical insult to osteochondral tissue [10]. IL-1β is a pro-inflammatory cytokine that is secreted by macrophages and modulates inflammatory response. This cytokine stimulates release of other inflammatory cytokines and can promote chondrocyte protease production. In addition, IL-1β is capable of activating COX-2.

Interleukin 1 receptor antagonist (IL-1Ra) is an important anti-inflammatory cytokine that modulates inflammatory response, competitively inhibiting the action of the pro-inflammatory cytokine IL-1β. Comparison of synovial fluid cytokine concentrations between knees undergoing surgical treatment of injury and contralateral asymptomatic knees have shown increased concentrations of IL-1Ra to be a negative predictor of severe cartilage injury [5]. Regarding the presence of IL-1Ra in ACL-injured knees, Kingery et al. [7] demonstrated a decrease in synovial fluid concentration of IL-1Ra at the time of ACL reconstruction, compared to samples taken a month earlier at the time of acute knee injury. Anil et al. [9] evaluated synovial fluid concentrations of IL-1Ra at the time of ACL reconstruction, and then again at the first postoperative visit. Postoperatively, synovial fluid concentrations of IL-1Ra were increased compared to samples taken prior to ligament reconstruction. Progressive articular cartilage injury is a concern after ACL injury, and this was evaluated by Markus et al. [11]. Synovial fluid concentrations of IL-Ra in the knee after ACL injury were significantly positively correlated with MRI findings of articular cartilage injury at a mean of 8 years postoperatively.

35.3.2 Tumor Necrosis Factor Alpha (TNF-α)

TNF-α is a pro-inflammatory cytokine released by monocytes, lymphocytes, and granulocytes. It is involved in the acute inflammatory response and is able to suppress proteoglycan synthesis and promote focal cartilage loss and degradation of cartilage matrix [12]. TNF-α receptor activation has been shown to promote cellular apoptosis, although there are multiple receptor types, and alternate TNF-α-specific receptor activation can modulate inflammatory responses that are involved in tissue regeneration.

According to Williams et al. [13], TNF-α has been shown to stimulate the release of collagenase by synovial cells and to be involved with resorption of cartilage and bone. Timmen et al. [14] compared femur fracture healing in wild-type mice to human TNF-α transgenic mice that develop polyarthritis. Fracture healing was evaluated after treatment with anti-TNF antibody, and in untreated controls. Increased TNF-α concentrations impaired callus formation and fracture healing, and TNF-α blockade in the transgenic mice led to similar fracture healing as in the wild-type mice.

Seguin and Bernier [15] examined the effects of TNF-α on rat chondrocytes. TNF-α was shown to inhibit type II collagen expression, disrupting the synthesis, organization, and turnover of articular cartilage. A systematic review of the inhibitory effects of TNF-α on articular cartilage by Chisari et al. [16] determined that TNF-α had the capacity to impair healing of chondral and osteochondral defects, potentially affecting the regenerative environment after cartilage repair surgery. It was suggested that use of a biologic agent to inhibit TNF-α may result in superior clinical outcomes after surgical treatment to repair chondral injury.

35.3.3 Macrophage Inflammatory Proteins (MIPs)

MIP-1α (also known as CCL3) and MIP-1β (also known as CCL4) are macrophage inflammatory proteins that act as chemokines. These molecules are released from macrophages and recruit monocytes, lymphocytes, and granulocytes. They promote the release of numerous pro-inflammatory cytokines and are also involved in bone resorption and remodeling.

Increased synovial fluid concentration of MIP-1β has been shown to be associated with high-grade chondral lesions in meniscus-injured knees, and concentrations have been shown to be increased in knees treated for meniscal injury in comparison to control knees [4]. Vangsness et al. [17] identified significant increases in MIP-1α concentration in knees with more advanced articular cartilage injury, even in the absence of radiographic differences.

Synovial fluid concentrations of MIP-1β have been shown to be increased at the time of acute ACL injury in comparison to paired sample analysis from synovial fluid taken at the time of ligament reconstruction [7]. Postoperatively, synovial concentration of MIP-1β has been shown to be decreased at the initial postoperative visit in comparison to samples obtained at the time of ACL reconstruction [9]. When examining chondral damage in the setting of ACL-injured knees, significantly greater concentrations of synovial fluid MIP-1β have been shown in cases of ACL injury, both with and without associated articular cartilage injury, when compared to controls [8].

35.3.4 Monocyte Chemotactic Proteins (MCPs)

Monocyte chemotactic proteins are pro-inflammatory chemokines that promote monocyte, macrophage, and other leukocyte migration to sites of tissue injury or other inflammatory processes. The role of MCP-1 (also known as CCL2) has been extensively studied in neoplastic processes, and this chemokine features prominently in the inflammatory response to injury.

Synovial fluid concentration of MCP-1 in the knee has been shown to be increased after acute ligament injury when compared to samples obtained a month later at the time of surgical intervention to reconstruct the ACL [7]. Synovial fluid analysis at the first postoperative visit after ACL reconstruction has shown that MCP-1 levels are increased compared to the samples analyzed at the time of surgery [9]. Markus et al. [11] found that increased synovial fluid concentration of MCP-1 at the time of ACL reconstruction was associated with greater severity of articular cartilage injury when assessed by MRI 6 to 10 years after surgery.

Evaluation of the synovial fluid cytokine profile in meniscus-injured knees has demonstrated increased concentration of MCP-1 compared to contralateral knee controls [4]. Severity of chondral injury at the time of knee arthroscopy has been shown to be associated with increased MCP-1 concentration based on work by Cuellar et al. [5]. This study also reported a strong positive correlation of MCP-1 concentrations with patient-reported pain scores at final follow-up, and a positive correlation of MCP-1 with the severity of chondral injury based on ICRS grade. Interestingly, this research demonstrated that increased MCP-1 concentrations in the synovial fluid of asymptomatic control knees were associated with increased patient-reported pain affecting the control knee at final follow-up 1–2 years after surgery, according to the visual analogue scale (VAS). Regarding knee OA, synovial fluid concentration of MCP-1 has been shown to positively correlate with severity of radiographic degenerative change in patients undergoing total knee arthroplasty, when compared to normal control knees [6].

Concentrations of synovial fluid MCP-1 can be elevated in knees affected by increased grade of articular cartilage injury at the time of direct examination according to ICRS criteria, despite similar grading of radiographic change according to Kellgren-Lawrence criteria [17]. In vitro research has found that blockade of MCP-1 activity in knee OA leads to a significant reduction in macrophage activity and attenuation of articular cartilage injury and synovitis [18]. Notably, absence of the main MCP-1 receptor, CCR2, has been shown to enhance regeneration of articular cartilage in mice, whereas absence of MCP-1 did not demonstrate the same cartilage regeneration after injury [19]. Furthermore, MCP-1 has been shown to promote cartilage degradation and chondrocyte apoptosis, even in the presence of a CCR2 antagonist [20].

35.3.5 Matrix Metalloproteinases (MMPs) and Tissue Inhibitors of Metalloproteinases (TIMPs)

MMPs are proteases that degrade extracellular matrix, including collagen, elastin, and fibronectin. These molecules play a role in cellular recruitment and apoptosis at sites of tissue injury or inflammation. TIMPs (tissue inhibitors of metalloproteinases) are anti-inflammatory bioactive molecules that modulate inflammatory processes by inhibition of MMPs. They have been shown to play a chondroprotective role and can promote chondrogenesis. While both MMPs and TIMPs may be elevated at sites of tissue injury and inflammation, the ratio of these factors may change, reflecting various states of pro-inflammatory and anti-inflammatory response.

MMP-1, MMP-2, and MMP-3 concentrations have been shown to be significantly higher in the synovial fluid of OA patients compared with normal asymptomatic knees [6, 21]. Driban et al. [21] also demonstrated elevated TIMP-2 concentrations in the synovial fluid of OA-affected knees. When comparing synovial fluid samples at the time of knee arthroscopy in patients suffering from meniscal injury, chondral injury, and/or ACL injury, Cuellar et al. [5] reported a positive correlation of MMP-3 concentration and increased preoperative patient-reported pain.

At the first postoperative visit after isolated ACL reconstruction, Anil et al. [9] reported increased synovial fluid concentration of MMP-3, and decreased concentration of TIMP-2, compared to samples obtained on the day of surgery, prior to the procedure. According to Kingery et al. [7], synovial fluid analysis within days of acute knee ACL injury identified decreased concentration of MMP-3, and increased concentration of TIMP-1, compared to samples analyzed a month later at the time of surgery. Clair et al. [4] identified significantly increased concentrations of knee synovial fluid MMP-3 in those treated for meniscus injury, compared to contralateral control knees. Additionally, in cases of knee synovitis, there was an associated decreased concentration of TIMP-1, and an increase in the ratio of MMP-3 to TIMP-1 [4]. In those suffering from ACL injury in isolation, or in association with chondral injury, concentrations of MMP-3 and TIMP-1 have been shown to be elevated compared to control knees [8]. This study also demonstrated decreased concentration of synovial fluid TIMP-2 in ACL-injured knees without associated cartilage injury, when compared to asymptomatic control knees.

35.3.6 RANTES

RANTES ("regulated on activation, normal T cell expressed and secreted"; also known as CCL5) is a chemokine released by T lymphocytes and monocytes. It is a chemotactic factor for leukocytes, and promotes activation of these cells and migration to sites of injury and inflammation. This chemokine exerts greater influence several days after insult and can function to maintain an inflammatory response. There is an association of RANTES with increased concentrations of pro-inflammatory cytokines such as IL-6 and TNF-α. RANTES also promotes activation of MMP-1 and collagen degradation.

RANTES concentration has been shown to be significantly elevated in the synovial fluid of patients suffering from knee OA compared with normal knee controls [6]. In cases of ACL-injured knees, increased synovial fluid concentration of RANTES has been identified approximately a month after injury, compared to synovial fluid analysis performed less than a week after acute injury [7]. Synovial fluid concentration of RANTES is not necessarily predictable based solely on the presence of inflammation and injury. When examining synovial fluid from control knees and from knees affected by ACL tear, meniscus tear, or articular cartilage injury, Cuellar et al. [5] reported significantly increased concentrations of RANTES only in those knees with isolated articular cartilage injury. This research also reported that increased synovial fluid concentration of RANTES at the time of surgical treatment of knee meniscus, chondral, and/or ACL injury was a strong predictor of improvement of patient-reported pain at final follow-up.

35.3.7 Vascular Endothelial Growth Factor (VEGF)

VEGF is a growth factor involved in cellular signaling to promote angiogenesis. This protein stimulates migration and proliferation of endothelial cells and is a chemotactic factor for leukocytes.

Increased synovial fluid concentrations of VEGF have been shown to be associated with high-grade knee chondral injury in those treated for meniscus tear [4]. In those treated with ACL reconstruction, synovial fluid samples taken at the time of surgery have demonstrated significantly decreased VEGF concentration compared to samples obtained a month previous, at the time of acute injury [7]. After surgery, synovial fluid concentration of VEGF has been shown to be increased at the first postoperative visit compared to the concentration of fluid taken at the time of ACL reconstruction [9]. Postoperative knee MRI examination up to 10 years after ACL reconstruction has demonstrated a significant positive correlation of higher-grade articular cartilage injury with increased synovial fluid concentration of VEGF at the time of surgery [11].

In asymptomatic knees used as controls at the time of knee arthroscopy, increased synovial fluid concentration of VEGF has been associated with a significant increase in knee pain up to 2 years later, according to patient-reported outcome assessment [5].

35.3.8 Basic Fibroblast Growth Factor (bFGF)

Basic FGF (also known as FGF-2) is a cell-signaling growth factor from a large family of growth factors known as fibroblast growth factors. This growth factor is expressed in vascular smooth muscle cells and endothelial cells. bFGF is involved in tissue generation in embryonic development, and at later stages of maturation, has an ongoing role in tissue differentiation and development, including angiogenesis and chondrogenesis. bFGF is known to play a prominent role in wound healing and the formation of granulation tissue.

Synovial fluid analysis from samples obtained days after acute knee injury has been shown to contain decreased concentrations of bFGF compared to fluid samples obtained a month later, at the time of surgery [7]. Kaplan et al. [8] reported decreased synovial fluid concentration of bFGF in knees affected by coexisting ACL and chondral injury compared to contralateral asymptomatic knees.

35.3.9 Platelet-Derived Growth Factor (PDGF)

PDGF is a growth factor glycoprotein dimer involved in tissue differentiation during embryonic development. In later stages of development, PDGF continues to play a role in cellular proliferation, migration, and differentiation. It acts as a mitogen and chemokine, capable of promoting tissue growth as well as angiogenesis. PDGF-BB is a specific dimer of PDGF that is capable of retarding cartilage degradation that can occur in the presence of IL-1β.

Cuellar et al. [5] examined concentrations of PDGF-BB in the synovial fluid of knees undergoing surgical treatment of meniscal injury, ACL injury, and/or chondral injury, in addition to contralateral asymptomatic knees. PDGF-BB concentrations were greatest in contralateral asymptomatic knees, and significantly decreased in all cases of knee injury. Concentrations of this growth factor were especially low in those knees affected by more severe chondral injury. Increased concentration of PDGF-BB at the time of surgical intervention was a strong predictor of reduced patient-reported pain at the time of final follow-up 1–2 years after surgery.

35.4 Overview of Biomarkers and Future Directions

A greater understanding of the interaction between pro-inflammatory and anti-inflammatory biomarkers, in the setting of degenerative conditions and following acute injury, is expected to facilitate development of targeted therapies to manipulate regenerative processes and optimize long-term durability of articular cartilage and joint health.

It has been shown that synovial fluid pro-inflammatory cytokine concentrations increase after joint injury or surgery. For instance, there is alteration of knee synovial fluid biomarker constituents to a profile that is more indicative of OA in cases of ACL-injured knees, with or without surgical ligament reconstruction [22]. Several bioactive molecules remain elevated for years after joint injury and have been linked to the pathogenesis of progressive articular cartilage injury and osteoarthritis [23]. These biomarkers make attractive therapeutic targets to alter pathologic pathways of joint degeneration, whether idiopathic or instigated by injury.

There are some well-known therapies that affect synovial biomarkers, however, altering the natural history of degenerative joint arthropathy is a much greater challenge. Research examining human synovium has demonstrated that NSAIDs downregulate expression of COX-2 mRNA in the synovial membrane, decrease synovial macrophage infiltration, and decrease expression of the pro-inflammatory cytokines TNF-α and IL-1β [24]. NSAIDs have been shown to decrease IL-1β-induced synthesis of pro-inflammatory cytokines in cultured human chondrocytes and decrease expression of TNF-α and IL-1β in human articular cartilage [3]. Percutaneous NSAID administration successfully suppressed the pro-inflammatory cytokines IL-6 and TNF-α in a rat model involving osteoarthritis (OA) [25]. Clinical research in humans undergoing surgical treatment of OA has demonstrated that preoperative oral NSAID administration can significantly decrease synovial fluid concentrations of pro-inflammatory IL-6, VEGF, and TNF-α, compared to controls [26].

There is potential for biologic therapy to modify catabolic processes in hyperglycemic environments seen in diabetes mellitus. Hyperglycemia and insulin resistance are associated with development of osteoarthritis, likely resulting from hyperglycemia-induced overabundance of pro-inflammatory cytokines leading to oxidative stress [27, 28]. In vitro research demonstrated increased MCP and MMP expression in chondrocytes within a hyperglycemic environment, whereas treatment with an MCP inhibitor led to decreased MMP and upregulation of collagen 2 [29].

Traumatic joint injury, as well as surgical insult, can lead to long-term derangement of the cytokine profile in synovial fluid and periarticular tissues. Alteration in synovial fluid biomarkers after surgical treatment of knee injury may promote and advance tissue degradation, and adversely affect clinical outcomes [30]. Acute therapeutic intervention to target specific biomarkers after joint injury, or to directly modulate activity of resident leukocytes such as synovial macrophages [31], has the potential to moderate cellular apoptosis and turnover, and to minimize degradative effects on supportive matrix that result from inflammatory responses.

Cytokine profiles change over the course of articular cartilage injury progression, with certain cytokines having greater influence earlier in the disease course [32]. The timing of intervention for specific biomarkers to modulate inflammatory responses may vary, depending on disease stage. There is also opportunity to develop therapeutics that are capable of treating a range of inflammatory disorders. There are numerous pro-inflammatory synovial fluid biomarkers found to be common in synovial fluid analyzed from patients suffering from osteoarthritis and rheumatoid arthritis [33], suggesting that successful targeting of a single inflammatory mediator may have broad clinical implications.

An improved understanding of the interaction and function of biomarkers will enable the development of more potent treatments and supportive care, potentially with the capability of substantially altering the natural history of articular cartilage injury and joint degeneration.

References

1. Cohen S, Bigazzi PE, Yoshida T. Commentary. Similarities of T cell function in cell-mediated immunity and antibody production. Cell Immunol. 1974;12(1):150–9.
2. Cuéllar VG, Strauss EJ. Molecular biomarkers of knee pathology. Bull Hosp Joint Dis. 2017;75(1):30–6.

3. Álvarez-Soria MA, Herrero-Beaumont G, Moreno-Rubio J, Calvo E, Santillana J, Egido J, et al. Long-term NSAID treatment directly decreases COX-2 and mPGES-1 production in the articular cartilage of patients with osteoarthritis. Osteoarthr Cartil. 2008;16(12):1484–93.
4. Clair AJ, Kingery MT, Anil U, Kenny L, Kirsch T, Strauss EJ. Alterations in synovial fluid biomarker levels in knees with meniscal injury as compared with asymptomatic contralateral knees. Am J Sports Med. 2019;47(4):847–56.
5. Cuéllar VG, Cuéllar JM, Kirsch T, Strauss EJ. Correlation of synovial fluid biomarkers with cartilage pathology and associated outcomes in knee arthroscopy. Arthroscopy. 2016;32(3):475–85. Available from: https://doi.org/10.1016/j.arthro.2015.08.033.
6. Monibi F, Roller BL, Stoker A, Garner B, Bal S, Cook JL. Identification of synovial fluid biomarkers for knee osteoarthritis and correlation with radiographic assessment. J Knee Surg. 2016;29(3):242–7.
7. Kingery MT, Anil U, Berlinberg EJ, Clair AJ, Kenny L, Strauss EJ. Changes in the synovial fluid cytokine profile of the knee between an acute anterior cruciate ligament injury and surgical reconstruction. Am J Sports Med. 2022;50(2):451–60.
8. Kaplan DJ, Cuellar VG, Jazrawi LM, Strauss EJ. Biomarker changes in anterior cruciate ligament–deficient knees compared with healthy controls. Arthroscopy. 2017;33(5):1053–61. Available from: https://doi.org/10.1016/j.arthro.2016.11.019.
9. Anil U, Jejurikar N, Kenny L, Strauss EJ. Changes in synovial fluid biomarker concentration before and after ACL reconstruction. Bull Hosp Jt Dis. 2019;77(3):189–93. Available from: http://www.ncbi.nlm.nih.gov/pubmed/31487484.
10. Dwivedi G, Flaman L, Alaybeyoglu B, Struglics A, Frank EH, Chubinskya S, et al. Inflammatory cytokines and mechanical injury induce post-traumatic osteoarthritis-like changes in a human cartilage-bone-synovium microphysiological system. Arthritis Res Ther. 2022;24(1):198.
11. Markus DH, Hurley ET, Mojica ES, Anil U, Kanakamedala A, Avila A, et al. Concentration of synovial fluid biomarkers on the day of anterior cruciate ligament (ACL)-reconstruction predict size and depth of cartilage lesions on 5-year follow-up. Knee Surg Sports Traumatol Arthrosc. 2023;31(5):1753–60. Available from: https://doi.org/10.1007/s00167-022-07045-9.
12. Westacott CI, Barakat AF, Wood L, Perry MJ, Neison P, Bisbinas I, et al. Tumor necrosis factor alpha can contribute to focal loss of cartilage in osteoarthritis. Osteoarthr Cartil. 2000;8(3):213–21.
13. Williams RO, Feldmann M, Maini RN. Cartilage destruction and bone erosion in arthritis: the role of tumour necrosis factor alpha. Ann Rheum Dis [Internet]. 2000;59 Suppl 1(Suppl 1):i75–80. Available from: http://www.ncbi.nlm.nih.gov/pubmed/11053094.
14. Timmen M, Hidding H, Wieskötter B, Baum W, Pap T, Raschke MJ, et al. Influence of antiTNF-alpha antibody treatment on fracture healing under chronic inflammation. BMC Musculoskelet Disord. 2014;15(1):184.
15. Séguin CA, Bernier SM. TNFα suppresses link protein and type II collagen expression in chondrocytes: role of MEK1/2 and NF-κB signaling pathways. J Cell Physiol. 2003;197(3):356–69.
16. Chisari E, Yaghmour KM, Khan WS. The effects of TNF-alpha inhibition on cartilage: a systematic review of preclinical studies. Osteoarthritis Cartilage. W.B. Saunders Ltd. 2020;28:708–18.
17. Thomas Vangsness J, Burke WS, Narvy SJ, MacPhee RD, Fedenko AN. Human knee synovial fluid cytokines correlated with grade of knee osteoarthritis: a pilot study. Bull NYU Hosp Jt Dis. 2011;69(2):122–7.
18. Raghu H, Lepus CM, Wang Q, Wong HH, Lingampalli N, Oliviero F, et al. CCL2/CCR2, but not CCL5/CCR5, mediates monocyte recruitment, inflammation and cartilage destruction in osteoarthritis. Ann Rheum Dis. 2017;76(5):914–22.
19. Jablonski CL, Leonard C, Salo P, Krawetz RJ. CCL2 but not CCR2 is required for spontaneous articular cartilage regeneration post-injury. J Orthop Res. 2019;37(12):2561–74.
20. Xu YK, Ke Y, Wang B, Lin JH. The role of MCP-1-CCR2 ligand-receptor axis in chondrocyte degradation and disease progress in knee osteoarthritis. Biol Res. 2015;48:64.
21. Driban JB, Balasubramanian E, Amin M, Sitler MR, Ziskin MC, Barbe MF. The potential of multiple synovial-fluid protein-concentration analyses in the assessment of knee osteoarthritis. J Sport Rehabil [Internet]. 2010;19(4):411–21. Available from: https://www.ncbi.nlm.nih.gov/pmc/articles/PMC3624763/pdf/nihms412728.pdf.
22. Harkey MS, Luc BA, Golightly YM, Thomas AC, Driban JB, Hackney AC, et al. Osteoarthritis-related biomarkers following anterior cruciate ligament injury and reconstruction: a systematic review [Internet]. Osteoarthritis Cartilage. Elsevier Ltd. 2015;23:1–12. Available from: https://doi.org/10.1016/j.joca.2014.09.004.
23. Markus DH, Berlinberg EJ, Strauss EJ. Current state of synovial fluid biomarkers in sports medicine. JBJS Rev [Internet]. 2021;9(8):1–7. Available from: https://journals.lww.com/10.2106/JBJS.RVW.21.00024.
24. Álvarez-Soria MA, Largo R, Santillana J, Sánchez-Pernaute O, Calvo E, Hernández M, et al. Long term NSAID treatment inhibits COX-2 synthesis in the knee synovial membrane of patients with osteoarthritis: differential proinflammatory cytokine profile between celecoxib and aceclofenac. Ann Rheum Dis. 2006;65(8):998–1005.
25. Orita S, Ishikawa T, Miyagi M, Ochiai N, Inoue G, Eguchi Y, et al. Percutaneously absorbed NSAIDs attenuate local production of proinflammatory cytokines and suppress the expression of c-Fos in the

spinal cord of a rodent model of knee osteoarthritis. J Orthop Sci. 2012;17(1):77–86.
26. Gallelli L, Galasso O, Falcone D, Southworth S, Greco M, Ventura V, et al. The effects of nonsteroidal anti-inflammatory drugs on clinical outcomes, synovial fluid cytokine concentration and signal transduction pathways in knee osteoarthritis. A randomized open label trial. Osteoarthr Cartil. 2013;21(9):1400–8.
27. De Roover A, Escribano-Núñez A, Monteagudo S, Lories R. Fundamentals of osteoarthritis: inflammatory mediators in osteoarthritis. Osteoarthr Cartil. 2023;31(10):1303–11.
28. Du X, Liu Z-Y, Tao X-X, Mei Y-L, Zhou D-Q, Cheng K, et al. Research progress on the pathogenesis of knee osteoarthritis. Orthop Surg. 2023;15:2213–24.
29. Quincey A, Mohan S, Edderkaoui B. Monocyte chemotactic proteins mediate the effects of hyperglycemia in chondrocytes: in vitro studies. Life. 2022;12(6):836.
30. Tollefson LV, Kennedy MI, Tagliero AJ, Malinowski K, Chahla J, Moatshe G, et al. Inflammatory synovial biomarkers and state of the tibiofemoral joint in the post-surgical settings: a narrative review. Ann Jt [Internet]. 2024;9:6–6. Available from: https://aoj.amegroups.com/article/view/8366/html.
31. Chen Y, Jiang W, Yong H, He M, Yang Y, Deng Z, et al. Macrophages in osteoarthritis: pathophysiology and therapeutics. Am J Transl Res [Internet]. 2020;12(1):261–8. Available from: http://www.ncbi.nlm.nih.gov/pubmed/32051751.
32. Li L, Li Z, Li Y, Hu X, Zhang Y, Fan P. Profiling of inflammatory mediators in the synovial fluid related to pain in knee osteoarthritis. BMC Musculoskelet Disord. 2020;21(1):99.
33. Meehan RT, Regan EA, Hoffman ED, Wolf ML, Gill MT, Crooks JL, et al. Synovial fluid cytokines, chemokines and mmp levels in osteoarthritis patients with knee pain display a profile similar to many rheumatoid arthritis patients. J Clin Med. 2021;10(21):5027.